Bird Families of the World

A series of authoritative, illustrated handbooks, of which this is the sixth volume to be published

Series editors
C. M. PERRINS Chief editor
W. J. BOCK
J. KIKKAWA

THE AUTHORS Clifford Frith is a self-employed ornithologist, wildlife photographer, and publisher. Early ornithological positions included the British Museum of Natural History and the Royal Society of London. Interests include the study of tropical Australasian birds, with particular reference to birds of paradise and bowerbirds and their behaviour, nesting biology and systematics. He has also worked on various avian and other animal groups in the Indian Ocean, Southeast Asia, and the tropical Pacific. He is, jointly with his wife Dawn, an Honorary Research Fellow of the Queensland Museum and recipient of the 1996 D. L. Serventy Medal of the Royal Australasian Ornithologists' Union for original contributions to ornithology. Together they produced *Cape York Peninsula—a Natural History* (1996). He has observed 22 bird of paradise species in the wild, eight more in captivity, and others on film. He lives in the Wet Tropics rainforests of north Queensland, habitat of Victoria's Riflebird. Bruce Beehler, who has had a life-long interest in birds, has conducted research in ornithology, tropical ecology, and conservation biology in India and the southwest Pacific. He observed his first bird of paradise in the wild in 1975 and subsequently conducted doctoral and postdoctoral research on this family, mainly in Papua New Guinea. Senior author of *Birds of New Guinea* (Princeton), Beehler has also authored *A Naturalist in New Guinea* (University of Texas) and several other books on birds. He edited *A Biodiversity Analysis for Papua New Guinea* (Biodiversity Support Program). Research interests have focussed on avian mating systems, avian frugivory, and biogeography. He has observed 31 species of bird of paradise in the wild. He resides with his wife Carol and three children in Bethesda, Maryland, and is a research associate in the Department of Vertebrate Zoology, National Museum of Natural History, Washington, DC.

THE ARTIST William T. Cooper is acknowledged as one of the greatest and most productive of internationally acclaimed bird artists. In acknowledgement of his work he was awarded the Gold Medal for distinction in natural history art by the American Academy of Natural Sciences in 1994 and appointed an Officer of the Order of Australia (A.O.) in 1995. His numerous ornithologically and botanically accurate and artistically beautiful watercolours illustrate much admired works including *A Portfolio of Australian Birds*, *Parrots of the World*, *The Birds of Paradise and Bower Birds*, *Parrots of Australia*, *Kingfishers and Related Birds* (six volumes), and *Turacos*. In order to prepare the plates for his own work on the birds of paradise he travelled extensively in Papua New Guinea and observed and drew 22 species in the wild. He lives in the tropical rainforests of north Queensland, Australia.

Bird Families of the World

1. The Hornbills
   ALAN KEMP

2. The Penguins
   TONY D. WILLIAMS

3. The Megapodes
   DARRYL N. JONES, RENÉ W. R. J. DEKKER, AND CEES S. ROSELAAR

4. Fairy-wrens and Grasswrens
   IAN ROWLEY AND ELEANOR RUSSELL

5. The Auks
   ANTHONY J. GASTON AND IAN L. JONES

6. The Birds of Paradise
   CLIFFORD B. FRITH AND BRUCE M. BEEHLER

Bird Families of the World

The Birds of Paradise
Paradisaeidae

CLIFFORD B. FRITH
Honorary Research Fellow
Queensland Museum, Brisbane

BRUCE M. BEEHLER
Research Associate
Smithsonian Institution, Washington

Illustrated by
WILLIAM T. COOPER

Sonogram production by
MICHAEL McGUIRE

Oxford   New York   Tokyo
OXFORD UNIVERSITY PRESS
1998

Oxford University Press, Great Clarendon Street, Oxford OX2 6DP

Oxford New York
Athens Auckland Bangkok Bogota Bombay
Buenos Aires Calcutta Cape Town Dar es Salaam
Delhi Florence Hong Kong Istanbul Karachi
Kuala Lumpur Madras Madrid Melbourne
Mexico City Nairobi Paris Singapore
Taipei Tokyo Toronto Warsaw
and associated companies in
Berlin Ibadan

Oxford is a trade mark of Oxford University Press

Published in the United States
by Oxford University Press Inc., New York

© Clifford B. Frith and Bruce M. Beehler, 1998

All rights reserved. No part of this publication may be reproduced, stored in a retrieval system, or transmitted, in any form or by any means, without the prior permission in writing of Oxford University Press. Within the UK, exceptions are allowed in respect of any fair dealing for the purpose of research or private study, or criticism or review, as permitted under the Copyright, Designs and Patents Act, 1988, or in the case of reprographic reproduction in accordance with the terms of licences issued by the Copyright Licensing Agency. Enquiries concerning reproduction outside those terms and in other countries should be sent to the Rights Department, Oxford University Press, at the address above.

This book is sold subject to the condition that it shall not, by way of trade or otherwise, be lent, re-sold, hired out, or otherwise circulated without the publisher's prior consent in any form of binding or cover other than that in which it is published and without a similar condition including this condition being imposed on the subsequent purchaser.

A catalogue record for this book is available from the British Library

Library of Congress Cataloging in Publication Data
Frith, Clifford B.
The birds of paradise: Paradisaeidae / Clifford B. Frith, Bruce M. Beehler; illustrated by William T. Cooper.
(Bird families of the world; 6)
Includes bibliographical references and index.
1. Birds of paradise. I. Beehler, Bruce Mcp. II. Cooper, William T., 1934– . III. Title. IV. Series.
QL696.P26F75 1998   598.8′65–dc21   97-52151
ISBN 0 19 854853 2

Typeset by EXPO Holdings, Malaysia

Printed in Great Britain by Butler and Tanner Ltd, Frome and London

The authors dedicate this book to:

Dawn W. Frith

who contributed immeasurably to this work in the field, museum, and study in addition to providing CBF with ultimate companionship and support throughout his studies of birds of paradise and much more

Carol Beehler

who honeymooned on the Wara Sii, weathered the rains on the Puwani, and bathed baby Grace in the milky pools of the Narirogo

Mary LeCroy

who learned the birds of paradise under the tutelage of Tom Gilliard; who subsequently advised and informed the work of virtually every ornithologist studying the avifauna of New Guinea; and who herself has made original contributions to our knowledge of the Paradisaeidae

and the memory of E. Thomas Gilliard (1912–1965)

whose groundbreaking field and museum studies significantly advanced our understanding of the biology of the birds of paradise

Foreword

by Sir David Attenborough

You might, with reason, suppose that birds of paradise were given their name as a poetic compliment to their incomparable beauty. But not so. The plumes of the birds, which for at least two thousand years have been arriving in ports of Southeast Asia, were in antiquity valued so highly there that they became incorporated into royal regalia. Such skins lacked both wings and legs for they were regularly cut off in order to emphasise the glory of the plumes by the hunters who collected the birds in the forests of New Guinea two thousand miles away to the east. When traders selling such skins in the Moluccas, who themselves had never seen the living creatures, were asked how a bird could perch or fly without any limbs, they replied that these creatures were *bolong diuata*, the birds of the gods. They floated wingless, high in the sky feeding on dew, and only came into human hands when they died and fell from paradise to earth.

These stories accompanied the specimens when, during the sixteenth century, they reached Europe. There, literal-minded naturalists, who could accept the notion of celestial suspension without question, were puzzled by the question of procreation. Where could a bird nest, high in the sky? Some solemnly declared that the female bird of paradise laid her eggs in a hollow in the middle of the male's back and sat on him to incubate them. Even as late as the seventeenth century, some European artists were still painting pictures of the Garden of Eden with birds sailing through the skies of paradise—without either legs or wings.

But the connection with paradise is, in one way, truer than the tellers of the original fables could have known. The vast island of New Guinea is indeed a paradise—for birds. The tropical forests that cover it are rich in food of all kinds, but unlike most other forests, they have very few vegetarian or insect-eating placental mammals. There are no monkeys to grab the fruits, no squirrels to gnaw the nuts. So a female bird of paradise might well be able to collect food with such ease that she can rear her own chicks unaided. Not only that, but there are no large mammalian carnivores either, so a male bird is not dangerously encumbered if he develops huge plumes, nor is it too risky for him to dance on the ground while displaying them. How these two characteristics of the forests have led to the birds of paradise becoming so pre-eminently beautiful will be clear from the pages that follow.

New Guinea is also, however, uniquely wild. Even today, travel through its mountains is very arduous and there are still areas that are unexplored by outsiders. During the nineteenth century and well into the twentieth, a great number of its birds were known to science only by dead specimens. European millionaire collectors and aristocratic ladies of fashion were as bedazzled by the birds as the kings of the orient had been and paid huge prices for unusual specimens. Scholars, working in their cabinets of curiosities and struggling to make sense of the astonishing but crumpled and bedraggled feathers they wonderingly unpacked, did their best to imagine how such improbable plumes could be used in display. Understandably, they made mistakes in their interpretations and some of the spectacular plates that were drawn to illustrate their monographs were wildly inaccurate.

The authors and the illustrator of this book, however, are no cabinet naturalists. They have collectively spent many years travelling through New Guinea, sitting in hides in the drenching rain waiting for birds to appear, climbing trees in search of nests, meticulously observing and noting the details of the birds' behaviour. Between them, they have recorded many details of the life cycle of these birds for the first time. Yet there are still species whose nests have not yet been described and whose displays have yet to be scientifically recorded. Even so, this book is certainly the most complete account yet written and the most comprehensively illustrated of this amazing family of birds and is likely to remain the authoritative work of reference for a long time to come. Its appearance will be a cause for celebration by those who are already fascinated by these wonderful birds, and a revelation to those who have not, as yet, fallen under their spell.

David Attenborough

Preface and acknowledgements

The authors of this work were born on opposite sides of the North Atlantic Ocean, and grew up watching titmice and chickadees, thrushes and robins, firecrests and kinglets, orioles and blackbirds, crossbills and grosbeaks, respectively. As our ornithological interests grew we were amazed to learn of the existence of the birds of paradise—a family of astounding beauty and mystery ensconced in an obscure corner of the globe—the southwest Pacific. We independently set about getting to study them in the museum, in the zoological park and in the field. For CF, that work began in 1968, and for BB it started in 1975. We came to know of each other, naturally, because of our mutual obsession, but we corresponded for years before actually meeting face-to-face at Ambua Lodge, overlooking the marvellous Tari Basin, Papua New Guinea. By that time we were already 'old friends'. That's one of the wonders of a career in the natural sciences.

The idea of a joint work on birds of paradise was first raised by CF in a letter to BB in early 1989. As plans for creation of the Oxford University Press *Bird Families of the World* series unfolded, BB suggested that our book plan be offered to Oxford, which we happily did. We now know how much work was hidden behind such a proposal. BB can attest to the fact that this long effort was mostly CF's, not so much because of lack of desire by BB, but simply because of competing commitments.

In order to gather as much of the unpublished and widely-dispersed biological information together as possible for this work CF made a world study tour of larger institutional collections (see below). This world tour, of five months duration, provided the baseline dataset that made such a work possible. As with the museum tour, in many other tasks that faced us in the production of the manuscript and art for this book, we were aided by many people generous with their time and knowledge, whom we would like to thank here. In addition, a number of colleagues who have performed fieldwork on birds of paradise have made unpublished information available to us. As a result, this volume contains a good deal that has not been previously brought together in a review of the birds of paradise. If we have overlooked anyone (almost an inevitability) who has helped in some way we do apologise. Because of the need to economise space, we have dispensed with titles and positions. Abbreviations used elsewhere (see Abbreviations) are used here.

First, we owe the greatest debt of thanks to Dawn Frith for expert long-term collaboration and unwavering assistance with numerous and often tedious aspects of data gathering, its initial analysis, and valued discussion. In addition to accompanying CF on the world tour of museums to document paradisaeid collections, she undertook the sorting and synthesis of these museum data and myriad measurements into usable form.

We also particularly thank William T. (Bill) Cooper for his magnificent artwork, knowledge, patience, and interest. We thank Mike McGuire for his skilled work generously freely given to the production of the sonograms. Sir David Attenborough's life-long personal interest in the birds of paradise has stimulated the present authors and many others to appreciate these birds and the fragile natural world they inhabit, and we deeply appreciate his Foreword. In addition, he has kindly provided copies of rare literature, artwork, and most valuable video footage of bird behaviour for our analysis.

For scientific guidance, as well as generous permission to study ornithological collections in their care we sincerely thank the following persons and institutions: Dean Amadon, Chris Blake, Walter Bock, Joel Cracraft, Mary LeCroy, Manny Levine, Lester Short and The Frank M. Chapman Memorial Fund, AMNH, New York; Robert Prys-Jones, Michael Walters, and Peter Coulston, Ornithology, BMNH, Tring; Storrs Olson, Gary Graves, Richard Zusi, Phil Angle, Jim Dean, and Carla Dove, USNM, Washington, D.C.; Ned Johnson, MVZ, University of California, Berkeley; David Willard, FMNH, Chicago; Raymond A. Paynter Jr., MCZ, Cambridge; Allen Allison and Carla Kishinami, BPBM, Honolulu; James Dick and Brad Millen, Royal Ontario Museum, Toronto; Frank Gill, ANSP, Philadelphia; James M. Loughlin, CMNH, Pittsburgh; Fred C. Sibley, PMNH, Yale University, New Haven; Gene K. Hess, Delaware Museum of Natural History, Wilmington; Grey Budney, Cornell Laboratory of Ornithology, Ithaca; Mohammad Amir and Darjono, MZB, Bogor; Frank Bonaccorso, Ilaiah Bigilale, and Paul Wanga, National Museum and Art Gallery of Papua New Guinea, Port Moresby; Clem Fisher, National Museums and Galleries, Liverpool; Josefina Barreiro, Museo Nacional de Ciencias Naturales, Madrid; Giuliano Doria, MCSN, Genoa; René Dekker and Peter Van Dam, RMNH, Leiden; Per Ericson, Gunnar Johansson, Göran Frisk, SMNH, Stockholm; Jon Fjeldså, Zoologisk Museum Københavns Universitet, Copenhagen; Eric Pasquet, MHN, Paris; Siegfried Eck, SMT, Dresden; Josef H. Reichholf, ZSMM, Munich; Claudio Pulcher, Museo Regionale di Scienze Naturali, Torino; Josefina Barreiro, Museo Nacional de Ciencias Naturales, Madrid. Prof. B. Stephan, ZMB. R. van den Elzen, ZFMK, Bonn for kindly forwarding skins to Frankfurt. Claus König, SMNS, Stuttgart, for kind hospitality in addition to access to collections. D. S. Peters, Prof. Dr. Steinbacher, Martina Küsters, and Karin Böhm, ZSMF, Frankfurt; H. Hoerschelmann, ZIZM, Hamburg for kindly sending skins to Frankfurt; Ernst Bauernfeind, NMW, Wien for kindly providing comprehensive label data.

Richard Schodde, Ian Mason, and John Wombey, CSIRO, Canberra; Walter Boles and Wayne Longmore, AM, Sydney; Les Christidis and Rory O'Brian, MV, Melbourne; Philippa Horton, SAM, Adelaide; Steve Van Dyck, Glen Ingram, and Carden Wallace, QM, Brisbane; Glen Storr, Ron Johnstone, WAM, Perth; David Parer, Dionee Gilmore, Richard Campbell, ABC, Melbourne; Richard Major, Australian Nest Record Scheme; Barry Baker, Belinda Dettmann, ABBBS, Canberra; Stephen Marchant and Ian Rowley as editors of *The Emu*.

For kindly permitting access to data held by the ABBBS, obtained during their efforts in banding live birds of paradise, we thank R. and M. Mackay, W. Peckover, S. and M. Pruett-Jones, the late H. Bell, P. Driscoll, K. and L. Fisher, T. and I. Weston, T. Pratt, J. Hardy, Townsville Bird Group, E. Lindgren, L. Filewood, L. Lamothe, A. Jones, R. Campbell, and others.

The following people, institutions, governments, and publications kindly provided permission to reproduce here text or illustrations, as credited in the body of the work: Walter Bock, The British Library National Sound Archive, Mary Clench, Brian Coates and Dove Publications, Christopher Healey, Stephen and Melinda Pruett-Jones, American Ornithologists' Union (*The Auk*), British Ornithologists' Union (*Ibis*), Cooper Ornithological Society (*The Condor*), P. Paiva, Papua New Guinea Philatelic Bureau, Republic of Indonesia, Royal Australasian Ornithologists' Union (*Emu*), Springer-Verlag (*Behavioural Ecology and Sociobiology*), and the University of California Press.

Brain Coates provided valuable discussions and shared his considerable expertise. Margit

Preface and acknowledgements xi

Cianelli, Derek Goodwin, Norbert and Christine Lenz, and Ralph Keller kindly translated several papers from German. Anders and Dagny Mhry kindly translated Bergman's book *Mina Paradisfåglar* from Swedish. Marion Buchanan kindly transcribed ROM bird of paradise specimen data prior to CF's world study tour.

Other people who helped by providing unpublished data, film, literature, recordings, map production, translations, observations, thoughts, ideas, discussions, encouragement and support, and enthusiam in various ways include: David Attenborough, Magnus Aurivillius, Demianus Bagali, (K.) David Bishop, Walter Boles, Jeffery Boswell, Les Christidis, Margit Cianelli, Brian Coates, Dan Cole, Bill and Wendy Cooper, Ann Datta, Jared Diamond, John Dumbacher, Dawn Frith, Errol Fuller, David Gibbs, David Gillison, Derek Goodwin, Andreé Griffin, Steve Hamilton, Colin Harrison, Anthony Hiller, Glenn Holmes, Paul Jepson, Dick Kimber, Richard Kirby, Mark Laska, Mary LeCroy, Roy and Margaret Mackay, Richard Major, Justin Marshall, Michael McGuire, Helmut Michi, Ken de la Motte and Khin May Nyunt of Jurong Birdpark, Elizabeth Nash, Mark Norrie of Bali Bird Park, David Parer, Bill Peckover, Yan Persolessy, Graeme Phipps, Gary Opit, Leslie Overstreet, Mike Potts, Bent Poulson, Michael Poulsen, Thane Pratt, Steve and Melinda Pruett-Jones, David Purmiasa, Paul Reddish, Pauline Reilly, Dieter Rinke, Ian Rowley, Eleanor Russell, Richard Schodde, Peter Shanahan, David Snow, Raimund Spect, the late Ray Swaby, Keith Turner, Klaus Uhlenhut, Carlo Violani, Michael Walters, Effie Warr, Norman Wettenhall, Richard Whiteside, John and Junell Young.

The following persons kindly read and commented on the chapters of the work indicated: David Bishop, Dawn Frith, Mary LeCroy, and Angela Turner all chapters; Stan Breeden chapter 2; Bill and Wendy Cooper 2; Errol Fuller 4; Derek Goodwin 1, 2; Graham Harrington 1, 2; Amy Jansen 1, 2, 3, 8; Roy Mackay 1, 2, 8; Thane Pratt 3–7; Jared Diamond 3, 5, 7, 8; Stephen Pruett-Jones 4, 5; Pamela Swadling 2, 8; Kathie Way 1–8.

The following people kindly read one or more of the species accounts: Brian Coates, Jared Diamond, Glenn Holmes, Paul Jepson, Andy Mack, Michael Poulsen, Thane Pratt, Stephen Pruett-Jones, and Richard Whiteside. David Attenborough, Jeffery Boswell, Richard Kirby, Mike Potts, and Paul Reddish of the British Broadcasting Corporation, Natural History Unit, Bristol, John and Junell Young and Thomas Schultze-Westrum provided copies of scientifically-valuable film and sound of birds of paradise. Cornell Laboratory of Ornithology, Mike McGuire, Michael Poulsen, Christoper Roberts, S. W. Smith, David Stewart, Richard Whiteside, and the BBC Natural History Unit kindly provided recordings of wild bird calls for sonogram production. Additional recordings were obtained from Richard Ranft, National Sound Archive, The British Library. The numerous quality sound recordings of David Bishop, generously freely provided, were particularly valuable.

People who kindly provided hospitality, field assistance and company, interest, and help in other ways include: Joan Airey, Allen Allison, Akia Aruah, David Attenborough, Peter and Jan Barter, Carol Beehler, David Bishop, Ellie Brown, Ian and Barbara Burrows, Brian and Del Coates, Bill and Wendy Cooper, Jeff and Barbara Davies, Sam Dibella, James and Rosemary Dingwall, Joe Dumoi, Pauline Duncan, Anne Dybka, David and Ruth Evans, Brian Finch, Peggy, Paul and Luisa Frith, Peter and Daphne Fullagar, Andrew and Trish Gillison, Rodney Goga, Derek Goodwin, Andreé Griffin, Colin Harrison, Mike and Helen Hopkins, Iamo Ila, Peter Kaestner, Ninga Kawa, Guy Kula, Norbert and Christine Lenz, David Lupke, Roy and Margaret Mackay, Stephen Marchant, Peter and Janet Marsack, Media Mediu, James Menzies, Eddy Mills, Paddy Osborne, Mary Pearl, Thane and Linda Pratt, Lee Rocke, Ian Rowley, Eleanor Russell, Agnes Safford, Harry Sakulas, Geoff Smith, Michael Lucas Simu, Joseph Tano,

Tom and Susan Tomasso, Andrew Taplin, Meg Taylor, Bill Timmis, David and Barbara Snow, and Michael Yamapao.

A special debt of thanks is due to Bob and Pam Bates of Ambua Lodge and Trans Niugini Tours for their interests in promoting and supporting studies of birds of paradise at Tari Gap, and the Frank M. Chapman Memorial Fund, Department of Ornithology, AMNH, which granted funds supporting work in New York. The Wildlife Conservation Society provided support to both CF and BB for field studies at Tari Gap, English Peaks, and the Lakekamu Basin. Conservation International, the MacArthur Foundation, the Smithsonian Institution, and the National Geographic Society supported field research by BB in various sites in PNG.

We thank the series editors Professors Christopher Perrins, Walter Bock, and Jiro Kikkawa for their constructive guidance and criticism. Judith May, formerly Senior Editor, and the staff of Oxford University Press have provided much appreciated encouragement and sympathetic support. We thank OUP for publishing this innovative series and for the opportunity to contribute to it.

Malanda, North Queensland   C. B. F.
Bethesda, Maryland   B. M. B.

Contents

List of colour plates	xvi
List of abbreviations	xvii
Plan of the book	xix
Diagrams of bird topography	xxv
Diagrams of bird of paradise structural morphology	xxvi
Map showing some more important locations mentioned in the text	xxviii
Boundaries of the regional maps used in the species accounts of Chapter 9	xxx

PART I *General chapters*

1	The incredible birds of paradise—an introduction	3
2	Discovery of the birds of paradise and history of their study	29
3	Evolution and biogeography of the birds of paradise	47
4	Ecology of the birds of paradise	78
5	Reproductive behaviour	102
6	Nesting biology and parental care	123
7	Birds of paradise in human tradition and culture	143
8	Conservation	154

PART II *Family, subfamily, genus, and species accounts*   169

9	Family **PARADISAEIDAE**	42 species in 17 genera	171
	Subfamily **Cnemophilinae**	3 species in 2 genera	174

Genus ***Cnemophilus***		175
(Subgenus *Loria*)		176
Loria's Bird of Paradise	*Cnemophilus loriae*	176
(Subgenus *Cnemophilus*)		183
Crested Bird of Paradise	*Cnemophilus macgregorii*	183
Genus ***Loboparadisea***		189
Yellow-breasted Bird of Paradise	*Loboparadisea sericea*	190

Subfamily **Paradisaeinae**	39 species in 15 genera	195
Genus ***Macgregoria***		197
Macgregor's Bird of Paradise	*Macgregoria pulchra*	198

Genus ***Lycocorax***		204
Paradise Crow	*Lycocorax pyrrhopterus*	205
Genus ***Manucodia***		210
(Subgenus *Manucodia*)		210
Glossy-mantled Manucode	*Manucodia atra*	211
Jobi Manucode	*Manucodia jobiensis*	217
Crinkle-collared Manucode	*Manucodia chalybata*	220
Curl-crested Manucode	*Manucodia comrii*	224
(Subgenus *Phonygammus*)		229
Trumpet Manucode	*Manucodia keraudrenii*	229
Genus ***Paradigalla***		241
Long-tailed Paradigalla	*Paradigalla carunculata*	242
Short-tailed Paradigalla	*Paradigalla brevicauda*	244
Genus ***Astrapia***		249
Arfak Astrapia	*Astrapia nigra*	250
Splendid Astrapia	*Astrapia splendidissima*	253
Ribbon-tailed Astrapia	*Astrapia mayeri*	257
Stephanie's Astrapia	*Astrapia stephaniae*	266
Huon Astrapia	*Astrapia rothschildi*	273
Genus ***Parotia***		277
Western Parotia	*Parotia sefilata*	277
Lawes' Parotia	*Parotia lawesii*	283
Wahnes' Parotia	*Parotia wahnesi*	292
Carola's Parotia	*Parotia carolae*	298
Genus ***Pteridophora***		304
King of Saxony Bird of Paradise	*Pteridophora alberti*	305
Genus ***Ptiloris***		314
(Subgenus *Craspedophora*)		315
Magnificent Riflebird	*Ptiloris magnificus*	315
(Subgenus *Ptiloris*)		327
Paradise Riflebird	*Ptiloris paradiseus*	327
Victoria's Riflebird	*Ptiloris victoriae*	334
Genus ***Lophorina***		345
Superb Bird of Paradise	*Lophorina superba*	345
Genus ***Epimachus***		356
Black Sicklebill	*Epimachus fastuosus*	357
Brown Sicklebill	*Epimachus meyeri*	366
Genus ***Drepanornis***		376
Buff-tailed Sicklebill	*Drepanornis albertisi*	377
Pale-billed Sicklebill	*Drepanornis bruijnii*	385

Genus ***Cicinnurus***		390
(Subgenus *Diphyllodes*)		391
Magnificent Bird of Paradise	*Cicinnurus magnificus*	391
Wilson's Bird of Paradise	*Cicinnurus respublica*	401
(Subgenus *Cicinnurus*)		407
King Bird of Paradise	*Cicinnurus regius*	407
Genus ***Semioptera***		417
Standardwing Bird of Paradise	*Semioptera wallacii*	417
Genus ***Seleucidis***		427
Twelve-wired Bird of Paradise	*Seleucidis melanoleuca*	428
Genus ***Paradisaea***		438
Lesser Bird of Paradise	*Paradisaea minor*	439
Greater Bird of Paradise	*Paradisaea apoda*	448
Raggiana Bird of Paradise	*Paradisaea raggiana*	456
Goldie's Bird of Paradise	*Paradisaea decora*	470
Red Bird of Paradise	*Paradisaea rubra*	475
Emperor Bird of Paradise	*Paradisaea guilielmi*	482
(Subgenus *Paradisornis*)		488
Blue Bird of Paradise	*Paradisaea rudolphi*	488

Appendices
1. The hybrid birds of paradise.   499
2. Exploration of Australasia and the study of birds of paradise—an annotated list.   521
3. Results of an examination of 4852 museum skin specimens, with a locality and date, of collections for obvious signs of active moult.   552
4. Summary of records of plant species recorded to be eaten by birds of paradise, from the literature and unpublished sources.   554
5. Some published recordings of bird of paradise vocalisations.   560
6. A brief guide to where and how best to study wild birds of paradise.   561
7. Gazetteer [of more important locations mentioned in the text with their geographical coordinates].   566

Glossary   573
Bibliography   580
Index   607

Colour plates

Colour plates fall between pages 174 and 175

Plate 1   Nineteenth century bird of paradise illustrations (1801–98)
Plate 2   Some bird of paradise habitats—subalpine to lowland forest
Plate 3   Nidification and moult of some birds of paradise
Plate 4   The wide-gaped birds of paradise
Plate 5   The sexually-monomorphic species of the Paradisaeinae
Plate 6   The astrapias
Plate 7   The parotias
Plate 8   The riflebirds and Twelve-wired Bird of Paradise
Plate 9   The sicklebills and sabretails
Plate 10 The King of Saxony, Superb, and sickle-tailed Birds of Paradise
Plate 11 The typical *Paradisaea* birds of paradise
Plate 12 The Standardwing and less typical *Paradisaea* birds of paradise
Plate 13 Some eggs of some bird of paradise genera
Plate 14 Ten presumed intergeneric and one intrageneric hybrid birds of paradise
Plate 15 Five presumed intergeneric and six intrageneric hybrid birds of paradise

Abbreviations

ABBBS	Australian Bird and Bat Banding Scheme		Industrial Research Organisation, Canberra
ABC	Australian Broadcasting Corporation	CYP	Cape York Peninsula, Queensland, Australia
ad	adult(s)		
AM	Australian Museum, Sydney	dia	diameter
AMA	Auckland Museum, Auckland	et al.	et alia, and all (additional authors)
AMNH	American Museum of Natural History, New York	♀	female
		♀♀	females
		ft	feet
ANSP	Academy of Natural Sciences, Philadelphia	FMNH	Field Museum of Natural History, Chicago
asl	above sea level		
Aust	Australia	g	gram(s)
BB	Bruce M. Beehler	ha	hectares
BBC	British Broadcasting Corporation	hr	hour(s)
		in litt.	unpublished information received in writing
BMNH	British Museum of Natural History, Tring (The Natural History Museum)		
		ins	inches
		IJ	Irian Jaya, Indonesia
BPBM	Bernice Pauahi Bishop Museum, Honolulu	I., Is	Island(s)
		imm	immature(s)
BRS	Baiyer River Sanctuary, Papua New Guinea	juv	juvenile(s)
		km	kilometre(s)
c.	circa—approximately	m	metre(s)
C	central	♂	male
cf.	confer (consult, see)	♂♂	males
CF	Clifford B. Frith	MCSN	Museo Civico di Storia Naturale, Genova
cm	centimetre(s)		
CMNH	Carnegie Museum of Natural History, Pittsburgh	MCZ	Museum of Comparative Zoology, Harvard University, Cambridge, Massachusetts
CSIRO	Commonwealth Scientific and		

Abbreviations

MHN	Museum National D'Histoire Naturelle, Paris	SAM	South Australian Museum, Adelaide
MV	Museum of Victoria, Melbourne	SD	standard deviation
MVZ	Museum of Vertebrate Zoology, Berkeley	sec	second(s)
MZB	Museum Zoologicum Bogoriense, Bogor	SMNH	Swedish Museum of Natural History
min	minute(s)	SMNS	Staatliches Museum für Naturkunde, Stuttgart
mm	millimetre(s)	SMT	Staatliches Museum für Tierkunde, Dresden
MRSN	Museo Regionale di Scienze Naturali, Torino	sp.	species (singular)
Mt./Mts	Mountain(s)	spp.	species (plural)
NG	New Guinea	subsp.	subspecies (singular)
NMW	Naturhistorisches Museum Wien, Vienna	subspp.	subspecies (plural)
		THL	total head length
		UK	United Kingdom
NSW	New South Wales, Australia	USA	United States of America
NYZS	New York Zoological Society (now Wildlife Conservation Society)	USNM	United States National Museum of Natural History, Washington
P1	innermost wing primary	WAM	Western Australian Museum, Perth
P10	outermost wing primary	YIO	Yamashina Institute of Ornithology, Tokyo
Penin	Peninsula	yr	Year(s)
PMNH	Peabody Museum, Yale University, New Haven	ZFMK	Zoologisches Forschungsinstitut und Museum Alexander Koenig, Bonn
PNG	Papua New Guinea		
QM	Queensland Museum, Brisbane	ZIZM	Zoologisches Institut und Zoologisches Museum, Hamburg
R.	River(s)		
Ra	Range(s)	ZMB	Museum für Naturkunde der Humboldt-Universität zu Berlin (Zoologisches Museum)
RAOU NRS	Royal Australasian Ornithologists' Union, Nest Records Scheme		
RMNH	Rijksmuseum van Natuurlijke Historie, Leiden (National Museum of Natural History)	ZSMF	Forschungsinstitut und Naturmuseum Senckenberg, Frankfurt
ROM	Royal Ontario Museum, Toronto	ZSMM	Zoologische Staatssammlung, Munchen

Plan of the book

The main aim of this book is to provide or direct the reader to as much information as possible about the birds of paradise in a single volume with particular emphasis on their biology. While the birds of paradise do provide extremely fertile ground for a variety of theoretical biological considerations, in this book we will focus on the birds and not on theory. This said, we attempt to review herein all such interesting aspects of the family with the aim of: (1) making the reader aware of the broader significance of the birds of paradise to general ornithological and biological studies and (2) providing reference to the key literature on these issues. In this regard, E. Thomas Gilliard's (1969) monograph provided the stimulus for the great deal learnt about these birds since its publication. We shall be pleased if the present volume proves as effective a stimulus for future researchers, naturalists, and nature lovers.

Owing to their beauty and the mystery surrounding them, birds of paradise became a focus of many natural history expeditions, which collected specimens for museum study. For this reason the museums of the world contain substantial numbers of bird of paradise skins. Many of the earlier specimens originated in the days of the plume trade and these specimens therefore lack data such as when and where collected, and thus are of little value in the context of this book. Specimens labelled with no more than a locality and a date of collection can be of considerable value, however, as information on distribution, geographical variation, and moult can be obtained. Better-documented specimens provide far more information (see below).

In our museum work, we passed over specimens whose labels indicated only 'New Guinea' or similarly imprecise locations. We did, however, record label data such as location and/or collector and date of collection, for distributional and historical use, even if the specimen was not measured. In dealing with museum skins, time permitted only looking for grossly obvious signs of moult in the wing, tail, and body plumage. Plumage was not scrutinised for body moult, this being noted only if readily apparent during handling and measuring the material in collections (see Appendix 3).

All available unpublished data for live birds of paradise caught and examined by ornithologists were obtained from the Australian Bird and Bat Banding Scheme (ABBBS) and from individual bird ringers. As a result, many new data, such as live bird weights, records of moult activity, and colour of bare parts, have been included in the species accounts. These are, however, almost exclusively for species and populations occurring within Papua New Guinea, where bird ringing has been carried out over 25 years or so.

A number of common names of birds of paradise are long, some excessively so, owing to the suffix 'Bird of Paradise'. In order to save space in the main body of text we do not use this. Thus we abbreviate to: 'the Blue', 'King of Saxony' or 'the Blue Bird' or 'King of Saxony Bird'. We use Wahnes', (instead of Wahnes's) and Lawes' (instead of Lawes's) Parotia as most recent authors do and because of their convenience. Also to save space, we refer only to a place name followed by the country abbreviation, omitting the wordy provinces as many of the more significant places are indicated on the maps (see page xxviii) or are listed in Appendix 7. Where we quote other authors, notably in display descriptions, we use the abbreviations standard in this volume, both to save space and for consistency.

PART I

PART I presents eight thematic chapters reviewing what is known of the birds of paradise and in places speculating on remarkable features of their appearance or life histories. *Chapter 1* aims broadly to introduce this family of birds to the reader. *Chapter 2* provides a brief history of discovery and study of the group, accompanied by references for further reading on the former, and providing more scrutiny of the latter, especially related to the more recent acquisition of biological knowledge of the birds.

Chapter 3 discusses the origins and evolution of the birds of paradise as well as their systematics. It also reviews some of the numerous theoretical considerations given to various aspects of the biology of the family. *Chapter 4* reviews the ecology of the birds, with special attention to diet, foraging, and bird–plant interactions.

Chapter 5 provides insights into how the elaborate adult male plumage structures of many species are used in display and how body movements enhance their presentation. The often complex, always highly ritualised, courtship displays are described and are discussed in the context of mating systems, sexual selection, display sites, and their significance to understanding relationships within the family. This discussion is much enhanced by the numerous text drawings of display to be found within the species accounts constituting *Chapter 9*.

The nesting biology is treated in *Chapter 6*—predominantly based on knowledge gained since the publication of Gilliard (1969). These new findings prove particularly exciting in relation to some of the insights they provide. While a book could be written on the broad and fascinating subject of *Chapter 7* (birds of paradise in human tradition and culture), we attempt to provide an overview with reference to the more significant literature. Finally, *Chapter 8* reviews the conservation status of the family and its member taxa as well as potential threats to some habitats and species. It represents a contemporary status report.

PART II

PART II or *Chapter 9* presents comprehensive species accounts in the systematic order of the family adopted for this work. The various aspects of the descriptions, biometrics, nomenclature, and biology of each species are presented in a uniform way as follows. Scientific names of birds other than birds of paradise are given only at their first mention.

Standard common and scientific names are followed by the author and reference to the original description. Reference is given to illustrations or maps related to the account. This is followed by a list of other names applied to the species. **Species introduction** briefly provides a thumb-nail introductory description of the species.

Description details what the species looks like, and follows the order Ad ♂, Ad ♀, Subad ♂, Imm ♂, Nestling–Juv. In the vast majority of species, 'immature' denotes males in wholly female plumage (as opposed to 'subadult males', which are in female plumage with some male plumage intruding). Total length of the bird from bill tip to tail tip is approximate only, based on specimens, and is for the **nominate subspecies only** in the case of polytypic species. Where it differs between the sexes it is also indicated for the female. Where a species has modified, elongate, central tail feathers or other plumes that extend beyond the normal tail structure the total length from bill tip to the tip of the modified plumes as they sit normally on the living bird (curved ones not being straightened) are also given (in parentheses).

Descriptive terms: such as nuchal bar, ear tufts, flank plumes, pectoral plumes, occipital plumes, etc. and those more typically used for bird topography are defined by example in the figures on pages xxvi–xxvii.

Plumages: note that there are quite possibly subtle differences between the plumages of immature male and female sicklebills, riflebirds, Superb Birds, and others to be discerned and described but such work was beyond the scope of the research performed for this book. Several recent works provide reasonably detailed, but nevertheless far from comprehensive, descriptions (e.g. the underwing and ventral surface of tail are not described) for most to all bird of paradise species (Rand and Gilliard 1967; Gilliard 1969; Cooper and Forshaw 1977; Coates 1990). The present work is therefore the place to present new thorough descriptions in 'handbook' style with respect to detail and objectivity. These descriptions are detailed and should provide answers to most researchers' questions about the appearance of each plumage beyond what appears in the colour plates.

Plumages and colour: plumage descriptions are of the **nominate subspecies** of each species and were assessed and described in good diffused natural light. To provide standardisation and objectivity to the plumage descriptions we refer to the numbers ascribed to colours by Smithe (1975) with his colour nomenclature indicated by capitalisation. We only use his names when appropriate and additionally use simple colour names that will be widely understood if only in general terms. If our own colour name is used, the number only of the colour in Smithe most like it appears in parenthesis. To avoid repetition, the colour number of Smithe is given only once in each plumage description unless a colour name has two or more Smithe numbers (i.e. Cinnamon and Buff). If only the one Cinnamon is referred to several times in the same text, however, only the number first cited applies. Greenish-blue means more blue than green, the second mentioned colour always being the dominant one. Structural characters readily apparent in the plates, such as a slightly-decurved bill, are not necessarily mentioned in the plumage descriptions. Colours of bill, iris, legs, and feet, etc. are given for female or age classes only if they differ significantly from those of adult males.

Decisions concerning the preparation of plumage descriptions and biometrics of some species are often extremely difficult. For example, it is useful to be able to compare descriptions and biometrics of adult male *Paradisaea* species with those of subadult males. The latter, however, include some individuals with fully-developed, elongate, central tail 'wires', some with such wires terminating in extensive, spatulate, vaned tips (see Plate 12) and others with elongate, narrow, but normally feather-vaned, pointed, central rectrices.

In describing structural characteristics of the species we pay special attention to those that are difficult to discern in the plates. Few aberrant plumages are known in the birds of paradise and consist of little more than a handful of specimens with part to near-complete albinism and a couple of individuals showing partial to complete replacement of normal adult male plumage with an odd 'fawn' one (Frith 1998). These are detailed following the descriptions of normal plumage.

Distribution outlines the geographical range of the species and complements a species distribution map. The maps illustrate each species' range by showing point localities (collections, sightings) for which we have geographical co-ordinates except for species confined to the islands (often printed too small for this kind of treatment). The islands involved are therefore shaded and emphasised by arrows. The reader should note that the island-dwelling birds involved can be expected to occur only within suitable altitudes and habitats. The point distributions are generated from this database of paradisaeid distributional records using a simple BasicCad mapping program in DesignCad 5.0. These are overlaid upon a base map adapted from MundoCart software mapfiles. Note that it was impossible for us to locate the co-ordinates for some known species localities. A purely-hypothetical species' range, the presence of suitable altitudes and habitats permitting, can be visually assessed from the plotted locations. We stress that these are merely based on the plotted data and that the species could only be expected within these hypothetical ranges where extensive suitable habitat occurs. Species distribution maps show only the relevant areas. Their geographical locations and limitations are shown on page xxx.

Systematics, nomenclature, subspecies weights, and measurements opens with a brief summary of the present systematic status of the species where applicable. Known hybridisation is indicated by listing other birds of paradise species involved only, as this topic is dealt with briefly at the end of Chapter 3 and fully in Appendix 1. The etymological derivation of the genus is dealt with under the first species of each genus only, and the species name only in species accounts thereafter.

The original species description citation and selected synonyms are given if applicable. Type specimens examined or located in the literature are detailed but where 'not located' is indicated this may mean no type was designated or that the type is an illustration and not a specimen. In listing subspecies in the systematics section of the species accounts we give the nominate subspecies first and then follow this in a geographical order (i.e. NW to SE). Any subspecific difference from the appearance of the nominate form is noted under the subspecies diagnosis. Subspecies range is also given.

Measurements and weights: measurements (by CF) are provided by sex and age group; we attempted to provide means and ranges based on a sample of 25, where available (see Frith and Frith 1997*b*). In the case of all manucode species, however, non-adult plumages are difficult to differentiate and thus are lumped. The size of some biometric samples for some taxa (e.g. *Cicinnurus regius*) are larger than 25 because of subspecific lumping subsequent to the data collection. All weights presented have been rounded-off to the nearest gram.

Wing length was measured as the straightened, flattened, maximised length (the wing arc). Tail length is the maximum length from the point of insertion of the longest tail feather into the skin to its tip. Central tail feather lengths are given only when these rectrices are 5 mm longer or shorter than the other rectrices. In manucodes the tip is often the tip of a narrow fine point to the centre of the feather vane. Total head length (THL) is the maximum distance from the tip of the upper mandible to the back of the skull (skull + bill length). This was taken from museum specimens only if the rear skull and upper mandible tip were undamaged. No skin considered inadequate was measured but THLs cannot be considered utterly accurate, owing to inevitable variation in skin preparation. Thus a small percentage of THL measurements may be slightly too short, owing to partial skull removal, rather than too long. Bill length is from the union of the bill with the foreskull to the tip of the upper mandible. Bill width was taken at the anterior extent of the nostril and should not be confused with the width of the gape.

Pectoral flank plume length is the average length of those plumes projecting beyond the tip of the normal tail (i.e. excluding the central pair of rectrices) as an approximate guide to relative plume development only. Measurements of modified central pairs of rectrices and of pectoral plumes are minimum lengths only as these may have been broken or worn, respectively.

As modified central tail feathers in adult males are typically and conspicuously elongate we present measurements of them in addition to those for the rest of the normal tail length, particularly as the proportionate length of the central pair to the remainder of the tail can be useful (Frith and Frith 1997*a*). This convention can present problems, however, as in the case of the Huon Astrapia in which the regular and gradual graduation of the tail means it is not possible to differentiate the central pair of tail feathers from the rest. In most cases no such problem exists. Measurements of specific plumes (i.e. occipital plumes of parotias) may be found in this section of the species account.

A discussion of the biometrics contained herein, with particular reference to subspecies, appears in Frith and Frith (1997*b*). In the latter paper the bird of paradise biometrics are comprehensively presented in tabular format, and are therefore more readily comparatively available for workers than in the present volume.

All egg maximum lengths and breadths were measured with digital callipers to the nearest 10th of a millimetre.

Abbreviations for New Guinea, Papua New Guinea, and Irian Jaya may be used (e.g. after a type locality) although at the time of the original description the political entity in question may have been what was then British, German, or Dutch New Guinea.

Habitats and habits details the environments that the species inhabit as well as miscellaneous information on behaviour that does not naturally fall into the other account sections. For a summary overview of habitat selection by the species, refer to Table 4.1. **Diet and foraging** reviews the data available on diet and foraging behaviour, supplemented by summary data in Table 4.2 and Appendix 4. Nestling diet is detailed under Nestling care. **Vocalisations and other sounds** details the known range of sounds produced by the species and their apparent significance. Where adequate tape recordings of vocalisations were available sonograms have been prepared and are presented in the species accounts. To avoid repetition and to enable **sonograms** to appear larger herein we do not indicate on each that numbers on the vertical (left) axis indicate frequency in kilohertz and on the horizontal (bottom) are time in seconds. Data files of vocalisations were made using Soundblaster Pro hardware (Creative Labs Inc.) on an IBM compatible PC running Microsoft Windows 95. Sonograms were made using Avisoft Sonagraph Pro software by Raimund Spect of Berlin whose generous support is appreciated. Corel Draw was used to remove most extraneous noise including vocalisations of other bird species. The digitising rate and other parameters were selected to optimise the sonogram displays (M. McGuire personal communication). **Mating system** is indicated and discussed, when adequately known. **Courtship behaviour** delineates advertisement and displays; these tend to be detailed, specific, and precise. As these are subsequently often misquoted by authors, or even misinterpreted, we make a point of quoting such text verbatim where we deem it appropriate in order to avoid misrepresentation and/or misinterpretation. **Breeding** descriptions follow the standard order Nest site, Nest, Eggs, Incubation, Nestling care, and Nestling development.

Annual cycle summarises all seasonal aspects of the biology of the species, with reference to timing of Display, Breeding, Egg-laying (see Table 6.5), and Moult (see Appendix 3). As so little is recorded about moult, we are bound to present recorded moult activity for each species as a whole. This is less than satisfactory, as a species distribution will often cover an extensive geographical and/or altitudinal range, but to reduce the sample further in any way would be fruitless. In species/subspecies not well represented in collections, the majority of specimens measured will have been examined for moult. In the case of species common in collections, such as Glossy-mantled Manucode, many (those not measured) will not have been examined for moult as time did not permit. Moult that was recorded is summarised in Appendix 3. **Status and conservation** treats the current conservation status of the species and any potential threats; these data are summarised in Table 8.1. **Knowledge lacking and research priorities** provides a brief list of the most interesting biological questions that remain unanswered about the species—subjects for future research. **Aviculture**: the cnemophilines excepted, the birds of paradise have great avicultural potential, as they are hardy birds that take well to captivity and are not difficult to breed given appropriate conditions. While always to be highly restricted and controlled, true aviculture (i.e. reproduction) of these birds in captivity should be encouraged, as much can be learned about them in so doing. Published and some unpublished aviculture experience is therefore summarised here. **Other**: is where we present odd information about a species not appropriate under the above headings.

Colour plates: the first colour plate shows some nineteenth century bird of paradise illustrations and the second some bird of paradise habitats. Plate 3 consists of photographs of some adult birds of paradise at their nests, eggs, and nestlings in the wild, and an adult male Emperor Bird in moult. The latter are presented not as examples of bird photography but have been selected for maximum biological information content. Thus they show aspects of biology not possible in the artist's colour plates or drawings of behaviour of the birds that may prove of interest or use to readers. The descriptions of juvenile or immature appearances in the captions for Plates 5 to 12 are broad generalisations only, usually based on the known appearance of same in at least one species of each genus, and not necessarily upon documentation.

Appendices: the subject of Appendix 1, the hybrid birds of paradise, provides an in-depth analysis of these little-known plumages, greatly aided by our ability to illustrate them all here and the recent publication of a major work on them (Fuller 1995). Appendix 2 summarises the history of ornithological exploration relevant to birds of paradise. Appendix 3 provides a summary of active moult observed in museum specimens. Appendix 4 is a table indicating plant species involved in observations of fruit eating by the birds of paradise. Appendix 5 presents a list of some published bird of paradise vocalisation recordings in various forms. Appendix 6 provides the reader with a broad guide of where and how to see birds of paradise. Appendix 7 is a gazetteer of more important locations mentioned in the text with their geographical co-ordinates (the map on page xxviii indicates the location of some more frequently-mentioned locations) and Appendix 8 provides a glossary of words and terms used.

Overlooked and new information: while every effort has been made to gather all published, and some previously-unpublished, information concerning the birds of paradise it is inevitable that some has been overlooked. If the reader is aware of any errors herein, published material we have omitted or overlooked, or unpublished knowledge of which we were unaware of, we would be most grateful for it being communicated to us directly or via the publisher.

Plan of the book xxv

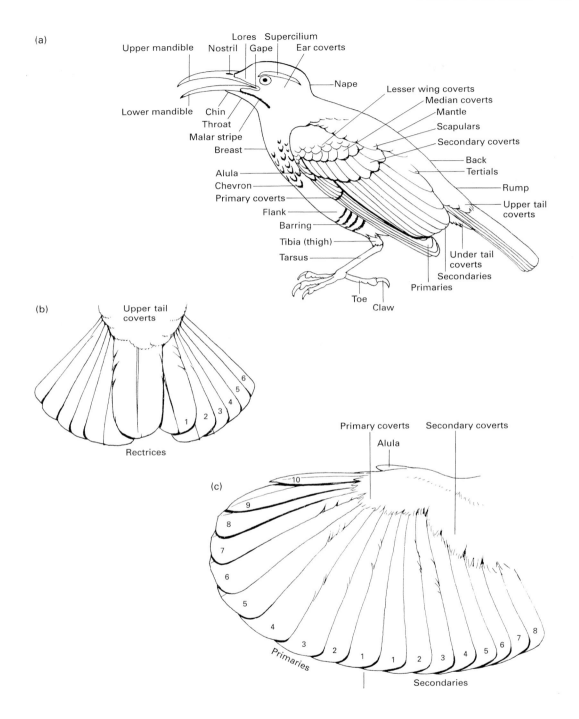

Selected topography and markings of a bird: (a) a female-plumage Victoria's Riflebird; (b) upper surface of a tail and (c) under surface of a wing of an adult male Victoria's Riflebird to show the tail feathers (retrices), uppertail coverts, primaries (P1–10), secondaries (S1–8), underwing coverts and the alula.

xxvi   Plan of the book

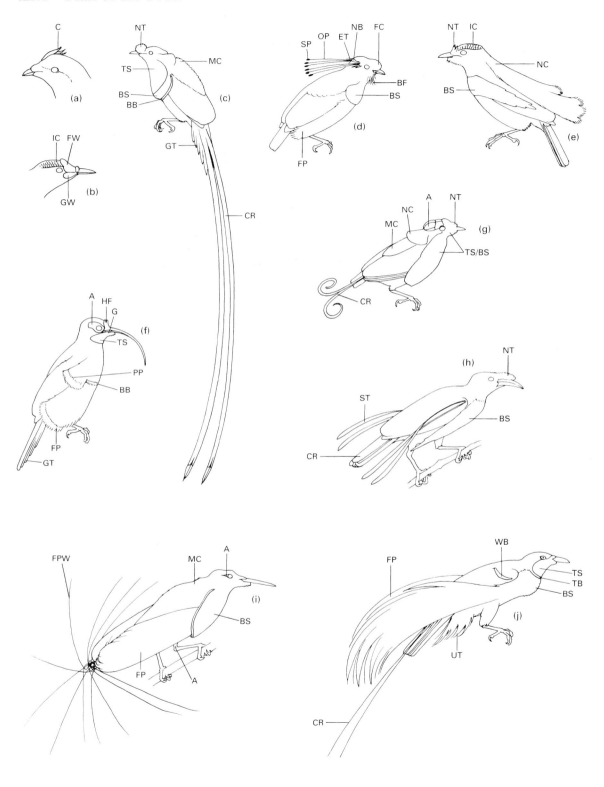

External structural features of adult male birds of paradise with particular reference to diverse yet homologous structures and their nomenclature (see also diagram on page xxv and plates 4–13): (a) Crested Bird of Paradise, (b) Long-tailed Paradigalla, (c) Ribbon-tailed Astrapia, (d) Carola's Parotia, (e) Superb Bird of Paradise, (f) Buff-tailed Sicklebill, (g) Wilson's Bird of Paradise, (h) Standardwing Bird of Paradise, (i) Twelve-wired Bird of Paradise, and (j) Raggiana Bird of Paradise.

Key to abbreviations:

A	= apterium (bare skin)	HF	= horn-like feathers
BB	= breast band	IC	= iridescent crown (discrete and sometimes of scale-like feathers)
BF	= hair-like 'beard' feathers		
BS	= breast-shield	MC	= mantle cape
C	= crest (of fine, central, sickle-shape feathers)	NB	= nuchal bar, or crest
CR	= central rectrices (elongated 'wires', 'sickles' and 'ribbons' or shortened and modified)	NC	= nape cape
		NT	= naral tuft
		OP	= occipital plumes
ET	= ear tuft	PB	= pectoral band
FC	= frontal crest	PP	= pectoral (fan) plumes
FP	= flank plumes	SP	= spatulate (tipped)
FPW	= flank plume wires	ST	= standard (modified lesser wing covert)
FW	= facial wattle	TB	= throat band
G	= gape	TS	= throat shield
GW	= gape wattle	UT	= undertail coverts
GT	= graduated tail	WB	= wing bar

Note: Pectoral (fan) plumes are derived from breast band/breast-shield edges.

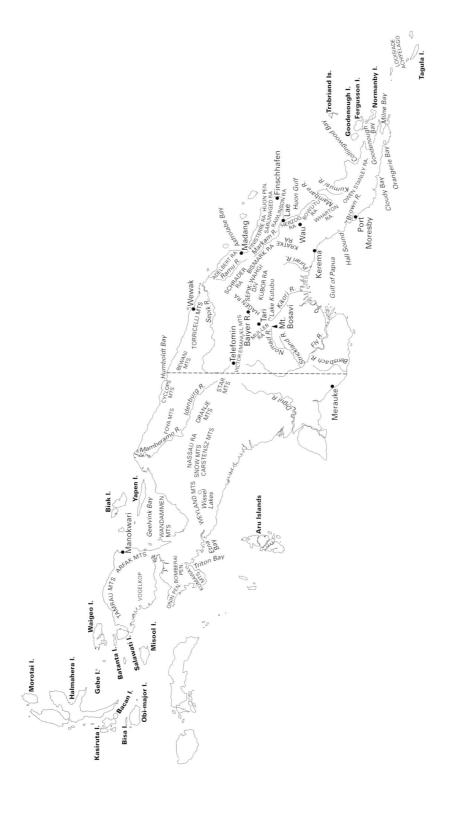

Map showing some more important locations mentioned in the text.

Map showing some more important locations mentioned in the text.

xxx   Plan of the book

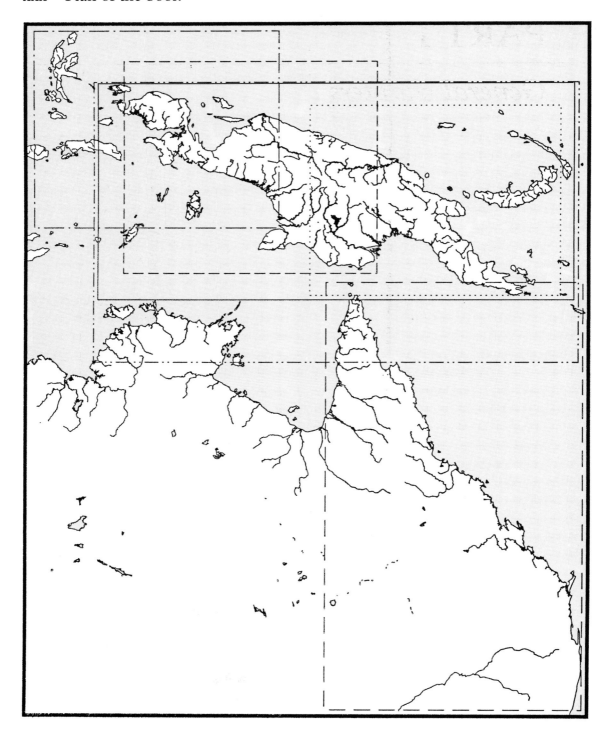

Boundaries of the regional maps used in the species accounts.

PART I

General chapters

1

The incredible birds of paradise—an introduction

In this chapter we provide a concise introduction to the bird of paradise family Paradisaeidae and also highlight the range of issues and topics that will appear in much greater detail in the chapters and species accounts that follow. This introductory discussion largely parallels that of Chapters 2–8, but touches only on the main points of interest about the group, its biology, and its history of discovery and study. In this manner the reader can gain a quick overview of the contents of this book, which will allow a focused follow-up on particular topics of interest.

The birds of paradise and their world

The birds of paradise have fascinated naturalists, scientists, and collectors for the four and a half centuries they have been known to the western world. By contrast, for uncounted millennia these extraordinary birds have served as the centrepiece of myth, ceremony, dress, and dance for a diversity of peoples who have long occupied New Guinea and its neighbouring islands. In this region, adult males of a number of birds of paradise have been traditionally hunted for their lovely nuptial plumes. These plumes are worn as decoration even today by the people of New Guinea, and they continue to be traded within that vast mountainous and forest-clad island and occasionally beyond.

Until the last half century, the West knew little of the life and habits of these birds in the wild, beyond what could be reported by the fieldworkers who collected them for museum and aviary. The rich repositories of knowledge of these birds and their ways were the barefoot hunter-naturalists who not only learned by keen and patient field observation, but also shared an orally-transmitted traditional natural history of their forest resources. Typically, this had been passed on from father to son around a smoky campfire on the trail or else in a men's house in some remote mountain hamlet. The pioneering work of western naturalists in the study of birds of paradise has been immeasurably aided by these indigenous naturalists, with their remarkable understanding of these birds and their environment. This was information shared freely, but it was only occasionally acknowledged by those who wrote of these birds.

It is ironic that, as we prepare this volume, measuring the considerable progress that has been made by western researchers to learn about the birds of paradise in recent decades, much of that Melanesian oral knowledge, and the oral tradition itself, is threatened by changing conditions in New Guinea today, especially because of the overpowering influence of mass culture from the West (not excluding Indonesia). Just as most readers would abhor the thought of the imminent extinction of any one of the remarkable bird species featured here, we would greatly lament the loss of any single traditional Melanesian culture and its oral tradition. For in each of these knowledge systems are many lifetimes of experience and information whose wealth cannot be measured in dollars, rupiah, or kina. Who knows what wonderful new cure or product is within the forests of New Guinea, waiting to be unleashed on the world at large? Part of the wonder of the birds of paradise is their wondrous forest homelands and the amazing people who guard these forest lands for all future generations.

History of discovery and study

The birds of paradise are famous for the wonderful nuptial plumes of the adult males. So bizarre and in some cases un-feather-like are the plumes of several birds of paradise that some early ornithologists were convinced that specimens of certain species were fakes—artefacts crafted to sell to the unwary scientist or naturalist as taxonomic novelties. The spectacular and ornate plumages of these birds in fact represent the extreme expression of the processes of sexual selection.

Notwithstanding the apparent close relationship among its member species, the range of colouration and feather modification is perhaps greater in this family than any other on earth, with the possible exception of the pheasants. Indeed it was the fabulous beauty and extraordinary structures of the first paradisaeidine specimens carried to the western world that so profoundly thrilled both natural historians and the public. The way in which the first bird of paradise specimens reached the western world is recounted in Chapter 2.

There are diverse reasons why the birds of paradise continue to attract the attention of both ornithologists and lay people (Table 1.1). In addition to their remarkable appearance and associated courtship behaviour, there is a certain romance regarding the remote and, even today, relatively-inaccessible habitats of these birds. Moreover, since first discovered by explorers from the western world they have been closely associated with the names of famous and colourful scientists, monarchs, aristocrats, and adventurers, including Charles Darwin, Alfred Russel Wallace, John Gould, Sir Joseph Banks, Queen Victoria, Crown Prince Rudolph of Austria, Prince Charles Lucien Bonaparte—a nephew of Napoleon, Count Tommaso Salvadori, Prince Luigi D'Albertis, the Hon. Walter (later Lord) Rothschild, Sir William Macgregor, Sir William Ingram, Sir Edward Hallstrom, Count Nils Gyldenstolpe, Prince K. S. Dharmakumarsinhji, Captain Neptune Blood, and Sir David Attenborough.

The history of the ornithological discovery and pursuit of birds of paradise in their homelands has been a truly international enterprise, having been directly influenced by the imperial aspirations and colonial intrigues of several European nations and their ornithologists. The partitioning of the great island of New Guinea in 1884 by the Netherlands, Germany, and Great Britain was a turning point in the history of the region, and set the stage for culturally disparate efforts to explore the island and study its natural history. By the 1920s and 30s, New Guinea was in the hands of the Dutch, British, and Australians, all of whom initiated expeditions to study the natural history of their colonial possessions. During this period, the Americans made contributions to the field study of the birds of paradise, and birds in general, as several wealthy benefactors such as Harry Payne Whitney and Richard Archbold financed a series of major zoological collecting expeditions there. Other naturalists involved in the study of the birds of paradise included Frenchmen, Germans, Indians, Indonesians, Irianese, Italians, Japanese, New Zealanders, Papua New Guineans, Portuguese, Spanish, and Swedes.

———

The first significant compilation of knowledge about birds of paradise was by the Frenchmen Vieillot and Audebert published in 1802, but it contained little more than illustrations of 11 species. Since then, monographs on the family have been produced by Lesson, Gould and Sharpe, Elliot, Iredale, Gilliard, and Cooper and Forshaw (see Bibliography). The last au-

Table 1.1 Characteristics found in the birds of paradise.

Great size variation in the family (16–96 cm in total length)
Diverse examples of extraordinary nuptial feather structures and structural colours
Cryptic and conservative female plumages
Gross skull shape modification to accommodate muscles for display plumes
Diverse facial and body skin specialisations (wattles, apteria, eye-patches)
Examples of grossly-elongated and modified tail feathering in males
Distinct juvenile plumages as well as female-like plumages in young males
A great range in sexual dimorphism in both size and plumage
Mechanical sound production by males in display
Examples of dramatic tracheal modification and position (*Manucodia*)
Extreme morphological radiation in both bill size and shape
Examples of reversed (female larger) sexual bill dimorphism
Island- and mountain-isolated monotypic genera
Lowland- to subalpine-confined monotypic genera
Complex altitudinal distribution of genera and species
Propensity to interspecific hybridisation (among the polygynous species)
Grossly-delayed male plumage maturation (and sexual bimaturism)
Extreme longevity (for passerines)
Diversity in mating systems: monogamy to lek polygyny
Complex terrestrial, arboreal, and aerial courtship displays
Examples of court-clearing and resource-accumulating behaviours in males
Example of tactile courtship display
Dietary diversity ranging from specialised insectivory to total frugivory
Examples of complex plant/bird interactions
Important participation in mixed-species foraging flocking
Examples of intra- and interspecific aggressive food-plant defence
Crow-like use of the feet to manipulate foods
Open-bill probing
Apparent relationship between plumage colour and diet
Very small clutch size
Diversity of both nest form and placement (site)
Examples of use of sloughed snake skin as nest 'decoration'
Nesting association with aggressive passerine of another family
Brief to very long passerine incubation periods
Brief to very long passerine nestling periods
Examples of apparent distastefulness

thoritative synthesis of the family was that of E. Thomas Gilliard (1969). Stimulated by Gilliard's contribution, a new generation of researchers has vigorously pursued study of the biology and systematics of these birds, and yet there still exists the slim possibility that not all species of bird of paradise have been discovered by science (see Beehler 1991a; Fuller 1995). This should provide impetus for the next generation of explorers and field students.

During the past decade, the two new intergeneric hybrid combinations have been discovered. In addition, a previously-unknown plumage has been described for two species. The nesting biology of nine species has been described for the first time. A nest and egg of the Standardwing was first reported during the writing of this text. Comparative studies of the feeding ecology of almost a quarter of the species have been carried out in the wild. These have provided important insights into the relationship between foraging ecology and mating behaviour. We now have some knowledge of the life history of virtually all bird of paradise genera (*Loboparadisea* and to a lesser extent *Lycocorax* being the exceptions) although there is still much that remains to be learned.

The Paradisaeidae

The family of the birds of paradise is considered by us to include 42 species, in 17 genera, of higher (oscine) passerine birds. The birds of paradise are confined to part of the region known as Australasia. Within this area 38 species live on New Guinea and its satellite islands. Two species are endemic to the northern Moluccan Islands of Indonesia. Two riflebird species are endemic to the narrow tracts of wet coastal forests of eastern Australia. Finally, the predominantly New Guinea-dwelling Trumpet Manucode and Magnificent Riflebird also inhabit the northeast tip of Australia (Fig. 1.1). Unlike the members of numerous other families of perching birds, the birds of paradise do not inhabit a wide range of habitats, but are predominantly confined to rainforests and a few other densely-vegetated habitats. While four species do occur in Australia none has adapted to the vast areas of woodland, savanna, and desert that predominate there.

General consensus has been that the birds of paradise are, for want of a better expression, elaborate crows. Despite the contrary views of several avian anatomists, ornithologists long treated the birds of paradise as most closely related to the bowerbirds (Ptilonorhynchidae). Surprisingly, recent DNA–DNA hybridisation studies indicated that the two groups are, in fact, not closely related. The higher-level relationships of the birds of paradise remain somewhat cloudy. On the other hand, the composition of the family, as recognised by taxonomists, has long remained relatively stable, the only doubts being with regard to the three wide-gaped species for which Bock (1963) erected the subfamily Cnemophilinae (then including Macgregor's Bird of Paradise). That several modern ornithologists thought it possible that these distinctive wide-mouthed birds of paradise might be closer to the bowerbirds than to typical birds of paradise emphasises just how little was known about them.

The anatomy of birds of paradise is broadly typical of the oscinine passerines, and is generally crow-like as is reflected in their average size, shape, strength, demeanour, powerful bills and feet, and harsh vocalisations. Conspicuous features of the typical paradisaeidine skull are a small or absent lacrymal and subsequently large ectethmoid plate solidly fused with the frontal bone and a short orbital process of the quadrate with expanded distal tip (Bock 1963; see Fig. 9.1).

Unlike the typical birds of paradise that form the subfamily Paradisaeinae, the skull of the wide-gaped birds lacks nasal ossification and the lacrymal, and the head of the orbital processes of the quadrate is expanded. The three wide-gaped birds are atypical in being far less active, weaker-billed, and weaker-footed birds (Table 1.2). This may relate to their specialised, exclusively-frugivorous, diets. Nevertheless, a number of characters shared by both subgroups support the hypothesis that they merit placement in the Paradisaeidae.

The Paradisaeinae, or 'true' birds of paradise, includes several older groups of apparently obscure affinities (e.g. *Macgregoria*, *Paradigalla*) as well as what appear to be newer lineages that have recently radiated explosively (e.g. *Parotia* and *Paradisaea*). The systematics of the birds of paradise will remain a source of contention until researchers find a suite of systematic characters that clearly resolves affinities of both the older and newer radiations. One taxonomic study suggests that the little-known lesser Melampitta *Melampitta*

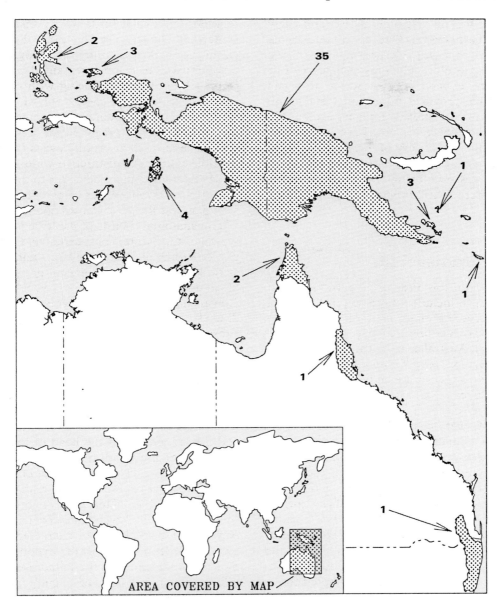

1.1 Map showing the world distribution of the bird of paradise family Paradisaeidae (shaded areas). Numbers indicate species present in specific areas.

lugubris, traditionally placed in the babbler family Timaliidae, is the basal lineage of the bird of paradise family. While we accept that this species may be more closely related to the Paradisaeidae than was previously believed, we do not treat it as paradisaeid in this work for reasons that are elaborated in Chapter 3.

Body size and proportions

Because of its enormously-elongated and broadened tail and relatively-long neck and bill, the adult male Black Sicklebill appears at first glance to be the largest bird of paradise. At an average adult male weight of 277 g,

Table 1.2 Some characters that vary between subgroups of the Paradiseaidae.

Character (reference—see below)	Cnemophilinae	Macgregoria	Manucodia	Paradisaeinae
Skull—nasal region (1)	Not ossified	Not ossified	Ossified	Ossified
Skull—maxillopalatines (1)	Long, expanded	Long, expanded	Short, blunt	Short, blunt
Skull—ectethmoid (1)	Large, 'entire'	Large, 'semiwinged'	Larger, 'winged'	Larger, 'winged'
Skull—lacrymal (1)	Absent	Absent	Small to absent	Medium to absent
Skull—lateral edge of palatine (1)	Not thickened	Somewhat thickened	Thickened	Thickened
Skull—transpalatine process (1)	Short, expanded	Long, pointed	Short, blunt	Short, blunt
Skull—postorbital & zygomatic processes (1)	Poorly developed	Well developed	Well developed	Well developed
Skull—mandibular rami (1)	Meet only at a narrow symphysis	Fused into a weak wedge	Fused into a strong wedge	Fused into a strong wedge
Skull—pseudotemporal process (1)	Absent	Absent	Moderately developed	Moderately developed
Skull—posterior wall of articular cavity (1)	Slightly developed	Moderately developed	Well developed	Well developed
Feather tract—dorsal pterylosis, number of rows (2)	13	10	7–8	7–11
Feather tract—dorsal pterylosis, basal gap (2)	Absent	Absent	Present	Present
Appearance of adult sexes (3)	Dimorphic	Monomorphic	Monomorphic	Dimorphic (+ monomorphic)
Adult female plumage (3)	No ventral barring	No ventral barring	No ventral barring	Ventral barring (+ none)
Juvenile plumage (4)	Unlike adults	Like both adults	Like both adults	Like adult female (+ like both adults)
Tips of rectrices (3, 5)	No hair-like tips	Hair-like tips	Hair-like tips	Hair-like tips
Modification of wing primaries (3)	None	Slight	None	Great to slight
Legs and feet (3)	Weak	Powerful	Powerful	Powerful
Bill gape width (3)	Wide	Narrow	Narrow	Narrow
Sexual dimorphism in size (3, 5)	Slight	Great	Slight	Great
Diet (6)	Obligate frugivorous	Predominantly frugivorous	Predominantly frugivorous	Omnivorous
Mating system (7)	Polygyny	Monogamy	Monogamy	Polygyny
Nest structure (3, 8)	Domed on few sticks	Bulky open supported cup, no sticks	Shallow sparse suspended cup, no sticks	Bulky open supported cup, no sticks

References: (1) Bock (1963); (2) Clench (1992); (3) present work; (4) C. Frith (1987, 1996), Frith and Frith (1992b, 1993b,c, 1995b); (5) Frith and Frith (1997b); (6) Beehler (1989); (7) Beehler (1987b, 1989b); (8) Gilliard (1969), Frith and Harrison (1970), Cooper and Forshaw (1977), Frith (1970, 1971), Frith and Frith (1990c, 1992b, 1993b,c,d, 1994b, 1995b).

however, it is considerably lighter than the more compact and bulky Curl-crested Manucode (up to 448 g or more). Larger adult male Greater Birds might weigh as much as an adult male Black Sicklebill but, surprisingly, no weights are presently available for the former. The lightest species is the King Bird, at *c.* 50 g (individual females weigh as little as 38 g) approximately one-tenth of the weight of an adult male Curl-crested Manucode (Fig. 1.2, Table 1.3).

Body dimensions such as wing and tarsus length provide useful gauges for morphological comparison. The latter is a more valid measurement for comparing birds of various ages as the tarsus quickly grows close to maximum size early in life. Means, ranges, and sample sizes for the species are provided in Table 1.3 (additional measurements appear in the species accounts). The differences in relative sizes and proportions of species forming a group of closely-related birds might provide insights into various aspects of their ecology and, to a limited degree, their relationships. For example, birds with proportionately-long wings usually use them for sustained flight in contrast to those with shorter broader ones that are more sedentary or terrestrial in habits. Not surprisingly, some of the island-isolated species (i.e. Curl-crested Manucode) have shorter wings proportionate to their weight than mainland manucodes (Fig. 1.3). Likewise birds with proportionately-longer tarsi are typically more terrestrial and those shorter in the leg tend to be more arboreal.

An analysis of the relationship between mean wing length and body weight shows the cnemophilinine genera (*Cnemophilus* and *Loboparadisea*) to be typically shorter-winged than other genera as a group. Within the subfamily Paradisaeinae, the Paradise Crow and the manucodes are generally longer-winged birds and Macgregor's Bird moderately long-winged.

1.2 Outline figures of adult male birds of paradise of selected species to indicate the diversity of size and shape (see also the diagrams on pages xxv and xxvi and Plates 4–13). Left to right are: King Bird, Yellow-breasted Bird, King of Saxony Bird, Lawes' Parotia, Standardwing Bird, Stephanie's Astrapia, Raggiana Bird, Macgregor's Bird, Black Sicklebill, and Curl-crested Manucode.

Table 1.3 Biometrics of adult male and female of each bird of paradise species (data amalgamated from biometrics in species accounts).

Species	Sex	Wing length (mm)		Tail* length (mm)		Tail centrals** (mm)		Tarsus length (mm)		Bill length (mm)		Weight (g)	
		Mean	Range	Mean	Range	Mean	Range	Mean	Range	Mean	Range	Mean	Range
Loria's Bird *Cnemophilus loriae*	♂	104	97–109	74	66–81			37	32–42	26	21–29	85	75–101
	♀	103	97–109	76	69–88			37	34–40	26	24–28	81	60–96
Crested Bird *Cnemophilus macgregorii*	♂	114	107–118	91	86–96			41	38–46	29	24–32	100	90–120
	♀	110	101–115	90	80–101			40	36–43	26	23–29	93	79–125
Yellow-breasted Bird *Loboparadisea sericea*	♂	95	90–100	57	52–61			31	29–34	21	19–24	64	50–75
	♀	97	91–101	59	55–63			34	30–33	22	21–23	73	71–75
Macgregor's Bird *Macgregoria pulchra*	♂	199	184–211	150	128–165			63	55–66	40	37–44	279	242–357
	♀	177	164–187	134	120–152			57	53–61	38	35–42	206	190–230
Paradise Crow *Lycocorax pyrrhopterus*	♂	198	175–224	145	127–161			44	39–51	50	42–56	302	242–370
	♀	192	176–210	140	130–153			43	39–49	47	41–54	267	218–316
Glossy-mantled Manucode *Manucodia atra*	♂	190	160–211	160	133–183			41	35–47	41	34–47	238	170–315
	♀	182	159–205	151	127–172			40	35–44	39	34–44	208	155–252
Jobi Manucode *Manucodia jobiensis*	♂	176	161–194	133	119–147			37	34–41	39	35–42	228	212–257
	♀	169	157–177	127	113–141			36	34–39	37	35–39	180	150–205
Crinkle-collared Manucode *Manucodia chalybata*	♂	173	161–183	141	128–152			38	33–41	40	38–44	221	150–265
	♀	167	153–182	136	124–154			37	34–41	38	34–43	215	160–289
Curl-crested Manucode *Manucodia comrii*	♂	237	213–257	167	148–182			50	47–53	57	52–61	448	
	♀	225	209–240	160	147–172			49	46–52	53	50–57	418	
Trumpet Manucode *Manucodia keraudrenii*	♂	167	144–194	130	110–143			36	28–41	34	30–38	171	130–240
	♀	160	139–186	124	108–141			35	30–40	33	29–37	151	130–182
Long-tailed Paradigalla *Paradigalla carunculata*	♂	186	180–201	132	122–137	160	132–170	49	48–50	43	38–45		
	♀	165	157–175	125	118–131	132	125–138	46	41–50	42	39–44	170	
Short-tailed Paradigalla *Paradigalla brevicauda*	♂	158	151–168	53	44–88			44	42–46	44	40–49	173	160–184
	♀	150	144–154	68	46–96			42	39–44	44	42–48	163	155–170
Arfak Astrapia *Astrapia nigra*	♂	185	179–193	399	369–437	569	518–756	42	39–43	41	40–42		
	♀	170	157–182	258	233–274	310	290–332	40	37–43	41	39–43		
Splendid Astrapia *Astrapia splendidissima*	♂	137	131–145	174	131–207	217	193–243	38	34–40	40	38–43	138	120–151
	♀	135	128–145	187	166–224	211	189–249	37	33–40	40	36–43	125	108–151
Ribbon-tailed Astrapia *Astrapia mayeri*	♂	179	173–185	111	97–126	893	657–1017	41	38–43	33	29–35	146	134–164
	♀	156	150–163	152	122–178	309	260–374	39	36–42	33	31–35	132	102–157

Table 1.3 continued

Species	Sex	Wing length (mm)		Tail* length (mm)		Tail centrals** (mm)		Tarsus length (mm)		Bill length (mm)		Weight (g)	
		Mean	Range	Mean	Range	Mean	Range	Mean	Range	Mean	Range	Mean	Range
Stephanie's Astrapia	♂	169	156–182	144	107–185	637	565–727	42	38–45	38	34–41	155	144–169
Astrapia stephaniae	♀	154	144–165	193	151–230	322	268–362	40	36–43	38	34–42	137	123–159
Huon Astrapia	♂	188	182–194	352	271–387	443	367–486	42	38–44	39	37–41	207	186–225
Astrapia rothschildi	♀	164	152–180	220	208–233	256	240–283	40	39–44	39	37–41	159	143–200
Western Parotia	♂	166	161–170	130	125–137			53	50–57	35	30–40	192	175–205
Parotia sefilata	♀	152	141–163	129	122–134			48	44–54	35	30–37	172	140–185
Lawes' Parotia	♂	155	148–163	80	73–84			50	46–53	34	29–39	167	153–195
Parotia lawesii	♀	148	141–159	98	92–107			46	40–53	33	26–36	144	122–169
Wahnes' Parotia	♂	161	157–166	187	159–198	213	201–225	51	49–53	31	29–33	171	170–172
Parotia wahnesi	♀	149	142–157	153	140–164	166	161–175	46	44–52	31	28–33	146	144–154
Carola's Parotia	♂	155	149–161	77	73–86			50	47–53	34	31–39	205	
Parotia carolae	♀	144	133–154	92	86–97			46	43–52	36	29–39	138	110–163
King of Saxony Bird	♂	126	119–133	86	79–97			32	30–35	24	22–27	87	80–95
Pteridophora alberti	♀	115	108–128	84	77–91			31	28–34	24	22–26	77	68–88
Magnificent Riflebird	♂	188	174–200	101	92–110	95	83–108	39	36–43	57	51–64	194	143–230
Ptiloris magnificus	♀	153	139–165	98	88–113	97	86–115	35	31–44	50	45–57	128	94–185
Paradise Riflebird	♂	160	153–165	98	92–103	89	84–95	33	30–36	54	49–62	141	134–155
Ptiloris paradiseus	♀	144	137–156	91	86–99			32	30–36	59	51–66	104	86–122
Victoria's Riflebird	♂	139	133–148	81	77–85			32	30–35	43	40–47	105	91–119
Ptiloris victoriae	♀	125	116–136	78	72–85			30	28–33	45	40–49	86	77–96
Superb Bird	♂	136	126–149	92	82–109			32	28–35	30	28–34	87	60–105
Lophorina superba	♀	120	110–133	84	72–110			30	27–37	30	26–33	67	54–85
Black Sicklebill	♂	201	188–232	402	340–539	735	593–946	51	47–55	78	70–86	277	250–318
Epimachus fastuosus	♀	171	152–193	226	171–280	298	214–387	47	43–55	75	66–89	195	160–255
Brown Sicklebill	♂	182	171–206	252	218–282	706	522–822	50	46–54	84	78–90	234	144–310
Epimachus meyeri	♀	157	142–185	201	151–250	294	213–359	46	39–53	82	66–91	166	140–202
Buff-tailed Sicklebill	♂	153	145–161	129	118–141			34	32–36	76	69–83	112	103–125
Drepanornis albertisi	♀	148	143–156	124	115–133			34	32–37	79	68–91	111	92–138

Table 1.3 continued

Species	Sex	Wing length (mm)		Tail* length (mm)		Tail centrals** (mm)		Tarsus length (mm)		Bill length (mm)		Weight (g)	
		Mean	Range	Mean	Range	Mean	Range	Mean	Range	Mean	Range	Mean	Range
Pale-billed Sicklebill	♂	160	153–165	110	103–116			33	31–35	79	74–83	161	160–164
Drepanornis bruijnii	♀	155	148–162	111	107–116			32	30–34	75	68–79	146	144–149
Magnificent Bird	♂	115	105–121	39	33–44	275	213–321	32	29–34	30	26–35	97	75–119
Cicinnurus magnificus	♀	110	104–125	60	53–69			30	27–33	30	26–33	82	62–113
Wilson's Bird	♂	98	94–101	39	35–41	140	121–150	27	26–28	25	23–28	61	53–67
Cicinnurus respublica	♀	97	93–101	52	49–59			27	25–29	25	23–28	56	52–60
King Bird	♂	100	93–110	32	28–38	158	134–182	26	22–29	26	23–31	53	43–65
Cicinnurus regius	♀	99	93–107	58	51–63			26	25–28	27	23–30	50	38–58
Standardwing Bird	♂	157	152–162	82	71–93	70	60–93	42	39–44	45	40–48	163	152–174
Semioptera wallacii	♀	144	135–154	85	78–94	86	76–96	39	36–43	43	41–47	135	126–143
Twelve-wired Bird	♂	173	163–185	69	61–86			40	31–44	69	63–75	193	170–217
Seleucidis melanoleuca	♀	162	152–171	103	93–113			37	34–40	63	56–69	171	160–188
Lesser Bird	♂	191	180–210	130	116–144	487	353–641	45	39–49	39	34–43	254	183–300
Paradisaea minor	♀	163	152–175	114	103–126	106	93–129	40	37–47	37	35–41	165	141–210
Greater Bird	♂	217	200–240	154	140–175	612	480–584	49	45–55	42	38–48		
Paradisaea apoda	♀	180	159–215	133	118–158	125	105–153	43	39–53	40	37–45	172	170–173
Raggiana Bird	♂	186	160–198	133	124–154	455	358–527	42	39–46	39	35–41	270	234–300
Paradisaea raggiana	♀	162	148–182	119	108–132	111	100–128	38	34–44	37	35–40	173	133–220
Goldie's Bird	♂	181	177–185	137	128–143	431	368–536	41	39–43	37	35–40	237	
Paradisaea decora	♀	160	157–165	128	123–131	109	104–112	37	36–39	36	34–39		
Red Bird	♂	176	169–186	119	114–125	563	478–765	42	38–44	36	31–39	201	158–224
Paradisaea rubra	♀	159	151–174	114	105–121	106	102–117	39	36–44	35	33–39	158	115–208
Emperor Bird	♂	177	172–188	114	107–121	557	455–693	45	42–47	41	37–43	256	250–265
Paradisaea guilielmi	♀	158	151–171	106	101–117	104	95–113	40	39–45	40	38–42		
Blue Bird	♂	157	151–163	82	76–91	414	341–458	39	36–42	41	39–46	176	158–189
Paradisaea rudolphi	♀	149	142–159	93	86–102	94	88–101	38	36–41	41	38–43	153	124–166

* Tail length is the maximum length of the tail except in birds with the central pair of feathers more than 5 mm longer or (rarely) 5 mm shorter than the rest in which case the additional measurements ** are given.
** In most cases in which only a single tail measurement is given the difference between central rectrices and the rest if < 1–2 mm.

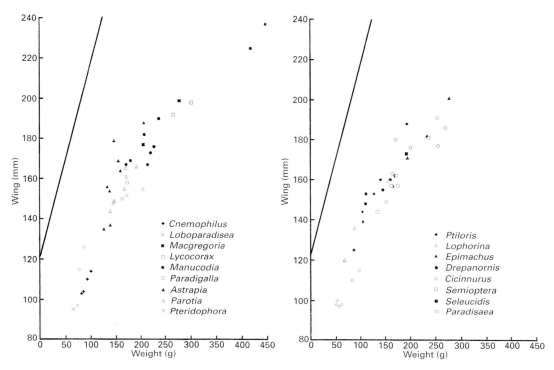

1.3 The relationship between mean wing length and body weight in the birds of paradise (divided into two diagrams for clarity only). Both sexes of all species for which weights were available are plotted. The diagonal line tracks the regression for birds in general (Greenwalt 1962). After Snow (1982).

The astrapias and riflebirds tend, on average, to have proportionately-shorter wings. The parotias (*Parotia*) are of average wing length for their mass. The King of Saxony Bird (*Pteridophora*), Superb Bird (*Lophorina*), and the sickletails (*Cicinnurus*) are relatively short-winged birds. The courtship behaviour of all the species of the Cnemophilinae remains unknown but of the remaining, typical, birds of paradise there is an apparent trend (Fig. 1.3) towards relatively-shorter wings in most groups (particularly their males) that display on or close to the ground (*Pteridophora, Lophorina, Cicinnurus*).

Juvenile and immature individuals are generally smaller than subadults and adults of their species as measured by wing length or weight. Museum specimens indicate that younger individuals in many, and quite probably all, species have discernibly-narrower tail feathers (rectrices) that are more pointed distally than those of older ones (Gyldenstolpe 1955b: 301 and personal observation). Future studies of plumages and bare parts of live birds, particularly of long-term individually-marked ones, are necessary to permit better-resolved ageing and sexing of both live birds in the hand and museum specimens.

Sexual dimorphism in size and proportions

The mean wing length of most adult female birds of paradise averages 85–95% that of males but the exceptions to this rule include the wide-gaped birds (96–102%), the manucodes (95–97%), Splendid Astrapia (99%), the two smaller (*Drepanornis*) sicklebills (97%), and the three sickletails (96–99%). Sexual dimorphism in wing length of the monomorphic and monogamous Macgregor's

Bird is relatively great (female mean wing length being 89% that of the male) and is twice as great as in all five manucodes and the three sexually-dimorphic cnemophilines. Sexual size dimorphism, as expressed by female/male wing length, is greatest in the Magnificent Riflebird (81%), Greater and Lesser Birds (83 and 85%, respectively), and the two larger (*Epimachus*) sicklebills (85–86%). These polygynous species show little difference in their sexual size dimorphism notwithstanding that they include both solitary- and lek-displaying males.

Data for wing length suggest that sexual dimorphism in body size tends to be reduced in smaller species of polytypic genera. This is so in the paradigallas, astrapias, sicklebills, riflebirds, and sickletails and is particularly clearly demonstrated within the seven plumed birds of *Paradisaea*. The Blue Bird, the only non-lekking species of *Paradisaea*, is the least sexually dimorphic in size and appearance and is overall smaller than its congeners (Table 1.4).

Sexual dimorphism in bill size

Bill-size sexual dimorphism in birds is of interest because it can indicate ecological isolation between the sexes (Darwin 1871; Lack 1971; Selander 1972). In such cases the foraging sites and diet of each sex may differ sufficiently to lessen intersexual competition for limited food resources, such as wood-dwelling arthropods (Selander 1966; Jamieson and Spencer 1996; Moorhouse 1996; Frith 1997). Figures for adult female mean bill length and width as a percentage of those of adult males show (Table 1.4) that, in general, the bills of adult male birds of paradise average c. 5% larger than those of adult females. Relative sexual dimorphism in bill size is, with few notable exceptions, typically consistent within the species of a genus or closely-related genera (e.g. *Macgregoria*, *Lycocorax*, and *Manucodia*; *Paradigalla*, *Parotia*, and *Pteridophora*; *Cicinnurus*; and *Paradisaea*—Table 1.4).

Within the Paradisaeinae, sexual dimorphism in bill size is most diverse in the riflebirds. In one species the male has the longer bill (typical of the family as a whole), whereas in the other two the female's is longer (reverse dimorphism). It is also reversed to a slight degree in the Buff-tailed Sicklebill. These birds all probe and excavate tree crevices, bark, and dead wood—a specialised foraging technique closely correlated with some of the most marked examples of bill sexual dimorphism in birds of other families (see Jamieson and Spencer 1996; Moorhouse 1996; Frith 1997). This may well also explain the (normal, male larger) bill sexual dimorphism in the Twelve-wired Bird (Table 1.4), but at present little is known of its foraging and diet.

In the genera *Macgregoria*, *Lycocorax*, *Manucodia*, *Ptiloris*, *Epimachus*, and *Paradisaea* female bills average narrower than in males whereas in *Paradigalla*, *Astrapia*, *Parotia*, *Pteridophora*, *Lophorina*, *Drepanornis*, *Cicinnurus*, and *Seleucidis* they average broader than in males (Table 1.4). The large samples notwithstanding, these measurements are, however, of small features and these tentatively-observed, but potentially ecologically-significant, patterns require further investigation.

The species accounts of Chapter 9 include detailed descriptions of the plumage of both adult sexes and of younger birds of the nominate subspecies. These also contain a brief diagnosis and full biometrics for all subspecies we accept in the present work (Table 1.5).

Plumage sexual dimorphism

Some birds of paradise exhibit dramatic sexual dimorphism in plumage colour (Plates 4 and 6–12), whereas in other lineages the sexual differences in plumage are modest or minimal (Plate 5). All sexually-dimorphic (dichromatic) birds of paradise are known or are thought to be polygynous. The converse, however, is not true, as some of the monomorphic birds of paradise are now known to be polygynous.

In the polygynous and sexually-dichromatic birds of paradise that have been studied, adult male plumage characters develop over a number

Table 1.4 Various proportions of some measurements of both sexes and of sexual dimorphism in others in bird of paradise species.

Species	Mean tail* length as a % of mean wing length	Mean tarsus length as a % of mean wing length	Mean bill length as a % of mean wing length	Mean bill length as a % of mean tarsus length	Adult female mean bill length as a % of adult male mean bill length	Adult female mean bill width as a % of adult male mean bill width	Adult female mean tarsus length as a % of adult male mean tarsus length	Adult female mean wing length as a % of adult male mean wing length
Loria's Bird *Cnemophilus loriae*	71 74 73	36 36	25 25	70 70	99	100	100	99
Crested Bird *Cnemophilus macgregorii*	80 82 81	36 36	25 24	71 65	91	98	98	96
Yellow-breasted Bird *Loboparadisea sericea*	60 61 61	33 33	22 23	68 65	105	110	110	102
Macgregor's Bird *Macgregoria pulchra*	75 76 76	32 32	20 21	63 67	94	93	90	89
Paradise Crow *Lycocorax pyrrhopterus*	73 73 73	22 22	25 24	114 109	95	96	98	97
Glossy-mantled Manucode *Manucodia atra*	84 83 84	22 22	22 21	100 98	95	96	98	96
Jobi Manucode *Manucodia jobiensis*	76 75 76	21 21	22 22	105 103	95	99	97	96
Crinkle-collared Manucode *Manucodia chalybata*	82 81 82	22 22	23 23	105 103	95	96	97	97
Curl-crested Manucode *Manucodia comrii*	70 71 71	21 22	24 24	114 108	94	97	98	95

Table 1.4 continued

Species	Mean tail* length as a % of mean wing length	Mean tarsus length as a % of mean wing length	Mean bill length as a % of mean wing length	Mean bill length as a % of mean tarsus length	Adult female mean bill length as a % of adult male mean bill length	Adult female mean bill width as a % of adult male mean bill width	Adult female mean tarsus length as a % of adult male mean tarsus length	Adult female mean wing length as a % of adult male mean wing length
Trumpet Manucode *Manucodia keraudrenii*	78 78 78	22 22	20 21	94 94	95	97	97	96
Long-tailed Paradigalla *Paradigalla carunculata*	86 80 83	26 28	23 25	88 91	98	113	94	89
Short-tailed Paradigalla *Paradigalla brevicauda*	34 45 40	28 28	28 29	100 105	101	105	96	95
Arfak Astrapia *Astrapia nigra*	308 (216)** 182 (152) 245 (184)	23 24	22 24	98 103	100	106	95	92
Splendid Astrapia *Astrapia splendidissima*	158 (127) 156 (139) 157 (133)	28 27	29 30	105 108	100	103	97	99
Ribbon-tailed Astrapia *Astrapia mayeri*	499 (62) 195 (97) 347 (80)	23 25	18 21	81 85	100	106	95	87
Stephanie's Astrapia *Astrapia stephaniae*	377 (85) 208 (125) 293 (105)	25 26	22 25	90 95	100	108	95	91
Huon Astrapia *Astrapia rothschildi*	236 (187) 156 (133) 196 (160)	22 24	21 24	93 98	100	103	95	88
Western Parotia *Parotia sefilata*	78 85 82	32 32	21 23	66 73	99	107	91	92

Table 1.4 continued

Species	Mean tail* length as a % of mean wing length	Mean tarsus length as a % of mean wing length	Mean bill length as a % of mean wing length	Mean bill length as a % of mean tarsus length	Adult female mean bill length as a % of adult male mean bill length	Adult female mean bill width as a % of adult male mean bill width	Adult female mean tarsus length as a % of adult male mean tarsus length	Adult female mean wing length as a % of adult male mean wing length
Lawes' Parotia *Parotia lawesii*	52 / 66 / 59	32 / 31	22 / 22	68 / 72	99	106	92	95
Wahnes' Parotia *Parotia wahnesi*	132 (116)** / 111 (103) / 122 (110)	32 / 31	19 / 21	61 / 67	101	108	90	93
Carola's Parotia *Parotia carolae*	50 / 64 / 57	32 / 32	22 / 25	68 / 78	106	106	92	93
King of Saxony Bird *Pteridophora alberti*	68 / 73 / 71	25 / 27	19 / 21	75 / 77	99	111	97	91
Magnificent Riflebird *Ptiloris magnificus*	54 / 64 / 59	21 / 23	30 / 33	146 / 143	89	100	90	81
Paradise Riflebird *Ptiloris paradiseus*	61 / 63 / 62	21 / 22	34 / 41	164 / 184	108	95	97	90
Victoria's Riflebird *Ptiloris victoriae*	58 / 62 / 60	23 / 24	31 / 36	134 / 150	104	98	94	90
Superb Bird *Lophorina superba*	68 / 70 / 69	24 / 25	22 / 25	94 / 100	99	104	94	88
Black Sicklebill *Epimachus fastuosus*	366 (200)** / 174 (132) / 270 (166)	25 / 27	39 / 44	153 / 160	97	100	92	85

Table 1.4 continued

Species	Mean tail* length as a % of mean wing length	Mean tarsus length as a % of mean wing length	Mean bill length as a % of mean wing length	Mean bill length as a % of mean tarsus length	Adult female mean bill length as a % of adult male mean bill length	Adult female mean bill width as a % of adult male mean bill width	Adult female mean tarsus length as a % of adult male mean tarsus length	Adult female mean wing length as a % of adult mean mean wing length
Brown Sicklebill *Epimachus meyeri*	388 (138)** 187 (128) 288 (133)	27 29	46 52	168 178	97	97	92	86
Buff-tailed Sicklebill *Drepanornis albertisi*	84 84 84	22 23	50 53	224 232	104	111	100	97
Pale-billed Sicklebill *Drepanornis bruijnii*	69 72 71	21 21	49 48	239 234	95	108	97	97
Magnificent Bird *Cicinnurus magnificus*	239 (34)** 55 147 (44)	28 27	26 27	94 100	100	103	94	96
Wilson's Bird *Cicinnurus respublica*	143 (40) 54 99 (48)	28 28	26 26	93 93	100	117	100	99
King Bird *Cicinnurus regius*	158 (32) 59 108 (46)	26 26	26 27	100 104	101	108	100	99
Standardwing Bird *Semioptera wallacii*	52 59 56	27 27	29 30	107 110	97	101	93	92
Twelve-wired Bird *Seleucidis melanoleuca*	40 64 52	23 23	40 39	173 170	92	103	93	94
Lesser Bird *Paradisaea minor*	255 (68)** 70 163 (69)	24 25	20 23	87 93	96	96	89	85

Table 1.4 *continued*

Species	Mean tail* length as a % of mean wing length	Mean tarsus length as a % of mean wing length	Mean bill length as a % of mean wing length	Mean bill length as a % of mean tarsus length	Adult female mean bill length as a % of adult male mean bill length	Adult female mean bill width as a % of adult male mean bill width	Adult female mean tarsus length as a % of adult male mean tarsus length	Adult female mean wing length as a % of adult mean wing length
Greater Bird *Paradisaea apoda*	282 (71)** 69 (74) 176 (73)	23 24	19 22	86 93	95	96	88	83
Raggiana Bird *Paradisaea raggiana*	244 (72) 73 159 (73)	23 23	21 23	93 97	96	94	90	87
Goldie's Bird *Paradisaea decora*	238 (76) 80 159 (78)	23 23	20 23	90 97	98	100	90	88
Red Bird *Paradisaea rubra*	320 (68) 72 196 (70)	24 25	20 22	86 90	98	99	93	90
Emperor Bird *Paradisaea guilielmi*	315 (64) 67 191 (66)	25 25	23 25	91 100	98	101	89	89
Blue Bird *Paradisaea rudolphi*	264 (52) 63 164 (57)	25 26	26 28	105 108	98	101	97	95

* The three figures given vertically for each species are for males, females, and the sexes combined, respectively.
** Mean total length of tail is given and (in parenthesis) that excluding the central pair of feathers if they are more than 5 mm longer or shorter than the rest.

Table 1.5 A comprehensive list of bird of paradise taxa included in the present work

Family	Paradisaeidae	Birds of paradise
Subfamily:	Cnemophilinae	WIDE-GAPED BIRDS of PARADISE
Genus:	*Cnemophilus*	
(Subgenus	*Loria*)	
C. l. loriae		Loria's Bird of Paradise
C. l. amethystina		
C. l. inexpectata		
(Subgenus	*Cnemophilus*)	
C. m. macgregorii		Crested Bird of Paradise
C. m. sanguineus		
Genus:	*Loboparadisea*	
L. s. sericea		Yellow-breasted Bird of Paradise
L. s. aurora		
Subfamily:	Paradisaeinae	TYPICAL BIRDS of PARADISE
Genus:	*Macgregoria*	
M. p. pulchra		Macgregor's Bird of Paradise
M. p. carolinae		
Genus:	*Lycocorax*	
L. p. pyrrhopterus		Paradise Crow
L. p. obiensis		
L. p. morotensis		
Genus:	*Manucodia*	
(Subgenus	*Manucodia*)	
M. a. atra		Glossy-mantled Manucode
M. a. subaltera		
M. a. altera		
M. jobiensis		Jobi Manucode
M. chalybata		Crinkle-collared Manucode
M. c. comrii		Curl-crested Manucode
M. c. trobriandi		
(Subgenus	*Phonygammus*)	
M. k. keraudrenii		Trumpet Manucode
M. k. aruensis		
M. k. jamesii		
M. k. neumanni		
M. k. adelberti		
M. k. diamondi		
M. k. purpureoviolacea		
M. k. hunsteini		
M. k. gouldii		
Genus:	*Paradigalla*	
P. carunculata		Long-tailed Paradigalla
P. brevicauda		Short-tailed Paradigalla
Genus:	*Astrapia*	
A. nigra		Arfak Astrapia
A. s. splendidissima		Splendid Astrapia
A. s. helios		
A. mayeri		Ribbon-tailed Astrapia
A. s. stephaniae		Stephanie's Astrapia
A. s. feminina		
A. rothschildi		Huon Astrapia

Table 1.5 *continued*

Genus:	*Parotia*	
P. sefilata		Western Parotia
P. l. lawesii		Lawes' Parotia
P. l. helenae		
P. wahnesi		Wahnes' Parotia
P. c. carolae		Carola's Parotia
P. c. meeki		
P. c. chalcothorax		
P. c. berlepschi		
P. c. clelandiae		
P. c. chrysenia		
Genus:	*Pteridophora*	
P. alberti		King of Saxony Bird of Paradise
Genus:	*Ptiloris*	
(Subgenus	*Craspedophora*)	
P. m. magnificus		Magnificent Riflebird
P. m. intercedens		
P. m. alberti		
(Subgenus	*Ptiloris*)	
P. paradiseus		Paradise Riflebird
P. victoriae		Victoria's Riflebird
Genus:	*Lophorina*	
L. s. superba		Superb Bird of Paradise
L. s. niedda		
L. s. minor		
L. s. latipennis		
L. s. feminina		
L. s. sphinx		
Genus:	*Epimachus*	
E. f. fastuosus		Black Sicklebill
E. f. atratus		
E. f. ultimus		
E. m. meyeri		Brown Sicklebill
E. m. albicans		
E. m. bloodi		
Genus:	*Drepanornis*	
D. a. albertisi		Buff-tailed Sicklebill
D. a. cervinicauda		
D. bruijnii		Pale-billed Sicklebill
Genus:	*Cicinnurus*	
(Subgenus	*Diphyllodes*)	
C. m. magnificus		Magnificent Bird of Paradise
C. m. chrysopterus		
C. m. hunsteini		
C. respublica		Wilson's Bird of Paradise
(Subgenus	*Cicinnurus*)	
C. r. regius		King Bird of Paradise
C. r. coccineifrons		
Genus	*Semioptera*	
S. w. wallacii		Standardwing Bird of Paradise
S. w. halmaherae		

Table 1.5 *continued*

Genus	*Seleucidis*	
S. m. melanoleuca		Twelve-wired Bird of Paradise
S. m. auripennis		
Genus:	*Paradisaea*	
(Subgenus	*Paradisaea*)	
P. m. minor		Lesser Bird of Paradise
P. m. jobiensis		
P. m. finschi		
P. a. apoda		Greater Bird of Paradise
P. a. novaeguineae		
P. r. raggiana		Raggiana Bird of Paradise
P. r. salvadorii		
P. r. intermedia		
P. r. augustaevictoriae		
P. decora		Goldie's Bird of Paradise
P. rubra		Red Bird of Paradise
P. guilielmi		Emperor Bird of Paradise
(Subgenus	*Paradisornis*)	
P. r. rudolphi		Blue Bird of Paradise
P. r. margaritae		

of years. Females, by contrast, develop their drab and cryptic adult plumage within a year of fledging. The youngest males are cloaked in plumage that so closely approximates that of the female they cannot be visually sexed. We would suggest that the slow male plumage progression from cryptic to gaudy may be a response to the severe competition among males for access to females as mates. It may also, at least in part, have to do with the evolution of plumage stages that reflect distinct life-history and alternative mate-acquisition strategies by those males that have not achieved full status in the mating hierarchy. The very prolonged development of full nuptial plumage is also strong evidence for considerable longevity in these species.

Feather form and function

Birds of paradise uniformly exhibit 10 primaries and 12 tail feathers. In the male plumage of many species, these feathers have been highly modified for nuptial display. This is particularly notable in the central pair of rectrices of males, which may become, even within a single genus, longer or shorter and more ornate with increasing age (Table 1.6). While most birds of paradise have forward-pointing feathering covering their nostrils, like their corvid cousins, few have thick, stiff, rictal bristles as do the true crows (*Corvus*).

No other avian family exhibits the degree of diversity of feather structure and colour found in the Paradisaeidae (Fig. 1.2, Plates 4–12). Long after the revelations of the existence of these birds, the numerous detailed descriptions and illustrations of their plumes, the function of most of their unique feathers remained a mystery. It was evident to the earliest ornithologists, however, that the specialised plumage of the birds of paradise was for courtship. What we know of these displays and how individual plumes are incorporated into them is reviewed and illustrated in Chapters 5 and 9.

Some feather modifications in adult males are unusual in not being directly exhibited in courtship. These appear to be primarily for the production of sound or to support or strengthen nuptial plumage that is presented

Table 1.6 Type of development of central pair and other tail feathers in male birds of paradise as they age.

	Species	Central pair of tail feathers gets longer (+) or shorter (−) with increasing age	Remaining tail feathers get longer (+) or shorter (−) with increasing age
Loria's Bird	*Cnemophilus loriae*	−	−
Crested Bird	*Cnemophilus macgregorii*	−	−
Yellow-breasted Bird	*Loboparadisea sericea*	−	−
Macgregor's Bird	*Macgregoria pulchra*	?	?
Paradise Crow	*Lycocorax pyrrhopterus*	?	?
Glossy-mantled Manucode	*Manucodia atra*	+	+
Jobi Manucode	*Manucodia jobiensis*	+	+
Crinkle-collared Manucode	*Manucodia chalybata*	+	+
Curl-crested Manucode	*Manucodia comrii*	+	+
Trumpet Manucode	*Manucodia keraudrenii*	+	+
Long-tailed Paradigalla	*Paradigalla carunculata*	++	+
Short-tailed Paradigalla	*Paradigalla brevicauda*	−	−
Arfak Astrapia	*Astrapia nigra*	++	++
Splendid Astrapia	*Astrapia splendidissima*	−	−
Ribbon-tailed Astrapia	*Astrapia mayeri*	++	−
Stephanie's Astrapia	*Astrapia stephaniae*	++	−
Huon Astrapia	*Astrapia rothschildi*	++	++
Western Parotia	*Parotia sefilata*	−	−
Lawes' Parotia	*Parotia lawesii*	−	−
Wahnes' Parotia	*Parotia wahnesi*	++	++
Carola's Parotia	*Parotia carolae*	−	−
King of Saxony Bird	*Pteridiphora alberti*	−	−
Magnificent Riflebird	*Ptiloris magnificus*	−	−
Paradise Riflebird	*Ptiloris paradiseus*	−	+
Victoria's Riflebird	*Ptiloris victoriae*	−	0
Superb Bird	*Lophorina superba*	+	+
Black Sicklebill	*Epimachus fastuosus*	++	+
Brown Sicklebill	*Epimachus meyeri*	++	+
Buff-tailed Sicklebill	*Drepanornis albertisi*	−?	−?
Pale-billed Sicklebill	*Drepanornis bruijnii*	−	−
Magnificent Bird	*Cicinnurus magnificus*	++	−
Wilson's Bird	*Cicinnurus respublica*	++	−
King Bird	*Cicinnurus regius*	++	−
Standardwing Bird	*Semioptera wallacii*	−	−
Twelve-wired Bird	*Seleucidis melanoleuca*	−	−
Lesser Bird	*Paradisaea minor*	++	+
Greater Bird	*Paradisaea apoda*	++	+
Raggiana Bird	*Paradisaea raggiana*	++	+
Goldie's Bird	*Paradisaea decora*	++	−
Red Bird	*Paradisaea rubra*	++	−
Emperor Bird	*Paradisaea guilielmi*	++	−
Blue Bird	*Paradisaea rudolphi*	++	−

Note: +(+) and − indicate proportionately greater increase or decrease in length, respectively.
? = inadequate sample of specimens for determination or for a fully confident one.
0 = tail does not increase or decrease in length with age.

in display (see Chapter 3). In addition to normally-pigmented plumages, it appears likely that an ephemeral yellow pigment (possibly obtained or derived from certain favoured fruits) is found in the plumes of the Twelve-wired Bird of Paradise. This colour rapidly fades in captive individuals and museum skins. The male nuptial plumage of many other birds of paradise includes a kaleidoscope of brilliantly-iridescent, structural colours produced by light refraction (see Dorst 1973a,b; Dorst *et al.* 1974 and plates therein).

Skull shape modification

Some remarkable and unique modifications of skull shape occur in a few species. While these apparently facilitate the insertion of muscle masses required for the movement of ornate head plumes or the erection of display plumage by adult males, the skull modification occurs in both sexes. The skull of parotias is broadly flattened on the medial and posterior sections of the crown (Fig. 1.4). This is a peculiar modification that accommodates the muscles required to manipulate the nuchal bar, or crest. In both the Magnificent and Superb Bird the rear of the skull is abruptly flattened or square ended (Fig. 1.4). It can be no coincidence that adult males of both species wear an extensive cape of much-elongated feathers, originating from the skin of the lower nape which they elevate and hold flat against the back of the head in display.

Other body colour

Aside from their plumage specialisations, the birds of paradise are further enhanced by bright colouring of legs, eyes, facial, narial, and gap wattles, the bare skin of the face and crown, and even the interior skin of the mouth and palate. While the courtship functions of areas of bare skin are discussed in Chapter 5 and described in the species accounts of Chapter 9, the broader significance of these unusual characters is reviewed here. Adult males of all three cnemophilines have brightly-

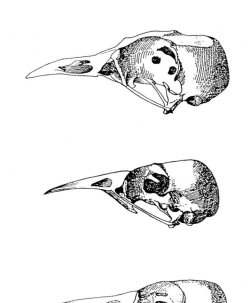

1.4 Drawings (not to scale and from top to bottom) of a skull of Lawes' Parotia (note 'dished' crown), Superb and Magnificent Birds of Paradise (note 'square' back to skulls): see text. From Stonor (1938).

coloured mouths. In Loria's Bird this is extended to form a small gape flange; in the Yellow-breasted Bird it is highly elaborated into bulbous narial wattles (Plate 4). As wattles in Yellow-breasted Birds are peculiar to adult males, they are probably the result of sexual selection and their size and colour may vary with age. By contrast, the colourful, wafer-like, orbital wattles of Macgregor's Bird and the mouth and facial wattles of the paradigallas occur in both sexes at all ages, including hatchlings. Their significance is unknown but may relate at least in part to species recognition.

All four sicklebills, three riflebirds, the Superb Bird, and the three sickletails exhibit a bright yellow or pale green mouth (and in some a small coloured gape). This is brighter and more obvious in adult males and is used in display. This may also be true of the pale mouth colour of the wide-gaped cnemophilines. The adult male Twelve-wired Bird lacks gape wattling but exhibits a brightly-coloured palate to females in courtship display. Likewise, the mouth interior of the adult King of Saxony Bird is a brilliant aqua-green which is exhibited during advertisement song and display. The extensive and pigmented bare facial and crown skin of the two smaller sicklebills and Wilson's Bird, respectively, occurs in both sexes. It is little different in colour between the sexes of the sicklebills but is obviously brighter in male than in female Wilson's Birds.

Iris colour is quite variable in the family, probably because it is such an important signal distinguishing age, sex, and species. It is common to observe a graded series of iris colours with age in some species (Magnificent Bird, some manucodes) and this may be widespread in the family. Most adults of both sexes share the same iris colour, the exception being the Black Sicklebill in which the male's iris is red, the female's brown. The two species of parotias that show geographical overlap exhibit distinct iris colour—Lawes' Parotia is blue and yellow, whereas Carola's is all yellow, lacking the prominent blue inner ring to the iris (Plate 7).

Bill colour is typically blackish or black in both sexes, with a few exceptions. Pale and/or colourful leg colour (i.e. not black, blackish or a nondescript dark colour) is confined to Macgregor's Bird, the sickletails, Standardwing, Twelve-wired Bird, and most of the plumed birds (Plates 5 and 10–12).

Vocalisations and associated adaptations

Most vocalisations produced by birds of paradise are loud harsh sounds, frequently referred to as corvid-like. Some exceptions are dramatic departures from this norm. The latter vary from weak and rather pathetic-sounding higher-pitched notes (Loria's Bird and some manucodes) to an explosive machine-gun-like staccato series (Brown Sicklebill), a complex and 'fizzing' radio-static-like noise (King of Saxony Bird), loud clear 'popping' notes (Emperor Bird), and an almost indescribable sound like that of an electric motor humming (Blue Bird in inverted courtship display). Note also that for all but a few monogamous species, the loud and typical vocalisations of birds of paradise are invariably given only by the males; the females are virtually silent. For full descriptions and sonograms of vocalisations, see the individual species accounts of Chapter 9.

The grossly-elongated and subcutaneously-coiled trachea of the manucodes (see Figs 1.5 and 9.9), greater in males than females, is unique among passerines, though similar modifications are found in a variety of non-passerine birds. This sexual dimorphism in trachea is correlated with distinctly-different vocalisations between the sexes (Frith 1994b). Forbes (1882a) noted that the Twelve-wired Bird exhibits a minor tracheal modification that may relate to male vocalisation.

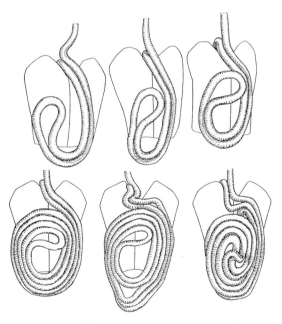

1.5 The modified subcutaneous trachea of six male Trumpet Manucodes showing (top left to bottom right) development with increasing age. The trachea at bottom right, from a bird captive at NYZS for > 13 years, was c. 750 mm long. From Clench (1978), with kind permission.

Ecology

Birds of paradise have evolved to inhabit closed humid tropical and subtropical forest. There are a number of lowland species, but most species inhabit montane forest. The only species that inhabits savanna as well as forest is the Glossy-mantled Manucode. One of the most interesting aspects of the distribution of birds of paradise in New Guinea is the manner in which species occupy different elevation zones, with the largest number concentrated between 1000 and 2000 m.

Birds of paradise show a range of diets, but most species take a combination of plant and animal material—mainly fruit and arthropods (and in some cases small frogs and skinks). Some species are mainly insectivorous and others apparently nearly wholly frugivorous. Those species that depend more on fruit tend to range more widely in search of food, whereas the insectivores are more sedentary. Insectivorous birds of paradise tend to forage in the manner of the creepers and woodpeckers.

Because there are so many kinds of fruit available in the forest it is difficult to identify taxonomic specialisations by individual bird species. A single bird of paradise might consume fruit from several hundred plant species in a single year. A few generalisations are worth noting. Some manucodes are apparently fig specialists; Macgregor's Bird appears to inhabit only environments where a particular species of subalpine conifer (*Dacrycarpus compactus*) is common, and may depend on its fruit when available; and many of the polygynous paradisaeinine species consume a range of morphologically-specialised fruits that must be plucked from a dehiscent woody husk. Some field research suggests that particular dietary specialisations are correlated with specific mating systems and this is discussed in Chapter 4.

Mating systems

The plumed *Paradisaea* birds of paradise typify that most extreme of mating systems known as lek behaviour, in which males gather at a communal site to call, display, and compete for the right to inseminate visiting females. We say 'extreme' because there are few sexual reproductive systems more remarkable in the animal world, and because mating skew (where one or only a few males perform most or all of the copulations) is extreme. Thus it is not surprising that the birds of paradise are prized subjects of study of mating behaviour and are particularly good for comparative studies. Different genera exhibit a range of behaviours, from pair-bonded monogamy to a number of court-based polygynous systems. More remarkable is that one can find on a typical forest plot in montane New Guinea a suite of bird of paradise species that exhibit the range of mating systems found within the family.

Whereas one thinks primarily of display and mating when treating mating systems, there are a number of equally-significant components to consider—sexual dimorphism, male–male interactions, female mate-choice, inter-court spacing of displaying males, territoriality, the relationship between diet and mating behaviour, delayed male plumage maturation, alternative mating strategies, and the like. Suffice it to say, the birds of paradise offer fertile ground for the study of all of these phenomena, and researchers have barely scratched the surface. A few species have been well studied, but the fundamental theoretical questions have only been addressed in passing. Here, then, is a remarkable opportunity for future research, especially as the New Guinea region becomes better developed and more accessible to the researcher with novel field technologies.

Nesting

The reproductive 'pay-off' of the efforts birds put into their mating systems depends upon the successful rearing of offspring that will in turn reproduce. For most birds of paradise, this is entirely the responsibility of the female who builds the nest, incubates, broods, and provisions the nestling until fledging and independence. Only the monogamous manucodes, the Paradise Crow (unconfirmed), and Macgregor's Bird exhibit biparental care at the nest. The remaining large majority of species that are known or presumed to be polygynous exhibit female-only nesting.

Known nests of the wide-gaped birds are domed structures of orchid stems with a lush moss and fern exterior, with a few sticks in the foundation, placed on the ground or in low vegetation. By contrast, nests of the typical bird of paradise are large, deep, bulky, stickless open cups of leaves, vine and orchid tendrils and stems, with moss and fern fronds on their exterior, supported in a tree fork. Those of manucodes are quite different in being shallower and sparser and suspended between horizontally-forking branches, similar to the nests of drongos and Old World orioles.

Bird of paradise eggs are typically elliptical ovate and pinkish to buff with long, broad, brush-stroke-like markings of browns, greys, and lavender- or purplish-grey (Plate 13). The clutch consists of one or two, rarely three, eggs. Hatchlings are naked or almost so, with only the slightest of down, and their skin characteristically becomes dark to almost blackish after several days (Plate 3). The nesting of a number of species remains to be detailed, and most are relatively poorly documented, especially with respect to the details of incubation, rearing, and nestling diet and development.

Birds of paradise and traditional cultures

A remarkable aspect of the birds of paradise is their importance to local Melanesian cultures, especially on the island of New Guinea. The birds commonly form important components of myth and ritual. Adult males are traditionally hunted for their plumes, which are worn for adornment both as traditional battle dress and for ceremonies and festivals ('sing-sings'). The trade in skins and plumes of the male birds of paradise has long been important both locally and, over the last several centuries, over an ever-wider area. The plume trade reached its zenith in the years around 1900, when tens of thousands of skins were commercially exported to Europe and the United States of America to supply the millinery trade.

One of the great pleasures of an ornithologist working in New Guinea is experiencing first hand the deep knowledge of local field naturalists who inhabit the forests where birds of paradise are commonplace. These are a great boon to the field researcher both as a source of information about the habits of the birds, and also as mentors on methodologies and lore essential to successful field investigations.

Conservation

The depredations on birds of paradise caused by the commercial plume trade nearly a century ago provided an important lesson about the resilience of the populations of some birds of paradise. Because only the older males of these species produce the desirable plumes, the culling for the plume trade never seriously threatened these populations. In the absence of the old males, there was a large pool of younger males to mate with the females. Given the polygynous mating systems of these plumed species, a single male is capable of inseminating dozens or scores of females in a season—a natural protection against the impact of commercial hunting of plumed males. This all said, it is of course theoretically possible that a local long-term lack of adult-plumaged males, due to culling, may have repercussions for sexual selection by females for mates.

No bird of paradise is known to have become extinct in historical time. Today, most species of birds of paradise are not immediately threatened with extinction, although threats over the near term and longer term can be readily identified—large-scale forest disturbance from industrial timber operations, regional development, human population growth, and deforestation from subsistence agriculture. These sorts of pressures present the greatest threats to small or circumscribed populations. Regionally- or locally-endemic species are, without question, most at risk.

Global climate change, about which there is considerable uncertainty, might pose the greatest threat of all, depending on its severity. Most conservation biologists today would agree that the goal for a region like Australasia is to conserve large tracts of natural habitat with entire biotas intact. Because they are relatively well known, birds of paradise can serve as useful indicators of this biodiversity, and reserves established to conserve particular groupings of birds of paradise will be valuable for the preservation of much of the terrestrial flora and fauna.

Hybridisation

Perhaps there is no other bird family that has produced as many wild interspecific hybrid crosses as have the birds of paradise. Our analysis notes 21 distinct combinations from the wild (see Table 3.5 and Appendix 1). Neither the cnemophilines nor manucodes are represented. All hybrids are ascribed to parental forms that are known or presumed to employ polygynous courtship display. We assume that hybridisation in these forms is relatively more frequent than in other avian taxa for several reasons: (1) the close phylogenetic affinities of many of the hybridising species, (2) the ephemeral nature of the pair bond (if any) that is formed prior to mating, (3) the presumed indiscriminate nature of mate selection by males on a display site, and (4) the imperfect nature of female mate-selection behaviours when faced with the stimulus provided by the remarkable mate-attraction behaviours of heterospecific males.

2

Discovery of the birds of paradise and history of their study

The history of the discovery and study of birds of paradise is as colourful as the birds themselves. It has included the lives (and sometimes the deaths) of monarchs and other eminent persons, famous and infamous explorers, traders and collectors, and some of the great evolutionists, zoologists, and ornithologists. Recent decades have seen all manner of people spend significant parts of their lives involved with these magnificent birds. American, Australian, and British self-made, and some knighted, millionaires, academics interested in the theoretical potential of the birds, and enthusiastic but near penniless young birdwatchers have suffered the extreme discomforts of New Guinea for a glimpse of avian paradise. Errol Flynn, before gaining Hollywood fame, had an eye to the bird of paradise plume trade as a possible back-up venture when in New Guinea in search of gold (Flynn 1960; Fuller 1995).

Much detail of the discovery, history, and mythology of the birds of paradise has been documented in earlier works (Lesson 1834–5; Salvadori 1880–2; Elliot 1873; D'Albertis and Salvadori 1879; Sharpe 1891–8; Iredale 1950; Stresemann 1954; Gilliard 1969). A full account of the history of the discovery of the birds, detailing the itineraries, successes, and failures of the colourful people involved would require another volume of the present size. The purpose of this chapter is to provide a compact overview while concentrating more on the acquisition of biological knowledge of the birds.

Trade skins of birds of paradise—the role of their plumes in ornithology

To this day the dried skins of adult male birds of paradise figure conspicuously in the economy of the peoples of New Guinea, particularly in the lives of the primarily-agricultural highlanders (Attenborough 1960; Majnep and Bulmer 1977; Healey 1990). Men hunt, trade, and lend these skins which are used as personal adornment (Fig. 2.1), symbolic of wealth and status. They were in many instances part of the 'bridewealth' a man must pay to obtain a wife (see Chapter 7). Thus, given that there is evidence of people living in New Guinea for at least 50 000 years, plume trading has quite probably existed for tens of thousands of years.

A comprehensive review of the long history of the plume trade has recently been published (Swadling 1996), and forms the basis of some of the following discussion. It is believed that dried 'trade skins' of adult males of the more elaborately-plumaged birds of paradise (particularly *Cicinnurus* and *Paradisaea* spp.) may well have been traded between New Guinea and the eastern Indonesian Archipelago, the Philippines and the SE Asian mainland for as long as 5000 years. They were certainly valued items of decoration in Asia more than 2000 years ago. Small wonder, then, that the first seafaring visitors to reach New Guinea from the West were offered bird of paradise skins as gifts and as trade goods (Swadling 1996).

2.1 Wigmen of the Huli People, Southern Highlands, Papua New Guinea at a 'sing-sing'. Adult male plumes of eight bird of paradise species (Ribbon-tailed and Stephanie's Astrapia, Lawes' Parotia, King of Saxony and Superb Bird, Black Sicklebill, Lesser and Raggiana Birds) are visible. Photo: CF.

Elcano, successor to the Portuguese sea captain Magellan as leader of the great Spanish voyage of exploration, reached the Moluccan island of Tidore in November of 1521. Here the Sultan of Batchian presented the expedition with several dried bird of paradise skins for the King of Spain. These were highly-prized gifts and were described by the local people as 'Birds of the Gods'. The birds were referred to in their island homes as *bolon diuata* (Swadling 1996) or as *Mamuco-Diata*, a variation of *Manucodiata* or *Manuccodiata*. The latter is translated from the Malay into 'Birds of the Gods'—origin of the modern name 'manucode'.

One of these skins was sent to Spain where Massimiliano Transilvano, the young private secretary to Karl the Fifth, saw it and wrote to his father the Archbishop of Salzburg. The letter was published in Cologne in January 1523 and contained news of the wonderful birds and the belief that they never alight but float in the air until they die and fall to earth (Stresemann 1954). Similar stories circulated in the West as more paradisaeid skins, prepared New Guinea style with their legs and feet removed, were sent back to Europe. An artefact of preparation, these 'footless' skins led to the belief that the birds were indeed footless and that they never alighted but fed upon dew or the fresh air on which they perpetually floated (Fig. 2.2). The 'perpetual flight' myth persisted until the end of the sixteenth century. During the early 1600s intact skins, complete with legs and feet, had, however, arrived to resolve this issue and to confirm that the birds of paradise were after all mere (albeit spectacular) avian mortals with large crow-like feet. Linnaeus (1758) alluded to this story in giving the specific name *apoda* (= footless or legless) to the Greater Bird of Paradise.

A period of scientific enlightenment followed spanning c. 1600–1825, during which much was written about the birds of paradise, although very little was actually known about them. A dozen or so species were made known to western science from skins reaching Europe (see Tables 2.1, 2.2 and Stresemann 1954). French naturalist Pierre Sonnerat published

Discovery of the birds of paradise and history of their study

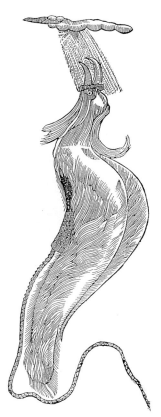

2.2 An adult male plumed bird of paradise depicted as they were once thought to live—by passively floating in air feeding and/or drinking upon dew/rain. From a woodcut in Gesner (1669).

[Golden birds, or birds that have metallic reflections] illustrated by the paintings of Audebert. These two works were the first monographs dealing primarily, or in substantial part, with the birds of paradise.

The first half of the nineteenth century was an exciting time for the study of birds of paradise. Intrepid and intensely-ambitious naturalists sailed to the Moluccan Islands, New Guinea, and Australia in search of them. Frenchman René Lesson, who joined a voyage of exploration aboard the 'Coquille', was the first western naturalist to see a wild bird of paradise in the forest. During late July 1824, upon encountering a male Lesser Bird of Paradise, he wrote 'The gun remained idle in my hand for I was too astonished to shoot … It was like a meteor whose body, cutting through the air, leaves a long trail of light' (Lesson in Stresemann 1954). He was clearly deeply moved by the experience and his great interest in the birds eventually expressed itself in the publication of his small, but important, treatise entitled *Histoire naturelle des oiseaux de paradis et des epimaques* [The Natural History of the Birds of Paradise and the Sicklebills] finely illustrated in colour by Prêtre and Oudart, in Paris during 1834–5. Given the near complete lack of biological knowledge of the group at the time, the title of his book now seems ambitious.

Lesson's experiences with live birds of paradise were the first to reach the scientific community of the western world. While in Sydney, Australia, he obtained a specimen of the Paradise Riflebird and in so doing became the first to record a bird of paradise living outside New Guinea and its adjacent islands (Lesson 1830, 1834–5).

The narrative of the voyage of H.M.S. *Rattlesnake* includes reference to the collection of some birds of paradise. John Macgillivray was the naturalist aboard the *Rattlesnake* and he collected many zoological novelties on Cape York Peninsula and elsewhere in north-east Australia as the ship's crew surveyed the coast under Captain Owen Stanley (Macgillivray 1852). His finds included the records of the Trumpet

his *Voyage à la Nouvelle Guinée* in 1776. While the title of this book is somewhat misleading, as Sonnerat did not actually reach the New Guinea mainland, it does contain descriptions and illustrations of six un-named birds of paradise. In 1801 the French ornithologist F. Levaillant started to publish the series of parts constituting his *Histoire naturelle des oiseaux de paradis* magnificently illustrated in colour by Jacques Barraband (Plate 1). In addition to being well ahead of his time in finely depicting birds with feather texture, depth, life-like postures, and animated eyes, Barraband also enjoyed the favour of Napoleon. In 1802 L. J. B. Vieillot, another French ornithologist, published *Oiseaux dorés, ou à reflets méttalliques*

Manucode and Magnificent Riflebird in Australia (already known from New Guinea) and discovery of Victoria's Riflebird (Gould 1850).

These observations were eclipsed by those of the great English explorer, naturalist, and evolutionist Alfred Russel Wallace. His wonderful accounts of discoveries and observations in his book *The Malay Archipelago* (Wallace 1869) created worldwide interest in the birds of paradise. Having seen Greater Birds displaying in the Aru Islands he wrote 'When seen in this attitude, the Bird of Paradise really deserves its name, and must be ranked as one of the most beautiful and most wonderful of living things' (Fig. 2.3). No mean accolade this, from a man that had by then seen many of the most remarkable parts of the tropical world and countless living things few others had seen. His descriptions of living birds of paradise attracted the attention of leading zoologists. Not least of these was Charles Darwin, for whom the birds of paradise were to serve as important examples in his great treatise on sexual selection (Darwin 1871). He noted that 'Lesson says that birds of paradise, so remarkable for their sexual differences, are polygamous, but Mr. Wallace doubts whether he had sufficient evidence'. Wallace was indeed correct in as much as Lesson did lack evidence, but Lesson was in fact correct about the sexually-dimorphic birds of paradise being polygamous.

A number of other early collectors of birds of paradise followed Wallace's example by carefully documenting their explorations, acquisitions, and observations. Amongst the first of these was the Dutch collector C. B. H. von Rosenberg, whose 1875 book illustrated nine species, including a head portrait of the Buff-tailed Sicklebill. The remarkable explorations of the Fly River by naturalist and adventurer Luigi D'Albertis are documented in *New Guinea: What I Did and What I Saw*—a remarkably-colourful narrative that is as informative, entertaining, and readable today as it was when published (D'Albertis 1880). Recent attempts to cast these explorations in a darker light (Goode 1977) are misguided, as D'Albertis was risking his life in an unknown river system in the heart of lands guarded by hostile populations. These threats never prevented the courageous D'Albertis from pursuing his first love—natural history. He was, in fact, the first to observe birds of paradise joining mixed-species foraging flocks and he also discovered and documented hybridisation between the Greater and Raggiana Birds. Writing of his experiences on the Fly River, he wrote (p. 241) 'Now that the sun has set, I hear the notes of the tallegallus, gowra, and many other birds unknown to me. A little before I had heard the long and cadenced notes of the *Seleucides*, while the *Paradisaea Raggiana* gaily repeated its own note. There are enough birds to make a naturalist happy; how many would rejoice to be with me tonight!'

The account of Captain J. Moresby's explorations (1876: 127–8), after whose father Port Moresby, today's capital of Papua New Guinea, is named, includes the following experience while at anchorage off Somerset, Cape York on 18 January 1873: 'Whilst here we fell in with a lonely waif of society, named Cockerill, who has betaken himself to live in a tiny vessel of about eight tons, and accompanied only by his son and two natives, cruises about these seas as a naturalist, and seems to be happy enough in his own way. His boat was laden with specimens of beautiful birds; and from the Aru Islands, 500 miles west of Somerset, which he had just left, he had brought some boxes full of the Great Bird of Paradise, and the still more exquisite King Bird of Paradise, of which he kindly gave me a specimen.'

The account of the cruise of the 429 ton yacht 'Marchesa' by Englishman F. Guillemard (1886) is not only very entertaining, but also contains several significant ornithological records. Among other observations, Guillemard was able to describe the growth and development of the unnatural-looking and plastic-like central rectrices of the Red Bird (see Plate 12). By far the most ornithologically important publication of this period was, however, the monumental, three volume *Ornitologia della Papuasia e delle Molucche* (Ornithology of Papuasia and the Moluccas) by the Italian

2.3 The illustration of 'Natives of Aru shooting the Great Bird of Paradise', at what is clearly a true lek, which first appeared in A. R. Wallace's book *The Malay Archipelago* (1869).

explorer, collector, and ornithologist Tommaso Salvadori (1880–2). This represented the first comprehensive handbook to the avifauna of the region and included accounts of all birds of paradise then known, with details of scientific collections of them.

Intense public interest generated by the extraordinary stories associated with the discovery of the birds of paradise in these remote corners of the globe did not abate. Each newly-discovered species, while clearly a member of this group, proved strikingly different in dazzling ways. The scientific specimens on which western knowledge rested were accessible only to a select few—cabinet naturalists and curators of museums housing natural history collections. A growing public interest in them therefore led to the publication of folios depicting them in the form of the magnificent hand-coloured works of Elliot (1873), Gould and Sharpe (1875–88) and Sharpe (1891–8).

By the end of the nineteenth century, growing interest by the public and scientific community had already resulted in considerable demand for specimens both for the museum and for the millinery trade. A subsequent demand was to be for the live bird trade.

Birds of paradise for the plume trade

The trade in feathers and plumes was long-established before it came to include birds of paradise (Doughty 1975), but once they were offered on the market their bewitching splendour ensured their popularity. Shortly after the turn of the century, the number of birds of paradise taken in New Guinea for the plume trade was staggering. Some 155 000 bird of paradise skins (all adult males and predominantly of *Paradisaea* species) were sold through the London Auction Sales alone during 1904–8 (Doughty 1975). In 1912 a single British vessel docked with 28 300 bird of paradise skins aboard (Bergman 1968). It is estimated that some 80 000 bird of paradise skins, worth at that time approximately £113 000, left New Guinea in 1913 (Swadling 1996)—all to satisfy a market created by hat styles for women that featured plumes and other feathers. These massive cullings focused exclusively on adult-plumaged males of only a few of the more widespread and common species, and therefore appear to have had no long-term effect upon local populations or upon species (see Chapter 8).

Several movements to restrict or stop such plume trade were eventually implemented. As early as 1869 the Sea Birds' Preservation Act of Christopher Sykes, instigated to control the heavy harvesting of Yorkshire seabirds for millinery plumage, was passed in the UK (Doughty 1975). While this had the potential to raise concern for the protection of other bird groups, initial steps were not taken in this direction until the 1885 establishment of the UK Plumage League (subsequently to become the Royal Society for the Protection of Birds) and the 1886 formation of the Audubon Society in the United States. In 1908 commercial hunting of, and trading in, bird of paradise plumes was declared illegal in Papua—what was, until 1906, British New Guinea (Swadling 1996). While the flow of plumes from Papua, the southeastern segment of the island, was slowed, their export from elsewhere in New Guinea continued apace, and the trade represented a major component of the economies of German (northeastern) and Dutch (western) New Guinea. The histories of the administrations of various parts of New Guinea are outlined in Fig. 2.4.

An example of the kind of illegal trade that continued in Papua is recorded in the Australian Archives: in October 1916 several German residents of the Mandated Territory were charged with plume hunting and trading. Of particular note were the brothers Joseph and Hermann Pieper who shot numerous birds of paradise to smuggle into Dutch New Guinea. In addition to being caught with several hundred bird of paradise skins, a letter by them was confiscated which referred to over 500 more skins they had hidden (R. Kimber *in litt.*). At this period the Australian newspaper columnist Captain W. M. B. Ogilvy wrote that 'Numbers of white men, Chinese, and Malays have been killed, cooked and devoured by cannibals whilst out shooting this beautiful and valuable bird [*Paradisaea* species], whose price before the war averaged anything between £3 and £7 per head. Some of the hunters shoot as many as 700 head in one season and obtained an average price of £3.10/- for them. It is profitable business' (Ogilvy 1916).

In Great Britain, more effective legislation, instigated in 1917, became the 1921 Import-

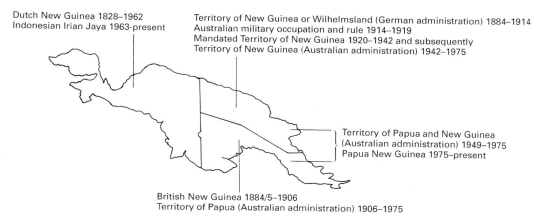

2.4 Outline map showing the history of colonisation and administration of mainland New Guinea.

ation of Plumage (Prohibition) Bill. Public outrage and the closing of American and Australian markets, due to 1913 legislation banning importation of plumes, caused the German Government to decree a one-year trial ban on plume hunting in German New Guinea in 1914. Plume hunting recommenced in what became, after the First World War, the Mandated Territory of New Guinea (see Fig. 2.4) but was halted there in 1922 by Australian legislation. In the face of increasing domestic and international pressure, the Dutch introduced progressively-stricter laws protecting live birds of paradise during the 1920s. This culminated in comprehensive anti-plume trade legislation in 1931 (see Swadling 1996). It has been optimistically stated that no birds of paradise have been collected for commercial purposes anywhere in New Guinea since 1924 (Gilliard 1969), but plumes do continue to be sought for relatively small markets in Indonesia and other parts of SE Asia (P. Swadling *in litt.*).

No birds of paradise now leave Papua New Guinea legally, except under permit and only for scientific and non-commercial purposes. The domestic and traditional trade does, however, continue within the country where, contrary to a law prohibiting the use of firearms to hunt these birds, shotguns are widely employed (see Chapter 8). In addition to the domestic hunting of birds in Irian Jaya, it is unfortunately true that there is a demand for plumes throughout Indonesia and a limited export trade exists that also includes live birds (see Chapter 8) notwithstanding that the Indonesian Government retained the Dutch laws prohibiting it. In 1990, moreover, Indonesia implemented legislation prohibiting persons hunting and trading in birds of paradise or transporting them about the Indonesian Republic (Swadling 1996).

Birds of paradise in the context of the international trade of plumes to the West are dealt with in considerable detail by several authors (see Doughty 1975; Everett 1978; Healey 1990; Fuller 1995; Swadling 1996 and references therein).

Birds of paradise for the museum cabinet

The international plume trade did certainly benefit the museum and private 'cabinet' ornithologists of Europe and America. As the trade progressed, ornithologists quickly realised the potential for discovering new bird of paradise species among the thousands of skins arriving in the major ports and cities. In the context of the prevailing descriptive ornithology of the time, the desire among European naturalists to discover, describe, and name new species (especially of birds and butterflies) was considerable. The extra prestige of naming a new bird of paradise would have been great indeed. Wealthier institutions and private collectors were quickly approached by those in the plume trade whose experienced eye noted anything out of the ordinary arriving at the trading centres. Much higher prices could be obtained for such specimens.

During the three centuries spanning the 1520s to the 1820s about a dozen valid species of birds of paradise were made known to the world from trade skins, reaching scientists via the plume trade (Table 2.1). While these species included the Yellow-breasted Bird, Paradise Crow, Glossy-mantled, and Crinkle-collared Manucodes, many others subsequently proved to be the remarkable products of interspecific hybridisation (see Appendix 1).

René Lesson's participation in the epic voyage of the 'Coquille' began the long period of what was to become intense scientific collecting of birds of paradise for museum collections. This reached its zenith from the 1860s to the 1930s (see Appendix 2). By the turn of the century the plume collectors were joined in the forests of New Guinea and associated islands by enthusiastic private collectors and naturalists eager to acquire and observe birds of paradise.

Larger institutions sent individuals, small teams, or major expeditions to New Guinea to obtain bird specimens for their collections. They also regularly bought from dealers resident in the area. Foremost among these various field-based efforts (Appendix 2) in terms of significant collections of birds of paradise were: C. E. H. von Rosenberg (during 1858–70), A. B. Meyer (1873), A. A. Bruijn and his hunters (1871–85), A. Goldie (1876–1884), C. Hunstein (1878–88), H. A. Bernstein (1861–3), W. Doherty (1896–7), L. D'Albertis and O. Beccari (1872–7), Sir William Macgregor (1889–97), A. S. Anthony (1895–9), C. Wahnes (1905–9), J. C. Frost (1908–34), W. Goodfellow (1904–9, 1925–6), The British Ornithologists' Union Expedition (1909–11), A. S. Meek (1894–1916), J. Bürgers (1912–3), the A. F. F. Wollaston Expedition (1912–3), the Netherlands expeditions (1909–13), A. E. and H. Pratt (1902–3, 1920–1), and H. Stevens (1932–3).

This topic cannot be concluded without reference to Lord Walter Rothschild, who invested much of his extraordinary wealth, energy, and intellect to the study of zoology. His father built for him what became one of the greatest of all private natural history museums. It was opened to the public at Tring in Hertfordshire, England in 1892 and was to house the most comprehensive collection of birds of paradise (and of many other birds, mammals, and insects) in private hands. Because the birds of paradise were very high on his list of scientific interests, Rothschild purchased large numbers of them. He employed some of the most adventurous and efficient collectors of the day to seek them out and ship them back to his museum in Tring. Sadly, a series of misfortunes led Rothschild to sell off a large segment of his bird collection, and the museum itself was shuttered (Rothschild 1983). Rothschild tried hard not to include his beloved birds of paradise in the sale but to no avail—they now reside in the Whitney Wing of the American Museum of Natural History in New York where they are freely available for study. Only some of Rothschild's prolific contributions are listed in the Bibliography, where it is made clear he enjoyed making significant bird of paradise discoveries from about 1895 to 1930.

This lengthy period of actively seeking new kinds of birds of paradise (Table 2.1) was followed by one of taxonomic reassessment. Numerous so-called species were determined merely to represent geographical forms (subspecies) of widespread and variable species. This rationalisation accounts in part for the often long and confusing list of previously-used scientific names (synonyms) of some species. With improved taxonomic perceptions more attention started to be given to the biology of the birds. In this respect the collections made by Ernst Mayr (in 1928–9), the

Table 2.1 Chronology of some major discoveries and studies of the birds of paradise.

1521	Trade skins of *Paradisaea* species obtained at Tidore, Moluccas by Elcano of Magellan's voyage (Pigafetta 1524–34)
1555	Lesser Bird first illustrated (Belon 1555)
1605	King Bird first illustrated (Clusius 1605)
c. 1610	Twelve-wired Bird first illustrated by Jacob Hoefnagel (in Stresemann 1954)
1726	Black Sicklebill, Greater Bird, and probably the Glossy-mantled Manucode first mentioned (Valentijn)
1734	Arfak Astrapia first illustrated (Seba 1734)
1750	Greater Bird first illustrated (Edwards 1750)
1758	King and Greater Birds described from plume trade skins [= Aru Is., Mayr 1941] (Linnaeus 1758)
1774–5	Magnificent, Superb Bird, Western Parotia, and Crinkle-collared Manucode first illustrated by Daubenton (1774) and then described by Montbeillard (1775)
1776	Black Sicklebill first illustrated (Sonnerat 1776)
1781	Crinkle-collared Manucode, Superb and Magnificent Birds, and Western Parotia more formally described from trade skins (Pennant 1781)
1783	Black Sicklebill named from plume trade skin [= Arfak Mountains, Hartert 1930] (Hermann 1783)
1788	Arfak Astrapia described from plume trade skin [= Arfak Mts., Beccari 1872] (Gmelin 1788)
1800	Twelve-wired Bird [= Salawati or Vogelkop, Mayr 1941] and Red Bird formally described from trade skin [= Waigeo I., Mayr 1941] (Daudin 1800)
1801	Magnificent Riflebird illustrated from plume trade skin by Levaillant (Levaillant 1801–6)
1801–6	*L'Histoire naturelle des oiseaux de Paradis* ..., illustrated by Barraband, published (Levaillant 1801–6)
1809	Lesser Bird described as *P. minor* from plume trade skin [= Dorey, Vogelkop, Mayr 1941] (Shaw 1809)
1824	Trumpet Manucode collected and live King and Lesser Birds seen at Dorey, by Rene Lesson (Lesson and Garnot 1826)
1824	Glossy-mantled Manucode collected at Dorey, Vogelkop (Lesson 1830)
1824	Paradise Riflebird first specimen purchased in Sydney, NSW by Lesson (Swainson 1825)
1834–5	*Histoire naturelle des oiseaux de paradis et des epimaques*, illustrated by Pretre and Ouchard, published (Lesson 1834)
1835	Long-tailed Paradigalla described from plume trade skin [= Arfak Mts., Beccari 1875] (Lesson 1835)
1848	Victoria's Riflebird collected by J. Macgillivray, Barnard Is., N. Queensland (Gould 1850)
1850	Wilson's Bird described from plume trade skin [= Waigeo I., Mayr 1941] (Bonaparte 1850)
1851	Paradise Crow described (Bonaparte 1851)
1858	Standardwing Bird discovered by Alfred Russel Wallace on Bacan I. (Gray 1859)
1858	Magnificent Riflebird first encountered in the wild near Dorey, Vogelkop by Allen for A. R. Wallace (1869)
1863	Wilson's Bird first encountered in the wild on Waigeo I. by Bernstein (1864)
1869	*The Malay Archipelago* published (Wallace 1869)
1870	Western Parotia, Superb Bird, and Arfak Astrapia found to live in Arfak Mountains by Rosenberg's collectors (Rosenberg 1875)
1872	Buff-tailed Sicklebill discovered in Arfak Mountains by D'Albertis (Sclater 1873)
1873	Raggiana Bird first collected by corvette 'Vettor Risani' crew at Orangerie Bay and obtained by D'Albertis (Sclater 1873)
1873	*A Monograph of the Paradisaeidae*, illustrated by Joseph Wolf, published (Elliot 1873)

38 The Birds of Paradise

Table 2.1 *continued*

1874	Curl-crested Manucode discovered on Fergusson I. by Dr. P. Comrie (Sclater 1876, Mayr 1941)
1875	Volume I of John Gould and Bowdler Sharpe's *The Birds of New Guinea and the Adjacent Papuan Islands*, including all known birds of paradise illustrated by hand coloured lithographs, published (Gould and Sharpe 1875)
1876	Jobi Manucode discovered on Jobi I. [= Japen I.] by Beccari and L. Laglaize for Bruijn (Salvadori 1876)
1880	Pale-billed Sicklebill discovered at eastern coast of Geelvink Bay (Oustalet 1880)
1880–3	*Ornitologia della Papuasia e delle Molucche* published (Salvadori 1880–3)
1882	Goldie's Bird discovered by A. Goldie on Fergusson Island (Salvin and Godman 1883)
1883	Lawes' Parotia discovered [= Owen Stanley Mts.] by Hunstein (Ramsey 1885)
1884	Brown Sicklebill, Stephanie's Astrapia, and Blue Bird discovered by Hunstein in Owen Stanley Range (Finsch and Meyer 1885)
1888	Emperor Bird described from plume trade skin [= Sattelberg Mts., Mayr 1941] (Cabanis 1888)
1889	Crested Bird discovered at Mt. Knutsford, Owen Stanley Mountains by W. Macgregor (De Vis 1890)
1891–8	*Monograph of the Paradisaeidae and Ptilonorhynchidae* published (Sharpe 1891–8)
1893	Loria's Bird discovered in Owen Stanley Range by L. Loria (Salvadori 1894)
1894	Carola's Parotia and King of Saxony Bird described from Weyland Mountains [Mayr 1941] (Meyer 1894*a,b*)
1895	Splendid Astrapia described from plume trade skin, probably from Weyland Mountains [Mayr 1941] (Rothschild 1895)
1896	Macgregor's Bird discovered on Mt. Scratchley, Wharton Ra. by A. Guilianetti with W. Macgregor (De Vis 1897)
1896	Yellow-breasted Bird from plume trade skin [= Weyland Mts., Mayr 1941] (Rothschild 1896)
1898	*Family Paradisaeidae* published (in *Das Tierreich, Aves*) in Berlin (Rothschild 1898)
1906	Wahnes' Parotia and Huon Astrapia discovered in Rawlinson Mountains by C. Wahnes (Rothschild 1906; Foerster 1906)
1911	Short-tailed Paradigalla discovered on Mt. Goliath by A. S. Meek (Rothschild and Hartert 1911)
1910–1	Yellow-breasted, King of Saxony Birds, and Splendid Astrapia first encountered live in the wild by A. S. Meek (Rothschild and Hartert 1911)
1930	*Welche Paradiesvogelarten der Literatur sind hybriden Ursprungs* published (Stresemann 1930*a*)
1938	Ribbon-tailed Astrapia discovered by F. Shaw Mayer in Wabag area (Stonor 1939)
1950	*Birds of Paradise and Bower Birds*, illustrated by Lilian Medland, published (Iredale 1950)
1954	*Die Entdeckungsgeschichte der Paradiesvögel [History of the discovery of birds of paradise]* published (Stresemann 1954)
1969	*Birds of Paradise and Bower Birds* published (Gilliard 1969)
1977	*The Birds of Paradise and Bower Birds* published (Cooper and Forshaw 1977)

well-organised and highly successful Archbold Expeditions of 1933–9, the various expeditions of E. Thomas and Margaret Gilliard from 1948 to the 1960s and those of Fred Shaw Mayer (1928–49), S. Dillon Ripley (1937–60), Jared Diamond (1965–8), Richard Schodde and colleagues of the Australian CSIRO, Canberra (1966–90), Charles Sibley and the Alpha Helix Expedition (1969), and the Bernice P. Bishop Museum in Honololu (1967–89) stand out as highly significant (see Appendix 2).

Specimens taken by these latter collectors and institutions were far more valuable to museums and scientists because they were carefully prepared and documented. The early bird of paradise skins reaching Europe and America had been only crudely prepared in the traditional way by the people of New Guinea. This involved little more than removing flesh and bone (skull included), legs, and often also the wings prior to drying over a wood fire, which caused significant shrinkage and often some discoloration. While the shrinkage was desirable to plume traders, as it greatly exaggerated the plumes, the skins resulting from this treatment left a great deal to be desired by ornithologists, who require properly-prepared entire skins.

Museums around the world contain, in total, several million bird study skins, the BMNH and AMNH alone each house in excess of a million and 800 000, respectively (Knox and Walters 1992; M. LeCroy *in litt.*). During a recent tour of perhaps three-quarters of the larger collections containing birds of paradise, the senior author examined more than 6000 bird of paradise study skins bearing at least a location and date of collection on their labels. Many thousands of earlier study skins and all plume trade skins lack even this basic information and were therefore not examined in detail. Study skins provide a fundamental dataset critical to the study of geographical distribution, speciation, biometrics, plumages, and evolution. It has largely been such specimens that have enabled ornithologists to define genera, species, and subspecies of birds, their sexual and other variation. They also serve as a permanent record of the existence of a particular population at a fixed point on earth at a specific time—a permanent record that becomes more important with the advent of major changes in global environments and with wholesale extinctions of species. The importance of museum collections is today rarely acknowledged, but remains significant to the study of ornithology (see Collar 1997).

Living birds of paradise for aviaries

Several early explorers and zoologists noted how easy some birds of paradise are to maintain in captivity, some individual birds taking food from the hand within a day of capture (Wallace 1869; Guillemard 1886; Horsbrugh 1909; Goodfellow 1910). En route to England by sea from New Guinea, Wallace purchased two adult male Lesser Birds of Paradise at Singapore and fed them during the subsequent voyage of several months on cockroaches, rice, and fruit (Wallace 1869). They subsequently lived for several years in the London Zoo.

Concurrent with the period of collecting birds of paradise for plumes and museum specimens was an extensive period of collecting live birds and their subsequent maintenance in European and American aviaries. In 1908 the London Zoo housed 10 species (Seth-Smith 1923a). Walter Goodfellow (1910) noted that a Mr Brook of Hoddam Castle, Dumfries, Scotland, had no fewer than 19 species of birds of paradise in captivity. The first recorded attempt at captive breeding took place in the Hoddam Castle aviaries where a Stephanie's Astrapia laid three single-egg clutches but failed to incubate them (Seth-Smith 1923a). By 1935 some 22 paradisaeid species had been maintained in the New York Zoological Society (Crandall 1935). Little effort was, however, made to breed these birds while they remained relatively easy to replace. Attitudes changed dramatically after the Second World War as birds became extremely expensive and difficult to obtain. By this time only major zoological gardens or wealthy individuals could afford to own and keep these ultimate avian showpieces.

Sir William Ingram, founding editor of the popular Illustrated London News, was a well-connected Englishman with a deep personal commitment to wildlife conservation and aviculture. While his concern for the survival of the Greater Bird of Paradise was based on a misunderstanding of its conservation status, he was prepared to spend large sums of money in an effort to establish an independent population of this species free of hunting pressure. During 1909 Sir William had 44 Greater Birds translocated from the Aru Islands to the small (1.6 km long) uninhabited West Indian island of Little Tobago, which he had purchased for this purpose. By 1912 the number of Greater Birds introduced from the Aru Islands to Little Tobago was at least 47 and possibly 51 (Dinsmore 1970a,b). In a popular account of the Little Tobago population, Gilliard wrote of c. 35 individuals remaining in 1958 and in the text of his book stated all those remaining were at least third generation Little Tobago birds (Gilliard 1958,1969). By July 1966 a population of some seven birds remained (Dinsmore 1970a,b), however, and evidence is that the population was extirpated by 1980 (J. Terborgh personal communication). Before its demise, however, more useful knowledge on this translocated population had been gained than has been obtained in its native New Guinea thanks to the efforts and considerable expenditure of Ingram (C. Ingram 1913, 1956; W. Ingram 1911, 1917, 1918) and the field work of Gilliard (1958, 1969) and Dinsmore (1969, 1970a,b).

Based in Sydney, the Australian industrialist Edward Hallstrom developed a considerable interest in zoology and New Guinea. He raised the national and international profile of Taronga Park Zoo in Sydney and kept a personal collection of birds of paradise there. Hallstrom also maintained large private aviaries with birds of paradise in suburban Sydney and at a sheep farm at Nondugl in the Eastern Highlands of what is now Papua New Guinea, and successfully bred several species. The original manager at Nondugl was Captain Neptune Blood, a zoological collector and naturalist closely associated with the birds of paradise.

In 1953 Hallstrom employed Fred Shaw Mayer, a first-class field collector, as his Nondugl manager and aviculturist. In 1967 the Nondugl birds of paradise and other captive animals were moved to a new facility at Baiyer River. At this new park, exclusively dedicated to live zoological collections, Shaw Mayer continued to encourage captive breeding, with considerable success (Bishop and Frith 1979; Peckover and George 1992). The park was subsequently known as Baiyer River Bird of Paradise Sanctuary or simply Baiyer River Sanctuary, and became a Mecca for ornithologists and dignitaries visiting the New Guinea highlands and wishing to see living birds of paradise. In addition, many observations, films, and photographs of captive and wild birds were made in the aviaries and the surrounding natural forest within the park's grounds.

Other than the instances of captive breeding at Nondugl and Baiyer River, most of which were unfortunately not usefully documented, few birds of paradise were bred in captivity by institutions until the 1970s. Those that were successfully bred were mostly the result of inordinate care and avicultural skill, and accounts of them are therefore of value to present-day aviculturists and ornithologists. The first real success and documentation was achieved in the private aviaries of Prince K. S. Dharmakumarsinhji (1943) at Bhavnagar, India, where a nestling Raggiana Bird was raised to independence by its female parent (originally collected by Fred Shaw Mayer). A brood of two Raggiana Birds, one of each sex, was bred and raised at Taronga Park Zoo in Sydney in 1958 involving birds shipped into Australia before the 1948 import ban (Hallstrom 1959, 1962; Muller 1974).

One of the most notable and best documented captive bird of paradise breedings is that of King Birds in Sweden by the explorer and ornithologist Sten Bergman (1956, 1957d, 1968) who spent much time in New Guinea and personally collected birds for his aviary.

American aviculturist A. Isenberg had a female Red Bird nest and raise her nestling to 10 days old (Isenberg 1961, 1962). Another private American aviculturist was Edward M. Boehm. He made himself a wealthy man as a sculptor and ceramicist whose work, primarily of birds, became widely known and very popular. He built a magnificent complex of aviaries over four acres in rural Washington Crossing, near Trenton, New Jersey, and filled it with exotic birds including some birds of paradise. He obtained the services of the highly-experienced English aviculturist

Charles Everitt and together they bred the Ribbon-tailed Astrapia (Boehm 1967), Stephanie's Astrapia (Boehm 1967; Yealland 1969), and Magnificent Bird (Boehm 1967; Everitt 1965, 1973)—see Chapter 9 species accounts for details.

Before the 1970s the private owners of captive birds of paradise were not successful in breeding them or were not publishing the results of their successes. By this time birds were difficult to obtain legally. The few zoological gardens still holding birds thus realised the need to breed them and this led to more informative publications. These papers placed more emphasis on recording the biology of the birds. The first such publication reported the nesting of a female Superb Bird who raised a single nestling to fledging at Chester Zoo in England (Timmis 1970).

At Baiyer River Sanctuary, Papua New Guinea, during 1975–83 several captive Raggiana Bird breeding attempts were successful as was a breeding of the Magnificent Bird, but no more than this was published (Anon 1981, 1984).

Other than a well-documented successful 1979 breeding of Raggiana Birds in the Hong Kong Zoo (Searle 1980), more recently-documented captive breedings have been achieved only in America. Here the practical and scientific approach has been impressively meticulous. Raggiana (Rimlinger 1984), Lesser (Hundgen et al. 1990; Laska et al. 1992), and Red Birds (Todd and Berry 1980; Hundgen et al. 1991; Worth et al. 1991) have reproduced in captivity and in the process considerable knowledge was gained by the observant zoo researchers. We are also aware of a number of other successful, undocumented captive breedings of various birds of paradise, notably of Superb and Magnificent Birds at Honolulu (P. Luscomb in litt.) and the first of the Twelve-wired Bird at Jurong Birdpark, Singapore (K. M. Nyunt personal communication).

The majority of public zoological gardens or aviaries were until recently primarily concerned with exhibiting birds and not with breeding them. Adequate aviary space could not usually be made available to birds to permit regular reproduction. Should the future wild status of any bird of paradise species become critical to the point that birds are taken into captivity for breeding and subsequent reintroduction into the wild, well-documented successful captive breeding data would prove invaluable. Thus captive birds and carefully-prepared reports of their breeding might have the potential to play a significant role in the future conservation of birds of paradise. The conservation value of making any knowledge gained from captive populations publicly available is considerable, as indeed is the opportunity for the public to see these magnificent creatures in life. Another real value of captive birds of paradise today is their considerable potential for advancing our understanding of birds otherwise difficult to study in the wild. Moreover, knowledge of their reproductive behaviour, nesting biology, physiology, and other aspects gained from captives can prove of real scientific value (Crandall 1936; Bergman 1968; Gilliard 1969; Frith 1976, 1981; Frith and Frith 1981; LeCroy 1981; Coates 1990; Johnsgard 1994 and references therein). This is further demonstrated by the amount of such data obtained from captives to be found in the species accounts of Chapter 9.

Birds of paradise for study
Courtship displays

The difficult and costly enterprise of breeding birds of paradise in the West has paid considerable dividends in our knowledge, especially of the behaviour of species. Given the elaborate and beautiful nature of bird of paradise courtship displays it is inevitably these that attract the attention of the keepers of captive birds. In this regard, the early descriptions of displays performed in captivity by Ogilvie-Grant (1905), Stresemann (1931), Crandall (1932, 1936, 1937a,b, 1938, 1940, 1946a,b), Crandall and Leister (1937), Friedmann (1934, 1935), Manson-Bahr (1935), Morrison-Scott (1936), Seth-Smith (1936), Stonor

(1940), and Bergman (1956, 1957*b,c,d*) are excellent. These pioneering observations stimulated subsequent and more systematic descriptions of courtships and matings (Everitt 1962; Frith 1968, 1974, 1976, 1981, 1992; Dinsmore 1970*a*; Timmis 1972; Frith and Coles 1976; Frith and Frith 1981, 1997*c*; LeCroy 1981; Beehler 1983*d*, 1987*a*; 1988; Beehler and Beehler 1986; Coates 1990; Frith and Beehler 1997; Frith and Cooper 1996). The latter studies have permitted meaningful comparative observations and some conclusions of taxonomic significance.

Observing birds of paradise displaying and mating in their wild habitats involves considerable investments in time, physical effort, and money. Alfred Russel Wallace first described in broad terms the lekking displays of both the Greater Bird and the Standardwing (Wallace 1862, 1869) and these caused great excitement and interest. Of Greater Birds on the Aru Islands he wrote 'The adults frequent the very loftiest trees, and are shy and wary, and so strong and tenacious of life, that I know no bird of its size so difficult to kill. It is in a state of constant activity, flying from tree to tree, scarcely resting still a moment on the same branch, and, at the slightest alarm, flying swiftly away among the tree tops' (Wallace 1857).

Seventy years would pass before naturalists provided detailed descriptions of the courtship displays performed by wild birds of paradise. The first of these was the detailed observations of courtship-chasing Macgregor's Birds and the finely-illustrated court-based display and mating of the Magnificent Bird by Rand (1940*a,b*).

The difficulties of making useful observations of paradisaeid display in the wild are emphasised by the limited number of similar studies published since Rand's initial papers, these being of the Black Sicklebill (Ripley 1957), Greater Bird (Dinsmore 1970*a,b*), King of Saxony Bird (Beach 1975; Frith and Frith 1997*c*), Victoria's Riflebird (Sharland 1977; Frith and Cooper 1996), Emperor Bird (Draffan 1978), Stephanie's Astrapia (Healey 1978*b*), Goldie's Bird (LeCroy *et al.* 1980), Lesser Bird (Beehler 1983*d*), Pale-billed Sicklebill (Beehler and Beehler 1986), Buff-tailed Sicklebill (Beehler 1987*a*), Superb Bird (Frith and Frith 1988), Raggiana Bird (Beehler 1988), Lawes' Parotia (Pruett-Jones and Pruett-Jones 1990), Standardwing (Bishop 1992; Frith 1992), and Twelve-wired Bird (Frith and Beehler 1997).

Biological studies of wild birds of paradise

In the introduction to his monograph of the birds of paradise and bowerbirds, Bowdler Sharpe (1891) correctly observed that good biological data on wild bird populations, invaluable for systematic analyses, were badly needed to advance understanding of the evolution of these lineages. He continued 'but that we shall ever discover them can scarcely be expected, for the aim of every ordinary collector in the present day seems to be, not to furnish us with details of the nesting-habits of the Birds of Paradise, but to see how many of these beautiful creatures he can procure for the decoration of the hats of the women of Europe and America'. He expressed his frustration as an ornithologist by writing. 'I cannot draw a line between the Paradisaeidae and Ptilonorhynchidae, [the bowerbird family] simply from lack of information as to the habits of many species'. Sharpe would certainly take much pleasure in both the status of the birds in the wild and our knowledge of them today, although there remains very much more to learn about them.

The year 1950 saw the start of two decades of remarkable growth in our knowledge and understanding of bird of paradise biology (Table 2.2, Appendix 2), mainly resulting from the field and museum work of E. Thomas Gilliard. Originally stimulated by the exciting ornithological discoveries made in New Guinea by the American Museum of Natural History's Archbold Expeditions (Archbold and Rand 1940, see Appendix 2), Gilliard was to spend the rest of his life as a student of New

Guinea ornithology and of the birds of paradise (and bowerbirds) in particular. Tragically, his repeated long-term need for anti-malarial medication while in New Guinea apparently had serious consequences. He died of a heart attack in New York on 26 January 1965 aged 53, before the publication of his monograph (Murphy and Amadon 1966; Gilliard 1969).

Gilliard's work in the field as a meticulous collector and observer of free-flying birds, his studies in the museum, and his prodigious output of publications, were equally remarkable. The sum of his works, embodied in his posthumous book, provided both the first serious synthesis of the biology of the birds of paradise and a theoretical framework for their evolution, systematics, mating systems, and associated behaviour. The importance of this monograph (Gilliard 1969) was that it brought into sharp focus the great significance of the birds of paradise to biologists of many persuasions. Here was a single family of songbirds that exhibited the extremes of monogamous and polygynous mating systems, sexual dimorphism, adaptive radiation in bill shape and size, courtship behaviour, degrees of subspeciation, and even hybridisation. The family continues to provide fertile research in sociobiology, behavioural ecology, systematic, adaptive morphology, molecular evolution, historical biogeography, foraging ecology, and more.

The name of Mary LeCroy of the Department of Ornithology at New York's AMNH is synonymous with that of both E. Thomas Gilliard and the birds of paradise. Having assisted Gilliard in his museum work on New Guinea birds in general and the birds of paradise in particular she then undertook not only the writing up and production of a number of substantial ornithological contributions but also of seeing Gilliard's monograph through to publication after his death (Gilliard and LeCroy 1961, 1966, 1967, 1968, 1970; Gilliard 1969). Since that time she has provided continuous support to ornithologists working in the New Guinea region and has shown particular interest in anyone's work on the birds of paradise. She has also contributed in no small measure with her own field and museum studies of the Paradisaeidae (LeCroy 1981, 1983, unpublished data; LeCroy *et al.* 1980).

While almost nothing has been learned of the Greater Bird in the New Guinea wilds (other than that males form leks), much of the daily activities, vocalisations, and courtship displays of the population introduced onto Little Tobago Island (see above) has been documented. Some male display postures were initially described and photographed by Gilliard (1958, 1969) who visited the island with bird photographer Frederick Truslow in Feb–Mar 1958. It was the study of this population by James Dinsmore during Oct–June 1966, however, that documented much of the present knowledge of Greater Birds. He had seen the potential for a serious study of one of the most fascinating birds of paradise without even leaving the Americas and he made the best of this unique opportunity (Dinsmore 1967, 1969, 1970*a,b*). This done, it remained for other young, fit, and enthusiastic biologists to follow in the impressive footsteps of Thomas Gilliard in the distant and demanding forests of New Guinea.

Jared Diamond is an American academic whose professional field is membrane physiology, but whose intellectual pursuits and publications include the avifauna of New Guinea, particularly the birds of paradise. He has found time and energy to visit many remote parts of New Guinea for various ornithological studies since 1963. His fieldwork has contributed greatly to the knowledge and understanding of the evolution and life history of the birds of paradise. His most significant contributions with respect to bird of paradise studies are undoubtedly his *Avifauna of the Eastern Highlands of New Guinea*, in which he reassessed paradisaeid systematics, and a review of the biology of the family (Diamond 1972, 1986*a*).

Since Diamond, a variety of students, including the present two authors, have studied the birds of paradise in the wild and in captivity.

Table 2.2 Bird of paradise species known to authors (= +) at the time of their works.

	Levaillant 1801–6	Lesson 1834–5	Elliot 1873	Gould 1840–8, 1869, 1888	Sharpe 1891–8	Gilliard 1969
PARADISAEIDAE						
CNEMOPHILINAE						
Loria's Bird					+	+
Crested Bird					+	+
Yellow-breasted Bird					+	+
PARADISAEINAE						
Macgregor's Bird					+	+
Paradise Crow				+	+	+
Glossy-mantled Manucode			+	+	+	+
Jobi Manucode					+	+
Crinkle-collared Manucode			+	+	+	+
Curl-crested Manucode				+	+	+
Trumpet Manucode			+	+	+	+
Long-tailed Paradigalla			+	+	+	+
Short-tailed Paradigalla						+
Arfak Astrapia	+	+	+	+	+	+
Splendid Astrapia					+	+
Ribbon-tailed Astrapia						+
Stephanie's Astrapia					+	+
Huon Astrapia						+
Western Parotia	+	+	+	+	+	+
Lawes' Parotia				+	+	+
Wahnes' Parotia						+
Carola's Parotia					+	+
King of Saxony Bird					+	+
Magnificent Riflebird		+	+	+	+	+
Paradise Riflebird		+	+	+	+	+
Victoria's Riflebird			+	+	+	+
Superb Bird	+	+	+	+	+	+
Black Sicklebill		+	+	+	+	+
Brown Sicklebill					+	+
Buff-tailed Sicklebill				+	+	+
Pale-billed Sicklebill				+	+	+
Magnificent Bird	+	+	+	+	+	+
Wilson's Bird			+	+	+	+
King Bird	+	+	+	+	+	+
Standardwing Bird			+	+	+	+
Twelve-wired Bird	+	+	+	+	+	+
Lesser Bird	+	+	+	+	+	+
Greater Bird	+	+	+	+	+	+
Raggiana Bird			+	+	+	+
Goldie's Bird				+	+	+
Red Bird	+	+	+	+	+	+
Emperor Bird					+	+
Blue Bird					+	+
Number of taxa known	9	12	21	25	38	42

These include the late Harry Bell, David Bishop, Brian Coates, Robert Draffan, Dawn Frith, David Gillison, Don Hadden, Chris Healey, Navu Kwapena, Andy Mack, Roy and Margaret Mackay, William Peckover, Thane K. Pratt, Stephen and Melinda Pruett-Jones, Richard Whiteside, and Debra Wright (see Bibliography).

Images of birds of paradise

For more than 150 years after the first footless and wingless trade skins of plumed birds of paradise to be seen by the western world arrived in Spain, their great novelty was expressed on canvases of the period. Fully-plumed male birds were repeatedly depicted, initially as wingless, darting, and meteor-like celestial creatures and subsequently as birds borne on their own wings across oil-paint skies above all manner of scenes, by numerous European masters (e.g. see Jackson 1993, 1994).

It is true to say that while the mythology surrounding the birds of paradise has been laid to rest, the facts that have come to light concerning their biology, behaviour, and appearance are no less fascinating. As a result of this and their intrinsic beauty the demand for images of them continues to grow.

The earlier bird of paradise monographs were mainly a showcase for beautiful illustrations. Such works are now eagerly sought as collector's items. The splendid and ornithologically-informative hand-coloured works of Daniel G. Elliot (1873) and Richard Bowdler Sharpe (1891–8) were even at the time of their appearance expensive publications issued in parts to subscribers. These fine and rare books were illustrated by some of the greatest of bird artists, including John Gould, Joseph Wolf, William Hart, and J. G. Keulemans. Painted images were transferred onto slabs of stone used to produce lithographic prints which were hand-coloured by crafts people using the artists' originals as reference. Today collectors of antiquarian books pay prices in excess of £35 000 for these magnificent volumes. Not to be overlooked in this context is the first volume of John Gould's great five-volume work on the *Birds of New Guinea*, which reviews and sumptuously illustrates 23 bird of paradise species and two hybrid forms (Gould and Sharpe 1875–88).

In 1977 yet another large-format book on the birds of paradise and bowerbirds was issued, illustrated by Australian artist William T. Cooper and accompanied by species accounts by Australian ornithologist Joseph M. Forshaw. This was the first monograph to appear that dedicated a full plate to each paradisaeid species since the grand hand-coloured folios of a century earlier. Cooper's illustrations reflect his knowledge of living birds and show the birds more accurately and life-like than in any preceding work (see also Plates 4–12 herein). While the creators of this beautiful book stated it primarily represented a companion work to Gilliard's monograph (which had been intended to be fully illustrated in colour) it contained much new information and met with great demand. It quickly went out of print.

The chronology of publication of the more significant and/or finely-illustrated books on birds of paradise appears in Table 2.1 and those (presently acknowledged) species presented in the more important of them are listed in Table 2.2. The former indicates a steady continuing interest in, or market for, such books particularly as it does not include smaller books on the family that deal with only part of the group (e.g. Bergman 1968; Peckover 1990).

Recent improvements in film and camera technology have now made it possible for the world at large to experience, at home, the wonder of birds of paradise in display in the forest. The Australian Broadcasting Corporation sponsored a film on birds of paradise in the wild that was the product of film maker David Parer (1981). For years this remained the best cinematography of the birds in life. It was to be eclipsed by an even more remarkable work that appeared in 1996—a labour of love and long-held ambition by Sir David Attenborough. This was spectacular colour film of 11 species of birds of paradise in the wild. It showed species never before filmed in the wild including previously-undescribed courtship displays (Attenborough 1996).

Papers describing displays of Victoria's Riflebird (Frith and Cooper 1996), Standard-wing (Frith 1992), Twelve-wired (Frith and Beehler 1997), and King of Saxony Bird (Frith and Frith 1997*c*) are in part the result of analysis of film or video tape of wild bird activity obtained for Attenborough's film. This technology, especially developments in video recording, cameras and tape, is revolutionising the study of animal behaviour and should facilitate another quantum leap in future studies of bird of paradise courtship and nesting (e.g. Frith and Frith 1995*b*).

Where from here?

We should in no way imply from this review that there is nothing left to discover or study about the birds of paradise in the wild. Nothing could be further from the truth. Certainly only a handful, at most, of new paradisaeid taxa (probably subspecies only) remain lurking, undescribed, in New Guinea's mountainous strongholds. But there are many other mysteries awaiting solutions in the fields of distributions, behaviour, ecology, systematics, physiology, and evolution. There are a myriad of questions with regard to species distributions (e.g. Crested and Blue Birds), displays (e.g. Crested, Yellow-breasted), interactions between distinct populations (e.g. Trumpet Manucode, Magnificent Riflebird), nesting (e.g. Yellow-breasted, Standardwing), seasonal movements (e.g. Macgregor's Bird), longevity, male plumage acquisition, foraging ecology, and mate-choice that await resolution by a new generation of field students. We attempt to highlight the questions we find most interesting or most pressing in the species accounts in Chapter 9. The more one studies the birds of paradise, the more it becomes clear how little is actually known about them.

3

Evolution and biogeography of the birds of paradise

As have most traditional treatments (Mayr 1941*a*; Bock 1963; Gilliard 1969; Diamond 1972; Cooper and Forshaw 1977), we consider the birds of paradise monophyletic—a single evolutionary lineage constituting the family Paradisaeidae. As may become clear from our systematic analysis that appears later in this chapter, proof of the evolutionary cohesiveness (e.g. monophyly) of the various genera here treated as birds of paradise is built on a rather slim dataset, and yet it remains the most parsimonious treatment of this assemblage of genera. Resolution of some of the thornier questions related to the systematics of the group may come once molecular analyses (see, for example, Christidis and Schodde 1992; Nunn and Cracraft 1996) are performed on a complete range of the species in the family. In particular, we believe keys to this systematic puzzle are the proper positioning of the genera *Cnemophilus*, *Loboparadisea*, *Macgregoria*, *Paradigalla*, and *Astrapia*, taxa not included in the molecular analyses of Sibley and Ahlquist (1990), Christidis and Schodde (1992), or Nunn and Cracraft (1996), for lack of tissue samples. Having made this admission, it should now become clear that some aspects of a discussion of origins and evolution of the group must remain speculative, and this we fully acknowledge.

Starting at the most general level, we know that the birds of paradise are passerine (or 'perching') birds—a group that represents *c.* 60% of all bird species. The approximately 9–10 000 species of bird are divisible into passerine and non-passerine lineages. The passerine group is subdivided into a more primitive group (the suboscine assemblage) and a more advanced group (the oscines), based, in part, on construction of the avian syrinx. Birds of paradise belong to the latter group.

Systematic relationships within the oscine assemblage, which is thought to be the product of a relatively-recent evolutionary radiation, have long been a source of scholarly debate—possibly no more so than during the past two decades. An examination of these debates about oscine systematics is impractical here; for that, the reader is referred to Sibley and Ahlquist (1990), the most useful single review of avian systematics currently available.

Of course, it was Sibley and Ahlquist who revolutionised the world's view of the systematics of birds with their reassessment of the entire class Aves using DNA–DNA hybridisation as a probe of inter-taxonomic relationships. Although today their methodology has largely been superseded by analyses based on actual sequencing of segments of the genetic code, Sibley and Ahlquist's work provided the first molecular assessment of the origins of the birds of paradise and their current sister-groups. This is just one very important example in a long line of systematic analyses produced, over the decades, by the students of the Paradisaeidae.

Origins of the birds of paradise

Linnaeus, who formalised the field of scientific taxonomy in his oft-revised *Systema Naturae*, was perhaps the first to classify the birds of paradise (1758: 110). His treatment placed *Paradisaea apoda* and *Paradisaea regia* (today's *Cicinnurus regius*) in the order 'Aves Picae' or crow-like birds, which this chapter will show to be a remarkably-prescient assessment. Gmelin (1788: 399–402) maintained this group's position in the Aves Picae, but noted that the Picae was, in fact, a 'grab-bag' of unrelated birds with crow-like features. Lesson (1831: 335–9), in his *Traité D'Ornithologie*, treated the birds of paradise as a distinct family in the passerine order, following the corvine genera *Corvus* and *Nucifraga*.

Linnaeus's and Lesson's works were followed by a series of updates to Linnaeus's original attempt to order the birds of the world (see Sibley and Ahlquist 1990: 184–245), all of which are linear sequences that provide, at best, an imperfect rendering of branching relationships (see Table 3.3). This sort of treatment would continue, with ever-greater elaboration of defining characters. During much of this time, very little was known of the biology of birds of paradise other than what could be divined from a museum study skin.

Naturalists and systematists remained fascinated with the birds of paradise and their origins, and rarely failed to make some sort of comment on relationships. Alfred Russel Wallace (1869: 317) believed the Paradise Crow to be a starling, but Guillemard (1886: 237) thought it a 'miniature crow approaching the Paradisaeidae'. Even more perceptively Gray (1870) treated the Paradise Crow as a manucode, using the binomial *Manucodia pyrrhoptera*. It would be the likes of Wallace, D'Albertis, and other pioneering field naturalists who would eventually provide field observations and more reliable specimen material that could serve as grist for the systematic mill.

Conservative treatment placed the birds of paradise near the widespread crows, orioles, and drongos, as well as the strictly Australasian bowerbirds and affiliated families. In a break from this tradition, Stresemann (1934) considered the Australasian butcherbirds to be close relatives of the birds of paradise. Following this treatment, the butcherbirds and closely-related endemic Australasian families continued to be treated as close relatives of the birds of paradise.

The birds of paradise received their first in-depth systematic assessment by Stonor (1937, 1938) who studied cranial anatomy, musculature, and the pattern of feather distribution on the body (pterylography) in search of clues to relationships. Most notably, given Sharpe's (1899–1909) placement of bowerbirds and some birds of paradise in a single family, Stonor (1937) reported 'I am unable to find a single feature which warrants inclusion of the Ptilonorhynchidae with the Paradisaeidae.' Stonor further noted (1938: 417) 'The limits of the family are well-defined, although their precise relationships are uncertain, and it is difficult to go further than stress their similarity to the Corvidae …' No less an authority on avian systematics than Dean Amadon (1944) thought the birds of paradise should be included in the Corvidae.

Beecher (1953) developed a phylogeny (or family tree) of the oscines based on morphology, musculature, and details of the bony structure of the skull. Beecher placed the birds of paradise on a major branch arising from the bulbuls, which was shared with the New Zealand wattlebirds (Callaeidae) and the bowerbirds (Ptilonorhynchidae). The crows and orioles he placed on a separate branch that originated from the bulbuls. Berger's (1956) study of the anatomy of the Red Bird of Paradise produced conflicting data on relationships, about which the author noted 'When further evidence is available to show that the Birds of Paradise do or do not belong to the corvine assemblage, it seems likely that the judgment will be made on a relatively few anatomical features'.

Bock (1963), in his influential systematic study of the birds of paradise and bowerbirds, hypothesised that the starlings were the most likely sister group to the Paradisaeidae. Mayr and Greenway (1962) placed the birds of paradise between the bowerbirds and the crows, whereas Gilliard (1969) did not name a possible sister group, but considered the ancestors of these birds to be 'doubtless forest loving, arboreal birds whose reproductive processes were patterned on monogamy'. Diamond's linear sequence (1972) was based on that of Mayr (1941a), which had the immediate relatives postulated by Mayr and Greenway (1962) reversed, by placing the birds of paradise adjacent to the crows, but followed by the bowerbirds.

Families most frequently cited as closely allied with the birds of paradise include (other than the bowerbirds) the crows, starlings, Old World orioles, drongos, Australian mudnest builders (Grallinidae), bell-magpies and butcherbirds (Cracticidae), and New Zealand wattlebirds (Callaeidae) (Table 3.1). Here we follow results of the DNA–DNA hybridisation work of Sibley and Ahlquist (1990) with regard to the placement of birds of paradise, and we largely follow that treatment in the discussion below.

Sibley and Ahlquist treat the birds of paradise as a tribe within the 'corvine assemblage'—a large radiation of passerines of corvine stock, which today has a global distribution. This corvine assemblage, which they classified as the Parvorder Corvida, was subsumed in the Suborder Passeri. Within the Parvorder Corvida are three superfamilies—the Menuroidea (lyrebirds and scrub birds), Meliphagidoidea (honeyeaters), and the Corvoidea (a large group including Australian robins, shrikes, logrunners, and crow relatives). Sibley and Ahlquist hypothesise the Parvorder Corvida probably originated in Africa or Eurasia, with early representatives of the lineage dispersing to Australia during the Tertiary Period where a massive radiation occurred.

The Superfamily Corvoidea includes the 'old endemics' of Australia and New Guinea (Sibley and Ahlquist 1990: 614), plus lineages that are presumed to have originated in Australia and dispersed and radiated on other continents (crows, magpies, jays, orioles, cuckoo-shrikes, leafbirds, etc.). Within this lineage, Sibley and Ahlquist's family Corvidae (much more inclusive than traditional treatments) totals some 650 species, including many Australian endemics, plus the more widespread drongos, monarchs, fantails, orioles, etc.

Within the Corvidae is Sibley and Ahlquist's subfamily Corvinae, which includes six lineages traditionally treated as families in their own right—crows and jays, birds of paradise, wood-swallows, butcherbirds, orioles, and cuckoo-shrikes. The phylogeny of this group produced by Sibley and Ahlquist (1990: Fig. 378) shows the crows and jays to be the sister group to the other five, and that the birds of paradise are the sister group to the orioles, cuckoo-shrikes, wood-swallows, and butcherbirds (including two obscure genera—the shieldbills [*Peltops*] of New Guinea and the monotypic Bristlehead [*Pityriasis*] of Borneo. The implication of Sibley and Ahlquist's phylogeny is that the four sister groups to the birds of paradise are equally closely related and thus it is not possible to pinpoint a single 'closest' sister-lineage from among these (see Sibley and Ahlquist 1990: Fig. 378).

That being said, we need to highlight an anomaly within which may lie a well-defined sister-lineage to the birds of paradise. The Lesser Melampitta *Melampitta lugubris*, a little known, but common and widespread, terrestrial insectivore endemic to the mountains of New Guinea, appears in Sibley and Ahlquist's phylogeny as the basal bird of paradise (see Sibley and Ahlquist 1987). For a number of reasons we have not followed this treatment, but we do concede that the Lesser Melampitta, though not a paradisaeid, may indeed represent a closely-related sister form to the birds of paradise. If that is corroborated by further research, then the birds of paradise may in fact have a narrowly-defined outgroup—one that is even less well known than the birds of paradise. The reason we voice misgivings about the hypothesised melampitta–paradisaeid relationship is that it is based on a single tissue sample dating from the 1960s, and there are many more features that distinguish the Lesser Melampitta from the Paradisaeidae than link it to the family.

The Lesser Melampitta is an obscure, diminutive, pitta-like ground-dweller. It is a sexually-monomorphic, glossy-black oscine never before thought to be remotely close to birds of paradise. It was almost completely unknown in life when Sibley and Ahlquist (1987) stated that it was a bird of paradise. Subsequent field study indicated that nothing about this bird's nesting behaviour would corroborate a relationship with the birds of paradise (Frith and Frith 1990a). The view that the Lesser Melampitta's nest is similar to that of Loria's and the Crested Bird (Sibley 1996) is correct only to the extent that all three are domed, as are nests of numerous passerine groups and species. Nests of cnemophilines are otherwise conspicuously different in most respects, with the materials and form of lining being typical of birds of paradise but those of the melampitta not being so (Frith and Frith 1990a). The Lesser Melampitta's egg and nestling morphology are most unlike those of birds of paradise as is almost everything about the sociobiology of this bird. It is age and sexually dimorphic in iris colour. Within the birds of paradise, sexual dimorphism in iris colour is known only in the Black Sicklebill. Thus while we accept that this odd and taxonomically-interesting bird would appear to be more closely related to the birds of paradise than ever considered possible, we can see no justification for its inclusion in the Paradisaeidae. We concur with Sibley's former view (cited in Diamond 1986a) that the Lesser Melampitta 'appears distantly related to' birds of paradise and as such it is not considered further in this book except as an outgroup for our systematic analysis.

As a final point of clarification, the phantom-like Greater Melampitta *Melampitta gigantea* is almost certainly not closely related to the Lesser Melampitta, but instead may prove to be a terrestrial pitohui—the Papuan lineage recently highlighted by the discovery that their skin and feathers carry an alkaloidal toxin that serves as a chemical defence from predators or ectoparasites (Dumbacher *et al.* 1992).

Thus we are left with the somewhat unsatisfying hypothesis (supported by data from Sibley and Ahlquist) that the Lesser Melampitta is the closest living relative of the birds of paradise. We say unsatisfying because we would like to see further systematic analysis of the Lesser Melampitta, and because this little bird is, itself, a mystery in terms of phylogenetic affinities.

The Lesser Melampitta aside, we accept the hypothesis that at the next level of distance, the Paradisaeidae is the sister-group to the oriole/cuckoo-shrike/wood-swallow/butcherbird radiation. At the next higher level, the crows and allies are the sister group to the paradisaeid + oriole/butcherbird lineage.

There seems little doubt that the Paradisaeidae originated from the same stock as the crows and their allies. But it is difficult, indeed, to enumerate shared derived features that link them all without at the same time linking these taxa to other groups. Such are the challenges of higher level oscinine systematics.

Zoogeographically, most of the sister groups to the Paradisaeidae have more extensive geographical ranges westward into southern Asia (wood-swallows) or even to Africa (cuckoo-shrikes, orioles), while the crows, of course, span the globe. Only the butcherbirds show a concordant distribution with the Paradisaeidae, with the sole cracticid outlier being the peculiar Bristlehead of Borneo.

We next review the range of avian lineages that have been nominated as close relatives of the birds of paradise.

True crows

The birds of paradise share a number of morphological and life-history characters with the crows. While it is true that only a single species of the Corvidae has ventral barring in the plumage (the immature of the Crested Jay *Platylophus galericulatus* of the Malay Peninsula, Sumatra, Java, and Borneo) the majority of the few unbarred Paradisaeidae are quite crow-like. Barring, which is widespread in the birds of paradise, may have arisen because of selection pressures favouring female crypticity given that they alone carry out nesting duties. Other characters of some or many corvine genera are typical of the Paradisaeidae, including naked or near-naked nestlings that quickly become dark-skinned, fine structurally-modified feathering projecting forward and over the nostrils, powerfully 'crow-like' legs and feet, glossy and iridescent plumage, erectile feathering above and behind each eye, pale or brightly-pigmented iris colour, whitish or pale blue bill colour, long and sometimes spatulate-tipped central rectrices, narial tufts, and elongated, plush, or curled crown feathering. Abnormal fawn/grey plumaged birds occur in both groups (see below).

Behaviour found in members of both the Paradisaeidae and the crows and jays that might be of taxonomic significance includes regurgitation of meals to nestlings, open-bill probing, use of a foot or feet to hold food to a substrate (Kramer 1930), a hopping gait in the main but the ability to walk at times, sunning and a characteristic sideways posture when doing so, a predilection for bathing and preening, and even a fondness for hanging completely inverted, and finally, in the Asia/Pacific region, the tendency of many of the species to join mixed-species foraging flocks (that typically include a drongo *Dicrurus* sp). While crows typically use wooden sticks in their nest foundation, only the cnemophilines (to a limited degree) do so within the Paradisaeidae.

Starlings

Starlings are strong-footed and strong-billed, generally crow-like birds, that do open-bill probing but do not use their feet for feeding, do not have feathered nostrils, and are predominantly hole nesters. Authorities have tended to affiliate starlings with birds of paradise as a result of similarities in skull features (Bock 1963; Borecky 1977; Cracraft 1981). Starlings exhibit plumage features that show similarities to those of some birds of paradise, but otherwise appear unrelated.

Orioles

Orioles typically build tidy, sparse nests suspended, hammock-like, in a branch fork like manucodes and the Paradise Crow. The orioles, relatively weak-footed, with a very short tarsus, show relatively few obvious shared characteristics with the birds of paradise. These include a tendency for bare skin around the eye, red iris colour, strong bill, and sexual dimorphism.

Wood-swallows

The wood-swallows are superficially quite unlike any of the birds of paradise, mainly because they have evolved a specialised lifestyle of aerial insect-foraging that certainly requires a body form specific to the task—compact body, short neck, pointed, triangular wings, and very short tarsus. The species tend to be quite sociable with a lot of inter-individual contact—features unknown in the birds of paradise.

Australian butcherbirds

The butcherbird lineage exhibits several paradisaeid traits, including its zoogeographical distribution, preference for forest and woodland, glossy black plumage, slender skull with nasal ossification, the strong and stout bill, strong legs (used to hold food by at least *Strepera*), and far-carrying vocalisations. The group also exhibits characters distinct from the Paradisaeidae: the primarily black-and-white plumage, twig nest, peculiar flight pattern of species of *Cracticus*, habit of impaling or wedging prey items on thorn or fork of a branch, and pugnacious defence of the nest.

New Zealand wattlebirds

The New Zealand wattlebirds use their feet to hold food, forage by open-bill probing, and possess facial wattles like some birds of paradise. Moreover the extinct Huia *Heteralocha acutirostris* exhibits the rare 'reversed' sexual dimorphism in bill length (longer in females) and shape found in some birds of paradise (Frith 1997). A serious objection to the notion of the wattlebirds being particularly close to birds of paradise was zoogeographical (Schodde 1976) but this has recently been negated by research arguing that the probably extinct New Zealand Thrush *Turnagra capensis* was most closely related to bowerbirds (Olson *et al.* 1983; Christidis *et al.* 1996), even though widely separated from them geographically.

The Silktail

The Silktail *Lamprolia victoriae* is an odd flycatcher-sized passerine peculiar to two islands in Fiji, in eastern Melanesia. Because it looks superficially like a diminutive bird of paradise it has been repeatedly considered a possible close relative (Mayr 1941*b*). Recent studies have provided evidence that the Silktail is a part of the myiagrine/monarchine flycatcher radiation (Heather 1977; Olson 1980; Frith *et al.* 1989; Sibley and Ahlquist 1990).

Bowerbirds

The bowerbirds were long believed to be the sister group of the birds of paradise (Mayr 1941*a*; Bock 1963; Gilliard 1969; Diamond 1972; Schodde 1976; see Table 3.1). The reasons for the postulated close relationship are readily apparent and understandable. Both families are confined to the Australasian region with the vast majority of member species living in rainforests of New Guinea and tropical Australia. Species of both families are typically highly sexually dimorphic, predominantly or partly frugivorous, and possess bizarre polygynous mating behaviours. Adult males of a number of species in both groups are ornately-and colourfully-plumaged, and the dull-coloured females share a patterned ventral plumage. Several species in each group clear foliage and ground litter from an area of forest floor to create a court for nuptial display. The assumed close relationship between birds of paradise and bowerbirds seemed so clear to ornithologists studying museum specimens that a number of authorities went so far as to combine the Ptilonorhynchidae with the Paradisaeidae (Lesson 1834–5; Elliot 1873; Sharpe 1891–8; Rothschild 1898; Stresemann 1934; Iredale 1950; Gilliard 1969; Schodde 1975, 1976; Cracraft 1981).

This long and uncontested systematic affiliation of birds of paradise and bowerbirds was not supported in any way by the first systematic studies of skull structures (Pycraft 1907). Also studying skull morphology, with additional feather tract, plumage, and egg characters, Stonor (1937) concluded that the bowerbirds 'constitute a singularly complete and isolated family of the acromyodian [of the Menurae and the oscine] passerine birds and show no special relationship to any other'.

Stonor's studies of skull osteology were subsequently reviewed and expanded by Bock (1963) who also examined the mandibular musculature of members of the two families. He fully agreed with Stonor's finding that the birds of paradise and the bowerbirds represent two distinct families but also concluded that the 'Paradisaeidae and the Ptilonorhynchidae are assumed to be closely related, which is supported, although not fully proven, by the cranial evidence'. Bock's main thesis encouraged subsequent ornithologists to continue to view bowerbirds and birds of paradise as sister-lineages. Moreover, they thereafter continued to be linked collectively in reviews and monographs (Gilliard 1969; Cooper and Forshaw 1977; Everett 1978; Diamond 1986*a*). Although Bock's work has been very influential, a recent unpublished assessment by BB of the key characters used in Bock's osteological analysis indicates that, given individual variation in skeletal form, such an analysis cannot be fully rendered until adequate sample sizes of skeletal material are collected. Note that many species are represented in osteological collections by only one or a handful of specimens. Thus more field collecting and additional study of skeletal characters are needed.

Gilliard was deeply interested in the life histories of wild, free-living birds of paradise and bowerbirds, but most of his field work was constrained by obligations to collect and prepare large collections of specimens for his museum while in New Guinea. Nonetheless, his unique knowledge of these birds, summarised in his posthumous monograph (Gilliard 1969), was to stimulate a generation of field investigators able to focus on the ecology, courtship, and nesting behaviour of members of both groups. Findings from these post-Gilliard field studies supported Stonor's (1937) assertion that the birds of paradise and bowerbirds are not closely related. Birds of paradise are, unlike bowerbirds, typically crow-like in their use of feet for feeding (the three wide-gaped species and possibly Macgregor's Bird excepted), in the 'open-bill probing' during foraging, and in the habit of regurgitating food to their offspring. There are, moreover, fundamental differences in the nests, eggs, and nestlings between the two groups. Finally, the results of the DNA–DNA hybridisation studies also indicated a distant relationship between the birds of paradise and the bowerbirds. Thus field and laboratory studies fully endorse the view that the two groups are not sister-lineages.

This may not spell the end of the controversy over the relationship between the birds of paradise and bowerbirds. The protein electrophoretic analysis carried out by Christidis and Schodde (1991, 1992) suggests a close relationship between the two. In the face of such diametrically-opposing views, we nonetheless see the weight of evidence from living birds as presently tipping the scales in favour of a greater rather than lesser distance between the two groups.

Systematics of the Paradisaeidae

In the scientific literature, the evolutionary relationships of taxa within the birds of paradise remain the subject of debate (Diamond 1972; Schodde 1976; Christidis and Schodde 1992; Nunn and Cracraft 1996). For use in this monograph we offer our own assessment, based on a cladistic analysis of a paradisaeid character set based on anatomy, external morphology, plumage, behaviour, and ecology. We consider our treatment to be both conservative and provisional, but it nevertheless is based on an explicit dataset and transparent analysis. We fully acknowledge the diversity of scholarly opinion on paradisaeid systematics and after the discussion of our own analysis, we provide a review of the other treatments that have been published in the recent past, with discussion of areas of concordance and incongruence.

Our phylogenetic reconstruction of the Paradisaeidae is based on 52 characters (Table 3.2) generated from our review of the biology of the family plus three outgroup taxa. Drawn from a much larger pool of possible characters, the 52 selected for our cladistic analysis we deemed appropriate because they exhibited a level of variation that resolved taxonomic relationships at the generic and subfamilial levels. The more invariant characters were useful for resolving basal branching of the phylogeny, whereas the more variable characters resolved branching in the upper branches. We make no special claims about our character matrix other than it includes characters from a broad suite of features. We have not included molecular data mainly because we question our ability to integrate properly that data-rich set with our information-rich matrix. Our character matrix and analysis includes no character weighting. The characters are coded numerically (1 or 0 for presence/absence, or ordinally for multistate characters). Character polarity of the dataset was treated as unordered in the parsimony analysis, mainly because we lacked

Table 3.1 The systematic placement of the Paradisaeidae relative to that of immediately-adjacent passerine taxa viewed as their most close relatives by previous authorities.

Gray 1869, 1870	Sharpe 1877	Sharpe 1891	Sharpe 1901–9	Stresemann 1934	Wetmore 1960	Mayr and Greenway 1956, 1962	Cracraft 1981	Bock 1982	Sibley et al. 1988	Sibley and Ahlquist 1990
Oriolidae[1] + 9 families	Oriolidae	Ptilonorhynchidae	Ptilonorhynchidae[4]	Sturnidae	Dicruridae	Sturnidae	Grallinidae	Sturnidae	Ptilonorhynchidae + 9 families to	Ptilonorhynchidae + 10 families to
Corvidae[2]	Dicruridae + 4 families	Sturnidae	Corvidae	Oriolidae	Oriolidae	Oriolidae	Corvidae	Dicruridae	Corvidae	Corvidae
Paradisaeidae	Ptilinorhynchidae	Dicruridae		Dicruridae	Corvidae	Dicrururidae	Craticidae	Corvidae	Corvinae	Corvinae
Sturnidae[3]		Oriolidae		Cracticidae	Cracticidae	Callaeidae	Sturnidae	Cracticidae	Corvini	Paradisaeini[6]
		Corvidae		Corvidae	Grallinidae	Artamidae	Callaeidae	Grallinidae	Paradisaeini[6]	Oriolini
		Paradisaeidae		Paradisaeidae[5]	Ptilonorhynchidae	Cracticidae	Paradisaeidae[5]	Callaeidae	Artamini	Artamini
					Paradisaeidae	Ptilonorhynchidae		Ptilonorhynchidae	Oriolini	Dicrurinae
						Paradisaeidae		Paradisaeidae	Dicrurinae	Aegithininae
						Corvidae			Rhipidurini	Malaconotinae
									Dicrurini	Callaeaidae
									Monarchini	Picathartidae
									+ 2 subfamilies to Callaeaidae	
									+ 3 families to Sturnidae	Sturnidae[7]

1 = includes Ptilonorhynchidae.
2 = includes some Cracticidae and Callaeidae.
3 = includes some of the Paradisaeidae (manucodes, astrapias) and Callaeidae.
4 = includes *Loria*, *Loboparadisea*, and *Macgregoria*.
5 = includes Ptilonorhynchidae.
6 = includes *Melampitta*.
7 = Sibley and Ahlquist (1990) place Sturnidae far-removed, in the Parvorder Passerida.

confidence in unambiguously assigning the direction of character-state transformations for most characters.

Because of limitations of our character set, we included only 21 taxa in our paradisaeid analysis—representing all generic and subgeneric lineages (see Table 3.2(a)). Because of this we made no effort to resolve branching patterns within most species-groups (e.g. *Manucodia*, *Parotia*, *Astrapia*, among others). These are left as unresolved 'polychotomies' in our phylogeny—each with more than two branches emanating from a single node (Fig. 3.1). Our selection of outgroups was based on the phylogeny of Sibley and Ahlquist (1990), the outgroup taxa selected being Lesser Melampitta, and the genera *Coracina* (Campephagidae) and *Artamus* (Artamidae). This dataset was analysed using the computer algorithm PAUP version 2.4 (Swofford 1985), with options set for global branch-swapping and a search for multiple equally-parsimonious trees. The computer runs produced 27 equally-parsimonious trees (each of 220 steps, and a consensus index of 0.725). The equally-parsimonious trees showed minimal significant structural variation. From these 27, we then generated a strict consensus tree (Fig. 3.1) using the CONTREE subprogram of Swofford (1986). From this consensus tree we generated the sequence of genera that is followed in this book. Species sequences within each genus follow geography (taxa listed ordinally from NW to SE) as well as inferred sister-species relationships.

We treated the Lesser Melampitta as an outgroup on most runs, but also carried out runs

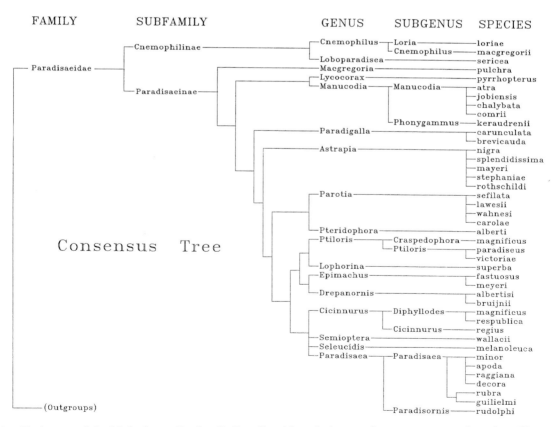

3.1 Phylogeny of the bird of paradise family Paradisaeidae—being a strict consensus tree based on 52 characters using the CONTREE subprogram of Swofford (1986). See text and Table 3.2.

Table 3.2a Matrix of 52 characters (at top, left to right) and the character states (see 3.2(b)) used to construct the bird of paradise phylogeny used herein and ascribed to the avian outgroups and the bird of paradise taxa (at left, top to bottom).

Genus/subgenus	1	2	3	4	5	6	7	8	9	10	11	12	13	14	15	16	17	18	19	20	21	22	23	24	25	26	27	28	29	30	31	32	33	34	35	36	37	38	39	40	41	42	43	44	45	46	47	48	49	50	51	52
Artamus	1	0	0	0	0	2	9	0	0	9	1	9	0	0	0	0	0	1	2	0	0	0	0	0	0	0	0	0	0	0	0	0	0	0	0	0	0	0	0	0	0	0	0	0	0	0	0	0	0	0	1	0
Coracina	0	0	0	0	0	9	9	0	0	1	1	9	0	0	1	0	0	0	1	0	0	0	0	0	0	0	0	0	0	0	0	0	0	0	0	0	0	0	0	0	0	0	0	0	0	0	0	0	0	0	2	0
Melampitta	0	0	0	0	0	0	1	0	0	1	1	0	1	0	0	0	0	1	3	0	0	1	0	0	0	0	0	0	0	0	0	0	0	0	0	0	0	0	0	0	0	0	1	0	0	0	0	0	0	0	0	0
Cnemophilus	0	0	1	0	1	0	0	0	1	1	0	1	2	0	0	1	1	1	0	0	0	0	1	1	0	0	0	0	0	0	0	0	0	0	0	0	0	0	0	0	2	0	0	0	0	1	0	0	0	1	3	0
Loboparadisea	0	0	1	0	1	0	0	0	9	9	9	9	9	0	0	1	1	1	2	1	0	0	0	9	1	1	0	1	0	0	0	0	0	0	0	0	0	0	9	0	9	0	0	0	9	0	0	0	0	1	3	9
Macgregoria	2	0	0	0	2	2	1	1	0	0	9	0	2	0	0	1	0	0	0	0	0	0	1	0	0	1	0	1	0	0	0	0	0	0	0	0	0	0	0	0	2	0	0	0	0	0	0	0	0	3	0	0
Lycocorax	2	1	0	1	2	2	1	1	2	0	0	9	3	0	0	1	0	0	0	0	1	1	0	0	1	1	0	9	0	0	0	0	0	0	0	0	0	0	0	0	2	0	0	0	0	0	9	0	0	3	3	9
Manucodia	2	1	0	1	2	2	1	1	2	0	0	1	3	0	0	0	0	0	1	1	0	9	0	1	1	0	0	1	0	0	0	0	0	0	0	0	0	0	0	0	2	0	0	0	0	0	0	0	0	3	0	0
Phonygammus	2	1	0	1	2	2	1	1	2	0	0	1	3	0	0	0	0	0	0	1	0	0	0	1	0	0	1	0	0	0	0	0	0	0	0	0	0	0	1	0	2	0	0	0	0	0	0	0	0	3	0	0
Paradigalla	2	1	0	0	2	2	1	1	0	0	0	1	2	0	0	0	0	0	0	0	0	0	0	1	1	0	0	0	0	0	0	0	0	0	0	0	0	0	9	0	2	0	0	0	9	0	0	0	0	0	4	1
Astrapia	2	1	0	0	2	2	1	1	0	0	0	1	2	2	0	0	2	0	0	0	0	1	0	1	1	0	1	0	0	1	0	2	1	1	0	1	1	0	0	0	2	0	0	0	9	0	0	0	0	0	4	1
Parotia	2	1	0	0	2	2	1	0	0	0	0	9	2	2	1	1	2	0	1	0	1	0	1	0	0	1	0	0	1	2	1	2	0	3	0	1	1	0	0	0	2	1	0	0	1	0	0	0	0	0	4	1
Pteridophora	2	1	0	0	2	2	1	0	0	0	0	9	2	2	0	0	2	0	1	0	0	0	1	0	0	1	0	1	1	0	1	2	0	0	3	0	1	1	9	0	2	0	0	0	0	0	0	0	0	0	4	1
Craspedophora	2	1	0	0	2	2	1	0	0	0	0	0	1	2	2	0	1	2	0	1	0	1	0	1	0	0	1	0	1	0	0	3	0	3	0	1	1	9	0	1	2	0	0	0	1	0	0	0	0	0	4	1
Ptiloris	2	1	0	0	2	2	1	0	0	0	0	1	2	2	1	1	2	0	1	0	0	0	0	0	0	1	0	0	1	0	1	1	0	0	2	0	1	1	1	0	2	0	0	0	1	0	0	0	0	0	4	1
Lophorina	2	1	0	0	2	2	1	0	0	0	0	9	2	2	1	1	2	0	1	0	0	0	0	0	0	1	0	1	0	0	0	0	2	0	0	2	0	1	1	1	2	0	0	0	1	0	0	0	0	0	4	1
Epimachus	2	1	0	2	2	2	1	0	0	0	0	1	9	2	0	1	2	0	1	0	0	0	0	0	0	1	0	0	1	0	3	0	0	2	0	2	1	0	1	1	2	0	0	0	1	0	0	1	0	0	4	1
Drepanornis	2	1	0	2	2	2	1	0	0	0	0	9	9	2	0	1	2	0	0	0	0	0	0	1	0	1	0	1	0	3	1	0	3	1	0	2	1	1	0	0	2	0	0	0	1	0	0	0	0	0	4	1
Diphyllodes	2	1	0	2	2	2	1	0	0	0	0	0	1	2	2	0	1	2	0	0	0	0	0	1	0	1	1	0	0	0	0	2	1	0	0	2	0	1	1	0	2	0	0	0	1	0	0	1	0	0	4	1
Cicinnurus	2	1	0	0	2	2	1	0	0	0	0	0	1	2	2	0	1	2	0	1	0	0	1	1	0	1	1	1	0	0	0	0	4	4	2	1	1	1	0	0	2	0	1	1	1	1	1	1	0	0	4	1
Semioptera	2	1	0	0	2	2	1	0	0	0	0	0	1	2	2	0	1	2	0	1	0	0	1	1	0	1	1	0	1	0	0	1	2	0	4	4	2	1	1	0	2	0	1	1	1	1	1	1	0	0	4	1
Seleucidis	2	1	0	0	2	2	1	0	0	0	0	9	9	0	0	1	2	0	1	0	0	2	1	0	1	0	0	0	0	0	0	0	0	6	2	1	1	9	0	0	1	0	1	0	1	1	0	4	1	0	4	1
Paradisaea	2	1	0	0	2	2	1	0	0	0	0	1	2	2	0	1	2	0	0	0	0	0	0	0	0	1	0	1	0	1	0	0	0	4	2	0	2	1	1	1	2	0	0	0	1	0	0	1	0	0	4	1
Paradisornis	2	1	0	0	2	2	9	0	0	0	0	1	2	2	0	0	2	0	0	0	0	0	1	1	0	1	0	1	0	5	0	0	0	3	4	0	1	1	9	0	2	0	0	1	1	0	0	0	0	0	4	1

Note: A number 9 indicates no data. See text.

Table 3.2b The 52 characters used in the matrix (see 3.2(a)) defined (at left, top to bottom) and the six ascribed character states (at top, left to right). See text.

	Character description	State Zero	State One	State Two	State Three	State Four	State Five	State Six
1	Skull: nasal ossification	Not ossified/weakly ossified	Moderate ossif./amphirhinal	Fully ossified				
2	Skull: lateral edge of palatine	Not flanged	Flanged					
3	Gape Depth	Typical	Deepened					
4	Bill morphology	Unspecialised, various	Crowlike, very strong	Sicklebilled				
5	Foot anatomy	Gracile/weak	Moderately robust	Robust-very robust				
6	Pterylography: Pteryla Spinalis pars dorsalis (rows)	13 rows	12 rows	<11 rows				
7	Pterylography: Pteryla Spinalis (number of feathers)	>220	<210					
8	Terminal rachis-protrusion of rectrix	Absent	Present					
9	Nest type	Open, supported cup	Domed, supported cup	Suspended cup				
10	Incubation period	<23 days	>26 days					
11	Hatching-condition	Naked	Downy					
12	Nestling feeding	In bill	Regurgitated					
13	Nest-materials	Typical/unspecialised	Moss-dominated	With orchids and/or ferns	Nest lined with wood chips			
14	Ventral barring of female plumage	Absent	Black and white	Brown and black				
15	Pale superciliary in female plumage	Absent	Present					
16	Female wing colour	Lacking brown or ochre	With brown or ochre					
17	Juvenile plumage (pattern)	Dull version of ad. male	Distinct from adult	Like adult female				
18	Juvenile plumage (colour)	Not distinct from adult	Entirely grey	Brown, streaked	Brown and black			
19	Mantle plumage	Sexes alike	Sexes unlike					
20	Supra-orbital feather ridging	Absent	Present					
21	Female plumage	Cryptic	Cock-feathered					
22	Leg colour	Lacking bright pigmentation	Brightly pigmented/blue	Brightly pigmented/pink	Brightly pigmented/orange			
23	Male mouth colour (skin)	Unspecialised	Specialised/brightly coloured					
24	Gape/facial wattling	Absent	Present					
25	Special contour feathers: silky/glossy/iridescent	Absent	Present					
26	Adult male with plumage, tracheal, or wattle modification	Absent	Present					
27	Male mantle plumage	Red absent	Red present					
28	Male delayed maturation	Absent	Present					
29	Male occipital flags	Absent	Present					
30	Ornamental flank plumes	Absent	Short, non-erectile	Short, erectile	Short, terminal iridescence	Large, colourful, non-erectile	Large, colourful, erectile	

Table 3.2b continued

	Character description	State Zero	State One	State Two	State Three	State Four	State Five	State Six
31	Male gular erectile display plumage	Absent	Present					
32	Feathers of metallic breast shield	Absent	*Ptiloris*-type	*Parotia*-type				
33	Erectile pectoral fan plumes	Absent	Cryptic	Iridescent-tipped				
34	Specialised neck feathering	Absent	Metallic	Velvety ear-tufts				
35	Elongation of central rectrices	Absent	Greatly elongated	Greatly elongated, sabrelike	Greatly elongated, wirelike	Elongated, wirelike, recurved		
36	Specialised crown feathering	Absent	Present, met, extens., unbounded	Met, confined to crown	Met, reduced to nuchal bar	Non met, velvety	Non met, stubbly yellow	Flattened
37	Iridescent green ventral display plumage	Absent	Smooth, metallic	Velvety	Scaly			
38	Male elongated erectile body plumes	Absent	Present					
39	Polygynous court-type advertisement and display	Absent	Present					
40	Wing sound produced by male in flight	Absent	Present					
41	Display-use of wing sound	Absent	By movement of primaries	Produced by carpel percussion				
42	*Paradisaea* display characters	Absent	Present					
43	Flight pattern	Typical songbird	Primarily non-flying	Undulating flap and glide				
44	Head-waggle display	Absent	Present					
45	Halo display (occipital wire plumes held over head)	Absent	Present					
46	Paradisaeinine advertisement vocal (cawing series)	Absent	Present					
47	Male advertisement vocalisation	Absent/irregular/weak	Persistent, regular, loud					
48	Frontal display	Absent	Present					
49	Vertical display flight over canopy	Absent	Present					
50	Obligate frugivory	Absent	Present					
51	Primary foraging method	Terrestrial litter search	Aerial	Hover-snatch/glean	General upright branch glean	Scansorial trunk glean		
52	Foot/food-manipulation	Not used	Used					

Note: A number 9 in Table 3.2(a) indicates no data. extens. = extensive; met. = metallic.

of the program in which this taxon was treated as an in-group taxon; these, however, produced trees that allied *Melampitta* with the outgroup taxa rather than the birds of paradise. For this and other reasons (see above) we do not treat the Lesser Melampitta as a bird of paradise. Our dataset and systematic analysis do not resolve the question of whether the Lesser Melampitta is the sole closest sister form to the birds of paradise.

The phylogeny

The two genera of wide-gaped birds of paradise *Loboparadisea* and *Cnemophilus* appear as a basal clade, and, following tradition, we treat this group as a subfamily. There is no dispute about the taxonomic cohesion of this small clade, but it is not robustly allied by our character analysis to the main paradisaeid radiation and we strongly recommend further study of its affinities. Characters that link the Cnemophilinae with the Paradisaeinae include the following: egg and nestling characters, nestling feeding method, nest materials, female wing coloration, specialised male mouth colouring, presence of specialised contour feathers in adult males, adult males possessing specialised secondary modifications for mate attraction, evidence of delayed plumage maturation in males, polygynous display at traditional perches, the distinctive paradisaeid flight pattern, and paradisaeid-type male advertisement vocalisation. For these reasons it is useful to continue to associate the cnemophilines with the typical birds of paradise until clear evidence arises that unequivocally contradicts this association. Given the group's morphological uniformity, its restricted distribution in montane New Guinea, and the characters that it shares with the Paradisaeidae, we maintain this is the most parsimonious resolution of the relationship.

We thus recognise two paradisaeid subfamilies, the small and obscure clade of wide-gaped species—the Cnemophilinae, and the speciose and diverse true birds of paradise—the Paradisaeinae. Within the latter much larger clade, there are several lineages, two very well defined, and three of uncertain position. The two well-defined groups are the manucodiine lineage (*Manucodia* and *Lycocorax*) and the plumed birds of paradise (including *Pteridophora*, *Parotia*, *Ptiloris*, *Lophorina*, *Epimachus*, *Drepanornis*, *Cicinnurus*, *Semioptera*, *Seleucidis*, and *Paradisaea*). Here we use the term 'plumed bird of paradise' to indicate the cluster of polygynous taxa that all have specialised display plumes. This includes four clades—the flagbirds, the riflebirds, the sicklebills, and the sickletail/*Paradisaea* cluster. The lineages of uncertain affinities are *Macgregoria*, *Paradigalla*, and *Astrapia*. *Macgregoria* and *Paradigalla* appear basal in the lineage, whereas *Astrapia* is, by our treatment, the sister group to the plumed birds of paradise, and is arguably a member of this plumed group by morphology.

Our analysis places *Macgregoria pulchra* as a basal outlier to the Paradisaeinae, as it shares characteristics with the Cnemophilinae and the true birds of paradise. Bock (1963) argued for placing it in the Cnemophilinae, but Gilliard (1969) and subsequent treatments placed it at the base of the Paradisaeinae. The relative branching position of *Macgregoria* and *Manucodia* remains weakly resolved, but the greater number of shared cnemophilinine characters lies with *Macgregoria*. This monomorphic and monogamous high-elevation specialist is, with its remnant distribution atop the central range and its many unique characters, a true relict.

Characters that link *Macgregoria* to the Cnemophilinae include the unflanged lateral edge of the palatine (Bock 1963) and a basal gap in the dorsal feather tracts (Clench 1992). Characters linking *Macgregoria* to the Paradisaeinae include the ossified nasal septum, typical (male larger) sexual size dimorphism, mechanical sound production by the wings, the typical narrow gape, strong feet, aspects of feather tracts (Fig. 3.2), nest form, briefer incubation period, and absence of a distinct juvenile plumage.

The manucodiinine lineage is very distinct, and verges on constituting a subfamily. The group is well defined by a range of characters,

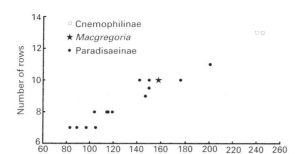

3.2 Mean number of feathers versus mean number of rows in the *pars dorsalis* feather tract of Paradisaeidae genera examined by Clench (1992). Star = *Macgregoria*, squares = the two cnemophilinine genera, solid circles = the other Paradisaeinine genera (*Paradigalla* and *Drepanornis* were not studied). After Clench (1985), with kind permission.

especially in plumage and tracheal morphology. Data from Sibley and Ahlquist (1985) indicate the manucodes and plumed paradisaeinine lineages may have branched c. 18–20 million years ago. The remarkable thing about the manucodes is that whereas there is a whole suite of characters defining them as distinct from the plumed lineage (Table 3.2), there are also characters that unambiguously link the two. Anyone who has studied manucodes and species of *Paradisaea* in the wild will note the characters they share, from the floppy undulating flight, to the body morphology and the strong legs, feet, and claws. And, once again, the geographical distribution of the manucodes and the plumed lineage is almost exactly concordant—from the Moluccas southeastward to the humid forests of eastern Australia.

The two paradigallas comprise another lineage of obscure affinities. They are little-studied and poorly-understood species, and this certainly limits our ability to place them evolutionarily. Nesting habits are paradisaeinine, but the sombre black plumage pattern shared by the sexes and the egg morphology link them to the basal groups, such as *Manucodia* and *Macgregoria*. By habit, general plumage pattern, bill morphology, and iridescence, *Paradigalla* appears closest to *Astrapia*, where our phylogeny puts it.

Astrapia is a genus of five species of primarily black-plumaged but sexually-dichromatic birds. Their main remarkable feature is the prodigiously-elongated tail of the adult male of each species. They also exhibit blackish to black dorsal female plumage and distinctive vocalisations. The dominant black coloration for both sexes links the *Astrapia* superspecies to the basal paradisaeinine taxa, but the brown-and-black ventral barring in females of three of the five species and the egg colour pattern ally the lineage with the plumed lineage. *Astrapia*, too, exhibits specialised erectile plumage on the head and mantle that is shared by the plumed group. More study of *Astrapia* is merited.

The 25 species of plumed birds of paradise are apparently the product of a single radiation (Gilliard 1969; Schodde 1976; Sibley and Ahlquist 1990; Christidis and Schodde 1992; Nunn and Cracraft 1996). Our character set linked the flagbirds and riflebirds as sister forms to a larger assemblage that, within itself, placed the sicklebills as a sister group to a cluster that includes the sickletails (*Cicinnurus* and its subgenus *Diphyllodes*) and the *Paradisaea* clade (*Seleucidis*, *Semioptera*, *Paradisaea*). This rather coarse resolution, we believe, is a product of the recent explosive radiation of the plumed birds of paradise.

The four clades within the main paradisaeinine lineage appear well defined by plumage and morphology. The 'flagbirds' include only *Parotia* and *Pteridophora*, the two genera with erectile occipital plumes. Although seemingly-distinct taxa, these two also share the presence in the males of a small iridescent throat patch (almost vestigial in *Pteridophora*), the velvety-black dorsal plumage, and some distinctive courtship behaviour. The 'riflebirds' include the monotypic genus *Lophorina* plus the superspecies of the three true riflebirds of the genus *Ptiloris*. As with the previous clade, this cluster is a somewhat novel combination, mainly because of the aberrant nature of *Lophorina*.

As a group, the riflebirds share the very distinctive metallic breast-shield, the habit of displaying on a log or stump, and the static and upright main components of display. Note that the advertisement song notes of *Lophorina superba* are nearly identical to those of *Ptiloris victoriae* and *P. paradiseus*. Some courtship display components of *Lophorina* are equivalent to those of *P. magnificus*.

The four sicklebills, although obviously linked by the remarkable bill morphology, are, in fact, two closely-related but distinct lineages, each with a number of defining characters. The group as a whole is defined by the unique sickle-shaped bill, the male pectoral plumes that are erected to frame the face in display, and the mantle and upper wing colour of females. Within the sicklebills the *Epimachus* 'sabretails' are defined by the conspicuous long and pointed tail, the iridescent mantle of the male, and prominent sexual dichromatism. The short-tailed *Drepanornis* sicklebills have a shorter rounded tail, a facial apterium, a brown iris in both sexes, and the hen-type dorsal plumage in both sexes.

The final clade within the plumed group includes several distinct sublineages, so is the most complex of any single clade within the Paradisaeinae, with four genera and four additional subgenera—eight distinct lineages above the species level. The group is linked by the presence of several shared characters—iridescent green and brilliant red and/or yellow in the display plumage, *Paradisaea*-type advertisement vocalisations, *Paradisaea*-type display postures, and the general absence or limited extent of black in the plumage of the adult males.

Within this *Paradisaea*-based clade, the diminutive sickletails include three species in the genus *Cicinnurus*, two of these comprising the subgenus *Diphyllodes*. All three have the central pair of tail feathers modified into remarkable recurved wires, iridescent green ventral display plumage, red in the dorsal display plumes, and bright pastel blue legs. The *Paradisaea* cluster is composed of the typical plumed birds of paradise (*minor, apoda, raggiana*, and *decora*), plus three aberrant species (*rubra, guilielmi*, and *rudolphi*), the last of which we accord subgeneric status. The genus *Paradisaea* is defined by the presence of both filamentous flank plumes and the unrecurved wire or tape-like central tail feathers.

Two additional odd monotypic genera complete the *Paradisaea* clade—*Semioptera* and *Seleucidis*. Each belongs in this group, but each has sufficiently-discordant character attributes to merit its own genus. *Semioptera wallacii* inhabits the northern Moluccas and almost certainly is a sister form of the sickletails and *Paradisaea* group (the green ventral plumage of males, and the fact that the sickletails and *Paradisaea* are the only higher paradisaeines on the adjacent west Papuan islands). *Semioptera* is very aberrant and its weirdness is probably a product of founder effects (Graves 1995). *Seleucidis melanoleuca* is less aberrant than *Semioptera*, but its correct systematic position is difficult to assign firmly because it possesses two suites of characters, one of which links it to the riflebirds (female plumage, bill morphology, predominantly black male plumage, etc.), the other of which links it to the *Paradisaea* lineage (filamentous yellow flank plumes, *Paradisaea*-type displays, *Paradisaea*-type advertisement vocalisations, pale leg and iris colour, etc.). We posit that these last cited characters are derived characters, whereas the three preceding characters are primitive, and thus of lesser systematic value. Schodde (1976) reached similar conclusions.

Paradisaea ('*Paradisornis*') *rudolphi* merits a comment. All recent monographs have treated this species as a member of the genus *Paradisaea*. We, on the other hand, only followed suit with the greatest reluctance, for even though this species is clearly derived from the *Paradisaea* sublineage (by vocalisation, erectile flank plumes, elongated central tail feathers, etc.), it possesses so few shared characters and so many unique character states (cock-plumed female, black in both male and female plumage, blue wings and flank plumage, broken white eye-ring, inverted display, absence of green in male display plumage, etc.)

that it almost certainly merits being treated as a genus of its own. We have not done so mainly because this would force us to treat the less aberrant *Paradisaea rubra* and *P. guilielmi* as distinct genera.

Other phylogenies

Anyone who has worked in systematics knows that phylogenies attract attention and criticism, and the literature is often home to phylogenies produced by different workers, often using differing methodologies. There is no 'final' phylogeny. This is because the recovery and reconstruction of the branching pattern of the evolutionary tree is not a simple, transparent process. Published phylogenies are hypotheses of the evolutionary process, nothing more. Although we posit that the phylogeny that we present in Fig. 3.1 represents our best current approximation of paradisaeid evolution, we believe it is useful to review the range of available phylogenies that have appeared in the scientific literature over the last five decades (Fig. 3.3). In some cases we have had to construct these based on imperfect information, as earlier treatments did not present graphic or unambiguous trees (see Table 3.3), and most have been simplified to exhibit only major branching. Our effort to present this has

Table 3.3 Order and taxa of linear systematic lists of the birds of paradise taxa by previous authors.

Elliot 1873	*Rothschild 1898*	*Stonor 1938*	*Mayr 1942*	*Iredale 1950*	*Mayr and Greenway 19*
PARADISAEIDAE	PARADISAEIDAE	PARADISAEIDAE	PARADISAEIDAE	PARADISAEIDAE	PARADISA
PARADISAEINAE	*Loboparadisea*	*Lycocorax*	[*Lycocorax*]	PAROTINAE	CNEMOPH
Paradisaea	*Cnemophilus*	*Manucodia*	*Manucodia*	*Ptiloris*	*Loria*
Manucodia	*Loria*	*Phonygammus*	*Phonygammus*	*Craspedophora*	*Loboparadise*
Astrapia	*Paradigalla*	*Macgregoria*	*Macgregoria*	*Lophorina*	*Cnemophilus*
Parotia	*Macgregoria*	*Paradigalla*	*Paradigalla*	*Parotia*	PARADISA
Lophorina	*Parotia*	*Loria*	*Drepanornis*	*Seleucidis*	*Macgregoria*
Cicinnurus	*Lophorina*	*Loboparadisea*	*Epimachus*	*Paradigalla*	*Lycocorax*
Paradigalla	*Pteridophora*	*Parotia*	*Astrapia*	*Macgregoria*	*Manucodia*
Semioptera	*Ptiloris*	*Craspedophora* [*Ptiloris*]	*Parotia*	ASTRAPINAE	*Phonygammu*
EPIMACHINAE	*Drepanornis*	*Lophorina*	*Lophorina*	*Astrapia*	*Ptiloris*
Epimachus	*Seleucidis*	*Semioptera*	*Craspedophora*	EPIMACHINAE	*Semioptera*
Seleucidis	*Astrapia*	*Pteridophora*	[*Ptiloris*]	*Epimachus*	*Seleucidis*
Ptiloris	*Cicinnurus*	*Epimachus* [*Drepanornis*]	[*Semioptera*]	*Drepanornis*	*Paradigalla*
	Diphyllodes	*Seleucidis*	*Seleucidis*	CICINNURINAE	*Drepanornis*
	Semioptera	*Astrapia*	*Diphyllodes*	*Cicinnurus*	*Epimachus*
	Paradisaea	*Paradisaea*	*Cicinnurus*	*Diphyllodes*	*Astrapia*
	Manucodia	*Cicinnurus* [*Diphyllodes*]	*Paradisaea*	PARADISEINAE	*Lophorina*
	Phonygammus		*Pteridophora*	*Paradisaea*	*Parotia*
	Lycocorax		*Loria*	MANUCODIINAE	*Pteridophora*
			Loboparadisea	*Lycocorax*	*Cicinnurus*
			PTILONORHYNCHIDAE	*Manucodia*	*Diphyllodes*
			Cnemophilus	*Phonygammus*	*Paradisaea*
				UNDETERMINED FAMILY	
				Pteridophora	
				Loria	
				Loboparadisea	
				Semioptera	
				PTILONORHYNCHIDAE	
				Cnemophilus	

Stonor (1938) presented a dendrogram, not a linear list, and as the above is derived from it other interpretations are possible.
Mayr (1942) did not deal with extra-New Guinea taxa and so these are inserted (in parenthesis) where Mayr, in Mayr and Greenway (1962), latter place
Gilliard (1969) is the order the author presented on page 47, not that in which species accounts appear in that work.
Diamond (1972) suggested a gross reduction of genera (indicated in parenthesis).
Cooper and Forshaw (1977) followed the order taxa appear in the species accounts of Gilliard (1969: 20) as opposed to his subsequent view (Gillia 1969: 47).
Beehler and Finch (1985) did not list subfamilies or deal with extralimital *Lycocorax* and *Semioptera* which we have added. They accepted only thos reduced genera of Diamond (1972) as indicated in parenthesis.
In this work we acknowledge the subgenera indicated in square parenthesis in the right hand column.

been aided by the review provided by Nunn and Cracraft (1996) and our figure is an adaptation of a figure in their publication. We comment on the most recent treatments in the following section.

Diamond (1972) carried out a major generic revision of the family, based on a non-cladistic assessment using plumage similarity to guide formation of taxa. No tree was presented. Instead Diamond offered a novel lineage list and a novel generic formulation, mainly by collapsing small genera into larger ones. This was an attempt by Diamond to highlight the recent nature of the radiation of Paradisaeinae and to link closely-related forms. Without pro-viding an explicit phylogeny, however, this simply reduced the recoverable systematic information available in his taxonomy. Basally, Diamond's treatment is quite similar to ours, except for his placement of *Semioptera*. Diamond also places the sicklebills as sister forms to the main paradisaeinine radiation, and places *Astrapia* in the heart of this latter clade, which we do not. Finally, Diamond's treatment differs from ours by placing *Seleucidis* with the riflebirds rather than the genus *Paradisaea*.

Schodde (1976) provided, like Diamond, a non-cladistic assessment, but with an explicit but non-traditional family tree. Schodde's

1969	Diamond 1972	Cooper and Forshaw 1977	Beehler and Finch 1985	Cracraft 1992	Sibley and Monroe 1996	Frith and Beehler (this volume)
ISAEIDAE	PARADISAEIDAE	PARADISAEIDAE	PARADISAEIDAE	PARADISAEIDAE	PARADISAEINI	PARADISAEIDAE
oria	CNEMOPHILINAE	CNEMOPHILINAE	CNEMOPHILINAE	CNEMOPHILINAE	*Melampitta*	CNEMOPHILINAE
x	*Cnemophilus* [*Loria*]	*Loria*	*Cnemophilus* [*Loria*]	*Loria*	*Loboparadisea*	*Cnemophilus* [*Loria*]
dia	*Loboparadisea*	*Loboparadisea*	*Loboparadisea*	*Loboparadisea*	*Cnemophilus*	*Loboparadisea*
mmus	PARADISAEINAE	*Cnemophilus*	PARADISAEINAE	*Cnemophilus*	*Macgregoria*	PARADISAEINAE
	Macgregoria	PARADISAEINAE	*Macgregoria*	*Macgregoria*	*Lycocorax*	*Macgregoria*
era	*Lycocorax*	*Macgregoria*	*Lycocorax*	PARADISAEINAE	*Manucodia*	*Lycocorax*
is	*Manucodia*	*Lycocorax*	*Manucodia*	*Lycocorax*	*Semioptera*	*Manucodia*
ulla	[*Phonygammus*]	*Manucodia*	[*Phonygammus*]	*Manucodia*	*Paradigalla*	[*Phonygammus*]
rnis	*Semioptera*	*Phonygammus*	*Paradigalla*	*Phonygammus*	*Epimachus*	*Paradigalla*
us	*Paradigalla*	*Ptiloris*	*Ptiloris*	*Ptiloris*	*Lophorina*	*Astrapia*
	Epimachus	*Semioptera*	[*Semioptera*]	*Semioptera*	*Parotia*	*Parotia*
a	[*Drepanornis*]	*Seleucidis*	*Seleucidis*	*Seleucidis*	*Ptiloris*	*Pteridophora*
	Lophorina [*Parotia*,	*Paradigalla*	*Epimachus*	*Paradigalla*	*Cicinnurus*	*Ptiloris*
hora	*Ptiloris*, *Cicinnurus*	*Drepanornis*	[*Drepanornis*]	*Drepanornis*	*Astrapia*	[*Craspedophora*]
us	*Astrapia*, *Pteridophora*,	*Epimachus*	*Astrapia*	*Epimachus*	*Pteridophora*	*Lophorina*
des	*Seleucides*]	*Astrapia*	*Lophorina*	*Astrapia*	*Seleucidis*	*Epimachus*
ea	*Paradisaea*	*Lophorina*	*Parotia*	*Lophorina*	*Paradisaea*	*Drepanornis*
hilus		*Parotia*	*Pteridophora*	*Parotia*		*Cicinnurus*
		Pteridophora	*Cicinnurus*	*Pteridophora*		[*Diphyllodes*]
adisea		*Cicinnurus*	[*Diphyllodes*]	*Cicinnurus*		*Semioptera*
		Diphyllodes	*Paradisaea*	*Diphyllodes*		*Seleucidis*
		Paradisaea		*Paradisaea*		*Paradisaea*
						[*Paradisornis*]

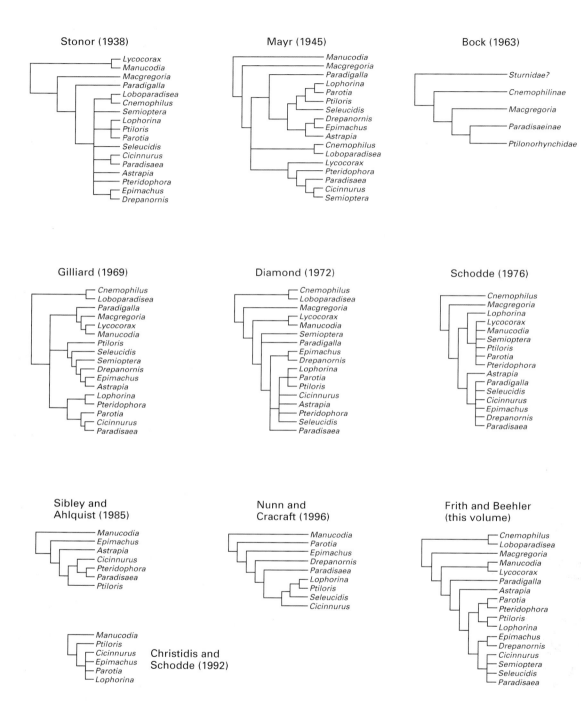

3.3 Phylogeny of the bird of paradise family Paradisaeidae acknowledged by previous authors (levels of fusion arbitrary).

treatment, as do most traditional assessments (Fig. 3.3), places the Cnemophilinae and *Macgregoria* as basal lineages. Assuming we have interpreted his tree correctly, it appears that he has delineated two paradisaeinine lineages, one with the manucodes, riflebirds, *Semioptera*, and *Pteridophora*, the other with *Astrapia*, *Paradigalla*, *Seleucidis*, the sicklebills, the sickletails, and *Paradisaea*.

Sibley and Ahlquist (1985) provide an explicit but partial phylogeny based on DNA–DNA hybridisation studies. This phylogeny places the sicklebills and *Astrapia* basal to the upper paradisaeinine clade, *Paradisaea* and *Pteridophora* as sister taxa (excluding the sickletails), and the riflebirds sister to this latter more inclusive lineage.

Christidis and Schodde (1992), using protein data from *Manucodia* (*Phonygammus*), *Cicinnurus* (+ *Diphyllodes*), *Drepanornis*, *Parotia*, *Lophorina*, and *Ptiloris*, is most distinct from our treatment in placing *Ptiloris* basal to a clade that includes *Cicinnurus*, *Parotia*, *Epimachus*, and *Lophorina*. The dissociation of the riflebirds from *Lophorina* and the flagbirds is a curiosity.

Nunn and Cracraft (1996) provided a first partial phylogeny based on base-pair sequences of mitochondrial DNA. As with those of Sibley and Ahlquist and Christidis and Schodde, their analysis was constrained by lack of material from a number of key genera (the Cnemophilinae, *Macgregoria*, *Paradigalla*, *Astrapia*, *Pteridophora*, and *Semioptera*). Their dataset produced a phylogeny distinct from ours in several important ways. *Parotia* and the sicklebills are placed basally in the Paradisaeinae; the two sicklebill lineages are separated; and *Paradisaea* is placed in a lineage basal to a clade that includes the sickletails, *Seleucidis*, and the riflebirds. Their tree agrees with ours in sister placement of *Ptiloris* and *Lophorina*.

In sum, this remarkable diversity of results reinforces three points. First, there is considerable agreement about the general placement of basal taxa, but this is possibly no more than a product of custom rather than data analysis. There is too little information to address this objectively at this stage, so tradition prevails. Second, there is little agreement about the phylogeny of the upper half of the tree—intergeneric relationships remain poorly resolved in the Paradisaeinae. Third, this latter radiation is, in all likelihood, a recent one, with rapid character evolution being driven by sexual selection and the effects of court and lek mating systems. We are not confident that the systematic uncertainties exemplified here will be fully resolved in the near future.

A final reckoning may perhaps be coming, however, with the advent of DNA sequencing (e.g. Nunn and Cracraft 1996), mainly dependent upon the acquisition of fresh field samples of DNA-bearing tissue from the key genera (especially *Melampitta*, *Cnemophilus*, *Macgregoria*, *Lycocorax*, *Paradigalla*) plus the delineation of genes that have evolved rapidly enough to track the radiation that all agree has been in large part a remarkably-recent phenomenon (Gilliard 1969; Schodde 1976; Christidis and Schodde 1992, 1993).

Treatment at the species level

Compare the paradisaeid species lists of Mayr (1941*a*, and in Mayr and Greenway 1962), Gilliard (1969), Diamond (1972), Beehler and Finch (1985), and Coates (1990) and one will find negligible discordance of species lists, with the total in the family ranging from 40 to 43 species (Table 1.5). The two main points of contention in these 'traditional' lists are (1) whether the major taxa in the *Paradisaea apoda* superspecies group are accorded full species status, and (2) whether *Parotia lawesii helenae* is indeed a subspecies (as treated herein) or should be considered a full species of *Parotia*. These are rather minor issues, indeed. In other words, by employing the biological species concept, ornithologists over the last 55 years have established a remarkable stability of species definition within this group. Strongly-polytypic taxa (e.g. *Manucodia keraudrenii*, *Lophorina superba*, *Parotia carolae*, *Paradisaea raggiana*) are commonplace in the family, but there has been little contention about the appropriate limits of these more complex taxa. We follow this traditional treatment, based on the biological species concept.

In a break with tradition, Cracraft (1992) reassessed the species of birds of paradise using the phylogenetic species concept, in which each taxonomically-diagnosable population is accorded species status. With this method, the list of 'species' of birds of paradise rose dramatically to *c.* 90. Cracraft argues that this is necessary to permit an objective phylogenetic assessment of the group. We believe that the reason cited by Cracraft for requiring this sort of treatment is both unnecessary to his goal of phylogeny construction and taxonomically arbitrary in many instances (see Snow 1997). First, the taxa (whether traditionally called 'species' or 'subspecies') had already been well defined 20–50 years ago and have already been readily accessible to objective systematic analysis. They have names and they possess codable characters, so any systematist can assess relationships of all taxa by whatever means deemed appropriate. Second, a considerable number of geographical forms (deemed 'subspecies' of polytypic species under traditional treatment) are poorly defined and possess characters that vary clinally with the named taxa simply representing distinct endpoints on geographical clines. The phylogenetic species concept has no mechanism for accommodating continuously-varying characters, and must use the same 'arbitrary' process that is a principal criticism of the more traditional methods used in applying the biological species concept.

Although we agree with the concept of basal taxa, we believe, in the case of the birds of paradise, that there is no substantial utility or necessity in reforming species limits because neither is such a process required to derive the basal taxa, nor are there sufficient data available on geographical variation at the 'subspecific' level to apply the phylogenetic species concept objectively at this time. Many of the 'species' delineated by Cracraft (1992) would fail Cracraft's own test of being objectively definable by characters that are expressed without variation throughout the population.

Given the extreme morphological divergences between related bird of paradise species or superspecies, the typically trivial differences between most bird of paradise subspecies (*Lophorina* excepted) are striking. This might, for example, reflect strong sexual selection against divergence in reproductively connected populations (J. Diamond *in litt.*).

Biogeography and speciation

The avian family Paradisaeidae has one of the most restricted distributions of any of the bird families of the Australasian region (Fig. 1.1). Although there are a number of species-poor 'relict' groups with more circumscribed distributions (e.g. scrub birds, lyrebirds, etc.), there are few species-rich families with ranges more limited than the birds of paradise. The 19 bowerbirds inhabit New Guinea, Australia, and (if *Turnagra* is included) New Zealand. The 25 fairy-wrens and the 15 quail-thrushes are confined to Australia and New Guinea. No Australasian group with as many species as the birds of paradise inhabits as small a geographical range.

The current disposition of the species of birds of paradise (Fig. 1.1) strongly supports the thesis that the group radiated in New Guinea. All of the major generic radiations (e.g. *Manucodia, Astrapia, Parotia, Paradisaea*) are either endemic or largely confined to New Guinea, and, in addition, the 'aberrant', 'old endemic' or 'relictual' forms (e.g. *Cnemophilus, Loboparadisea, Macgregoria, Paradigalla, Pteridophora*) are entirely confined to the uplands of New Guinea. No lineages have dispersed across the Vitiaz Straits to the Bismarck Archipelago, and the outliers in Australia and the Moluccas are either local representatives of New Guinea lineages (*Manucodia, Ptiloris*) or else are apparently recently evolved from New Guinean lineages (*Semioptera*, from the *Cicinnurus/Paradisaea* ancestor, *Lycocorax* from the *Manucodia* ancestor). No major lineages or relictual forms are found off the island of New Guinea.

Noting that New Guinea and Australia share the same tectonic plate, and are separated only by a shallow and geologically-ephemeral water barrier (Torres Strait/Arafura Sea) that has been breached most recently between 40 000 and *c.* 8000 years ago (Kikkawa *et al.* 1981; Heatwole 1987), it may seem anomalous that the family Paradisaeidae is not more 'Australian'. An explanation for this is that, unlike the Ptilonorhynchidae, the Paradisaeidae has never evolved any dry-habitat lineages, and that most of Australia today is arid. Thus we see today that the bowerbirds range virtually throughout Australia, whereas the birds of paradise inhabit only the relictual patches of humid forest in the east and northeast. This is yet another reason for believing that the birds of paradise evolved in the wet upland forests of the New Guinea's central mountain cordillera that over recent geological time never gave way to open habitats and woodland during even the driest climatic periods (Kikkawa *et al.* 1981).

The Australian distributions of the riflebirds and the Trumpet Manucode appear to be the products of either very recent (Magnificent Riflebird, Trumpet Manucode) or moderately recent (the two smaller riflebirds) geographical speciation events (Beehler and Swaby 1991). We postulate two sequential colonisations during periods when the Torres Strait was dry and when closed forest extended from New Guinea southward into eastern Australia. The first colonisation was by an ancestral riflebird that differentiated geographically into the three current forms. The second colonisation was by the Magnificent Riflebird (*contra* Christidis and Schodde 1992) and Trumpet Manucode (only to Cape York Peninsula)—recently enough that these isolates have not yet achieved distinct species status.

The Moluccan outpost of the Paradisaeidae is a bit of an anomaly, about which we know little, in part because the two species are relatively poorly known (especially the Paradise Crow). Still, we believe the evidence is that each of these is the product of a single overwater dispersal event, both older than any of the Australian colonisations. We see no evidence for any significant evolutionary radiation in the Moluccan paradisaeids—just insular subspeciation that is, by far, strongest in the Crow (which we believe to be a product of an older dispersal event). The remarkable morphological divergence of the Standardwing male and female is deceptive and we believe is the product of strong sexual selection and the effects of a very small founding population, which may include a mutation known as a non-eumelanic schizochroism that produces a plumage lacking in eumelanin (Graves 1995).

To return to our hypothesis that the family evolved on the island of New Guinea, it is worth noting that in no location off the New Guinea mainland and its adjacent land-bridge islands can one today count more than three sympatric bird of paradise species, whereas virtually any mainland forested site supports no fewer than five and as many as nine sympatric species. The great, tropical, physiographically-diverse island of New Guinea appears to be an evolutionary engine for biological diversity and species radiations.

Subfamilial and generic patterns

Each lineage within the birds of paradise, once it has diverged from its sister-lineage, takes on a history of its own that is no longer tied to its relatives. These are, thus, nested subsets of differentiation that tell a series of biogeographical stories—some unique and some remarkably concordant with chronologically-comparable lineages.

The Cnemophilinae is a lineage with a distribution narrowly confined to the highlands of the central mountainous spine (or cordillera) of New Guinea—from the Weyland Range in the west to the Owen Stanley Range in the southeast. By contrast, the species-rich Paradisaeinae entirely encompass the range of the family—all outliers are members of this larger subfamily. Note that the range of the putative sister-form to the family as a whole, the Lesser Melampitta, is much like that of the Cnemophilinae (but also includes the two main outlier mountain groups of the Huon Peninsula and the Vogelkop). And, finally, the apparently relictual Macgregor's Bird of Paradise inhabits a range that nearly identically approximates that of the Cnemophilinae—but is slightly more restricted because of its specialisation on high altitude forest. Thus the three most basal paradisaeid lineages are strictly Papuan-montane (i.e. confined to the uplands of New Guinea) without exception—further evidence of the family's restricted evolutionary geography.

One other generic lineage is restricted to the New Guinea's central cordillera—that of *Pteridophora*. All other genera range more widely. For instance, the genera *Paradigalla*, *Astrapia*, *Parotia*, *Lophorina*, and *Epimachus* are mainly distributed on the central cordillera, but also have representative populations on the various outlier ranges: mountains of the Vogelkop and Huon Peninsula, plus the north coastal ranges in some instances (*Lophorina*, *Epimachus*). Of the 17 genera we recognise in the family, 10 are restricted to the highland regions of New Guinea.

The genera *Manucodia*, *Ptiloris*, *Cicinnurus*, *Seleucidis*, and *Paradisaea* are primarily lowland-dwelling, and not surprisingly have broader distributions that encompass the island of New Guinea, some key fringing islands, and, in the case of *Manucodia* and *Ptiloris*, include eastern Australia. In most cases these genera show nearly continuous and widespread distributions over the New Guinea lowlands, but *Seleucidis* is exceptional in being absent from the easternmost section of the northern watershed (apparently because that area lacks flat expanses of lowland habitat because of the presence of coastal mountain ranges).

Both *Manucodia* and *Paradisaea* have dispersed across deep water barriers, albeit ones probably considerably narrowed by Pleistocene (1.6 million to 10 000 years ago) lowering of sea level. *Manucodia* populations are today found on the three main islands of the D'Entrecasteaux Archipelago, the Trobriand Islands, and Tagula Island of the Louisiade Archipelago (all in southeastern insular PNG), and *Paradisaea* inhabits the upland forests of two islands of the D'Entrecasteaux group and Waigeo and Batanta Islands of the Raja Ampat group in the far west of Irian Jaya. *Cicinnurus respublica* is confined to Waigeo and Batanta also. Finally, the two Moluccan paradisaeids are geographical isolates of these same two lineages that have elsewhere demonstrated the ability to disperse across water.

Distributional patterns of species

Generic distributions are composites of species distributions. These can be broken down into species and species-group distributions, which create geographical patterns that may help inform us of the modes of speciation that produced these groups (Fig. 3.4). The distributions of most species are readily classifiable into one of three categories: widespread (16 spp.), regional isolate (18 spp.), and insular isolate (6 spp.): see Table 3.4. Only the Pale-billed Sicklebill and Blue Bird fall into none of these categories, and can best be classified as having 'unexplained' distributions—restricted ones that fit no pattern.

Widespread distributions are those that occupy most or all of the habitat available to the species. Thus the range of montane species like Loria's Bird extends the length of the central cordillera. Or, for instance, the distribution of a lowland species like the King Bird extends throughout the lowlands of mainland New Guinea and onto a number of the land-bridge islands.

The regional isolate distributions exemplify the effects of geographical speciation in a widespread ancestral population. A montane example would be the parotias, with regional isolates on the Vogelkop (*P. sefilata*), the western and central cordillera (*P. carolae*), Huon Peninsula (*P. wahnesi*), and eastern cordillera (*P. lawesii*). A lowland example would be the *Paradisaea apoda* superspecies, with mainland

Evolution and biogeography of the birds of paradise 69

forms in the southwest (*P. apoda*), west and northwest (*P. minor*), and east (*P. raggiana*).

The widespread distributions of some species are of interest, when contrasted with the regional isolate distributions, because of the curious absence of speciation over their extensive ranges. It will be interesting to speculate why some widespread populations have differentiated into a series of regional forms, whereas others have not. One can assume that age of the distribution might have something to do with it, as well as dispersal ability of local populations.

Insular species distributions are different from mainland New Guinean regional isolates in several respects (aside from the trivial fact that they involve islands). First, bathymetric data suggest that every insular species isolate colonised its insular range via active over-water dispersal. Second, all but one appear to give evidence that they are older and more divergent than typical New Guinean regional isolates from the mainland. Only Wilson's and Goldie's Birds are unambiguous recent sister forms of the widespread Magnificent and Raggiana Birds, respectively. The Curl-crested Manucode is grossly differentiated from its postulated sister form (Crinkle-collared Manucode), the Red Bird is nothing like its Vogelkop sister form, the Lesser Bird, and the two Moluccan endemics are divergent enough to merit full generic status.

Speciation

Historical biogeography is a kind of distributional archaeology that, among other things,

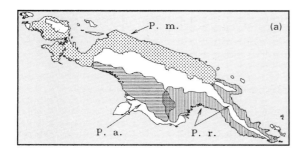

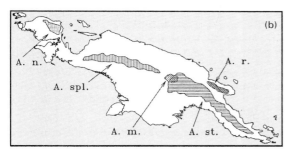

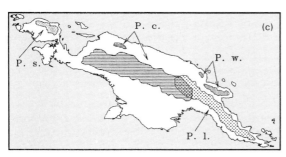

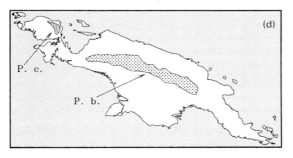

3.4 Sample geographical distributions of paradisaeid species-groups on the New Guinea mainland. (a) *Paradisaea apoda* superspecies, P. a. = *Paradisaea apoda*, P. m. = *P. minor*, P. r. = *P. raggiana*. This exemplifies a lowland ring-species distribution. (b) A. n. = *Astrapia nigra*, A. spl. = *A. splendidissima*, A. m. = *A. mayeri*, A. st. = *A. stephaniae*, A. r = *A. rothschildi*. This distribution, as well as the examples that follow (c,d) exemplify patterns that are created by speciation in montane lineages, hypothetically produced by the 'drop-out' phenomenon (see Fig. 3.5). (c) P. s. = *Parotia sefilata*, P. c. = *P. carolae*, P. l. = *P. lawesii*, P. w. = *P. wahnesi*. (d) P. c. = *Paradigalla carunculata*, P. b. = *P. brevicauda*. The drop-out phenomenon has produced species groups of 5, 4, and 2 species taxa in these latter three examples.

Table 3.4 Types of bird of paradise geographical distributions

Species	Widespread	Regional isolate	Insular isolate	Anomalous
Loria's Bird	+			
Crested Bird	+			
Yellow-breasted Bird	+			
Macgregor's Bird	+			
Paradise Crow			+	
Glossy-mantled Manucode	+			
Jobi Manucode		+		
Crinkle-collared Manucode	+			
Curl-crested Manucode			+	
Trumpet Manucode	+			
Long-tailed Paradigalla		+		
Short-tailed Paradigalla		+		
Arfak Astrapia		+		
Splendid Astrapia		+		
Ribbon-tailed Astrapia		+		
Stephanie's Astrapia		+		
Huon Astrapia		+		
Western Parotia		+		
Lawes' Parotia		+		
Wahnes' Parotia		+		
Carola's Parotia		+		
King of Saxony Bird	+			
Magnificent Riflebird	+			
Paradise Riflebird		+		
Victoria's Riflebird		+		
Superb Bird	+			
Black Sicklebill	+			
Brown Sicklebill	+			
Buff-tailed Sicklebill	+			
Pale-billed Sicklebill				+
Magnificent Bird	+			
Wilson's Bird			+	
King Bird	+			
Standardwing Bird			+	
Twelve-wired Bird	+			
Lesser Bird		+		
Greater Bird		+		
Raggiana Bird		+		
Goldie's Bird			+	
Red Bird			+	
Emperor Bird		+		
Blue Bird				+
Distribution type totals	16	18	6	2

uses current species patterns to reconstruct speciation events that transpired in the distant past. For birds of paradise the patterns are few and are, in most instances, remarkably straightforward, allowing us to envision repetition of past events in different lineages, and also providing apparent snap-shot views of geographical differentiation at different stages in the process. We recognise five distinct patterns of differentiation, as follows.

Lowland differentiation.—This is exemplified by the ring-species concept of Rensch (1933), which was applied by Mayr (1942) and others, and is typified by the pattern created by the geographical differentiation of the *Paradisaea apoda* superspecies (Fig. 3.4). In this process, it is postulated that a widespread parent population occupies the lowlands that ring a central mountain range, forming a distribution that roughly approximates a doughnut shape. Periodically, as climate changes, the natural creation of habitat discontinuities produces distributional barriers in one or more sections of the 'doughnut', and the geographically-segmented populations diverge into distinct subspecies or species. The postulated 'barriers' to distribution are created by habitat shifts or range constrictions that, in times of climatic regression (e.g. desiccation, cooling, etc.) form patches of unfavourable habitat that reduce or prevent gene flow between neighbouring populations in the favourable habitat patches. With time, these regional isolates may diverge genetically to a degree that when they expand their ranges and recontact their neighbour the two maintain their own identities instead of merging back together and reamalgamating into a single freely-breeding population.

For the current distributions of birds of paradise, we cite what we believe exemplify the various stages in this differentiation process. Stage one: King Bird of Paradise, inhabiting the entire 'doughnut' of lowland New Guinea, with two poorly-differentiated races. Stage two: Magnificent Riflebird, with vocally very distinct (near-species) populations in (1) the eastern and (2) central/western lowlands. Stage three: the three-species example of the *Paradisaea apoda* superspecies. Stage four: the manucode species-group having achieved full lowland sympatry in some species (e.g. Glossy-mantled and Jobi Manucodes—with the expansion of the range of the Glossy-mantled to include the entire lowlands of mainland New Guinea).

Cordilleran differentiation.—This is the process highlighted in Diamond's (1972, 1973) 'drop-out' model of speciation (Fig. 3.5). In this process, a single parental form has a distribution that includes the entire central cordillera. Subsequently, because of local extinction of intermediate populations and minimal dispersal, regional isolates achieve effective isolation and differentiate into two or more distinct cordilleran isolate populations. With time and further ecological differentiation, the cordilleran isolates successfully invade each other's ranges and achieve sympatry. Stage one in this process is exemplified by Loria's Bird, with minor geographical variation over the length of the cordillera. A second stage is presented by the distribution of Macgregor's Bird, a species with central cordilleran populations that are only slightly differentiated and yet separated by an enormous east–west gap. Stage three is best exemplified by the cordilleran populations of astrapia (Splendid in west, Ribbon-tailed in central, and Stephanie's in central and east) with little or no sympatry. A fourth stage of this process is offered by the cordilleran populations of Carola's and Lawes' Parotias, with significant overlap in a central segment of the cordillera, but also with significant areas of allopatry in the west and east. The fifth and final stage is illustrated by the cordilleran populations of the *Epimachus* sicklebills, with sympatry established over most of the cordillera through displacement of the two species' ranges into complementary elevational zones (the Brown Sicklebill higher, the Black lower).

One of the most intriguing examples of avian intraspecific variation is that the geographical forms (subspecies) of both Carola's Parotia and the Superb Bird show extremely striking parallel variation in facial plumage

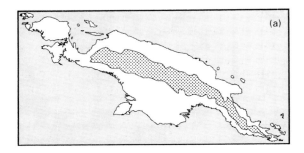
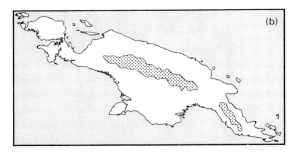
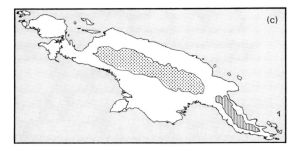
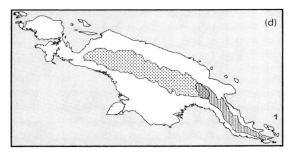

3.5 Diamond's 'drop-out' model of montane speciation in New Guinea. Diamond (1972) hypothesised a process by which species populations on a continuous montane cordillera differentiate geographically. This is illustrated in the four panels. (a) Initially, a single contiguous and undifferentiated population inhabits the entire cordillera. (b) Because of some environmental regression, the once-continuous population is broken into two allopatric subpopulations. (c) During this period of allopatry, the eastern population differentiates into a morphologically-distinct form. (d) In a period of environmental improvement, the two populations re-expand and meet in limited sympatry. Note that this is a hypothetical time-series.

pattern and dorsal coloration (Diamond 1972, 1982). So alike are some forms of the two species that, differences in size notwithstanding, they have been confused by experts (see Frith and Frith 1996b). Is this an example of a smaller and less aggressive bird mimicking a larger and more aggressive one in order to gain advantage in competitive foraging situations? This phenomenon deserves study.

Differentiation via land-bridge.—A third mode of geographical differentiation involves the classic pattern of localised isolation (or 'vicariance process') produced by the rise and fall of sea level that periodically produces a land-bridge. In this, during a period of low sea level a widespread form occupies the lowlands of 'greater' New Guinea that include a 'bridged' fringing island connected to the big island; subsequent sea level rise isolates a population on the newly-reisolated island; this population subsequently differentiates in isolation.

In the birds of paradise of the New Guinea region this process has produced no full species, but has produced a number of insular subspecies. The sole example of land-bridge speciation involves Torres Strait and the riflebirds (Beehler and Swaby 1991). As discussed earlier in this chapter, there is clear evidence that there have been two 'bridging' events in this example: the first and older event produced a small Australian lineage (the parent to the current Victoria's and Paradise Riflebirds) versus a large, New Guinean species (the parent to the current Magnificent). The second bridging and isolation produced subspecies of the Magnificent Riflebird (*P. m. alberti*, *P. m. magnificus*) on either side of the Strait.

Over-water differentiation.—Over-water differentiation follows chance dispersal across a permanent water barrier. Although this is evidently a rare event in the birds of paradise, it has produced six insular species—the Paradise Crow, Curl-crested Manucode, Standardwing, and Wilson's, Goldie's, and Red Birds. Apparently over-water dispersal is a highly-effective pathway to speciation in birds of paradise. Only three species have over-water isolates that have not achieved species status—the insular populations of the Glossy-mantled and Trumpet Manucodes on Tagula Island and the D'Entrecasteux Archipelago and Gebe, respectively, and the Trobriand populations of the Curl-crested Manucode.

Time-scale of speciation.—How long does paradisaeid speciation take? A 'natural experiment' is offered by bird of paradise populations extant on land-bridge islands, all of which were connected to the mainland as recently as 10 000 years ago. None of these land-bridge populations has produced a really strongly-differentiated subspecies over that period of time. Thus 10 000 years appears to be much too short a period for paradisaeid speciation. At the other end of the scale, Sibley and Alhquist (1985) used molecular clock calibrations in DNA–DNA hybridisation studies of the corvine assemblage to calculate that the manucodes diverged from fellow paradisaeinines 18–20 mya [million years ago], and that the upper paradisaeinine genera diverged anywhere from 4 to 5.4 mya. Extrapolation of additional DNA data from Sibley and Alhquist indicate divergence between *Cicinnurus* and *Diphyllodes* took place about 1.5 mya. Although all of these calculations are certainly subject to considerable 'fudge factors' we can at least use them to postulate that fission of sister species forms takes considerably less than 1.5 million years—but how much less is not known.

Anomalies.—A minor anomaly in paradisaeid distribution is the absence of four widespread cordilleran birds of paradise from the mountains of the southeastern peninsula (Yellow-breasted Bird, Short-tailed Paradigalla, King of Saxony Bird, and Black Sicklebill). The southeastern terminus of the distributions of these birds (except for Yellow-breasted) is the Kratke Range. It is apparent there is some sort of distributional barrier southeast of the Kratke Mountains—certainly a reduction of the height of the central cordillera at this point is a possible explanation for the limits for the King of Saxony Bird, but does not explain the limitation on a middle elevation species like the Black Sicklebill. The Yellow-breasted Bird ranges as far as Wau but no further, for which we have no explanation.

The ranges of the Blue Bird of Paradise and, to a lesser extent, the Crested Bird are equally anomalous. Both are widespread on the cordillera but do not extend westward to its terminus at the Weyland Range. The Blue Bird does not cross the border into Irian Jaya, and the Crested Bird ranges no farther than central Irian Jaya. An explanation for these truncated ranges is not readily forthcoming, but accords with Diamond's postulated 'drop-out' phenomenon, in which local populations of sedentary forest species are extirpated without rapid recolonisation. Alternatively, these could be species of eastern and western origin that have not yet spread to the opposite end of the island since the last rise in mountains/life zones (Schodde and Hitchcock 1972).

The most anomalous range in the family is that of the Pale-billed Sicklebill, confined to the lowlands of northwestern New Guinea. It shows no apparent relation to its putative sister species, the Buff-tailed Sicklebill, which patchily inhabits the highlands from the far northwest to the far southeast of New Guinea. One hypothesis (M. LeCroy *in litt.*) is that this species differentiated in isolation in the foothills of the north coastal ranges of present-day Irian Jaya during a period of high sea level that flooded the basins of the Idenburg and Sepik Rivers.

Ecological biogeography

Species distributions are defined by their two components, often tightly interrelated—geography and ecology. We have focused on geography in the preceding section, but shall examine the ecological component here.

The main ecological factor in species distribution is habitat selection (see Chapter 4). For most species that inhabit forests of the equatorial tropics, this is determined by altitude. The sharply-delineated and very specific distribution of humid tropical habitats at different elevations has a direct impact on the geographical distribution of birds of paradise. Elevation or altitude thus presents a climatological and ecological limitation, but also offers closely-related species the opportunity to avoid competition while establishing geographical sympatry—by occupying adjacent elevational zones without significant direct contact between populations.

Schodde and Hitchcock (1972) and Diamond (1972) hypothesised that elevation is perhaps the most important ecological sorting mechanism that permits the adaptive radiation of birds of paradise (see Fig. 4.6). We agree. Geography permits the morphological divergence of sister forms, but demands no adaptive or ecological divergence. For true biological diversification to occur, sister forms must acquire a degree of ecological separation that permits sympatry. Elevational specialisation (which is an analogue to habitat specificity) is a ready pathway to ecological divergence for recently-differentiated sister forms.

We thus see the pattern in the two *Epimachus* sicklebills, in which the Black inhabits lower montane forest, the Brown mid- and upper-montane forest, with hundreds of kilometres of range interface but no significant physical overlap over this entire distribution. Both species inhabit remarkably narrow cordilleran ranges—'ribbons' of habitat often little more than a few kilometres wide and hundreds of kilometres long.

We see what appears to be an earlier stage in the process with Ribbon-tailed and Stephanie's Astrapias; there is slight elevational displacement of distributions (the Ribbon-tailed at higher elevations) and slight geographical overlap (measured in tens of kilometres).

The Crinkle-collared and Jobi Manucodes, although offering a considerably less transparent distributional tableau, provide yet another example of the importance of altitude. The Jobi Manucode is adapted to lowland forest, whereas the nearly-identical Crinkle-collared is adapted to upland forest, with complete geographical overlap on the broad scale but apparently little or no actual sharing of forest.

More divergent species-pairs include the Emperor Bird and its lowland counterparts (Lesser and Raggiana Birds) that in some places on the hills of the Huon Peninsula achieve microsympatry, and the Blue Bird and its two lowland counterparts (Lesser and Raggiana Birds), which are found in local sympatry (with minor elevational overlap) over virtually all of the range of the Blue Bird. These last two examples highlight evolution of new lineages—a later stage of the differentiation process, where the two 'paired' forms are sufficiently different in morphology and habit that they no longer pose any competitive threat to each other, and the resultant coexistence creates heightened local species richness.

Hybridisation

A phenomenon covered in some detail in Appendix 1 is the remarkable degree of interspecific and intergeneric hybridisation among the birds of paradise (Table 3.5). It is now known that only the polygynous species have been implicated in such a pattern. This is not a surprise considering the promiscuous nature of the mating system of the males of polygynous species—they are 'mating machines' with apparently little or no discrimination in mate-choice. Presumably the females, which are the more discriminating sex, may mate with males of the wrong species in situations where conspecific mates are not readily available (at geographical or elevation limits of a species' distribution—see also Appendix 1). Also note that it is not uncommon to find display sites of different species close to each other in the

Table 3.5 Systematic cross reference to birds of paradise (exclusively Paradisaeinae) known or assumed to be involved in wild interspecific hybridisations.

		1	2	3	5	6	7	8	9	11	13	14	15	16
1	Long-tailed Paradigalla *Paradigalla carunculata*													
2	Arfak Astrapia *Astrapia nigra*													
3	Ribbon-tailed Astrapia *A. mayeri*													
4	Stephanie's Astrapia *A. stephaniae*			X										
5	Western Parotia *Parotia sefilata*	X?												
6	Lawes' Parotia *P. lawesii*													
7	Carola's Parotia *P. carolae*													
8	Magnificent Riflebird *Ptiloris magnificus*													
9	Superb Bird *Lophorina superba*	X?				X		X	X					
10	Black Sicklebill *Epimachus fastuosus*	X?	X?							X*				
11	Magnificent Bird *Cicinnurus magnificus*									X				
12	King Bird *C. regius*										X			
13	Twelve-wired Bird *Seleucidis melanoleuca*							X						
14	Lesser Bird *Paradisaea minor*							X*			X	X		
15	Greater Bird *P. apoda*													
16	Raggiana Bird *P. raggiana*											X	X	
17	Emperor Bird *P. guilielmi*											X		X
18	Blue Bird *P. rudolphi*					X								X

Note: Some cause for doubt exists about those putative parent combinations indicated by a question mark and also, but to lesser degree, those marked by an asterisk.

forest, and it may happen that the seductive power of a male of one species, in combination with the female of another species' predisposition to respond to a display stimulus, could lead to the occasional interspecific cross. In spite of the spectacular by-products of the interspecific hybridisation (Plates 14, 15), we remain uncertain as to the evolutionary significance of this remarkable tendency. In our systematic studies we do not see modern species that show strong evidence of hybrid origin or introgression (as we see, for instance, in certain populations of the New Guinean montane honeyeaters *Melidectes belfordi* and *rufocrissalis*). Other views about the systematic significance of hybridisation in the Paradisaeidae have, however, been expressed (see Schodde 1976; Christidis and Schodde 1993). Further discussion of this subject is presented in Appendix 1.

Some issues for further study

Several issues remain unresolved with regard to the geographical distribution of birds of paradise. These offer ideal study topics for future field researchers, and their resolution may provide us with a better understanding of vertebrate speciation in the New Guinean region, as well as the evolution of the Paradisaeidae.

Papuan vocal dialects of the Magnificent Riflebird.—As highlighted by Beehler and Swaby (1991), the eastern and central/western Papuan populations of the Magnificent Riflebird possess distinct vocalisations—a remarkably-loud upslurred series of 2–4 whistles in the central/west, versus 2–3 guttural growled notes in the east. Given the importance of male advertisement vocalisation in court-displaying polygynous species, Beehler and Swaby postulated that these vocal dialects might signal that these two populations have reached semi-species status. Apparent zones of contact occur in the south near the Purari River and in the north near the Sepik/Ramu deltas. We would like to know if there are zones where the two song-types coexist, and, if so, is there assortative mating and effective sympatry, or do these types hybridise and produce offspring with intermediate song types? Or, are there, instead, offspring that produce both song-types? There is a report of the western song-type from deep into the eastern population on the northern scarp of the Huon Peninsula, hinting either at the latter or else potential for coexistence.

Splendid/Ribbon-tailed contact.—The contact/hybrid zone between Stephanie's and Ribbon-tailed Astrapias is relatively well documented, but further west, nothing is known of the easternmost edge of the Splendid Astrapia versus the westernmost edge of the Ribbon-tailed Astrapia range. Given the considerable morphological divergence between these latter two, it will be interesting to learn whether they meet each other in places, and whether they freely hybridise, as the Ribbon-tail does with Stephanie's.

Westernmost range of the Blue Bird.—The strangely-truncated western portion of the distribution of the Blue Bird is an anomaly. Does the species, in fact, range westward into Irian Jaya (as has recently been found for the Crested Bird)? If not, what is the barrier that keeps the Blue Bird from ranging further westward along the continuous cordillera?

Paradisaea Hybrid Zones.—The contact zones between the three members of the *Paradisaea apoda* superspecies support considerable hybridisation. LeCroy and Peckover (personal communication) have found mixed leks of Greater and Raggiana in the Bensbach region of southwest Papua New Guinea. What is the nature of these hybrid interactions? Are these contact zones stable and is mating random or assortative?

Glossy-mantled/Crinkle-collared/Jobi Manucode.—The manucodes in the lowland and lower hill forest zone present a considerable field identification problem for ornithologists. This problem, in fact, masks a deeper problem in ecological biogeography—what is the local distribution of these birds in zones of contact, and how do they sort out ecologically and microgeographically? Is there hybridisation? Are there pre-mating behavioural isolating mechanisms that prevent this? If they do not coexist, in what way do they exclude each other from their adjacent ranges?

Trumpet Manucode elevational populations.—The Trumpet Manucode includes lowland races and upland races. In some areas (e.g. south slope of southeast New Guinea) the lowland form exists just downhill from the highland form, and the two have distinct vocal dialects, with only the male of the lowland form giving the high, clear *ee-yoh!* song. Are there, in fact, more than one species of Trumpet Manucode? What is their geographical relationship?

Carola's Parotia and Superb Bird parallel geographical variation.—The parallel variation in adult female plumage across the various subspecies of these two species is a remarkable phenomenon. A detailed study of this would be additionally rewarding because (1) the considerable geographical variation in both species is inadequately understood and (2) the status and home of *Parotia carolae berlepschi* and of birds on the Foya Mountains (possibly *berlepschi*) require determination (see species accounts in Chapter 9).

4

Ecology of the birds of paradise

A remarkable feature that defines the birds of paradise is the group's near-obligate dependence upon the tropical closed humid forests of Australasia, entirely east of Wallace's Line. The dependence on humid forest is nowhere more obvious than in Australia, where one finds that continent's four paradisaeid species restricted to small geographical enclaves of remnant rainforest. It can be safely said that the birds of paradise have evolved a close dependence on rainforest. Also, as will be discussed in this chapter, there is evidence that some Australasian rainforest trees appear to depend upon birds of paradise for important aspects of their maintenance and regeneration—seed dispersal in particular.

The reliance of birds of paradise on the rainforests of New Guinea, the Moluccas, and Queensland is a factor that, for decades after their discovery, made them virtually inaccessible to naturalists and researchers. Today, this affinity for the rainforest gives the birds of paradise a certain cachet. In addition, it makes the presence of these birds one useful measure of a specific rainforest's ecological 'quality.'

To find such a forest rich in birds of paradise requires knowledge of the birds' ecology and their habitat preferences. In this chapter we discuss these and we attempt to address the origins and maintenance of this rainforest affiliation. We also examine how habitat and diet may have influenced the evolution of the group.

The environment of the birds of paradise

Although most species of birds of paradise occur in rainforest, the Paradisaeidae is not a single-habitat family. First of all, rainforest is a rather inexact term that subsumes a range of tropical humid forest environments, some lowland and some montane. Second, there are some paradisaeid species that regularly use non-forest environments—forest-edge, gardens, savanna, and subalpine woodlands, for instance. But these examples are, in fact, exceptional. There is no single species of bird of paradise that does not depend upon closed humid forest over most of its range. Still, given the structural and floristic variation in closed forest habitats in Australasia, one can find plenty of variety in the types of environments used by these birds.

On a global scale, two critical parameters can be invoked to define the environment of the birds of paradise—latitude and rainfall. The birds are almost exclusively tropical and subtropical, with but a single species (the Paradise Riflebird) ranging southward across the Tropic of Capricorn to just beyond 32° South Latitude in the Hunter River region north of Sydney, Australia.

Rainfall is equally important in determining the distribution of birds of paradise. In this case it is minimum annual precipitation that indirectly limits the distribution of the birds by controlling the distribution of humid closed forest. There is some correlation between length of annual dry seasons and annual precipitation, but the latter is a more readily available measure for our purposes, and one finds that both closed forest and birds of paradise are generally not found in areas that receive less than approximately 1500 mm of rainfall per annum.

There is one exception to the low rainfall/closed forest limit—the Glossy-mantled Manucode (Peckover and Filewood 1976; Bell 1982d). This widespread species commonly inhabits open savanna woodland in the three major rain shadow zones in New Guinea (the Trans-Fly, Port Moresby, and Raba Raba in the far southeast), as well as lowland rainforest. It is notable that its relative, the Curl-crested Manucode, is often found perched and singing in open, anthropogenic, savanna-like formations in the D'Entrecasteaux Archipelago, and is also abundant in the interior of closed forest. Other common and widespread lowland species such as the Raggiana Bird of Paradise traverse open savanna, but do not regularly use this open habitat, instead confining their activities to forest and forest-edge that abuts the savanna.

Apparently, there is no upper rainfall limit to the distribution of birds of paradise. A number of sites in New Guinea and in northern Queensland receive in excess of 5 m of rain a year (mostly along the southern scarp of the Papuan central cordillera, and on the eastern slopes of North Queensland's high ranges) and in these regions paradisaeid communities flourish. Thus it is likely that the presence or absence of forest, not rainfall per se, is limiting the distribution of birds of paradise. And it is not minimum temperature, per se, but a prolonged cold season, that is also critical—probably by limiting the distribution of tropical or subtropical humid forest. We say this because Macgregor's Bird is a denizen of the cordilleran subalpine forests that receive frost regularly through the year.

Habitat selection

There is evidence for habitat selection among paradisaeid species, but because of our rather rudimentary understanding (or definition of) 'habitats' in tropical forests of Australasia, it becomes a somewhat subjective analysis. In Table 4.1 we present a full listing of our current knowledge of habitats used by the various birds of paradise. For purposes of this analysis, we confine 'use' to be those habitats where a particular species will nest. The various forest types are adapted from the system of Johns (1977).

Most birds of paradise are habitat specialists, especially the montane species (Table 4.1). Fourteen species are found only in a single environment, as defined by this classification, and only possibly eight species span more than three habitats—mostly lowland species (such as the Raggiana Bird) that range up into the hills. The most restricted species are found in the midmontane forest, it seems, although defining 'midmontane' forest using any measure besides altitude would be a challenge. Still, there is something about the midmontane forests (between 1200 and 2500 m—e.g. Johns 1977) that birds of paradise find particularly benign (Pruett-Jones and Pruett-Jones 1986). For it is in the middle elevations that one can find the greatest concentrations of species—a study site at 1480 m on Mount Missim supported nine resident paradisaeids and an additional vagrant species. Pruett-Jones and Pruett-Jones (1986) demonstrated that the birds of paradise have narrower elevational ranges than typical forest birds in New Guinea, another measure of habitat specialisation.

What of particular specialisations? Macgregor's Bird appears to have the most restricted habitat requirements of any in the family—subalpine woodland dominated by two gymnospermous tree species—*Dacrycarpus compactus* and *Libocedrus papuanus*. Preliminary fieldwork indicates its distribution is limited by the distribution of the species of the

Table 4.1 Habitat* selection by the birds of paradise.

Species	Rainforest	Swamp forest	Savanna	Hill forest	Midmontane forest	Upper montane forest	Subalpine forest	Garden areas	Mangrove forest	Subtropical forest	Habitats per species
Loria's Bird						X	X				2
Crested Bird						X	X				2
Yellow-breasted Bird					X						1
Macgregor's Bird							X				1
Paradise Crow	X			X	X				X?		3–4
Glossy-mantled Manucode	X	X	X					X	X		5
Jobi Manucode	X	X		X							3
Crinkle-collared Manucode				X	X						2
Curl-crested Manucode	X		X	X	X	X		X	X?		6–7
Trumpet Manucode	X			X	X				X?		3–4
Long-tailed Paradigalla					X						1
Short-tailed Paradigalla					X			X?			1–2
Arfak Astrapia					X	X	X				3
Splendid Astrapia					X	X	X				3
Ribbon-tailed Astrapia						X	X				2
Stephanie's Astrapia					X	X	X				3
Huon Astrapia					X	X	X				3
Western Parotia					X						1
Lawes' Parotia					X						1
Wahnes' Parotia					X						1
Carola's Parotia					X						1
King of Saxony Bird					X	X					2
Magnificent Riflebird	X	X?		X	X				X?		3–5
Paradise Riflebird										X	1
Victoria's Riflebird	X			X				X	X?		3–4
Superb Bird					X	X					2
Black Sicklebill					X						1
Brown Sicklebill					X	X					2
Buff-tailed Sicklebill					X						1
Pale-billed Sicklebill	X										1
Magnificent Bird	X?			X	X						2–3

Table 4.1 continued

Species	Rainforest	Swamp forest	Savanna	Hill forest	Midmontane forest	Upper montane forest	Subalpine forest	Garden areas	Mangrove forest	Subtropical forest	Habitats per species
Wilson's Bird	X?			X	X?						1–2
King Bird	X										1
Standardwing Bird	X			X				X?			2–3
Twelve-wired Bird	X	X									2
Lesser Bird	X	X		X				X?			3–4
Greater Bird	X			X							2
Raggiana Bird	X			X	X			X			4
Goldie's Bird				X				X?			1–2
Red Bird	X										1
Emperor Bird				X							1
Blue Bird					X			X			2
Habitat species counts	15–17	4–5	2	15	25–26	11	7	1–6	1	1	

*See Johns (1977) for habitat definitions. A question mark indicates the bird has been recorded in the habitat but to what extent it typically does so is unknown.

Dacrycarpus tree (Beehler 1991a,b). The Twelve-wired Bird of Paradise was long thought to be dependent upon swamp forest (e.g. Gilliard 1969), but recent fieldwork in a range of lowland forest sites (BB) has shown it to be present in virtually all lowland forests, although it does show a particular affinity to swamp forest.

The manucodes appear to be habitat generalists, and certain lowland forests in New Guinea support as many as three species. We do not know how these manucodes differ ecologically in sympatry. All manucodes appear to be highly frugivorous, some specialising on figs. The Glossy-mantled Manucode inhabits savanna, gallery forest, forest-edge, closed forest, swamp forest, and hill forest. The Trumpet Manucode inhabits lowland forest, hill forest, and midmontane forest to an elevation of 2000 metres, Likewise, the island endemic Curl-crested Manucode inhabits all the woodland and forest habitats of its range—from coastal mangroves and savanna woodland upward into stunted mossy forest of the higher mountains. These two probably have the widest elevational ranges of any paradisaeid species (Pruett-Jones and Pruett-Jones 1986).

Some species are very patchily distributed—being present in some geographical locales but absent elsewhere. The Yellow-breasted Bird provides an extreme example of patchiness. In some sites it is common but in many others is absent. Is this some undiscovered form of habitat selection? Again, we do not know, in part, because the Yellow-breasted Bird remains one of the least-known birds of paradise.

Diamond (1972) reported finding sexual segregation of birds of paradise by elevation in the mountains of New Guinea, with the adult males inhabiting the highest segment of the species range, and the females and immature males at the lowest elevation. Diamond notes that Stein (1936) reported the same phenomenon in the Weyland Mountains of western New Guinea. The detailed analysis of this phenomenon by Pruett-Jones and Pruett-Jones (1986) for Lawes' Parotia indicated that, at least for that species, both adult female and male distributions were centred on the same elevation, and that immature males ranged widely above and below this presumed 'optimum' habitat for the adult birds.

Diet

Birds of paradise show a range of diets, which is not surprising given the remarkable range of body sizes (Tables 1.2 and 1.3) and bill morphologies (Fig. 4.1). Collectively, the various species have been recorded taking fruits of several hundred plant species (Appendix 4), various insects, spiders, and other arthropods, frogs, lizards, and even leaves, flowers, and apparently nectar (Gilliard 1969; Coates 1990). Contrary to what is written in a number of earlier reviews of the family, the birds of paradise are not known to digest seeds and thus are not known to be seed predators, but instead serve as important seed dispersers, about which more will be said in a later section. Victoria's Riflebirds have been observed to eat mixed 'canary' seed placed at bird feeders for estrildid finches but it is not known if these are digested (CF).

Although some species are dietary specialists, which confine themselves primarily to eating arthropods or fruit, the typical bird of paradise consumes a mix (Table 4.2). It is notable how little we do know about paradisaeid diets. Our generalisations have to be based on a few field studies (e.g. Schodde 1976; Beehler 1983a; Pratt 1983).

There are good data on the diets of 25 species, and fragmentary data on all species but the Wilson's Bird of Paradise. All of the species for which data exist are known to take fruit. Thus the most general point to be made about diet is that birds of paradise are frugivores first and foremost. Arthropods comprise the second major component of paradisaeid diets. Only the three Cnemophilinae are thought to refrain entirely from consuming arthropods. So the second generalisation is that after fruit, arthropods constitute a varying but often important component of paradisaeid

4.1 Heads of selected female birds of paradise to indicate the diversity of bill size, shape, and structure adapted to foraging (see also diagrams on pages xxvi and 9 and Plates 4–13): (a) Loria's Bird, (b) Macgregor's Bird, (c) Curl-crested Manucode, (d) Short-tailed Paradigalla, (e) Lawes' Parotia, (f) King of Saxony Bird, (g) Paradise Riflebird, (h) Buff-tailed Sicklebill, (i) King Bird, (j) Twelve-wired Bird, (k) Standardwing, and (l) Raggiana Bird.

diets. After fruit and arthropods, only very minor constituents can be appended. Five species (Short-tailed Paradigalla, Brown Sicklebill, Splendid and Ribbon-tailed Astrapia, and Victoria's Riflebird) are known to consume frogs and/or skinks, and probably most of the larger species do this from time to time, given the availability of small members of the herpetofauna. Several species have been reported to consume leaves in captivity, so this may be an uncommon but regular activity in the wild. Two species (Trumpet Manucode and Twelve-wired Bird of Paradise) have been implicated in nectar-feeding, but because of unspecialised bill and tongue structure, such an activity may be opportunistic, at best. Specialised nectar-feeders require long, extensible tongues, something that birds of paradise do not have. They thus would be required to get the nectar by plucking open the inflorescence and consuming the nectary (which is a common phenomenon among birds but has not been explicitly reported for birds of paradise).

Range of diet

As measured by proportion of fruit versus arthropods, there is a wide range of dietary

Table 4.2 Major dietary types of wild birds of paradise.

Species	Major dietary types					Fruit types		
	Fruit	Arthropods	Herpetofauna	Leaves	Nectar	Capsular	Drupe	Fig
Loria's Bird	+++++					+++++		
Crested Bird	+++++					+++++		
Yellow-breasted Bird	++++	+				+++++		
Macgregor's Bird	+++++	*	?			+++++		
Paradise Crow	*	*?				?	*	?
Glossy-mantled Manucode	++++	+			?	+	+	+++
Jobi Manucode	*	*				?	*	?
Crinkle-collared Manucode	++++	+				+	+	+++
Curl-crested Manucode	++++	?	?			?	++	?
Trumpet Manucode	++++	+			*	+	+	+++
Long-tailed Paradigalla	*	*?	?			?	*	?
Short-tailed Paradigalla	+++	++	+			?	*	?
Arfak Astrapia	*	*				*	*	?
Splendid Astrapia	++++	+	+			?	*	?
Ribbon-tailed Astrapia	++++	+	+			+++	++	?
Stephanie's Astrapia	++++	+				++	+++	?
Huon Astrapia	?	?				?	?	?
Western Parotia	*	?				*	?	*
Lawes' Parotia	++++	+				+	+++	+
Wahnes' Parotia	*	*		*		?	?	?
Carola's Parotia	*	*		*		?	?	?
King of Saxony Bird	++++	+				?	+++	?
Magnificent Riflebird	++	+++	?			+++	+	+
Paradise Riflebird	*	*	?			*	*	*
Victoria's Riflebird	++	+++	*			+++	*	*
Superb Bird	+	+++				+++	+	+
Black Sicklebill	++	+++				+++	?	?
Brown Sicklebill	++	+++	+			?	+++	?
Buff-tailed Sicklebill	+	++++				+++	?	+
Pale-billed Sicklebill	+++	++				+	+++	+
Magnificent Bird	+++	++				+	+++	+
Wilson's Bird	?	?				?	?	?
King Bird	*	*				*	*	*
Standardwing Bird	*	*				?	?	?
Twelve-wired Bird	+++	++			+	*	*	*
Lesser Bird	+++	++				*	*?	*
Greater Bird	*	*				?	?	?
Raggiana Bird	++++	+		*		+++	+	+
Goldie's Bird	*	*				?	?	?
Red Bird	*	*				?	?	*
Emperor Bird	*	*				*	*	*
Blue Bird	++++	+				++	++	+

+ = takes this food type, in proportion to number of +s.
* = take this food type, but data inadequate to determine relative preference.
? = presumed to take this food type but no direct observations.

preference among the birds of paradise. Whereas the three wide-gaped species are entirely frugivorous, the great majority of species split their diet between fruit and arthropods, and a few other species (e.g. Buff-tailed Sicklebill, Superb Bird of Paradise) are primarily insectivorous (Table 4.2). It is notable that there are no examples of purely insectivorous paradisaeids, and even the primary insectivores are known to take a range of fruit types typical of the more frugivorous species. There is an interesting functional relationship linking adaptations for insect-foraging to specialised fruit-foraging that we shall discuss in a later section. A final note on the 'purely' frugivorous Cnemophilinae is worth making. There is considerable uncertainty about the absolute nature of their frugivory. The Yellow-breasted Bird is associated with a single record of insectivory, and the Crested Bird is recorded taking tiny shelled snails. In this respect, it is interesting to note that Snow (1982) recorded that one of the wholly frugivorous bellbirds (Cotingidae) was recorded eating snail shells 'probably for their calcium content'. We believe it is important to study closely the 'extreme' frugivory of the Cnemophilinae because it may tell us a great deal about dietary specialisation and its nutritional, energetic, and developmental limitations.

Knowledge about diet is useful to the fieldworker for two reasons. It can inform one how to look for the bird in the field (search for the frugivore at favoured fruit trees, the insectivore in the forest canopy or middle story bark-gleaning and/or rotten wood probing). It also correlates with social behaviour. The frugivores tend to be more sociable, the insectivores more solitary and territorial (Beehler 1983a, 1985a, 1987b).

Insectivory

Three types of insectivory have been recognised in the birds of paradise—bark-gleaning, dead wood/foliage probing/tearing, and generalised twig-and-foliage gleaning. The first two are widespread among most paradisaeines and the latter seems to be the predominant habit among the manucodes. Bark-gleaning subsumes a range of specific foraging methods, but overall requires that the forager hop or creep about scansorially on the bark of a trunk or limb, searching for prey on or under the bark surface or in bark fissures. It is common to see birds of paradise on near-vertical trunks of trees, hitching about, peering at cracks in the bark, prising at loose bark, and even removing bark and penetrating rotting wood in search of insect larvae or adult beetles hidden therein. Another very common behaviour is for the foraging bird to move along the top of a large horizontal branch and crane its neck down to peer for prey on the bark of the side and underside of the branch. In the upper montane forests where moss and epiphytes tend to overgrow all woody surfaces, the most common paradisaeid behaviour is to search for prey by probing and pulling apart moss clumps.

The riflebirds are remarkable in their woodpecker-like activities, using their powerful chisel-like bills to chip and dig into dead bark and wood in order to find and dislodge prey. Riflebirds commonly 'open-bill probe' or 'gape' into wood. This is an advanced technique involving the probing of the closed bill into a substrate, such as soft rotten wood or dead and curled leaves and accumulated foliage, and then opening the bill to expose prey (Harrison 1964). It presumably requires powerful jaw muscles and a strong bill. In the northern hemisphere, one of the best known species to perform this behaviour commonly is the Common Starling *Sturnus vulgaris*. The Buff-tailed Sicklebill exhibits another highly-specialised insect-foraging behaviour, sometimes using its narrow sickle-shaped bill to probe wormholes and knotholes. In some instances it opens its bill wide and uses only the lower mandible to probe for and spear prey in holes in dead wood. This is remarkable to witness, given both the size of the bird's bill and the peculiar sight of the bill being gaped with the maxilla out of use altogether. It is analogous to the foraging habit of the Hawaiian

Nukupuu honeycreeper *Hemignathus lucidus* (Drepanidinae)—although the latter uses the maxilla for probing (T. Pratt *in litt.*).

The manucodes forage for arthropods by upright twig-and-foliage gleaning—not nearly so remarkable as the scansorial methods, mainly because many arboreal perching birds that feed on insects employ this method. It requires no specialised physical ability. But given the large size of the manucodes, it is a little out of the ordinary—most of the twig/foliage gleaners being considerably smaller.

The types of arthropods consumed by birds of paradise include caterpillars, beetles and beetle grubs, katydids and their relatives, and ants (Table 4.3). Our tabulation is necessarily very incomplete, given the very few data obtained from museum collections as well as field treatments of mist-netted birds. The latter requires holding trapped birds in a darkened cage for a half-hour and collecting the excreta. These samples can be suspended in ethanol and later examined microscopically to identify the arthropod parts (Beehler 1983*a*).

Frugivory

Frugivory is by far the most important dietary pattern of birds of paradise. Indeed, it has been said by more than one authority that frugivory has provided a key ecological potential in the evolution of the remarkable polygynous lifestyle that characterises the birds of paradise (Snow 1976*b*; Snow and Snow 1979; Ricklefs 1980).

In spite of its complexity, fruit-eating by birds of paradise can be summarised in a few lines. Typical paradisaeid foragers feed solitarily by plucking and consuming fruit in fruiting trees that grow in the middle and upper stories of the forest. Such birds often join foraging aggregations that form at fruiting trees, but usually interact very little with other foragers and usually depart the tree immediately after consuming a relatively small number of the fruits (Pratt and Stiles 1985).

The birds of paradise are important frugivores in New Guinea and are perhaps the single most important group of vertebrate seed dispersers on the island (Beehler 1989*a*). Birds of paradise are good at manipulating complex fruits, are acrobatic foragers, their guts do not damage seeds, and they scatter seeds throughout the forest while they spend some of each day away from fruit trees foraging for arthropods. By contrast, the parrots tend mostly to be seed predators, as are many of the pigeons. The true frugivore pigeons (*Ducula*, *Ptilinopus*) differ by being far less capable than birds of paradise of manipulating many kinds of fruits, and are also less adept at dispersing seeds efficiently because of their tendency to loaf for long periods in fruit trees and at resting sites—and thus creating a clumped 'seed rain' (Pratt and Stiles 1985). The other significant seed dispersers on New Guinea are the cassowaries and the fruit bats.

Perhaps the least understood and most complex aspect of frugivory is fruit-choice. Except in some marginal habitats (e.g. the highest altitudes) most of the forests where birds of paradise live support hundreds of species of fruit-bearing trees, vines, lianas, and shrubs. A number of studies now show that the birds of paradise consume fruit of only a small number of the available plant species (Beehler 1983*a,b*; Pratt 1983; Pratt and Stiles 1985; Brown and Hopkins 1998). Why do they consume certain species and ignore others?

Size certainly has some bearing on fruit selection. Birds of paradise generally consume small to medium-sized fruits, whose edible portion is approximately 1 cm in diameter. The birds commonly take smaller fruits, but only rarely consume very large fruits. Typical forests in northern Queensland and New Guinea support fruits of a large range of sizes, from a few millimetres in diameter to as much as 10 cm. The larger fruits are consumed by, among others, cassowaries, imperial pigeons, hornbills, and flying foxes (Pratt and Stiles 1985).

Size considerations aside, there are a number of fruit morphologies, only some of which are preferred by the birds of paradise.

Ecology of the birds of paradise 87

Three general fruit morphologies (Beehler 1983b) have been recognised for comparative purposes—capsule, drupe/berry, and fig (Fig. 4.2). These are defined by the placement of the seed or seeds and the degree of structural protection of the edible part (the diaspore). Capsular fruits are structurally protected—situated in a leathery or woody husk that encloses it either partially or entirely. The diaspores of capsular fruits are varied but tend

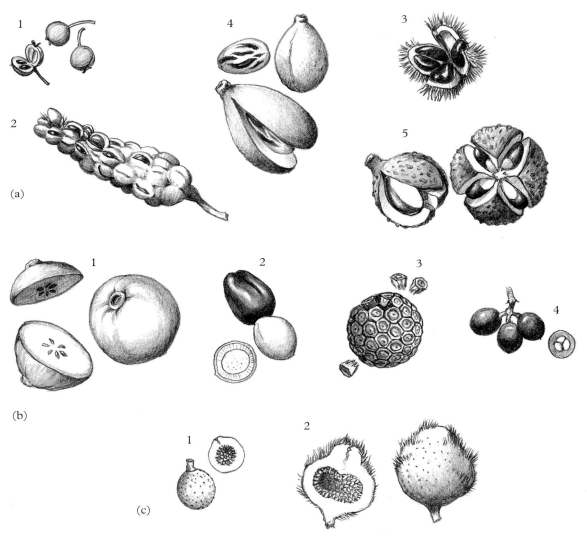

4.2 Examples of the main structural fruit types eaten by birds of paradise: (a) complex capsular fruits, requiring skillful and strong manipulation by bill if not also with feet and often involving strong acrobatic agility (see Fig. 4.4)—such as (1) *Omalanthus novoguineensis*, (2) *Elmerrillia papuana*, (3) *Sloanea sogerensis*, (4) *Myristica subaluata* and (5) *Dysoxylum* sp.; (b) simple berries or drupes, easily plucked and swallowed entire—such as (1) *Endospermum labios*, (2) *Podocarpus nereifolius*, (3) *Schefflera 'chimbuensis'*, (4) *Cissus hypoglauca*; (c) figs, easily plucked and swallowed entire if small but larger ones requiring pecking to pieces *in situ* or plucking and taking elsewhere to hold to a perch with foot to tear apart with the bill (see Fig. 4.3)—such as (1) *Ficus* sp. and (2) *Ficus odoardi*.

Table 4.3 Summary of the known invertebrate dietary components of birds of paradise.

Species	Consumes arthropods	Lepidoptera larvae	Lepidoptera imago	Coleoptera	Orthoptera+	Hymenoptera	Dictyoptera	Dermaptera	Diptera	Hemiptera	Odonata	Araneida	Chilopoda	Myriapoda	Annelida
Loria's Bird															
Crested Bird															
Yellow-breasted Bird	?														
Magregor's Bird	?														
Paradise Crow															
Glossy-mantled Manucode	X	?*													X?
Jobi Manucode	X														
Crinkle-collared Manucode	X				X							X			
Curl-crested Manucode	X														
Trumpet Manucode	X			X	X				X	X		X			
Long-tailed Paradigalla	?														
Short-tailed Paradigalla	X	X		X	X							X			X
Arfak Astrapia															
Splendid Astrapia	X														
Ribbon-tailed Astrapia	X			X											
Stephanie's Astrapia	X	X		X					X		X	X	X		
Huon Astrapia	X?														
Western Parotia	?														
Lawes' Parotia	X			X				X					X		
Wahnes' Parotia	X														
Carola's Parotia	X				X										
King of Saxony Bird	X	X		X	X	X		X				X			
Magnificent Riflebird	X	X	X	X	X	X	X	X				X		X	
Paradise Riflebird	X	X		X	X	X	X	X				X			
Victoria's Riflebird	X	X?		X	X							X	X	X	
Superb Bird	X	X**		X	X	X	X	X							
Black Sicklebill	X						X								
Brown Sicklebill	X			X											
Buff-tailed Sicklebill	X	X		X	X	X	X	X				X	X		
Pale-billed Sicklebill	X			X	X	X	X					X			X?
Magnificent Bird	X	X		X	X	X	X		X	X		X			
Wilson's Bird	X														
King Bird	X														
Standardwing Bird	X														
Twelve-wired Bird	X														
Lesser Bird	X														
Greater Bird	X				X										
Raggiana Bird	X		X	X	X	X	X					X			
Goldie's Bird															
Red Bird	X?														
Emperor Bird	X														
Blue Bird	X				X	X	X					X			

References include: Schodde (1976); Beehler (1983, 1987); Frith and Frith (1992b, 1993b,d, 1995b, 1997e); Barker and Vestjens (1990).
+ here includes Phasmatodea.
* see von Rosenberg (1875).
** see Hicks and Hicks (1988).

to be arillate—a seed with an edible aril appended (Fig. 4.2(a)). These are characteristically the most complex fruits and in a number of instances form an important part of the diet of birds of paradise. Drupe/berry fruits are usually simple structures (Fig. 4.2(b)), with the stone (testa) surrounded by the pericarp or edible portion (as in a cherry). Figs comprise a single genus (*Ficus*) in the family Moraceae. Although they show remarkable morphological variety in the Australasian region, the figs consumed by birds of paradise tend to be relatively uniform—more or less spherical, the outer layer edible pulp, the inner layer an array of tiny seeds, the centre hollow (Fig. 4.2(c)).

Of the three general types, the figs tend to be consumed by all frugivorous bird species. This is because figs are structurally unprotected and the seeds are tiny. Thus even small birds can consume larger figs by pecking off only a part of the fruit and swallowing this portion (Fig. 4.3). Consumers of the drupe/berry class are usually limited by the size of the seed. Small-seeded species are popular with a wide range of birds, whereas the large-seeded species are taken only by the species with gapes large enough to swallow them. Finally, the capsular species are typically consumed only by a restricted subset of the avifauna—those species capable of physically extracting the fruit from its protective capsule. These fruit are structurally protected in order to deter seed predation by insects, parrots, and mammals, among others. A number of birds of paradise excel at foraging on capsular/structurally-protected fruit, and thus serve as important dispersers of their seeds. Birds of paradise take fruits from all categories, but show a preference for the figs and capsular types (Table 4.2).

Fruit foraging

Most typical birds of paradise forage for fruit in a stereotyped fashion. The typical individual enters the feeding tree warily, forages nervously but with discrimination, testing each fruit prior to consumption, and then leaves after several minutes (Pratt and Stiles 1983). Birds of paradise harvest fruit while perched and almost never take items in flight, whereas the latter technique is characteristic of the prominent neotropical fruit-eaters, the cotingids and manakins (Snow 1976b). On many occasions a single individual bird of paradise will visit the same tree several times in a morning, but each visit tends to be brief. Typical foraging-bout lengths for sample species range from three to five minutes, which is brief when compared to bouts of the fruit-pigeons (Pratt and Stiles 1983). The mean number of fruit taken per bout also tends to be low (5–10) for most paradisaeines (Beehler 1983a,b). The birds rarely rest or preen in the feeding tree (Pratt and Stiles 1983). To perform such activities they instead move to adjacent vegetation. In addition, the birds prefer to forage in a tree that already has foraging birds in it (Beehler 1983b), although this generalisation does not hold for small fruiting plants.

Once a foraging group assembles in a fruit tree the foraging is kinetic (Beehler 1980). More study is needed to determine the degree to which interference competition and dominance hierarchies form at fruiting trees visited by birds of paradise (Pratt 1984). The bustle

4.3 Trumpet Manucode fruit-foraging by holding a large *Ficus hispida* fig to the perch with a foot and eating a piece from it.

of activity is punctuated by rapid evacuations of the tree, in what are presumed to be predator-escape manoeuvres (Howe 1979). The wary foragers flush from the tree by dropping nearly vertically towards the ground then veering into the nearest dense vegetation. Because bird-eating hawks were only rarely spotted after such evacuations, it is believed that many of these occurrences are 'false alarms' caused by a single, overly-wary, individual fleeing and thereby flushing the entire group. Avoidance of predation is the likely force motivating this foraging style and is probably a conservative force suppressing the acquisition of interspecific differences in foraging styles among the species (see Howe 1979).

Fruit preferences

Because we still have only a rudimentary understanding of frugivory in birds of paradise or of the New Guinean rainforest flora, it is difficult to generalise about favoured food plants of the various species. The families Moraceae, Araliaceae, Euphorbiaceae, Elaocarpaceae, Meliaceae, Myristicaeae, Pandanaceae, and Urticaceae appear particularly important in paradisaeid diets (Beehler 1989*a*). Some plant genera appear to be widespread and popular with birds of paradise in virtually all forest habitats (especially figs, *Schefflera* species, and nutmegs and genera of Meliaceae).

Although most birds of paradise forage for fruit in a stereotyped fashion, their preference for specific fruit types differs considerably. This can be measured by the proportional representation of the different fruit types in the diets of the birds. A dataset of 1187 fruit foraging records from Mount Missim (Beehler 1983*b*) shows this rather well. The Trumpet and Crinkle-collared Manucodes both took mostly figs (80% and 93% of foraging records, respectively) and only small quantities of capsular or drupe/berry fruit. By contrast, four species took more capsular fruit than any other class (Magnificent Riflebird 71%, Superb Bird 69%, Magnificent Bird 58%, and Raggiana Bird 49%). Finally, Lawes' Parotia took more drupe/berry fruit (57%) than any other category. The most specialised food plants used by the birds of paradise are the capsular species, in which the foraging bird must extract the edible 'reward' from its protective husk (Fig. 4.4(a,b)). It appears that considerable dexterity is needed to accomplish this, and some capsular food plants appear to be fed upon only by certain birds of paradise. The complex fruit of the genus *Pandanus* appear to be the preferred foods of the Twelve-wired Bird of Paradise (Fig. 4.5), the *Epimachus* sabretails, and, to a lesser degree, the riflebirds. It is interesting to note that there are some structurally-

(a)

(b)

4.4 (a) Adult male Lawes' Parotia fruit-foraging by holding a complex *Myristica subaluata* fruit to the perch with a foot to extract the aril; and (b) a female-plumaged Raggiana Bird acrobatically removing a fruit from a *Dysoxylum pettigrewianum* capsule.

Ecology of the birds of paradise

4.5 Adult male Twelve-wired Bird clinging onto a bunch of ripe *Pandanus lauterbachi* fruits to pick and swallow pieces repeatedly from one of them.

protected figs that are visited primarily by birds of paradise (e.g. *Ficus odoardi*)— these offer a taxonomic exception that supports our contention about the special access to fruit that birds of paradise have. While requiring additional study, the accumulating fruit-foraging data for the King of Saxony Bird indicate the possibility of considerable specialisation upon the fruits of false figs *Timonius* spp. (Rubiaceae).

One final example of apparent fruit-specialisation is worth mentioning. Compelling circumstantial evidence suggests that Macgregor's Bird has established a dietary specialisation on the resin-rich fleshy 'cones' of the subalpine tree *Dacrycarpus compactus*. The bird appears to live only in habitats where this tree is common. In one site in the highlands of eastern Papua New Guinea, *Dacrycarpus* is the single dominant tree species forming the forest canopy. A most remarkable feature of the plant that impacts *Macgregoria* is that the fruiting cycle appears to be very long, with local populations of tree going for long periods (in excess of a year) without carrying ripe fruit. In one field study site, the birds entirely departed to an unknown location during this long period when *Dacrycarpus* was not fruiting.

Two final points are worth considering. It is evident there is no tight bird–plant mutualism, and, for the plant, the major biological event of its life history seems to be the spectacular synchronous wind-pollination that occurs every two years. Finally, the bird consumes a range of other fruits when *Dacrycarpus* in not available. Clearly, more study of this system is warranted (Beehler 1991*b*).

Handling and processing foods

From numerous published anecdotal records and our own observations it would appear likely that all genera of typical birds of paradise (Paradisaeinae) have the ability to use a foot or the feet to hold down food while feeding (Figs 4.3,4). The only genera (or subgenera) for which a species is not recorded doing so are *Macgregoria*, *Lycocorax*, *Semioptera*, *Seleucidis*, *Paradigalla*, and *Drepanornis*. Nevertheless, there is no reason to doubt they do so. We are not so confident, however, about such behaviour being performed by the obligate frugivorous members of the Cnemophilinae. It is quite possible this may be a character separating the two subfamilies.

It has been pointed out that the use of the feet by birds of paradise in the same way as the corvids suggests a close relationship (Kramer 1930). Some casual experiments illustrate just how characteristically-frequent foot use is in the Paradisaeinae (*contra* Kramer 1930). By presenting captive birds at Baiyer River Sanctuary with a live cockroach, grasshopper, or cricket and timing from the instant the insect was snatched in the bird's bill until it was placed beneath a foot for feeding the following results were obtained: Crinkle-collared Manucode 19 sec (8–29 sec; $n = 3$), Curl-crested Manucode 7 sec (6–9 sec; $n = 3$), Brown Sicklebill 6 sec (2–12 sec; $n = 12$), Stephanie's Astrapia 14 sec (5–22 sec; $n = 2$), Wahnes' Parotia 7 sec (4–11 sec; $n = 5$), Superb Bird 12 sec (3–36 sec; $n = 4$), Raggiana Bird 5 sec (2–12 sec; $n = 6$), Lesser Bird 6 sec (2–11 sec; $n = 5$), Emperor Bird 16 sec ($n = 1$), Blue Bird 10 sec (3–17 sec; $n = 3$) (CF). Foot use in birds of paradise is quite sophisticated. In another example, a wild adult male Victoria's Riflebird that visited a garden to perch upon the edge of a deep box full of mealworm culture would lift and gather the edges of up to four layers of sack-cloth pieces beneath one foot to expose and then eat the larvae (W. Cooper personal communication).

In representative species of at least seven genera of Paradisaeinae and one of the Cnemophilinae (*Cnemophilus*), adults have been seen to regurgitate meals to nestlings. This method of provisioning young is likely to be common to all species. Likewise, adults of representative species of at least eight genera of Paradisaeinae and one of Cnemophilinae (*Loboparadisea*) have been seen regurgitating fruits and either billing and reswallowing them or else dropping the seeds. This behaviour is probably typical of the entire family. The seeds dropped in and around terrestrial display courts of the Magnificent Bird and the parotias attract terrestrial granivorous pigeons on a regular basis. In this interesting ecological interaction, the pigeons get food, while the male bird of paradise gets court-cleaning services.

Crandall (1932: 87) reported that captive adult male Lawes' Parotia, Superb, and Raggiana Birds regurgitated 'casts' in the form of 'sacs, sometimes containing food detritus but usually empty. Their shape, size, and heavily corrugated folds led to the supposition that they were formed by a shedding of the inner lining of the gizzard.' While histological examination apparently confirmed this opinion, no other such reports are known to us.

Adult versus nestling diet

Nestling diet has been an important question with regard to the evolution of polygyny in altricial birds (Snow 1976*b*). The generally-accepted notion is that for male emancipation from nesting duties to arise (something that permits male promiscuity and the evolution of lek behaviour), it must be possible for solitary females to raise the offspring without help. In most insect-eaters both sexes attend, and in many cases both provide food to, the nestlings. Most typical birds of paradise initially feed their nestling arthropods (Frith and Frith 1992*b*, 1993*b*, 1995*b*; Davis and Beehler 1994, and references therein). After the nestling grows to a certain age, it is then fed predominantly fruit or a mix of fruit and insects. The assumption is that the arthropods provide the necessary calcium for bone formation, structural proteins to provide the critical constituents for tissue, organ, and feather production, and lipid for energy needed for growth. Once the nestling achieves a certain size, dietary needs can probably be provided by a selection of fruit, some of which are relatively rich in protein and lipid (Beehler and Dumbacher 1996).

One of the unexpected reversals of an initial study of the behavioural ecology of birds of paradise (Beehler 1983*a*) is the nesting ecology of the Trumpet Manucode (Beehler 1985*a*). This is a monogamous species in which both sexes attend the nest. It had provisionally been assumed that in monogamous birds of paradise both parents were needed to provision arthropods to the nestlings. Not so. It was found that these frugivorous specialists feed their nestlings a diet that is virtually all figs. It appears that in order to give the nestlings the necessary suite of macronutrient building-blocks for growth, massive amounts of nutritionally-poor figs are needed. Here is a situation where the nestlings apparently receive an oversupply of sugar/carbohydrate but obtain insufficient lipid and protein, which can be found in good quantities in arthropod prey. How the manucode nestlings handle the oversupply of carbohydrate is unknown. Presumably it is passed through the gut without being absorbed. The hypothesis (Drew 1988) that some fruits may be 'protein enhanced' by infestation by arthropods or their larvae does not appear to apply to the figs taken by birds of paradise—a range of these fig species from New Guinea has been analysed for nutrition, and none showed either moderate or high levels of protein or lipid (BB; Beehler and Dumbacher 1996).

The generalisation is also broken by the Cnemophilinae, in which both parent and nestling subsist entirely on fruit (Frith and Frith 1993c). But note that the nestling period is much longer in the one wide-gaped species that has been studied attending a nestling, possibly in part an adaptation to dietary limitations.

The fact that we understand little about the nutritional components of the diet of birds of paradise becomes clear in the rearing of these species in zoological gardens. Zoos invariably have had problems with haemochromatosis, a syndrome in which the birds sequester too much iron from their iron-rich diet in captivity and often die from liver failure (Frankenhuis et al. 1989). Zoos thus must feed birds of paradise an especially low-iron diet. We believe this tells us that the dietary landscape for birds of paradise in the wild is iron-poor. As a result, the birds have evolved to be especially efficient in sequestering and storing iron. This peculiar fact may offer us a useful insight into the dietary ecology of birds of paradise in the wild.

Bird–plant interactions

Anyone who has observed birds foraging in fruit trees in the tropics knows that it is a rather complex and messy affair—often including large assemblages of true frugivores and seed predators, and including chases, interspecific aggression, and predator alerts (Terborgh and Diamond 1970; Howe 1979; Beehler 1980; Pratt 1984). Fruit-foraging and seed dispersal are not as ecologically prone to strict specialisation as the classic pollinator/flower interaction for a range of reasons (Wheelwright and Orians 1982). This generalisation holds in the Neotropics as well as Australasia, which is particularly rich in fruit-eating birds (Beehler 1988). A typical fruiting fig might attract as many as 20 species of birds plus a selection of nocturnally-foraging frugivorous fruit-bats. Plants producing large quantities of small and easily-consumed fruit are visited by large suites of vertebrate foragers (both large and small), whereas those producing larger fruit are visited very predominantly by the larger species.

It is thus of considerable interest that researchers who have studied fruit-foraging in New Guinea have found that certain food plants in the forest appear to be visited only, or primarily, by birds of paradise (Table 4.4). Field studies on Mount Missim by Beehler (1983a, 1988) and Pratt (1983) and field studies on the Sogeri Plateau by Beehler and Dumbacher (1990) have provided a number of examples of what appear to be restricted foraging assemblages. All of these examples are of plants producing relatively-small quantities of highly nutritious and structurally-protected fruits (e.g. Meliaceae: *Chisocheton* spp., *Dysoxylum* spp.; Araliaceae: *Gastonia spectabilis*; Myristicaceae: *Myristica* spp.).

Specialised fruit-foraging systems are rare. In New Guinea they seem in part to be the product of the very restricted mammalian fauna (with few seed predators, and notably the absence of squirrels and monkeys). The paradisaeid examples from New Guinea are among the most species-specific fruit-foraging systems known (Howe 1984). Whether they represent some co-adaptation or mutualism remains to be seen, however. First of all, the examples are never of the single plant/single bird systems hypothesised by mutualism theory (Schemske 1983). Instead, the specialisation appears one-sided: the plant depends largely or exclusively on birds of paradise for dispersal of its seeds, but the birds of paradise (usually a suite of species) never restrict themselves to a single plant species. Mutualism or not, this is an intriguing phenomenon, and we believe it indicates how important some species of birds of paradise are to rainforest seed dispersal in New Guinea.

Why has this phenomenon been recorded with the birds of paradise in New Guinea and not with other animal groups in other parts of the world? This paradisaeid/food plant syndrome may have arisen because of the ecological opportunities produced by the general impoverishment of placental mammals. Birds have evolved to be primary dispersers of rainforest seeds, and birds of paradise have evolved anatomical and physical characteristics that

Table 4.4 Food plants visited predominantly by birds of paradise (BOP).

Plant taxon	Family	Number of feeding records	Number of bird species recorded	Number of bird families recorded	Number of BOP species	% of visits by BOP	Numerically dominant BOP species (% of records)
Chisocheton cf. *weinlandii*	Meliaceae	91	9	1	9	100%	Superb Bird (25%)
Chisocheton lasiocarpus	Meliaceae	199	3	1	3	100%	Raggiana Bird (85%)
Dysoxylum cf. *macrothyrsum*	Meliaceae	50	5	1	5	100%	Raggiana Bird (42%)
Dysoxylum pettigrewianum	Meliaceae	395	3	1	3	100%	Raggiana Bird (97%)
Myristica 'varirata'	Myristicaeae	44	6	2	3	84%	Raggiana Bird (77%)
Gastonia spectabilis	Araliaceae	327	7	1	7	100%	Lawes' Parotia (38%)
Schefflera pachystyla	Araliaceae	55	5	1	5	100%	Lawes' Parotia (76%)
Omalanthus novoguineensis	Euphorbiaceae	278	9	4	7	97%	Magnificent Bird (38%)
Ficus odoardi	Moraceae	118	8	2	7	94%	Trumpet Manucode (47%)
Ficus gul	Moraceae	244	13	6	7	86%	Trumpet Manucode (47%)

References: Beehler (1983); Beehler and Dumbacher (1996).

permit them skilfully to manipulate and open structurally-protected fruits. Thus the plants have evolved fruit structures that limit access mainly to the birds of paradise, who serve as their primary or sole seed dispersers. For a number of reasons detailed elsewhere (Pratt 1983; Pratt and Stiles 1985; Beehler 1988, 1989a,b; Beehler and Dumbacher 1996; and above), birds of paradise serve as excellent seed dispersers.

An interesting related aspect of fruit-foraging by birds of paradise is the defence of specific fruit resources by pugnacious individuals (Pratt 1984). In this, an individual will stake out a small and relatively-compact fruit resource (a single branch or vine, or a small tree) and spend much of each day there, foraging, loafing, and aggressively driving away other birds that visit to forage. This is unusual for frugivores and fruit resources, and occurs only when the resource is defensible and, presumably, worthy of defence. Not surprisingly, it has the effect of reducing the number of species that use the resource.

Foraging ecology
Preferred foraging zone

When foraging, birds of paradise typically spend most time in the upper levels of the forest vegetation, although they do descend considerably lower when they join mixed-species flocks. In lowland rainforest it is typical to see foraging birds of paradise at heights between 20 and 30 m above the ground. Our guess is that this is where most of the desirable fruit is found as well as the types of substrates that are home to favoured arthropod prey. It seems there are more epiphytes, dead limbs, dry stubs, vine tangles, and growing vegetation above 20 m. This predilection for the forest canopy has two practical impacts. The field observer must concentrate on the canopy, and those attempting to mist-net the birds are required to make special efforts to capture them (ridge-crest netting, putting nets into the upper stages of the forest using pullied rigs).

One exception is female-plumaged Magnificent Birds which Jared Diamond (personal communication) regularly caught in mist nets within 2 m of the ground.

Mixed flocks of brown and black birds

For fieldworkers in New Guinea, one of the most thrilling first experiences is to encounter a large mixed-species foraging flock of birds in the lowland rainforest. One can hear it coming from a distance—the varied notes of the Rusty Pitohuis *Pitohui ferrugineus* (Pachycephalidae) and Rufous Babblers *Pomatostomus isidorei* (Pomatostomatidae) and the metallic whistles of the drongos—and often the scold or *tup* notes of the attendant birds of paradise. Birds of paradise are almost invariably an important component of any mixed flock in lowland or hill forest, what Diamond (1987) has called the 'brown-and-black flocks'. In particular, one can expect manucodes, the Magnificent Riflebird, the King and Twelve-wired Birds, *Paradisaea* species, and even a sicklebill species. Some other (of at least 16) bird of paradise species recorded as part of such a flock are Stephanie's Astrapia, Western and Lawes' Parotias, Superb, and Magnificent Birds.

Most of the flock members tend to stay relatively low (from 3 to 10 m in height) and travel fairly quickly. Often the male birds of paradise call regularly, as if to keep the flock together. Most of the birds, in fact, are predominantly brown or black. All of the female and some of the male birds of paradise follow this rule, as do most of the non-paradisaeid members of the flock (see below).

The mixed-flock phenomenon is important for several reasons. Flocking is an important component of the social behaviour of lowland forest birds of paradise that cuts across sexual lines. Ecologically, it promotes rapid movement through the forest and combines insect-eating and fruit-eating in a way that, we suspect, has important ramifications for seed dispersal and forest regeneration.

Mixed flocks also occur in the uplands, but are less prominent, in part because of the less species-rich bird communities. It is common to see birds of paradise in mixed flocks in the middle elevations and at higher elevations. At 1400 m on Mount Missim one finds the Superb, Lawes' Parotia, Magnificent Riflebird, Raggiana, and two manucode species joining mixed flocks (mostly as insectivores). At higher altitudes it is common to find several Stephanie's Astrapias and Brown Sicklebills foraging together for arthropods.

What are the adaptive advantages of these mixed-species assemblages? Two factors, presumably acting in tandem, seem plausible forces promoting the phenomenon. One is as an anti-predator ploy. With many sharp eyes at work (including the 'sentinel' drongo species) it is safer for individuals to focus more on looking for food rather than for predators. The other factor is the 'beater' effect, in which any mobile arthropod that escapes the grasp of one bird will, in all likelihood, be noticed and captured by a nearby flock-member (Diamond 1987).

Population biology

It is safe to say we know little about the population biology of birds of paradise. We do know that in the polygynous Magnificent Bird of Paradise, although the ratio of court-holding males to females is strongly skewed in favour of females, the sex ratio across an entire local population is 1 : 1 (Beehler 1983*a*), as predicted by Fisher (1958). Thus the functional sex ratio is skewed as a product of the mating system (Emlen and Oring 1977), in large part because of the large proportion of (non breeding) immature males wearing female plumage in any population of polygynous, sexually-dimorphic bird species. The species for which data are available appear to be typical of medium to large tropical forest birds. They have low reproductive rates, very small clutch sizes (1–2), high rates of nest predation, and high adult survivorship (Frith and Frith 1990*c*, 1992*b*, 1993*b*,*c*,*d*, 1994*b*, 1995*b*, 1998*a*).

Longevity

Evidence of the potential longevity of birds of paradise is seen in the extremes of delayed plumage maturation of males. For the few species for which there are data (mainly based on captive birds), the plumed males of polygynous species require from 4 to 7, or possibly more, years to gain full adult plumage. This sort of plumage development pattern could not evolve in a short-lived species. Known maximum ages of birds of paradise are few, especially for wild birds. Captive individuals have been recorded living in excess of 30 years. A wild Buff-tailed Sicklebill, first banded as a fully-plumed male on a territory on Mount Missim in 1978, was observed still on its territory in 1987. Although no less than 9 years old, it is much more likely it was a minimum of 15 years old. A wild Lawes' Parotia caught as an adult female was recaptured at the point of original capture more than 8 years later (Anon 1989), a wild male Victoria's Riflebird lived at least 15 years before a domestic cat killed it, and a Raggiana Bird almost 17 years (see species accounts).

Heterochrony

Heterochrony is unequal development in a species. Heterochrony in the birds of paradise is exemplified by the female gaining adult plumage within a year, versus the male achieving full adult plumage over a number of years, and passing through a series of intermediate plumages entirely absent in the female. Male plumage heterochrony in polygynous birds of paradise can be remarkably complex. This is a topic in need of further study. The interesting thing about graded heterochrony is that it may be an important signalling system among male birds of paradise. Not surprisingly, this graded heterochrony is probably the most complex in the lek-displaying *Paradisaea* birds of paradise. As far as is known, in these species the male first assumes the female plumage, and may retain this for several years. The first plumage indicator that the bird is a male is when it produces the elongated and narrowly pointed central tail feathers (which in the next moult may have paddle-shaped tips) and subsequently the green throat and yellow cap. The truly wire-like central rectrices follow, and also the dark velvety breast cushion and short flank plumes. In each subsequent year the flank plumes become longer, until they reach an 'adult' length. Still, after this standard is reached (9–10 years?) it is exceeded in later years by even longer tail wires, and longer and fuller flank plumes. All of these plumage traits are probably signals that figure in the social and sexual behaviour of polygynous birds of paradise.

Heterochrony also occurs in at least some of the monogamous manucodes (the Trumpet Manucode in particular)—in the form of the male's coiled trachea. We have no knowledge of the number of years required to develop each of the tracheal 'loops' but the degree of tracheal variation found in specimens of the Trumpet Manucode suggests it is likely to be 5 years at a minimum (see Fig. 1.5).

Age at first breeding

We know virtually nothing of this except what (very little) has been learned from captive individuals. Our assumption is that females are probably capable of breeding in their first or second year, whereas males of all species probably do not breed until at least 2 years old. For some of the polygynous plumed birds of paradise we know female-plumaged males are capable of siring offspring in captivity (Delacour in Gilliard 1969; Laska *et al.* 1992), and in the wild female-plumaged males have been seen to mount females in a lek of Goldie's Bird (LeCroy *et al.* 1980), but we suspect that males typically do not reproduce until after they obtain full adult plumage—several to many years later than females. Mating system theory supports the notion that those species with the most extreme mating systems will produce the longest delay of first reproduction in males, and plumage heterochrony supports this. Five-year-old males in subadult plumage are signalling that they are not yet prepared to enter fully the mating competition.

Predation and other causes of mortality

Most birds of paradise are large and aggressive and well armed with their powerful bill and claws. They face many natural enemies, including snakes and hawks (mainly *Accipiter* species). Snakes probably prey mainly on eggs, nestlings, and the occasional female brooding at night or male at its roost. New Guinea's Brown Tree Snake *Boiga irregularis* is a notorious bird predator and one presumes that pythons will also consume roosting birds of paradise. There are a few records of evidence of natural predation but these are not particularly informative. A Macgregor's Bird was recorded by Rand (1940*a*) in the gut of a Spotted Marsh Harrier *Circus spilonotus*. Victoria's Riflebird was said to be taken by the Lesser Sooty Owl *Tyto multipunctata* (Schodde and Mason 1980) and one was taken by a Grey Goshawk *Accipiter novaehollandiae* (CF). Opit (*in litt.*) located the carcass of a recently-killed Buff-tailed Sicklebill on Mount Missim. A Lesser Bird was eaten by a Doria's Hawk *Megatriorchis doriae* (Rand and Gilliard 1967). The behaviour of the birds, however, speaks volumes. On leks and in fruiting trees, birds of paradise are exceedingly skittish and wary of anything hawk-like. From this behaviour, it is evident that hawk predation is a factor in the lives of these birds. Bird-eating raptors are particularly well represented in forest bird communities in New Guinea. Probably all seven species of New Guinea's forest-dwelling goshawks and sparrowhawks (*Accipiter* spp.) take birds.

Aside from predation, one must assume that these birds succumb to parasites and disease, although little is known of these for birds of paradise. Varghese (1977) described two novel parasitic protozoans inhabiting the gut of Raggiana Birds, and the ectoparasitological field surveys of mammals and birds conducted under the auspices of the Bernice P. Bishop

Museum collected a range of taxa from birds of paradise. A study of paradisaeid blood parasite loads of 10 species of birds of paradise showed that 83% of a sample of 216 birds carried some form of blood parasite, including the following identified genera: *Haemoproteus*, *Plasmodium*, *Leucocytozoon*, *Trypanosoma*, as well as unidentified microfilaria (Pruett-Jones *et al.* 1990). The *Haemoproteus/Plasmodium* malarial protozoa were the most abundant in the samples, followed by the microfilaria. The authors did not comment on the relationship of parasite loads to mortality, but they did establish an inverse relationship between courtship activity and parasite load. Perhaps most interesting of all is that they demonstrated a statistically-significant positive relationship between plumage 'showiness' and parasite loading at the species level—the showy polygynous species tend to carry more parasites than the less showy monogamous species. This confirms part of the Hamilton and Zuk (1982) hypothesis that argues that the bright plumage of males evolved as a signal of heritable true fitness to females—with only the healthy males being able to produce showy and beautiful plumage and those with heavy parasite loads being easily recognised and rejected by females as potential mates.

Nesting ecology

With respect to nesting ecology, several interrelated factors seem to be potentially relevant to the birds of paradise: (1) the small size of the clutch, (2) nest crypticity, (3) the threat of nest predation, (4) the need to feed nestlings efficiently, and (5) female-only nest care for most species (Snow and Snow 1979; Willis 1972). Clutch sizes of bird species in New Guinea tend to be small, compared to northern temperate datasets (BB), but paradisaeid clutches are even smaller than those of typical New Guinean passerines. It has been argued that, because nest predation in the tropics is high, individual investment per nesting attempt should be low (hence a small clutch is best) but with several nesting attempts each season. In the case of upland breeding species other factors may, however, limit clutch size, such as the colder and more frequently humid ambient conditions and associated relative scarcity of arthropod foods. Male emancipation from nesting duties presumably evolves after clutch size has been reduced, so it is perhaps only a factor in the maintenance of small clutch size in polygynous species.

The nest should be cryptic to avoid attracting predators and fewer nestlings require fewer visits to the nest. Under these conditions, presumably one parent is sufficient to provision nestlings, especially if the parent can efficiently feed the nestlings on regurgitated fruit pulp. Most nests of birds of paradise are exceedingly cryptic, and placed out of the way of the traditional arboreal pathways of potential predators (usually in vegetation that is isolated from surrounding vegetation). In addition, they are hidden in the vegetation in a manner that makes discovery (at least by human searchers) very difficult indeed.

If there are real links between nesting ecology and the evolution of polygyny, one might expect consistent differences in nesting habits between monogamous and polygynous birds of paradise. One might expect that monogamous species build better-protected or invulnerable nests, or have larger clutches, or both. This has not been found for the birds of paradise, so it leaves us without a rigorous explanation for the very small clutch sizes of typical birds of paradise. Nonetheless, one assumes that two adults at a nest could provide additional nest protection against some predators, and this alone may be an important benefit of biparental care. Conversely, activity of only a single parent would reduce the probability of predation at the nest.

Ranging habits

Most birds of paradise appear to be sedentary forest-dwellers with relatively small permanent home ranges. This generalisation is based on detailed field data from a few key species (Beehler 1983a, 1988, 1991a,b; Pruett-Jones and Pruett-Jones 1988a). The primary insectivores (such as the sicklebills and riflebirds) have compact home ranges that appear to be exclusive male foraging territories (Beehler and Pruett-Jones 1983; Beehler 1987a; CF), whereas the home ranges of primary frugivores appear to be larger and non-exclusive (Beehler 1988; Pruett-Jones 1985). The seven-day range of a plumed male Buff-tailed Sicklebill (an insectivore) was 11 ha, whereas a seven-day range of a male Raggiana Bird (frugivore) was 44 ha. For the sicklebill, no more than one adult male was observed on the study area, and singing males were widely dispersed. By contrast, male Raggianas forage together, feed in the same feeding trees, and display together in communal leks.

Distributional ecology

Many birds in New Guinea appear to sort out ecologically along an elevational gradient (Fig. 4.6). This sorting operates among closely-related species of birds of paradise (Diamond 1972; Pruett-Jones and Pruett-Jones 1986). The King and Magnificent Birds occupy only marginally-overlapping ecological zones, with the King in the lowlands and hills and the Magnificent in the upper hills and lower montane zone. Two upland sicklebills, the Brown and the Black, exhibit a similar pattern, with some minor overlap in midmontane forest, but with the Black exclusively at lower elevations and the Brown at higher elevations. The Blue and Raggiana birds of paradise exhibit a similar pattern, with the Blue above the Raggiana with considerable marginal overlap.

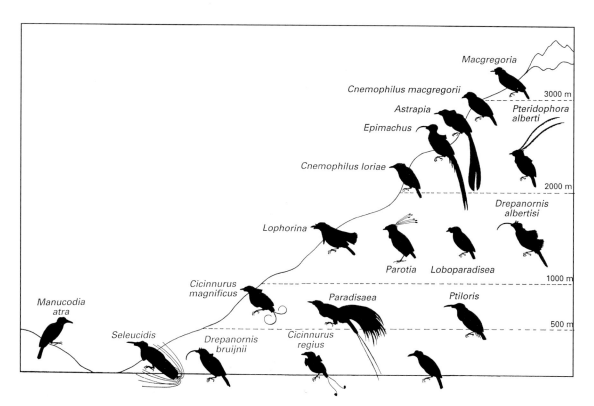

4.6 Diagramatic representation of a profile of the New Guinea central cordillera with the main altitudinally-restricted distribution of various bird of paradise taxa indicated by the named silhouettes.

The same occurs in the Huon Peninsula with the Emperor and Raggiana Birds. The Emperor inhabits the uplands, the Raggiana the lowlands, and populations meet in upper hill forests. The very closely-related Ribbon-tailed and Stephanie's Astrapias exhibit elevational segregation, but there is considerable hybridisation where the two meet.

Birds of paradise and rainforest regeneration

Perhaps our most important discovery about the ecology of birds of paradise is the role they play in rainforest seed dispersal (Beehler 1989*b*). On the island of New Guinea, the Paradisaeidae supports more species of seed-dispersing frugivores (34) than any other vertebrate family—the pigeons are second with 24 species of primary frugivores. Thirty-one species of honeyeaters (Meliphagidae) take some fruit, but are a much less significant dispersal group because of small size and lack of dietary diversity. As noted earlier in this chapter, certain plants appear to rely exclusively on birds of paradise for seed dispersal, and many other plants rely on the birds of paradise as major dispersers. Dispersal by birds of paradise involves hundreds of species of plants from scores of botanical families. The degree to which the birds of paradise influence rainforest composition and regeneration is a complex question that has not yet been addressed. Understanding how natural vertebrate-assisted forest regeneration operates could provide important clues as to how to regulate timber extraction to promote rapid forest renewal.

Annual cycles

The annual cycles of birds of paradise are poorly understood, mainly because there is a great range of seasonal and climatic regimes over their distributions. On the tropical island of New Guinea, most regions have an end-of-calendar-year wet season, whereas other sections (centred, for instance, on Lae, Wewak, Kerema) have a middle-of-the-year wet season. Some other regions in New Guinea are virtually aseasonal, with significant rainfall every month of the year. These three general patterns are a product of the interaction of the major trade/monsoon wind systems in concert with the peculiarities of local physiography. In general, birds of paradise display during the dry season and nest in the beginning of the wet season (Diamond 1972). For most of New Guinea, then, display peaks from April to October, with nesting from November to January. Plumage moult typically follows nesting (February–March).

The Tari Basin has an equable rainfall regime with no real wet or dry season (MacAlpine *et al.* 1975). In this region birds of paradise nest at the end of the year as in the typical parts of New Guinea. What is not clear is whether these annual avian cycles are reversed in the parts of New Guinea where the climatic seasonality is reversed. It is notable that in the Crater Mountain Reserve, in an area of high and equably-distributed rainfall, paradisaeid breeding is possibly bimodal, with nesting records both from April–May and October–November (A. Mack personal communication). Annual seasonal cycles are a phenomenon that merit greater scrutiny by fieldworkers. The issue of periodic El Niño droughts also adds a layer of complexity and unpredictability to the issue, and that, too, needs to be better understood, especially in terms of annual and supra-annual fruiting cycles (M. G. Hopkins personal communication; Beehler 1991*a,b*).

Plumage mimicry, pitohuis, and distasteful birds

The various medium-sized to large bird species typically constituting New Guinea mixed-species foraging flocks are generally patterned in brown and black and it has been hypothesised this colour scheme facilitates flock social cohesion and also confuses potential predators (Diamond 1987). In this regard it should be noted that the majority of individuals of polygynous bird of paradise species participating in these flocks are in the predominantly brown female plumage. These facts led to the notion that 'social mimicry' is involved in the similar plumage patterns and coloration across the taxonomically-diverse flock-member species (Moynihan 1968; Diamond 1987). It was suggested that in New Guinea mixed-species flocks, this primarily involves mimicry of the general appearance of the Rufous Babbler *Pomatostomus isidorei* and various pitohuis *Pitohui* species—by far the most typical flock leaders. This novel idea preceded the discovery that the Hooded Pitohui *Pitohui dichrous* is chemically defended by toxic feathers and skin (Dumbacher *et al.* 1992).

At a point when the New Guinea zoological collector A. E. Pratt and his associates were extremely hungry they 'were compelled to try [unspecified] bird-of-paradise soup: it was truly abominable, and after the first spoonful we got no further' (Pratt 1906: 258). Earlier in the same book Pratt wrote of tasting a roasted Buff-tailed Sicklebill, which also proved to be inedible. Chalmers and Gill (1885) also record boiling a Raggiana Bird of Paradise for several hours but it proved 'as tough as leather, and the soup not much to our taste'. It is therefore probable that some birds of paradise are unpalatable to at least some predators some of the time. This discovery opens the possibility that mimicry may be operating, a notion we discuss at greater length below.

The above records of distastefulness of some birds of paradise may take on greater significance since publication of the details of the toxicity in the Hooded Pitohui. Two questions devolve from this apparent example of chemical defence in a forest bird in New Guinea. First, might some birds of paradise actually be chemically defended? Second, might the brown-and-black plumages of certain birds of paradise be the product either of Batesian or Müllerian mimicry (Barnard 1979)? Certainly the brown-and-black bird flock phenomenon may include mimicry that is driven by avian toxicities—the example of the plumage variability of the Variable Pitohui *Pitohui kirhocephalus* appears to include components of Müllerian mimicry (where the much less toxic form *P. k. meridionalis* exhibits a plumage nearly identical to the highly-toxic Hooded Pitohui). As noted above, the females of many birds of paradise are dorsally uniformly coloured much like the more typical flock leaders. Of particular note in this regard is that while female plumage in the *Paradisaea* species is unlike that of any other family members, it consists of more contrasting areas of colour like the Hooded and Variable Pitohui. We believe the various speculative threads we have spun here merit additional examination.

5

Reproductive behaviour

The birds of paradise have long figured as prominent examples of sexual dimorphism and unusual mating behaviour. Charles Darwin referred to them on 12 occasions in *The Descent of Man and Selection in Relation to Sex* (1871). Most of Darwin's knowledge of this bird family was borrowed from the writings of his colleague and fellow evolutionist, Alfred Russel Wallace. So taken was Wallace by the beautiful plumages and bizarre courtship behaviour of the birds of paradise that he spent months searching for them in the field (with more disappointment than success), and he included them in the subtitle of his most enduring work, *The Malay Archipelago* (1869).

For Darwin, the fantastic male plumages of birds of paradise exemplified the end product of a phenomenon he named 'sexual selection'—the often fierce competition to gain mates and to reproduce. This is an evolutionary process Darwin believed to be distinct from natural selection. He used sexual selection to explain the development of morphological and behavioural characteristics in the one sex (the male in birds of paradise) that competes for reproductive access to members of the other sex. It is this process that produces, via evolutionary mechanisms, the prominent sexual differences so common in the animal world, for example the gaudy and colourful look of the male peafowl (peacock) versus the more cryptic and unadorned plumage of the peahen.

Sexual selection and sexual dimorphism

Sexual dimorphism is one outcome of sexual selection—and the one that makes birds of paradise so alluring to birdwatchers and naturalists. We focus on the 'beauty' of birds of paradise, but, in fact, the extravagant beauty is confined to the males of each species. The scales of morphological evolution have been heavily tipped in the favour of males, and all the extreme characters and spectacular colours have been channelled into male morphology. It is not so much that naturalists adore birds of paradise—as it is that they adore the adult male birds of paradise.

The birds of paradise offer some of the best examples of extreme differences between males and females. This is most obvious in several male physical characteristics: size, plumage coloration and the presence/absence of ornate nuptial plumes, and vocalisation. It is also evident in sexual differences in behaviour related to reproduction—mate-selection, competition for mates, and activities related to nesting and care for offspring.

Size dimorphism

Virtually all male birds of paradise are larger than their female counterparts, even among the monogamous and monochromatic species (the aberrant Yellow-breasted Bird is the sole exception). This size difference, best exemplified by comparative weights (Table 1.3) indicates that sexual selection operates in all species of birds of paradise—the physical disparity being a product of differing selection pressures faced by males versus females of each species. The male/female weight ratios vary considerably, from as high as 1.56 in the Raggiana Bird to as low as 0.87 in the Yellow-breasted Bird. In birds of paradise, males are the displaying sex, females the choosy sex, and it is safe to assume that to females, bigger is better. Any male display, no matter how simple, will be more impressive if given by a larger individual. The anomalous Yellow-breasted Bird is the curious counter-example in the family, in which the male is smaller than the female. Because this is perhaps the least known and least understood species of the family, we cannot yet speculate as to the reasons for this sex-size reversal.

Plumage dimorphism

Only the manucodes (Paradise Crow included), paradigallas, and Macgregor's Birds are sexually monochromatic, the remaining genera exhibiting strong plumage dimorphism (Plates 4–12). Even a species with the most exaggeratedly-plumed male is balanced by a female whose plumage borders on drab. The dull female plumages are products of selection that rewards the most cryptic individuals—after all, each must construct a nest and tend nestlings without male assistance in an environment inhabited by predators.

Vocal dimorphism

Birds of paradise are well known for their loud, distinctive (but rarely pleasant) vocalisations. Among the monogamous species, both male and female vocalise regularly, giving sex-specific calls. By contrast, among the polygynous species, all of the loud broadcast vocalisations are male advertisement calls, the females being virtually mute (but see Superb Bird account, Chapter 9). For while the male's call can carry more than a kilometre, female vocalisations are restricted to minor *sotto voce* growls and single notes that carry no more than tens of metres. The relatively-silent nature of females in polygynous species has been difficult to document in the field because many birds in female plumage are, in fact, young males.

Sexual differences in reproductive investment

Although the monogamous bird of paradise species form pair bonds and forage together, the lives of male and female polygynous birds of paradise are more divergent. In polygynous species, the males display and the females nest, and the two sexes come together briefly to copulate. The degree to which they socialise away from the display sites remains incompletely answered, but appears to be confined to little more than opportunistic foraging associations.

The sexes in polygynous species lead separate lives, following their own pathways to fitness. A male seeks to rise in the mating hierarchy and then to mate with as many females as possible once he has achieved high status. A female struggles to produce one or two offspring a year. As a result, a male will focus on competing with other males and on gaining a favoured display site, whereas a female will focus on selecting the best mate, but will invest most time and effort in nesting and caring for her offspring. A court-holding adult male spends much of his time on or near his court, prepared to copulate at very short notice. A female tends to range more widely, becoming fixed to a specific location only when nesting. Because of differing nutritional needs and time budgets, it is likely that females and males face differing foraging demands, with females seeking more arthropods during the nesting season, and males seeking food that permits minimum search and handling times (fruit), so that they can forage quickly and then return to their display site. Virtually every behaviour differs between male and female of the polygynous species.

Mate-choice

The extremes of sexual dimorphism are apparently related to the degree of reproductive competition among males, as evidenced by the type of mating system employed. Sexual dimorphism is slight in the crow-like and monogamous manucodes, in which male and female co-operate in raising the offspring. By contrast, sexual dimorphism is very pronounced in the polygynous Raggiana Bird of Paradise and Lawes' Parotia, in which the males gather into lek assemblages. In the Raggiana, the males of a display group share a single canopy tree, the traditional lek site. Males and females interact mainly or solely at the lek. In lekking species, a very small percentage of the males in the population carry out most of the mating in any single mating season. In Lawes' Parotia, leks are based on variably-clustered terrestrial courts, but the same sort of mating system operates, with considerable mating skew among males (Pruett-Jones and Pruett-Jones 1990).

Much of the recent debate about sexual selection focuses on the nature of interactions among males and between males and females. Are the remarkable nuptial plumes of the male birds of paradise primarily 'badges' of dominance among males ascending a mating hierarchy or advertisement to the females of male genetic quality and vigour? Do females discriminate between males based on physical characteristics such as size and plumage? Are females free to 'choose' their mate or do males (at least in the lekking species) in some way limit their choice?

Based on data from birds of paradise and the neotropical manakins (Pipridae), Beehler and Foster (1988) suggested that competitive interactions between males as well as female choice influence these mating systems. Likewise, studies of mating behaviour of the Satin Bowerbird *Ptilonorhynchus violaceus* by Gerald Borgia (1986) indicated that both female discrimination and aggression between males are critical to the understanding of sexual selection in this polygynous court system. Pruett-Jones and Pruett-Jones (1990) stressed that, for Lawes' Parotia, female choice is of predominant importance in the operation of that lek mating system.

Mate-choice has a significant impact on sexual selection and sexual dimorphism when mating diverges from random pairing of individuals and instead includes only a small proportion of the males in the population. This allows for a rapid selection of traits exhibited by these favoured males, and sets up a situation that can permit 'runaway' selection—the ever-stronger selection for more and more extreme male traits. Field studies of mate selection in Lesser and Raggiana Birds (and to a lesser extent Lawes' Parotia) have documented strongly-skewed mating—in which few males perform the vast majority of matings (Beehler 1983*d*, 1988; Pruett-Jones and Pruett-Jones 1990). In the Lesser Bird, among eight lek males, the alpha male obtained 25 of 26 recorded copulations. In the Raggiana, the most successful of four lek males obtained 15 of 35 copulations. In Lawes' Parotia, season-long observations showed than only 30–40% of the males successfully mated (Pruett-Jones and Pruett-Jones 1990). It is interesting to note that the lek with the highest mating skew had the most lek males.

Studies of paradisaeid species exhibiting lek and exploded lek behaviour (see Court display, below) have provided minimal evidence of significant morphological differences among adult males (Beehler 1988; Pruett-Jones and Pruett-Jones 1990). Any existing physical differences were either subtle and visible only to the females and not to the human observers (very possible), or else the females were responding to other factors, such as position in the lek or male behaviour. Pruett-Jones and Pruett-Jones (1990) found that it was the likelihood of a male displaying to visiting females that was the most important factor correlated with male mating success. We believe that in sedentary, long-lived birds such as these polygynous birds of paradise, females recognise males as identifiable individuals. The Pruett-Jones' (1990) examples of females remating with the same males over two seasons supports this notion, unless females select for the same position (not same male) on the lek.

Mating systems

The various types of mating systems employed by the birds of paradise can be classified by male–female interactions and by the kinds of dispersion of the males (Table 5.1). The major dichotomy is between monogamous and polygynous systems. We review the range of paradisaeid mating systems in the following sections.

Monogamous mating systems

The behaviour of three monogamous birds of paradise have been observed in the field. Rand (1940a) and Beehler (1983c, 1991a,b) have studied Macgregor's Bird, Rand (1938) reported on the Glossy-mantled Manucode, and Beehler (1985a) on the Trumpet Manucode. In these monogamous species, the pair bond appeared to be maintained year-round, and both sexes attended the nest and provisioned the nestlings.

Perfunctory male–female courtship has been observed in Macgregor's Bird and the manucodes. In Macgregor's this involves little more than chases with attendant vocalisation, often with more than one male in the chase. With the manucodes, the male has been seen to give a brief and simple posturing display to the female in the canopy of the forest, in which the male opens the wings and erects the tail and neck hackles in a striking pose (Trumpet Manucode).

In both Macgregor's Bird and the manucodes, males and females forage together and maintain contact while foraging by counter-calling. In at least one manucode (Trumpet) the male and female give sex-specific call notes. This is apparently the case also in Macgregor's Bird.

A final characteristic of the monogamous birds of paradise is that none appears to be territorial. All are highly frugivorous and wide-ranging, with little opportunity to defend a foraging space. Trumpet Manucodes habitually travel in small parties (family groups?) and as many as 10 individuals have been observed foraging in a fruiting fig at the same time (Beehler 1981).

Polygynous mating systems

Court display

It is the polygynous court-and lek-type mating systems that are of particular interest to students of the birds of paradise, mainly because they exhibit all sorts of peculiarities and specialisations. All of the polygynous paradisaeid species studied to date exhibit 'court' behaviour, in which a male occupies a mating area (Beehler 1987b)—which can be as small as a perch in a treetop or a larger all-purpose territory with a display site or series of display sites therein. Court behaviour is unusual because it is based on male self-advertisement rather than male defence of resources required by the female (Emlen and Oring 1977). The male advertises himself to any and all females whom he can attract to his court. In court behaviour, the only interaction between male and female is courtship and subsequent copulation, the male providing no special access to a foraging territory or nesting site, nor does he assist with nesting. Polygynous court display ranges from solitary and territorial (e.g. Buff-tailed Sicklebill) to communal lekking (e.g. Raggiana Bird), with a range of intermediate manifestations (e.g. Lawes' Parotia). See Table 5.1.

The most common type of mating system in the birds of paradise is solitary, non-territorial (e.g. Magnificent Bird), occurring in at least 10 species. The solitary-court species themselves vary considerably in dispersion and behaviour. Some species establish terrestrial display grounds that are cleared of leaves and debris (e.g. Magnificent Bird) whereas others display from a tree stub or fallen tree bough (e.g. Magnificent Riflebird) that requires little or no tending or modification. Some display high above the ground (e.g. Twelve-wired Bird), whereas most display in the understorey or terrestrially. Some appear to defend an exclusive all-purpose territory within which lies the display court (e.g. Buff-tailed Sicklebill).

Lek display

In most court-displaying species the males are arrayed through the forest in a regular fashion, but in some the males are clustered in space. The phenomenon of male clustering for display has at times been called 'arena' behaviour (Mayr 1945; Gilliard 1969), but we use the more widely-accepted term 'lek.' Those systems in which the males are loosely clustered are termed 'exploded leks' whereas those in which males are tightly clustered are termed 'leks' or 'true leks.' The operational definition of an exploded lek is a spatially-clustered array of male courts in which males are in vocal but not immediate visual contact. In true lek dispersion the males are in visual contact. At least three species are known or suspected to employ the exploded lek dispersion—two species of parotia (Lawes' and Carola's) and the King Bird of Paradise. Males of the King Bird are in some instances arrayed in dispersed pairs, in which the paired males occupy perches *c.* 70 m apart. But there is a range of variation in this dispersion, with many males displaying in solitude, to as many as four males clustered. In Lawes' Parotia, a similar range of variation occurs, from solitary to clustered (Pruett-Jones and Pruett-Jones 1990). In true lek dispersing species, one also finds leks with as few as one to as many as 40 or more males (Goodfellow 1908; Beehler 1988).

The lek mating system represents the most extreme form of polygynous mating system known in the animal world. The Paradisaeidae aside, the Australasian region is home to only two additional lek-breeding bird species, the bizarre ground-dwelling parrot called the Kakapo *Strigops habroptilus* of New Zealand (Merton *et al.* 1984) and the Tooth-billed Bowerbird *Scenopoeetes dentirostris* of Australia (Frith and Frith 1994c, 1995a). At least seven species of birds of paradise exhibit true lek behaviour (Stephanie's Astrapia, the Standardwing, and five *Paradisaea* species). In all of these species males display in a tight cluster (inter-male distance 0.5–5 m) in the canopy branches of one or more forest trees (a single tree is typical), with display perches of small diameter (*c.* 2–4 cm), steeply-sloping limbs.

Leks of these species tend to be distantly spaced, with nearest-neighbour leks ranging from several hundred metres to a kilometre or more (Beehler 1988). In the best-studied lek species, the Raggiana Bird, lek dispersion varies considerably from site to site. At Varirata National Park, Papua New Guinea, where the species is abundant, Beehler and co-workers (unpubl. data) located six leks in a tract of hill forest and savanna about 2 km in diameter. A similar-sized tract of lowland alluvial forest in the Lakekamu Basin supported only a single Raggiana lek.

Paradisaeid leks tend to be long-lived or 'traditional'. One Raggiana lek at Varirata National Park was active for at least two decades (*c.* 1969–89) but was then gradually abandoned. A Lesser Bird lek on the grounds of the Baiyer River Sanctuary was active in the same tree for at least a decade or more, but wind damage to the tree caused most of the males to shift to a tree about 200 m away. Most males displayed in the new tree, but some individuals continued to use the old one.

How a lek is originally established is not known, but the fact that we have documented leks that contain a single male supports the notion that an independent male may found a lek and eventually attract subordinates.

Beehler and co-workers (unpubl. data) have documented temporary leks in the Raggiana Bird. The best-monitored such lek was found in a fruiting nutmeg tree nearly midway between two permanent leks. This temporary lek attracted plumed males on successive mornings for more than a week before being abandoned. Males were seen to call and display in this tree, which produced fruit that was consumed by the birds. In a similar phenomenon, Raggiana males were seen to display sporadically in a fruiting fig tree on Mount Missim, another food-plant tree favoured by the species. No mating was seen at either lek, although female-plumaged birds were seen to visit them.

Further highlighting the complexity of the lek system, we have observed individual unplumed males displaying opportunistically to presumed females in the forest away from any known lek. In addition, on Mount Missim, Beehler and co-workers radio-tracked a plumed male Raggiana Bird that occupied no lek, although it visited at least one of the active leks (Beehler 1988).

Male maturity and mating strategy

It was Delacour (1963: 232) who first noted that 'plumeless ... non adult' males can fertilise females. He was shown by Shaw Mayer a nest containing newly-hatched chicks in an aviary that contained only three males, which were all in the (plumeless) female plumage of immaturity. This event, and the fact that some female-plumaged Goldie's Birds were observed to mount females in a lek (LeCroy et al. 1980; LeCroy 1981), led to the interesting speculation that males might retain female plumage as an alternative mating strategy (see also the Chapter 9 species account for Goldie's Bird). A captive female Lesser Bird, having lost her adult-plumaged male sire, was successfully fertilised by the only available mate, a younger male in female plumage. Notwithstanding that the captive female had no choice of mate, it was argued that certain natural conditions may result in selection of unplumed males by females (Laska et al. 1992). Thus the elaborate plumages, displays, courts, and leks of the birds of paradise reflect but a part of their highly-complex reproductive behaviour and biology.

These observations support the hypothesis that males in lekking species may attempt a range of mate-acquisition behaviour in response to the highly-competitive and strongly-skewed mating regime. The long delay in full acquisition of the nuptial plumage of the plumed *Paradisaea* species accords with this theory. An immature male 'floats' in the population for several years dressed in the cryptic female plumage. Only after this floating period does the young male begin the slow, staged process of acquiring male nuptial plumage. The floating period allows the novice male to study the social/sexual environment while 'camouflaged' as a female. The staged acquisition of adult male plumes permits a gradual apprenticeship in the intensely-competitive male mating hierarchy.

The ecology and evolution of mating systems

Two questions worth asking about lek behaviour in the birds of paradise are: why does lek behaviour occur in this lineage; and why does it occur in a few species but not in most others? The evolution of lek behaviour involves two behavioural shifts from the typical monogamous condition, and we believe that ecological factors may weigh heavily in this evolutionary transition. The first shift is from monogamous pairs to unpaired polygynous males that attend courts spaced regularly through the forest. The behavioural shift from monogamy to court-based polygyny requires that males be 'emancipated' from duties at the nest. The single prominent ecological factor that correlates with polygynous court behaviour in birds of paradise is the reliance on a suite of food plants that produce nutritious fruit on a predictable schedule (Beehler 1985a).

The second shift is to unpaired polygynous males on courts that are clustered spatially. Hints about the causes for these behaviour shifts can be gleaned from the work on analogous forest songbirds in the New World and on field studies of African bats.

Frugivory, insectivory, and paradisaeid lek behaviour

David and Barbara Snow pioneered in the study of lek behaviour by their detailed work on a number of species of cotingas (Cotingidae) and manakins (Pipridae) in Central and South

America. These two rainforest-dwelling avian families are remarkably similar to the birds of paradise in both plumage and behaviour, in spite of being quite unrelated to them. The Snows showed that there is a correlation between lek behaviour and frugivorous specialisation in these neotropical bird families. They postulated that the economic and nutritional advantages of frugivory may be the major ecological force promoting the evolution of the polygynous mating behaviour within the cotingas and manakins (Snow and Snow 1979; Snow 1976b).

What is the socio-ecological basis for the lek clustering exhibited by birds of paradise? Based on studies of polygynous bat species in Africa, Jack Bradbury has postulated that wide-ranging females, with overlapping home ranges, produce a heightened potential for polygynous mating (Bradbury 1981). A male stationed at any single point might have the opportunity to encounter large numbers of potential mates, especially if the male chooses a site where a maximum number of females overlap in their movements. Only then would there be any perceivable benefit to clustering by males. Females may, in turn, favour certain segments of the forest because of an abundance of food plants or other ecological benefits (Bradbury and Gibson 1983). The competition among males for these favoured sites encourages clustering. Older and dominant males tend to control these sites, and younger and subordinate males tend to associate with these experienced birds.

Bradbury's models predict a positive relationship between the number of females that use a site and the number of males that cluster at it. Bradbury supposed that this transition in female ranging habit would occur because of a dietary shift to hard-to-find or 'trap-lined' food plants. For birds of paradise, the trap-lined foods are the specialised nutritious food plants (lipid and protein rich), whereas the hard-to-find varieties are typified by the figs (water and carbohydrate rich). Monogamous Trumpet Manucodes depend on hard-to-find fruit. They have large, overlapping home ranges, but male emancipation is prevented because both parents are needed to provision the nestlings with the nutritionally-poor figs. In contrast, the diet of the lekking Raggiana Bird of Paradise includes significant proportions of both fruit types. The hard-to-find foods promote wide ranging among females and opportunistic access to rich lodes of food to lekking males. The trap-lined foods allow the lone female to provision her nestlings, and also provide a predictable and steady source of food for males stationed at a lek (Beehler 1987b, 1989a). Thus the data obtained for the birds of paradise support the hypothesis that the lekking species have larger ranges with greater range overlap than non-lekking, solitary species—but there are exceptions.

The example of the Buff-tailed Sicklebill suggests that the socio-ecological scenarios outlined above do need to be accepted with considerable caution. The sicklebill is polygynous and yet is primarily insectivorous (Beehler 1987a). The male defends a foraging territory, which is typical of insectivorous forest birds but unlike most other birds of paradise. How can this anomaly be interpreted? Although predominantly insectivorous, both male and female sicklebills have been observed foraging for fruit—the nutritious specialised varieties but also the poorer figs. Field studies of the Pale-billed Sicklebill (sister species of the Buff-tailed) showed that it has a predominantly fruit diet typical of the other polygynous birds of paradise (Beehler and Beehler 1986). It is suggested that this genus of sicklebills has evolved from typical frugivorous stock, but that the bill specialisation (Fig. 4.1, Plate 9) has permitted the group secondarily to become insectivorous specialists.

Following the Snows' lead (see above), recent ecological studies of birds of paradise have focused on the importance of dietary ecology and specialised frugivory in the evolution of lek behaviour. The remarkable diversification in bill size and shape in the Paradisaeidae indicates a range of diets and foraging methods that tallies closely with subsequent field research on diet (see Chapter 4).

Thus some species of birds of paradise are nearly wholly insectivorous, some have a diet with equal proportions of fruit and insects, and some are nearly exclusively frugivorous.

We gain an insight on constraints and opportunities to mating system evolution by comparing the divergent ecologies of two birds of paradise. The monogamous Trumpet Manucode relies on figs, a fruit resource that is locally concentrated and easy to harvest, but unpredictable in space and time, very patchy, and nutritionally poor. For the manucode, each few weeks bring a new search, far and wide, for a fig tree bearing ripe fruit. Adults take very few arthropods and feed little or no arthropod prey to the nestlings.

In contrast, the favourite food plants of the polygynous Magnificent Bird produce specialised capsular fruit. These plants are numerically more abundant; they produce nutritionally-rich fruit annually, with interplant synchrony. The predictable nature of the fruit diet of the Magnificent Bird permits a male attending a fixed display court to visit and feed efficiently at a range of food plants as they ripen predictably each year. The long fruiting season of individual plants means fruit is available over a long period. Females are able to provision the nestlings with a combination of nutritious fruit and arthropods.

These contemporary studies show a close correlation between specialised frugivory and the evolution of the complex polygynous mating systems in the Paradisaeidae. What is uncertain is the degree of co-evolution that has taken place between bird and plant. In some environments, the birds of paradise are exclusive dispersers of some species of Papuan rainforest trees (Beehler and Dumbacher 1996), and may exemplify the tight frugivore/food-plant relationship rarely observed in nature (see Chapter 3). Radiations of bill morphology and male plumage within the family have been great, but are not paralleled by significant genetic divergence (Sibley and Ahlquist 1990). The morphological radiation appears to be the result of two forces—extreme sexual selection, and dietary and foraging shifts among closely-related species. The mechanism by which significant morphological differentiation can occur despite limited genetic change requires investigation.

Courtship display sites and types

Courtship display in birds of paradise is the process by which the normally-aggressive behaviours between individuals are broken down by a series of predictable and ritualised displays by the male that, if successful, encourage the female to submit to mating. The remarkable poses and displays of the male, be he solitary at his court or communal in a lek, must attract females and prove his solicitous intention towards them. Given the highly charged and competitive nature of paradisaeid mating systems, this male–female interaction is no simple union. Close study of courtship behaviour can provide a window into the functioning of a mating system.

Location of display site

In some birds of paradise, like the monogamous Trumpet Manucode, the male displays to the female at a random canopy perch (Table 5.1). Most species, however, carry out courtship displays at a traditional (i.e. perenially used) site. These range from branches in the canopy or understorey to vertical tree trunk stumps, fallen logs, and terrestrial dance grounds cleared by the birds. In polygynous species, the adult males use traditional perches for at least some combination of advertisement, display, and mating. A different site may be used for each component and multiple perches or sites may be used for single components, as well. In some instances there is evidence that these sites or perches are defended; in others a territory around the sites is apparently defended as well (Buff-tailed Sicklebill).

The most common type of traditional display sites are, not surprisingly, branches or limbs of living trees (< 5 cm and up to 20 cm

Table 5.1 Known and assumed (based on known sister forms) mating strategy, courtship display and site, main components, and visual characters of bird of paradise courtship displays.

Bird of paradise genera and species	Mating strategy	Male display strategy	Display site type	Main display movements	Main visual display character(s) presented
Cnemophilus					
Loria's Bird	Polygyny	Solitary	Tree branches	Static + ? inverted	Mouth & gape + part opened wings
Crested Bird	Polygyny	?	?	?	crest ?
Loboparadisea					
Yellow-breasted Bird	?	?	?	?	?
Macgregoria					
Macgregor's Bird	Monogamy	? Solitary	Tree branches	Chase	?
Lycocorax					
Paradise Crow	Monogamy?	Solitary?	?	?	?
Manucodia					
Glossy-mantled Manucode	Monogamy	Solitary	Tree branches	Chase then static	Part opened wings
Jobi Manucode	Monogamy	Solitary	Tree branches	Chase then static?	Part opened wings?
Crinkle-collared Manucode	Monogamy	Solitary	Tree branches	Chase then static?	Part opened wings?
Curl-crested Manucode	Monogamy	Solitary	Tree branches	Chase then static?	Part opened wings?
Trumpet Manucode	Monogamy	Solitary	Tree branches	Chase then static	Part opened wings
Paradigalla					
Long-tailed Paradigalla	Polygyny	Solitary?	?	?	?
Short-tailed Paradigalla	Polygyny	Solitary?	?	?	?
Astrapia					
Arfak Astrapia	Polygyny	?	Tree limbs?	?	?
Splendid Astrapia	Polygny	?Lek	Tree limbs?	?	?
Ribbon-tailed Astrapia	Polygny	?Lek	Tree limbs	Limb hopping	Central rectrices
Stephanie's Astrapia	Polygny	Lek	Tree limbs	Limb hopping	Central rectrices
Huon Astrapia	Polygny	Solitary?	Tree limbs	Static + inverted	Central rectrices + head plumage
Parotia					
Western Parotia	Polygyny	?	Cleared court + horizontal perches	Dance + static + leg flex head waggle	Flank plumes, pectoral shield, mantle cape, nuchal crest, & occipital plumes
Lawes' Parotia	Polygyny	Exploded lek	Cleared court + horizontal perches	Dance + static + leg flex head waggle	Flank plumes, pectoral shield, mantle cape, nuchal crest, & occipital plumes
Wahnes' Parotia	Polygyny	?	Cleared court + horizontal perches	Dance + static + leg flex head waggle	Flank plumes, pectoral shield, mantle cape, nuchal crest, & occipital plumes
Carola's Parotia	Polyggyny	Exploded lek	Cleared court + horizontal perches	Dance + static + leg flex head waggle	Flank plumes, pectoral shield, mantle cape, nuchal crest, & occipital plumes

Table 5.1 continued

Bird of paradise genera and species	Mating strategy	Male display strategy	Display site type	Main display movements	Main visual display character(s) presented
Pteridophora					
King of Saxony Bird	Polygyny	?Solitary	Tree branches + low supple vine stem	Static + leg flex + perch-bounce + dance + head waggle	Mantle cape, pectoral shield, occipital plumes, wings, mouth
Ptiloris					
Magnificent Riflebird	Polygyny	Solitary	Stout vine stem or tree limb	Static + dance + leg flex + head sway + ? wing clap	Wings, pectoral shield, mouth
Paradise Riflebird	Polygyny	Solitary	Tree limbs	Static + dance + leg flex + head sway + wing clap	Wings, pectoral shield, mouth
Victoria's Riflebird	Polygyny	Solitary	Atop vertical tree trunk stump	Static + leg flex head sway + wing clap	Wings, pectoral shield, mouth
Lophorina					
Superb Bird	Polygyny	Solitary	Low horizontal tree log or ground	Static + dance + wing clicks + leg flex?	Nape cape, pectoral shield, fore-crown, mouth
Epimachus					
Black Sicklebill	Polygyny	Solitary	Atop vertical tree trunk stump	Static + leg flex to lean to horizontal + sway	Pectoral, flank & tail plumes, mouth?
Brown Sicklebill	Polygyny	Solitary	Tree branches	Static + leg flex to lean + sway	Pectoral, flank & tail plumes, mouth
Drepanornis					
Buff-tailed Sicklebill	Polygyny	Solitary	Tree branches	Static + inverted	Pectoral, flank & head plumes, mouth?
Pale-billed Sicklebill	Polygyny	Solitary	Tree branches	Static + ?	Pectoral & flank plumes, tailed fanned, mouth
Cicinnurus					
Magnificent Bird	Polygyny	Solitary	Cleared court + vertical perches	Static + lean + dance	Nape cape, pectoral shield, central rectrices, mouth
Wilson's Bird	Polygny	Solitary	Cleared court + vertical perches	Static + lean + dance	Nape cape, pectoral shield, bald head, central rectrices, mouth
King Bird	Polygny	Exploded lek?	Tree branches + vines thereon	Dance + inverted + flight	Wings, pectoral fans, relict flank plumes, central rectrices, mouth
Semioptera					
Standardwing	Polygyny	Lek	Tree limbs, branches + space above	Limb-hopping + static + flight	Wings, pectoral shield, wing 'standards'
Seleucidis					
Twelve-wired Bird	Polygyny	Solitary	Atop vertical tree trunk stump	Dance + hen-pecking	Wings, flank plumes & wires, pectoral shield, bare thighs, mouth

Table 5.1 continued

Bird of paradise genera and species	Mating strategy	Male display strategy	Display site type	Main display movements	Main visual display character(s) presented
Paradisaea					
Lesser Bird	Polygyny	Lek	Higher tree limbs	Dance + static + hen-pecking + wing clap	Wings, flank plumes, central rectrices
Greater Bird	Polygyny	Lek	Higher tree limbs	Dance + static + hen-pecking + wing clap	Wings, flank plumes, pectoral cushion, central rectrices
Raggiana Bird	Polygyny	Lek	Higher tree limbs	Dance + static + pecking + wing clap + inverted	Wings, flank plumes, pectoral cushion, central rectrices
Goldie's Bird	Polygyny	Lek	Higher tree limbs	Dance + static + pecking + wing clap + inverted	Wings, flank plumes, central rectrices
Red Bird	Polygyny	Lek	Higher tree limbs	Dance + static + pecking + wing clap + inverted	Wings, flank plumes, pectoral cushion, central rectrices
Emperor Bird	Polygyny	Lek	Higher tree limbs	Dance + static + pecking + wing clap + inverted	Wings, flank plumes, abdomen decoration, pectoral shield, central rectrices
Blue Bird	Polygyny	Solitary	Lower branches, vines, & grasses	Static & inverted	Flank plumes, abdomen decoration, eye feathering, central rectrices

Notes: ? = unknown.
? after a character = assumed but requires confirmation.
? before a character = assumed but there remains some doubt.
dance = hops and/or steps over ground and/or on perch.
Pectoral cushion, pectoral shield, and pectoral fans are homologous.
See also Table 3.5.

diameter, respectively), which are used by at least 25 species (Table 5.1). The display site of some species, exemplified by the stump-top display perch of Victoria's Riflebird, the Black Sicklebill, and the Twelve-wired Bird, dictates that the male perform a static courtship display or one otherwise limited to hopping up and down. A stout horizontal vine stem or tree limb permits the linear hopping or dancing display of the Magnificent and Paradise Riflebirds, whereas a large fallen log or the ground permits the semicircular hopping dance of the male Superb Bird (see Chapter 9).

Display site modification

The males of most polygynous birds of paradise are known to modify their display site. This most commonly takes the form of simple leaf-removal, whereby the male plucks leaves from the immediate vegetation. At least some astrapias, parotias, the three sickletails, the Standardwing, and the plumed *Paradisaea* birds remove foliage from their traditional display sites. This 'gardening' may serve one or more functions, as follows: (1) removing foliage that otherwise might hinder display; (2) providing a better view of the male for females; (3) improving the displaying male's view of approaching females or predators; (4) better illuminating the male and his display by admitting more light; and (5) providing a visual marker for conspecifics of the location of the display site.

Males of all four parotias and the Magnificent and Wilson's Birds create terrestrial display courts, or cleared sites. The courts of the latter two species encompass the lower portion and bases of several vertical sapling stems, from which foliage is removed and which are often well worn by repeated use. Males also daily clean the forest floor beneath the vertical display perches of fallen leaves and other debris. This exposes a uniformly-cleared area of bare ground and roots that contrasts with the surrounding forest litter (Fig. 5.1).

The creation and careful maintenance of a cleared terrestrial court area is probably performed for one or more of the reasons cited above. Other reasons for this more intensive/

5.1 A cleared terrestrial court of an adult male Magnificent Bird of Paradise (on centre-left of court). Note vertical (and the worn leaning) court sapling perches. Photo: courtesy of B. J. Coates, with kind permission.

extensive clearing may be involved, however. Female Magnificent and Wilson's Birds typically watch a courting male from directly above him on one of the vertical court perches. The male is thus presenting his plumage to the female against an uncluttered and contrasting plain prepared court floor devoid of visual distractions. Moreover, the relative extent and cleanliness of a male's court floor might provide females with a measure of his time/energy expenditure in display site maintenance. This could in turn inform females about the relative experience and quality of the court owner. In the Satin Bowerbird, males with the best quality and larger numbers of decorations are older individuals (successful long-term survivors) that obtain most copulations (Borgia 1985, 1986).

The parotias have made the quantum leap from merely clearing a court to decorating it with specific items. Such court items include pieces of natural chalk, sloughed snake skin, and mammal dung. Female parotias will carry such items away from courts during the nesting season. While no direct observation has been made of what females do with these items, chalk and mammal dung may be ingested for their mineral and calcium content, respectively, and snake skin may be used in nest construction or 'decoration' (Pruett-Jones and Pruett-Jones 1988b). Note also that Majnep and Bulmer (1977) indicated that Majnep had once encountered a Lawes' Parotia nest with pig dung decorating the nest rim. Thus these things, obtained and laid upon the court by males as resources, are possibly significant to the nutrition, egg formation, and the nest of females. Or they may merely provide a female with much the same (symbolic) information about a male's fitness as do bowerbirds' bower decorations (Borgia 1986).

The terrestrial display courts of birds of paradise can be likened to analogues of the bowers of bowerbirds. As it is now believed that the two groups are not closely related, these similarities in court creation are viewed as convergent male courtship strategies. Similar male display site modification (but not decoration) is also seen in the manakins (Pipridae) and the cotingas of the Neotropics (Snow 1976b).

Advertisement of display sites

Most males advertise their presence on a display site by producing loud and regularly-repeated vocalisations, described in detail in the species accounts (Chapter 9). In the manucodes and the Twelve-wired Bird, tracheal modifications (see Figs 1.5, 9.9 and 9.11) are thought to enhance the far-carrying quality of calls by significantly lowering their pitch (Clench 1978; Frith 1994b). Vocalisations are the single most important means for male birds of paradise to broadcast their presence to females. Not surprisingly, paradisaeid vocal repertoires have radiated profusely, and exhibit a remarkable diversity of sound types, patterns, and qualities that merits a study of its own.

Diverse non-vocal sounds produced by birds of paradise in display could also act as auditory advertisement. Non-vocal sounds include wing-beating, bill rattling, primary-swishing and wing-snapping, plus wing-sounds in flight.

The flight of at least some individual manucodes, Macgregor's Birds, paradigallas, and of adult male astrapias, parotias, riflebirds, sicklebills, and sickletails, produces conspicuous sounds. Individual males of some polygynous species are able to control the volume of wing-noise in flight. Most of these birds have blackish to jet black, glossy, dense flight feathers which produce a dry rustling sound as they move against each other, even when a wing is merely opened and closed. Adult-plumaged male Magnificent and Wilson's Birds produce a characteristic clear and sharp clacking sound as they fly about their court. How this mechanical sound is produced remains to be determined. The mildly-emarginated to highly-modified outer primaries of adult males of several genera (Fig. 5.2) might contribute to wing-noise in flight but this remains to be investigated.

Courting male Standardwings bring their wing carpal joints sharply together repeatedly

Reproductive behaviour

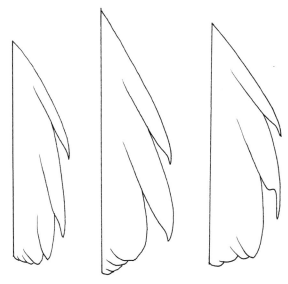

5.2 Outline representations of the modified two outermost primaries of (from left to right) adult male Superb Bird, Magnificent Riflebird, and Lawes' Parotia. See also diagram on page xxv and plates 4–13.

in a powerful movement to produce a series of loud, sharp cracks. This behaviour appears analogous to wing carpal joint beating performed by inverted courting males of some *Paradisaea* species which produce a softer but clearly-audible thud (Frith 1992).

Finally, conspicuous areas of brightly- or contrastingly-coloured plumage are used to advertise the male on his display perch visually. The long, white central rectrices of the adult male Ribbon-tailed Astrapia can be seen by females at a considerable distance as males fly or hop about. The brilliant yellow flank plumes of Twelve-wired Birds and all species of *Paradisaea* are made widely visible by adult males raising their wings to expose them.

Sociality of courtship
Solitary courtship display

Males of most bird of paradise species court females in solitude (Table 5.1). Solitary courts are the product of the avoidance of contact with rivals by their owning males. In this mating system, the pattern of regular dispersion of males on their display sites produces the greatest possible distance between displaying males. Thus solitary and communal (i.e. lekking) displays have a fundamental difference that, in all likelihood, influences the observed behaviour in these two types of system (see below).

Displays by males of solitary species are not universally different from those of communally-displaying males, but many of the solitary species do clearly employ more static displays. Males of solitary species do not typically perform communal displays, although rival males and young males commonly visit display sites both when the court-holder is absent and when he is present (when brief communal displays can occur). In solitary court systems males tend to be far less active and to display less frequently in the absence of females. The average solitary male spends more time in vocal advertisement and court maintenance, and less time in display and male–female interactions (Beehler 1983a).

Communal courtship display

Of the 42 bird of paradise species, a potential maximum of 13 perform communal courtship displays. Males of eight species are confirmed to form true leks, and an additional three species apparently form exploded ones (Table 5.1). Additional species probably exhibit these types of mating organisation but confirming data are lacking.

Males of the King Bird and the lekking *Paradisaea* species court females with conspicuous use of expansively-opened wings. This is in marked contrast to their non-lekking congeners (Magnificent, Wilson's and Blue Birds) that display solitarily and close to the forest floor without opening or moving their wings. This generalisation about wing-movement in lek species fails, however, with regard to the astrapias and wing-movement is performed by the solitary riflebirds. It is thus difficult to define unambiguous behaviour patterns that differentiate the displays of solitary and communal species.

One type of display that is unique to communal systems is the 'convergence display'—in which dispersed males quickly and noisily return to, or 'converge' on, their lek perches upon the arrival of a female. In the typical species of *Paradisaea* and the Standardwing, this is the most striking display in their behavioural repertoire—colourful, kinetic, noisy, and exciting to the observer (Beehler 1988; Bishop 1992). One can identify a 'convergence' event at a lek just from the sound of the vocalisations, which can be heard from several hundred metres distant. None of this is a surprise, as members of the communal lek group of males stand to benefit from the attraction and excitement of fertile females—the more that are attracted to the lek, the greater the opportunity to copulate for the lek males. Convergence display is also known from Carola's Parotia, and, in fact, a version of this kind of display is given from time to time by the solitarily-displaying Twelve-wired Bird.

Co-operative courtship by birds is usually performed only by communally-displaying species. It typically involves two or more males performing a synchronised communal display routine to a visiting female. This type of display is unusual, and rare in nature, because females will usually mate with only one of the displaying males—providing no incentive to the 'losing' male. Such remarkable displays are performed by several neotropical manakins (Snow 1976b). Males of some lekking birds of paradise (*Semioptera*, *Paradisaea*) may carry out some co-operative courtship displays but currently there is no proven instance of it. The most likely candidate for co-operative display is the Emperor Bird, in which closely-perching, paired males give co-ordinated duetting calls and carry out synchronised inverted displays that may constitute co-operative behaviour (Coates 1990). In only a few species other than lekking ones, the Western Parotia being a notable example, do promiscuous adult males appear to tolerate other males at their display sites when courting. Male King Birds of Paradise appear to associate in pairs (Beehler and Pruett-Jones 1983), but the nature of this association is unknown.

Much display activity of lekking males may maintain a male–male dominance hierarchy that could limit the choice of potential mates by females to one or more males occupying the central lek position (LeCroy 1981). One main point of evidence for this hypothesis was the proposed presence of specific male–male displays in the genus *Paradisaea*. In a study of lek behaviour in the Raggiana Bird, Beehler (1988) found no evidence of specific male–male displays distinct from the displays that are given to visiting females. Follow-up studies of the Raggiana Bird (Beehler, Dumbacher, and Laska unpublished data) have, however, documented considerable physical male–male aggression in the lek that may serve the purpose of hierarchy formation and maintenance.

Male–male aggression in the lek

The lek is an extraordinary phenomenon because it brings together in close association highly-competitive males that are each seeking to maximise their fitness through copulation with a limited population of females, each of which probably produces no more than one or two offspring a year. How do these males in the lek avoid physical conflict with their competitors? Actually, they often do not. Male–male fights in leks are not uncommon, and some level of fighting is probably necessary to establish the type of 'truce' that in these systems is better known as a dominance hierarchy or pecking order (Beehler and Foster 1988). Bill wiping (stropping the bill on a perch) is frequently performed by males of lekking species but is seen far less frequently in solitary ones. This may reflect a situation opposite to that involving gaping (commonly performed by several solitarily-displaying species) as bill wiping is generally a submissive or displacement behaviour in other birds (Feare 1984; Goodwin 1986).

Male–male fighting has been observed in Stephanie's Astrapia, the Standardwing, and several *Paradisaea* plumed bird species. Fighting males have been observed locked together in combat, tumbling out of the canopy to the ground. Seeing this reinforces our perception of the competitive nature of the system.

One irony of leks is that the newer examples tend to have fewer males, more aggression, and fewer visiting females, whereas the older leks tend to have more males, less evident aggression, and more visiting females (Beehler and Foster 1988). Aggression does not necessarily benefit visiting females, but is necessary to establish lek hierarchies. Once the hierarchy is established, aggression is reduced and females can visit without an excess of harassment by the lek males. Less-dominant males in a lek may, however, be forced to play the role of 'expendable' peripheral individuals that are more prone to predator attack. Females may select to mate with centrally-located older/more dominant males (see above) where the vulnerable act of copulation would be protected by a periphery of subdominant/younger males.

Courtship display signals, postures, and movements

Reports of the nineteenth century naturalists indicated little more than that, in the wild, male birds of paradise called loudly and exhibited their elaborate and colourful plumage by peculiar, if not random and contorted, posturing. It was not until the 1920s to 1940s, mainly from observations of captive birds, that it became clear that highly-specific postures were performed by males of each species (Winterbottom 1928a,b; Friedmann 1934; Crandall 1921–46). For the most part, observers described various display postures but did not discern oft-repeated sequences of them. Ogilvie-Grant (1905) and Winterbottom (1928a) provided early exceptions.

Courtship display appears to be more complex in the more ornately-plumed birds of paradise (Table 5.1). Other species perform courtship displays that are static and simple or involve simple chases or limb hopping (Table 5.1; see also Chapter 9 text and figures). It is evident that males of the more derived genera (parotias and genera higher in the family) perform more discrete courtship body movements in a highly-ritualised and little-varying progressive sequence. Sexual selection has produced complex male vocalisations and plumages as well as increasing the complexity of courtship choreography. We provide a review of courtship and its various components below.

Vocalisations during courtship display

As noted earlier, males of most if not all polygynous species sing loudly and regularly to advertise their presence and the location of their display site. Males of lekking species call loudly and more excitedly when assembling during their 'convergence' displays. Males of few species vocalise while directing display at a female, but if they do then their calls are muted.

A male Black Sicklebill has been observed to vocalise during courtship while male Pale-billed Sicklebills have not (the latter gives a bill rattle). While parotias display in silence, male King of Saxony Birds produce a continuous hissing whisper-song of varied notes while performing courtship perch-bouncing and precopulatory display. In the latter display context the sickletails also softly vocalise. While displays performed by male *Paradisaea* to females are relatively quiet, some brief soft notes are given and may be accompanied by sharp clicking presently thought to be produced by the mandibles (also true of the Magnificent Bird). Remarkable among the *Paradisaea* is the unique mechanical-sounding buzzing given by the male Blue Bird in fully-inverted display. Male Standardwings display to females while producing vocalisations characteristic of both *Cicinnurus* and *Paradisaea* but with the addition of high-pitched twitterings and also a sharp snapping sound produced by the carpal joints being brought together.

We must briefly allude to the question of vocal mimicry during courtship. A male Raggiana Bird raised by hand from a nestling lived many years at the Baiyer River Sanctuary and mimicked several words and the whistling of local people. This it appeared to do in the context of 'courting' people, upon which it was imprinted. No reports of vocal mimicry by wild birds of paradise exist. In reviewing recordings of bird calls for the sonograms in this work, however, the vocalisations of a courtship-displaying male King of Saxony and a Wilson's Bird appeared to incorporate some avian vocal mimicry of the calls of some smaller passerine birds, notably *Sericornis* scrubwrens (CF). As vocal avian mimicry is characteristic of courting males of the polygynous bowerbirds (Frith and Frith 1993*e*, 1994*c*; Frith and McGuire 1996; Frith *et al.* 1996 and references therein) which show much convergent associated behaviour with the birds of paradise, this phenomenon in the Paradisaeidae merits investigation.

Use of plumage and colour in display

Numerous observations demonstrate that the elaborate nuptial plumage characters of adult male birds of paradise are specifically presented to females in a stereotyped manner during courtship. Thus astrapias limb-hop, fly, and invert in a way best designed to emphasise their long tails; parotias form a skirt of their flank plumes and erect occipital and nuchal plumes; King of Saxony Birds manipulate occipital plumes, raise their mantle cape, and make best use of their contrasting wing-patch colour by trembling their wings; riflebirds fully open their highly-modifed wings; the Superb Bird raises its enormous nape cape and breast-shield; and sicklebills erect their pectoral fan and flank plumes and fan their long tail. For every plume form there is a clear function.

The birds of paradise include diverse species that use their wings conspicuously in courtship and other species that do not. Wings are used in display by manucodes, which partly open and raise them, and by the King and Standardwing Birds and most *Paradisaea* species, which partly or fully open them in a motionless pose and also flutter or flay them and/or beat their carpal joints together. Their wing fluttering is like the similar and probably analogous wing fluttering, waving, or flailing of true corvids (Goodwin 1986), starlings (Ellis 1966; Feare 1984), and other higher passerines.

The riflebirds, King, Standardwing, Twelve-wired Bird, and the *Paradisaea* species (except the Blue Bird) typically hold their opened wings out to either side of the body or up at an angle above the back as a 'wing pose' courtship component. Only in lekking males of the typical *Paradisaea* species is this followed by a communal static 'flower' pose with lowered head. During this females might have the opportunity to inspect and compare several males in the same static posture. This flower pose specifically exposes and presents the males' flank plumes to best advantage. It has developed a step further in the lekking male Emperor Bird into a completely hanging-inverted posture. The solitarily-displaying male Blue Bird similarly hangs inverted but its wings are held tightly closed, unlike all of its lekking immediate relatives.

In the riflebirds, wing use in vigorous male courtship is much elaborated and has resulted in remarkable modifications of entire wing and individual flight feather shape and texture (see also Surprise elements of courtship, below). The primary feathers of adult male riflebirds produce loud and dramatic noises during parts of their courtship display.

During each of the more expansive hops of the courtship dance of adult male Superb Birds the wings are rapidly flicked open and closed to produce a loud sharp, dry tick sound. As the female cannot see the courting male's wing-movements (Fig. 9.46(d)) this is an exclusively auditory signal. The actual physical source of this remarkable and explosive sound is presently unknown.

In other polygynous species, solitary courtship-displaying males do not typically use their wings (the King Bird is an exception)—a general pattern exemplified by the lack of wing use in the Blue Bird (Table 5.1 and see above).

How the bizarre occipital plumes of adult male King of Saxony Birds were put to use remained a tantalisingly-fascinating question ornithologists long puzzled over (Fig. 9.32 provides the answer). Only recently has some understanding of the use of such singular characters as the pectoral fan plumes of sicklebills (see Figs 9.48, 51, 54 and 57), pale wing patches of King of Saxony Birds, whitish primaries of Standardwings (see Fig. 9.68), and plume wires and bare thighs of Twelve-wired Birds (see Fig. 9.71) been gained (Beehler and Beehler 1986; Frith 1992; Frith and Beehler 1997; Frith and Frith 1997c). But the way in which characters unique to adult males of several other species are used remain complete mysteries. For example, what is the function, if any, of the delicate crest of the male Crested Bird, the fleshy facial lobes of the male Yellow-breasted Bird and the paradigallas, the twisted central rectrices of the Curl-crested Manucode, the orange facial wattles of Macgregor's Birds, or the various kinds of grossly-elongated central rectrices in *Paradisaea* species?

The tail is used in limited ways by courting males, notwithstanding the considerable diversity of overall shape and individual feather modifications. In most species a courting male's tail is not held or manipulated in a particular way, but in the graduated-tailed sicklebills and Wahnes' Parotia it is repeatedly fanned. In the longer-tailed Western and Wahnes' Parotias it is, perforce, held to one side during terrestrial dancing movements. Only in courting riflebirds, the Superb Bird, and all three sickletails is the male's tail conspicuously cocked upward, between 45 and 90°. In the typical *Paradisaea* the elongated undertail coverts of adult males are unusually held well away from the tail during parts of the display.

Males of many species with a brightly-coloured mouth open the bill widely to present this visual signal in courtship and, of course, they expose it whenever calling. Advertisement and courtship gaping is typical of adult male Loria's Bird and the obvious external, whitish, fleshy gape flanges in this species might function to enhance this mouth display. Little is known of mouth-gaping in the other cnemophilines, the paradigallas, or the astrapias. Gaping is probably not an important display element in the dark-mouthed and monogamous Macgregor's Bird and the manucodes. Male King of Saxony Birds, riflebirds, Superb Birds, sicklebills, sickletails, and Twelve-wired Birds all typically expose their coloured mouths in courtship. Adult males of all riflebirds and the larger two sicklebills have small but visible fleshy gape flanges that are as bright yellow as their mouths. Parotias, the Standardwing, and the *Paradisaea* birds appear not to gape in display.

Most family members have dull blue-grey to blackish legs and feet while in the *Paradisaea* birds they are certainly paler but are not colourful. Bright pigmentation of the legs and feet is confined to the Standardwing (bright yellow), Twelve-wired Bird (deep coral pink), and the three sickletails (blue). The bare head skin of Wilson's Bird is also brightly-pigmented blue. As these bare parts tend to be slightly brighter in males than females they may be related to courtship display. The Twelve-wired Bird also has (uniquely within the family) bare and pigmented thigh skin that features in display (Frith and Beehler 1997). These are far from forgone conclusions, however, and are not necessarily supported by variation in other characters.

It has been known for many years that bird feathers reflect ultraviolet (UV), a part of the spectrum to which the human observer is blind. Many birds are also able to perceive UV light due to their having a fourth cone type (humans have only three) specifically sensitive in the UVb range 350–400 nm. With further investigation, UV vision in birds is likely to be the norm rather than an exceptional find. These facts suggested the intriguing possibility that birds may use colours which we can not see to communicate information, such as sexual state or territorial dominance to others of the same or opposite sex. Convincing support for this idea has recently come with the finding that the mate choice in Zebra Finches *Taeniopygia guttata* and Common Starlings *Sturnus vulgaris* relies heavily on the UV reflected from the plumage of the males.

The iridescent feathers of the Common Starling are coloured by the same phenomenon (constructive interference) as those of the metallic blue, green and copper coloured feathers seen in the birds of paradise. The colours of the feathers used by birds of paradise in their stunning displays are currently being charaterised (J. Marshall personal communication) and interestingly a strong and indeed the most constant component of the colours is found in the UV region of their spectral reflectance. It is almost certain that the brilliant displays of the birds of paradise which we see look quite different and possibly even more colourful to the birds themselves. Additionally it may be that a vital part of the message contained in the display, the UV colours, are lost to our eyes altogether (J. Marshall *in litt.*).

Body movements during display

Leg 'flexing' causes a statically-perched or standing bird's body to rise and fall, or leap vertically up and down, but in some cases to sway and/or rotate to one side or side to side. Courting males of all parotias and the King of Saxony Bird, all three riflebirds and the Superb Bird, and the two larger sicklebills are known to perform this type of ritualised movement. In the sabretailed sicklebills this involves leaning to one side, until horizontal, and there further flexing the legs to sway and/or slightly rotate the body (see Figs 9.48 and 9.51(a)). This action may be performed by riflebirds but they also, and more typically, slowly raise and lower the body vertically on flexing legs (see Figs 9.38 and 9.41). The Magnificent Riflebird and Superb Bird elaborate upon this, however, to perform repeated, bouncing, upward leaps (see Figs 9.36 and 9.46).

All parotias incorporate leg flexing movements, both slow and rhythmic and rapidly and sharply repeated (see Figs 9.23 and 9.27). Parotias have also developed ground stepping to a sophisticated level of leg and foot movement coordination. This enables males to perform dance choreography incorporating rapid and mincing forward, backward, and sideways steps more complex than other family members are capable of (Frith and Frith 1981). Leg flexing by courting male King of Saxony Birds is like that of parotias but is much more vigorous in order to bounce a display vine (see Fig. 9.32).

Courting male King Birds invert themselves from perches in a remarkable manner that might involve an unusual form of leg flexing. Not only does the male hang suspended but he also elongates his body with his plumage tightly sleeked, while swinging, rigidly straight, by the legs rapidly from side to side like a pendulum (Fig. 9.66(e)). Inverted displays are also given by Loria's Bird, the Huon Astrapia, Buff-tailed Sicklebill, Emperor and Blue Birds (Beehler and Beehler 1986; J. Hicks and R. Hicks 1988; Coates 1990; T. Pratt personal communication). In these species, the courting male hangs inverted from the display perch while the female watches at close range (Fig. 5.3). Inverted displays are usually performed late in a display sequence and immediately prior to copulation.

The *Paradisaea* type of inverted display has developed further in the Emperor and Blue Birds to become their main feature of courtship. To this end, adult males feature ornate abdomen plumage specifically presented to the female perched directly above the hanging male (Fig. 5.3). In the Emperor Bird the (usually concealed) deep yellow bases to its pure white flank plumes are exposed and in the Blue Bird it is contrasting adjacent areas of jet black and deep crimson plumage that are presented (see Fig. 5.3). While adult male Goldie's Birds also wear a striking black ventral marking, they are not presently known to invert in display.

An amazing aspect of the stunning Blue Bird's display is that wild males may remain fully inverted and vocalising for prolonged

Reproductive behaviour   121

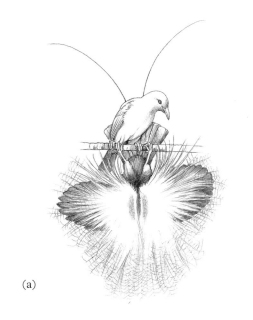

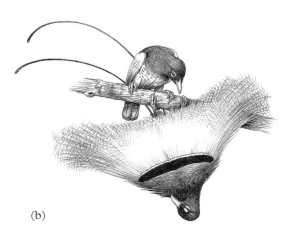

5.3 Courting adult male Emperor (a) and Blue (b) Birds of Paradise in inverted display that specifically presents dorsal plumage markings to females observing from directly above.

periods of time. Inverted males appear to be in an almost trance-like state, often oblivious to distractions. Certainly this is a vulnerable position for a male to assume, and one can speculate that this is a mechanism for reducing aggression between the male and a visiting female, as well as an additional means for the male to exhibit his physical prowess and plumage finery.

While male Lesser, Greater, Raggiana, and Goldie's Birds do not hang completely and statically inverted they do lean downward on a sloping limb with the head well below the feet and plumes erected in a cascade. These postures are not, however, truly inverted, as are those of the Emperor and Blue Birds. Has the Blue Bird, then, developed its inverted display into a solitary one independently or is it derived from the progressively downward-leaning displays of the lekking Lesser, Greater, Raggiana, Red, and Emperor Birds, respectively?

Courtship display flights are rare and the few described remain, for the most part, inadequately characterised. Extensive and deeply-undulating flights of male Ribbon-tailed Astrapias exhibit the elongated central rectrices in a striking way. An adult male Black Sicklebill was seen once performing a spectacular flight display by diving deeply into a gorge from a high tree limb, then, when nearing the vegetation below, soaring upwards back to its perch (Ripley 1957). Another single anecdotal observation involved an adult male Victoria's Riflebird posturing in such a way as to enhance and draw attention to its iridescent, short central tail feathers. This suggests the central rectrices might function in some kind of invitation flight to the male's display site (Frith and Cooper 1996). The loud wing-noise given by male riflebirds in flight would presumably much enhance such a signal.

Male King of Saxony Birds advertise in the forest canopy but subsequently lead a visiting female down to a display vine a few metres from the ground with a descending invitation flight. An adult male King Bird has been seen to fly vertically upwards from the forest canopy before dropping vertically in a display flight. Males are also known to drop from their display perch in another form of display descent or 'flight' (see Chapter 9). Standardwing Birds perform a dramatic display flight above the canopy that ends with a slow descending display on vibrating wing tips (Bishop 1992).

Finally, males of several species produce, and appear to control, sound made by their wings, and their flights about traditional display perches may therefore represent an audible form of display flight (but also see Advertisement of display sites, above).

Male Twelve-wired Birds perform a tactile courtship display by brushing recurved pectoral plume wires across the face of a visiting female (see Fig. 9.71(c,d)). The feather shafts involved are worn only by adult males. This observation provides a clue to the possible function of similarly-bare plume shafts in the case of the wire-like central pair of rectrices of most *Paradisaea* species. These feathers have lost their visual significance in all but the Red and Blue Birds (see Plates 11 and 12). One ornithologist wrote in frustration of these particular rectrices 'I for one am quite unable to see what possible use they can be in display, since they are almost invisible except at very close quarters' (Stonor 1940). Their function may thus be tactile.

Surprise elements of courtship

Courtship display by promiscuous male birds displaying solitarily at a traditional site typically includes an element(s) of vigorous advance at, and/or surprise to, the female. These include dramatic and sudden changes of size and appearance, attitude, and/or sound of the male. This is especially true of the parotias, King of Saxony Bird, riflebirds, Superb Bird, sicklebills, and the sickletails. Males of some other species, including the Twelve-wired Bird and the *Paradisaea*, approach closely in pre-copulation display and repeatedly peck at the female. Courting male riflebirds and, interestingly, also the typical *Paradisaea* species follow an open-wings display by closely confronting a female to swing side-to-side rapidly and repeatedly before her while vigorously embracing her with alternately-cupped wings immediately prior to copulation. It is not known if this similarity merely reflects behavioural convergence or is indicative of relationship.

These intimidating display actions frequently result in the visiting female retreating or departing, which would appear counter-productive to the male. Alternatively these pseudo-aggressive actions by males might function to frighten or shock visiting females in order to test just how ready/determined they are to remain and be courted/mated. If the male's abrupt change to a more assertive/aggressive behaviour causes an unready female to leave, the male saves wasting more effort on an insufficiently-receptive audience. Thus such sudden changes in size/appearance and/or behaviour before a potential mate might have selective value as part of a male's mating strategy. Conversely, and potentially most importantly, females may in fact actively select males exhibiting such overtly aggressive display characteristics, as they must presumably seek to maximise the reproductive potential of their male offspring. Thus they might select not only for male nuptial plumage characters but also for traits such as assertive and aggressive posturing and movements that would prove advantageous in male–male interactions.

It is clear that visual avian displays, particularly the more stereotyped elements of intense courtship, can clearly reflect phylogenetic relationships (Prum 1990). They therefore represent important systematic characters, and particularly so within groups of closely-related species (Irwin 1996) such as the typical birds of paradise. While we do use such elements of courtship display in this way, within a much broader set of useful characters (Chapter 3; Table 3.2(b)), systematic study of the colourful and phylogenetically-informative bird of paradise displays has barely begun.

6

Nesting biology and parental care

The nesting biology of the birds of paradise has not been reviewed previously, in large part due to limited knowledge of the subject. Notwithstanding the fact that the greatest ornithologists of their time eagerly encouraged collectors to seek out the nests and eggs of birds of paradise, few were found until the twentieth century. When Bowdler Sharpe wrote his monograph (1891–8) on the birds of paradise and bowerbirds, all he could include on the nesting biology of the former group were brief descriptions of the nest and eggs of four species and the eggs of two additional ones. Nothing was known of the incubation and nestling periods, parental care, nestling diet, or growth and development of even a single species (Table 6.1).

By the time Gilliard completed the text for his 1969 monograph, representatives of the nests of 23 species and the eggs of 25 species had been described. Even then, four-and-a-half centuries after the first bird of paradise specimens had reached Europe and one-and-a-half centuries after the first monograph on the family appeared (Levaillant 1801–6), the incubation and nestling periods of only one species were known accurately. These were for the King Bird, and they were recorded as a result of a single 1956 captive breeding in Sweden (Table 6.1). The incubation period of the Glossy-mantled Manucode was estimated by Austin Rand (1938) to be more than 14 and less than 18 days, and remains our only knowledge of it today.

By 1977, when Cooper and Forshaw's companion volume to Gilliard's monograph was published, an incubation period for four additional species and a nestling period for three had been established—all from single captive breeding events in the UK or the USA. In addition, the nest and nestling of the Short-tailed Paradigalla were described from specimens stored but unexamined since their collection in New Guinea in 1913 (Frith 1970). Since then there has been a virtual revolution in the knowledge of the nesting biology of the family (Table 6.1).

Most advances in knowledge of some species of plumed birds (*Paradisaea* spp.) over the past two decades have been the result of captive breedings in the USA (Tables 6.1 and 6.2). A Wahnes' Parotia egg laid in captivity in Papua New Guinea was recently described (Mackay 1990). Observations have been made of a wild nesting Raggiana Bird (Davis and Beehler 1994) and the Blue Bird (Pruett-Jones and Pruett-Jones 1988a; Mack 1992), which has also been studied by Richard Whiteside (1998). The remaining studies of birds nesting in the wild (Table 6.1) have been made primarily by the senior author and Dawn Frith. We still know relatively little, however. For example, do nesting females of polygynous species defend any area beyond a few metres of the nest against conspecifics? Do both sexes assist with nest construction in any monogamous species? What type of nest does the Yellow-breasted Bird construct?

Table 6.1 History of bird of paradise species known to authors without (=+) and with (see Key below) knowledge of their nesting biology.

	Gould 1840–48 1869 & 1888	Sharpe 1891–98	Gilliard 1969	Cooper and Forshaw 1977	Frith and Beehler present work	Notes*
PARADISAEIDAE CNEMOPHILINAE						
Loria's Bird		+	+	+	N C I	
Crested Bird		+	N	N	N C F	The egg in Gilliard (1969) was misidentified (Frith and Frith 1993c)
Yellow-breasted Bird		+	+	+	+	
PARADISAEINAE						
Macgregor's Bird	+	+	N C	N C	N C	
Paradise Crow	+	+	N C	N C	N C	
Glossy-mantled Manucode	+	+	N C I	N C I	N C I F	F = captive
Jobi Manucode	+	+	N C	N C	N C	
Crinkle-collared Manucode	+	+	N C	N C	N C	
Curl-crested Manucode	+	N C	N C	N C	N C	
Trumpet Manucode	+	+	N C	N C	N C	
Long-tailed Paradigalla	+	+	+	+	+	
Short-tailed Paradigalla			+	N	N C I F	
Arfak Astrapia	+	+	+	+	+	
Splendid Astrapia		+	+	+	+	
Ribbon-tailed Astrapia			N	N I	N I C F	I = captive data
Stephanie's Astrapia		+	N C	N C I F	N C I F	I and F = captive data
Huon Astrapia			C	C N	C N	
Western Parotia	+	+	+	+	+	
Lawes' Parotia	+	+	N C	N C	N C	
Wahnes' Parotia		+	+	+	C	Captive data
Carola's Parotia		+	+	+	+	
King of Saxony Bird		+	+	+	N C I	
Magnificent Riflebird	+	N C	N C	N C	N C	
Paradise Riflebird	+	N C	N C	N C	N C	Nest in Elliot (1873) erroneously identified
Victoria's Riflebird	+	N C	N C	N C	N C I F	
Superb Bird	+	+	N C	N C I F	N C I F	I and F = captive data

Table 6.1 continued

	Gould 1840–48 1869 & 1888	Sharpe 1891–98	Gilliard 1969	Cooper and Forshaw 1977	Frith and Beehler present work	Notes*
Black Sicklebill	+	+	+	+	+	
Brown Sicklebill		+	NC	NC	NC	
Buff-tailed Sicklebill	NC	NC	NC	NC	NC	
Pale-billed Sicklebill	+	+	+	+	+	
Magnificent Bird	+	+	NC	NCIF	NCIF	I and F = captive data
Wilson's Bird	+	+	+	+	+	
King Bird	+	+	NCIF	NCIF	NCIF	I and F = captive data
Standardwing Bird	+	+	+	+	NC	
Twelve-wired Bird	+	+	NC	NC	NCIF	CIF = captive data
Lesser Bird	+	+	NC	NC	NCI	I = captive data
Greater Bird	+	C	C	C	C	
Raggiana Bird	+	C	NC	NC	NCIF	I and F = captive data
Goldie's Bird	+	+	+	+	+	
Red Bird	+	+	C	C	CIF	I and F = captive data
Emperor bird		+	C	CN	CN	
Blue bird		+	NC	NC	NCI	
Number of taxa known	25	38	42	42	42	

Key: N = nest description, C = clutch size and/or egg description, I = incubation period, F = nestling (or fledging) period, diet and development.
* See Tables 6.2 and 6.3 for references.

One value of Table 6.1 is that it shows at a glance the gaps in our knowledge of basic nesting biology of the birds of paradise. We know nothing of the monotypic genus *Loboparadisea*, little of *Manucodia* species, and next to nothing except brief and in some cases inadequate nest and egg descriptions for the genera *Lycocorax*, *Parotia*, *Epimachus*, *Semioptera*, and *Seleucidis*. Nothing whatever is known of 10 species, including some of those longest-known to Western science (e.g. Arfak Astrapia, Black Sicklebill, and Wilson's Bird). Of those for which we do know something, our knowledge is limited to nothing more than descriptions of nests and eggs for 17 species (Table 6.1).

Despite these limitations to our understanding of paradisaeid nesting biology it is clear that great advances have been made in the past three decades (Tables 6.1 and 6.2). As a result of this, some interesting facts and initial patterns are emerging that are presented here for the first time. We hope this will stimulate future exploration and research in the wild, and encourage the systematic observation and recording of reproductive events in aviculture.

Nest sites

All species of birds of paradise for which nests are known are solitary nesters. It is highly unlikely that any species ever nests gregariously, colonially, or with the involvement of 'helpers' (*cf.* Skutch 1987). As the sites of nests collected by earlier explorers or ornithologists were not well documented, and relatively few nests have been found by modern biologists, information on nest sites is limited. Still, we know enough to state that there are at least three general types of nest placement.

First is that of the Cnemophilinae: of six active or recently-disused Loria's Bird nests found in one patch of forest, five were placed upon exposed, low, damp rocks covered by lush mosses and ferns and one upon a moss-covered tree trunk, at an average height of 1.5 m above ground. The unique nest site of Loria's Bird appears to be selected, at least to some degree, to enhance crypticity. Three nests of the Crested Bird (Loria's closest relative) have been found in low tree branches, two on a tree trunk, and one atop a tree stump. The mosses and ferns of the immediate nest site area perfectly match the external nest materials of these two species, making their structures extremely difficult to discern even at close quarters and with knowledge of the presence of a nest.

Nests of the Paradisaeinae have all been found in tree branches with the single exception of the King Bird, whose single known wild nest was a cupped structure placed in a tree crevice. It remains to be seen if this is always the case. Nests of manucodes and the Paradise Crow are, uniquely within the family, slung hammock-like from a horizontal branch fork by their upper outer rim. Their nest placement is, thus, much like that of drongos (Dicruridae) and Old World orioles (Oriolidae).

Most nests of Trumpet Manucodes, Short-tailed Paradigallas, Ribbon-tailed Astrapias, and Victoria's Riflebirds have been found in the densely-foliaged and small crowns of a sapling or other plants. These plants were usually growing within a small gap in the forest (sky and not forest canopy directly above) somewhat isolated from other rainforest vegetation. Nesting in such a situation may represent a means of reducing chances of potential predation upon egg, nestling, or the brooding adult by tree-climbing predators (e.g. cuscuses [Phalangeridae], possums [Pseudocheiridae], rats [Muridae], and reptiles).

While records remain few for firm conclusions, information (see Chapter 9) suggests that at least one Riflebird (the Magnificent, but also possibly Victoria's) and Sicklebill (the Brown), the Superb Bird, and the Twelve-wired Bird specialise in nesting (as well as foraging—see Chapter 5) in pandans (*Pandanus* spp.), palms, or in plants of similar architecture (personal observations).

Several authors have reported that Loria's and Macgregor's Birds, Short-tailed Paradigallas, Ribbon-tailed Astrapias, Paradise, and Victoria's Riflebirds not uncommonly build a nest upon, immediately adjacent to, or near their nest of the previous season(s). Thus the females of some birds of paradise appear to nest 'traditionally', a scenario that has recently proved true of the females of polygynous bowerbirds (Frith and Frith 1994a, 1997d). This raises the possibility that such traditionally-nesting females may mate with the same male over several years (see Chapter 5).

In a broader context of nest sites, we briefly mention here the fascinating nesting association between the Trumpet Manucode and Black Butcherbird *Cracticus quoyi* (Cracticidae) in the rainforests of Cape York Peninsula (Barnard 1911; Frith and Frith 1993d). The assumption has been made that the manucodes seek to nest adjacent to the pugnacious and predacious butcherbirds (efficient nest predators) to gain some kind of protective advantage from the latter's presence, but this remains to be demonstrated as does the manucode's ability to prevent the butcherbirds from predating their own nest contents. The Trumpet Manucode's means of nest defence would thus be a 'protective nesting association with a formidable species' in the classification of bird nesting associations by Collias and Collias (1984).

Nests

Nests of birds of paradise are constructed in three architectural forms: domed; suspended, shallow, open cup; and supported, bulky, open cup. The nests of the cnemophilines Loria's and Crested Birds are dense and substantial, roughly-spherical, domed structures predominantly composed of long slender orchid stems overlaid with fresh mosses and ferns that incorporate a token 'foundation' of relatively-few, stout, short, woody sticks (Plate 3, Fig. 9.3). They are quite distinct from the open cup-or bowl-shaped nests of the typical birds of paradise (i.e. excluding the manucodes), but have in common with them the use of orchid stems, mosses, and fern fronds. The use of significant quantities of mosses by the highland-nesting cnemophilines is noteworthy as American Robins *Turdus migratorius* (Turdidae) use more moss in those nests built in colder microenvironments (Horvath in Collias and Collias 1984). The use of fresh fern fronds on the nest exterior is probably for camouflage—to 'confuse or deceive predators' (Collias and Collias 1984). At this time we eagerly await a description of the nest and egg of the Yellow-breasted Bird. Is the nest of this little-known species domed like those of the other two cnemophilines?

While the known domed nests of the relatively-weak (see Chapter 9) cnemophilines are built in relatively-exposed sites and are highly cryptic, the bulky open cup nests of the Paradisaeinae are usually well concealed in dense vegetation and difficult for human searchers to locate. Manucode nests are, however, relatively less cryptic and usually more easily seen.

Within the typical birds of paradise there exists a clear dichotomy of basic nest structural type. The manucodes construct a sparse and relatively-shallow open cup (suspended oriole-like between horizontal forking branches), predominantly of vine tendrils (Plate 3), and in some cases including leaves and rotten wood pieces within the egg cup. There is some inconsistency in nest descriptions for the Paradise Crow, but the nest found by Heinrich (1956) was said to be 'like an oriole nest' and recently-seen ones fit this discription (Frith and Poulsen in press).

Nests of manucodes and Paradise Crows thus differ considerably from those of the other paradisaeines. That the manucodes are monogamous whereas all others are not (atypically high-altitude Macgregor's Bird excepted) may be no coincidence. In monogamous birds one parent can be in nest attendance at all times. The unique manucode nest type could thus reflect this situation with respect to a number of ecological factors (e.g. greater nest attendance and/or predation levels) but clearer understanding must await future field researches.

The more typical paradisaeines construct a dense and bulky open cup (supported below by predominantly upright-forked branches), which is composed of orchid stems, leaves, fresh mosses, and ferns (Plate 3, Figs 9.16, 17, 20). This nest type, which rarely includes wood debris in the egg cup, is remarkably consistent across all genera.

The Standardwing is the only paradisaeinine genus (*Semioptera*) for which a representative nest remains to be described in detail, although one was recently discovered (Anon 1995; Frith and Poulsen in press). We now know the Standardwing's nest and eggs are at least superficially similar to those of the other typical birds of paradise, but further details are awaited.

A few early ornithologists and one or two authors of contemporary popular accounts have casually mentioned that some of the typical birds of paradise include woody sticks in their nests. In examining preserved nests, active nests of 11 species in the wild, and photographs of additional ones, we have failed to see a single woody stick incorporated into that of a paradisaeinine bird of paradise. Moreover, a careful reading of published nest descriptions that include reference to sticks suggests that they are probably inaccurate. They certainly do not suggest that any species typically uses sticks, if at all.

Nest-building

Females of polygynous birds of paradise construct their nest unaided. Most remarkably, no detailed observations of nest-building by any bird of paradise have been published. It is not known if both sexes share in this duty among the monogamous species. In the domed-nesting Loria's and Crested Birds it would be interesting to learn if some of the sticks incorporated into the lower nest are the first materials placed onto an intended nest site as true foundations or have a different significance. Observations published to date of birds nesting in the wild have unfortunately all resulted from the discovery of a completed nest structure. Surprisingly, no published accounts of captive breeding provide details of female nest-building behaviour, other than that it may involve just 'a matter of days' (Rimlinger 1984) or a week (Hundgen *et al.* 1990) prior to egg-laying. Most observers recorded nothing more than 'the nest was constructed' (Dharmakumarsinhji 1943; Everitt 1965; Timmis 1968, 1970, 1972; Muller 1974; Searle 1980; Todd and Berry 1980; Laska *et al.* 1992; Cooper 1995). Nest-building behaviour by birds of paradise remains an untouched field of study.

Nest 'decoration'

The remarkable fact that female Victoria's and Paradise Riflebirds 'decorate' the rim of their nests with pieces of sloughed snake skin has long been known. In two instances such snake skin was added to the nest after the clutch had been laid, but in another instance, several large snake skin pieces were added to a nest before egg-laying. Thus a female riflebird will apparently incorporate into the nest any snake skin it finds, no matter the timing, presumably because this item is a rare one to find. Several authors have suggested that placement of snake skin on a nest may deter some potential predators of nest contents. So ingrained is this behaviour in Victoria's Riflebird, that females today, in the absence of snake skins, will use strips of clear or semi-opaque plastic if made available. Egg collectors used this knowledge in order to attract nest-building females. They would then follow the birds as they flew off to their nest site with strips of plastic and so discover the nest location (J. Young personal communication).

Recent intensive studies of the behaviour of male Lawes' Parotias revealed that they 'decorate' their terrestrial court area with shed snake skin, mammal dung, chalk pieces, mammal fur, feathers, and bone fragments in that order of preference (Pruett-Jones and Pruett-Jones 1988*b*). Female visitors to the courts remove such items during the nesting season only, and although nests were not studied, the behaviour of the females led the Pruett-Jones's to suspect they were using the snake skin as nest-lining material. The Pruett-Jones's cited a Lawes' Parotia nest lacking snake skin described by Hartert (1910) and cited by Gilliard (1969), but this nest was probably wrongly identified (see species account for Lawes' Parotia). If females do incorporate snake skin into their nest it is more likely to be as nest rim 'decoration' (*cf.* Victoria's Riflebird) and not as nest (egg cup) lining. More on the relation between this court 'decoration' and the nest appears in the Lawes' Parotia species account of Chapter 9.

Eggs

Clutch size

The usual bird of paradise clutch size is one or two, with three eggs occurring only rarely. A clutch size of one or two eggs is perhaps to be expected in tropical polygynous species because, with male emancipation from nesting duties, only the female is available to incubate and provision the young. A clutch of two eggs (rarely three) is typical of the monogamous manucodes. That said, it must be pointed out that sample sizes for clutches remain few for most species. Given the little we do know, however, it is likely that polygynous species will prove to have a modal clutch size of one, whereas the monogamous manucodes will have a modal clutch size of two. While records of active nests of the presumed monogamous Paradise Crow remain few, all indicate a single-egg clutch.

The odd three-egg clutch does occur rarely in lowland polygynous species such as the riflebirds. In the highland species for which clutches have been recorded they are of a single egg only. While this may reflect colder climates where sparser animal food is available (Mani 1990) numerous other factors might play a part in the evolution of clutch size (Lack 1968).

A larger clutch may require a larger nest, thus making it more visible to predators. Resultant losses might therefore cancel out any advantage gained from larger clutch size. A single nestling might be easier for a female parent to protect than a larger brood during rains in tropical rainforest-dwelling, polygynous, passerine species (Snow, B. in Goodwin 1983: 23). While possibly having an influence on clutch size in open-cup nesting birds of paradise this would probably be less true of the wide-gaped birds which lay their single-egg clutches in bulky and more protective domed structures.

A seemingly logical explanation for small clutch size in tropical rainforest-dwelling birds is that it reflects an inability on the part of parent birds to provide adequate food for a larger number of nestlings (Lack 1968). This limitation would be expected to have a far greater influence upon polygynous species in which the female alone provisions her offspring. Observers of tropical rainforest passerines have, however, noted that species in which both sexes feed nestlings do not on average raise more young from a nesting attempt than do closely-related species in which only the female provisions the brood. Indeed some species are considered to be quite capable of raising more young each year than they have been observed doing (Skutch 1949, 1976; Snow 1976*b*). It has been noted, moreover, that little evidence for the hypothesised limiting nature of food availability actually exists (Perrins 1985).

Another idea explaining small clutch size in tropical rainforest birds is that it is an adaptation to potentially high predation levels upon nesting attempts. A small clutch is more easily and quickly replaced than a larger one and this may therefore improve the chances of renesting within a breeding season in the event of failure through predation. Thus, this is a strategy literally involving birds not putting all of their eggs in one basket.

What little we know of bird of paradise demographics suggests that, at least in males of the polygynous species, adult survival rates are markedly high and that reproductive rates are low. Emphasising this fact is that the males of such species require in excess of 5 years or more to acquire their first nuptial plumage. High predation levels plus low food availabilty may favour birds investing in many small clutches over a long lifetime—so that the loss of one clutch does not represent a large proportion of the parent's lifetime investment in reproduction.

Numerous additional variables, including latitude, length of daylight, habitat type, diet, body size, parent age, egg size, nest site and form, and social structure of the nesting birds (to name but some) might influence the evolution of clutch size.

The bottom line with regard to selection for clutch size must be, however, that any successful strategy should ultimately be reflected by what is the optimal clutch size for the species. This is the clutch size producing the largest number of surviving offspring to a parent throughout its life span (Perrins 1985).

Egg shape and colour

Bird of paradise eggs are typically elliptical ovate (or long oval), with the odd one tending towards oblong oval (or elliptical). The few known eggs of the Cnemophilinae are pale pinkish with most markings consisting of reddish-brown fine spotting, predominating about the larger end. Macgregor's Bird and the manucodes' eggs vary in ground colour from whitish and pinkish to buff, and they are variably heavily spotted to blotched with browns to blacks throughout but more densely about their larger end. The buff to beige Paradise Crow egg is unique within the family in being more pointed at its smaller end and in being scrawled and scribbled throughout with swirling, fine, blackish lines. It is a little surprising that the egg of this species is so different from those of its close relatives the manucodes.

Of the remaining lower typical (Paradisaeinae) species, the paradigalla egg is much like the more spotted (not blotched) eggs of the manucodes. Those of astrapias may be similar but are more usually marked with longer broad blotches tending to be more elongated down the egg length. In the higher paradisaeines, eggs vary in ground colour but are typically all characteristically marked with broad, brush-stroke-like, elongate markings down their length. This gives them a most attractive 'hand-painted' appearance (see Plate 13).

Attractive though bird of paradise eggs are, particularly those of the Paradisaeinae, it is most probable that their colour and markings have evolved to enhance crypticity—to make eggs left uncovered in the nest less conspicuous to predators searching visually from above. Blotches and streaks break up the outline of eggs (Lack 1965). There is no point in cryptically-coloured and marked eggs, however, if they are usually constantly concealed as in monogamous species (e.g. pigeons) and hole or domed-nesting birds (e.g. parrots and logrunners). Such considerations might account for the paler and/or less cryptically-marked eggs found within the monogamous and domed-nesting species of birds of paradise. Numerous other factors could, however, also be involved (Skutch 1976).

Egg weight

Egg weight as a proportion of the laying bird's body weight provides a relative measure of female reproductive investment. A high egg-to-female-body-mass ratio is of interest because it may indicate a strategy to ensure offspring are relatively advanced at hatching. This may be an attempt to shorten the length of the nestling stage, and thus reduce the risk of nestling predation.

The few data available for birds of paradise laying single-egg clutches (Loria's Bird, Short-tailed Paradigalla, and Ribbon-tailed Astrapia) suggest their egg represents between 10 and 15% of mean adult female weight. Each egg of a Victoria's Riflebird clutch of two represented 12% of average adult female weight (Fig. 6.1). Thirteen freshly-laid eggs of a captive Red Bird averaged 9.1 g (Worth *et al.* 1991) which represents 6% of a mean adult female weight of 157 g ($n = 5$, LeCroy 1981). These eggs were all products of a series of two-egg clutches. With the exception of the captive-laid eggs of the latter species, these relative egg weights are large for passerines (Lack 1968).

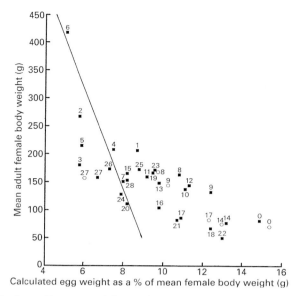

6.1 Approximated fresh bird of paradise egg weights, calculated from mean egg measurements (Bergtold 1929; Preston 1974), plotted (solid squares) as a proportion of mean adult female body weight for species for which data are available (sample sizes appear in species accounts). Also plotted (open circles) are the actual fresh egg weights as a proportion of mean adult female body weight, for the six species for which these are known. The diagonal line shows the general relationship between proportionate egg weight and unspecified (presumably mean of adults of both sexes) body weight for passerine birds (*cf.* Lack 1968).

Numbers above, below, or beside symbols indicate the species, in systematic order, as follow: 0 = *Cnemophilus loriae* (1), 1 = *Macgregoria pulchra* (1), 2 = *Lycocorax pyrrhopterus* (1), 3 = *Manucodia atra* (2), 4 = *M. jobiensis* (2), 5 = *M. chalybata* (1–2), 6 = *M. comrii* (1–2), 7 = *M. keraudrenii* (1–2), 8 = *Paradigalla brevicauda* (1), 9 = *Astrapia mayeri* (1), 10 = *A. stephaniae* (1), 11 = *A. rothschildi* (1), 12 = *Parotia lawesii* (1), 13 = *P. wahnesi* (1–2), 14 = *Pteridiphora alberti* (1), 15 = *Ptiloris magnificus* (2), 16 = *P. paradiseus* (2), 17 = *P. victoriae* (2), 18 = *Lophorina superba* (1–2), 19 = *Epimachus meyeri* (1), 20 = *Drepanornis albertisi* (2), 21 = *Cicinnurus maginficus* (2), 22 = *C. regius* (2), 23 = *Seleucidis melanoleuca* (1), 24 = *Paradisaea minor* (1–2), 25 = *P. apoda* (1), 26 = *P. raggiana* (1–2), 27 = *P. rubra* (1–2), 28 = *P. rudolphi* (1). Numbers in parenthesis after a species name indicate known or approximated clutch size.

Note: the plotting (open circles) of five actual proportionate fresh egg weights (Frith and Frith 1990*c*, 1992*b*, 1994*b*, 1995*b*; Worth *et al.* 1991) indicate that those calculated from egg measurements are reasonably accurate. The larger disparity between the actual and calculated plot for *Astrapia mayeri* probably reflects the fact that the eggs concerned were not relatively freshly laid (see Frith and Frith 1993*b*).

For example, egg weight as a percentage of female weight ranges from only 4.2 to 5.7% in the monogamous corvine magpies *Pica* spp. (Birkhead 1991). Conversely, it is particularly high in the polygynous, Neotropical, passerine fruit-crows (Cotingidae) (Snow 1982) and the bowerbirds (Frith and Frith 1994*a*, 1998*b*).

Fresh bird of paradise egg weights are few. It is therefore worthwhile to calculate estimates of them for species for which both egg measurements and female weights are available (Fig. 6.1) as this can test the impression given by the few actual data. The volume (V) of an egg can be calculated (Preston 1974) from its length (l) and breadth (b) measurements ($V = \pi/6 \times l \times b^2$) and assuming a specific gravity of 1.05 g/cc (Bergtold 1929) for fresh egg content the fresh weight can be thus estimated. Calculated paradisaeid egg weights are a relatively high proportion of adult female body weight in most species regardless of mean clutch size, but are particularly high in those laying single-egg clutches (Fig. 6.1). Estimated fresh egg weights plotted as a proportion of female weight do agree closely with the few data obtained from actual fresh weights (Fig. 6.1).

Notwithstanding variables such as uniparental versus biparental care, nest type, and exclusively-frugivorous (*Cnemophilus*) through mixed diets to predominantly-arthropod ones, a general correlation between increasing altitude and longer incubation periods is indicated (Table 6.2). Far more data are required before significant patterns can be discerned. We note that meagre available data do broadly suggest, however, that higher-altitude species have a higher proportionate single egg weight as follows: lowland Red Bird 6%, lowland to hill-forest Victoria's Riflebird 12%, highland Short-tailed Paradigalla and Ribbon-tailed Astrapia 10%, King of Saxony Bird 13% (Fig. 6.1).

It is clear that egg weight as a proportion of adult female body weight in monogamous species of birds of paradise is typically closer to the general relationship pertaining in passerine birds. In polygynous species, the egg weight is, however, typically proportionately higher (Fig. 6.1). Uniparental care by females and their single-egg clutches may partly account for the high proportionate egg weights in polygynous birds of paradise (and cotingas and bowerbirds). This tentative observation of a general pattern is only made in the hope of stimulating the gathering of further pertinent information from wild birds.

Egg-laying intervals

From our limited dataset, it is apparent that most multi-egg clutches of birds of paradise are laid on successive days. The eggs of a wild clutch of Victoria's Riflebird (Frith and Frith 1995*b*), two clutches of a captive Magnificent Bird, a clutch of a captive King Bird and a clutch of a captive Raggiana Bird were all laid on consecutive days (although the evidence for the King Bird is circumstantial; all clutches were of two eggs). A single captive Red Bird, stimulated to lay again by the removal of her initial clutches, may have exhibited atypical laying intervals as a result. Over two breeding seasons she laid eight two-egg clutches and while it is reported the eggs of a clutch were laid 'in intervals of about 36 hours' the same authors indicate that eggs of five clutches were laid on alternate days and those of two on consecutive days. (Worth *et al.* 1991). Contrary to birds of paradise, evidence suggests that bowerbirds typically lay eggs on alternate days (Marchant 1986; Frith 1994*d*).

Egg incubation and hatching

A 5-hour observation over 2 days at a nest of the monogamous Glossy-mantled Manucode suggested to the observer that only one parent, assumed to be female, incubated. The possibility that the male was too nervous to incubate was, however, noted (Rand 1938). By contrast, both sexes of the Trumpet Manucode certainly do incubate (Frith and Frith 1993*d*); (see Plate 3). No observations of incubation by Paradise Crows have been made. In the monogamous Macgregor's Bird, only the female is said to incubate (Rand 1940*a*). This requires confirmation, as single-parent incubation would be an unusual strategy in a monogamous species nesting in such a cold habitat. In all polygynous species observed nesting (Tables 6.2–4) only the female incubates the clutch.

Known incubation periods for birds of paradise vary from 14 to *c.* 26 days (Table 6.2). Exceptional is Loria's Bird which inhabits montane forest, builds a domed nest, and has a heavier egg proportionate to female weight (Fig. 6.1). It is not surprising that this species has a longer incubation period (*c.* 26 days) than the average for seven lowland to hill-forest nesting species that build an open cup (*c.* 18 days).

The typical or normal breeding bird spends 60–80% of its day incubating (Skutch 1976: 181), this being true of both species in which only the female and species in which both sexes incubate. The four paradisaeid species where nesting data are available fall within this pattern (Table 6.3).

Table 6.2 Summary of known bird of paradise incubation and nestling periods.

Species	Altitude in metres*	Clutch size	Incubation period (days)	Nestling period (days)	Nesting location	Reference
Loria's Bird	2000–2400	1	26 +/-1	—	PNG	Frith and Frith 1994b
Crested Bird	2400–3500	1	—	> 30	PNG	Frith and Frith 1993c
Glossy-mantled Manucode	0–1000	1–2	> 14–< 18	—	IJ	Rand 1938
Short-tailed Paradigalla	1600–2400	1	> 20	25	PNG	Frith and Frith 1992b
Ribbon-tailed Astrapia	2400–3400	1	21	22–26	(c) USA	Boehm 1967, Everitt 1973
				> 23 & 25–29	PNG	Frith and Frith 1993b
Stephanie's Astrapia	2000–2600	1	21–22	22–26	(c) USA	Boehm 1967
			—	27	(c) UK	Yealland 1969
King of Saxony Bird	1500–2900	1	> 23	—	PNG	Frith and Frith 1990c
Victoria's Riflebird	0–1000	1–3	18–19	4	AUST	Frith and Frith 1995b
Superb Bird	1500–1800	1–2	18–19	18	(c) UK	Timmis 1970
Magnificent Bird	0–1500	2	18–19	17–18	(c) USA	Everitt 1965
King Bird	0–1000	2	17	14	(c) SWE	Bergman 1957
Twelve-wired Bird	0–200	1	20	c. 21	(c) SING	Nyunt, unpubl. data
Lesser Bird	0–1000	1–2	18	19	(c) USA	Hundgen et al. 1990
Raggiana Bird	0–1500	1–2	20	20	(c) AUST	Muller 1974
			—	17	(c) HK	Searle 1980
			18–19	17	(c) USA	Rimlinger 1984
Red Bird	0–600	1–2	14–18	15–20	(c) USA	Todd and Berry 1980
			x = 16.4	x = 17.5	(c) USA	Worth et al. 1991
			n = 5	n = 4		
Blue Bird	1300–1800	1	> 18	—	PNG	Pruett-Jones and Pruett-Jones 1988a

Key: (c) denotes a captive breeding in: PNG = Papua New Guinea, AUST = Australia, IJ = Irian Jaya, Indonesia, USA = United States of America, UK = United Kingdom, HK = Hong Kong, SING = Singapore, SWE = Sweden. * = wild birds only.

Table 6.3 Female parental investment in incubating the clutch* in some birds of paradise.

Species	Total observation time (hr)	% time spent incubating	Mean length of incubation bout (mins)	Mean no. incubation visits per hr	% time spent absent	Mean duration of absences (mins)	Reference
Short-tailed Paradigalla	30	58	17	2	42	13	Frith and Frith 1992b
Ribbon-tailed Astrapia	19	66	16	2.4	34	9	Frith and Frith 1993b
King of Saxony Bird	20	58	9	4.1	42	7	Frith and Frith 1990c
Victoria's Riflebird	128	71	44	1	29	18	Frith and Frith 1995b
Raggiana Bird	59	78	?	1.3	20	?	Davies and Beehler 1995
Blue Bird	38	64	31	—	36	19	Mack 1992
Mean		66	23	2.2	34	13	

* Single egg clutches save for Victoria's Riflebird clutch of two. Note that sample observations vary as to age of the egg(s).

At a Victoria's Riflebird nest, the female removed the shells of two freshly-hatched young slowly and carefully from amongst her hatchlings before flying off with them to return briefly thereafter (Frith and Frith 1995b). This observation, and fresh egg shells of this species located on the rainforest floor (S. Breeden personal communication; R. Whitford personal communication; and personal observation.) indicate that egg shells are not necessarily eaten by this species (a habit typical of crows and their allies).

Nestlings, their growth, and development

A most striking characteristic common to nestlings of birds of paradise is that they hatch naked or with only the sparsest of down, their dorsal skin already pigmented dark blue-grey to blackish-grey or else becoming so within a few days.

Hatchlings of Macgregor's Bird and the paradigallas are known to have well-developed facial and mandibular wattles typical of their respective adults. Nestlings and juveniles of the relatively long-billed genera *Ptiloris*, *Seleucidis*, *Epimachus*, and *Lophorina* have shorter, broader, and blunter bills that grow longer, narrower, and finer as they develop. It is not known if this pattern indicates that young of the longer-billed species are parent-dependent for speciality foods for a longer period than short-billed species.

Nestling growth of the few birds of paradise species studied indicates that their eyes first open at about 6 days old, but hand-raised Red Bird nestlings are recorded to do so at an average of 11 (9–14) days (but this may well refer to fully- rather than part-opened eyes), when some body contour feather tracts are visibly in pin. At about 8–10 days these feathers have just burst from the pin tips, and the primary coverts and primaries burst from pin at about 2 weeks old.

Growth curves for wild nestling birds of paradise are available for so few broods that we present them all without synthesis (Fig. 6.2). They show that the larger species grow at a fast rate and that all growth curves are virtually linear over much of their length, except for the down-turn just prior to departure from the nest. Then nestlings regularly lose some weight because of the energetic cost of rapid growth of the tail and flight feathers. As with incubation periods, nestling periods are generally longer in higher-altitude species—these varying within the family from 14 to more than 30 days (Table 6.2).

At 25 days old, captive King Birds had a tail about 3 cm long and the legs and feet blue but paler than those of their mother. At 30 days their tail was almost as long as their mother's and when almost 60 days old they moulted their head and underpart feathers with those of the back which then became a pure pale grey colour.

Nestling captive Raggiana Birds left the nest at 20 days when described as approximately three quarters the size of their parent (Muller 1974). At 10 weeks they were the size of their mother and by 4 months had attained the golden crown and nape of the female. A captive nestling Red Bird stretched and preened at 15 days old and left the nest the next day when capable of fluttering flight. At 24 days its tail was about 13 mm long, when the neck was sparsely feathered but by 28 days old it had golden-brown feathers like those of its female parent. It was truly flying a week later (Todd and Berry 1980).

Nestling care and diet, and post-fledgling care

Nestling care

Care of nestlings in birds of paradise is in many respects typical of tropical rainforest-dwelling passerines. In all birds of paradise for which nest sanitation behaviour has been noted the parent(s) swallows nestling faeces throughout all or most of nestling life. Only when the nestlings are near nest departure do parents carry faeces and seeds away from the nest.

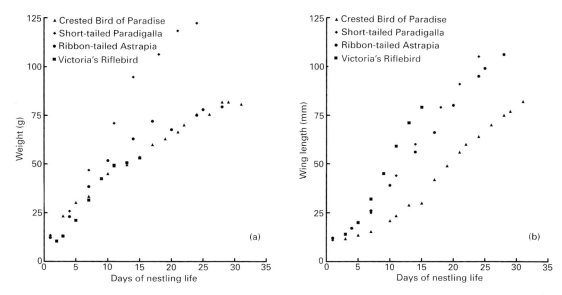

6.2 Composite growth curves for wild single nestling broods of four polygynous bird of paradise species showing increase with age in (a) weight and (b) wing length to within 1–2 days of fledging (after Frith and Frith 1992b, 1993b,c, 1995b).

An anti-predator 'freezing' behaviour has been recorded in Victoria's Riflebird and the Raggiana Bird. If, in approaching the nest, a potential nest predator is seen or heard, the returning parent will simply 'freeze' on its perch. A female of the former species was seen to freeze on a perch 1–2 m below her nest and nestling with a large insect in her bill tip for 9 minutes in response to a Spotted Catbird *Ailuroedus melanotis*, a well-known predator of other birds' nest contents, calling close by. Under normal circumstances, typical birds of paradise approach the nest site silently and via several perches as they attempt to satisfy themselves that no danger is about. An incubating or brooding bird starts to turn the head about at an increasing rate as it appears to assure itself it is safe to depart the nest. It then silently raises itself, hops very briefly onto the nest rim and drops quickly downward and then away from the nest site.

Several incubating or brooding female Ribbon-tailed Astrapias were noted to ignore other species up to the size of a large honeyeater (Meliphagidae) within a few metres of their nests but to chase off immediately any conspecific venturing into even the more general area (Frith and Frith 1993b). Similar behaviour is recorded for a nesting Raggiana Bird (Davis and Beehler 1994). A nesting Blue Bird was seen to be aggressive towards conspecifics and other birds of paradise, notably the Trumpet Manucode and the Superb Bird, but only near the nest and not in a conventionally territorial way (Pruett-Jones and Pruett-Jones 1988a).

The few measures of the percentage of time a wild female bird of paradise spends brooding a single nestling varies from *c.* 14% to 48% (Table 6.4). It must be emphasised that these data are from single nests for few species making it difficult to generalise. A most conspicuous point is, however, that the domed nesting, polygynous Crested Bird spent far less time brooding young, suggesting that the more cryptic, protective, and presumably better-insulated domed nest enabled the parent to be less attentive notwithstanding its relatively cold nesting habitat.

The rate at which single-nestling broods of polygynous species are fed by the female

Table 6.4 Female parental investment in brooding and feeding a single young in some birds of paradise.

Species	Total observation time (hr)	% time spent brooding	Mean length of brooding bout (mins)	Mean no. brooding visits per hr	No. nestling meals per hr	% nestling diet animal foods	% time spent absent	Mean duration of absences (mins)	Reference
Crested Bird	54	14	10	0.8	2.8	0	78	15	Frith and Frith 1993c
Short-tailed Paradigalla	57	27	13	1.3	2.1	65	73	18	Frith and Frith 1992b
Ribbon-tailed Astrapia	51	18	7	1.5	1.8	33	76	9	Frith and Frith 1993b
Victoria's Riflebird*	149	21	19	0.7	1.7	80–90	74	24	Frith and Frith 1995b
Raggiana Bird	112	48	?	1.5	1.3	>75	52	?	Davis and Beehler 1994
Mean		26	12	1.2	1.9	—	71	17	

* Initially a brood of two but one young died on its third day of life. Note that sample observations vary as to age of the nestling.

parent has been recorded for five species. It varies from about one to three meals per hour (Table 6.4). Of these, the Crested Bird has by far the highest feeding rate and this is doubtless a reflection of the low protein diet of exclusively fruit provided to its nestlings, which is relatively easy to harvest quickly in the habitat (*cf.* Snow 1976*b*).

During *c*. 26 hours of observations of parental care of a two-nestling Trumpet Manucode brood over 4 days, 87 nestling-feeding visits were made by the parents at an average of 3.4 each hour (see species account of Chapter 9). More studies of additional monogamous species are required.

Nestlings are fed by parents regurgitating fruit and/or animal items with the exception that some species will carry the first item (usually an animal prey) to be offered to the young in the bill. This is as in crows, but is quite distinct from the behaviour of the Lesser Melampitta and all bowerbirds, which do not regurgitate food items but instead carry them in the bill and/or mouth (Frith and Frith 1990*a*, 1994*a*).

Nestling diet

Female Crested Birds raise their offspring exclusively upon regurgitated simple fruits (see Chapter 4). Parents of the paradisaeine species undoubtedly discriminate in selection of food items for their young based on their age. Meals brought to hatchlings and young up to 3–4 days old are so small as to be unidentifiable to the observer. A captive female Red Bird first fed her offspring only fruit, by regurgitation, for 5 days and then changed to live food (insects) only thereafter. Parents reduce larger animal and fruit food items in size to be able to swallow and regurgitate them and so make them small enough for their offspring to eat. This they do by holding the prey or fruit onto a perch by one foot (rarely both feet) to tear it apart with the bill. Even within the confines of small aviaries, female King Birds and *Paradisaea* spp. are noted to hold a grasshopper onto a perch by the foot, remove the legs and wings, and swallow the edible pieces before then flying to the nest to regurgitate these to their nestlings. See also Table 6.4.

Post-fledgling care

The care of dependent young once they have left the nest is little known. Only for the Short-tailed Paradigalla is a period of dependence well documented. A ringed female was seen to feed her fully-grown but duller-plumaged juvenile 108 days after it left the nest; this behaviour apparently ceased a week later (Frith and Frith 1992*b*). Female-plumaged Ribbon-tailed Astrapias and Brown Sicklebills have been seen being followed by a juvenile that would flutter its primaries and give soft begging notes to solicit a meal. A female Victoria's Riflebird was seen to feed her begging juvenile 61 days after it had left the nest (W. Cooper personal communication).

A captive fledgling Raggiana Bird one week out of the nest was sheltered beneath one extended wing of its mother at dusk. It took fruit by itself at 43 days old but was still being fed by its mother, by regurgitation, at 75 days old when it looked much like its parent (Searle 1980).

During intensive work with nesting female Loria's and Crested Birds, a Short-tailed Paradigalla, Ribbon-tailed Astrapias, and Victoria's Riflebirds, mobbing, distraction, or other active anti-predator behaviour was rarely seen. On one occasion a female Victoria's Riflebird, returning to her week-old nestling, flew directly and powerfully at a Spotted Catbird (Ptilonorhynchidae) 20 m from her and immediately displaced and chased it (Frith and Frith 1995*b*). In most instances, a sitting female would simply disappear from the nest as a person approached, but if surprised upon returning to her nest by a person nearby, the female might hop about at a distance and give a few soft, low-pitched, scold notes before retreating (Frith and Frith 1995*b*). Clearly some birds of paradise do, however, perform a distraction display and aggressive nest defence, as exemplified by the captive female King Bird that both vigorously distracted and physically attacked its keeper (Bergman 1956).

Brood parasites

There exists a single record of the Common Koel *Eudynamys scolopacea* (Cuculidae) laying an egg into an active nest of Victoria's Riflebird—the only such occurrence of nest parasitism recorded for any bird of paradise (Brooker and Brooker 1989).

Lord Rothschild (1930b) reported the claim of his collector W. Frost that a Trumpet Manucode laid an egg in the nest of a Greater Bird—acting as a brood parasite. A reading of Rothschild suggests that Frost was encouraged in this belief by often seeing the manucode apparently following female-plumaged Greater Birds. What we now know of paradisaeid participation in mixed-species foraging flocks (see Chapters 4 and 9) suggests that this was probably what Frost witnessed. It was subsequently suggested that the manucode may simply use nests newly built by Greater Birds to nest in (Harrison and Walters 1973). Brood parasitism or intergeneric nest use is unlikely in view of the distinct nest form, incubation regime, and nestling diet of the Trumpet Manucode (see species accounts).

Climatic cycles and annual breeding seasons

The annual cycles followed by birds of paradise are poorly understood (see Chapter 5). The 'breeding season' repeatedly mentioned in ornithological literature has often been inadequately defined. We use this term to refer to time when a species is nesting. We have re-examined all literature and bird, nest, and egg specimens for the birds of paradise in order to define objectively the breeding seasons, species by species (Table 6.5). In addition to presenting records of nests found under construction and containing eggs or nestlings (the nesting season) we also indicate records of mating or birds seen nest-building or of birds found to have been in 'breeding condition' as indicated by significantly-enlarged (> *c.* 60%) gonads. In regard to 'breeding condition', a female with enlarged oocytes or a conspicuous brood patch is good evidence of nesting. By contrast, a male may show enlarged gonads throughout much of the display season, which is far longer than the nesting season. In general, we believe much of the data on gonadal condition are of only limited use for this analysis.

Combining breeding data for the more extensively-distributed species inevitably leads to delineation of unnaturally long nesting seasons—because it will combine populations that are following differing cycles. Also we believe this bias will be exacerbated by inter-annual variation as well, if data from a range of years are combined. Despite these biases, the summary of breeding and nesting seasonality (Table 6.5) provides some broad patterns. It is clear that far more breeding takes place between August and January than from February to July, with March to June being the least productive period, as was found to be true near Port Moresby in a 2-year study (Bell 1982b). Thus general breeding activity over New Guinea as a whole falls during the seasonally-heaviest rains of January to April and then picks up again as rains recommence after the driest period of May to August (McAlpine *et al.* 1983). This broadly concurs with Gilliard's (1969: 5) generalisation that the peak of nesting by the birds of paradise in New Guinea is during the early part of his 'wet season' of October–April. It is also in agreement with Alexander Skutch who wrote (1976: 60) 'In the rainier parts of the tropics, the principle breeding season starts as the dry season passes into the wet and reaches its peak early in the rainy season, when the greatest number of species have the greatest number of nests. Some birds start a few weeks before the rains, some a few weeks after they begin. The majority avoid nesting in the driest or the wettest months.'

This breeding season appears to coincide with a period of abundance of fruit and arthropod prey. This is consistent with the statement

Table 6.5 Summary of knowledge of gonad activity, mating, and nesting seasonality of birds of paradise over their entire ranges.

PARADISAEIDAE	Jan	Feb	Mar	April	May	June	July	Aug	Sept	Oct	Nov	Dec
CNEMOPHILINAE												
Loria's Bird	EN	N		m			m			m	B	B
Crested Bird	N				m	m	m f	E	N	P	E	E
Yellow-breasted Bird					m	m		m		m		
PARADISAEINAE												
Macgregor's Bird		J		m			E	BEN	EJ		m	
Paradise Crow	mE	A	E	N		mE	f	m	m f	m	m	E
Glossy-mantled Manucode	E	m	E		A	m		EN	EN	EN	B	A f m
Jobi Manucode	m	m	m	m	m	m	m	m				E
Crinkle-collared Manucode	EN	m	m		f	m	B	E	EN	m	m	
Curl-crested Manucode			E			E	EN	E	EN	N	B	
Trumpet Manucode	EN	m	m	m	E	f	m	N	m	E	CE	EN
Long-tailed Paradigalla	N											E
Short-tailed Paradigalla*	E	BN		E	EN	EN	E	N	BE	E	B	E
Arfak Astrapia							J	J				
Splendid Astrapia			B	m	m		m	J	m	E	N	
Ribbon-tailed Astrapia	EN	EN	N	m	E	EN	N	N	E	BEJ	BEN	EN
Stephanie's Astrapia	m	m		m	N	EN	NJ	f	B	E	E	BN
Huon Astrapia			m f									
Western Parotia	m						m	m	m			
Lawes' Parotia	m	mB			m	E	NJ	m f	m f	m f	EN	EN
Wahnes' Parotia	E++	f		BN							m	
Carola's Parotia	m			m	m	m	m	m	m f	m f	m	
King of Saxony Bird	E	N		C	m	m	N	m	J	f	f	CBJ
Magnificent Riflebird	E	E	m	m f	E	E	fB	m f	EJ	E	BNE	E
Paradise Riflebird	E	J	J						E	EN	BEN	EN
Victoria's Riflebird	EJ	E					m f	E	BEN	EN	BEJ	E
Superb Bird	EN	J	EJ	BN	m	E	N	m f	m fN	m fN	BE	N
Black Sicklebill	J	EJ		m	m		m	m	m		B	
Brown Sicklebill	fJ	m		E	N	E	EN	ENJ	N	E	N	fJ
Buff-tailed Sicklebill				m	m			m	m	m	J	
Pale-billed Sicklebill						m f	m	m		m f		m

Table 6.5 continued

PARADISAEIDAE	Jan	Feb	Mar	April	May	June	July	Aug	Sept	Oct	Nov	Dec
Magnificent Bird		m f	E	m	m f	m f	E	E	E	B E N	E N	E N
Wilson's Bird					m	m f				m		
King Bird	m	m	E	N	J	B	m	m f	m	C	m	m
Standardwing Bird				m	C E	m N				m		
Twelve-wired Bird	B E	B E			E	m	m	E	m f	E N	m	m
Lesser Bird	E++	m f		m	m	m	m f	B	m f	N	m f	m
Greater Bird	N J		E		E			E	E N			E N
Raggiana Bird	m f	m f	m	m	N	B N	C B N	C B E N	C E N	C B E N J	B E J	E
Goldie's Bird									m			
Red Bird					m	m			m f	m f		m
Emperor Bird		m						E	E	m	m	
Blue Bird	E	N J	m E++	E	N		E	E N	E	E N	E N	E
Totals	19	16	11	7	14	14	19	24	27	25	22	22

Key: m and f = male and female found in breeding condition (enlarged gonads), m is not indicated where better data are available; C = copulation and B = nest building seen; P = female with brood patch examined; E = egg(s) in nest; N = nestling present; J = juvenile recently out of nest; A = nest attended but contents unknown. * = data are from the one female and area over several years (see text): ++ = captive laying in Papua New Guinea, excluded from totals.

Note: males with enlarged gonads are not as indicative of breeding as are females because males display and have enlarged gonads for a greater period than the nesting season. For this reason they are excluded from the totals.

Sources include: the literature and world-wide museum collections of birds and eggs and unpublished data provided as are summarised and cited in the species accounts.

that the early (Neotropical) wet season is most rich in fruits and insects (Skutch 1976). Data for leaf-litter invertebrates, mobile insects, and tree fruits of the tropical northeastern Australian upland rainforest habitat of Victoria's Riflebird also fully agree with this (Frith and Frith 1985, 1990, 1994c). Also data on fruit availability on Mount Missim indicated that fruit was most abundant during the paradisaeid breeding season (Beehler 1983a).

The relatively brief breeding seasons suggested for some species by the data in Table 6.5 may be a reflection of limited data on selected species combined with seasonal biases of data collection. Data for better known species such as the Short-tailed Paradigalla, Ribbon-tailed Astrapia, Magnificent Riflebird, Superb, Magnificent, King, Greater, and Raggiana Birds indicate that, over a period of years, nesting attempts occur in virtually every month of the year—to be expected with periodic El Niño effects, renesting, and the like.

The simple fact is, to understand properly paradisaeid breeding seasonality, one must carry out a multi-year single-site study that delineates inter-year variation and single-year patterns. We suspect that peak breeding at a single site over a single year is probably temporally quite discrete—a few months at most, but that the compilation of data from various sites and various years blurs this pattern.

Bird breeding seasons in the tropics are hypothesised to become contracted at higher altitudes (Skutch 1976). While the limited data available show that the nesting season of highland birds such as the Loria's, Crested, and Macgregor's Birds might be briefer than some, there are other highland species, such as the Short-tailed Paradigalla, Ribbon-tailed Astrapia, and Brown Sicklebill that apparently have prolonged seasons. A complication is that the former three species feed their nestlings an exclusively—or predominantly—fruit diet that may impose a more restricted seasonality on nesting.

Results of the findings of egg collectors and of recent studies of several nesting birds of paradise suggest that it is unlikely that females of any polygynous species raise more than a single brood in a breeding season (but see Victoria's Riflebird account, Chapter 9). This is not surprising in long-lived birds in which the female must raise her offspring alone. However, evidence does exist to show that such females will attempt several renestings in a season as a result of nest failure.

It is clear from the above that while all too little is known about bird of paradise nesting biology some broad patterns are tentatively indicated. The dichotomy between the domed nesting cnemophilines and open nesting paradisaeines on the one hand, and that between the sparse and shallow suspended nests contrasting with the bulky and deep supported ones within the latter group on the other, suggest topics for comparative studies.

Egg colour and markings, and egg weight proportionate to that of adult females, tentatively suggest a dichotomy between the monogamous and polygynous species. Preliminary observations of limited datasets suggest that higher-altitude species lay smaller clutches. These require longer periods of incubation and their resultant nestlings remain longer in the nest than those of lowland species. The annual nesting season of a species at any single location is probably brief but at this time knowledge does not permit firm conclusions. That said, the future is a challenging and exciting one for students conducting field investigations into the nesting biology of the Paradisaeidae.

7

Birds of paradise in human tradition and culture

Ships' crews of the earliest intrepid voyages to reach the shores of New Guinea were the first to witness the degree to which the indigenous cultures venerated and used birds of paradise. As exploration of the region and observation of local cultures increased it became apparent that the birds of paradise represented an important source of wealth and custom, as well as spiritual totems; their displays, behaviour, and forest life history served as a source of myth and religion.

It is undoubtedly true that the peoples of New Guinea have admired and utilised birds of paradise for many thousands of years. In the more remote villages, affiliation with the birds remains undiminished today. In some instances, traditions have died or been irrevocably altered—in many cases by the overpowering effects of evangelical Christianity. At the social and political level, what were once traditional or customary affiliations have been transformed—the most graphic examples being a bird of paradise now serving as a symbol on a national flag, on a can of beer, or packet of coffee.

A remarkably strong aesthetic sense and desire for personal adornment by the people of New Guinea was clearly demonstrated to the first western visitors to the island. Explorers to the island found abundant evidence of complex architecture, elaborately-decorated sea-going outriggers and war canoes, exquisitely-carved everyday artifacts, and the intricately-artistic scarring of body skin. Conspicuous about the necks and decorating the upper chest of many people, particularly the highlanders, were large, crescent-shaped, and polished mother-of-pearl pieces of the gold-lip 'kina' shell and necklaces of dog teeth or bright yellow orchid stems. Other than these crafted items, natural products that could provide colourful body decoration were then, and remain today, few. They consist of bands of more strikingly-marked mammal fur, brightly-iridescent green elytra (wing cases) of cetoniine beetles, and of course, the plumes of birds of paradise.

Plumes, personal adornment, and wealth

The early explorers were, if diplomatic and patient enough, privileged by being shown the men's highly-prized plumes. These were the carefully-accumulated dried skins of the plumed and long-tailed adult male birds of paradise hunted and/or traded by older initiated men. Once prepared and dried, they were, if not for everyday wear, carefully placed between the flattened, broad, or woven fronds of a palm or pandanus tree, within a length of bamboo tube or other safe-keep. These were then stored in the smoky rafters of the men's house. Here they were safe from the depredation of insects, light, and damp, the prying eyes of the women, or the grip of curious children. The brilliant colours and delicate structures of these highly-treasured plumes were exposed to the detrimental light and elements only during the most important of cultural events such as a sing-sing. Bird of paradise plumes do not retain their pristine colour and structure for many decades under tropical conditions and may remain socially acceptable for such use for as few as four or five events of major social significance (Majnep and Bulmer 1977).

A sing-sing is the New Guinean equivalent of a ball—a social gathering of the community that enables one and all to exhibit wealth (by personal adornment), social status, dancing ability (physical prowess), musical ability, wit, and other social graces. One of many functions of a sing-sing is that it provides a forum for single males to exhibit publicly their hoard of, or access to, bird of paradise plumes. These may be owned by the individual bachelor, may be loaned to him by his relatives or clansmen, or may be temporarily hired by him in exchange for goods or return favours, etc. Because of the long tradition of trading plumes within New Guinea a man may own and wear the plumes of species not found in or near his own homeland.

In most parts of New Guinea, a single man wishing to obtain a wife must provide the woman's family with a negotiated 'bridewealth', which traditionally may have consisted primarily of plumes, shells, cassowaries, and some other goods, but now increasingly involves other considerations such as pigs, cash, and other assets. In the Wahgi Valley of Papua New Guinea during 1925–74 the average number of bird of paradise plumes included in bridewealth exchanges was 18 (Heaney 1982). At a traditional sing-sing predating western influences, a fit and handsome young man could dance and chant until all his peers were exhausted and voiceless, but it may have been to little avail without an adequate bride price of bird of paradise plumes. In this the men closely associated themselves and their day-to-day lives with the male birds. They knew only too well, from their own observations in the forest or from folklore, that the male birds rarely, if ever, manage to mate with a female until and unless fully plumaged. Moreover, the males must display properly and with levels of competence and skill critically demanded by the females if they are to mate successfully.

Men adorning themselves with plumes certainly do so at least in part because they relate closely to the masculinity and vigour of the courtship displays of the male birds. Aesthetic values vary from place to place and do change over time, as with fashion elsewhere. For example Majnep (in Majnep and Bulmer 1977) writes that whereas his ancestors used Stephanie's Astrapia and Black Sicklebill plumes as the central feature of their sing-sing head-dresses, today the well-dressed man of that area employs the plumes of the Lesser Bird as a centrepiece. The quantity of plumes a man wears does not necessarily reflect his real status, however. It may do just that, particularly if he owns all he wears, but it may also simply demonstrate the socially-useful and desirable fact that he maintains good connections with kith and kin (who have rented or loaned him plumes). On the other hand in some Papua New Guinea societies the younger men tend to own larger numbers of more bird of paradise species because they participate in more sing-sings. Women may don plumes and dance but are mostly married individuals and wear the plumes belonging to their spouses (Healey 1990).

Because of the great value of bird of paradise plumes, the men of New Guinea jealously guard their knowledge of display areas, where the birds can be predictably culled. In the case of species in which numerous adult males may congregate at a traditional lek, year after year or decade after decade (as with some plumed *Paradisaea* species or Stephanie's Astrapia), ownership of this 'resource' is much like a valuable holding of stock that can be passed on from generation to generation. The birds are acknowledged as such, and traditionally the harvest of birds was and (in remote areas) is carefully monitored so that sufficient fully-plumaged adult males are left to maintain the lek site and to reproduce there. In some areas female-plumaged birds are not hunted at all or such activity is carefully restricted and bag limits on the number of male birds taken in a season are applied (Healey 1990). For much of interest concerning the use and significance of bird of paradise plumes, hunting, and the development and structure of trade in them by the Maring People of Papua New Guinea see Healey (1990).

Where there are denser human populations, the land is traditionally rigidly owned by clans, families, and individuals (there is minimal government- or publicly-owned land in Papua New Guinea). Ownership is in fact highly complex, involving various resource rights. In such areas the rightful ownership of any bird of paradise display lek, court or tree, or nest contents is widely acknowledged and is usually respected, but some poaching certainly takes place. If a man is found to have poached a land-owner's bird, a local investigation and discussion usually take place and, unless skilfully eloquent, the poacher is usually made to hand over the bird and/or adequate compensation. If he expended considerable energy in reaching the bird, building a blind, and waiting for the kill, these efforts are taken in consideration and he may be required to pay all or part of the estimated value of the bird while being permitted to keep it. In forests more remote from habitation it is easier to obtain birds surreptitiously without drawing the attention of the rightful owner(s). New Guinea land tenure and hunting rights have been refined to the present highly-territorial status quo by peoples that have diverged into hundreds of distinctive linguistic and cultural groups over thousands of years of geographical and cultural isolation.

Which species of birds of paradise are in vogue as decoration varies with time and of course with geography. The skins of all of the New Guinea mainland species of *Paradisaea* are popular and those of the rare Blue Bird are particularly sought by some groups who pay high prices for them. Elsewhere the central pair of tail feathers of the Ribbon-tailed and Stephanie's Astrapias (Fig. 2.1) and, more rarely, the two *Epimachus* sabretails are favoured, if not the height of fashion. Plumes of other species of birds of paradise are used only occasionally compared with those mentioned above. In parts of the highlands the unique occipital plumes of adult King of Saxony Birds and the breast-shield of Superb Birds are in most demand. The former are used for head and facial decoration (Fig. 7.1) and the latter specifically and almost ubiquitously as the central adornment to the fascinating human-hair wigs of the Huli People of the Southern Highlands of Papua New Guinea (Fig. 7.2). The occipital plumes of the King of Saxony Bird are particularly difficult to obtain because, while female-plumaged birds are numerous and are attracted to fruiting trees, vines, and fruit baits set in the forest, adult males more rarely visit them (Majnep and Bulmer 1977). The latter spend most of their time calling from uppermost forest branches, only coming lower to feed and to court females for brief spells.

A novel avian counterpoint to the human acquisition of the rare and bizarre King of Saxony plumes was recently found at the Tari Gap, Papua New Guinea. Here males of the rare Archbold's Bowerbird *Archboldia papuensis* construct and maintain large mat-like bowers, which they decorate with fern fronds, orchid stems, glossy black beetle wings, pieces of fungus, tree resin, and, apparently as important as any other decoration type, the occipital plumes of the King of Saxony Bird (Frith and Frith 1990b; Frith *et al.* 1996). While no Huli man was found who even knew of this bowerbird, let alone its decoration-hoarding habits, a single young boy was eventually located who had been cashing in on his knowledge of this avian behaviour for several seasons. He had been secretly and systematically revisiting several bowers he had discovered for himself and was harvesting the plumes. These he then sold to men at the Tari markets and from here they were traded on to men of the Mount Hagen area where the popularity of this bird's plumes assures higher prices.

The vogue for the wearing of plumes of birds of paradise as hat decorations by women of Europe and America from the 1880s through the 1920s is detailed elsewhere (see

146 The Birds of Paradise

7.1 Men of Mendi, Southern Highlands, Papua New Guinea prepare for a 'sing-sing' in which King of Saxony Bird plumes feature conspicuously in their head-dresses. Photo: CF.

Chapter 2). After the 1920s, few people outside Papua New Guinea and Indonesia wore bird of paradise plumes. A colourful and conspicuous exception remained in Nepal until recently, however, in the long-continued use of the yellow flank plumes of the Greater Bird in ornate headgear worn by men of the Court of the Royal Family. The 1957 coronation of King Mahendra incorporated the particularly long plumes of several of these birds (see Ripley 1950; Swadling 1996).

An article published in the New York Times of 2 May 1956 relates an interesting story regarding bird of paradise plumes, the ornithologist E. Thomas Gilliard, international

7.2 A Huli 'wigman' of the Tari Valley, Southern Highlands, Papua New Guinea. The traditional centrepiece of the human hair wig is the breast-shield of an adult male Superb Bird—augmented here by lorikeet tails and Sulphur-crested Cockatoo crest feathers atop the Superb Birds cape with Raggiana Bird flank plumes above. The necklace is of cassowary wing quills. Photo: CF.

diplomacy, and the Nepalese Royal Family (Rosenthal 1956). Gilliard was well aware of the fact that the kings, prime ministers, and generals of the Nepalese Royal Court had traditionally worn the plumes of birds of paradise in head-dress for centuries. While visiting New Delhi in India, he learned that since the long-standing ban on international traffic in bird of paradise products the plumes of the Nepalese Royal Household had inevitably deteriorated. The impending coronation of Mahendra, ninth King of the Shah Dynasty, made Gilliard think of a package of immaculate bird of paradise plumes seized long ago by USA customs officials and now stored in the basement of the American Museum of Natural History (Gillard 1957). He suggested to the American Embassy in New Delhi that these plumes be obtained and sent as an addition to the formal gift being made for the King's coronation (an autographed picture of President Eisenhower). The plumes were duly recovered from the Museum and President Eisenhower's personal representative, Mrs Robert Low Bacon, took them to Kathmandu and the King. This most appropriate and timely of gifts was warmly received.

Hunting birds of paradise

Birdwatchers spending time in the immediate environs of villages in the uplands of New Guinea quickly come to notice that birds other than domestic fowl are extremely difficult to approach and observe. A vital part of growing up for every young New Guinean boy is opportunistic hunting with a slingshot or a hurled stone. Nutritional protein is limited in rural communities in New Guinea, and birds of any size, eggs, and nestlings are eagerly collected and consumed. By age 8 or 10 boys have graduated to their first bow and arrows, and the hunting life becomes ingrained in most rural youth.

Hunting of birds and other wildlife remains a lifelong pursuit, however. The bows of older men are made of the dense and heavy wood of the black palm. Arrow shafts are made of straight lengths of narrow 'pit pit' cane *Saccharum* sp. Those for general bird hunting are multi-pronged with sharp slivers of bamboo.

Birds of greater significance such as cassowaries (for food, feathers, and bones) and larger pigeons (for food) and birds of paradise (for plumes) are often actively hunted by stalking or by more sophisticated means. In species where the traditional display site is accessible, hunters construct a hide (blind) of foliage upon a frame of supple sapling trunks or branches tied together with vines. Within this they wait with bow and arrow at-the-ready and pre-aimed at the display perch. If the hide is close enough a blunt-or club-tipped arrow specifically made for shooting birds of paradise is used to kill or stun. This avoids soiling the bird's plumage with blood. To enable the hunter to have his arrow ready at the required alignment and fire it with minimum movement a tube of bamboo or tree bark may be attached to the hide and pre-directed at the perch (Fig. 7.3).

In cases where a desired species cannot be approached, reached, or successfully killed at his traditional perch the hunter must employ techniques involving a more profound knowledge of the bird's habits and behaviour. The hunter may know of a tree bearing fruit favoured as food by the species, or of a forest stream or pool frequented by the species for drinking and bathing (particularly on hot days in the dry season). If so, he can build a blind close by and also cleverly construct a convenient perch to attract the bird to a spot where it can be readily shot (Fig. 7.3). Alternatively, a bunch of enticingly-attractive ripe wild bananas, pandanus, or other fruit might be tied in front of a hide constructed in the vegetation beneath the desired bird's display perches.

Traditional methods other than the bow and arrow may be employed by the hunter concealed within his hide. A trap baited with some favoured food and with a spring-loaded or string-pulled door might be used near a display perch or beside a forest pool. A simple

7.3 A hide, or blind, of foliage at a forest stream pool with a bamboo tube and branch attached to the front of it. The white arrows show the hide and bamboo tube. Birds visiting the pool are shot by bow and arrow via the bamboo tube as they perch upon the convenient branch provided. Photo: courtesy of C. Healey, with kind permission.

noose laid upon, or a drop trap or net suspended above, a favoured perch or fruit bait, or upon the ground of a display court may be employed. These will catch the bird when quickly activated by the hidden hunter. Early European bird collectors who spent long periods in the field with Papuan hunters write of men waiting patiently and silently for up to two days to shoot or capture a bird of paradise from a blind. This seemingly excessive effort is understandable when balanced against the huge material compensation offered in return for the bird. Remember that the forest societies of New Guinea were (and in many cases still are) rich in communal values but very poor in material possessions. The small material rewards offered by traders to native plume hunters were, to them, equivalent to a 'king's ransom'.

The above sit-and-wait hunting techniques are time-consuming and tedious but are much preferred for the killing of plumed birds of paradise. This is because the hunter can immediately retrieve his valuable acquisition and avoid loss to poachers, wild dogs, native predators, or the possible damage to the plumes by the environment. When opportunity permits, a nestling or young bird may be held and made to call in alarm so that adults attracted by them might be shot. Alarm calls may also be imitated by the more knowledgeable hunter in order to attract birds into range. Finally, bird-lime (sticky tree-sap usually from the bark of a *Ficus* fig) may be applied to favoured perches in display or food trees; the hunters then return to the 'limed' perches to harvest their catches. This method is not usually applied when the aim is to acquire plumes, however, as it can soil the birds' plumage.

Some sophisticated and highly effective means of capturing birds are easily implemented by the knowledgeable. During a brief stroll and discussion with the senior author, Papua New Guinean naturalist Joseph Tano plucked a terrestrial fern frond in passing in response to a question about the preparation of nooses for trapping birds. As he walked he stripped the almost horse-hair-fine, shiny, smooth, central frond stem of the leafy pinna, made a tiny slit in its thicker end with his fingernail and threaded the other end through it. The result was a small and inconspicuous noose that, once anchored to a perch or fruit bait, would slip tightly closed about a bird's toes and claws. A version of this very technique is one commonly used today by falconers and raptor biologists to catch birds of prey but involves nylon nooses placed atop a cage containing a live prey animal to attract the predator.

An impressive technique for obtaining Twelve-wired Birds has been recounted in the literature (Wallace 1869; Bergman 1959). Local men that know where birds roost at night in the heavily-foliaged crowns of sapling trees wait for a night lacking moonlight. Having located a sleeping bird and, noting its height above ground, they fell, clean, and sharpen a suitable tree trunk close by. This is then quietly and carefully carried to the roost tree and speared into the ground immediately adjacent to it. With his companions holding the pole upright the catcher climbs until level with the sleeping bird, then is gently swung towards the sleeping bird until he can grab it.

Elsewhere, men locate the roost sites of Twelve-wired Birds by looking for forest floor accumulations of their bright droppings which are typically reddened by pandanus fruit. The hunters conceal themselves near the roost tree to note which branch is used as the bird goes to roost and then return at night to climb gently to the branch to place a cloth over the sleeping bird (Guillemard 1886). The Kalam people of Papua New Guinea similarly erect timber ladders beside pandanus trees in which they have observed sicklebills roosting so that they can return at night to climb to and grab the sleeping birds with minimum disturbance (Majnep and Bulmer 1977).

The Biagge people of the upper Mambare Valley, Papua New Guinea, deploy a hand-woven net (much like a locally-manufactured version of a Japanese mist net) across the traditional escape route of male Raggiana Birds in their lek tree. They then pull on a rope attached to branches atop the tree (simulating a predator attack), which frightens the males into evacuating the lek. At the same time another person deploys the net, and the birds, following their escape route, fly right into it (BB).

Needless to say, the availability of shotguns is rapidly replacing traditional hunting techniques except in the remotest parts of New Guinea. An unfortunate effect of what has been the relatively instantaneous modernisation of New Guinea is that modern travel networks and technology have far out-paced the ability of resident landowners to protect and defend what is traditionally and rightfully theirs. Thus groups of people driving the highways with shotguns can simply stop in a forested area to shoot a few birds and be on their way with little fear of witnesses and reprisals. In this way the most important traditionally-vested interest in the sustainable harvesting of birds is broken down by poachers with no such long-term interest. This sad state of affairs is to some extent presently and thankfully confined to the geographically-linear distributions of coastlines, rivers, and interior roads.

Although shotguns are a boon to hunters in search of wild pigs, a most regrettable aspect of the use of shotguns for plume hunting is that it greatly devalues the need for people to have to learn the natural history of the birds. Thus it is ironically the case that while an increasing number of people of the developed world are eager to learn more about the ways of wild birds of paradise, the traditional landowners may be losing their close association with the birds and their habits.

Mythology, magic, poison, and the unknown

Mythology, if not magic, was already attached to the first few birds of paradise specimens arriving in the western world in the 1520s. The Portuguese sailors brought with them to Europe the story that these birds never alight but instead float about in the air, like ethereal fairies, until death. The scientist of the day had to provide a theoretical framework to support such a novel avian lifestyle, including explanations of how they reproduced and fed themselves. The suggestions that were made now seem utterly absurd but were long the subject of hot scientific debate among the ornithologists of the day. Here are two examples. The birds fed upon dew and mated as they floated in the breezes. Males had a depression in the back into which the females laid their eggs and the males then used their elongated wire- or thread-like central rectrices to entwine their mate as she dutifully sat upon his back incubating her clutch. It was also suggested that the wire-like central rectrices served to anchor these floating birds to a tree limb, in the manner that a mooring line does a hot air balloon or airship, as they rested in the air.

But people had created myths about these birds thousands of years before the remains of Magellan's fleet reached Tidore. Men of the Kalam people of the Kaironk Valley, Schrader Range, Papua New Guinea, hold that if one dreams of the long-tailed birds of paradise (astrapias and sicklebills) then one dreams of women, whose souls change into these birds. They unflatteringly observe that these birds are black and brown, the colours of women, whereas the red and green of the parrots are the colours of men. The plumes of these birds of paradise are also thought to be like women (the flank plumes being like women's skirts) because when they are new they are beautiful, 'but after a season or two they are faded and tattered, just as a young woman may look beautiful but after she has had one or two children her beauty has gone'. But the feathers of the parrots stay good for several years (Majnep and Bulmer 1977).

Another Kalam story relates to how a different bird of paradise keeps its plumage neat and tidy. There is apparently some debate as to where female and female-plumaged male Lesser Birds roost. One view is that they do so inside the dry stems of giant ferns which have bent over to be roughly horizontal. The birds are supposed to clear out thoroughly the inside of a stem and then enter it from one end to sleep and exit at the other and so avoid disturbing or soiling their feathers. It is also said that they use hollow dry bamboo stems in the same manner. Some Kalam people consider this erroneous and believe the birds sleep in low trees, including pandanus, and remove twigs and foliage immediately about their roost site 'so their feathers don't catch on these, and so that insects will not hide there and attack their plumes' (Majnep and Bulmer 1977). The latter clearly refers to adult males and the story possibly derives from people observing that males on their leks defoliate their perch areas (see Chapter 5). In view of this it is just possible that birds might clear the immediate area of their roost perch, but it is unclear as to why they would do so.

A Kaironk Valley story delightfully links the court-clearing parotias with the maypole-bower-building Macgregor's Bowerbird *Amblyornis macgregoriae*. It recounts that male Macgregor's Bowerbirds first decorate their bowers with flowers and fruits and then invite parotias to come and dance there and be presented with these attractive decorations as gifts. In return the parotias invite the bowerbirds to their courts of cleared ground but there insultingly offer only pig dung as gifts (Majnep and Bulmer 1977). This story is based on the actual differences between the respective bird's bower and court decorations (see Lawes' Parotia, Chapter 9).

New Guinean men may perform magic rituals or speak incantations prior to and during the hunting of birds of paradise in order to keep themselves safe, make their arrows fly true, or their traps work successfully. For example, Kalam men wishing to shoot adult male Stephanie's Astrapias at a lek first shoot the stem of a particular reed over the lek tree with a bow. This assists ghosts or spirits to get rid of troublesome goblins (*masalai*) and prevents the thoughts of any of the hunters' relatives knowing of the hunt and so following or distracting him. Much magic is involved in the hunting of the now particularly important Raggiana and Lesser Birds (Majnep and Bulmer 1977).

Huli Wigmen of the Tari Valley, Papua New Guinea, enhance the performance of their kundu drums (see Fig. 2.1) with bird of paradise feathers. A man takes a small bundle of delicate flank feathers of the adult male Blue Bird and carefully burns them while hollowing a tree limb with fire to form the drum. This is believed to impart to the drum the clear and penetrating sound of the bird's voice (Parer 1981).

The birds as modern symbols or as featured on products

Apparently, birds of paradise have been sustainably exploited in New Guinea for millennia. This was followed by the growth of a regional, then international, trade in the birds, for their plumes. Later still, the beauty and mystique of these birds were exploited by the publication for profit of folios of paintings of them by ornithological entrepreneurs Daniel Elliot and John Gould.

For many years the postal services and governments mints of Papua New Guinea and Indonesia have issued sets of lovely postage stamps and several coins depicting birds of paradise (Fig. 7.4) that have proved highly popular with philatelists and coin collectors around the world. The stamps contribute directly and indirectly to national economies

7.4 (a) Bird of paradise first day cover set of Papua New Guinea postage stamps, designed by W. T. Cooper, depicting (left to right) Stephanie's and Ribbon-tailed Astrapias, Goldie's Bird, and a Carola's Parotia; (b) set of eight forged and officially unsanctioned 'State of Oman' postage stamps of tropical birds including (from top left to bottom right) the Superb (2), Lesser (3), Goldie's (5) and Magnificent (6) Birds of Paradise (from Gould and Sharpe 1875–88). Reproduced with kind permission of the Papua New Guinea postal service and the knowledge of the Sultanate of Oman postal service.

and to awareness of the birds, as do other products depicting them. In addition, some other countries, often very distant from Papua New Guinea and Indonesia, have encouraged philatelic sales of their own postage stamps by using images of these exotic birds. The lovely images these birds provide have even led unscrupulous people to forge postage stamps. The supposed postage stamps of the State of Oman shown in Fig. 7.4 are actually forged fakes that have nothing to do with the Sultanate of Oman or its postal service (Q. bin Yousref *in litt*).

Most recently, the birds of paradise have come to be used as symbols in a range of regional enterprises. Papua New Guinea produces a great diversity of publications and consumer goods that boast the beauty of their birds of paradise as both a national and commercial logo. The species by far the most frequently used in this promotional way is the Raggiana Bird which is endemic to Papua New Guinea. The National Emblem of Papua New Guinea and its National Assembly is a stiffly-stylised, adult-plumaged male Raggiana Bird of Paradise perched in display atop the centre of a horizontal kundu drum and ceremonial spear (Fig. 7.5). The National Flag is divided into an upper red and a lower black triangle with the red triangle decorated by a stylised, soaring, adult male, plumed bird of paradise and the black one with the stars of the Southern Cross.

The handsomely-designed Indonesian 20 000 rupiah bank note (Fig. 7.6) is decorated with an adult male Red Bird (holding its plumes in a somewhat atypical position). The 50 rupiah coin depicts an adult male Greater Bird in display.

The corporate symbols of Air Niugini, the Papua New Guinea national air carrier, is a stylised plumed bird of paradise, and this and images of actual birds may be seen almost anywhere associated with airports and air travel in Papua New Guinea. The most globally-exposed, novel, and widely-admired representation of the Papua New Guinea national symbol in recent years must certainly have been

(a)

(b)

7.5 (a) The flag and (b) National Emblem of Papua New Guinea, first published in the Papua New Guinea Government Gazette 1 July 1971.

the skilfully designed exterior livery of at least one of Air Niugini's larger aircraft (Fig. 7.7).

While Papua New Guinea consumer products bearing images of birds of paradise are far too numerous to enumerate here, the range of vacuum-packed export coffee is worthy of note as it depicts several appropriate paradisaeid species particularly well—we say appropriate because both the Blue and Raggiana Birds are endemic to Papua New Guinea and can be found in or around midmontane coffee plantations (the third species being the Lesser Bird).

In the 'far north' of Queensland during the 1990s Victoria's Riflebird has become an 'exploited' symbol repeatedly featured by local councils, tourist, and other commercial enterprises. As a stylised logo the adult male in high intensity courtship display appears on the doors of taxicabs of the coastal town of Ingham (between Townsville and Cairns) and a number of other businesses in the area use

Birds of paradise in human tradition and culture

7.6 The Indonesian 20 000 rupiah bank note depicting an adult male Red Bird which is endemic to the islands of Waigeo and Batanta, Irian Jaya. Reproduced with kind permission of the Indonesian Government.

7.7 An airbus of Air Niugini, international air carrier of Papua New Guinea, painted as an adult male Raggiana Bird of Paradise. Photo: CF.

similar images of this Australian endemic bird of paradise.

Birds of paradise in contemporary and future human culture

Methods employed in the hunting of birds of paradise are inevitably moving quickly away from the traditional, and the use of plumes as bridewealth is certainly diminishing. There is every reason to believe, however, that the long-formed strong and close cultural ties between the people of New Guinea and these stunning birds will continue in some form, although western material culture continues to penetrate even into the remotest valleys. The demand for plumes certainly continues as many are required for sing-sings on a grand scale, such as the annual gatherings of tribes at Goroka and Mount Hagen, as well as for smaller and local performances for tourists.

For many young New Guineans, birds of paradise will best be known, however, as a logo on a beer can (South Pacific Beer), the mascot of a rugby team (the Kumuls), or perhaps one day as a symbol on a future baseball cap. The serious question that remains is to what extent can the ancient traditional customs that link humankind with the birds of paradise be maintained in today's world? That is a question for the social scientists of the next few decades. The survival and documentation of the associated oral knowledge of New Guinea's rainforest biodiversity, traditional medicines, and customary uses of the plants and animals, will also be an issue that should be addressed before it is too late. It is of more than esoteric interest, as the global society broadens its search for novel products and better practices on a planet crowded with people and evermore impoverished of forest and wildlife.

8

Conservation

During the first decade of the twentieth century, hundreds of thousands of skins of adult male birds of paradise were exported from the island of New Guinea. They were destined to adorn the hats of European and American women and head-dresses of sultans and other Asian aristocracy (Gilliard 1969; Swadling 1996). The millinery trade was booming, no respectable person appeared in public without a hat, and no fashionable woman wished to be seen on the street without a hat sporting aigrettes, plumes, or even whole birds (especially the brilliantly-metallic male hummingbirds of the neotropics).

It was during this period, and largely because of the profligacy of the millinery business, that the bird conservation movement was born. After considerable grass-roots agitation, laws were enacted to ban the collection, sale, and exchange of most kinds of wild birds, and since 1924, the birds of paradise have been formally protected from commercial exploitation throughout their range. This was a marvellous achievement and is testament to the remarkable appreciation of birds that had arisen in the West.

Today, all birds of paradise are formally protected by the fauna protection acts in Papua New Guinea, Australia, and Indonesia, and the Convention on the International Trade of Endangered Species (CITES) worldwide. That being said, it is becoming evident that the future of the birds of paradise is probably less certain today that it was a century ago, when wholesale commercial slaughter was the norm. This is because of large-scale resource development and growing human populations in the two tropical countries that are home to most of the species of birds of paradise. In the following pages we address the conservation of the Paradisaeidae as an ongoing challenge. To do this we treat the threats, the particular species that are vulnerable, and opportunities for action. The good news is that no single species appears to be under direct threat of extinction at this time. But at the same time we should not be complacent, given the growing pace of change, especially on the island of New Guinea.

Threats to birds of paradise

Populations of birds of paradise potentially face threats from a familiar range of culprits—habitat loss, poaching and culling, introduction of exotic predators, human population growth, and (over the longer term) global climate change. Although no single force may be sufficient to jeopardise any single species seriously, in combination, they may constitute a serious impact. We shall consider the threats separately and collectively, in an effort to predict situations worthy of concern, either now or in the future.

Habitat loss

Throughout the tropical world, habitat loss threatens the very survival of tens of thousands of species of plants, mammals, birds, and fishes, not to mention the millions of lower forms of life, most of which have not yet been enumerated or described. It is thus natural that we put 'habitat loss' as the main threat—at least the most obvious one. For the birds of paradise, most of which are habitat specialists by virtually any measure (see Chapter 3), habitat loss is a real issue, especially for the species with the most restricted ranges or most specific habitat requirements.

Some uncertainties remain about the nature of habitat specialisation in birds of paradise, and, indeed, about the stability or fixity of what we call tropical forest habitats. In particular, researchers are attempting to come to grips with the degree of disturbance (both natural and human-caused) that tropical forests in Australasia have suffered over the last several thousand years. In addition, there is considerable interest in the ability of these forests to recover from disturbance—the thought being that if the forests can recover from disturbance, then so can the birds. One must assume that the birds and forests have evolved together, with the birds adapting to the rhythms and cycles of the forest.

It is instructive to examine the highly-disjunct and dissected paradisaeid populations that inhabit eastern Australia. The four species of birds of paradise that inhabit the continental island of Australia occur in a broken string of populations along the east coast from Cape York, in the far north, southward to the Sydney area of New South Wales. Whereas the populations of the two species found at the isolated and undeveloped northern tip of Cape York Peninsula (Trumpet Manucode and Magnificent Riflebird) remain unaffected by habitat loss, those of the more southerly two species (Victoria's and Paradise Riflebirds) have declined.

The world distribution of the Paradise Riflebird (see page 329) coincides with that of eastern Australia's most intensely-settled and developed area. As a consequence, the subtropical lowland habitat of this southernmost of all birds of paradise has been grossly reduced by clearing of rainforest. The species is now largely restricted to small pockets of forested uplands (Blakers et al. 1984).

Victoria's Riflebird of the 'wet tropics' of northeastern Queensland has also suffered significant loss of range owing to the clearing of tropical rainforest, notably in the lowlands and on the Atherton Tableland. The recent protection of remaining tracts of rainforest within the Wet Tropics World Heritage Area has, however, greatly increased the chances for long-term survival of the species.

Both Paradise and Victoria's Riflebirds will attack cultivated fruit crops and to this day numerous birds (but not significant to species) continue to be killed in agricultural control programmes. Permits can be obtained for the destruction of these species when they are deemed to pose serious threats to fruit crops. In practice, most birds are killed illegally by farmers on small properties adjacent to rainforest.

Both the Trumpet Manucode and Magnificent Riflebird populations on Cape York Peninsula represent distinct endemic subspecies of species widely distributed in New Guinea (see pages 230 and 317) and therefore the future of the species is not dependent upon these isolated populations. As distinctive subspecies, considered to be full species by Cracraft (1992), these two Australian populations occupy diminutive ranges (Beehler and Swaby 1991; Frith 1994a), far smaller than their two endemic Australian small riflebird relatives, and as such are vulnerable to any clearing of tropical rainforest on Cape York Peninsula (D. Frith and C. Frith 1996).

In Papua New Guinea, large-scale human-caused habitat loss has been minor until recently, and large-scale habitat conversion did not commence until the rapid growth of foreign national logging and agricultural operations in the mid-1980s. In fact, the main cause of forest loss in New Guinea was and is swidden (slash-and-burn) agriculture. The key distinction is that swidden clearing tends to be patchy and small-scale. Only in situations of high population density (as occurs in the Chimbu Province of Papua New Guinea) does it pose a threat to entire forest tracts. Now in 1998, in a disturbing turn of events, major portions of Papua New Guinea's lowland and hill forest environments are being scheduled for industrial-scale selective logging. The rate of extraction has been growing and has been assessed to be unsustainable. At current rates, all of the accessible forests will be cut-over in the next decade. The more remote forests will certainly be logged in the first or second decade of the twenty-first century. In addition, the pace of habitat conversion from other (non-logging) activities will seriously impact every province of the country. Predicting the where and when is impossible, but if current trends continue, this outcome is inevitable. It is also sobering to note that as early as the 1970s, local informants in the Kaironk Valley, Papua New Guinea, recognised and commented on the reduction of populations of Stephanie's Astrapia and the Brown Sicklebill (Majnep and Bulmer 1977)—in the absence of any significant western influences aside from some mission activity, presumably.

In Irian Jaya, the pace of large-scale development has been slower than in Papua New Guinea, mainly because of the Indonesian Government's focus on other provinces as well as its desire to manage development through Indonesian initiatives (or in close partnerships, as exemplified by the vast P.T. Freeport mine operation in the western Snow Mountains). Given the startling pace of economic development in Indonesia as of this writing, the large-scale 'opening' of Irian Jaya to development of all sorts cannot be far behind. The current proposal to dam the Mamberamo River in a large hydro-electric scheme (and thereby flooding a large section of the vast Lake Plain region) shows that development in various forms may soon be taking shape in Indonesia's easternmost province. Irian's provincial planning office has proposed extensive additional transmigration settlements in the southeast and west, and most of the lowland forest tracts have been leased as timber concessions. There is thus reason for concern for the future of the habitats of Irian Jaya's lowland and hill forest birds of paradise.

Two paradisaeid species, the Paradise Crow and Standardwing, inhabit only the Northern Moluccan archipelago (Halmahera, Bacan, Morotai, Obi, and Kasiruta). These islands are isolated and, for the most part, rugged, offering only minor opportunities for development. But given the size of Indonesia's populace and the entrepreneurial spirit of its people, developers will certainly be knocking on the door in the Moluccas as well.

What, then, will happen to the bird of paradise when the habitats are exploited by loggers, miners, oil-palm agriculturalists, commercial rattan-harvesters, and settlers from more populous regions? Some species, exemplified by the Greater Bird superspecies, will probably suffer little or not at all, whereas the future of some of the rarer or less widespread species may become threatened. This is because of effects of scale and demography.

For example, the Raggiana Bird inhabits a range that includes nearly two-thirds of mainland Papua New Guinea, from sea level to 1500 m elevation. This includes a range of original and secondary forest types. The species is known to nest in forest but also in isolated shade trees near villages; it is known to forage in remnant trees in areas dominated by cultivation. Given this breadth of habitat-use, it seems highly probable that the species will survive the type of habitat alteration that we foresee for Papua New Guinea over the next 20–30 years.

By contrast, a species such as Wahnes' Parotia, which is confined to a sliver of forest habitat between 1100 and 1700 m in mountains of the Huon and the Adelbert Ranges, is considerably more vulnerable, especially because it appears to require the interior of undisturbed forest. If the type of logging pressures that we have witnessed in Madang Province (where the Adelbert Range is located) are indicative of the future, the forests that today support Wahnes' Parotia should not be considered safe just because they are largely unthreatened today. The experimental helicopter-logging operation that was initiated in the hill forests of the eastern Huon Peninsula in 1990 showed that even the more rugged terrain can be exploited, given suitably-high timber prices. The helicopter-logging can be managed in a manner that is relatively benign ecologically, but it does not have to be so, given different management prerogatives.

We thus see habitat loss as a long-term threat to birds of paradise and all of the larger forest vertebrates. One unanswered question that remains relates to time lags in disturbance effects. What are the prospects of the isolated populations of riflebirds in eastern Australia? They may appear healthy today simply because of time lags in population decline or impacts related to forest fragmentation (heightened impacts of brood parasites, nest predators, exotic pests and diseases, fire etc.). It is not always safe to predict the future based on limited information about the impact of recent changes.

Hunting and the trade

Adult male birds of paradise have been hunted for their plumes for hundreds of years and the demand for skins continues in East Asia—mainly as exotic collectibles for regional sale.

The indigenous domestic trade of skins for traditional use appears to be minimal (or at least for the most part undocumented) in Irian Jaya, but it continues in Papua New Guinea, where annual or periodic 'sing-sings' both for traditional purposes and to satisfy tourist demand has created a market for the skins. Traditional trade and exchange of birds of paradise is legal in Papua New Guinea; commercial sale is illegal. In spite of this, a black market trade continues, especially between the 'producer areas' in highlands fringe areas (where human populations are low and paradisaeid populations are high) and the 'consumer areas' in the populous central highlands. In some instances, the plumes or skins are simply offered for sale in town markets (illegal but overlooked by the police). In other instances there appears to be an underground trade that involves recruitment of lowland collectors by travelling highland entrepreneurs who visit source areas to stimulate the commerce. It is strictly cash for plumes. The sale of paradise skins can represent an important source of cash to isolated rural villages in Papua New Guinea. In many cases the local hunters are quite willing to sell their paradise skins for cash, for the price paid is usually well above the local market price (if there is a market at all) and in some instances the species being cropped are not locally used for adornment (called *bilas* in Papua New Guinea) in traditional celebrations and thus have no special cultural significance.

In the early twentieth century, as the millinery demand for skins declined, a much more restricted clientele was clamouring for live birds of paradise to enliven personal, royal, or public menageries around the world. This has remained a fairly small trade, although it has perhaps grown in the last two decades. Some of it is legitimate, but the major portion, especially today, is an illicit and little-documented black market business that operates primarily between Irian Jaya and distributors in several of the major cities arrayed between Jakarta, Singapore, and Hong Kong. Much of the big money is in the international trafficking of these birds, live, but there is probably a larger volume of trade within the confines of the Indonesian archipelago—where the cage bird trade thrives. Commerce of plumes and live birds in Irian Jaya is apparently an institutionalised underground operation managed (as personal income-enhancing side-businesses) by civil servants, the military, and private entrepreneurs (Rumbiak 1984). As happens in Papua New Guinea, the laws and regulations are simply ignored, but this occurs on a larger scale and on an inter-regional (and international) basis.

What is the impact of the current domestic and international trade in both live and dead birds of paradise? Although it is difficult to measure directly, it appears that the assessment made by Ernst Mayr half a century ago still holds—that the negative impacts are apparently negligible to the long-term survival of birds of paradise. As stated by Gilliard, Mayr, and A. E. Pratt long before them (cited in Gilliard 1969), so long as large tracts of suitable forest remain, there is little chance that collecting for the cage bird and plume trade will create more than a minor and very localised threat to paradisaeid populations.

There may be negative secondary effects, however. Roadside hunting/poaching of birds of paradise in some highland areas of Papua New Guinea has much reduced the visibility of plumed males from these now 'famous' sites (exemplified by the Tari/Ambua/Doma Peaks area). While there is no indication that this is much more than a local phenomenon of no demographic significance, it can pose a significant threat to the nature tourism or eco-tourism trade that, in turn, has ramifications for conservation, at least in Papua New Guinea. Being able to show Papua New Guinean decision-makers and wealthy foreign tourists beautiful birds of paradise will mean that birds of paradise have a commercial value as wild and free-ranging living things. In addition, the knowledge and appreciation of these creatures will spread to more influential people around the world. This knowledge can translate into power for preservation. Look at how the knowledge and appreciation for the Tiger and the Mountain Gorilla have fostered their conservation. Thus if key local or international decision-makers never see the wonders of birds of paradise, will they ever be moved to take action to save them?

Introduction of exotic predators

Feral cats, dogs, and rats pose serious problems to native populations of birds around the world, especially to those on islands. These three introduced mammals have had significant impacts in the Pacific as a whole, mainly in the form of extinctions of flightless and terrestrial bird species (Greenway 1967). Flannery (1995*a*,*b*) has more recently documented apparent island extinctions of mammals for the Melanesian islands north and east of New Guinea, brought about by the same factor, and certainly the same is well known for the Australian fauna. At this time no bird of paradise appears directly threatened by these exotic predators, but certainly terrestrial-displaying species might be vulnerable, and, in the future, new introductions to New Guinea (e.g. of macaques or other primates) could cause wholesale disruption of forest animal communities. Witness the havoc that has been brought about within the native bird community of Guam by the introduced Brown Tree Snake *Boiga irregularis* (Fritts 1988).

Human overpopulation

What will the island of New Guinea look like a century or two from now? A key to this will be the rate of human population increase over the next decades. Today the rate of increase in Irian Jaya is *c.* 2.5%, whereas in Papua New Guinea it is almost 3%. For Papua New Guinea, this rate of increase means that its population of nearly four million will double in 30 years. In Papua New Guinea, population increase is strictly the product of local fecundity. By contrast, in Irian, it is further influenced by transmigration resettlement. Transmigration is a scheme by which the Ministry of Population shifts population from overcrowded parts of Indonesia (in the west) to uncrowded provinces (in the east).

No matter how the population increases, its net effect will be the same—a heightened rate of habitat conversion and less forest for birds of paradise. Over the short term (decades) this will have a limited impact on the birds. Over the long term (centuries) this will certainly pose the greatest threat of all—as it does to the wildlife and forests of the entire globe. Habitat loss should, in fact, be considered nothing but a discrete effect produced by human population increase. If we wish to conserve birds of paradise for our great great grandchildren, our most efficient route of action will be the regional stabilisation of human population—i.e. fewer grandchildren.

Global change

If we are to expand our view over an even greater span of time—to centuries adding up to millennia, then a major threat to the future of the birds of paradise must be human-caused global climate change—in particular, the anthropogenic 'greenhouse' effect. The greenhouse effect is being caused by the release into the atmosphere of carbon dioxide and other greenhouse gases by automobiles and industries. It is postulated to cause warming of the earth, with attendant climatological and environmental effects that are difficult to predict with much accuracy (e.g. drought, flood, sea level rise, and more subtle effects).

How would this threaten birds of paradise? The main effects would probably be three. First and foremost, there could be a rise in annual temperature of as many as several degrees centigrade, which over a short period might cause a radical shift in the composition and location of montane forest types that are home to so many birds of paradise. Particularly vulnerable would be the many species that inhabit narrow elevational zones in the middle altitudes. The disappearance of the high altitude forests might also cause the extinction of the high elevation specialists. Global temperature shifts have happened many times before, but never in such a rapid fashion as is currently predicted. One wonders if the bird populations could adapt to such a rapid environmental change.

Second, the rise in temperature could have a widespread effect on the location and intensity of regional rainfall. Whereas some areas might become more rainy, others are likely to become drier, with the potential that humid forest is replaced by open woodland or savannas. This might threaten lowland species of birds of paradise.

Third, and linked to the second effect, there might be major changes in the El Niño/Southern Oscillation—a periodic pattern of sea water warming that at this time dominates major weather patterns over the entire tropical Pacific region every 4–6 years. Today New Guinea lies in the heart of the El Niño region. Disappearance or shift of the El Niño, or the far more likely growing frequency of it, might cause a shift in the pathway of major Pacific typhoons that can cause catastrophic damage to tropical forests.

It is difficult to envisage what the state of the world will be in two centuries time. But if we recall that humankind has been resident in Australasia for more than 50 000 years (White and O'Connell 1982), and that the evolutionary age of the birds of paradise perhaps exceeds 20 million years (Sibley and Ahlquist 1990), then a century or two is indeed a relative blink of the eye. Given that the global effects that we think we are seeing today took several hundred years to bring about, we need to think today about taking remedial actions whose benefits may require decades or centuries to manifest themselves fully. Does our global society have the patience and forbearance for such prescriptive solutions?

Factors in combination

Threatening synergies might be created when rapid population growth, large-scale development, and the expanded subsistence needs of a hungry human population are combined in a single society. It is certainly these sorts of combinations of threat that the birds of paradise will face either sooner or later, depending upon the pace of change. Predictions of wholesale species extinctions, over and above the background rate (extinction being a natural process), are now heard over and over in the literature of conservation biology (Ehrlich and Ehrlich 1981; Myers and Simon 1994; Lawton and May 1995). Might these events include birds of paradise? Probably so. Just which species are high on the list of potential losses is discussed in the next section.

Species in trouble

The most recent red data book for animals (Baillie and Groombridge 1996), published by the IUCN Species Survival Commission, lists six paradisaeid species as vulnerable: Macgregor's, Blue, and Goldie's Birds, the Ribbon-tailed Astrapia, Black Sicklebill, and

Wahnes' Parotia. Thus the international conservation community does acknowledge potential threats to birds of paradise, and has been examining threats species-by-species. We revise, update, and expand on that analysis here, mainly by looking at each species with regard to types of vulnerability (Table 8.1). These factors are the flip side of the conservation coin—those that complement the threats described in the preceding section. In its simplest form: environmental threat + species vulnerability = potential for extinction/extirpation.

Types of vulnerability

For birds of paradise, the most relevant types of vulnerability relate to circumscription of range, overall rarity, patchiness of distribution, and single-nation status. Range vulnerability is produced by a geographically-circumscribed species distribution. With a small range, some geographically-localised environmental catastrophe (volcanic eruption, forest clearance, sea level rise) might wipe out an entire species population. Overall rarity denotes a very low population density even in the heart of a species' range. Thus a species may be geographically widespread but at the same time have a global population that is smaller than a species with a very restricted geographical range. Patchiness of distribution is a geographically-sporadic form of rarity, in which local populations may vary from abundant to absent, in no pattern that is currently understood. Suffice it to say it is evidence of some latent weakness in a species' ability to maintain itself over a large range. Finally, single-nation status is a political category that addresses the vulnerability of a species whose future is inextricably tied to the future of a single nation. The Blue Bird of Paradise is a single-nation endemic, and its fate depends on the policies of a single nation—thus there is no margin for error.

Vulnerable species

Given the unwanted attentions visited on the birds of paradise by local hunters, plume traders, and nearly a dozen bird-eating raptor species, it is a pleasure to report that we can nominate only three species of the 42 that might be classified as currently vulnerable (Table 8.1). These are, from most to least threatened: Macgregor's Bird, Black Sicklebill, and Blue Bird. Certainly none of these three would qualify as endangered or near extinction. Thus, overall, our current assessment of the present state of the family Paradisaeidae is cautiously optimistic, at least over the short term. But let's focus on the bad news, briefly.

Macgregor's Bird is vulnerable for six reasons. Although it ranges from Irian Jaya to southeastern Papua New Guinea, its current global range probably encompasses less territory than a thousand square kilometres—perhaps even less. Worse, this global range consists of a series of discrete populations, most of which appear to be entirely isolated from neighbouring populations. The lack of morphological differentiation between these populations could at least suggest they have not been isolated for long, but that we don't know. The species is absent from a large section of suitable habitat in the middle of its range—evidence of past population extirpation (Barker and Croft 1977). Its reproductive cycle appears to be tied to the unpredictable fruit production of a favoured food plant (Beehler 1991a,b). It is large, edible, and hunted as game in at least some parts of its range (Beehler 1981). And finally, it inhabits a subalpine environment that may be warmed out of existence in the next century or two. These points, in sum, provide a cogent argument for the vulnerability of Macgregor's Bird.

At more than a metre from bill tip to tail tip, the adult male Black Sicklebill is the longest bird of paradise. It sports a magnificent pair of ornamental tail plumes that make splendid adornments to a traditional head-dress. It apparently must compete with its more successful cousin, the Brown Sicklebill, for habitat throughout the length of its range (Diamond 1972). These are some of the reasons we consider it to be a vulnerable species. Other reasons can be appended:

Table 8.1 Vulnerability of the birds of paradise.

Species	Range restriction	Habitat restriction	Population rarity	Geographical patchiness	Political endemic to	Hunted for plumes?	Hunted as game?	Vulnerability rating
Loria's Bird	—	++	—	—	—	—	—	Low
Crested Bird	—	+++	—	—	—	—	—	Low
Yellow-breasted Bird	+++	++	—	+++	—	—	—	Medium
Macgregor's Bird	++	+++	—	+++	M	—	+?	High
Paradise Crow	—	—	—	—	—	—	—	Low
Glossy-mantled Manucode	—	+	—	—	—	—	—	Low
Jobi Manucode	—	+	—	—	—	—	—	Low
Crinkle-collared Manucode	++	+	—	—	—	—	—	Low
Curl-crested Manucode	—	—	—	—	PNG	—	+?	Low
Trumpet Manucode	—	—	—	—	—	—	—	Low
Long-tailed Paradigalla	+	++	+?	+?	IJ	—	—	Medium
Short-tailed Paradigalla	+	++	?	+?	—	?	—	Low
Arfak Astrapia	+++	++	+	—	IJ	—	-	Medium
Splendid Astrapia	+	+++	?	—	—	—	—	Low
Ribbon-tailed Astrapia	+++	++	+	—	PNG	+	—	Medium
Stephanie's Astrapia	+	+++	?	—	PNG	+++	—	Low
Huon Astrapia	+++	+++	+	—	PNG	?	—	Medium
Western Parotia	+++	+++	?	—	IJ	—	—	Medium
Lawes' Parotia	+	+++	?	—	—	—	—	Low
Wahnes' Parotia	+++	++	+	—	PNG	—	—	Medium
Carola's Parotia	+	++	+	—	—	—	—	Low
King of Saxony Bird	—	++	—	—	—	++	—	Low
Magnificent Riflebird	—	—	—	—	—	—	—	Low
Paradise Riflebird	++	++	++	+	A	—	—	Medium
Victoria's Riflebird	+++	++	++	—	A	—	—	Medium
Superb Bird	—	+	—	—	—	+	—	Low
Black Sicklebill	+	+++	++	+?	—	++	+?	High
Brown Sicklebill	—	+	++	—	—	—	—	Low
Buff-tailed Sicklebill	—	++	?	—	—	—	—	Low
Pale-billed Sicklebill	+	++	—	—	—	—	—	Low
Magnificent Bird	—	+	—	—	—	—	—	Low
Wilson's Bird	+++	++	?	—	IJ	—	—	Medium
King Bird	—	—	—	—	—	—	—	Low
Standardwing Bird	++	++	+	—	M	+?	—	Medium
Twelve-wired Bird	—	+	—	—	—	+++	?	Low
Lesser Bird	—	—	—	—	—	—	—	Low

Table 8.1 (continued).

Species	Range restriction	Habitat restriction	Population rarity	Geographical patchiness	Political endemic to	Hunted for plumes?	Hunted as game?	Vulnerability rating
Greater Bird	—	—	—	—	—	?	—	Low
Raggiana Bird	—	—	—	—	PNG	+++	—	Low
Goldie's Bird	+++	++	+?	—	PNG	?	—	Medium
Red Bird	+++	?	—	—	IJ	—	—	Medium
Emperor Bird	+++	++	—	—	PNG	—	—	Medium
Blue Bird	+	++	+	+	PNG	+	—	High

Key: +s indicate the degree to which species accords with category. A = Australia; IJ = Irian Jaya; M = Moluccas; PNG = Papua New Guinea.

(1) the species has a very low density; (2) it inhabits only a very narrow elevational zone (mainly from 1800 to 2150 m); (3) it is probably hunted for the stewpot as well as for its plumes; (4) it is apparently patchily distributed, being absent from seemingly-suitable habitat; (5) its prime range lies entirely within the altitudinal zone favoured for human habitation and cultivation; and (6) it shows evidence of an attenuated geographical range (being absent from the eastern half of Papua New Guinea). The only factor weighing in advantage of the persistence of this species is the existence of large tracts of steep, ever-wet, and relatively-inaccessible habitat on the fringes of the New Guinean central cordillera. When one flies over sections of these outlying areas (many with little or no current human presence) one can again take hope for the future of this magnificent species.

The nearly mythical Blue Bird of Paradise suffers for many of the same reasons as the Black Sicklebill. The Blue Bird has an attenuated cordilleran range—its absence from Irian Jaya is a biogeographical mystery. It is hunted for its plumes. Its elevational range (mainly 1400–1800 m) is entirely within the zone most favoured for local human habitation and cultivation. It is patchily distributed (it is common on the south side of the Kokoda Trail and absent from the north slope) and it occurs in low population densities. This being said, most would agree that it is not as vulnerable as the Black Sicklebill, for several reasons. It is a smaller, less conspicuous species with higher population densities. It uses secondary forest and forest-edge and thus appears to be fairly compatible with human activities. It is not hunted as game (except incidentally). And its plumes are not as valuable as those of the Sicklebill. Still, the futures of the Blue Bird and the Black Sicklebill are probably closely linked. Majnep and Bulmer (1977: 72) write of the historical disappearance of the Blue Bird from the Kaironk valley, indicating that range reduction can occur even without western development and high population pressures.

Species to watch

Aside from the three prominently-vulnerable species discussed above, there are 14 species that will bear watching over the next decade or two. These include island forms (Standardwing, Wilson's, Red, and Goldie's Birds), forms from isolated mountain massifs (Arfak, Ribbon-tailed and Huon Astrapia, Western Parotia, and Emperor Bird), two species that inhabit remnant dissected habitats in Australia (Paradise and Victoria's Riflebirds), and an aberrant species with a widespread but patchy range (Yellow-breasted Bird). As a final footnote, Peckover (1990) expressed real concern for *Manucodia comrii trobriandi* populations, owing to habitat loss associated with remarkably-rapid human population growth on the Trobriand Islands off southeastern Papua New Guinea. These flat coral islands may also be threatened by sea level rise from global climate change.

None of these species is currently showing signs of threat, but because of their restricted distributions or the 'weakness' of their population (e.g. Yellow-breasted Bird), they may be vulnerable to major environmental changes. Before that time comes it is advisable that we make some effort to understand better the biology of these species, through field studies. It might then be important to develop conservation contingency plans for each. Finally, it may also be worthwhile to develop captive breeding programmes for them, with viable captive populations in a range of zoos and breeding facilities around the world (this has been shown to be possible to some limited degree for the Superb, Lesser, Raggiana, and Red Birds).

Taking action to conserve birds of paradise

We believe there already exist practical routes to fostering the long-term survival of all species of birds of paradise in the wild. Much of what appears below has been proven with other wildlife in other regions of the world. The key to success is not so much in the method as in its long-term implementation, and field implementation can be devilishly hard for a number of reasons. Thus we believe any progress with the conservation of birds of paradise will require significant progress over and above the rather paltry local successes that have been accomplished over the last two decades. We see three necessary activities to effect the conservation of birds of paradise, as follows: (1) creation of a system of large, locally-managed forest reserves; (2) appropriate grass-roots awareness programmes for local resource-owners; and (3) the full implementation of all current legislation protecting birds of paradise. We discuss each below.

Forest reserves

As argued by Beehler (1985b), we believe the birds of paradise should be used as a 'flagship' family for focusing habitat conservation efforts regionally and environmentally. In other words, we believe that by seeking to design a comprehensive system of conservation interventions to preserve healthy populations of all species of birds of paradise, we can thereby effectively conserve a major segment of the Australasian humid forest biota. The reserves we envision in New Guinea would be large multi-use areas supporting resident human populations.

It is notable that a range of experts, in a series of papers, has offered a listing of priority sites for habitat reserves appropriate to conservation of birds of paradise (Schodde 1973; Beehler 1985b; Diamond 1986b). This has led to some preliminary reserve designation in eastern Indonesia (Diamond 1986b), but practically not at all in Papua New Guinea (Collins *et al.* 1991). The synthesis of habitat conservation needs provided in the Papua New Guinea 'Conservation Needs Assessment' (Beehler 1993) was based at least partially on data on the distribution of birds of paradise, and thus provides an outline for paradisaeid conservation within Papua New Guinea. A 1997 biodiversity conservation priority-setting workshop for Irian Jaya (Burnett 1997) expanded on, but did not fundamentally alter, the original array of proposed Irianese reserves that grew out of Diamond's original analysis. The Northern Moluccas have been subjected to a similar planning exercise sponsored by the Indonesian government.

We have an array of proposals for biodiversity conservation in Papua New Guinea, Irian Jaya, and the Northern Moluccas, plus a system already in place for Australia. The question, then, remains how to foster timely implementation of the reserve system plans for Papua New Guinea and Indonesia. Because of the fundamentally-different systems of government and cultures dominating, the routes to success will certainly differ.

Indonesia, with its very centralised government and its government claim on virtually all wilderness lands, creates reserves by fiat. Superficially, this appears efficient and straightforward, and yet Indonesia's failure to defend the preservation of these reserves strongly against degradation by large development schemes currently renders their long-term protection uncertain. Some measure of local autonomy in the Northern Moluccas and Irian Jaya may help the matter—these eastern provinces have typically been viewed as resource reserves and open lands for resettlement of colonists from the crowded western Indonesian islands.

Irian Jaya today is in an ideal position to create a network of large forest reserves (Diamond 1986b). The critical hurdle to the long-term success of habitat conservation in Irian is the issue of traditional unregistered tenure over forest land. For just as the rural residents of Papua New Guinea recognise ownership of Papua New Guinea's rural lands, the culturally-similar Irianese residents 'own' vast tracts of land appropriate for the conservation of populations of birds of paradise. Unless these people are brought into the habitat conservation process, they will become a disenfranchised constituency that may eventually undermine conservation goals (as has happened repeatedly around the world). Resource managers in Irian Jaya must, then, be forward-looking in ongoing attempts at creating a reserve system. The Irianese living on the lands in question must be given the knowledge and incentive to become productive collaborators in the conservation process.

Papua New Guinea, which has been very forward-looking about conservation legislation, has very little to show for this attitude in the way of a significant network of habitat reserves appropriate for conservation of birds of paradise. And it is the issue of local landowners that has hindered this. In some ways then, Papua New Guinea is a useful test case for examination by Irianese conservation authorities, for it exemplifies a contrasting counter-example in which the local populace inhabiting and/or claiming wilderness tracts has been the main excuse used by national authorities to avoid creation of conservation reserves. The Papua New Guinea government argues that since 97% of the land area of that nation is traditionally owned by the rural populace, it is impossible for the national government to establish national reserves. Papua New Guinea authorities believe that local sovereignty must rule, and that any conservation must be locally owned and operated. This is the core of the 'Wildlife Management Area' concept (Paine 1991). The unspoken assumption is that it is thus in the hands of the rural landowners to take the initiative to create large-scale reserves of habitat for the purposes of conservation.

Thus in the examples of Papua New Guinea and Irian Jaya, we see two extremes in attitude that have hindered conservation—one very decentralised and the other very centralised. The first has failed to create a system out of deference to the rural populace. The other has created a system without consultation with the local decision-makers. In both instances, successful long-term conservation of habitat can happen only if there is the political will to make difficult and far-sighted choices in tandem with a consensus-building initiative with local landowners. The latter almost certainly must include some form of appropriate economic compensation to those who must give up certain rights to land development where habitat reserves are to be established.

Grassroots education and outreach

Any future local or national initiatives for conservation education at the grassroots level might benefit from the following conventions.

(1) The outreach should be made by local Melanesian field officers and their main goal should be to incorporate any conservation ideas into the context of the local culture. Outside and top-down efforts usually fail for lack of appropriateness.

(2) The basic needs of the local community must be recognised as primary, and any conservation goal must be subordinated to these and must grow out of these.

(3) One goal should be to educate the local resource-owners about how special their wildlife is when measured on a global perspective—giving the local people well-deserved pride in what they have.

(4) Another goal must be to help foster heightened communication about prudent land-use planning among neighbouring landowner groups in order to foster networks for land-use and conservation. The local decision-makers must be encouraged to think beyond their own valley and to see themselves as critical components in a regional ecosystem potentially threatened by uncoordinated development.

Enforcement of legislation

As mentioned earlier in this chapter, all birds of paradise are legally protected throughout their range—in theory, that is. Laws are on the books in Indonesia, Papua New Guinea, and Australia that fully protect birds of paradise from commercial exploitation and export. It is unfortunate that these laws are routinely ignored, and only rarely invoked, typically under unusual circumstances. In spite of Ernst Mayr's convincing argument that heavy cropping of plumed birds has minimal effect on paradisaeid populations, we believe that strict enforcement of the laws protecting birds of paradise is indispensable. It reinforces the rule of law that all civilised societies must maintain, and it spotlights the clear aesthetic value that society and the world at large place on the continued existence of these marvellous creatures. These aesthetic considerations are important, for they underlie all other efforts to protect the birds and their habitats. The laws are fair (they allow for traditional/customary use), and thus their enforcement places no undue burden on society or governments. Birds of paradise deserve the protection that forward-looking legislators envisioned for them decades ago.

Aviculture

Although not conservation, per se, the maintenance, in captivity, of birds of paradise, has proven valuable in learning about aspects of behaviour and biology of these birds difficult to determine in the wild. At some point in the future, aviculture may constitute an important tool in the restoration of certain paradisaeid populations. Although birds of paradise have been kept in captivity in the West for at least 200 years, much remains to be learned about care and breeding of them in the aviary.

Understanding the social structure, ecology, and behaviour of birds in the wild has provided new insights into the probable causes of previous captive breeding failures and appropriate actions are being taken to prevent them (e.g. Hundgen *et al.* 1991).

The results of a review of post mortems of birds of paradise that died in various zoos were that most had died from excessive iron accumulation in the liver (Assink and Frankenhuis 1981; Frankenhuis *et al.* 1989). This condition, known as haemochromatosis, is now easily prevented by a low iron (< 80 mg/kg) daily diet (Worth *et al.* 1991). The control of iron content in the diet is critically important for the successful maintenance of captive birds of paradise and technology has today provided infinitely better-balanced commercially-available foods for captive birds.

A tragedy of present Australian quarantine regulations is that they prevent Australians and people visiting Australia from seeing the non-Australian birds of paradise in life in their zoological gardens. Moreover, species from these areas are being bred and much learnt about them in the USA, Europe, Indonesia, Singapore, Hong Kong and elsewhere, while Australians are denied such opportunities for no justifiable reason given that millions of wild birds migrate between New Guinea and Australia annually. There is very real conservation value in maintaining general public awareness of the existence of these wonderful creatures. While we emphatically do not support anything remotely like free and easy import of birds of paradise (or any other birds) into Australia (or elsewhere) we do strongly feel that all would benefit from the careful resumption of import for legitimate educational and scientific purposes. In this regard we wholeheartedly agree with Gilliard (1969: 39) who observed that 'since we know that the annual killing of more than 80 000 males—and these were mostly of three species—did not threaten any of the birds of paradise, I would strongly advocate that small numbers of live birds be collected each year for the zoos of the world, to be used for educational purposes.'

A final word

We believe the birds of paradise merit more attention than they receive today, and we thus suggest that a Trust for the Birds of Paradise be established with its chief purpose being to raise global awareness of the need to conserve the group. We strongly advocate that the first line of defence is the maintenance of wild populations in managed tracts of natural habitat (as outlined above), we also suggest that a secondary goal is a scientifically-planned network of captive-bred populations of all paradisaeid species, distributed through the world's zoos, for the purposes of education, aesthetic appreciation of the birds, scientific research appropriate to captive populations, and raising awareness of the world's populace of the very existence of these birds. It is perhaps remarkable to note that there were many more species of live birds of paradise in zoological gardens in the 1930s than there are today.

PART II
Species accounts

Family PARADISAEIDAE: the birds of paradise

Paradisaeidae Vigors 1825, *Transactions of the Linnean Society of London* **14**, 395–517.

The family Paradisaeidae comprises **42 species** of oscinine songbirds with a global distribution restricted to the forests of eastern Australia, New Guinea and its satellite islands, and the north Moluccas (Fig. 1.1, Plates 4–12). All spp. inhabit humid (mostly tropical) forest. A few spp. also range into woodland savanna, and one ranges southward into the Australian subtropics. New Guinea is the centre of generic diversity and sp. richness, and this may be where the group evolved.

Of the 42 spp. 38 are found on New Guinea and its fringing islands. Four spp. (the three riflebirds and the Trumpet Manucode) are found in Australia (two of the former are endemic). A final two isolates—*Semioptera* and *Lycocorax* (both monotypic genera)—are endemic to the northern Moluccas, a relatively short distance WNW of New Guinea's western satellite islands. Only two groups within the family show evidence of an ability to colonise across deep water barriers—the manucodes (*Manucodia/Lycocorax*) and the plumed birds and their immediate relatives (a clade that includes *Cicinnurus*, *Semioptera*, *Seleucidis*, and *Paradisaea*). The group as a whole is morphologically and behaviourally diverse and thus difficult to define by a few unambiguous characters. The family as we delineate it exhibits the following defining **characters:** (1) hatchlings lack down and become black skinned, (2) both subfamilies have spp. that exhibit polygynous court display, with persistent calling by ♂♂ from a fixed post or series of calling posts, (3) plumage with iridescence and a range of bright plumages not found elsewhere in the corvine assemblage, (4) compact and robust body form, (5) distinctive flight, with broad-winged, undulating flap-and-glide, (6) remarkable sound production by the wings in flight by the ♂♂ of many genera, (7) nests that do not typically involve sticks but feature supple long stems of epiphytic orchids and vines.

Size varies greatly, from 15 cm length/50 g weight for the King Bird (comparable to a Common Starling), to 44 cm length/450 g weight for the Curl-crested Manucode (size of a crow, see Fig. 1.2). The ad ♂ Black Sicklebill, with its long sabre-like tail feathers, measures 110 cm long. In the family, ♂♂ are larger and heavier than ♀♀ except in the aberrant Yellow-breasted Bird, in which ♀♀ are slightly larger and heavier. As a result of a strong morphological radiation, the **bills** of the various genera vary from short to long, slim to stout, and straight to strongly decurved. Many of the spp. (e.g. *Macgregoria*, *Paradigalla*, *Astrapia*, *Cicinnurus*) have small, slim bills (see Fig. 4.1). Figure 9.1 shows a *Paradisaea* skull as typifying that of the Paradisaeinae.

There is a marked difference in bill and leg morphology between the two major lineages in the family. The small and relatively fine-tipped weak bills of the 'wide-gaped' birds of paradise (Cnemophilinae) have very wide gapes as an adaptation to an exclusively fruit diet such as is found in some Neotropical passerine bird groups (Snow 1976*b*). In addition, these birds also have relatively fine and weak legs and feet, in marked contrast to the paradisaeinine 'true' birds of paradise, including Macgregor's Bird, which have extremely-powerful legs and feet and use them to cling acrobatically to limbs, vines, and tree trunks, and to hold a food item to a perch whilst dismembering it.

In the majority of genera the **wings** are rounded and in ♂♂ of several genera some outer primaries are slightly to highly modified in shape, probably for mechanical sound production (see Fig. 5.2). The **tarsus length** as a proportion of wing length is a conservative 21–29% in most spp. but in *Macgregoria* and *Parotia* is 31–32% and in the cnemophilines

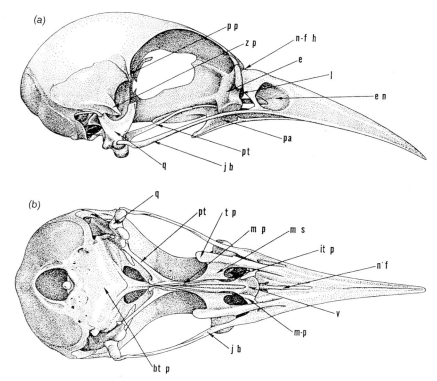

9.1 Profile and ventral view of a skull of *Paradisaea* as typifying that of the Paradisaeinae. Abbreviations: bt p = basitemporal plate, e = ectethmoid, e n = external naris, it p = inter-palatine process, j b = jugal bar, l = lacrymal, m p = mediopalatine process, m-p = maxillopalatine, m s = medial shelf of the palatine, n f = nasal floor, n-f h = nasal-frontal hinge, pa = palatine, pp = postorbital process, pt = pterygoid, q = quadrate, t p = transpalatine process, v = vomer, z p = zygomatic process. From Bock (1963), with kind permission.

33–36%. In manucodes the **trachea** is greatly elongated, coiled, and displaced to sit subcutaneously on top of the pectoral muscles (Forbes 1882*b*; Ogilvie-Grant 1904; Clench 1978; Frith 1994*b*—see Figs 1.5 and 9.9,11). The trachea of the ad ♂ *Seleucidis* is uniquely modified in a less dramatic way (Forbes 1882*b*). Several spp. (e.g. *Drepanornis*, *Cicinnurus*) have **modified bare skin**; both sexes of Macgregor's Bird have extensive circumorbital yellow eye wattles; and the paradigallas a fleshy yellow forehead wattle growing from the base of the maxilla and a blue or blue-and-red wattles from the base of the mandible. Uniquely, the ad ♂ Yellow-breasted Bird has a pale green mandibular wattle that forms a bulbous, fleshy, swollen structure atop the maxilla. Also uniquely, the thighs of ad ♂ Twelve-wired Birds are bare and pigmented deep coral pink as the legs and these play a role in courtship display.

The **monogamous** Macgregor's Bird, the manucodes, the Paradise Crow, and the two **polygynous** paradigallas are basically all-black in plumage and are **sexually monomorphic** with greater or lesser degrees of blue and/or green iridescence to their feathering. The remaining 33 spp. are all polygynous and **sexually dimorphic** to an extreme degree, with ad ♂♂ being adorned with colourful and beautiful, often extraordinarily-bizarre, ornately-elaborate plumage. The highly-modified, erectile head plumes of *Parotia* and *Pteridophora* are paired with cranial modifications to facilitate the large muscle masses required to move them (Stonor 1937). The only crested sp. is *Cnemophilus macgregorii*, in which both sexes wear a diminutive sagittal crest of a few filamentous sickle-shaped feathers.

♀♀ of these polygynous spp. are drably coloured in subdued browns and dull yellows (*Paradisaea* spp.) or are brown and/or rufous above and dully paler below marked with darker barring to give an overall cryptic appearance. **Males** take at least 5–7 yr to acquire fully ad plumage (conceivably longer, if the presence of ad ♂♂ inhibits younger birds), while limited evidence suggests that ♀♀ breed when younger (probably usually after their first year or two).

Ad ♂♂ of polygynous spp. attend and maintain one or more traditional arboreal or terrestrial **display perch(es)** for solitary or communal courtship display. Perhaps related to this kind of behaviour, the **vocalisations** of ♂♂ are mostly harsh, loud, strident notes, crow-like, but also some screeches, bell-and bugle-like notes, and rapid bursts or powerful notes sounding like a pneumatic jack-hammer. The curled and elongated trachea of ad ♂ manucodes produces low tremulous calls that are unique in the group.

Ad ♂♂ of some promiscuous spp., notably the predominantly-insectivorous ones, defend all-purpose territories, whereas most other spp. are entirely non-territorial or defend nothing more than their display perch, display site, or nest (in monogamous spp.).

Nests, **eggs**, and clutch **incubation** are detailed in Chapter 6.

The two **subfamilies** recognised in our treatment—the Cnemophilinae (wide-gaped fruit-eaters) and the Paradisaeinae (the typical birds of paradise)—are very distinct and share few unambiguous derived characters that prove they are of the same lineage (Table 1.2). The powerfully-and dextrous-footed, narrow-mouthed, omnivorous, open cup-nesting, typical birds of paradise constitute the subfamily Paradisaeinae in which we here include *Macgregoria* (see Chapter 3). In this subfamily juv birds resemble a duller version of the ♀/imm ♂ plumage. In pterylography, all members of the Paradisaeinae except *Macgregoria* feature a basal gap in the *Pteryla Spinalis*, *pars dorsalis* feather tracts which occur in 7–11 rows (Clench 1992). By contrast, the three weak-and non-dextrous-footed, wide-gaped, sluggish, exclusively-frugivorous, and domed-nest-building spp. exhibit a distinct juv plumage, lack a basal gap in back pterylography, and constitute the subfamily Cnemophilinae.

As defined by general biology, ecology, nidification, and anatomy, the sexually-monomorphic and monogamous manucodes form a very distinct clade within the Paradisaeinae (Bock 1963; Sibley and Ahlquist 1985; Beehler 1989*b*; Christidis and Schodde 1992; Frith 1994*b*). Future studies may indicate that these differences justify a separate subfamily for them, but in many respects (flight pattern, foraging behaviour, diet, gross body form) they appear to be typical (see Table 1.2). Our limited knowledge of *Lycocorax* indicates it is a member of the manucode lineage.

Fourteen intergeneric and seven intrageneric wild **hybrid crosses** (see Table 3.5 and Appendix 1) have been documented (see Plates 14–15), dramatically emphasising the close genetic relationships between the spp. of the Paradisaeinae, notwithstanding the extreme morphological diversity of ad ♂♂.

Subfamily CNEMOPHILINAE: *the wide-gaped birds of paradise*

Cnemophilinae Bock 1963, *Condor* **65**, 102.

The **Wide-gaped birds of paradise** include **three species** (Loria's, Crested, and Yellow-breasted, see Plate 4) confined to the uplands of mainland NG. Bock (1963) defined the subfamily by a series of cranial osteological characters. Whilst they possess some basic cranial characters of the typical birds of paradise, the cnemophilines lack those typical of the bowerbirds (Ptilonorhynchidae) to which they had long been assigned. Although Bock believed the cnemophilines were clearly birds of paradise, he provided few clear characters demonstrating the relationship. **Characters** defining the Cnemophilinae are a weak and broad-gaped bill and an unossified nasal region; Bock (1963) also noted large ectethmoid plate with a small dorsolateral foramen, large maxillopalatine, palatines without a thickened lateral edge, broad transpalatine process, and a weak mandible in which the two rami are not fused into a strong anterior wedge (these characters merit further study for their utility in defining this group). Feet are relatively small, weak, and non-manipulative. All members of the Cnemophilinae lack a basal gap in the feathers of their *Pteryla Spinalis, pars dorsalis* feather tracts which occur in 13 rows (Clench 1992), see Fig. 3.2.

In describing the ad ♂ type specimen of *Cnemophilus macgregorii* De Vis (1890) assigned the sp. to the bowerbirds, thinking it systematically between *Amblyornis* and *Sericulus* because of similar bright plumage coloration and structure. Sclater (1891) noted some characters he felt approached those of *Cicinnurus*, however, and so placed it in the birds of paradise. Ornithologists long searched for the bower of this supposed bowerbird until Mayr (in Mayr and Gilliard 1954), who together with Erwin Stresemann already suspected it to be a bird of paradise (*in litt.* to Marshall 1954), found anatomical characters linking it with the birds of paradise. This finding was supported by the osteological studies of Bock (1963) who concluded that whilst Loria's, the Crested, and the Yellow-breasted show basic features of the birds of paradise, lacking those of the bowerbirds, they exhibited characters that supported giving the group subfamilial status. No interspecific hybrids are known for any spp. of Cnemophilinae.

Notwithstanding these conclusions, several ornithologists subsequently resurrected the notion that the Crested was perhaps a bowerbird (Gilliard 1969; Beehler 1989*b*). Recently-acquired knowledge of the nest, egg, and nestling appearance of this bird does, however, support the view that the cnemophilines are birds of paradise and certainly demonstrates that they are not bowerbirds. Whilst this supports the hypothesis that the cnemophilinine birds are more closely related to the birds of paradise than to the bowerbirds, we admit the possibility that the cnemophilines may, in fact, not be birds of paradise. The fact that there is no other readily-available sister group for the cnemophilines militates against removing them from the assemblage at this time. It is certainly convenient to treat these three spp. here, and we believe this is their proper place. As a final point, it is interesting that the Kalam people of Papua New Guinea consider the ad ♂ Crested Bird to be a bowerbird (Majnep and Bulmer 1977).

See Subfamily Paradisaeinae, page 195, for a discussion of why we exclude *Macgregoria* from the Cnemophilinae.

Ad ♂♂ of *Cnemophilus* spp. are < 5% larger than ♀♀ but in *Loboparadisea* ♂♂ are fractionally smaller than ♀♀ (Table 1.3). **Bill** short, being slightly shorter than or the same length as, the head, weak and broad at the gape. **Tail** moderately graduated. **Wing** short

Plates

Plate 1
Nineteenth century bird of paradise illustrations (1801–98)

1. Red Bird of Paradise by Barraband (from Levaillant 1801–6) showing the artist's ability to give life-like shape, form, and texture to a bird he did not know in life. If not for unnatural perches, fashionable at the time, Barraband's paintings would compare favourably with the best of today's works.

2. Twelve-wired Bird of Paradise by Barraband (from Levaillant 1801–6) showing a lack of knowledge of a (unique) bird's shape and form in life (compare with Plate 8). Note: this species does not have black underparts, a point raising interesting questions (see Appendix 1; Twelve-wired Bird × Magnificent Riflebird).

3. Lesser Bird of Paradise by John Gould (from Gould and Sharpe 1875–88) accurately showing the appearance of the species in life and the fact that males gather to form leks.

4. Blue Bird of Paradise by John Gerrard Keulemans (from Sharpe 1891–8) showing how the artist imagined the male might have displayed to a female (compare with Fig. 9.86).

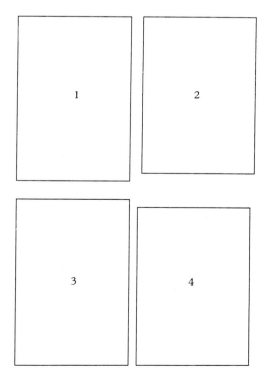

L'Oiseau de Paradis rouge. Pl. 6.

Le Nébuleux, étalant ses parures. Pl. 16.

PARADISEA PAPUANA.

PARADISORNIS RUDOLPHI, Finsch.

Plate 2
Some bird of paradise habitats—subalpine to lowland forest

1. Subalpine plateau with *Dacrycarpus compactus*-dominated woodland. Lake Habbema (at 3225 m) with Puncak Trikora, or Mt. Wilhelmina, in background, Irian Jaya. Habitat for Macgregor's Bird and upper limit of range of Splendid Astrapia. Photo: BB.

2. Aerial view of lower montane mixed forest with frost-induced (and fire-assisted) grassland. Western Tari Gap, Papua New Guinea at c. 2500–2800 m. Optimal habitat for Crested, Ribbon-tailed, King of Saxony, and Brown Sicklebill and suboptimal, upper limit, habitat for Loria's Bird. Photo: CF.

3. Interior of the previous lower montane mixed forest habitat—in which *Pandanus tectorius*, *Cyathea* tree ferns, epiphytic *Schefflera*, mosses, and other plants dominate. Note: observation tower (at Ribbon-tailed Astrapia nest) under construction. Photo: CF.

4. Aerial view of midmontane forest at 1600 m with typical Papua New Guinea (Central Province) village and its steep ridge-top 'bush' airstrip. Habitat for Trumpet Manucode, Buff-tailed Sicklebill, Superb Bird, Lawes' Parotia, Magnificent, and Blue Bird. Photo: BB.

5. Lower hill forest. Rouna Falls, Laloki River, Papua New Guinea at 500 m. Habitat for Glossy-mantled Manucode, Magnificent Riflebird, and Raggiana Bird. Photo: BB.

6. Lowland riverine forest. Nagore River, Papua New Guinea at 70 m. Habitat for Trumpet and Glossy-mantled Manucodes, Magnificent Riflebird, Twelve-wired, King, and Raggiana Birds. Photo: BB.

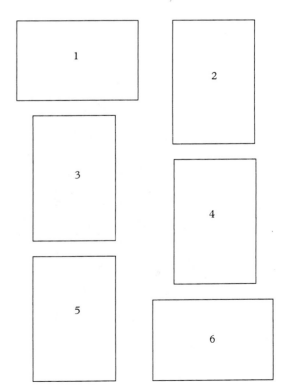

Plate 3
Nidification and moult of some birds of paradise

1. Adult female Crested Bird at her domed nest, 20 Jan 1989, Tari Gap, Papua New Guinea. Note nest material and contrastingly grey-plumaged nestling. Photo: CF.

2. Paradise Crow nest and hatchling, 19 Apr 1996, Gamkonora, Halmahera, Maluku, Indonesia. Note suspended lined tendril nest and naked, pale-skinned, hatchling. Photo: M. K. Poulsen.

3. Trumpet Manucode pair incubating two-egg clutch (bird at right relieving bird at left) in suspended unlined nest of curly vine tendrils, 20 Nov 1990, Iron Range, Cape York Peninsula, Australia. Photo: CF.

4. Hatchling Short-tailed Paradigalla, 13 Jan 1989, at 2100 m on Ambua Range, below Tari Gap on Mt. Hagen-Tari road, Papua New Guinea. Note well-developed facial wattles and dark skin. Photo: CF.

5. Adult female Ribbon-tailed Astrapia at nest and nestling, 4 Feb 1989, at c. 2650 m, Tari Gap, Papua New Guinea. Note nest materials, this particular female's male-like central tail feathers, that she is regurgitating food, and the blackish nestling plumage. Photo: CF.

6. Adult male Emperor Bird showing the severe moult typical of many adult males of the sexually-dimorphic polygynous bird of paradise species. Photo: CF.

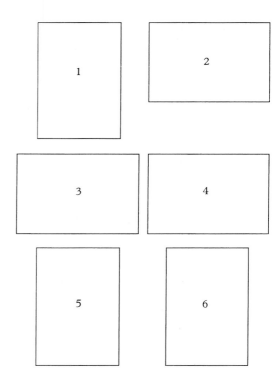

Plate 4
The wide-gaped birds of paradise

The three species comprising the subfamily Cnemophilinae. Juvenile plumage distinctly different from that of adult of both sexes.

1. Loria's Bird of Paradise
Cnemophilus loriae p. 176
(a) Adult ♂, (b) subadult ♂, (c) adult ♀, (d) juvenile. Immature similar to adult ♀.

2. Crested Bird of Paradise
Cnemophilus macgregorii p. 183
(a) *C. m. macgregorii* adult ♂, (b) adult ♀, (c) juvenile, (d) *C. m. sanguineus* adult ♂. Immature similar to adult.

3. Yellow-breasted Bird of Paradise
Loboparadisea sericea p. 190
(a) adult ♂, (b) adult ♀, (c) juvenile. Immature similar to adult ♀.

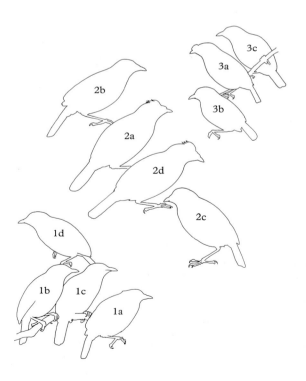

W.T. COOPER 1995

Plate 5
The sexually-monomorphic species of the Paradisaeinae

Monogamous (*Macgregoria* and *Manucodia*), presumed monogamous (*Lycocorax*), and polygynous (*Paradigalla*) birds of paradise. Juvenile similar to but duller than adult. No barring in plumage of birds of any age.

1. Macgregor's Bird of Paradise
Macgregoria pulchra p. 198
Adult. Juvenile and immature similar but slightly duller.

2. Paradise Crow
Lycocorax pyrrhopterus p. 205
(a) *L. p. pyrrhopterus* adult—concealed bases of primaries ochraceous-grey, (b) *L. p. obiensis* adult—concealed bases of primaries with only a trace of white, (c) *L. p. morotensis* adult—concealed bases of primaries extensively white. Juvenile and immature similar but slightly duller.

3. Glossy-mantled Manucode
Manucodia atra p. 211
Adult. Juvenile more dull sooty black, immature generally similar but slightly duller.

4. Jobi Manucode
Manucodia jobiensis p. 217
Adult. Juvenile more dull sooty black and immature generally similar but slightly duller.

5. Crinkle-collared Manucode
Manucodia chalybata p. 220
Adult. Juvenile more dull sooty black and immature generally similar but slightly duller.

6. Curl-crested Manucode
Manucodia comrii p. 224
Adult. Juvenile more dull sooty black and immature generally similar but slightly duller.

7. Trumpet Manucode
Manucodia keraudrenii p. 229
Adult. ♀♀ probably with less shaggy hackles than older ♂♂. Juvenile more dull sooty black and immature generally similar but slightly duller. For display posture see Fig. 9.15.

8. Long-tailed Paradigalla
Paradigalla carunculata p. 242
(a) Adult ♂. Adult ♀ like adult ♂ but slightly duller, (b) juvenile to immature—more sooty black and duller version of adults but tail *shorter* and facial wattles entirely yellow and paler than in adult.

9. Short-tailed Paradigalla
Paradigalla brevicauda p. 244
(a) Adult ♂. Adult ♀ like adult ♂ but slightly duller, (b) juvenile to immature—more sooty black and duller version of adults but tail *longer* and facial wattles entirely yellow and paler than in adult.

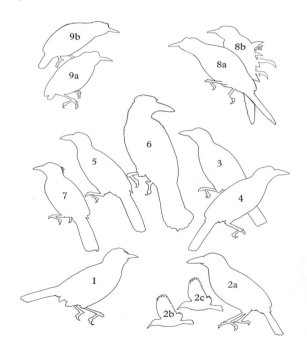

Plate 6
The astrapias

Adult ♂♂ ornately plumaged, promiscuous, and solitary or lek-displaying on traditional perches. Adult ♀, juvenile, and immature cryptically plumaged with underparts finely barred. Tail grossly elongated in both sexes, more so in adult ♂.

1. **Arfak Astrapia**
Astrapia nigra p. 250
(a) Adult ♂, (b) adult ♀. Juvenile to immature are similar to adult ♀.

2. **Splendid Astrapia**
Astrapia splendidissima p. 253
(a) Adult ♂, (b) adult ♀. Juvenile to immature are similar to adult ♀.

3. **Ribbon-tailed Astrapia**
Astrapia mayeri p. 257
(a) Adult ♂, (b) adult ♀. Adult ♀♀ have variable amount of white in central pair of tail feathers which may, rarely, be entirely white as in adult ♂ (see Plate 3). Juvenile to immature are similar to adult ♀.

4. **Stephanie's Astrapia**
Astrapia stephaniae p. 266
(a) Adult ♂, (b) adult ♀. Juvenile to immature are similar to adult ♀.

5. **Huon Astrapia**
Astrapia rothschildi p. 273
(a) Adult ♂, (b) adult ♀. Juvenile to immature are similar to adult ♀. For display posture see Fig. 9.21.

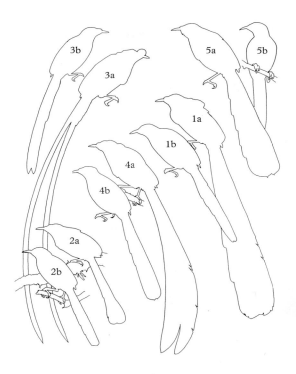

Plate 7
The parotias

Adult ♂♂ ornately-plumaged, promiscuous, and solitary or exploded-lek-displaying on cleared traditional ground courts. Adult ♀, juvenile, and immature cryptically plumaged with paler underparts barred.

1. Western Parotia
Parotia sefilata p. 277
(a) Adult ♂, (b) adult ♀. Juvenile to immature are similar to adult ♀. For display postures see Fig. 9.23.

2. Lawes' Parotia
Parotia lawesii p. 283
(a) *P. l. lawesii* adult ♂, (b) adult ♀ (c) *P. l. helenae* adult ♂. Juvenile to immature are similar to adult ♀. For display postures see Figs 9.26–7.

3. Wahnes's Parotia
Parotia wahnesi p. 292
(a) Adult ♂, (b) adult ♀. Juvenile to immature are similar to adult ♀. For display postures see Fig. 9.29.

4. Carola's Parotia
Parotia carolae p. 298
(a) Adult ♂, (b) subadult ♂ (c) adult ♀. Juvenile to immature are similar to adult ♀.

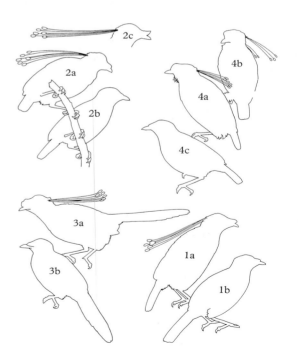

Plate 8
The riflebirds and Twelve-wired Bird of Paradise

Adult ♂♂ ornately plumaged, promiscuous, and solitary-displaying on traditional perches. Adult ♀, juvenile, and immature cryptically plumaged with paler underparts strongly marked with barring and/or chevrons.

1. Magnificent Riflebird
Ptiloris magnificus p. 315
(a) Adult ♂, (b) adult ♀. Juvenile to immature are similar to adult ♀. For display postures see Fig. 9.36.

2. Paradise Riflebird
Ptiloris paradiseus p. 327
(a) Adult ♂, (b) adult ♀. Juvenile to immature are similar to adult ♀. For display posture see Fig. 9.38.

3. Victoria's Riflebird
Ptiloris victoriae p. 334
(a) Adult ♂, (b) subadult ♂, (c) adult ♀. Juvenile to immature are similar to adult ♀. For display postures see Figs 9.40–2.

4. Twelve-wired Bird of Paradise
Seleucidis melanoleuca p. 428
(a) Adult ♂ (note bare colourful thighs), (b) subadult ♂, (c) adult ♀. Juvenile to immature are similar to adult ♀. For display postures see Fig. 9.71.

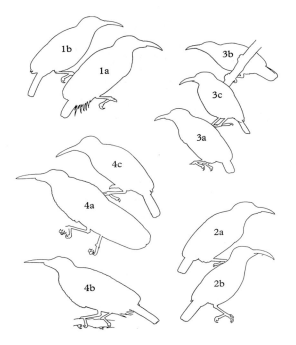

Plate 9
The sicklebills and sabretails

Adult ♂♂ ornately plumaged, promiscuous, and solitary-displaying on traditional perches. Adult ♀, juvenile, and immature cryptically plumaged with paler underparts strongly marked with barring and/or chevrons. Tails graduated.

1. Black Sicklebill
Epimachus fastuosus p. 357
(a) Adult ♂, (b) subadult ♂, (c) adult ♀. Juvenile to immature are similar to adult ♀. For display posture see Fig. 9.48

2. Brown Sicklebill
Epimachus meyeri p. 366
(a) adult ♂, (b) adult ♀. Juvenile to immature are similar to adult ♀. For display postures see Fig. 9.51.

3. Buff-tailed Sicklebill
Drepanornis albertisi p. 377
(a) Adult ♂, (b) adult ♀. Juvenile to immature are similar to adult ♀. For display posture see Fig. 9.54.

4. Pale-billed Sicklebill
Drepanornis bruijnii p. 385
(a) Adult ♂, (b) adult ♀. Juvenile to immature are similar to adult ♀. For display posture see Fig. 9.57.

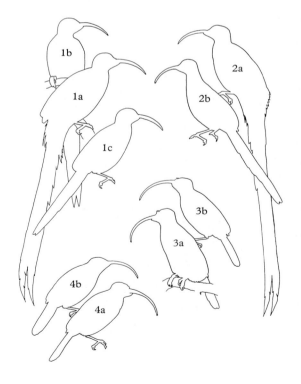

Plate 10
The King of Saxony, Superb, and sickle-tailed Birds of Paradise

Adult ♂♂ ornately plumaged, promiscuous, and solitary or lek (King Bird only) displaying on traditional perches (*Pteridophora* and *Cicinnurus regius*), perches and ground (*Lophorina*) or vertical perches on a cleared ground court (*C. magnificus* and *C. respublica*). Adult ♀, juvenile, and immature cryptically plumaged with paler underparts marked with bold chevrons (King of Saxony Bird only) or fine barring.

1. **King of Saxony Bird of Paradise**
Pteridophora alberti p. 305
(a) Adult ♂, (b) adult ♀. Juvenile to immature are similar to adult ♀. For display postures see Fig. 9.32.

2. **Superb Bird of Paradise**
Lophorina superba p. 345
(a) *L. s. superba* adult ♂, (b) adult ♀, (c) *L. s. feminina* adult ♀. Juvenile to immature are similar to adult ♀. For display postures see Fig. 9.46.

3. **Magnificent Bird of Paradise**
Cicinnurus magnifucus p. 391
(a) Adult ♂, (b) adult ♀. Juvenile to immature are similar to adult ♀. For display postures see Fig. 9.60.

4. **Wilson's Bird of Paradise**
Cicinnurus respublica p. 401
(a) Adult ♂, (b) adult ♀. Juvenile to immature are similar to adult ♀. For display postures see Fig. 9.62.

5. **King Bird of Paradise**
Cicinnurus regius p. 407
(a) Adult ♂, (b) subadult ♂, (c) adult ♀. Juvenile to immature are similar to adult ♀. For display postures see Fig. 9.66.

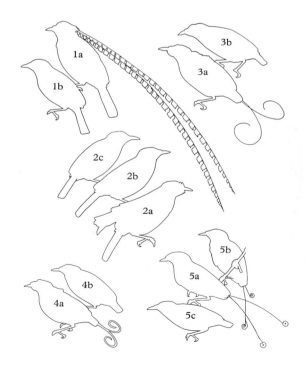

Plate 11
The typical *Paradisaea* birds of paradise

Adult ♂♂ ornately plumaged, promiscuous, and lek displaying on traditional perches. Adult ♀, juvenile, and immature drably plumaged with paler to white underparts but not marked with barring or chevrons.

1. **Lesser Bird of Paradise**
Paradisaea minor p. 439
(a) Adult ♂, (b) adult ♀. Juvenile to immature are similar to adult ♀. For display postures see Fig. 9.73.

2. **Greater Bird of Paradise**
Paradisaea apoda p. 448
(a) Adult ♂, (b) adult ♀. Juvenile to immature are similar to adult ♀. For display postures see Fig. 9.75.

3. **Raggiana Bird of Paradise**
Paradisaea raggiana p. 456
(a) *P. r. raggiana* adult ♂, (b) adult ♀; (c) *P. r. intermedia* adult ♂, (d) adult ♀; (e) *P. r. augustaevictoriae* adult ♂. Juvenile to immature are similar to adult ♀. For display postures see Fig. 9.77.

Note: hybrid combinations between some of the above forms produce adult ♂♂ showing various degrees of mixed plumage characters. See Plate 15 and Appendix 1.

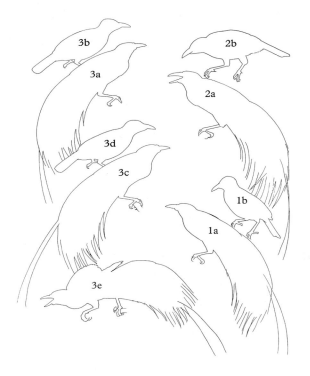

Plate 12
The Standardwing and the less typical *Paradisaea* birds of paradise

Adult ♂♂ ornately plumaged, promiscuous, and lek-displaying (solitary Blue Bird excepted) on traditional perches. Adult ♀, juvenile and immature drably plumaged (except Blue Bird) with paler underparts not marked with barring or chevrons (except Blue and Goldie's Birds).

1. **Standardwing Bird of Paradise**
Semioptera wallacii p. 417
(a) Adult ♂ (note central tail feather pair shortest), (b) adult ♀. Juvenile to immature are similar to adult ♀. For display postures see Fig. 9.68.

2. **Goldie's Bird of Paradise**
Paradisaea decora p. 470
(a) Adult ♂, (b) subadult ♂, (c) adult ♀. Juvenile to immature are similar to adult ♀.

3. **Red Bird of Paradise**
Paradisaea rubra p. 475
(a) Adult ♂, (b) adult ♀. Juvenile to immature are similar to adult ♀. For display postures see Fig. 9.82.

4. **Emperor Bird of Paradise**
Paradisaea guilielmi p. 482
(a) Adult ♂, (b) adult ♀. Juvenile to immature are similar to adult ♀. For display postures see Figs 5.3 and 9.84.

5. **Blue Bird of Paradise**
Paradisaea rudolphi p. 488
(a,b) Adult ♂ (note plumes dorsally rust-red), (c) adult ♀. Juvenile to immature are similar to adult ♀. For display postures see Figs 5.3 and 9.86.

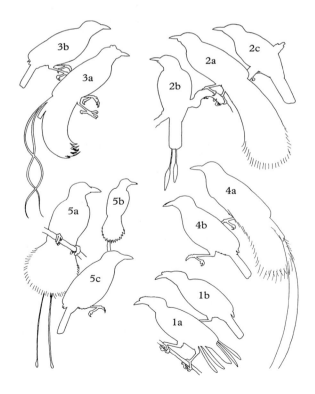

Plate 13
Some eggs of some bird of paradise genera

Numbers are of the egg figures from left to right and from top to bottom. No egg is presently available for illustration for the genera *Loboparadisea* and *Semioptera*.

1. *Cnemophilus*
Loria's Bird of Paradise *Cnemophilus loriae* p. 181
Painted from photographs taken in the wild of the only egg known and its near-complete remains now in BMNH.

2. *Macgregoria*
Macgregor's Bird of Paradise *Macgregoria pulchra* p. 203
Painted from the only known egg, in AMNH.

3. *Manucodia*
Curl-crested manucode *Manucodia comrii* p. 228
A typical egg derived from a sample of 37 in BMNH.

4. **Glossy-mantled Manucode** *Manucodia atra* p. 216
A typical egg derived from a total sample of 19 in BMNH (10), AMNH (3), and two each in MV, ZMB, and SMT.

5. **Trumpet Manucode** *Manucodia keraudrenii* p. 239
A typical egg derived from a total sample of 30 in BMNH (21), MV (8), and ZMB (1).

6. *Lycocorax*
Paradise Crow *Lycocorax pyrrhopterus* p. 209
A typical egg derived from a total sample of 6 in RMNH (3), BMNH (2), and ZMB (1).

7. *Paradigalla*
Short-tailed Paradigalla *Paradigalla brevicauda* p. 247
A typical egg derived from photographs of 2 eggs laid in the wild by same female.

8. *Astrapia*
Ribbon-tailed Astrapia *Astrapia mayeri* p. 263
A typical egg derived from photographs of 4 eggs laid in the wild by different females.

9. **Huon Astrapia** *Astrapia rothschildi* p. 276
A typical egg derived from a sample of 3 in BMNH.

10. *Parotia*
Lawes' Parotia *Parotia lawesii* p. 291
Painted from the only egg known, in BMNH.

11. *Pteridophora*
King of Saxony Bird of Paradise *Pteridophora alberti* p. 313
Painted from photographs taken in the wild of the only egg known.

12. *Lophorina*
Superb Bird of Paradise *Lophorina superba* p. 355
A typical egg derived from a total sample of 7 in BMNH (5) and ZMB (2).

13. *Drepanornis*
Buff-tailed Sicklebill *Drepanornis albertisi* p. 384
A typical egg derived from a sample of 3 in BMNH.

14. *Epimachus*
Brown Sicklebill *Epimachus meyeri* p. 374
Painted from the only two eggs known, one each in BMNH and AMNH.

15. *Seleucidis*
Twelve-wired Bird of Paradise *Seleucidis melanoleuca* p. 437
Painted from the only two eggs known, one each in BMNH and AMNH.

16. *Ptiloris*
Magnificent Riflebird *Ptiloris magnificus* p. 326
A typical egg derived from a total sample of 20 eggs in MV (13), BMNH (6), and ZMB (1).

17. **Victoria's Riflebird** *Ptiloris victoriae* p. 343
A typical egg derived from a total sample of 22 in MV (16) and BMNH (6).

18. *Cicinnurus*
Magnificent Bird of Paradise *Cicinnurus magnificus* p. 398
A typical egg derived from a sample of 6 in BMNH.

19. *Paradisaea*
Raggiana Bird of Paradise *Paradisaea raggiana* p. 466
A typical egg derived from a total sample of 19 in BMNH (13), AMNH (2), ZMB, and SMT (2).

20. **Lesser Bird of Paradise** *Paradisaea minor* p. 446
A typical egg derived from a total sample of 6 in BMNH (4), AMNH(1), and RMNH (1).

Note: sample sizes from which typical egg appearance were derived are not necessarily the total number of eggs of any given taxon in any particular collections(s).

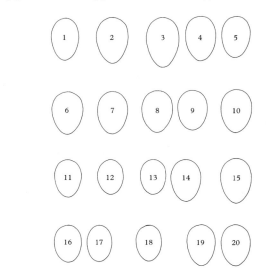

Plate 14
Ten presumed intergeneric and one intrageneric hybrid birds of paradise

1. **Arfak Astrapia × Black Sicklebill** p. 508
(*Astrapia nigra × Epimachus fastuosus*) adult ♂.
Astrapian Sicklebill (one specimen known), but see also 4 below.

2. **Ribbon-tailed × Stephanie's Astrapia** p. 504
(*Astrapia mayeri × Astrapia stephaniae*) adult ♂.
Barnes' Astrapia (at least 12 specimens known).

3. **Long-tailed Paradigalla × Black Sicklebill** p. 509
(*Paradigalla carunculata × Epimachus fastuosus*) adult ♂.
Not named (one specimen known).

4. **Black Sicklebill × Arfak Astrapia** p. 508
(*Epimachus fastuosus × Astrapia nigra*) adult ♂.
Elliot's Bird of Paradise (two specimens known), but see also 1 above.

5. **Long-tailed Paradigalla × Superb Bird of Paradise** p. 510
(*Paradigalla carunculata × Lophorina superba*) adult ♂.
Rothschild's Lobe-billed Bird of Paradise (three specimens known).

6. **Black Sicklebill × Superb Bird of Paradise** p. 511
(*Epimachus fastuosus atratus × Lophorina superba feminina*) adult ♂.
Bobairo Bird of Paradise (one specimen known).

7. **Western Parotia × Superb Bird of Paradise** p. 512
(*Parotia sefilata × Lophorina superba*) adult ♂.
Duivenbode's Six-wired Bird of Paradise (two specimens known).

8. **Long-tailed Paradigalla × Western Parotia** p. 513
(*Paradigalla carunculata × Parotia sefilata*) adult ♂.
Sharpe's Lobe-billed Riflebird (one specimen known).

9. **Superb × Magnificent Birds of Paradise** p. 513
(*Lophorina superba × Cicinnurus magnificus*) adult ♂.
Wilhemnina's Bird of Paradise (three specimens known).

10. **Carola's Parotia × Superb Bird of Paradise** p. 514
(*Parotia carolae × Lophorina superba*) adult ♀.
Stresemann's Bird of Paradise (one specimen known).

11. **Lawes' Parotia × Blue Bird of Paradise** p. 515
(*Parotia l. lawesii × Paradisaea rudolphi margarita*) adult ♀.
Schodde's Bird of Paradise (one specimen known).

Note: the rare specimens of most forms were not available to the artist and all but forms 2 and 11 were painted from numerous photographs of specimens. Iris and bare part colours of most hybrids are unknown so those of the species dominating the plumage morphology are shown here. These hybrids are presented in less life-like postures that birds in previous plates to remind the user of their significance. See Appendix 1.

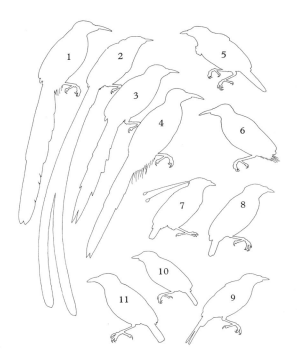

Plate 15
Five presumed intergeneric and six intrageneric hybrid birds of paradise

1. **Magnificent Riflebird × Superb Bird of Paradise** p. 515
(*Ptiloris magnificus × Lophorina superba*) adult ♂.
Duivenbode's Riflebird (three specimens known, one subsequently destroyed in WW II).

2. **Twelve-wired × Lesser Bird of Paradise** p. 516
(*Seleucidis melanoleuca × Paradisaea minor*) adult ♂.
Wonderful Bird of Paradise (five specimens known).

3. **Magnificent Riflebird × Lesser Bird of Paradise** p. 517
(*Ptiloris magnificus × Paradisaea minor*) adult ♂.
Bensbach's Bird of Paradise (one specimen known).

4. **Twelve-wired Bird of Paradise × Magnificent Riflebird** p. 517
(*Seleucidis melanoleuca × Ptiloris magnificus*) adult ♂.
Mantou's Riflebird (at least 13 specimens known).

5. **Magnificent × Lesser Birds of Paradise** p. 518
(*Cicinnurus magnificus × Paradisaea minor*) adult ♂.
Ruys' Bird of Paradise (one specimen known).

6. **Emperor × Raggiana Birds of Paradise** p. 504
(*Paradisaea guilielmi × Paradisaea raggiana augustaevictoriae*) adult ♂.
Frau Reichenow's Bird of paradise (at least six specimens known).
Note: the hybrid combination of Emperor × Lesser Birds (*Paradisaea guilielmi × Paradisaea minor finchi*), known from one specimen (and once named *P. duivenbodei*), is not illustrated. It is very similar to the present hybrid except that it has brown uppertail coverts not marked with straw yellow streaking and flank plumes more yellow that red, as would be expected.

7. **Raggiana × Blue Birds of Paradise** p. 505
(*Paradisaea raggiana salvadori × Paradisaea rudolphi margaritae*) adult ♂.
Captain Blood's Bird of Paradise (one specimen known).

8. **Greater × Raggiana Birds of Paradise** p. 506
(*Paradisaea apoda novaeguinea × Paradisaea raggiana salvadori*) adult ♂.
Lupton's Bird of Paradise (numerous specimens known).

9. **Raggiana × Lesser Birds of Paradise** p. 506
(*Paradisaea raggiana augustaevictoriae × Paradisaea minor finchi*) adult ♂.
Rothschild's Bird of Paradise (at least four specimens known).
Note: the hybrid combination of Raggiana × Lesser Birds (*Paradisaea raggiana salvadori × Paradisaea minor finchi*), known from several specimens, is not illustrated. It is very similar to the present hybrid except that it has slightly more reddish flank plumes.

10. **Magnificent × King Birds of Paradise** p. 507
(*Cicinnurus magnificus × Cicinnurus regius*) adult ♂.
King of Holland's Bird of Paradise (at least 25 specimens known).

11. **King × Magnificent Birds of Paradise** p. 507
(*Cicinnurus regius × Cicinnurus magnificus*) adult ♂.
Lyre-tailed King Bird of Paradise (at least three specimens known).

Note: the rare specimens of several forms were not available to the artist and all but forms 7, 9, and 10 were painted from numerous photographs of specimens. Iris and bare part colours of most hybrids are unknown so those of the species dominating the plumage morphology are shown here. These hybrids are presented in less lifelike postures that birds in previous plates to remind the user of their significance. See Appendix 1.

and distinctly rounded but with no modification of primary shape. **Tarsus** long, c. 35% of wing length, and relatively fine. Both sexes of *C. macgregorii* wear a unique, short, central, crown **crest** of a few filamentous sickle-shaped feathers; the ad ♂ *C. loriae* has a conspicuously-extensive pale gape; and the ad ♂ *L. sericea* exhibits unique, swollen, pale blue-green **wattles** above the base of the maxilla and nostrils (Plate 4).

All cnemophilinine spp. have a distinct **juv plumage** that differs from that of ad ♂♂, ad ♀♀, and imms (Plate 4). They are slow-moving birds unlike the animated and powerful members of Paradisaeinae. The two *Cnemophilus* spp. appear to be **obligate** (exclusive) **frugivores** and *Loboparadisea* may well prove to be so.

Thought until recently by several ornithologists to be monogamous *Cnemophilus* spp. are in fact **polygynous** (*Loboparadisea's* mating system is unknown). The *Cnemophilus* spp. were recently found to build a substantial domed **nest** of mosses and ferns lined with stems of epiphytic orchids, whilst that of *Loboparadisea* awaits discovery. These nests are unlike those of all known Paradisaeinae nests which are substantial cup-shaped structures of tree leaves built within an exterior of fresh green material and an egg-cup lining of fine supple stems of epiphytic orchids. **Eggs** are known only for *Cnemophilus loriae* and are most like those of *Macgregoria*, some *Manucodia* spp, and *Paradigalla*. **Nestlings** are naked and quickly become dark-skinned as in all known Paradisaeinine spp. (Twelve-wired Bird possibly excepted) and unlike the downy nestlings of melampittas and bowerbirds.

Genus *Cnemophilus* De Vis, 1890

Cnemophilus De Vis (1890). *Annual Report of British New Guinea 1888–1890*, p. 5 (dated 23 Aug 1889), Appendix C, 62. Type, by monotypy, *C. macgregorii* De Vis.

Synonym: *Loria* Salvadori 1894 *Annali del Museo Civico di Storia Naturale di Genova* ser. 2, **14**, 151.

The **two species** of *Cnemophilus* are unique among the birds of paradise in constructing domed nests. **Skull** of *Cnemophilus* includes some basic cranial features of the subfamily Paradisaeinae (see Fig. 9.1) but lacks those of the bowerbirds (Ptilonorhynchidae). It is short and bulbous like that of typical generalised passerines, being more like that of a thrush or starling than of a *Paradisaea* spp. As defined by Bock (1963) it features a generalised passerine palate, short straight internal processes of the mandible, and an expanded head of the orbital processes of the quadrate. There is no ossification of the nasal septum and the external naris, giving the bill a weak construction. The ectethmoid plate is large and entire, lacking the medial depression along the lateral edge that forms the 'winged' appearance found in *Paradisaea* (see Fig. 9.1). The ventral edge of the ectethmoid closely approaches the dorsal surface of the palatine and the jugal bar. Unique in the birds of paradise is a small foramen at the dorsolateral corner where the ectethmoid merges into the frontal bone. There is no lacrymal. The two rami of the lower jaw are fused only at the very tip of the bill and from there greatly diverge, thus emphasising the weakness of the wide-gaped bill (Bock 1963). The proximal culmen is broadly flattened in *C. macgregorii* (as in *Loboparadisea*) but is sharply keeled in *C. loriae*.

Compact birds of paradise with a relatively fine-tipped, weak **bill**, shorter than, or the same length as, the head but with a wide gape. **Tail** rounded, slightly graduated, and relatively long. **Wings** short, rounded, and the flight feathers deeply curved with no modification of primary shape. **Tarsus** length, at 36% of wing length, is proportionately the longest in the family. Tarsus smooth. Legs and feet are fine and relatively weak for birds of paradise. Sexually monomorphic in **size**, but ♂♂ slightly heavier than ♀♀, and with conspicuous variation in plumage related to sex and age.

Sexually dimorphic: ad ♂♂ black or blackish-brown ventrally; dorsally with brilliant iridescent orange or orange-yellow (*C. macgregorii*) or glossy velvety black with modified scale-like iridescent lore feathering (*C. loriae*). The latter sp. exhibits obvious whitish gape flanges. ♀♀ and subad ♂♂ olive-brown (*C. macgregorii*) or olive (*C. loriae*) throughout and rather bowerbird-like in general appearance. **Juvs** briefly wear a completely grey plumage. *C. macgregorii* is unique in the family Paradisaeidae in that ad of both sexes have a small, erectile, central, crown crest of filamentous sickle-shaped plumes. **Polygynous**, with assumed solitary court display by assumed promiscuous ♂♂. Nest attendance exclusively by ♀. **Nest** is placed on the ground or a few metres above it and is a substantial domed structure. Single-**egg** clutch, pinkish-buff spotted and finely blotched but lacking the longitudinal streaks typical of the Paradisaeinae (see Plate 13). No inter-or intrageneric **hybrids** known.

Loria's Bird of Paradise *Cnemophilus loriae* (Salvadori, 1894)

PLATE 4; FIGURES 9.2,3

Other names: Loria's Bird, Lady Macgregor's Bower-bird *Cnemophilus mariae*, Crow Bird-of-Paradise. *Arudimi* of Gimi people, Wahgi R., PNG (Diamond 1972). *Apenemeto* of the Gimi people, Crater Mt., PNG (D. Gillison *in litt.*). Kalam people call ♂ *Kabay mosb* or *Kabay sbtkep* and ♀ plumaged birds *Kabay gs* or *Kabay mosaj*, and call the sp. *Kabay kawslog* (Majnep and Bulmer 1977). *Mam-eh* of the Yali, IJ (BB). *Gatemup* and *Gagens* in Kubor Mts, PNG (W. T. Cooper *in litt.*). *Milome* of Victor Emanuel Mts (Gilliard and LeCroy 1961).

A small, compact, listless, and inconspicuous bird of midmontane forests of the central ranges. Round-winged, weak-billed, weak-footed, wide-gaped, and apparently exclusively frugivorous. Ad ♂ completely glossy blue-black, whereas the olive-green ♀ is easily mistakable for the larger, browner, ♀ Crested Bird. A distinct grey plumage appears to be briefly worn by juvs. Polytypic; 3 subspp.

Description

ADULT ♂: 22 cm, basal culmen sharply keeled (*contra* Crested Bird), gape wide, tail graduated. Entire plumage deep velvety black with sheens of rich silk-like purple (172A) and/or Magenta (2) in suitable light. Fine scale-like lores, forehead, and forecrown feathering are intensely iridescent, with metallic green-blues, and showing purple washes in suitable light. Exposed surfaces of tertials richly iridescent blue (270), appearing green-blue to blue washed purple (172) to Magenta in some lights. Bill shiny black, iris very dark brown, legs and feet dark olive brown to blackish. Conspicuous bare gape flanges and mouth interior cream yellow to yellowish white, sometimes with pale pink or green hue.

ADULT ♀: no bare gape flanges. Variably Greenish Olive (49) throughout upperparts with faint dusky feather edges, more so on crown, mantle, back, and upper breast, giving scalloped appearance. Upper wing and tail generally similarly coloured but with conspicuous Cinnamon-Brown (33) wash and this stronger on outer edges to wing flight feathers. Underparts paler greenish olive (150), more yellowish on belly. Averages only fractionally smaller than ♂.

SUBADULT ♂: ranging from as ad ♀, but with few feathers of ad ♂ plumage intruding, to as ad ♂ with few feathers of ♀ plumage remaining. Tail longer than ad ♂ ♂.

IMMATURE: as ad ♀, but rectrices more pointed (Gyldenstolpe 1955a) and longer than ad and subad ♂.

NESTLING–JUVENILE: hatchling naked, skin mid blue-grey dorsally and on legs, eyes, and forecrown. Crown, nape, and ventral body paler and yellowish fawn. Bill brownish grey with tiny white egg tooth, gape white, claws conspicuously white (Frith and Frith 1994b). Juv has a grey plumage, subsequently replaced by typical olive green ♀ plumage (Frith 1987). In view of a grey nestling–juv plumage also recently discovered in the Crested (Frith and Harrison 1989) it is likely this grey plumage of Loria's is briefly worn and may or may not occur in all subspp. (Frith 1987, 1996). Note: a colour photograph said to depict a juv Loria's in Peckover (1990: 6) is of a juv Yellow-breasted Bird of Paradise.

Distribution

The central ranges of NG from the Weyland Mts, IJ, southeastward to the southern Owen Stanley Ra of PNG (Mt. Dayman). Found from 1500 to 3000 m, but mostly 2000–2400 m. Present on Mt. Bosavi (T. Pratt *in litt.*). Absent from Vogelkop and Huon Penin.

Systematics, nomenclature, subspecies, weights, and measurements

SYSTEMATICS: a distinctive *Cnemophilus* sp. whose close relationship to the Crested is supported by similar juv and ♀ plumages and nest and egg morphology. First collected by Dr Lamberto Loria in 1893, being named Loria's Bird of Paradise *Loria loriae* by fellow Italian ornithologist Count Tommaso Salvadori in May 1894. Towards the end of 1894 De Vis, of the Queensland Museum, described specimens of the same taxon collected during Sir William Macgregor's expedition by W. E. Armit and R. E. Guise as Lady Macgregor's Bowerbird *Cnemophilus mariae*. Walter Rothschild obtained further specimens and from these concluded that *C. mariae* was a junior synonym of *L. loriae*. Diamond (1972) suggested returning Loria's to the genus *Cnemophilus*, established for the Crested Bird by De Vis in 1890. Since then the discovery of a nestling–juv grey plumage in both spp. and a Loria's nest that is all but identical to that of the Crested have provided support for Diamond's position. Lack of ornately-modified

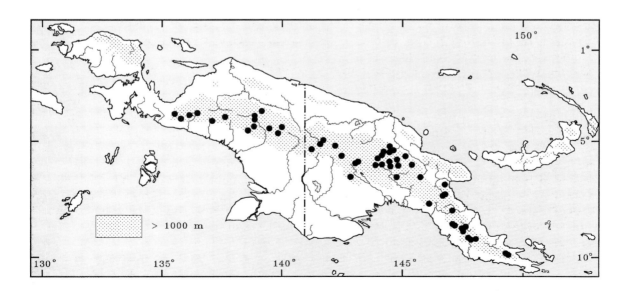

ad ♂ plumage, the uniform ♀ plumage, and the small stocky shape led early ornithologists to consider this a bowerbird. Stonor's anatomical studies (1937) indicated it to be a bird of paradise. Bock (1963) concurred with Stonor and by examining specimens of the Crested and Yellow-breasted birds realised these three spp., whilst birds of paradise, were sufficiently distinct to warrant subfamilial status, the Cnemophilinae. The recently-found nests and an egg of Loria's, and nests and nestlings of Crested, share characters with those of typical birds of paradise (Frith and Frith 1993c, 1994b). Hybridisation: unrecorded. Any successful hybridisation might be expected to be limited to its fellow cnemophilines, the Crested and Yellow-breasted Birds, both because of genetics and frugivorous nestling diet.

SPECIES NOMENCLATURE AND ETYMOLOGY

Cnemophilus loriae (Salvadori)
Loria loriae Salvadori, 1894. *Annali del Museo Civico di Storia Naturale di Genova*, ser. 2, **14**, 151. Moroka, Owen Stanley Ra.
Synonym: *Cnemophilus mariae* De Vis, 1894. *Annual Report for British New Guinea, 1893–96*, Appendix EE, p. 104. Mt. Maneao, southeast New Guinea.
Etymology: *Cnemophilus* = Gr. *knemos*, shoulder of a mountain or mountain slope; *philos*, loving, fond of. *loriae* = presumably in honour of the wife of the discoverer Dr Lamberto Loria.

SUBSPECIES, WEIGHTS, AND MEASUREMENTS

1. *C. l. loriae* (Salvadori, 1894). Type specimen MCG C.E. 22408.
Range: SE PNG, from the Herzog and Kuper Ra (E of the Watut/Tauri Gap) southeastward through the Owen Stanley Ra to Mt. Dayman. Exact western boundary of this subsp. has not been established, but presumably does not extend to the Kratke Ra.
Measurements: wing, ad ♂ ($n = 29$) 99–107 (103), imm ♂ ($n = 9$) 101–104 (102), ad ♀ ($n = 18$) 98–106 (102); tail, ad ♂ ($n = 29$) 71–81 (75), imm ♂ ($n = 9$) 71–81 (77), ad ♀ ($n = 18$) 72–84 (77); bill, ad ♂ ($n = 29$) 21–29 (27), imm ♂ ($n = 9$) 22–27 (26), ad ♀ ($n = 18$) 24–28 (26); tarsus, ad ♂ ($n = 28$) 34–42 (37), imm ♂ ($n = 9$) 35–38 (37), ad ♀ ($n = 18$) 34–40 (37). Weight: ad ♂ ($n = 7$) 76–94 (83), imm ♂ ($n = 3$) 81–100 (89), ad ♀ ($n = 8$) 78–96 (86).

2. *C. l. amethystina* (Stresemann, 1934). *Ornithologische Monatsberichte* **42**, 144. Schraderberg, Sepik Mountains. Type specimen ZMB 33. 1104.
Range: PNG's Western, Southern, and Eastern Highlands: the Schraderberg, Wahgi Divide, Bismarck, Kubor, Hagen, and Giluwe Ra. Presumably the population on Mt. Bosavi refers to this subsp. *C. l. inexpectata* is found to the W, whereas *C. l. loriae* occurs to the SE of this subsp.
Diagnosis: differs from other subspp. in the ad ♂'s deep violet purple (72) iridescent upper tertial surface. ♀ as other subspp., but mean tail length slightly longer.
Measurements: wing, ad ♂ ($n = 25$) 101–109 (105), imm ♂ ($n = 14$) 100–107 (103), ad ♀ ($n = 20$) 98–109 (103); tail, ad ♂ ($n = 25$) 73–81 (77), imm ♂ ($n = 14$) 75–86 (80), ad ♀ ($n = 20$) 73–88 (80); bill, ad ♂ ($n = 25$) 25–28 (26), imm ♂ ($n = 13$) 25–28 (26), ad ♀ ($n = 19$) 24–27 (26); tarsus, ad ♂ ($n = 25$) 32–39 (37), imm ♂ ($n = 14$) 35–39 (37), ad ♀ ($n = 20$) 34–40 (37). Weight: ad ♂ ($n = 5$) 80–101 (93), imm ♂ ($n = 6$) 66–100 (82), ad ♀ ($n = 7$) 69–95 (79).

3. *C. l. inexpectata* (Junge, 1939). *Nova Guinea*, new series **3**, 77. Bijenkorf, Oranje Mountains. Type specimen RMNH 412.
Range: western and central segment of the central ranges, including the Weyland, Nassau, Oranje, Hindenberg, and Victor Emanuel Mts.
Diagnosis: iridescence on tertials of ad ♂♂ much more green than those of nominate form or *amethystina*. ♀ as other subspp. Mean tail length slightly shorter than nominate form.
Measurements: wing, ad ♂ ($n = 25$) 97–107 (103), imm ♂ ($n = 13$) 100–109 (104), ad ♀

(n = 27) 97–108 (102); tail, ad ♂ (n = 25) 66–77 (71), imm ♂ (n = 13) 70–81 (76), ad ♀ (n = 28) 69–79 (73); bill, ad ♂ (n = 24) 25–29 (26), imm ♂ (n = 13) 25–28 (27), ad ♀ (n = 28) 24–28 (26); tarsus, ad ♂ (n = 24) 35–40 (38), imm ♂ (n = 12) 37–39 (38), ad ♀ (n = 28) 34–39 (36). Weight: ad ♂ (n = 4) 75–85 (79), imm ♂ (n = 1) (87), ad ♀ (n = 8) 60–95 (80).

Habitat and habits   Table 4.1

Midmontane forest, forest-edge, and second-growth. Observed to forage in all levels of forest vegetation, but perhaps most frequent at lower levels. That the sp. is regularly caught in terrestrial mist-nets supports this notion. Usually sits quietly or moves slowly about leafy substage feeding upon fruit (Rand 1942b). Single birds most often encountered but groups of 3–4 ♀-plumaged birds, sometimes with an ad ♂, seen in fruiting trees and sometimes with one or more other bird of paradise spp. (e.g. Stephanie's Astrapia, Lawes's Parotia). Usually quite close to the ground with a preference, like thrushes, for the rather large branches covered with moss along which they hop (Stein 1936; Melville 1979). An ad ♂ Loria's and ad ♂ King of Saxony were seen to call briefly from the same perch and the Loria's displace the King of Saxony (D. Bishop *in litt.*). Individuals on Mt. Bosavi, PNG, were described as 'fearless' (T. Pratt *in litt.*).

Diet and foraging   Table 4.2

Apparently wholly frugivorous (Schodde 1976; Beehler and Pruett-Jones 1983), with a diet of small or medium-sized fruits, harvested mainly in the middle and lower stories of the forest and forest-edge (Stein 1936; Rand 1942b; Gyldenstolpe 1955a; Ripley 1964). As many as 10 individuals in ♀ plumage have been reported foraging for fruit in shrubbery (Stein 1936). Fruits of 3 mm dia of *Sericolea pullei* are eaten (Hopkins 1992). Both ad ♂ and ♀-plumaged birds have been observed feeding in a *Planchonella* tree together, where one ♀-plumaged bird made on average, feeding visits of > 2 min and took an average of 1.5 fruits each visit (R. Hicks and J. Hicks 1988a). Ad ♂ visits averaged almost 3 min and 1.2 fruits eaten per visit. Fruits observed eaten are mostly simple drupes or berries, plucked and swallowed whole (see Chapter 4). Of five faecal samples obtained from mist-netted birds in Tari Gap all contained only fruit; four birds contained *Xanthomyrtus* sp., two *Pittosporum* sp., and one bird each contained remains of *Acronychia kaindiensis*, *Elaeocarpus* sp., *Timonius belensis*, *Riedelia* sp., *Drimys* sp., *Psychotria* sp., and *Symplocos* sp. See Appendix 4 for fruits known to be eaten. The apparent diet of only fruit does not accord with an interpretation of Loria's skull morphology which was thought indicative of a generalised insect and fruit eater (Bock 1963). Often seen on Ambua Ra, PNG in Jul–Aug feeding in fruiting trees with groups of 2–6 Tit Berrypeckers *Oreocharis arfaki* or occasionally a Short-tailed Paradigalla, but will chase off astrapias attempting to feed there (D. Bishop *in litt.*). Claims that large snails are broken open and eaten (Majnep and Bulmer 1977) seem unlikely given the weak bill and broad gape of this frugivore, but tiny snails possibly are eaten as they are by the Crested (Frith and Frith 1993c).

Vocalisations and   Figure 9.2
other sounds

Vocal repertoire poorly known, but includes, at minimum, a primary ♂ advertisement call plus an alarm note. The ♂ advertisement call is unusual and difficult to transcribe, apparently, for virtually every fieldworker has provided a different transcription. For sake of comprehensiveness we provide all here. Gilliard (1969: 81) reported 'its calls are powerful, highly ventriloquial bell-like tolls repeated at moderately long intervals'. BB reported that a loud musical *kerrng!* or *Herrng!* note delivered every *c.* 5 sec was heard on 19 Mar at *c.* 1700 m, S. Balim Gorge, IJ. CF reports that the advertisement song of the ad ♂ is a regularly-repeated ringing *weeep* or *pseep* note (Fig. 9.2) that is both ventriloquial and far-carrying, but is not powerful. It is penetrating and demands attention despite

Loria's Bird of Paradise *Cnemophilus loriae* (Salvadori, 1894)

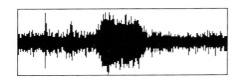

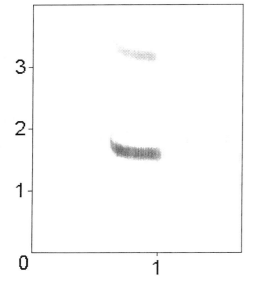

9.2 Sonogram from a recording of the ad ♂ Loria's Bird high-pitched *pseep* advertisement song note by K. D. Bishop, Ambua Ra, PNG in Aug 1988.

its apparent softness, and one suddenly becomes aware of it with the knowledge it has been audible for some time. Gregory (1995) reported that the ad ♂ produces a 'quiet, unobtrusive, rising and upward inflected *zheee* note'. An ad ♂ at 2100 m on the upper Tari Valley slopes, timed during seven singing periods between 1530 and 1730 hr (range 4–35 min), over 3 days between 27 Dec and 4 Jan, gave a note every 9 sec. After every 20–30 normal *weeep* notes this ♂ gave a distinctly-different, soft, tremulous, bleat-like *queeee* note. During the longest watch, of 35 min, ending at 1646 hr (when sunlight hit his perches), the ♂ gave 294 consecutive notes. This ♂ avoided direct sunlight by leaving his singing perches, but remained on them during overcast conditions, notwithstanding drizzle or light rain. During the call his bill was half opened with head and bill lifted sharply upward as the lower mandible was lowered, making the pale wide mouth interior obvious. The bird repeatedly hopped sharply and nimbly to face each direction on perches between singing, presumably to broadcast his song more widely (CF). Based on fieldwork in IJ, BB also reported another distinctive song that a Yali informant told him was produced by the ♂ Loria's: a 4–5 note series of upslurs on a rising scale, musical but coarse and querulous, with the cadence slowing over the series. Two ad ♂♂ chasing one another on Mt. Bosavi, PNG gave *ch-ch-ch* notes (T. Pratt *in litt.*). A ♀ disturbed from her nest and egg was usually silent but once gave a soft, low, rasping, scold note repeated 4–5 times (Frith and Frith 1994*b*). The regular clicking noise given by a displaying ♂ (see Courtship Behaviour) is presumably analogous to a similar sound produced by the Crested Bird. Vocalisations are apparently rarely given except by ♂♂ at singing perches. Stein (1936) reported an ad ♂ giving a 'screaming cry' as it flew away, presumably in alarm.

Mating system
That single ad ♂♂ each attend traditional singing perches and ♀♀ attend the nest alone indicates a polygynous mating system with promiscuous ♂♂, as is confirmed in many other strongly sexually-dimorphic birds of paradise exhibiting these characteristics.

Courtship behaviour
J. Hicks and R. Hicks (1988) observed a single brief behaviour they considered display on the Ambua Ra, PNG. A lone ad ♂ produced a 'regular clicking noise', probably with the bill, while hanging upside down from a perch 6 m high just below the forest canopy with quivering half opened wings for < 10 sec before flying off to feed. No ♂-♀ courtship interactions have been reported. Ad ♂♂ spend considerable periods of early mornings and late afternoons at traditional advertisement perches in the forest canopy. These are either bare and exposed or foliaged, and can be situ-

ated in the forest interior or edge. The ♂ advertises by monotonously-repeated song notes delivered with the mouth opened widely which exposes its whitish interior. A bird will appear to 'yawn' now and then, probably to expose its conspicuous contrastingly-pale mouth (CF). PNG naturalist Akia Aruah (personal communication) observed that a displaying ♂ lowered its head and raised its opened wings in display on singing perches. Gilliard (in Mayr and Gilliard 1954; Gilliard 1969) noted an ad ♂ singing mostly from 1600 hr in the afternoon and that its display perches were branches 21–27 m high and 25 mm in dia, in the canopy of undisturbed forest. Another calling ♂ used perches 25–35 mm in dia (CF). The plumage of ad ♂♂ suggests courtship might consist of the ♂ remaining relatively static near the ♀ while adopting simple postures that show off the highly-modified, iridescent lore feathering and perhaps also the pale gape flanges and/or mouth.

Breeding Table 6.5

Original reports of an active nest of this sp. by Sims (1956) and Loke (1957) were in fact of the Crested Bird. The first information on breeding by Loria's Bird was obtained by Frith and Frith (1994b).

NEST SITE: of nine nests located in an area of some 0.5 km² on the side of the Tari Valley (2150–2200 m), eight were placed on lushly-vegetated rock faces and one upon a moss-covered tree trunk on the steeply-sloping sides of a narrow torrent ravine, at an average of 1.5 m above ground. Appears to specialise in nesting upon well-vegetated, near-vertical rock faces in ravines within wet montane forest. Of the nine nests, five appeared to have been constructed and used in the immediate area of another used in a previous season, as may have two others elsewhere. Such evidence of apparent traditional nesting sites has been found in other promiscuous birds of paradise, and in some bowerbirds (Frith and Frith 1992b, 1993b, 1994a,b, 1995b, 1998b). See Chapter 6 for a discussion of this.

NEST: Majnep suggested a domed nest is built (in Majnep and Bulmer 1977). Frith and Frith (1994b) describe the nest as a substantial, globular, domed structure externally composed of much fresh moss and numerous filmy fern fronds with a horizontally-ovate entrance aperture in the front; inner nest chamber lined with a discrete basket of supple fresh stems of epiphytic orchids. In addition to the external filmy fern fronds, several nests had comb-tooth fern fronds like those used in the same manner by the Crested. As does the Crested, Loria's nest incorporates some 20–30 straight sticks into and upon the nest entrance perch and within the moss below it. Of five nests measured, average total height of the structure was 240 mm, width as viewed from the front 234 mm and depth from the front face of the nest to the outside of the rear wall 178 mm; nest entrance aperture was 101 mm wide and 77 mm high; inner nest chamber 119 mm high floor to ceiling, 128 mm wide wall to wall, and 105 mm deep. Nests *in situ* are extremely cryptic, their external fresh green mosses and ferns matching perfectly the lush plant life on adjacent rock or tree trunk surfaces (Fig. 9.3).

EGGS: pale pink-buff, spotted and blotched sparsely to densely over the entire egg with russet, rufous, tan-browns, and purple-grey

9.3 A recently-used Loria's Bird nest *in situ*. Note extremely cryptic nature of the structure. Photo: CF.

markings, these forming a denser band around the larger end. A complete, freshly-laid, slightly-glossy egg (Plate 13) measured 36.8 × 24.5 mm and weighed 11.3 g (12% of mean ad ♀ weight, n = 6). Clutch probably one egg.

INCUBATION: the above egg hatched on its 26th day of incubation (at 25 ± 1 days old).

NESTLING CARE: unknown but presumably ♀-only. Majnep (in Majnep and Bulmer 1977) stated both ad ♂ and ♀ attend the nestling but all subsequent data on ♂ and ♀ behaviour of this and the Crested Bird makes this likely to be erroneous. A ♀ returning to find a human at her nest containing a pipping egg made no attempt to mob, distract, or perform other anti-predator behaviour, but briefly hopped about immediate perches before departing.

NESTLING DEVELOPMENT: unknown. Two birds collected in the Ilaga Valley, Snow Mts, IJ in early Sept were said to be 'juveniles … not long out of the nest' and similar to ad ♀♀ in plumage colour (Ripley 1964). These are indeed typically ♀-like in plumage, if slightly duller and greyer, with some under-plumage revealed by preparation, but they are not juv.

Annual cycle   Table 6.5
DISPLAY: ad ♂♂ seen singing from advertisement perches on 11 Oct (J. Hicks and R. Hicks 1988). No calling recorded for Apr–May; singing noted during Mar, July, and Sept–Jan (Mayr and Gilliard 1954; Gilliard and LeCroy 1961; Gregory 1995; CF, BB). BREEDING: Nov–Feb. EGG-LAYING: the single known egg-laying date is 1 or 2 Jan near Tari, PNG. MOULT: 59 of 195 specimens were moulting, involving all months but predominantly Feb–Oct (Appendix 3).

Status and conservation   Table 8.1
Appears common and widespread throughout its range. Being inconspicuous except when foraging in fruiting trees, and silent save when ♂♂ are singing, this sp. is undoubtedly more abundant where found than observations suggest. For example, free-flying birds were not seen on Mt. Karimui but seven were mist-netted (Diamond 1972). Likewise, while no free-flying birds were seen during field studies in Tari Gap, five were mist-netted there at close to the upper altitudinal limit of the sp. (Frith and Frith 1993a). Thus, statements such as 'nowhere abundant' should be treated with suspicion. Ripley (1964) claimed ad ♂♂ were rare in the Ilaga forests of IJ owing to hunting for the plume trade, but the latter is unlikely. There is no apparent short-term threat to any of the three subspp. populations of Loria's Bird.

Knowledge lacking and research priorities
This poorly-studied sp. merits a number of field investigations, including descriptions of ♂ display, courtship and mating, ad diet and confirmation of potential total frugivory, and length of nestling period. Appearance of nestling and significance of grey juv plumage require clarification. Age at which ♂♂ acquire ad plumage and longevity in this (and all other) species are unknown. In addition it will be important to map dispersion of ♂♂ on display sites and confirm ♀-only nest attendance.

Aviculture
A ♂ was trapped by Goodfellow in 1909 and kept in the UK collection of a Mr Brook (Seth-Smith 1923a) but no information appears to have been documented. Internal parasites require immediate attention and birds adjust slowly to captivity, individuals preferring an aviary with dark areas to themselves or with few conspecifics only (P. Shanahan *in litt.*). Probably difficult to maintain in areas lacking a cool to cold wet climate and adequate suitable fruits particularly for ♀♀ raising young.

Crested Bird of Paradise *Cnemophilus macgregorii* De Vis, 1890

PLATES 3,4; FIGURES 6.2, 9.4

Other names: Sickle-crested, Multi-crested, Macgregor's, Black and Gold Bird-of-Paradise or Crested Golden Bird. *Wougle-bogamp* on Kubor Mts and *Wo-glia-bora* on Mt. Hagen (Mayr and Gilliard 1954). *Kwßb Ikañ-sek* of the Kalam people (Majnep and Bulmer 1977). *Keko* of the Gimi people, Crater Mt., PNG (D. Gillison *in litt.*).

A medium-sized, compact, and relatively listless bird of upper montane and subalpine forest and shrubbery of the main ranges. Round-winged, weak-billed, and wide-gaped; relatively weak-footed and exclusively frugivorous. The ad ♂ is strikingly patterned in orange and blackish, whereas the ♀ is uniformly brown-olive, rather bowerbird-like, and slightly bulkier than the very similar ♀-plumaged Loria's Bird. A distinct grey juv plumage appears to be worn briefly. Polytypic; 2 subspp.

Description

ADULT ♂: 24 cm, culmen broadly flat towards skull, gape wide, tail graduated. Forehead, upper ear coverts and entire upperparts brilliant, silky, flame yellow with orange wash (18), and iridescent white highlights, brightest on head and neck and becoming duller to Buff (24) on back and to Cinnamon (123A) on wings and dorsal surface of tail. Sickle-shaped feathers (4 to 6) of small, erectile (usually concealed), sagittal crest dark Buff to Cinnamon with golden iridescent gloss. Lores, lower ear coverts, a tiny narrow line above the central eye and entire underparts deep brownish black with rich coppery-bronze, dull iridescent sheen in suitable light. Variable amount of Cinnamon feathers on thighs and flanks. Underwing: coverts variably blackish grey, flight feathers Ground Cinnamon (239) with broad paler Buff (124), trailing edges. Ventral surface of tail pale Cinnamon-Brown (33) with paler central feather shafts. Bill dark brownish black, iris dark brown to bluish grey, legs and feet purplish brown to brown-black, mouth pinkish.

ADULT ♀: very slightly smaller than ♂. Head and entire upperparts Dark Brownish Olive (129) but with rich Cinnamon-Brown suffusion on back, rump, wings, and tail. Crest feathers the colour of the crown but much shorter and straighter than in ad ♂. Malar area, chin, throat, and breast a pale and grey-washed Raw Umber (123) becoming paler to creamy, slightly yellowish, Cinnamon (39) on lower breast and belly. Underwing: coverts Cinnamon, flight feathers grey-washed Olive-Brown (28). Thigh and undertail coverts as breast. Undertail Dark Brownish Olive with paler central feather shafts. Bill brownish black, iris dark brown-grey, legs and feet dark brownish to brownish black, mouth pale green.

SUBADULT ♂: variable—as for ad ♀ with few feathers of ad ♂ plumage intruding, initially on back and rump, to as ad ♂ with few feathers of ♀ plumage remaining.

IMMATURE ♂: as for ad ♀ with paler bill, legs, and feet, and iris brownish to brown-grey.

NESTLING–JUVENILE: newly-hatched nestling naked and becoming darker-skinned, as in almost all birds of paradise. Two birds collected as free-flying individuals (Kubor Ra, Mt. Hagen) wore a grey plumage, with brown in flight feathers, lacking all suggestion of yellow pigment (Frith and Harrison 1989). All three nestlings seen to date were in grey plumage with rufous in flight feathers (see Coates 1990: plate 427). A Mt. Hagen juv (wing length 104 mm) was in grey-green plumage, downy underneath, with pointed tips to rectrices (Mayr and Gilliard 1954; Frith 1996). There is little doubt that nestlings and juvs wear, if only briefly (first juv plumage), a grey plumage distinct from that of ♀♀ and imm ♂♂ (Sims 1956; Coates 1990; Frith and

Frith 1993c; Frith 1996). A *C. m. sanguineus* specimen (BMNH 1953.17.215) described as juv and indistinguishable from an ad ♀ (Sims 1956) appears imm to us.

Distribution

Confined to the highest forests of the central cordillera, from the central Owen Stanley Ra (Mt. Knutsford, Musgrave, Victoria, and Tafa) northwestward through PNG on the higher summits and into IJ apparently at least as far as Jabogema, N of Lake Habbema in the upper Ibele Valley, a tributary of the Balim (B. Poulson *in litt.*). Irian records are relatively recent, and indicate that the sp. had long been overlooked in spite of considerable zoological collection. These new records imply it should be looked for westward as far as the Weyland Ra and, in the SE of PNG, should be expected at least as far as Mt. Nisbet, and probably even as far as Mt. Suckling and Dayman. Occurs patchily in part owing to its restricted altitudinal range of 2100–3650 m (mainly 2600–3500 m).

Systematics, nomenclature, subspecies, weights, and measurements

SYSTEMATICS: a very distinctive sp., but the relationship to Loria's Bird is confirmed by the juv and ♀ plumages and nest and egg morphology. Originally described by De Vis as a bowerbird. A. P. Goodwin placed the sp. in *Xanthomelus [Sericulus]* in the belief that it was a regent bowerbird, to which ad ♂♂ bear a strong superficial resemblance. Hybridisation: unrecorded. The sp. is sympatric with Loria's at 2600 m on the slopes of the Hagen Ra and both occur between 2200 and 2650 m on the Ambua Ra (Cooper and Forshaw 1977; Frith and Frith 1992a, 1993a). Hybridisation is therefore possible. Hybridisation with Paradisaeinine spp. is unlikely, but should it occur, and the ♀ parent be a Crested, fledged young would be unlikely to survive owing to the nestling diet of only fruit provided by the parent.

SPECIES NOMENCLATURE AND ETYMOLOGY

Cnemophilus macgregorii De Vis, 1890. *Annual Report of British New Guinea 1888–89*, p. 62 Mt. Knutsford, Owen Stanley Mountains. See also De Vis (1891).

Synonym: *Xanthomelus macgregori* Goodwin, 1890 (Apr). *Ibis* 2, series 6 No. 6, 153. Mt. Musgrave, Owen Stanley Ra.

Etymology: *macgregorii* = after Sir William Macgregor, then Administrator of British NG,

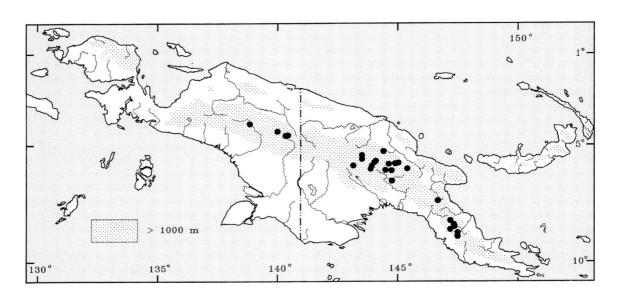

who first collected the bird on Mt. Knutsford in 1889. As an English name, we use Crested Bird of Paradise because this is the only crested sp. and the name is both brief and informative.

SUBSPECIES, WEIGHTS, AND MEASUREMENTS

1. *C. m. macgregorii* De Vis, 1890. Type specimen QM O. 19429.

Range: SE PNG northwestward at least to the Ekuti Divide, E of the Watut/Tauri Gap. Exact extent of range in the SE is unknown. The undiagnosed population from the Kratke Ra is reputed to be pale, like this subsp. (Gilliard 1969), and may merit inclusion in this subsp. or else may be intermediate between this and *C. m. sanguineus* to the W.

Measurements: wing, ad ♂, (n = 22) 107–117 (113), imm ♂ (n = 7) 107–116 (111), ad ♀ (n = 10) 104–114 (108); tail, ad ♂ (n = 22) 86–96 (92), imm ♂ (n = 7) 91–103 (94), ad ♀ (n = 9) 89– 101 (93); bill, ad ♂ (n = 24) 24–32 (29), imm ♂ (n = 6) 28–31 (30), ad ♀ (n = 10) 25–29 (27); tarsus, ad ♂ (n = 28) 38–46 (41), imm ♂ (n = 7) 40–42 (41), ad ♀ (n = 9) 36–43 (39). Weight: ad ♂, (n = 4) 94–104 (98), imm ♂ (n = 1) 81, ad ♀ (n = 1) 91.

2. *C. m. sanguineus* Iredale, 1948. *Australian Zoologist* **11**, 162. Kumdi, Mt. Hagen District. Type specimen AM O. 37683.

Synonym: *C. m. kuboriensis* Mayr and Gilliard, 1954. *Bulletin of the American Museum of Natural History* **103**, 361. Mt. Orata, Kubor Mountains. Type specimen AMNH 748584.

Diagnosis: ad ♂ easily distinguished from those of the nominate form by the considerably richer and reddish dorsal plumage. More orange-red (15) head, less red (16) on back to orange (17) rump; underparts with less copper-red suffusion. Mean wing length slightly longer, but tail shorter than that of nominate form. Characters supposedly distinguishing *C. m. kuboriensis* are variable, minor, and insufficient to warrant subspecific status (Diamond 1972; CF).

Range: known from the central and Eastern Highlands of PNG (Doma Peaks, Mt. Hagen, Giluwe, Karimui, Kubor Ra), W of the range of the nominate subsp. Exact delineation of distribution is impossible, pending diagnosis of the population in the Kratke Ra, as well as those that occur W of Tari into IJ. The only known specimens from IJ were preserved originally in fluid, which may have altered plumage colours. Clearly, fresh material from the Krakte Ra and IJ are required.

Measurements: wing, ad ♂ (n = 30) 110–118 (115), imm ♂ (n = 17) 108–119 (114), ad ♀ (n = 24) 101–115 (111); tail, ad ♂ (n = 29) 86–96 (90), imm ♂ (n = 16) 87–97 (90), ad ♀ (n = 24) 80–97 (89); bill, ad ♂ (n = 26) 26–31 (29), imm ♂ (n = 17) 26–30 (28), ad ♀ (n = 23) 23–28 (26); tarsus, ad ♂ (n = 30) 38–44 (41), imm ♂ (n = 17) 39–43 (41), ad ♀ (n = 24) 38–43 (40). Weight: ad ♂ (n = 6) 94–120 (103), imm ♂ (n = 2) 87–97 (92), ad ♀ (n = 10) 79–125 (93).

Habitat and habits   Table 4.1

Upper montane and subalpine forest and forest-edge, including second-growth, disturbed vegetation, and shrubbery. This obligate frugivore may be adapted to the fruiting plants of forest-edge vegetation, which forms a significant vegetation formation within its range (Frith and Frith 1993c). Predominantly frequents middle and lower forest strata. Foraging birds typically inconspicuous but when located are not particularly wary, whereas birds not feeding are often wary of humans. An ad ♂ and two ♀-plumaged birds were harassed and displaced from their feeding tree by a Common Smoky Honeyeater *Melipotes fumigatus* (Cooper in Cooper and Forshaw 1977). Usually encountered as lone birds unless an aggregation has formed at a fruiting tree, but two 'pairs' of birds once seen to remain together, c. 30–40 m apart, for c. 10 min (M. Laska *in litt.*).

Diet and foraging   Table 4.2

Of 19 faecal samples from ad birds at Tari Gap during Sept–Oct, all contained only fruit except that three samples contained tiny shelled molluscs < 3 mm dia. Fruit identified from these samples included *Zygogynum argentia*, *Timonius belensis*, *Riedelia* sp., *Rapanea* sp., *Acronychia kaindiensis*, *Schefflera* sp., *Xanthomyrtus* sp., *Ficus* sp., *Dimorphanthera alpinia*, *Alpinia* spp., *Garcinia* sp., *Elaeocarpus* sp., and *Syzygium* sp. (see also Appendix 4). Virtually all fruits in the diet are simple drupes or berries 3–12 mm dia (Diamond 1972) plucked and swallowed whole without manipulation by bill and/or feet (see Chapter 4). A free-flying ♀-plumaged individual ate 4 × 3 mm fruits of *Symplocos cochinchinensis* on 10 Oct in the Tari Gap, PNG (Hicks 1988b). Of six direct observations of feeding birds, five involved singletons and one a group consisting of two ad and one imm ♂. Several ♂- and ♀-plumaged birds may gather at one fruiting plant where other spp., including birds of paradise such as Stephanie's, may be feeding (Cooper and Forshaw 1977; Coates 1990). At one tree bearing small red to black berries in Tari Gap, PNG, 20–30 Crested Birds fed together, including up to nine ad ♂♂, with Loria's, Brown Sicklebills, and Ribbon-tails (J. Tano personal communication). May forage at 18–25 m in the forest but most activity has been observed between the ground and 12 m up, mostly closer to the ground, in dense forest-edge or second growth (Rand in Gilliard 1969). One bird said to have foraged in forest floor leaf litter (Hoyle 1975). Twenty-seven birds were caught in Tari Gap forests during 49 days of mist-netting with nets from ground to 3 m high, which indicates that birds do fly low in the vegetation.

Vocalisations and other sounds

Less vocal than most birds of paradise. Voice inadequately known. Harsh and rasping sounds are reported and an explosive muffled bark repeated at long intervals said to be similar to a Macgregor's Bowerbird *Amblyornis macgregoriae* call (Diamond 1972; Beehler et al. 1986). The latter is a loud RRRAK given every 4–5 min (G. Opit *in litt.*). Rand (in Mayr and Rand 1937) noted a low, harsh, hissing call, a loud clicking call repeated a number of times, and a loud call similar to that of two timbers rubbing together under stress. A prolonged squeak like the creaking of a rusty gate (presumably the latter call) and a rasping *aa-aah* or *haah* (Fig. 9.4) are reported (Cooper in Cooper and Forshaw 1977). An ad ♂ held captive and confined for 2 hr with a ♀-plumaged bird produced a low pleasant purring and the other bird replied with softer purring (Peckover 1990). A ♀ attending a nest and nestling gave a soft *wark, wark* as she approached them (Loke 1957). At Tari Gap a ♀ approaching her nest and nestling gave a repeated, soft, sharp *whit* and when disturbed there produced a soft churring-growl scold. The begging nestling produced a high-pitched, sharp, metallic-sounding, squeaky bleating (Frith and Frith 1993c). Flight wing-beats of ad ♂♂ produce a loud whirring noise. Whether ♂ has an advertisement vocalisation is unknown.

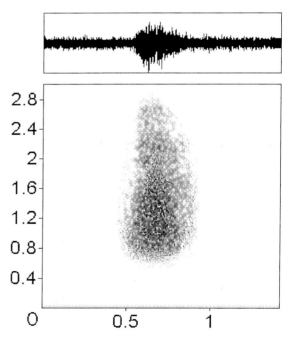

9.4 Sonogram from a recording of a Crested Bird, giving a short sharp harsh *haah* note, made near Okbap, Star Mts-IJ on 7 Apr 1993 by K. D. Bishop.

Mating system

Apparently polygynous, with ♀-only nest attendance (Loke 1957; Frith and Frith 1993c), which accords with the marked sexual dimorphism. An ad ♂ was observed carrying a stick in its bill (Clapp 1986) and another was shot (BMNH 1949.62.96) at a Macgregor's Bowerbird bower. A number of PNG informants indicated that ad ♂♂ visit Macgregor's Bowerbird bowers, while one specifically expressed the view they do not build a bower of their own (Bulmer in Majnep and Bulmer 1977). These events are inexplicable but the possibility that ♂♂ do somehow utilise woody sticks cannot be ruled out.

Courtship behaviour

In early July on Mt. Giluwe at 2440 m asl, an ad ♂ appeared to have a 'territory of a couple of hundred yards in dia [which he patrols] constantly using the same trees in succession' (Opit *in litt.*) but it is not yet clear if ♂♂ are territorial or maintain traditional display areas or perches. Courtship display unknown.

Breeding   Table 6.5

A 25 Sept photograph of a bird at the nest was first published as that of a Loria's Bird (Gyldenstolpe 1955a) but was subsequently correctly identified as a Crested (Loke 1957). The photograph shows a ♀-plumaged individual at a globular domed nest of mosses and ferns, a fact of great interest as no domed bird of paradise nest was then known.

NEST SITE: atop a decayed mossy tree stump, on the side of a mossy tree trunk or within the branches of a tree and associated vegetation at an average of 2.6 m (range 1.9–3.7 m, $n = 6$) above the ground (Sims 1956; Frith and Frith 1993c).

NEST: the nest photographed by Loke was of green ferns and moss with a lining of fern stalks (Sims 1956). Five additional nests were all roughly globular with a horizontally-ovate entrance hole in the middle of their front (Frith and Frith 1993c). The basic external structure was a substantial accumulation of green mosses heavily decorated or camouflaged on top, around the sides, and about and below the entrance perch with fresh green fern fronds with conspicuous 'comb-tooth' pinnae, probably of *Blechnum* or *Doodia* spp. A meagre nest foundation consisted of woody sticks 250–300 mm long and averaged 3.5 mm dia, mostly beneath the front of the nest but some extending to 70 mm in front of the nest entrance to form a 'ramp' overlaid with fern fronds. The nest chamber was lined exclusively with innumerable long, fine, supple, green-yellow, epiphytic orchid stems, probably *Glossorhyncha* spp., 1–2 mm in dia and *c.* 250–350 mm in length, some with two or three branching stems. The egg cup was lined with finer, supple, orchid stems up to the level of the entrance perch. Nests varied in their number of foundation sticks and number of conspicuously 'comb-toothed' fern fronds, one nest lacking them. Of five nests measured, average height was 220 mm, width 197 mm, and depth 175 mm. Nest entrance aperture 122 mm wide and 74 mm high and inner nest chamber 126 mm high, 107 mm wide, and 123 mm deep. Nests were extremely cryptic *in situ* (see Plate 3), their external fresh green plant material blending perfectly with surrounding vegetation (Frith and Frith 1993c). Rothschild (1898) described a nest and egg collected and ascribed to this sp. by A. S. Anthony. This nest was subsequently lost but, in the light of subsequent discoveries (Frith and Frith 1993c), it is now clear from the published measurements, the description of its materials, and the appearance of the egg collected with it (see below), that it could not have been that of the Crested.

EGGS: a complete clutch (probably typically one egg) remains to be found, but egg fragments found in a used nest clearly indicate a pale pinkish buff ground colour with broad purple-grey, grey-pink, and pale russet irregular blotches, similar to that of Loria's (Frith and Frith 1993c, 1994b; see Plate 13). The egg attributed to this bird since 1898 is unlike any known bird of paradise egg in shape, ground colour, and markings. Ernst Hartert (1910) pointed this out and Shane Parker expressed this view to E. Thomas Gilliard who subsequently examined the egg and accepted it as that of the Crested (Gilliard 1969). The measurements of the nest in which it was laid clearly indicate, however, an open cup structure far smaller than the substantial domed nest photographed by Loke. Moreover, this nest was lined with some (cassowary) feathers, a feature unknown in bird of paradise nests (Frith and Frith 1990c, 1992b, 1993b,c, 1994b, 1995b).

INCUBATION: unobserved but doubtless by ♀ only. In view of its close relationship to Loria's Bird and higher altitudinal range incubation is likely to be c. 26 days or more.

NESTLING CARE: exclusively by ♀ parent. Nestlings are fed exclusively regurgitated fruit (save the odd tiny shelled mollusc that may be unintentionally provisioned by the parent). The only identified fruit observed to be eaten by a nestling was of *Alpinia tephrochalmys*. Some seeds are regurgitated by the nestling, into or out of the nest. During 53 hr of observation at one nest containing a nestling > 4 days old, the ♀ averaged 3 visits per hr with a nestling meal. She spent an average of 22% of her time at the nest and an average of 4 min there at each visit. Brooding bouts averaged c. 1 per hr, but none occurred during the last 2 weeks of nestling life. Average brooding time was 10 min, and all brooding totalled an average of 14% of observation time. Parental absence from the nest averaged 16 min per bout. ♀♀ at two nests containing a nestling performed no mobbing, distraction, or other anti-predator behaviour upon returning to find a human at their nest (Frith and Frith 1993c).

NESTLING DEVELOPMENT: development slow, the nestling period of > 30 days being prolonged (Fig. 6.2), for any passerine of this size, possibly reflecting limitations of altitude where conditions are relatively cold and wet and food resources limited (see Chapter 6).

Annual cycle Table 6.5

DISPLAY: birds with enlarged gonads collected during June–Nov. Marshall (1954) stated the courting season is well established in June on Mt. Hagen and full sperm production by ♂♂ takes place July–Sept and mating before mid Nov. BREEDING: at least Aug–Jan. EGG-LAYING: assuming an incubation of c. 26 days (cf. Loria's, Frith and Frith 1994b) and a nestling period of 30 days (cf. other birds of paradise at similar altitudes) the egg laid in the nest photographed by Loke must have been laid c. 3 Aug (see below). The only two nests found since Loke's nest each contained a nestling and their eggs would have been laid on c. 28 Nov and 18 Dec. Thus, the known laying period is Aug–Dec in PNG central highlands. MOULT: 19 of 96 specimens were moulting during June–Dec, 89% of them moulting in July–Nov (Appendix 3). Of seven birds captured in Tari Gap, PNG in Sept, 57% were moulting and of 19 birds caught in Oct 58% were moulting (Frith and Frith 1993a).

Status and conservation Table 8.1

Wherever researchers have mist-netted in its habitat, the Crested Bird has been found to be common, but being otherwise inconspicuous and unobtrusive. Fairly common on Mt. Tafa and the Murray Pass (Rand in Gilliard 1969). Gilliard (1969: 90) collected 20 birds on Mt. Hagen, between 2400 and 3600 m. Kwapena (1985: 69) presented limited data suggesting that numbers of birds above 2600 m on Mt. Giluwe increased during the dry season, as birds moved to their upper altitudinal limit. Skins of ad ♂♂ are rarely used as human adornment. Near Goroka Gilliard (1969) found someone wearing one as a headdress. No immediate or long-term global threats to the bird or its habitat are apparent.

Knowledge lacking and research priorities
This sp. merits considerable attention, especially with regard to vocalisation, ♂ display, mating system, and nutrition and energetics. As a polygynous passerine in which both ad and nestling diet is apparently fruit-only, it would be interesting to determine how developing nestlings obtain adequate nutrition. A fuller understanding of its ecology and mating system is much desired as it will doubtless prove significant to theoretical considerations of the ecology and evolution of non-monogamous mating systems in tropical birds (see Chapter 5). The single observation of possible territoriality in ad ♂♂ requires confirmation and it is important to know if ♂♂ maintain a display area or perch(s) and how they advertise and court ♀♀.

Aviculture
Individuals have been kept at BRS (1200 m), near Mt. Hagen, PNG, considerably below their natural elevation. They did not do well in aviaries containing other birds possibly because of interference from fellow captives, the heat of lower altitude, and an inappropriate diet. Future attempts to keep and breed the sp. would doubtless result in valuable knowledge, but should not be attempted unless attention can be lavished upon birds in an appropriate location and climate.

Genus *Loboparadisea* Rothschild, 1896

Loboparadisea Rothschild, 1896. *Bulletin of the British Ornithologists' Club* 6, 15. Type, by monotypy, *L. sericea* Rothschild.

The Yellow-breasted Bird of Paradise, a little-known, slow-moving bird of midmontane forest, is the sole sp. in the **monotypic genus** *Loboparadisea*, which, together with *Cnemophilus*, constitute the subfamily Cnemophilinae. The **skull** and lower jaw are shorter and broader in *Loboparadisea* than in closely-related *Cnemophilus* but they are otherwise nearly identical save that the transpalatine process is broader in *Loboparadisea* (Bock 1963). **Bill** slightly shorter than the head and proximal culmen broadly flattened, as in *Cnemophilus macgregorii*.

Walter Rothschild correctly described this sp. as a bird of paradise but Bowdler Sharpe, and others, thought it a bowerbird (Sharpe 1891–8). We do not hold to the view of Schodde (1976) that this genus should be synonomised with *Cnemophilus*, given *Loboparadisea's* unique wattles, the pale underparts and rump of ad ♂♂, the ♂-like colourful appearance of the ♀, the ♀ being larger than the ♂, and the distinct rufous imm plumage. No **hybrids** known.

♀♀ **larger and heavier than** ♂♂. Ad plumage only **slightly sexually dimorphic**, the ad ♀ being only slightly paler than the ad ♂. Imm plumage uniquely rufous with dark ventral streaking (Plate 4). **Bill** relatively fine and weak, gape very wide. **Wings** short, rounded, and flight feathers deeply curved, primaries not modified in shape. **Tail** relatively very short (61% of wing length), rounded, and slightly graduated. In ad ♂, the base of the maxilla is decorated with a bulbous, swollen, aqua blue-green **wattle** covering and enclosing the nostrils, and mandible base and gape decorated by soft bare skin of this colour extending from the wattles. **Tarsus** smooth and its length 33% of wing length. Ad of both sexes, uniquely, have contrastingly rufous-coloured feathers on the thighs. Juv and imm birds wear brown and ventrally-mottled plumage distinctly different to that of both ad sexes and to the grey juv/imm plumage of *Cnemophilus* spp.

Mating system and nidification unknown.

Yellow-breasted Bird of Paradise *Loboparadisea sericea* Rothschild, 1896

PLATE 4; FIGURE 9.5

Other names: Wattle-billed Bird of Paradise, Wattled Bird of Paradise, Yellow-chested Bird-of-Paradise, The Shield-bill, Shield-billed Bower-bird, Shield-billed Bird of Paradise. *Jungam* of Victor Emanuel Mts (Gilliard and LeCroy 1961). *Fiotumo* of the Gimi people, Crater Mt., PNG (D. Gillison *in litt.*).

A short-tailed and compact little bird; inconspicuous, sluggish, quiet, and virtually unknown in the wild. The most distinctive sp. character is the silky-yellow plumage of the breast, belly, and rump, contrasting the rich honey-brown of remainder of the dorsal plumage. Imm is dark olive-brown above and the paler, more buff-brown, underparts streaked with darker markings. Imm looks much like a small ♀-plumaged Streaked Bowerbird *Amblyornis subalaris* (but the spp. never co-occur). Polytypic; 2 subspp.

Description

ADULT ♂: 17 cm, proximal culmen broadly flattened, gape wide, tail slightly graduated. Bilaterally-bifurcate swollen wattles atop maxilla and bare skin over mandible bases palest, chalky, Turquoise Green (64) or yellow. Sides of face, lores, and crown dark Raw Umber (23) with rich coppery-green sheen on crown. Nape, mantle, and upper back rich Raw Sienna (136), with darker greyish feather tipping, washed iridescent coppery yellow. Lower back and rump pale, silky, iridescent Sulphur Yellow (157). Upper wing, tail, and coverts Raw Sienna with darker primary, secondary, and tail tips. Malar area and entire underparts silky/glassy Sulphur Yellow. Underwing rich Cinnamon (39) with dark grey wash on flight feather tips. Thigh feathers Raw Umber. Bill blackish, iris dark brown, legs and feet blackish, mouth dull-coloured, lacking in bright pigment. Colour of naral wattle may be variable. Ogilvie-Grant (1915) noted that the naral wattles of live ad ♂♂ from the Utakwa R., Snow Mts, IJ were yellow and 'not dull blue with yellow tips as shown in Rothschild's plate'. Several ♂♂ trapped and photographed by BB on the Trauna Ridge above BRS, PNG, had pale green wattles.

ADULT ♀: slightly larger and averaging 14% heavier than ♂. Lacks wattles. Sides of face, crown, mantle, upper back, dorsal surface of tail, and coverts Cinnamon-Brown (33) with Amber (36) wash. Lower back and rump silky Sulphur Yellow mixed and washed with upper back colour. Upperwing Cinnamon-Brown strongly washed Amber, the latter colour almost pure on outer edges of flight feathers. Malar area and entire underparts as ad ♂ but mixed and washed with dilute Amber more so on flanks and undertail coverts. Underwing as for ad ♂. Thigh feathers Cinnamon-Brown. Ventral surface of tail Olive-Brown (28) washed Amber. Note: the ♀ plumage description of Rand and Gilliard (1967: 505) actually refers to the imm plumage.

IMMATURE: no wattles; wing and tail slightly larger than ad ♂. First-year birds with dark Olive-Brown upperparts strongly washed amber on wings and tail, lacking yellow, and underparts warm Cinnamon (123A) with dark breast feather edging forming broad greyish streaking. Belly pale greyish. Second-year plumage of both sexes as ad ♀ but darker below and less yellow above, ♂ with partially-developed wattles. Peckover (1990) includes a colour photograph of an individual in this plumage that is incorrectly identified as an imm ♂ Loria's Bird.

NESTLING: unknown. JUVENILE: the above first-year plumage may in fact be that of juvs but given the large number of specimens obtained wearing it and that both the other cnemophilinine spp., Loria's and the Crested

Yellow-breasted Bird of Paradise *Loboparadisea sericea* Rothschild, 1896

Birds, briefly wear a grey juv/imm plumage (rare in collections) it is just possible the Yellow-breasted briefly wears a distinctive juv plumage.

Distribution

Patchily distributed on the Central Ra of NG, from the Weyland Mts of IJ eastward as far as the Herzog Ra and Wau Valley, PNG, just east of the Watut/Tauri Gap. Has been recorded from 600 to 2000 m but mainly above 1200 m. More patchy and apparently scarcer in the E. Apparently absent from the Vogelkop, Huon Penin, and peninsular SE NG. Present on Mt. Bosavi (T. Pratt *in litt.*).

Systematics, nomenclature, subspecies, weights, and measurements

SYSTEMATICS: clearly the third member of the wide-gaped birds of paradise (Cnemophilinae). One authority has suggested the genus *Loboparadisea* be merged into the genus *Cnemophilus* (Schodde 1976). It is, however, a distinctive sp. best retained in its monotypic genus in view of its ad and juv/imm morphology and plumage (Frith and Frith 1994*b*).

Described as a bird of paradise from a specimen purchased at Kurudu I., just off the East coast of Yapen I.; this specimen was probably taken in the Weyland Mts by native trade collectors (Mayr 1941*b*: 179). Sharpe (1891–8) considered it a bowerbird. Ad ♀ larger than ♂ in the Weyland Mts but not on Mt. Karimui (Diamond 1972). Hybridisation: unrecorded.

SPECIES NOMENCLATURE AND ETYMOLOGY

Loboparadisea sericea Rothschild, 1896. *Bulletin British Ornithologists' Club* **6**, 16. Dutch New Guinea (restricted to Weyland Mts by Mayr 1941*a*).

Etymology: *Loboparadisea* = Gr. *lobos*, a lobe; *paradisea*, a genus of bird of paradise.

SUBSPECIES, WEIGHTS, AND MEASUREMENTS

1. *L. s. sericea* Rothschild, 1896. Type specimen AMNH 679048.

Range: the western two-thirds of the species range, from the Weyland Mts of IJ eastward at least to the Victor Emanuel and Kubor Ra and Mt. Karimui and Soliabeda (Soliabedo) of PNG.

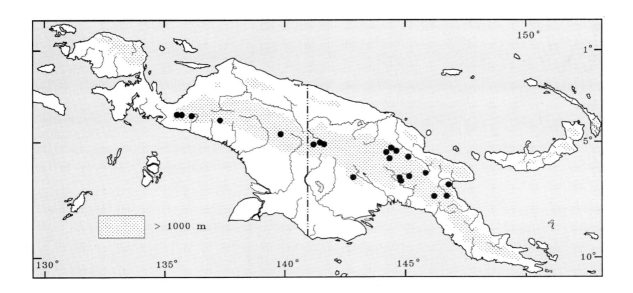

Measurements: wing, ad ♂ (n = 17) 90–98 (94), imm ♂ (n = 3) 96–98 (97), ad ♀ (n = 6) 96–101 (98), subad ♀ (n = 10) 92–99 (97); tail, ad ♂ (n = 16) 52–59 (55), imm ♂ (n = 4) 56–66 (60), ad ♀ (n = 6) 55–63 (58), subad ♀ (n = 10) 56–63 (60); bill, ad ♂ (n = 16) 19–23 (20), imm ♂ (n = 4) 20–21 (20), ad ♀ (n = 6) 21–22 (21), subad ♀ (n = 9) 20–21 (21); tarsus, ad ♂ (n = 16) 29–34 (31), imm ♂ (n = 4) 31–33 (32), ad ♀ (n = 6) 31–33 (32), subad ♀ (n = 10) 30–33 (31). Weight: ad ♂ (n = 2) 72 (72), imm ♂ (n = 1) 63, subad ♀ (n = 2) 75–77 (76). Measurements of imm include 1 subad ♂ and 4 subad ♀♀ as their measurements did not differ from those of imm.

2. *L. s. aurora* Mayr, 1930. *Ornithologische Monatsberichte* **38**, 147. Dawong, Herzog Mountains. Type specimen ZMB 30.1933.

Range: described from material from the Herzog Ra, E of the Watut/Tauri Gap, but also material from the upper Jimi R. is ascribed to this subsp. by Gilliard (1969). A photograph of a ♂ from the Trauna Ridge above BRS (Coates 1990, 428) shows the greenish crown attributable to *aurora* but also shows a strong contrast between crown and back, typical of the nominate subsp. Without explanation, Cracraft (1992) apparently rejects Diamond's (1972) attribution of birds from the Mt. Karimui, PNG area to the nominate form and considers them *L. s. aurora*. A single ad ♂ specimen in the BMNH, purchased from R. von Duivenbode in Dec 1910, is far darker and more chestnut brown above, particularly on nape, mantle, and upper back, than the presently-known forms and possibly represents an undescribed subsp. that inhabits some part of IJ.

Diagnosis: ad, on average, fractionally larger than those of nominate form, more so in tail length. Plumage of upperparts significantly brighter (paler), more brown-yellow (24) the crown far paler and more greenish (46), less brown, and underparts similar to nominate subsp. Mayr (1930*a*) noted the naral wattle to be 'pale blue'.

It is evident that these are poorly-defined subspp.; perhaps the pattern of variation is not consonant with any useful delineation of distinct and definable subspp. Perhaps colour of naral wattle is diagnostic, but this will require additional fresh material from the W and E.

Measurements: wing, ad ♂ (n = 16) 93–100 (97), imm ♂ (n = 2) 95–96 (96), ad ♀ (n = 3) 91–101 (97), subad ♀ (n = 7) 93–102 (99); tail, ad ♂ (n = 16) 56–61 (59), imm ♂ (n = 2) 62–64 (63), ad ♀ (n = 3) 62–63 (63), subad ♀ (n = 7) 59–69 (63); bill, ad ♂ (n = 16) 20–24 (21), imm ♂ (n = 2) 21–22 (21), ad ♀ (n = 3) 22–23 (23), subad ♀ (n = 7) 20–22 (21); tarsus, ad ♂ (n = 16) 29–33 (31), imm ♂ (n = 2) 29–32 (31), ad ♀ (n = 3) 30–33 (31), subad ♀ (n = 7) 30–33 (32). Weight: ad ♂ (n = 12) 50–75 (62), imm ♂ (n = 2) 68–72 (70), ad ♀ (n = 2) 71–75 (73), subad ♀ (n = 3) 60–77 (70). Measurements of imm include 1 subad ♀ as its measurements did not differ from those of imm.

Habitat and habits Table 4.1

Appears to be a bird of the interior of midmontane forest, in contrast to its fellow cnemophilines (Loria's and Crested) which commonly frequent forest-edges. A lone ad ♂ once flushed off a road (Gregory 1995). The scanty information is contradictory: several observers indicate birds predominantly inhabit the upper canopy level of the forests whilst others report it in the lower strata. Stein (1936) noted that birds in the Weyland Mts were never high in the forest. A subad ♂ collected for Gilliard (in Gilliard and LeCroy 1961) was said, however, to have been with a conspecific in the crown of a high forest tree. Diamond (1972) did not personally see or mist-net the sp. in the Eastern Highlands, which he took to support the statements of his local field assistants who told him they found birds in treetops. On the other hand an ad ♂ foraged over moss-covered vines and tree trunks *c.* 3 m from the ground in forest on S slopes of the Bismarck Ra (Cooper in Cooper and Forshaw 1977) and Beehler (1991*a*) netted birds at that level in forest. Quiet, lethargic, and wary bird of the lower forest where it moves slowly from branch to branch or sits for long periods between short rapid flights, usually alone but sometimes in parties of up to six (Rand and Gilliard 1967). Thane Pratt (*in litt.*) noted the sp. on Mt. Bosavi to be a very fast and unsystematic feeder dashing about in tree canopies and thought birds might pick only the ripest of fruits. Bishop (1987) saw and heard birds commonly in the forest sub-canopy at *c.* 1800 m on Mt. Pugent (Trauna Ridge), PNG during 7–9 Aug 1986, and BB observed individuals quietly foraging at this same locality, in mossy canopy branches, and mist-netted three individuals (all ♂) on a sharp ridge top. Bulmer informed Gilliard (1969: 85) that people of the Schrader Mts told him this bird was fairly common in the forest-edge and in bush fallow, but this was a misidentification of the Wattled Ploughbill *Eulacestoma nigropectus* of the Pachycephalidae (Majnep and Bulmer 1977).

Diet and foraging   Table 4.2

Diet almost exclusively fruit: simple drupes and berries plucked and swallowed whole; only a single bird has (of 6 examined) been found with arthropods in its stomach. Two birds from one location examined by Schodde (1976) contained only fruits as did a faecal sample of one from Trauna Ridge (BB). ♀-plumaged individuals were twice seen to visit and forage on soft and fleshy 8 mm dia *Ficus* fruits from a canopy vine in a large forest opening (BB). Beehler and Pruett-Jones (1983) tentatively considered it possible this bird is 100% frugivorous (but see above). That many attempts at maintaining birds in captivity have rapidly failed certainly suggests this sp. has specialised requirements. Single birds or feeding associations of up to 10 were noted in fruiting trees or undergrowth fruiting bushes (Stein 1936).

Vocalisations and   Figure 9.5
other sounds

The only known call is described as 'a series of loud, harsh, grating notes slightly upslurred, *sssh sssh sssh*, usually two notes followed by a brief pause and again two to three notes. As the series continues the notes become slower and stronger but lower-pitched than the Superb's (Bishop 1987). See Fig. 9.5. Bishop heard an ad ♂ calling and noted both sexes responded to conspecific vocalisations.

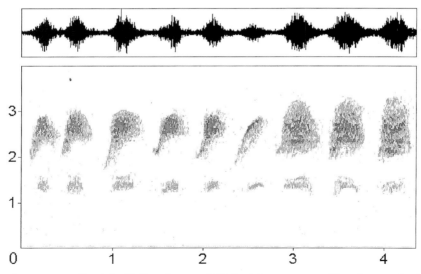

9.5 Sonogram from a recording of a Yellow-breasted Bird, giving a repeated harsh screeched note not unlike that of the Superb Bird, made at Trauna Ridge above Baiyer R. on 8 Aug 1986 by K. D. Bishop.

Mating system
Unknown but presumed to be polygyny. Clarification of mating organisation is eagerly awaited in view of the unusual disposition of plumages and the relatively small differences between the sexes, leading several ornithologists to speculate the sp. might be monogamous (Schodde 1976; LeCroy 1981).

Courtship behaviour
Unknown. That only ad ♂♂ are adorned with nasal wattles, and that they appear to require a number of years to attain them, suggests that these are significant secondary-sexual characters.

Breeding   Table 6.5
Inhabitants of the Weyland Mts stated that the sp. places its nest in branches in vegetation relatively close to the ground (Stein 1936).

NEST: unknown, but said to be an open moss structure (Stein 1936). EGGS: Stein (1936) stated a single-egg clutch was reported by a local informant.

INCUBATION, NESTLING CARE AND DEVELOPMENT: unknown.

Annual cycle   Table 6.5
DISPLAY: unknown. BREEDING: ♂♂ collected with much-enlarged gonads during May and Oct and with moderately-enlarged ones in June and Aug. MOULT: 22 of 53 specimens were moulting during Feb–Oct with most activity during Apr–Oct (Appendix 3).

Status and conservation   Table 8.1
Patchily distributed, but locally common. Too little is known of distribution, status, abundance, ecology, and general biology to comment usefully here. Difficult to assess significance of an apparently-patchy distribution as secretive habits may leave birds overlooked. Several ornithologists have, however, found it absent from seemingly-suitable forests within its range and it has been suggested that a patchy distribution may be dictated by availability of significant food plants. The lack or rarity of the sp. in certain areas that would appear suitable would suggest, however, that it deserves more attention and careful monitoring. A specimen was collected at Wau by Herbert Stevens in 1932 and it was found to be not uncommon on Mt. Kaindi on occasions during 1968–74 by Peter Shanahan (personal communication) who collected live individuals there. The sp. was not recorded in long-term ornithological studies in the area by Beehler, Pratt, the Pruett-Jones, and others (Beehler 1991a).

Knowledge lacking and research priorities
This sp. merits priority attention. One of the least known birds of paradise, this being especially significant because of its placement in a monotypic genus. Field students should make special efforts to determine the following: (1) mating organisation, (2) ♂ dispersion in forest (if court-displaying, as assumed to be), (3) courtship and mating, (4) details of diet of ad and nestlings, (5) reasons for patchiness of distribution, and (6) delineation of geographical variation.

Aviculture
Kept briefly at the BRS, PNG, where photographed by Mackay (1987; and in Coates 1990). A bird was maintained at the National Zoological Park, Washington (Clench 1992). Said to require immediate control measures for internal parasites, and not to adapt well to captivity, and to require a large shady aviary with low hiding places and a well-foliaged pool. Will not do well housed with other spp. (P. Shanahan *in litt.*).

Subfamily PARADISAEINAE: the typical birds of paradise

Paradisaeinae Vigors, 1825. *Transactions of the Linnean Society of London* **14**, 395–517. Typical, or narrow-gaped, birds of paradise.

The **39 species** presently constitute the **15 genera** *Macgregoria, Lycocorax, Manucodia, Paradigalla, Astrapia, Parotia, Pteridophora, Ptiloris, Lophorina, Epimachus, Drepanornis, Cicinnurus, Semioptera, Seleucidis,* and *Paradisaea* (Plates 5–12) of NG and offshore islands, several Moluccan Is of Indonesia, and wet coastal forests of E Australia. **Skull** morphology of this group is distinct from that of the cnemophilines, only the basal form *Macgregoria* sharing some of their features. Major Paradisaeinae skull characters are lacrymal small or absent and the enlarged ectethmoid plate solidly fused with the frontal bone. A short orbital process of the quadrate has an expanded distal tip (Bock 1963), see Fig. 9.1. **Bill** short, straight, relatively narrow, and stout to extremely long and narrow, laterally compressed, and greatly decurved (see Fig. 4.1). None is wide-gaped as in Cnemophilinae. **Tarsus** relatively stout and **feet** strong and highly manipulative.

A number of subfamilies within this group have been suggested (see Bock 1994) on the basis of external morphological differences (see Table 3.3), but growing knowledge of the birds suggests that, with the possible exception of the 'Manucodiinae' (see Genus *Manucodia*), these are inappropriate. Marked similarities in the general biology and the courtship behaviour of most members of this subfamily, in addition to numerous inter-and intrageneric **hybrids** (Chapter 3 and Appendix 1) support the notion that this subfamily is the product of a rapid and explosive radiation of no great age.

We again stress here that *Macgregoria* is very much the odd genus within this subfamily and its position there remains tentative and awaits confirmatory data. Anatomical studies found this unique high-altitude sp. to be in many respects intermediate between the typical birds of paradise, Paradisaeinae, and the three wide-gaped and weak-footed spp. constituting the subfamily Cnemophilinae (Bock 1963). Skull characteristics suggested to Bock that birds must use the bill as a probe and pincer, particularly using the 'open-bill probing' technique commonly performed by members of Paradisaeinae (but in fact not the Cnemophilinae). Bock tentatively concluded that on balance, *Macgregoria* should, however, be in the Cnemophilinae. As Gilliard (1969) placed *Macgregoria* in the Paradisaeinae it remained in that subfamily unquestioned, however, until Clench (1992) presented evidence of feather tract morphology, or pterylography, and interpreted it as indicative of a closer relationship to the cnemophilines. Unfortunately Clench was under the impression that the Cnemophilinae consists of exclusively monogamous spp., leading her to conclude that 'behavioural data also support the placement of *Macgregoria* in the Cnemophilinae.' We now know that at least two of the three latter subfamily members are very probably polygynous (Frith and Frith 1993c, 1994b).

Bock (1963) specifically made the observation that the lack of nasal ossification, a character given much weight in attributing *Macgregoria* to the Cnemophilinae, 'could be explained as an adaptation for fruit eating and can easily be derived from the typical cranial condition of the birds of paradise'. We do now know that Macgregor's Bird is a fruit-eating specialist (Beehler in Coates 1990; Beehler 1991b).

Bock concluded *Macgregoria* be best placed within the Cnemophilinae, but with considerable reservation because of several paradisaeinine characteristics (such as its strong relatively narrow bill). While following Bock in most respects Gilliard (1969: 47) considered Macgregor's Bird a typical bird of paradise and quite distant from the Cnemophilinae (which he saw as intermediate between the birds of paradise and the bowerbirds). All subsequent authors have followed Gilliard in restricting the cnemophilinae to the three wide-gaped spp., but have retained Bock's conclusion that the cnemophilines should be at the bottom of the bird of paradise family tree and that Macgregor's Bird be a basal to the typical birds of paradise of the subfamily Paradisaeinae.

We tentatively reject Clench's (1992) placement of *Macgregoria* within the Cnemophilinae and point out that a number of aspects of this bird's anatomy, biometrics, morphology, and biology argue more strongly for its placement in Paradisaeinae. It is a large, strong-billed and strong-footed, quick-moving bird that is significantly more sexually dimorphic in size than the cnemophilines but monomorphic in plumage and does not have a distinct juv plumage. It produces wing-noise in flight, showing slight primary feather shape modification, like many of the Paradisaeinae. It builds an open cup-shaped nest like those of the typical Paradisaeinae but unlike those of at least the two *Cnemophilus* spp. of Cnemophilinae.

The basal lineages that contain *Macgregoria* and *Manucodia* antedate the explosive radiation noted above. These birds are sexually monomorphic, apparently all pair-bonding and **monogamous** spp. that do not hybridise and do not have ventrally-barred juvs (as do other Paradisaeinae) or a juv plumage different to both ad sexes (as do the Cnemophilinae) but have plain young with plumage a dull version of the ad. The manucodes and, to a lesser extent, the Paradise Crow construct sparse, open, suspended **nests** unlike those of other Paradisaeinae in both structure and materials.

In all sexually-dimorphic paradisaeinine genera in which ad ♂ plumage is colourful and ornate the spp. are **polygynous** and the ♀ alone participates in nesting—building a bulky open nest predominantly of ferns, orchid and vine stems. Most **nestlings** are ventrally barred. In polygynous *Paradigalla* the sexes are, however, nearly identical and the nestlings are unbarred and approximate a dull version of the ad. All polygynous genera except *Pteridophora* and *Semioptera* are known to **hybridise** with at least one other genus, and often several genera, within the subfamily (see Appendix 1 and Plates 14–15).

Ad ♂♂ of the group exhibit remarkable feather modifications and plumage coloration, but no sp. is conventionally crested. It is conceivable, however, that the ornate head plumes of *Parotia* spp. might be homologous to the crest of *Cnemophilus*. Wings are not particularly rounded, as in the cnemophilines, except in ad ♂♂ only of some genera (e.g. *Parotia*, *Ptiloris*) modified for display purposes. ♀ and imm ♂ plumages in all polygynous genera are generally similar and cryptic, being brown to blackish brown above and off white to blackish brown below, with dark ventral barring; the one notable exception being monomorphic *Paradigalla*, in which juvs and imms are duller versions of the uniformly-black unbarred ad.

In the seven typical, or plumed, *Paradisaea* spp. most ♀ and imm plumages are distinctly different, however, five of the spp. being unbarred below and coloured uniformly chestnut (*P. apoda*) or with contrasting areas of chestnut-brown, blackish, yellow, and white. This exception is most striking in this otherwise uniform group of ventrally-barred ♀ plumages. The recent discovery of toxicity of the brownish and brown-and-black spp. of *Pitohui* (Pachycephalidae)— which are also important constituent spp. of mixed-spp. flocks—may offer a speculative explanation for this plumage shift in *Paradisaea*. Most members of the Paradisaeinae regularly take part in **mixed-species foraging flocks** involving pitohuis and other 'brown and black' spp. (D'Albertis 1880; Bell 1983; Diamond 1987; Coates 1990). Perhaps, therefore, ♀-plumaged *Paradisaea* spp. mimic the general appearance of typical mixed-spp. flocking (and/or the toxic) birds in order to participate in (and/or gain protection from visual hunting predators when joining) such flocks (CF).

Genus *Macgregoria* De Vis, 1897

Macgregoria De Vis, 1897. Ibis p. 251. Type, by monotypy, *M. pulchra* De Vis.

Macgregor's Bird of Paradise, the sole member of the genus *Macgregoria*, inhabits patches of subalpine woodland that cloak the heights of a selection of the major massifs of New Guinea's central cordillera. Its **affinities** have long puzzled avian systematists. The **skull** exhibits some features shared with both the Cnemophilinae and Paradisaeinae plus some others unique within the family (Bock 1963). See also Chapter 3, Subfamily Paradisaeinae (above) and Macgregror's Bird account (below). The culmen is sharply keeled. The ectethmoid plate is unlike *Cnemophilus* and is more like *Paradisaea*. Unlike *Cnemophilus* there is no foramen dorsolateral on the ectethmoid but a small one is located on the posterior face of the ectethmoid near its ventral border. The orbit, the somewhat elongate brain case, and the longer postorbital and zygomatic processes are more like *Paradisaea* than *Cnemophilus*. In addition, the posterior condyle of the quadrate projects backward as in *Paradisaea*. The mandible is much more like that of *Paradisaea* than *Cnemophilus* (Bock 1963). See Table 1.2.

Notwithstanding some similarities with the cnemophilines, there are also a good number of obvious and significant differences. If *Macgregoria* proved to be more correctly placed in the Cnemophilinae it would represent a cnemophiline that has diverged greatly from its far smaller, weaker, wider-gaped and weaker-footed, exclusively-frugivorous relatives to become a much more powerful paradisaeine-like bird with a broader, but predominantly-frugivorous, **diet** over a patchy high-altitude relict range. The facts that it is markedly **sexually dimorphic in size** (♂♂ larger), is narrow-gaped, has no distinct juv plumage, is **monogamous**, and builds an open cup-shaped **nest** suggest a significantly-closer relationship to the Paradisaeinae than to the Cnemophilinae. We therefore treat it as a basal lineage in the typical birds of paradise Paradisaeinae until more knowledge of its biology and/or genetics may permit more accurate placement in the phylogeny.

Bill relatively fine and nearly straight, being slightly depressed at the tip, but deeper, more pointed, and clearly much stronger than that of *Cnemophilus* and *Loboparadisea* but not as strong as in some Paradisaeinae. It is fractionally longer than the head. **Tail** slightly graduated, **wings** rounded, flight feathers slightly curved, P9 and P10 mildly emarginated on their inner vane. Tiny, hair-like, pointed, central tips to tail feathers. **Tarsus** length 32% of wing length, the lower tarsus scuted but less so higher up. Feet powerful.

An active bird, unlike the wide-gaped Cnemophilinae. Unique, large, hemicircular, orange-yellow orbital **wattles** originate from and attached only to circumorbital skin and are free and labile. These are slightly larger in ♂♂. Plumage entirely velvety black save for an extensive ochre wing patch on inner primaries (Plate 5). Body contour plumage underlaid with thick insulating down, presumably as an adaptation to high-altitude climate. Juv plumage is that of a dull ad complete with facial wattles (as in *Paradigalla*) and is not distinctive as in the cnemophilines.

Monogamous, pairs remaining together for at least one nesting and sharing the duties of raising offspring. Nest open and cup-shaped built into a tree fork, quite unlike the nest sites and domed nests known in the cnemophilines. Egg somewhat intermediate in appearance between those of the less typical Paradisaeinae and that of *Cnemophilus* (Plate 13).

Macgregor's Bird of Paradise *Macgregoria pulchra* De Vis, 1897

PLATE 5

Other names: Macgregor's Bird, Orange-wattled Bird of Paradise. Called *Kondimkait* by the people near Telefomin (Gilliard 1969). Called *Mo* by the Biagge people of Mt. Albert Edward area, PNG (Rand 1940*a*; Beehler 1991*a*). *Wunin* or *Genat* of the Western Dani, or Ilaga Valley, IJ and *Engabect* of the Uhunduni people (Ripley 1964). *Morkek* at Murray Pass, PNG (W. Cooper *in litt.*).

An unwary, black, crow-like bird of subalpine forest-edge, often conspicuous because of incessant vocalising, the bright eye wattles, and ochre wing patches. Encountered most often in pairs or small parties; characteristically flicks wings and tail. Polytypic; 2 subspp.

Description

ADULT ♂: 40 cm; ♀: 35–40 cm, tail slightly graduated, gape narrow. Entire head and upperparts jet to sooty black with slightly duller, Sepia (119) tertials, secondaries, and primary tips. Remainder of primaries extensively rich Cinnamon (123A). Erect feathering of fore-crown and lores dense, stiff, and plush. Contour plumage dense, soft, and silky in texture. Bases of contour feathers dark matte grey. Face conspicuously adorned with a large, fleshy, Orange-Yellow (18), hemicircular orbital wattle surrounding all but front of eye, and originating from the circumorbital eye-skin. Chin, throat, and underparts jet black with slightest brownish hue on belly, vent, and undertail coverts. Undertail brownish black (119). Underwing: coverts blackish brown (219), flight feathers black. Bill shiny black, iris reddish brown to red, legs and feet blue-grey. At any distance bird appears jet black with yellow facial wattles and extensive cinnamon wing patches.

IMMATURE: plumage browner than ad, softer and more fluffy, iris dark brown.

NESTLING–JUVENILE: on 22 Aug 1938 a nestling *c.* 12 days old at Lake Habbema was 'surprisingly small, with eyes not yet open' with a tarsus 35 mm and a bill 19 mm long. It was partly feathered dull black, with well-developed eye wattles 13 mm in dia, and with primaries just bursting from pin. It lacked ventral down but down was fairly plentiful on the ends of its crown, nape, back, rump, and humeral and femoral feather tracts and was present on tips of greater wing coverts, primary, secondary, and some dorsal surface of tail coverts. All down was sooty black. The eye wattles and other bare facial skin was dull yellow, the bill blackish streaked yellowish, a small egg tooth was white, the feet greenish yellow with greyish nails, and inside of mouth was orange-yellow (Rand 1940*a*). Nestling/juv plumage like that of ad but duller (unlike that of young *Cnemophilus* spp. and *Loboparadisea*, which are distinctly plumaged).

Distribution

Small remnant populations confined to the central high mountain cordillera of NG, being reported from the Snow and Oranje Mts of IJ, the Star Mts of the IJ/PNG border, and the central Owen Stanley Ra of E PNG, but apparently absent from the Central and Eastern Highlands of PNG. Has been recorded from 2700 to 4000 m, but is most commonly found between 3200 and 3500 m. A local informant considered reliable reported this sp. occurs in the highest forest of the Victor Emanuel Mts, near Telefomin (Gilliard and LeCroy 1961), but this is doubtful in view of their relatively-low elevation (the informant probably meant the Star Mts, where the bird does occur; Barker and Croft 1977). Erroneously reported at *c.* 1980 m on the Edie Creek road near Wau (Watson *et al.* 1962) and in view of description was probably a misidentification of *Melipotes fumigatus*, the Smoky Honeyeater (Meliphagidae). Cooper

Macgregor's Bird of Paradise *Macgregoria pulchra* De Vis, 1897

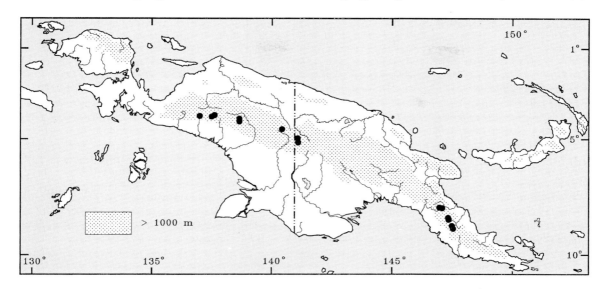

and Forshaw (1977) cited this erroneous record to support the claim that birds wander out of their favoured habitat (Beehler 1983c). Schodde *et al.* (1975) reported that Temple (1962) recorded seeing birds on the Kemabu Plateau in 1961 but he did not. It was Ripley (1964) that cited Temple's observation. J. Hope (in Hope *et al.* 1976) reports the sp. to be common in the Carstensz meadow, from 3600 to 3800 m.

Systematics, nomenclature, subspecies, weights, and measurements

SYSTEMATICS: a highly distinctive sp. constituting a monotypic genus of uncertain origin. By plumage and wattles, the genus *Paradigalla* appears to show the closest affinities. Described from three specimens collected by Amedeo Guilianetti for Sir William Macgregor, then Governor of British NG and the expedition leader, in May 1896 between the Mambare and the Vanapa R. Bowdler Sharpe (1891–8) considered *Macgregoria* 'undoubtedly closely allied to *Paradigalla*' and his view, given that it was based on external morphology only, was not unreasonable. Mathews (1930) placed *Macgregoria* in the bowerbirds (Ptilonorhynchidae). Iredale (1950) did not consider Macgregor's Bird a bird of paradise. Bock (1963) entertained the idea that *Macgregoria* possibly evolved from a fruit- and arthropod-eating *Paradigalla*-like ancestor into an exclusive frugivore and placed it in his newly-erected subfamily Cnemophilinae. This bird's position within the Paradisaeidae was reassessed in the light of its pterylography (feather tracts) by Clench (1992). Clench pointed out that *Macgregoria* shares the feature of no 'basal gap' between lines of feathers on the central back, or dorsal areas of feathers known as the *pars dorsalis* (saddle) and the lower area of back feathers, the *pars pelvica*, with the three cnemophilines. The three cnemophilines also show no sign of a 'basal gap' here whereas all genera of the Paradisaeinae do and Clench gave much weight to this in concluding that *Macgregoria* should be placed in the Cnemophilinae. A plotting of the number of feathers versus the number of rows in the *pars dorsalis* or back 'saddle' of bird of paradise genera demonstrates, however, that *Macgregoria* falls well within the typical birds of paradise and not the Cnemophilinae (Clench 1992), see Fig. 3.2. Furthermore the cnemophilines are not monogamous, like *Macgregoria*, as Clench was led to believe at the time and, as she pointed out, they are sexually dimorphic whereas *Macgregoria* is not. See also Chapter 3 and Genus *Macgregoria* (above). Hybridisation: unrecorded and not expected.

SPECIES NOMENCLATURE AND ETYMOLOGY

Macgregoria pulchra De Vis, 1897. *Ibis*, p. 251, pl. 7. Mt. Scratchley, south-eastern New Guinea. Type specimen BMNH 97.8.16.1. Etymology: *Macgregoria* = after Lady Mary Macgregor, wife of Sir William, Governor of British NG; *pulchra* = L. beautiful or lovely. Usage of this surname is the source of confusion, mainly because 'Macgregor' changed the spelling of his name from Mcgregor to MacGregor during his career in NG (Joyce 1971). We have decided, for purposes of stability, to follow Beehler *et al.* (1986) and Coates (1990).

SUBSPECIES, WEIGHTS, AND MEASUREMENTS

1. *M. p. pulchra* De Vis, 1897.

Range: confined to several of the highest summits of the central Owen Stanley Ra. To date, reported only from Mt. Strong/Chapman, Albert Edward, Scratchley (including English Peaks), Victoria, and Batchelor.

Measurements: wing, ad ♂ (n = 19) 187–211 (202), ad ♀ (n = 7) 175–187 (183); tail, ad ♂ (n = 18) 156–168 (161), ad ♀ (n = 7) 143–152 (147); bill, ad ♂ (n = 18) 37–42 (39), ad ♀ (n = 7) 35–42 (38); tarsus, ad ♂ (n = 19) 55–65 (61), ad ♀ (n = 6) 55–58 (56). Weight: ad ♂ (n = 6) 242–274 (256), ad ♀ (n = 3) 190–207 (198).

2. *M. p. carolinae* Junge, 1939. *Nova Guinea* (New Series) **3**, 82. Oranje Mts. Type specimen RMNH 255.

Range: Snow and Oranje Ra of IJ (Mt. Carstensz [Puncak Jaya] and Wilhelmina [Puncak Trikora], and Carstensz Meadow, Kemabu Plateau, and Lake Habbema), and the Star Mts of the IJ/PNG border region (Mt. Capella, Dokfuma meadow). Range of this subsp. is much more extensive and continuous than for the nominate subsp., because of the more extensive habitat in IJ.

Diagnosis: size proportions are significantly different from nominate form. Wing and tail smaller, but weight considerably greater. ♂ tail/wing ratio for nominate subsp. = 0.80, but for *carolinae* = 0.70.

Measurements: wing, ad ♂ (n = 10) 184–200 (192), ad ♀ (n = 10) 164–179 (172); tail, ad ♂ (n = 10) 128–141 (134), ad ♀ (n = 10) 120–133 (124); bill, ad ♂ (n = 10) 37–44 (42), ad ♀ (n = 10) 36–40 (38); tarsus, ad ♂ (n = 10) 62–66 (64), ad ♀ (n = 10) 53–61 (58). Weight: ad ♂ (n = 2) 340–357 (349), ad ♀ (n = 1) 230. For the sp. as a whole, ad ♀♀ mean wing length in 11% smaller and weight 26% lighter than those for ad ♂.

Habitat and habits Table 4.1

Subalpine forest interior and edge, and also glades where alpine grasslands are broken up by forest 'islands' of various sizes and shapes (Plate 2). Particularly closely associated with the podocarp tree *Dacrycarpus compactus*, which in many high-elevation areas is the numerically-and structurally-dominant woody plant in the habitat. When fruiting, this tree constitutes a very important component of the bird's diet. When *Dacrycarpus* is not producing ripe fruits birds will feed in low vegetation and on the ground. Birds on the ground are wary, more so than when perched, but some have been described as rather tame and curious. Easy to locate by flight wing-noise and rather sluggish foraging habits, and also by fairly-persistent vocalisations (BB). Birds exit tall trees by descending in a steep glide (D. Bishop *in litt.*). Observed as solitary individuals, as apparently-mated pairs moving about together, or as triplets presumably consisting of mated pairs with dependant imm (BB). Reported to behave at times more like a smaller sp., hopping actively about perches whilst often flicking wings and tail, the latter being raised, or cocked, above the wings. Members of bonded pairs roost together on open perches in groves of trees at the forest/alpine grassland edge and allopreen (Rand 1940a; Mayr and Rand 1937; Beehler 1983c, 1991a). Birds sometimes remain in the same immediate area for several hr. Flight heavy with very shallow and noisy wing-beats interrupted by short glides (Rand in Mayr and Rand 1937), consisting of *c.* 12 heavy strokes and then a short glide on outstretched wings. When wings are held set they produce a continuous *zing-g-g-g* reminiscent of a minor version of the sound produced by hornbills (Bucerotidae) in flight. Wing-beats produce a heavy rustling or low, loud, hollow, 'thopping' sound (Rand 1940*a*; BB).

Diet and foraging Table 4.2

Mostly forages in subalpine forest canopy for *Dacrycarpus compactus* fruits but when they are unavailable will forage at all vegetation levels for a variety of fruits. Observations and examinations of stomach contents and faecal samples show that fruits of *D. compactus* (33 records) are the favoured food. Birds may feed almost exclusively upon this plant when prolifically in fruit. These fruits are both periodically superabundant and annually unpredictable in availability. Some observers have noted that Macgregor's Bird ignores *Dacrycarpus* fruit when available, but this is probably because the fruit is unripe—fruits take about a year to ripen once they appear. On at least five occasions when field observers have watched this sp. at Lake Omha birds have foraged on fruit other than *Dacrycarpus*. Clearly dietary behaviour requires study. In the absence of *Dacrycarpus* fruits, birds will forage upon small fruits on sub-canopy trees and even upon tiny fruits of small prostrate shrublets at the forest/grassland ecotone; there the birds will descend to the ground and feed upon fruiting cushion-shrubs (Beehler 1991*a*). At Lake Omha, near English Peaks, one bird foraged at a low mat of fruiting cushion plants for 19 min (BB), and other individuals were observed to consume fruits of *Eurya brassii* (23 records), *Cladomyza acrosclera* (13), the low cushion plants *Styphelia suaveolens* (10) and *Coprosma divergens* (4), *Rapanea* sp. (3), and *Symplocos cochinchinensis* (2). For other fruits eaten see Appendix 4. Several ornithologists have watched birds actively foraging amongst moss and other epiphytic vegetation, and in foliage, as if seeking arthropods or small vertebrates—in a manner much like an astrapia (Clapp 1986; Safford and Smart 1996; BB in Coates 1990). Four birds foraging in a Star Mts, IJ forest interior on shrubs within 2 m of the ground were confiding and were 'pished' to within 3–4 m of an observer (D. Bishop *in litt.*). Animal prey no doubt constitutes a part of the diet. Ripley (1964) reported birds 'often feeding on flowers and berries' but provides no further details of flower feeding. This may, in fact, be the product of an informant's report, as Ripley was very ill during this period (*fide* M. L. Ripley). From a comparative study of skull morphology Bock (1963) concluded that Macgregor's Bird probably employs the advanced ('paradisaeine') method of foraging known as open-bill probing, whereby the closed bill is inserted into moss, soft wood, fruit, etc. and is then opened to provide better access to, and vision into, the space created.

Vocalisations and other sounds

The most commonly-given call is a rapidly-repeated, double, high-pitched, whistled *jeet* contact note. It is also given in alarm. Also repeatedly given is a *pseer* note, sometimes used in greeting (Beehler *et al.* 1986; Beehler in Coates 1990) or a *schweet schweet* at a rate of > 1/sec (BB). A low sharp *click* or *click-click* repeated several times, a low plaintive *quee*, and during flight chases a sharp but not loud *chick-chick-chick-chick* or *chick-chick chick-chick* call (Rand 1940*a*) or a weak but pleasant *psheer* 2–5 times (BB). Also a not very loud *cheu* note several times repeated (Rand in Mayr and Rand 1937). A pair give a nasal, slurred, lisping, and quiet *chiff* note when together. Of a disturbed pair the putative (larger) ♂ gave a loud, frightening *krahh!* and the putative ♀ a *jeet*. A *fwooiip!* note was given by a bird separated by 140 m from two associated individuals giving *scheet* and *schweet* notes (BB). When gliding on fully-extended wings, the widely-separated primary tips made a loud ripping *zing* (Rand 1940*a*). The wings in flight produce a whirring noise (Temple in Ripley

1964), 'musical ripping or whining sound' (Beehler in Coates 1990) or a remarkable rustling or zipping sound (Beehler 1991a) interspersed with silent glides. Recent observations confirm those of Rand (1940a) that at least some birds can control the wing-noise (BB).

Mating system
Monogamous, non-territorial. Three individuals that were radio-tagged in Aug 1986 had overlapping foraging ranges, and a week-long range for each was 10, 9, and 7 ha in extent (BB). Based on radio-tracking over a week, the home range of a pair was 12 ha (BB).

Courtship behaviour
Courtship flight and hopping chases are usually performed by a mated pair, but sometimes involve three and rarely up to six additional birds. These take place in the immediate area of a nest under construction. The chase consists of one bird following another, hopping through tree tops and rapidly travelling considerable distances on foot, and flying with wing-noise across open spaces back and forth through the forest. Sometimes when descending across open space birds glide on fully-extended wings. Frequently birds descend into undergrowth where the chase continues (Rand 1940a).

Breeding   Table 6.5
At Lake Omha, PNG, it appeared that nesting may often be synchronised with the local production of ripe fruit of *Dacrycarpus compactus*, which is not an annual event (BB). This said, a nest containing a single small chick was observed during a period when virtually no ripe *D. compactus* fruits were available. Two adult birds provisioned this nestling and defended an all-purpose territory around the nest by evicting other frugivorous birds from it (T. Pratt and E. Brown *in litt.*). A ♂ was observed accompanying a ♀ as she nest-built and on her feeding trips during her incubation of their clutch during Aug. A ♀ whose nest, egg, and mate were collected on 14 Aug had remated and had built one-third of a new nest on 21 Aug (Rand 1940a).

NEST SITE: only three described. One was 15 m high in multiple upright forks (10–14 mm dia) of a small branch in the crown of an isolated podocarp (probably a *Dacrycarpus*) tree rising above surrounding lower moss forest vegetation and the other *c.* 11 m high on a lateral 32 mm dia bough, supported by many smaller branches growing vertically from it of 9–13 mm dia, in the edge of a small clump of these trees near the forest at the edge of an extensive grass area (Rand 1940a). Significantly, Rand found four old nest structures, probably all of this sp., within 50 m of a courtship chase area. A nest he collected was replaced by a new one within 50 m of the original site. Another nest was 'high in the top of a canopy tree' (T. Pratt and E. Brown *in litt.*).

NEST: the 15-m-high nest was a bulky, firm, cup-shaped structure built externally mostly of moss, mixed with some herbaceous and woody vine or orchid stems and lichens. Only one or two short sticks, of up to 3 mm dia, were incorporated. Inside, a firm cup with walls of slender herbaceous and woody stems 10–20 mm thick at the top but very thin at the bottom and firmly lined with slender (up to 2 mm dia), unbranched, woody stems. The latter were presumably supple when used and possibly inflorescence stems. Within the cup were a considerable number of broad leaves ('phyllodes') of the conifer *Phyllocladus hypophyllus* and numerous leaves of *Nothofagus* beech trees. External structure 240 mm in dia and 190 mm in depth and inside cup dia 130 mm and 90 mm in depth. A second nest was similar but had a bulkier moss foundation and was lined with semi-woody stems and small oval leaves of an angiosperm shrub. It measured *c.* 315 mm dia and 190 mm deep externally and 135 mm dia and 90 mm deep inside (Rand 1940a and pers obs). These open-cup nests show no sign of even a rudimentary roof (*cf. Cnemophilus* spp.) but are generally similar to those of the Short-tailed Paradigalla (see Figs 9.16, 17).

EGGS: the only known clutch is one ovate egg with slightest gloss and a slightly-rough feel to its surface, Vinaceous Pink (221C) sparsely marked with soft spots of light brown and pale purplish grey, many overlaid with the colour of the shell to give secondary purplish black spots, the markings well distributed but more so about the larger end and with a slight concentration about the small end (Plate 13). It measures 39.9 × 28.6 mm and was apparently collected with the second nest described above with the ♀ parent (AMNH 342050).

INCUBATION: by ♀ only. During an hr observation, incubation bouts lasted *c.* 15 min and absences *c.* 10 min (Rand 1940*a*).

NESTLING CARE: Rand watched undisturbed activity at a nest containing a *c.* 1 day-old nestling for 88 min. Only the ♀ brooded, for periods of 7–14 min and whilst she fed the nestling only twice, the ♂ did so 16 times with absences from the nest of 1–13 min. The ♂ fed the nestling during brooding bouts, the ♀ merely hopping a few m from the nest to permit him access. Parents apparently regurgitated food to the young and ate its faecal sacs. At nestling age 11–12 days no brooding was seen (Rand 1940*a*).

NESTLING DEVELOPMENT: appears slow (see Description). Slow nestling development at high altitude with parents providing almost exclusively fruit is unsurprising (Frith and Frith 1993*c*, 1994*a*).

Annual cycle Table 6.5

DISPLAY: flight chases, considered courtship activity, seen on 11–13 Aug (Lake Habbema). BREEDING: apparently *c.* July–Feb. One or two ♂ birds only have been collected with enlarged gonads during Apr, Sept, and Nov. EGG-LAYING: given the above egg (see Eggs) was slightly incubated when collected on 14 Aug and the presence of a *c.* 12-day-old nestling on 22 Aug, and assuming and incubation period of 26–30 days (see Chapter 6), it can be said eggs are laid from at least mid July to mid Aug (Lake Habbema, IJ). MOULT: 12 of 35 specimens were moulting, all but one (Feb) during July–Nov (Appendix 3). It remains to be seen if individual birds do moult while breeding or if this is merely the result of combining data from various locations and periods.

Status and conservation Table 8.1

Birds are scattered in small numbers through the geographically-restricted and patchy habitat. The presently-known range of this bird is believed by several authors to be indicative of a declining sp. Many seemingly-suitable patches of appropriate habitat visited by ornithologists lack this bird. Birds may be fairly common in a given location at times and sparse or absent at others, however, their numbers apparently correlated with the availability of fruit of *Dacrycarpus compactus*. *Macgregoria* populations vacate the Lake Omha area for periods of as long as a year, possibly because their favoured tree spp. fail to produce fruit (BB). Reported as common on the main dome of Mt. Albert Edward in July 1933 (Mayr and Rand 1937) but was then not seen there by several ornithologists (Bell 1971; Barker and Croft 1977; P. Lambley in Safford and Smart 1996) and was considered possibly locally extinct. A group of three birds was, however, seen in Apr 1995 at 3080 m between the Neon (= Neowa) Basin and Woitape where the sp. was then considered rare by local people (Safford and Smart 1996). It has been suggested hunting pressure might cause the absence or paucity of birds in places (Barker and Croft 1977; Safford and Smart 1996). On the other hand, in the Okbap area, Star Mts, IJ the sp. was moderately common and confiding and a Ketengban informant reported the bird is not hunted but is revered and protected (J. Diamond and D. Bishop *in litt.*). Rand (in Mayr and Rand 1937) noted an abundance of fruit on Mt. Albert Edward but no fruit in the Murray Pass and suggested that such situations might cause the bird to make local movements. It is clear that at times many areas of adequate habitat lack fruiting *D. compactus* trees, and this may result in a form of relatively-localised nomadism/migration, causing the local temporary absence of birds. The grossly patchy distribution on the NG highlands is, however, highly suggestive of remnant isolated populations which may, by definition, leave the sp. vulnerable. Moreover, the periodically-superabundant but annually-unpredictable nature of fruit significant to this sp. dictates that it cannot be sedentary and that it probably suffers periodic and dramatic population fluctuations. This means that small areas of protected habitat will not guarantee long-term survival of the populations living therein.

Knowledge lacking and research priorities

Conservation biology studies are required, including an extensive and thorough survey of all suitable habitat at appropriate altitudes along NG's central cordillera. In particular, *Macgregoria*'s distribution in eastern IJ, its apparent absence in the extensive highlands of Enga Province, PNG, and its presence or absence on additional summits SE of Mt. Victoria, PNG need investigating. More biological and adequate molecular genetic data are required to settle the relationships of this sp. More study is needed to determine the nature of its dietary specialisation and nomadism. Determining the details of the bird–plant relationship might provide additional interesting dividends, as might studies of ecological relationships to other frugivores in the habitat with particular reference to dominance hierarchies. Finally, it is apparent that the fruit of *Dacrycarpus compactus* are high in aromatic compounds that may require special digestive capacity.

Aviculture

A live bird was taken to London Zoo by Fred Shaw Mayer on 29 Apr 1937. Unfortunately it was housed in the heated Tropical Bird House (Fisher 1938) and died 30 Aug 1938.

Other

The remains of a Macgregor's Bird were found in the stomach of a Spotted Marsh Harrier *Circus spilonotus* at Lake Habbema, Snow Mts on 19 Aug (Rand 1940a).

Genus *Lycocorax* Bonaparte, 1853

Lycocorax Bonaparte, 1853. *Comptes Rendus Hebdomadaires des Séance de L'Académie des Sciences, Paris* **37**, 829. Type, by original designation, *Corvus pyrrhopterus* Bonaparte.

The Paradise Crow is, in appearance, the most crow-like bird of paradise (Plate 5), and is the sole member of the genus *Lycocorax*. It is confined to the northern Moluccan islands of Indonesia. The **skull** is generally similar to that of *Manucodia* with a heavy, almost straight bill of medium length and width for the subfamily and, perhaps to slightly lesser degree, most characters of the skull are similar to those of *Paradisaea* (Bock 1963). ♂ fractionally larger than ♀, the proportions of **size** dimorphism being similar to those for *Manucodia*. The sp. is presumed to be most closely allied to the manucodes, which show a propensity for colonising islands. We hypothesise that *Lycocorax* is an island vicariant that evolved from what was a widespread form that gave rise to both *Lycocorax* and the genus *Manucodia*.

The elongate, powerful, crow-like **bill**, c. 30 mm longer than the head, is wider and heavier than in other birds of paradise but is similar to that of manucodes. There is no gape, as is the case for *Macgregoria* and *Manucodia*, and no wattles. Nostrils clearly exposed and not covered or approached by forward lore feathering. **Tail** slightly rounded (outermost feathers shortest) and of similar proportionate length to manucodes. Tiny hair-like points to central tips of tail feather (as in *Macgregoria*, *Manucodia*, and *Paradigalla*). No modification of primary shape. The large legs and feet are powerful and **tarsus** length is 22% of wing length, as in *Manucodia*. No *Manucodia*-like modification of the trachea is known.

A glossy blackish, dully brown-winged, bird in which the sexes are nearly identical (sexually monomorphic). Body plumage soft and silky in texture with only a slight greenish gloss, more so on upperparts, the sp. lacking ornate plumes and brightly-coloured, metallic-like, iridescent plumage. Presumed to be **monogamous** and pair-bonding.

Notwithstanding that **nests** are structurally more typical of other Paradisaeinae, they do show some manucode-like characters not found in other paradisaeid taxa. **Eggs** are distinctive but are more like those of manucodes than those of other paradisaeinine genera (Plate 13).

Paradise Crow *Lycocorax pyrrhopterus* (Bonaparte, 1851)

PLATES 3,5; FIGURES 9.6,7

Other names: Silky Crow, Brown-winged Bird of Paradise, Brown-winged Paradise-Crow, Obi Paradise-Crow, Morty Island Paradise-Crow. *Burung andjing* on Morotai I. (Hartert 1903).

A retiring, crow-like forest-dweller of the Moluccas. Frequents the interior of forest and appears to behave much like a Trumpet Manucode. Note brown wings at rest and pale underwing patches in flight. Obi I. population more confiding, noisier, and more commonly observed in non-forest habitats. Polytypic; 3 subspp.

Description

ADULT ♂: 34 cm, entire head slightly glossy blackish Dusky Brown (19), darkest on crown. Mantle, back, rump, and uppertail coverts similar but paler, greyer, and with a dull blue-grey gloss or sheen with slightest of green cast, more so on the mantle. Upperwing brown, being darker (119A) on coverts and tertials than the dark Cinnamon-Brown (33) flight feathers. Dorsal surface of tail darker than rest, being glossy blackish brown (119) the feathers with hair-fine central points to their tips. Underparts entirely as mantle but slightly paler and with only slightest of blue-green gloss. Lower belly, vent, and undertail coverts slightly browner (219) as is the glossy ventral surface of tail. Underwing uniform Olive-Brown (28), bases of primaries ochraceous grey, coverts darker, with pale, Tawny Olive (223D) central feather shafts. Bill shiny black, iris blood red.

ADULT ♀: as ad ♂, but slightly smaller.

IMMATURE: duller-coloured, less blackish below than ad, possibly with paler brown wings.

NESTLING: unknown. JUVENILE: recently fledged: crown, midback, and dorsal surface of tail as ad but rest brownish (closest to 223) with only slightest of blue-black gloss. Downy underparts matte brownish (closest to 223 but paler).

Distribution

Endemic to islands of the northern Moluccas (Maluku Utara), Indonesia (Halmahera, Kasiruta, Bacan, Obi, Bisa, Morotai, and Rau Is), from sea level to 1600 m.

Systematics, nomenclature, subspecies, weights, and measurements

SYSTEMATICS: a little-studied, highly-distinctive sp. comprising three insular subspp. The sole

Paradise Crow *Lycocorax pyrrhopterus* (Bonaparte, 1851)

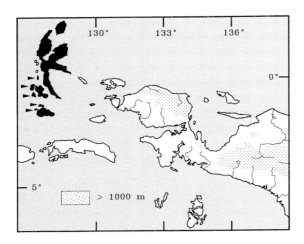

C[orvus] pyrrhopterus Bonaparte, 1851. *Conspectus Genera Avium* 1 (1850), 384. Gilolo and Halmahera. *Manucodia pyrrhoptera* Gray, 1870. *Hand list of birds* 2, 17.

First described as a crow *C[orvus] pyrrhopterus* by Bonaparte (1851) who thereafter (1853, *Comptes Rendus Hebdomadaires des Séance de L'Académie des Sciences, Paris* 37, 829) erected *Lycocorax*.

Lycocorax = Gr. *lycos* = wolf). Gr. *korax*, raven; *pyrrhopterus* = Gr. *pyrrhos*, flame-coloured, red; *pteros*, winged (red-winged).

SUBSPECIES, WEIGHTS, AND MEASUREMENTS

1. *L. p. pyrrhopterus* (Bonaparte, 1851). Type specimen not located.

Range: northern Moluccan islands of Halmahera, Bacan (= Batjan or Batchian) and Kasiruta.

Measurements: wing, ad ♂ (n = 28) 175–206 (190), ad ♀ (n = 27) 176–197 (186); tail, ad ♂ (n = 28) 127–160 (143), ad ♀ (n = 27) 131–153 (139); bill, ad ♂ (n = 27) 42–49 (47), ad ♀ (n = 27) 41–49 (45); tarsus, ad ♂ (n = 27) 39–44 (42), ad ♀ (n = 27) 39–45. Weight: ad ♂ (n = 8) 242–304 (269), ad ♀ (n = 7) 218–276 (250).

2. *L. p. obiensis* Bernstein, 1864. *Journal für Ornithologie* 12, 410. Obi I. Type specimen not located.

Range: northern Moluccan islands of Obi and Bisa.

Diagnosis: superficially like nominate form but much darker overall, more glossy blue-green, and darker and more blue-black on crown and dorsal surface of tail. Far more different from nominate than is *morotensis* and close to sp. status (Cracraft 1992; personal observations). Concealed bases of primaries with trace of white only (Plate 5). Lambert (1994) inexplicably writes of Obi birds having a whitish streak above and behind the eye.

Measurements: wing, ad ♂ (n = 26) 190–224 (204), ad ♀ (n = 19) 186–208 (198); tail, ad ♂ (n = 26) 135–156 (145), ad ♀ (n = 19) 130–152 (141); bill, ad ♂ (n = 26)

member of the genus *Lycocorax*. Wallace (1869: 317) considered this bird a starling. Guillemard (1886: 237), who collected it on Obi I. in 1883, perceptively thought it a 'miniature crow approaching the Paradisaeidae'. It is noteworthy that Gray (1870) treated it as a manucode. Iredale (1950) reluctantly included the Paradise Crow in his monograph of birds of paradise 'merely on traditional usage' but did place it with the manucodes in his subfamily Manucodiinae. Bock's anatomical studies led him to conclude confidently that the skull of the Paradise Crow is typical of the Paradisaeinae, being very like that of *Paradisaea*, but not showing characters particularly close to the manucodes. It is noteworthy, however, that *Lycocorax* exhibits the short and narrow non-fleshy-gaped mouth typical of the manucodes and not shared with other birds of paradise. We consider the Paradise Crow to be a sister form to the manucodes, and hypothesise that during a period of lowered sea levels, a parent form to both lineages ranged throughout NG and the northern Moluccas, its Moluccan and Papuan isolates subsequently differentiating into distinct lineages that today constitute *Lycocorax* and *Manucodia*. Hybridisation: unrecorded and not expected.

SPECIES NOMENCLATURE AND ETYMOLOGY

Lycocorax pyrrhopterus (Bonaparte)

49–56 (52), ad ♀ (*n* = 19) 45–54 (50); tarsus, ad ♂ (*n* = 26) 43–49 (46), ad ♀ (*n* = 19) 42–48 (45). Weight: ad ♂ (*n* = 9) 300–370 (332), ad ♀ (*n* = 5) 250–316 (291).

3. *L. p. morotensis* Schlegel, 1863. *Ibis* p. 119. Mortay [Morotai] I. Type specimen not located.

Range: northern Moluccan islands of Morotai and Rau.

Diagnosis: much like nominate form but paler, slightly more brownish above but slightly darker below. Concealed bases of primaries extensively white (Plate 5). Considerably larger than other two subspp.

Measurements: wing, ad ♂ (*n* = 4) 214–223 (219), ad ♀ (*n* = 3) 209–210 (209); tail, ad ♂ (*n* = 4) 152–161 (156), ad ♀ (*n* = 3) 143–147 (145); bill, ad ♂ (*n* = 4) 55–56 (56), ad ♀ (*n* = 3) 53–54 (53); tarsus, ad ♂ (*n* = 4) 48–51 (49), ad ♀ (*n* = 3) 45–49 (47). Weight: none.

Habitat and habits Table 4.1

Common in all habitats including fairly open agricultural land with scattered scrub and trees from sea level to at least 800 m on Obi I. On Halmahera, Bacan, and Kasiruta Is found in primary and tall secondary forest, forest-edge, scrubby gardens within forest, and in scrub at the boundary of forest and lightly-wooded cultivation. In rainforest interior frequents the canopy of moderately-high trees (Bernstein 1864*b*). Not typically seen in more open agricultural areas dominated by scrub (Lambert 1994). Rarely encountered in swamp forest and found in mangroves only where adjacent to swamp forest. Frequents vegetation from the canopy down to the substage. Perches in exposed free-standing trees only infrequently (M. Poulsen *in litt.*). Also recorded in coconut plantations and in orchards where it presumably comes for fruit crops. Scant literature suggests that on Halmahera, Bacan, and Kasiruta it is a shy and timid bird of denser foliage that nervously retreats from humans and is therefore difficult to view well (Heinrich 1956; Lambert 1994), but recent observations suggest it may be otherwise on Halmahera (Frith and Poulsen in press). Inconspicuous, secretive, and usually encountered singly or in twos but occasionally in groups of up to five and, rarely, with mixed-spp. foraging flocks (Bishop 1984; Coates and Bishop 1997). On Obi and Bisa Is it is, however, one of the most demonstrative and loud birds in the avifauna producing calls not given on Halmahera (Lambert 1994; D. Bishop *in litt.*). Flight fast, moderately direct, often interspersed with glides and producing an audible rustling, and is performed through forest midstorey and just above the canopy (Coates and Bishop 1997). Unlike the manucodes, does not make jerky body movements, but usually holds its tail slightly fanned when perched and flying. Birds often move from tree to tree in 'follow-the leader' fashion, with a pause between each bird's departure (Frith and Poulsen 1999)

Diet and foraging Table 4.2

Bernstein (1864) thought it feeds only on fruit. Heinrich (1956) found Pinang Palm *Pinanga* sp. fruit favoured, the finding of one seed of the latter in a nest with a nestling suggesting this fruit is fed to young. Several birds were seen by Lambert and Young (1989) to eat the flesh of large fruits of *Alpinia* sp. (probably *A. nutans*). Often seen on Obi I. feeding in fruiting trees with imperial pigeons and fruit doves (Linsley 1995). Birds forage for fruits and arthropods in dense upper-midstorey and canopy foliage (Coates and Bishop 1997).

Vocalisations and Figure 9.6
other sounds

Apparently gives several distinct vocalisations, some of which probably constitute ♂ advertisement, others serving as contact or alarm notes. There may be considerable inter-populational variation in vocalisation, as indicated by the following accounts. On Obi I., a *wuhk* or *wunk* note was reported by Bernstein (1864), which may be the same call described by Lambert (1994) as a very loud *who-up*, which was often answered by a loud *hwhoo*. On Halmahera, Heinrich (1956) recorded a call resembling a short, hoarse dog bark. The former call also transcribed as *woo-up*, the

Paradise Crow *Lycocorax pyrrhopterus* (Bonaparte, 1851)

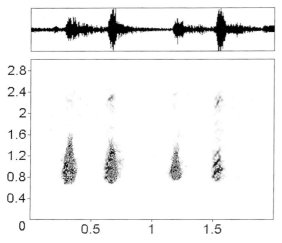

9.6 Sonogram from a recording of a Paradise Crow's *kreck rek* or *krek kek* advertisement notes by S. W. Smith on Halmahera I. in June 1993.

latter note rising and often followed by a harsh rasping *krek* or *rek* which is often given alone (Linsley 1995). The same latter distinctive vocalisation is described as a loud, two-note *wu-wnk* lasting c. 0.25 sec, the first note short and unmusical and the second nasal and hollow. This is repeated every 9–11 sec, possibly as advertisement or song (Coates and Bishop 1997). We transcribe a recording of the latter on Halmahera as a harsh, sharp, and rasping two-note *kreck rek* or *krek kek* (Fig. 9.6) given 10 times in 50 sec. A softer contact alarm call not unlike the *rrr* call of the coucal *Centropus celebensis* is reported by Heinrich (1956). On Halmahera, Ripley (1959) found a harsh rasping *tschak ... tschak* the common call, but also heard one ♂ give a low deep *om* apparently as a display call. The former call would appear to be the single, moderately-loud, dry, slightly upslurred, guttural bark-or croak-like *ekk* repeated at 4–8-sec intervals on Halmahera (Lambert and Young 1989; Coates and Bishop 1997) and considered the contact note. Also given is a short frog-like croak or bark *ech* (Coates and Bishop 1997). On Bisa I. the advertisement call is a loud, deep, resonant, disyllabic *OO-lip* or deeper *OO-lee* lasting c. 0.25 sec and produced with the quality of a woodwind instrument, of which the first syllable is deep and descending and the second high-pitched and rising (Coates and Bishop 1997). This call is repeated every c. 6–14 sec Flight wing-beats produce a rustling noise.

Mating system
Monogamy with biparental nest attendance is assumed, because the sexes are drably identical (Cooper and Forshaw 1977; LeCroy 1981). This requires confirmation, however, as ♀♀ of sexually-monomorphic Short-tailed Paradigallas attend the nest alone. Ripley (1959) collected a bird of each sex with enlarged gonads perched side by side on Halmahera on 7 Sept. The single-egg clutches and single-nestling broods recorded for this sp. would, if typical, be remarkably small for a lowland monogamous sp.

Courtship behaviour
Nothing known.

Breeding   Table 6.5
NEST SITE: 4–15 m above ground. Heinrich (1956) found a deep cup-shaped structure 'like an oriole nest' (i.e. suspended hammock-like in forked branches) containing a single chick at the end of Apr on Mt. Gamkanora, Halmahera, in forest. At 750 m on the same mountain a nest was found in a felled tree 10 m tall containing a single nestling 7 cm long on 19 Apr 1996 (M. Poulsen *in litt.*, see Plate 3). An Obi I. nest in primary forest at c. 500 m was at the edge of the lower forest canopy and appeared to be attached by pieces of vine or twine in such a way that it hung below the c. 5 cm dia branches (Lambert 1994).

NEST: Fred Shaw Mayer described a nest (BMNH 1941.1.2.98) bought by him from local collectors at Patani, SE Halmahera as a 'large basin shape nest built of roots, moss and lined with soft chips of wood' (Parker 1963). A similar second nest was collected on the same day. The Obi I. nest (above) was a large structure of dead leaves and moss. When

Paradise Crow *Lycocorax pyrrhopterus* (Bonaparte, 1851)

sitting in the nest the ad was obvious but the nest itself was well camouflaged by mosses on the tree trunk and branches (Lambert 1994). Another Obi I. nest collected by Bernstein on 7 Aug is a large, bulky, relatively-open bowl shape. The exterior is of lightweight, woody vine stems and tendrils, probably supple when incorporated, up to 4 (but mostly < 2.5) mm dia and these predominantly at the sides and rim. Exterior vines are sparse beneath the structure and predominantly < 1.5 mm dia there (Fig. 9.7) through which can be seen very predominantly lightweight, pale wood chips (up to 90 × 20 × 10 mm) with a couple of dead medium-sized leaves and a few pieces of paperbark-like bark. Some curled vine tendrils are in the outer rim and nest sides but there is no moss or orchid stems. The egg-cup lining is of coarse, springy, hair-like tendrils of 0.3 mm and less dia and *c.* 30 cm long. Beneath this moderately-dense lining can be seen the wood chips down to the bottom of the structure. Many external vine tendrils are > 30–60 cm long and encircle the nest sides and rim. A second generally-similar nest is 19 cm external dia, 9 cm total depth and 13.5 cm and 5.2 cm internal cup dia and depth, respectively. Nests are generally manucode-like but bulkier with numerous dry wood chips between lining and outer vine tendrils (CF). Many old nests were found on Halmahera, all 'placed well below the [forest] canopy, hanging in the fork of a branch' (M. Poulsen *in litt.*).

EGGS: six eggs each one representing a complete clutch average 37.8 (35.1–40.9) × 27.2 (24.8–29) mm, those from Obi I. being larger. Ground colour similar to Pale Horn (92) or Pale Pinkish Buff (121D) to Beige (219D) with sparse, fine, vein-like, scrawling markings over entire egg. Parker (1963) described a fresh single-egg clutch as 'pinkish-stone ground colour, marked all over in an irregular pattern with sparsely distributed lines of violet-brown and hairstreaks of pale lilac. Faint orange stains may be attributed to the wet woodchips'. As an egg from another nest had a much larger hole made in it by the preparator Parker concluded that it must have been more incubated, thus confirming a single-egg clutch. This second clutch was identical in colour and markings to the first, both being slightly glossy. Rothschild (1930b) described two eggs, which he implied came with the one nest. His measurements and descriptions make it clear, however, these are not the eggs, from different nests, described by Parker (1963) and are not Paradise Crow eggs. Field observations in 1995 confirm a single-egg clutch (D. Bagali *in litt.*). See Plate 13.

INCUBATION, NESTLING CARE, AND DEVELOPMENT: unknown.

9.7 (a) Nest and egg of Paradise Crow (RMNH collection) and (b) a view of the underside of the nest structure. Photo: CF.

Annual cycle Table 6.5
DISPLAY: nothing known. BREEDING: at least Dec–June and possibly from as early as July or Sept. ♂ specimens with enlarged gonads collected during Jan, June, Aug, Oct–Nov and a ♀ during July. EGG-LAYING: at least Dec–early June. MOULT: 59 of 103 specimens were moulting, involving all months of the year (Appendix 3).

Status and conservation Table 8.1
Locally common on Halmahera and moderately common to common on Obi and Bisa (Coates and Bishop 1997). Up to 10 individuals sighted daily on NW Obi I. during 11–23 Dec (Linsley 1995). An expedition performed systematic (variable circular plot) survey counts on Halmahera during July–Sept 1994, resulting in an average of 0.2 (range 0.1–0.4) birds per ha in habitats on both limestone and sedimentary rocks and 0.5 (range 0.3–0.8) birds per ha in montane habitats. BirdLife International field workers performed the same kind of counts on Halmahera during 1995–6 and found an average density of 0.4 (range 0.1–0.9) birds per ha in logged lowland forest on sedimentary rock (Frith and Poulsen in press). Given the differentiation of the Obi I. population (Cracraft 1992; CF) as a most marked subsp. its status deserves attention.

Knowledge lacking and research priorities
The following require investigation: (1) degree of insular differentiation, (2) evolution of vocal dialects, (3) mating organisation, (4) reproductive behaviour and display, (5) all aspects of nesting behaviour, and (6) diet of ad and nestling.

Aviculture
No known experiences.

Genus *Manucodia* Boddaert, 1783

Manucodia Boddaert, 1783. *Table des Planches Enluminées*, p. 39, no 634. Type, by monotypy, *M. chalybea* Boddaert = *Paradisea chalybata* Pennant (1781).
    Synonyms: *Eucorax* Sharpe, 1894. *Bulletin of the British Ornithologists' Club* **4**, 15. Type, by monotypy, *Manucodia comrii* Sclater.
    Phonygammus Lesson and Garnot, 1826. *Bulletin des Sciences Naturelles et de Géologie Férussac* **8**, 110. Type, by monotypy, *Barita Keraudrenii* Lesson and Garnot (originally described as a subgenus).

The **five species** of manucode (Plate 5) constitute the most homogeneous and most distinctive clade within the subfamily Paradisaeinae. Indeed, the lineage that includes the manucodes and the Paradise Crow may merit subfamily status. Manucodes are found throughout mainland NG and adjacent islands and the extreme NE of Australia. The members of *Manucodia* are in greater need of taxonomic revision than any other genus and also present the most difficult problems of field identification within the birds of paradise. Plumage descriptions for the *Manucodia atra, jobiensis, chalybata* complex must be considered to some degree tentative as this group requires an exhaustive review in order to define spp., subspp., sex, and age differences in plumage detail.

According to Bock (1963) the *Manucodia* **skull** is not generalised and primitive, but is, like its elongate and coiled trachea, highly specialised. The ectethmoid is a little larger than in *Lycocorax*, with its expanded foot resting on the jugal bar. The lacrymal is small, wedged into the dorsal corner between the ectethmoid and nasal, or is lacking. It is possible the manucodes represent a highly-specialised group derived from ancestral paradisaeinine stock (Stonor 1938; Sibley and Ahlquist 1990), but Bock (1963) considered this unlikely and thought their relatively-drab sexually-monomorphic plumage and **monogamous** pair-bonding possibly secondarily primitive.

♂♂ only fractionally larger than ♀♀. **Bill** is the length of the head or slightly to *c.* 20 mm longer (*M. comrii*), almost straight, curved on the culmen, wide and heavy and, as in *Lycocorax*, much like that of *Paradisaea*. The proximal culmen is broad and swollen in all spp. (but less so in the Jobi Manucode) except the Trumpet Manucode. Nostrils partially concealed by frontal feathers. There is no visible gape in the 'short-mouthed' *Manucodia* (and *Lycocorax*) spp., unlike all other birds of paradise in which the gape is clearly seen to extend to below or beyond the eye. No modification of primary shape. **Tail** long (71–84% of wing length), becoming longer with age in both sexes, and graduated (*atra, chalybata, comrii*) or slightly so (*jobiensis, keraudrenii*). **Wings** long and rounded. Legs and feet, strong, lower tarsus scuted becoming fainter to smooth higher up. **Tarsus** relatively short, 21–22% of wing length.

♂♂ exhibit a remarkable **tracheal modification**. The windpipe may be elongated into a simple loop as in *M. chalybata* (Fig. 9.9) or convoluted into a number of concentric loops as in *M. keraudrenii* (see Fig. 1.5). The modified trachea sits immediately beneath the breast skin and on top of the pectoral muscles (Clench 1978). It is believed that this elongation of the trachea is to lower the pitch of vocalisations so that they carry over greater distances, an adaptation found in a number of unrelated non-passerine groups (Frith 1994*b*).

An 'eye brow' tuft of slightly longer and dense feathering sits, at times more conspicuous than others, above each eye. Entire ad plumage highly glossed with purples, blues, and greens and some areas of some feathers in some spp. are tightly curled or crinkled. In *M. comrii* the central pair of tail feathers in both sexes curl through 180° at the tips to expose the ventral feather surface uppermost. Tail feathers of manucodes are tipped by a fine hair-like point (as in *Lycocorax*, *Macgregoria*, and *Paradigalla*). Previously placed in the monotypic genus *Phonygammus*, *M. keraudrenii* lacks curled or crinkled feathers but has conspicuous lanceolate 'ear tufts' and nape and throat feathers (Plate 5).

Monogamous and pair-bonding and, in *M. keraudrenii* at least, non-territorial (apparently a product of the highly-frugivorous diet). Unlike all other paradisaeinine spp., including *Lycocorax*, the **nest** is constructed predominantly of curled vine tendrils entwined to form a sparse basket-like structure suspended between tree branches at the upper rim level like nests of drongos *Dicrurus* spp. (Dicruridae). Like *Lycocorax* nests, however, wood chips are a typical component in some spp. **Eggs** are pale pinkish and are spotted and finely blotched like those of *Lycocorax* and *Paradigalla*—not longitudinally streaked as in eggs of most of the higher paradisaeinine spp. (Plate 13).

Glossy-mantled Manucode *Manucodia atra* (Lesson, 1830)

PLATE 5; FIGURE 9.8

Other names: Glossy Manucode, Black Manucode. *Quatbon* in middle Sepik, PNG (Gilliard and LeCroy 1966). *Yaga* of the Fair language at Bomakia and *Baliyo* of the Kombai language at Uni, IJ (Rumbiak 1994). *Kondawa* (Hollandia), *Quat Bove*, *Quali-Tienpa* (both Sepik R.), *Carray* (Batanta I., IJ) (specimen label data).

A large manucode that is relatively small-headed and long-tailed. Variably glossy-black, red-eyed, and lacking barring on the mantle. Flight appears floppy and ♂'s call is a mournful, high-pitched, drawn-out tone. The only bird of paradise to frequent savanna regularly. Polytypic; 3 subspp.

Description
ADULT ♂: 38–42 cm, tail graduated. Head, nape, and neck blue-black with highly-iridescent blue-green tipping to smooth feathering which may show plum-purple sheens in some lights. Feathers above each eye dense and plush but not elongated or erectile. Iridescent tipping on each feather of chin and throat forms an inverted V shape. Mantle, back, rump, upperwing, and dorsal surface of tail smooth-plumaged, blue-black with rich iridescent glosses of deep blue (70) and predominantly purples (1 and 172) to Magenta (2). Dorsal surface of tail appears subtly barred blackish in some light. Tiny hair-like central points to tail feathers. Breast as upperparts but with more blue-green iridescence which increases on belly, vent, and undertail coverts. Underwing: coverts blackish with iridescent blue-green edging, flight feathers brownish black (119) to blackish. Bill, legs, and feet black, iris blood red with a brown inner ring. Base of culmen broadened and flattened to a degree intermediate between that of the Jobi and Crinkle-collared Manucodes. The trachea is less modified than in other manucodes, forming a short single loop on the upper pectorals only (Forbes 1882b; Clench 1978; Frith 1994b).

ADULT ♀: 33–37 cm, slightly smaller than ad ♂. Similar to ad ♂ but greener, less purple, in iridescent glosses particularly on underparts.

NESTLING: a nestling that was c. 1 day old was naked with blackish-flesh skin, paler below, bill greyish flesh, gape white, mouth flesh, feet greyish flesh (Rand 1938).

IMMATURE–JUVENILE: juv downy and dull brownish black subsequently broken up by glossy feathers of imm plumage as they appear, initially more so on upperparts. Difficult to differentiate from imm *M. chalybata*. Iris orangy yellow to yellowish orange, probably changing with age. Tail on average slightly shorter than ad.

Distribution
Very widely distributed, throughout the lowlands of mainland NG and on the Western Papuan Is, Aru Is, and Samarai, Sariba, Mailu, Yule, and Tagula I. (Beehler *et al.* 1986; Coates 1990). From sea level into low hills, less commonly up to 1000 m (e.g. 850 m at Bernard Camp, IJ, and 1100 m, 5 km N of Wanuma, Adelbert Mts, PNG; specimen data from CF). Because of difficulties of field separation of *atra*, *jobiensis*, and *chalybata*, we lack detailed data on the relative elevational limits of these spp.

Systematics, nomenclature, subspecies, weights, and measurements
SYSTEMATICS: a sp. difficult to distinguish in the field from the Crinkle-collared and Jobi Manucodes. These three spp. constitute an assemblage deserving an integrated field, laboratory, and museum study, as, indeed, do the three poorly-defined subspp. of *M. atra*. The first bird of paradise discovered in its native habitat, by the Frenchman Lesson. Since a study by Gilliard (1956), resulting in the acceptance of the three following subspp., only Cracraft (1992) has reviewed the sp. In assessing plumage variation, from the phylogenetic sp. concept point of view, Cracraft found no diagnostic characters within *M. atra* populations. We agree with Gilliard, however, that it would be misleading to include all populations in a single taxon in the face of gross plumage differences apparent at the extremes of the range of this sp. and the clear differences in average size demonstrated here. We therefore follow Gilliard's treatment until additional studies clarify the nature of geographical variation in this sp. Hybridisation: unrecorded for any manucode sp. Even if it occurs (which we judge unlikely because of monogamous mating), it would be most difficult to detect from external morphology.

Glossy-mantled Manucode *Manucodia atra* (Lesson, 1830)

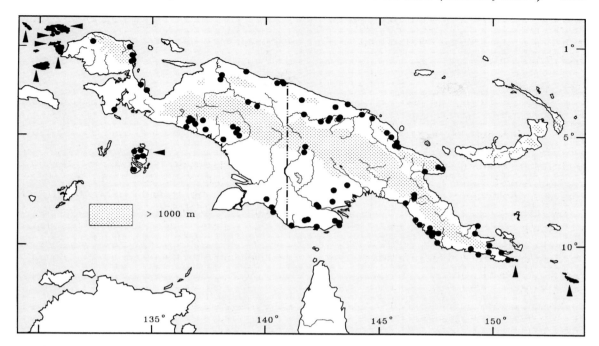

SPECIES NOMENCLATURE AND ETYMOLOGY

Manucodia atra (Lesson)

Phonygama ater Lesson, 1830. *Voyage of the Coquille, Zoology* **1**, 638. Dorey, northwestern New Guinea.

Etymology: *Manucodia* = from old Javanese (Malay) *manuk dewata* (birds of the gods); *atra* = black. Glossy-mantled Manucode is the preferred English name.

Note that we follow Filewood and Peckover (1978) and Beehler and Finch (1985) in treating *Manucodia* as feminine, which accounts for the spellings of scientific names herein (*contra* Gilliard 1969).

SUBSPECIES, WEIGHTS, AND MEASUREMENTS

1. *M. a. atra* (Lesson, 1830). Type specimen MHN 5505/136.

Range: lowlands of the western two-thirds of the NG mainland; easternmost extent of range of this subsp. is not well delineated, but according to Gilliard (1969: 98) in the N it ranges as far as the Huon Gulf, and in the S it extends to the Purari R.

Measurements: wing, ad ♂ ($n = 25$) 160–204 (181), imm ♂ ($n = 11$) 164–185 (173), ad ♀ ($n = 27$) 159–196 (175), imm ♀ ($n = 17$) 156–182 (167); tail, ad ♂ ($n = 25$) 133–170 (150), imm ♂ ($n = 11$) 123–158 (144), ad ♀ ($n = 27$) 127–166 (143), imm ♀ ($n = 17$) 125–155 (137); bill, ad ♂ ($n = 25$) 34–45 (39), imm ♂ ($n = 11$) 37–43 (39), ad ♀ ($n = 27$) 34–43 (38), imm ♀ ($n = 17$) 33–42 (36); tarsus, ad ♂ ($n = 25$) 35–42 (38), imm ♂ ($n = 11$) 36–41 (39), ad ♀ ($n = 27$) 35–41 (38), imm ♀ ($n = 17$) 32–40 (37). Weight: ad ♂ ($n = 17$) 170–315 (224), ad ♀ ($n = 19$) 155–252 (207), imm ♀ ($n = 3$) 153–200 (177).

2. *Manucodia a subaltera* Rothschild and Hartert, 1929. *Bulletin of the British Ornithologists' Club* **49**, 110. Dobbo, Aru Is. Type specimen AMNH 677266.

Range: disjunct and biogeographically inexplicable: Western Papuan Is (Misool, Salawati,

Batanta, Gemien, Gebe, Waigeo), Aru Is, and SE peninsular PNG, westward in southern watershed to Gulf of Papua, in northern watershed to Kumusi R. The disjunct disposition of this subsp. exemplifies the need for greater study of geographical variation in this sp.

Diagnosis: said to average much more purple and violet with oil green rare on ad (Gilliard 1969). On average, larger than nominate form, but smaller than *altera*, but biometrics in need of thorough analysis to delineate nature of contact between population of nominate form and of *subaltera* in SE PNG.

Measurements: wing, ad ♂ ($n = 29$) 175–211 (195), imm ♂ ($n = 7$) 172–199 (184), ad ♀ ($n = 26$) 173–205 (185), imm ♀ ($n = 6$) 177–200 (188); tail, ad ♂ ($n = 29$) 144–183 (165), imm ♂ ($n = 7$) 140–168 (155), ad ♀ ($n = 26$) 138–168 (154), imm ♀ ($n = 6$) 153–168 (160); bill, ad ♂ ($n = 29$) 39–44 (42), imm ♂ ($n = 7$) 38–43 (40), ad ♀ ($n = 25$) 37–41 (39), imm ♀ ($n = 6$) 37–43 (40); tarsus, ad ♂ ($n = 28$) 35–45 (41), imm ♂ ($n = 7$) 38–42 (40), ad ♀ ($n = 26$) 37–43 (40), imm ♀ ($n = 6$) 40–43 (42). Weight: ad ♂ ($n = 6$) 238–300 (275), imm ♂ ($n = 1$) 225, ad ♀ ($n = 5$) 190–245 (212).

3. *M. a. altera* Rothschild and Hartert, 1903. *Novitates Zoologicae* **10**, 84. Sudest I. Type specimen AMNH 677282.

Range: endemic to Tagula (= Sudest) I. of the Louisiade Archipelago of SE PNG.

Diagnosis: flanks and belly markedly more violet than nominate form; size considerably larger than other two subspp.; bill length of ad ♂ shows no overlap with that for the other two subspp.

Measurements: wing, ad ♂ ($n = 7$) 198–208 (204), imm ♂ ($n = 1$) 195, ad ♀ ($n = 11$) 182–201 (191); tail, ad ♂ ($n = 7$) 163–182 (172), imm ♂ ($n = 1$) 161, ad ♀ ($n = 11$) 152–172 (160); bill, ad ♂ ($n = 7$) 46–47 (46), imm ♂ ($n = 1$) 45, ad ♀ ($n = 11$) 41–44 (43); tarsus, ad ♂ ($n = 7$) 43–47 (45), imm ♂ ($n = 1$) 44, ad ♀ ($n = 11$) 41–44 (43). Weight: none.

Habitat and habits   Table 4.1

Rainforest, swamp forest, forest-edge, riverine and monsoon forests, woodlands, denser savanna woodlands, savanna, mangroves, gardens, and associated secondary growth. An active bird of clumsy appearance and movements. Usually encountered singly or in pairs, but occasionally in small feeding associations with conspecifics and/or other frugivorous spp. Moves actively and vigorously through foliage, seeking animal prey and fruiting and flowering trees and vines, rarely lingering in one place. Not known to fly over forest canopy. Will sing from exposed higher perches but mostly frequents the lower canopy and low dense vegetation. Two ♂♂, c. 350 m apart, were noted singing every 1–2 min from exposed canopy perches (BB). In giving its song, the ♂ raises itself on its legs and stretches its neck while expanding breast plumage and raising mantle and back feathers as the drooped wings are half opened—to return to normal perching posture (Coates 1990). Rand (1938) watched one bird hop towards and displace another, newly arrived in a tree top, whilst shaking its slightly-spread wings and tail. When nervous, birds sharply flick the tail (Rand 1938). Individuals have been observed chasing Figbirds *Sphecotheres viridis* (Oriolidae), potential competitors for fruits, from feeding trees (Coates 1990).

Diet and foraging   Table 4.2

Lesson (1834–5) stated that this manucode feeds on fruit of larger trees, and Wallace (1869), writing of Aru Is birds, wrote 'It is a very powerful and active bird; its legs are particularly strong, and it clings suspended to the smaller branches while devouring the fruits on which alone it appears to feed'. D'Albertis (1880) observed that it lives on fruits, especially figs but von Rosenberg (1875) reported insects and worms being eaten. Seen at Sii R., PNG at 70 m in mid July feeding on large, orange-red, pendant fruits of a *Piper* vine 10 m above ground (BB). See Appendix 4. Individuals visit flowers but it is not yet clear exactly what is eaten. Commonly joins mixed-spp. foraging flocks of black and/or brown plumaged spp. (Diamond 1987). One bird was seen feeding on fruits in company with at least four ♀-plumaged Raggiana Birds near Rouna Falls, PNG (Gilliard 1950). A single bird foraged in a ridgetop mixed-spp. flock with three ♀-plumaged Lesser Birds, a King Bird, Spangled Drongo *Dicrurus hottentotus*, Frilled Monarch *Arses telescophthalmus*, Grey Whistler *Pachycephala simplex*, Tawny-breasted Honeyeater *Xanthotis flaviventer*, and two Black Cuckoo-shrikes *Coracina melaena* at Gogol, PNG in late Oct (BB). Of 38 records of this species foraging in rainforest at Brown R., PNG, Bell (1982c) found that 11% were of birds in emergent trees (> 35 m high), 57% in the canopy (25–35 m), 21% in the sub-canopy (8–25 m), and 11% in the understorey (0–6 m high). Of 45 general (non-fruit) foraging records, involving 14 individual birds, observed by Bell (1984) 40% involved the bird feeding from 'debris' of dead foliage, 34% from branches, 13% from the air, 7% from tree trunks, and 6% from the ground. Foraging methods used were gleaning (94%) and hovering (6%). A flock of eight birds fed in a single fruiting tree in the Aru Is, early in Apr 1988 (D. Bishop *in litt.*). A manucode probably of this sp. was flushed from a mist-net where it had killed and pecked out the brain of a Spot-winged Monarch *Monarcha guttula* at 70 m asl in early Sept (BB).

Vocalisations and other sounds

Figure 9.8

The sp. (presumably the ♂) gives a high tone-like advertisement song, a simple contact note, plus an assortment of other miscellaneous call notes of uncertain significance. Finch (1983) described the song as a 'peculiar loud far carrying continuous note like something electrical,' and the call note as a *tuck tuck* given when feeding or in flight. A bird gave the song from the exposed upper branches of forest-edge trees at the edge of lake-side grasses of the upper Sepik R.—mournful, monotone, high-pitched, repeated, whistled notes of ringing quality, each lasting 1.5–2.5 sec and sounding like microphone electronic feedback (CF). This call is similar on the Aru Is (Fig. 9.8). At Milne Bay, PNG on 26 July, an individual sang a distinctive, single, pure, high tone call of 2–3 sec, monotonic like an electric tone that might be produced by an electric organ. The volume peaked in the middle of the note. When giving this call the bird raised its wings over its back, cocked its neck, and opened its beak (BB). Calls described by Hoogerwerf (1971) were 'a subdued *korrrrr-korrrrr* or a soft *kwek* and a hoarse *kek-kek-kek* and not very loud screeching'. A deep *chug* or *chook* was given by tail-flitting birds alarmed near their nest, and an undisturbed bird at the nest gave a low chattering call (Rand 1938). A parent bird alarmed at the nest uttered *ack, ack, ack* (Bergman 1961). Birds in flight produce a heavy, silken, swishing or rustling sound, like that of riflebirds but not so loud and harsh.

Mating system

Pair-bonding monogamy, both sexes confirmed brooding and provisioning nestlings, but apparently non-territorial.

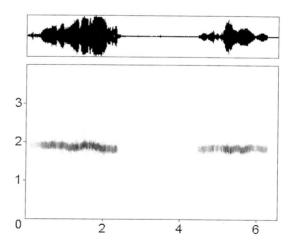

9.8 Sonogram from a recording of an ad ♂ Glossy-mantled Manucode advertisement song, the two even-toned notes sounding like microphone electronic feedback, by K. D. Bishop, Aru Is, Apr 1988.

Courtship behaviour

Said to consist of the ♂ slightly spreading and shaking his wings and tail and erecting his body feathers (Rand and Gilliard 1967). A bird near Tufi, PNG displayed at 0800 hr on 3 Jan; Wahlberg (1990) located it, in the crown of a tall tree in a forest clump amid wet savanna, by its characteristic song note. The bird 'threw its wings forward, so that the wings were level with its head, and then produced its long monotonous whistle'. This was immediately followed by a higher-pitched call in reply, assumed to be the ♀ replying to the assumed ♂ [believed to be another ♂ by BB]. The entire event lasted c. 15 sec and was then repeated once before the birds departed. It is not known if this represented courtship or was merely a calling display (between ♂♂).

Breeding   Table 6.5

NEST SITE: c. 4–12 m above ground in small trees in tall open rainforest, at or near the rainforest/open woodland or savanna edge, in savanna, beside a rainforest clearing with a break in the canopy, or in mangroves. Built between and among small forking horizontal branches (c. 10 mm dia), with some of the nest tendrils encircling branches to hold it firmly, with surrounding foliage.

NEST: a deep firm cup of slender, semi-woody to woody, dead and dry vine tendrils or stems, sometimes with a few green supple ones, coiled and interwoven into a neat cohesive whole. In the nest base, atop the first (outer) layer of woody stems, a good number of dead leaves (c. 160–170 × 30–60 mm) laid flat and atop these a quantity of dead, dry, rotten wood (as in Paradise Crow and Crinkle-collared Manucode). The egg-cup lining consists of a layer of blackish, semi-woody, supple stems, up to 1 mm dia, bent around inside the cup to form a rather rough finish. One nest measured 250 mm in dia and 140 mm deep externally and 130 mm in dia and 100 mm deep internally, the walls being 60 mm thick at one side and 20 mm thick and sparse enough to see through on the other (Rand 1938).

Another nest, of vine tendrils atop a few sticks of up to 2.5 mm dia, had an exterior of 240 × 175 mm and an interior, or cup, of 107 × 68 mm (Gilliard and LeCroy 1966)

EGGS: ovate with the shell fairly smooth and ranging from slightly glossy to dull (Plate 13). Of 19 eggs examined, 17 were measured (one clutch of 3, five of 2 and six of 1 egg), and averaged 39.2 (36–42.4) × 26.8 (24.6–29.6) mm. Ground colour dirty chalky to creamy white or very pale grey sometimes with faintest of Beige (219D) wash, spotted and blotched matte black through greys, grey-purples, and browns (CF).

INCUBATION: Rand (1938) observed a nest with a complete clutch of two eggs, less than a week old, for 5 hr over 2 consecutive days. A single bird, assumed to be ♀, incubated for periods averaging 30 min (range 21–51 min, $n = 5$) and was absent for on average of 26 min (range 5–45 min, $n = 6$). Thus incubation appeared to be performed by only the ♀. Rand acknowledged the possibility, however, that the ♂ was nervous of human activity and stayed away from the nest as a result. Incubation lasts > 14 and < 18 days (Rand 1938) and is likely to prove closer to the latter, which is brief for a bird of paradise.

NESTLING CARE: both sexes brood and feed nestlings, by regurgitating fruits, and eat nestling faeces.

NESTLING DEVELOPMENT: see Description above of a c. 1-day-old nestling. Another < 1 week old was naked and black, with wing feathers starting to grow. It was still in the nest when at least 19 days old (Bergman 1961).

Annual cycle   Table 6.5

DISPLAY: no information, except a 3 Jan record (Wahlberg 1990). BREEDING: at least Aug–Mar. For the sp. as a whole, ♂♂ have been recorded with much-enlarged gonads during June–Jan and with moderately-enlarged ones during Feb–May. A nestling a day or so old was seen on 29 Sept (Rand 1938). EGG LAYING: eggs collected or seen during Jan,

Mar, and Aug–Oct. MOULT: 65 of 160 specimens were moulting, involving all months but with Dec–June being the period of most, and July–Nov that of least, activity (Appendix 3).

Status and conservation   Table 8.1
Common though fairly inconspicuous and perhaps sparsely distributed throughout its range. Found to be common and widespread, usually in the canopy or sub-canopy of primary forest, mature secondary forest, and occasionally edges of gardens with remnant trees, on the Aru Is during 28 Mar–12 Apr 1988 (D. Bishop and J. Diamond *in litt.*). A widespread sp. clearly in no danger but the island endemic subsp. *altera* should be surveyed, especially given the amount of logging activity and forest clearing for gardens taking place on islands such as Tagula.

Knowledge lacking and research priorities
Intraspecific variation requires clarification (especially the validity to *M. a. subaltera* as currently delineated), as does behavioural and ecological isolation between this and its two sibling spp. relatives. The three are reputed to coexist in at least two localities known to us—Bernhard Camp, Idenburg R., IJ, and South Naru, Madang Province, PNG. It would be useful (though very difficult), to determine the nature of coexistence among the three; in particular it would be valuable to determine whether dietary differences distinguish the three. Are these spp. fig specialists like the Trumpet Manucode? Finally, how does tracheal looping of the ♂♂ vary among this group—both within the subspp. of *M. atra*, as well as within the spp.-group? The two better-studied manucodes exhibit sex-specific vocalisations and this can be expected for the present sp. Clarification of these will greatly aid study of the sp. in the field.

Aviculture
At least one bird was kept between 9 Mar 1881 and 11 Mar 1882 (Forbes 1882*b*). In 1904 Walter Goodfellow delivered a pair to 'Mrs. Johnstone' (Seth-Smith 1923*a*). The sp. was kept by the NYZS (Crandall 1935).

Jobi Manucode *Manucodia jobiensis* Salvadori, 1875 [1876]

PLATE 5

Other name: Allied Manucode

A secretive, little-known, and difficult-to-identify manucode all but identical to the Crinkle-collared, but with a relatively short tail that is less graduated than other manucodes, and a bill that is possibly blunter than that of the Crinkle-collared in areas outside the Sepik. Differences between this sp. and the latter are extremely subtle, as are their vocalisations, making identification of wild birds most difficult. Monotypic.

Description
ADULT ♂: 34 cm, tail only slightly graduated. Head and neck blue-black with highly-iridescent blue-green tipping to smooth feathering which may show plum purple gloss in some lights. No modification to feathers above eyes. Iridescent tipping contrasting with dark matte feather portions forms distinct barring on nape, chin, throat, and neck. Mantle, back, rump, and dorsal surface of tail coverts deep blue with rich, iridescent, violet purple (172) gloss and rich Magenta (2) sheens in some lights that are all conspicuously barred by broad matte black feather tipping. Dorsal surface of tail deep blue-black but rich and uniform, iridescently-glossed violet purple with Magenta sheens (without darker barring of Glossy-mantled). Tiny hair-like central points to terminus of each tail feather. Upper

Jobi Manucode *Manucodia jobiensis* Salvadori, 1875 [1876]

breast as mantle but slightly less purple, and lower breast, belly, and thighs progressively even less purple, more green-blue, in iridescence. Underwing: coverts Dusky Brown (19) with fine, iridescent, glossy, green-blue edges, flight feathers Dusky Brown. Vent and undertail coverts matte brownish black with subtle green-blue gloss. Ventral surface of tail blackish brown (119). Bill black, the base of the culmen only slightly broadened and flattened, far less so than other manucodes except the Trumpet. Iris blood red, legs and feet black. Subcutaneous, simple, single, looped trachea in ♂ to a similar degree of development as in *M. chalybata* (Forbes 1882b). See Fig. 9.9.

ADULT ♀: 31 cm, similar to, but averaging smaller and being slightly duller than, ad ♂. Lacks tracheal looping.

NESTLING: unknown.

JUVENILE–IMMATURE: downy upper and lower parts, crown and throat brownish black with crown more matte blackish to dull sooty black with upper wing and tail becoming progressively coloured and glossy with age. Lacks tracheal looping.

Distribution
Restricted to the northern lowlands, plus a small westernmost segment of the southern lowlands, of the main body of NG. In N, the sp. ranges from Rubi (SW shore of Geelvink Bay) eastward in N to the mouth of the Ramu R., PNG. In S, the sp. is confined to the isthmus of the 'Bird's Neck' (Etna Bay) eastward to the Mimika and Setekwa R. Ranges from sea level to 500 m (rarely to 750 m), being replaced above this by the Crinkle-collared Manucode. Not recorded from the Bomberai and Vokelkop Penin. In places sympatric with Glossy-mantled, Crinkle-collared, and Trumpet Manucodes.

Systematics, nomenclature, subspecies, weights, and measurements
SYSTEMATICS: poorly differentiated from Glossy-mantled and Crinkle-collared Manucodes, the group deserving intensive systematic study. Mainland birds are, on average, smaller than Yapen I. birds (Gilliard 1969; Frith and Frith 1997b) and possibly, on average, more intensely suffused with purplish (Cracraft 1992). As Yapen I. specimens are few ($n = 8$) and wing and tail lengths of mainland birds entirely overlap those of the island

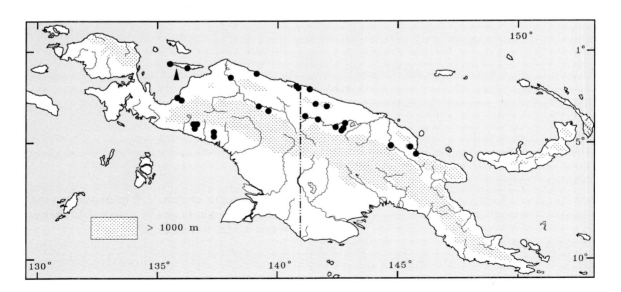

form, the size differences do not justify the retention of *rubiensis* and we therefore consider the sp. monotypic. Hybridisation: unrecorded.

SPECIES NOMENCLATURE AND ETYMOLOGY
Manucodia jobiensis Salvadori, 1875 [1876]. *Annali Museo Civico Storia Naturale di Genova* 7 (1875), 969. Wonapi, Jobi (= Yapen) I. Type specimen MCG C.E. 17230.

Synonym: *Manucodia rubiensis* A. B. Meyer, 1885. *Zeitschrift für die gesammte Ornithologie* 2, 374. Rubi, Geelvink Bay. Type specimen SMT C 1464 (8687), destroyed during WW II.

Etymology: *jobiensis* = after Jobi (Yapen) I., Geelvink Bay, IJ. The English name 'Jobi Manucode' is not ideal given the bird's extensive distribution beyond Jobi (= Yapen) I. Given the specific name *jobiensis*, the long usage of Jobi Manucode, and the problem of four other similar manucodes it is, however, best retained.

WEIGHTS AND MEASUREMENTS
Measurements: wing, ad ♂ ($n = 30$) 161–194 (176), imm ♂ ($n = 4$) 167–174 (170), ad ♀ ($n = 26$) 157–177 (169), imm ♀ ($n = 2$) 163–169 (166); tail, ad ♂ ($n = 29$) 119–147 (133), imm ♂ ($n = 4$) 123–132 (128), ad ♀ ($n = 26$) 113–141 (127), imm ♀ ($n = 2$) 117–120 (119); bill, ad ♂ ($n = 30$) 35–42 (39), imm ♂ ($n = 3$) 37–40 (38), ad ♀ ($n = 26$) 35–39 (37), imm ♀ ($n = 2$) 34–37 (36); tarsus, ad ♂ ($n = 29$) 34–41 (37), imm ♂ ($n = 4$) 36–37 (37), ad ♀ ($n = 26$) 34–39 (36), imm ♀ ($n = 2$) 35 (35). Weight: ad ♂ ($n = 8$) 212–257 (228), imm ♂ ($n = 1$) 197, ad ♀ ($n = 5$) 150–205 (180).

Habitat and habits — Table 4.1
Lowland rainforest, swamp forest, hill forest-edge. The sp. joins mixed-spp. foraging flocks. A ♀ was extremely nervous and difficult to observe near her nest (Grant in Ogilvie-Grant 1915*a*).

Diet and foraging — Table 4.2
Only unidentified fruits reported in the diet but it doubtless also eats arthropods, particularly in view of mixed-spp. feeding associations. Participates in mixed-spp. foraging flocks of predominantly black and/or brown birds; with other birds of paradise with or without Rusty Pitohuis *Pitohui ferrugineus* and Rufous Babblers *Pomatostomus isidorei* (Beehler and Beehler 1986). In the W Sepik Prov, PNG, in July–Aug a bird was seen searching for arthropods 1.5 m up in a shrub then flew down near ground and three individuals foraged for arthropods in the midstorey 15 m above ground (BB).

Vocalisations and other sounds
Little documented, for the simple reason that unless the singing bird is collected immediately after calling, it is difficult to ascribe a particular vocalisation to the sp. with certainty. Claude Grant (in Ogilvie-Grant 1915*a*) reported a 'long-drawn moaning note'. A series of 4–6 slowly-delivered and hollow *hoo* notes, much like those of the Crinkle-collared is considered the song and a harsh *chig* or *becheg* the call notes (Beehler *et al.* 1986). The behavioural contexts of vocalisations remain unknown, and some lingering doubt must remain concerning whether these vocalisations were indisputably from *M. jobiensis* (given the difficulty of field identification).

Mating system
Unknown but assumed to be pair-bonded monogamy, as known for the other manucodes.

Courtship behaviour
Unknown.

Breeding — Table 6.5
A ♀ was collected at her nest with two eggs on 28 Dec at the Mimika R, IJ.

NEST SITE: one known, 2.4 m above ground suspended between two horizontal branches (Ogilvie-Grant 1915*a*).

NEST: deep and cup-shaped, of roots, creepers, and leaves.

EGGS: the two eggs of the only known clutch are short oval, slightly pointed towards one end and almost glossless. Ground colour pale pinkish (between 121D and 219D), spotted with small dark brown dots and larger oblong lavender grey spots, chiefly about the larger end and thinly elsewhere. They measure 32.2 × 24.4 and 31.5 × 24.4 mm (CF).

INCUBATION, NESTLING CARE, AND DEVELOPMENT: unknown.

Annual cycle   Table 6.5
DISPLAY: unknown. BREEDING: birds have been collected with moderately-enlarged gonads in June–Aug and Dec–Jan and with much enlarged ones in Feb–Mar, May, and July (Table 6.5). EGG-LAYING: 28 Dec is the only laying date known. MOULT: 37 of 61 specimens show moult, involving wing, tail, or body, during all months (Appendix 3).

Status and conservation   Table 8.1
Said to be locally common, but the fact that this is a little-known sp. easily confused with the Crinkle-collared must make many records suspect.

Knowledge lacking and research priorities
This sp., and its distributions and ecological relationships with other manucodes, is of considerable interest. One must assume that at least certain of the vocalisations of *chalybata* and *jobiensis* are distinct and sp.-specific. Is the tracheal looping distinct, as well? The two better-studied manucodes exhibit sex-specific vocalisations and this can be expected for this sp. Clarification of these question will greatly aid study of the sp. in the field.

Aviculture
Not reported.

Crinkle-collared Manucode *Manucodia chalybata* (Pennant, 1781)

PLATE 5; FIGURES 9.9,10

Other names: Green-breasted Manucode, *Manucodia chalybatus*, Green Manucode, Crinkle-breasted Manucode. The Daribi of PNG call ad *Pagonabo* and imm *Tdgari* (Diamond 1972) and people at Lake Kutubu call ♂♂ *Wor* and *Kaliabudu* (Schodde and Hitchcock 1968). *Yagá* of the Fair language at Bomakia and *Bumburi* of the Kombai language at Uni, IJ (Rumbiak 1994). *Kaho* of the Gimi people, Crater Mt., PNG (D. Gillison *in litt.*).

A robust, long-tailed, glossy blue-black manucode; crow-like, with crinkled or frizzled, shiny green breast feathers and a prominent eyebrow tuft. The commonly-observed manucode of hill and lower montane forests, mainly because of its propensity for forest-edge. Monotypic.

Description
ADULT ♂: 36 cm, tail graduated. Head, neck, and nape blue-black with blue-green iridescent feather tipping forming inverted V shapes on chin, throat, neck, and nape. Feathers directly above each eye elongated and may be erected to form small supraorbital tufts. Mantle, back, rump, uppertail coverts, and dorsal surface of tail blue-black with rich iridescence predominantly of violet purple (172) with Magenta (2) sheens but also deep blues, particularly on mantle and alula. Tiny hair-like central points to rectrices. Conspicuous darker broad mantle barring formed by matte black feather tipping. Indistinct dark barring may infrequently be apparent on otherwise uniformly-glossy purple dorsal surface of tail in some lights. Lower throat, the neck, and upper breast feathering has fringed edging which creates an overall crinkled surface, the iridescence of which is conspicuously bronzed yellow-green. Lower breast and remaining underparts blue-black with overall violet purple gloss with Magenta sheens in some lights and darker, blackish, feather tipping giving a barred appearance that

Crinkle-collared Manucode *Manucodia chalybata* (Pennant, 1781)

is less conspicuous on the slightly more matte blackish vent and undertail covert feathering. Underwing: coverts blackish with deep blue iridescent edging, feathers blackish brown (119). Ventral surface of tail blackish, with slightest brownish wash and gloss. Bill black, iris deep red with dark brown inner ring (some individuals—subad?—reported with yellow-orange iris); legs and feet black. Older ♂♂ exhibit a single-looped, subcutaneous trachea (see Genus *Manucodia* and Fig. 9.9) that extends over *c*. two-thirds of the thorax.

ADULT ♀: 33 cm, like ad ♂ but averaging only fractionally (3%) smaller, generally duller below and being less purple, more green-blue, here. Trachea not looped.

SUBADULT ♂: similar to ad but less purple and less barred. Iridescent blue-green of throat and breast duller and belly more blackish than in ad. Lacks looped trachea.

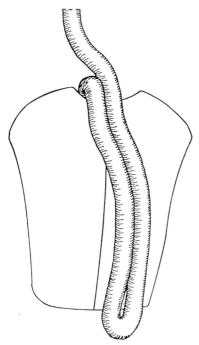

9.9 The single looped trachea of a male Crinkle-collared Manucode. From Clench (1978), with kind permission.

NESTLING: unknown.

JUVENILE–IMMATURE: like ad ♀ but breast and belly even greener (lacking purple) with matte blackish chin and throat. Iris greyish brown to dark brown. Difficult to differentiate from those of *M. atra*.

Distribution

Inhabits the hills and lower montane forest throughout the NG mainland, as well as the lowland forests of Misool I. (Western Papuan Is.), generally 600–1500 m; occasionally ranges from sea level as high as 1700 m.

Systematics, nomenclature, subspecies, weights, and measurements

SYSTEMATICS: a sp. indistinctly differentiated from the Glossy-mantled and Jobi Manucodes. Species distributions indicate these taxa are well established and probably relatively old. Without any explanation, Mayr (1941*a*) postulated that *M. chalybata* forms a superspecies with the insular *M. comrii*, an hypothesis that we find unconvincing, even though the two may indeed be sister-forms. Hybridisation: unrecorded.

SPECIES NOMENCLATURE AND ETYMOLOGY
Manucodia chalybata (Pennant)
Paradisea Chalybata Pennant, 1781. *Spec. Faunula Indica*, in *Forster's Indian Zoology*, p. 40. (based on Daubenton, *Table des Planches Enluminées*, pl. 634). New Guinea, restricted to Arfak Mts. Type is an illustration.
 Synonym: *Manucodia orientalis* Salvadori, 1896. *Annali del Museo Civico di Storia Naturale di Genova* ser. 2, **16**, 103. Type specimen not located.
 Etymology: *chalybata* = modern L. *chalybeatus*, steely, resembling same.

WEIGHTS AND MEASUREMENTS
Measurements: wing, ad ♂ ($n = 37$) 161–183 (173), imm ♂ ($n = 23$) 157–179 (169), ad ♀ ($n = 27$) 153–182 (167), imm ♀ ($n = 21$)

Crinkle-collared Manucode *Manucodia chalybata* (Pennant, 1781)

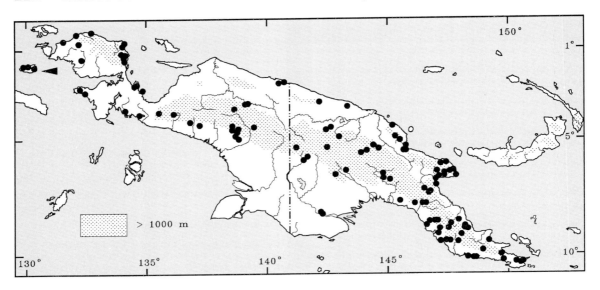

150–175 (161); tail, ad ♂ (*n* = 37) 128–152 (141), imm ♂ (*n* = 22) 120–144 (133), ad ♀ (*n* = 27) 124–154 (136), imm ♀ (*n* = 21) 119–138 (130); bill, ad ♂ (*n* = 36) 38–44 (40), imm ♂ (*n* = 23) 34–44 (40), ad ♀ (*n* = 26) 34–43 (38), imm ♀ (*n* = 21) 34–42 (37); tarsus ad ♂ (*n* = 34) 33–41 (38), imm ♂ (*n* = 23) 35–42 (38), ad ♀ (*n* = 27) 34–41 (37), imm ♀ (*n* = 21) 35–40 (38). Weight: ad ♂ (*n* = 36) 164–265 (223), imm ♂ (*n* = 2) 205–213 (209), ad ♀ (*n* = 21) 160–255 (209), imm ♀ (*n* = 5) 149–181 (172).

Habitat and habits Table 4.1

A bird of the hill forest interior and edge and midmontane rainforests where it mostly frequents the middle to canopy levels. Frequently observed, especially in vegetation of forest-edge, but individuals are nervous, wary, and flighty, and rarely perch in the open. Will flush at considerable distances. Normally perches with tail held down but when nervous crouches with body held low to perch and tail raised, or cocked, whilst making sudden jerky side-to-side movements and simultaneously flicking wings and tail, possibly as flight intention movements. Moves *c.* 10–20 m at a time. Commonly joins mixed-spp. foraging flocks. An individual radio-tracked on Mt. Missim in Nov 1980 covered a range of *c.* 45 ha during a week (BB). Flight appears floppy and awkward, much like that of *Paradisaea* spp. Wing-beats quick and convulsive each accompanied by a sharp jerking of the cocked tail and this being interspersed by short glides. A bird on Mt. Bosavi, PNG sunbathed after a brief shower by perching in a squatting posture on a horizontal branch and leaning to its right while stretching its left wing slightly and fanning its tail for several min (T. Pratt *in litt.*).

Diet and foraging Table 4.2

Predominantly frugivorous. Of 59 observed feeding instances (Beehler 1983*b*) 93% were upon figs *Ficus* spp., 5% upon complex or capsule fruits, and 2% upon simple drupe or berry fruits (see Chapter 4). Evidently specialises on figs; other fruits taken were those of *Chisocheton* cf. *weinlandii*, *Cissus aristata*, *Elmerrillia papuana* (*tsiampaca*), and a ginger (see also Appendix 4). Four faecal samples contained only fruit remains (BB). Often observed aggressively defending fruiting trees from other frugivorous birds even as they approach them (Pratt 1984; Mack and Wright 1996). Seen foraging actively for arthropod prey in riverine secondary growth in tangles

Crinkle-collared Manucode *Manucodia chalybata* (Pennant, 1781)

and among leaves of sapling stage 1–5 m up, even landing on ground several times, and often perched scansorially on vertical tree trunk (BB). Two birds and a single one were seen gleaning foliage and tangles of riverine shrubs and scrub of the sapling stage of forest 1–8 m above ground for arthropods (BB). Commonly joins mixed-spp. feeding flocks of predominantly black and/or brown birds (Diamond 1987).

Vocalisations and other sounds

Figure 9.10

Song of ♂ a slow series of up to eight haunting, low-pitched, hollow *hoo* notes of the same pitch. A response (presumed to be from the ♀), is a soft, ghostly, whistled series of descending notes *u-o-u-o-u-o-u-o*. The displaying ♂ gives a deep pigeon-like *ummmh* (like that of imperial pigeons *Ducula* spp.) or a deep hollow *whoo hoouw, hmmm* or *hmmm-hooo*. A deep hollow *hoouw* may be given singly or repeated at intervals of many sec during display, with the call fading as display subsides. A captive gave a repeated, unusual two-note call every *c.* 14 sec, the first a low, hollow, frogmouth-like humming followed after *c.* 1 sec by a louder, higher-pitched, *hmmm—hoo* lasting 2.75 sec—the whole being repeated many times. Sometimes only the first note was given (Coates 1990). The call note (Fig. 9.10) is variously transcribed as a *tuk, chenk, chook,* or *thtop* like that of other manucodes (Diamond 1972; Finch 1983; Beehler *et al.* 1986). A short, sharp, high-pitched *kok,* a sharp *tchich, kick* or *chack,* a nervous, high-pitched, nasal, mammal-like trill of *c.* 2 sec duration and short whining cries and a low pig-like grunting are also noted (Coates 1990). Also, ♀ gives a single *chengk!* note. Another quiet call is an inconspicuous nasal trill of *c.* 10 notes like a thinned-down version of a lamb bleeting (BB). In flight the plumage produces a silken swishing noise like, but not as loud as, that produced by the Magnificent Riflebird.

Mating system

Pair-bonding monogamy. Non-territorial and wide-ranging, as is typical in highly frugivorous bird spp.

Courtship behaviour

The ♂ chases a ♀ through foliage via numerous tree and vine perches, stopping now and then to display; may display repeatedly on a perch as the ♀ perches close to him. With each display call note (see above) the ♂ expands his breast and mantle feathering as he leans forward to stretch his neck and raise his head (Coates 1990).

Breeding

Table 6.5

NEST SITE: suspended between the forking branches of a tree limb 'like Orioles [*Oriolus* spp.] nests' (Hartert 1910).

NEST: built of 'brown wiry stalks, intermingled with leaves, and are lined with finer stalks and fibres' (Hartert 1910).

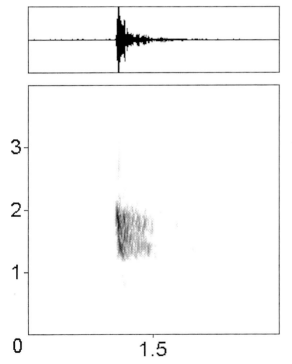

9.10 Sonogram from a recording of the Crinkle-collared Manucode *tuk* call note by B. M. Beehler in Varirata National Park, PNG, July 1989.

EGGS: rather roundish and chalky to creamy white, sometimes with slightest of pinkish wash, ground colour with brown and purplish markings. Four undated eggs, consisting of a two-egg and two single-egg clutches, averaged 35.4 (35–36.3) × 27.3 (26.9–28.1) mm (CF).

INCUBATION: unknown.

NESTLING CARE: Bulmer (in Gilliard 1969) appears to have collected both sexes at one nest containing two newly-hatched young; this suggests that both sexes provision young, but confirmation is required.

NESTLING DEVELOPMENT: unknown.

Annual cycle   Table 6.5
DISPLAY: captives displayed at BRS in Jan and Sept (Coates 1990), wild birds in Apr near Popondetta (Anon 1972), June and July near Port Moresby (Coates 1990), and Sept at Lake Kutubu (Schodde and Hitchcock 1968). BREEDING: nesting during at least July–Sept and Jan. EGG-LAYING: unknown. MOULT: 26 of 105 specimens were moulting, involving Jan–Apr and July–Nov with proportionately most occuring during Feb–Apr (Appendix 3).

Status and conservation   Table 8.1
Fairly common over much of its range, but secretive, making it difficult to census. Given the array of habitats it uses, the sp. is not under threat.

Knowledge lacking and research priorities
Nature and extent of fig specialisation needs study, as does the significance of the apparent ♂–♀ counter-calling. Where this sp. shares the forest with other spp. of manucodes (especially *jobiensis* and *atra*), the nature of ecological sorting of the spp. requires delineation.

Aviculture
Two live birds were delivered to London Zoo on 3 July 1908 by C. B. Horsbrugh and W. Stalker (Horsbrugh 1909). Kept by the NYZS (Crandall 1935). Restless in captivity, requiring a large well-lit aviary with shady areas, adequate flight area, ponds, and access to rain/sprinkler for bathing (P. Shanahan *in litt.*).

Curl-crested Manucode *Manucodia comrii* Sclater, 1876

PLATE 5; FIGURES 9.11–13

Other names: Curl-breasted Manucode. *Bulu oi* on Goodenough I. (BB). *Buli Buli* on Kiriwina I. (M. LeCroy *in litt.*).

A large, ungainly, and crow-like, glossy blue-black bird with peculiarly curled, frizzled, or crinkled feathering on head and breast. The tips of the central rectrices curl into a part spiral. Conspicuous and vocal in its habitat. Note floppy and undulating flight. Polytypic; 2 subspp.

Description
ADULT ♂: 43 cm, tail graduated. Head, neck, mantle, back, and rump deep blue-black, with crown feathering curled and that of neck crinkled, all tipped with strongly-iridescent yellowish green with deep blue and/or violet purple (172) and Magenta (2) sheens in some lights. Brighter iridescent tipping against black feather bases forms barring on the mantle and back in some lights. Upperwings blue-black, lesser coverts strongly-glossed iridescent green-blue tipped black and remainder intensely-iridescent deep violet-purple with strong Magenta sheen and feathers edged deep blue. Primaries blackish with dull, deep purplish blue sheen to outer vanes and tips. Dorsal surface of tail blue-black, intensely-iridescent deep purple with strong Magenta sheen throughout, black feather

shafts. Tiny hair-like central points to rectrices. Terminal third of central pair of tail feathers twisted to expose their underside. Breast and remaining underparts dark blue-black with contrasting strongly-iridescent broad feather tips deep purple/magenta with some deep blue to green-blue sheens. Thighs entirely glossed greenish blue. Underwing: lesser coverts glossed iridescent deep blue with purple/magenta sheens, greater coverts blackish brown with fine, iridescent, deep blue edges, flight feathers slightly glossy blackish. Undertail glossy black. Bill black, iris dull red or red-orange, legs and feet black. Ad ♂♂ develop a subcutaneous looped trachea, longer than in *M. atra*, *M. jobiensis*, and *M. chalybata*, that may extend the full length of the ventral body and double back up to the furcular (Beddard 1891). See Fig. 9.11. Rectrices arranged into a fairly-deep inverted V-shape rather than on one plane, the central pair at the apex.

ADULT ♀: like ad ♂ but averaging slightly (*c.* 5%) smaller.

NESTLING: unknown, but see Nestling development

JUVENILE–IMMATURE: two sibling birds 1–2 weeks out of the nest were well feathered except that their naked bluish black head skin was relieved only by a central thin line of matte black feathers along the central crown until reaching the feathered nape. The fluffy nape, neck, mantle, and wing covert feathers and the soft breast and flank feathers were matte black. Wing flight feathers were strongly glossed blue. Legs and bill were blackish, gape pale yellow, iris greyish brown, and mouth interior pinkish grey. Imm, like ad but plumage less glossy, predominantly dull matte blackish, throat and breast feathers not crinkled and curved, iris paler and more orange-brown at outer, and tiny gape pale pink. Mouth bright pink. See also Fig. 9.13.

Distribution

Inhabits Goodenough, Fergusson, Normanby, Wagifa, and Dobu Is of the D'Entrecasteaux Archipelago, and Kiriwina and Kaileuna Is of the Trobriand Is, SE PNG. On Goodenough ranges up to at least 2000 m (BB). Sympatric with Trumpet Manucode on Goodenough, Fergusson, and Normanby Is.

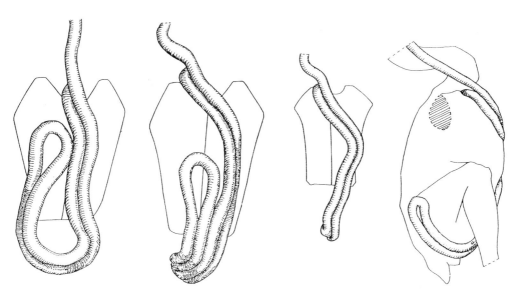

9.11 Examples of the elongated and looped trachea found in male Curl-crested Manucodes. From Clench (1978), with kind permission.

Curl-crested Manucode *Manucodia comrii* Sclater, 1876

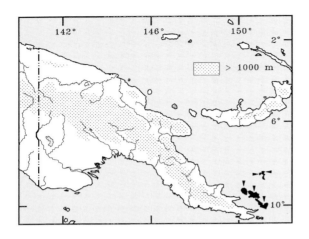

Systematics, nomenclature, subspecies, weights, and measurements

SYSTEMATICS: a distinctive island-isolated 'giant' manucode with two ill-defined subspp. Considered by Mayr (1941a) to form a superspecies with Crinkle-collared. Whereas *M. comrii* is likely an insular vicariant of the widespread *M. chalybata*, we question treating the pair as a superspecies, given the gross morphological disparities between the two. First collected on Fergusson I. in May 1874 by Dr P. Comrie, surgeon abroad the H.M.S. *Basilisk*. Hybridisation: unrecorded.

SPECIES NOMENCLATURE AND ETYMOLOGY

Manucodia comrii Sclater, 1876. *Proceedings of the Zoological Society of London* p. 459, pl. 42. Huon Gulf, in error for Fergusson I. Type specimen BMNH 87.11.20.373.

Etymology: *comrii* = named in honour of Dr P. Comrie, by P. L. Sclater.

SUBSPECIES, WEIGHTS, AND MEASUREMENTS

1. *M. c. comrii* Sclater, 1876.

Range: Goodenough, Wagifa, Fergusson, Dobu, and Normanby Is of the D'Entrecasteaux Archipelago, SE PNG.

Measurements: wing, ad ♂ ($n = 23$) 229–257 (244), imm ♂ ($n = 1$) 233, ad ♀ ($n = 13$) 223–240 (231), imm ♀ ($n = 5$) 207–224 (219); tail, ad ♂ ($n = 22$) 162–182 (173), imm ♂ ($n = 1$) 164, ad ♀ ($n = 13$) 158–170 (165), imm ♀ ($n = 5$) 149–171 (163); bill, ad ♂ ($n = 23$) 53–61 (58), imm ♂ ($n = 1$) 52, ad ♀ ($n = 13$) 51–57 (53), imm ♀ ($n = 5$) 50–55 (51); tarsus, ad ♂ ($n = 23$) 48–53 (51), imm ♂ ($n = 1$) 51, ad ♀ ($n = 13$) 46–51 (49), imm ♀ ($n = 5$) 47–49 (48). Weight: ad ♂ ($n = 1$) 448, ad ♀ ($n = 1$) 418; imm? ($n = 1$) 380.

2. *M. c. trobriandi* Mayr, 1936. *American Museum Novitates* **869**, 3. Kaileuna, Trobriand Is. Type specimen AMNH 224342.

Range: Kiriwina and Kaileuna Is of the Trobriand group, SE PNG.

Diagnosis: a thinly-differentiated subsp., diagnosed by the smaller wing and tail.

Measurements: wing, ad ♂ ($n = 14$) 213–235 (226), imm ♂ ($n = 1$) 204, ad ♀ ($n = 9$) 209–223 (217); tail, ad ♂ ($n = 14$) 148–169 (158), imm ♂ ($n = 1$) 149, ad ♀ ($n = 9$) 147–172 (154); bill, ad ♂ ($n = 14$) 52–57 (55), imm ♂ ($n = 1$) 50, ad ♀ ($n = 9$) 50–57 (53); tarsus, ad ♂ ($n = 14$) 47–52 (50), imm ♂ ($n = 1$) 45, ad ♀ ($n = 9$) 46–52 (48). Weight: none.

Habitat and habits   Table 4.1

Unwary and easy to observe. Inhabits all wooded habitats within its range: forest interior and edge, woodland, savanna, beach-crest casuarina woodland, hill and montane forest, secondary growth with larger trees about the forest-edge, neglected gardens, and occasionally in mangroves. Flies with floppy, loose-tailed, flight typical of manucodes. Frequents exposed perches and often will sit with tail cocked; folds tail like a North American grackle, Icteridae (BB). Tail flicking not pronounced, the tail is carried in an inverted V and in flight is opened and closed as the wings are (LeCroy *et al.* 1983). A flock of 4 and another of 6–7 silent birds moved as tight group in late July on Goodenough I. at *c.* 2000 m; birds were well spaced when stationary, but moved as a group (BB).

Curl-crested Manucode *Manucodia comrii* Sclater, 1876

Diet and foraging   Table 4.2

Forages alone, in pairs, or in parties. Stomach contents from birds on Goodenough I. contained fruit only, one specimen with large orange fruit 2–2.5 cm dia, the other with cranberry-like fruit (BB). About 12 birds seen feeding in one fruiting *Medusanthera laxiflora* tree with Goldie's Birds at *c.* 1400 m behind Nade, Fergusson I. (LeCroy *et al.* 1983).

Vocalisations and   Figure 9.12
other sounds

Very vocal, with a range of calls. Calling not uncommonly occurs from 1 hr before first light and throughout daylight (Beehler *et al.* 1986). A low-pitched resonant *ko-ko-ko-ko-ko* given all day and a less frequent metallic croak noted (Bell 1970). The usual call is a mournful, low-pitched, rolling series of notes *woodloodloodloodloodll* descending slightly (BB), probably equivalent to the low, bubbling call said to be typical and once heard given at dawn chorus (LeCroy *et al.* 1983). This is presumably the low even series of rapidly-repeated soft and rounded *oo* notes, almost pulse-like recorded by D. Gibbs (Fig. 9.12(a)). A briefer, high-pitched, and more musical, rapidally-descending call lasts *c.* 4 sec. Presumed ♂ gives low, haunting series while putative ♀ a prettier, more musical series, sharply descending (BB). The sexes occasionally give a rudimentary duet (like Crinkle-collared). Putative ♀ very rapidly gives a 4 sec. descending series of piping little notes, rather loud with whistled but throaty quality; much higher pitched than ♂ call. Voice on Goodenough I. in Mar was a ghostly, monotone whistle, louder then softer, very rapid series of notes all at the same pitch for 1.5–2 sec. A scold is a repeated, exclamatory, harsh, and nasal-sounding *ench* note (Fig. 9.12 (b)). Flying birds produce loud wing-noise (BB).

Mating system

Unknown but presumed to be pair-bonded monogamy, like other manucodes.

Courtship behaviour

Unknown.

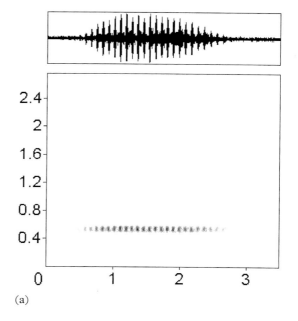

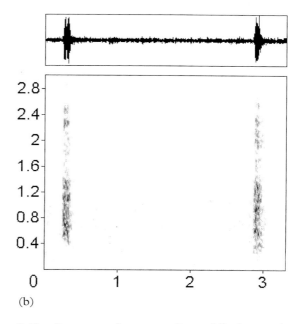

9.12 Sonograms from recordings of Curl-crested Manucode: (a) the low even series of rounded soft *oo* notes of the song and (b) the exclamatory, harsh, nasal-sounding, *ench* scold note. Recorded on Fergusson I. in Nov 1993 by D. Gibbs.

Breeding   Table 6.5

Reported to nest within 30 m of coast by Fred Shaw Mayer (in Gilliard 1969). A sibling pair of fledglings examined on 31 Oct 1985 had been taken from a D'Entrecasteaux Archipelago nest 1 or 2 weeks previously when they were about ready to leave it (CF in Coates 1990). A Kiriwina I. fledgling was examined on 27 Oct 1986 by M. LeCroy (*in litt.*)

NEST SITE: suspended in the fork of a tree branch like an oriole's *Oriolus* spp. nest, some vines of the structure holding it firmly to the fork. One nest placed at the extremity of a lower branch (7.6 m above ground) of a bread-fruit tree (Rickard in North 1892). A bird seen taking nest material *c*. 20 m up into a tree (LeCroy *et al.* 1983).

NEST: the nest found by Rickard was an open, loosely-made structure of vinelets and twigs. A Trobriand Is nest seen by A. S. Meek had a very thick base and was ornamented on the outside with large thick leaves. At the centre of the bottom layer of material were many pieces of rotten wood (as in Glossy-mantled Manucode and Paradise Crow nests).

EGGS: clutch of one ($n = 8$) or two ($n = 16$), 39 eggs averaging 44 (40.4–47.9) × 29.8 (20–31.4) mm (CF). See Plate 13. Ground colour Beige (219D) to Pale Pinkish Buff (121D) or like Pale Horn (92) with faintest wash of former marked with sparse to dense purple-grey spots and elongate blotches overlaid with deep brown and chestnut-purple ones, denser at larger end. May be streaked like typical *Paradisaea* eggs.

INCUBATION AND NESTLING CARE: unknown.

NESTLING DEVELOPMENT: nestlings/fledglings remain bare-headed, save for a fine central crown line of matte black feathers, for weeks after leaving the nest, the extensive, bare, blue-black, facial skin giving them an odd appearance (Fig. 9.13).

9.13 A juv Curl-crested Manucode several weeks after nest departure. Note persisting areas of bare facial skin. Photo: CF.

Annual cycle   Table 6.5

DISPLAY: unknown. BREEDING: at least June–Nov and Mar and, therefore, potentially any time of year. Nest building seen mid Nov (LeCroy *et al.* 1983). EGG-LAYING: Mar and June–Oct. The fact that Shaw Mayer collected 20 clutches (BMNH 1941.1.2.1–20) from the Faralulu district of W Fergusson I. coast at C. 300 m between 6 Aug and 28 Sept 1935 has remained uncited. Of these, all seven Aug clutches were of two eggs. Six two egg clutches were taken during 1–2 Sep, one-egg clutches were taken on 4, 9, 12, 17, and 22 Sep, and two-egg clutches on 12 and 28 Sep. Shaw Mayer collected five clutches of two eggs on 2 Sep, indicating the sp. nested conspicuously and, perhaps, densely in that area at that time. MOULT: 21 of 65 specimens were moulting, involving Apr–May and Nov–Dec. Specimens of Mar, June, Aug–Sept showed no moult. No samples were available for Jan–Feb, July, and Oct (Appendix 3).

Status and conservation   Table 8.1

Reported as abundant on Fergusson (Shaw Mayer in Gilliard 1969) and Goodenough Is (Bell 1970; BB) and as fairly common to abundant overall (Coates 1990). The collection of 20 clutches during Aug–Sept 1935 certainly indicates this sp. was then abundant on W Fergusson I. Very common at Sewa Bay, Normanby I. in Sept (Mackay and Mackay 1974). The small Trobriand Is subsp. said to be less plentiful and potentially threatened by habitat loss associated with rapid human population growth (Peckover 1990). This said, birds occur in over-grown gardens, which constitutes all habitat on Kiriwina I. where the sp. survives (M. LeCroy *in litt.*). Gilliard (1969) was informed that feathers of this manucode were used by Trobriand Islanders for ceremonial adornment. A Goodenough I. informant told Bell (1970) that this sp. is eaten, but the abundance and tameness of it on the islands suggests that birds are infrequently killed for any purpose and that such usage represents no threat to populations. It was commonly encountered in all habitats on Goodenough I. in 1976 and 1980 (BB).

Knowledge lacking and research priorities
Detailed knowledge of the diet would be of much interest, especially the degree to which it forages on figs vs. other plant taxa. Social behaviour and details of vocalisation are also in need of better study. This sp. would be an excellent focus for a study of a monogamous bird of paradise, mainly because of its abundance and confiding nature.

Aviculture
The two recently-fledged siblings seen by CF (above) were donated to the BRS where they proved easy to maintain and did well for at least 2 yr before that sanctuary closed.

Trumpet Manucode *Manucodia keraudrenii* (Lesson and Garnot, 1826)

PLATES 3,5; FIGURES 1.5, 4.3, 9.14, 15

Other names: Trumpetbird, Trumpet Bird, Manucode, *Phonygammus keraudrenii*. The northern Fore peoples' name is *Kaukábara* and that of the southern Fore *Kautyábo*, *Pagonábo* (ad) and *tágari* (imm) of the Daribi people (Diamond 1972). *Imonno* of the Fair language at Bomakia and *Mambaliyo* of the Kombai language at Uni, IJ (Rumbiak 1994).

A shy, canopy-dwelling, glossy blue-black manucode of forest interior; more often heard than seen. Aside from the distinctive vocalisations, the shaggy mane of nape and neck feathers is the most obvious field mark. Feeding aggregations of up to 10 birds may form in favoured fig tree. Polytypic; 9 subspp.

Description
ADULT ♂: 31 cm, tail only slightly graduated. Head, neck, and nape blue-black with rich iridescent green-blue with slight violet purple (172) sheen throughout, more so on nape, in some lights. Feathers of rear crown, nape, and lower neck elongated and finely pointed, those above the eyes forming conspicuous, erectile, occipital 'ear' tufts. Mantle, back, rump, uppertail coverts, and dorsal surface of tail smooth-plumaged and intensely-iridescent deep blue, with slightest yellow-green wash, heavily influenced by deep violet-purple sheen.

In some lights one or all colours may be visible. Upperwing intense, iridescent, deep violet purple with indistinct deep blue sheen and edging to coverts and tertials. Primaries blackish brown with narrow, iridescent deep blue outer edges. Upper breast and remaining underparts richly-iridescent oily green-blue with deep purplish sheens particularly on upper breast feathers elongated and finely narrowed into 'hackles'. Vent and undertail coverts duller and less iridescent. Undertail slightly glossy brownish black with tiny hair-like central points to feather tips. Bill, legs, and feet blackish to black; iris rich red with a fine inner ring of dark brown; mouth blackish. Convoluted trachea exhibits five or six coils under skin of breast.

ADULT ♀: 28 cm, as ad ♂ but on average slightly smaller, somewhat duller on underparts, and head and neck hackles on average

Trumpet Manucode *Manucodia keraudrenii* (Lesson and Garnot, 1826)

slightly shorter. Iris orange-red with thin brown inner ring. No looping of trachea.

IMMATURE–SUBADULT: generally dull blackish, particularly the head, with only slight bluish green iridescence on upperparts and breast but duller on wings and belly. Iridescent sheen on underparts of younger birds at least on some subspp., and particularly of *purpureoviolacea*, clearly greener. Iris greyish brown to brownish red to reddish orange (with increasing age). Brighter iridescent plumage and longer 'ear' tuft feathers increasingly acquired with age. No tracheal looping.

NESTLING–JUVENILE: a nestling of fledging age was dull sooty blackish brown on central to rear crown, mantle, and underparts with blackish lower mantle, upperwings and tail quite strongly-glossed, iridescent, deep blue-green. Sides of face and chin naked and blackish. Writing of the nominate form, Hartert (1930) noted that moulting specimens show that the juv plumage is a sort of raven black, burnished or graphite black. We found juv to be generally blackish with slight dull steel blue-green gloss but no purple on the back, wings, or tail. Head black, with scarcely any steel green gloss, dull, sooty, blackish brown down on central to rear crown, and sides of face and chin naked and blackish. Elongate feathers scarcely indicated. Underparts dull black, with almost no gloss except on throat.

Distribution

The most widespread of the birds of paradise. Inhabits NG and fringing islands, and CYP, Australia. Found throughout most mountain ranges of NG, at altitudes of 200–2000 m (mainly 900–1800 m), and in the lowlands of the Vogelkop of IJ and S and SE PNG, but not known from the Huon Penin or Cyclops Mts. Also inhabits the Aru Is of IJ, the Australian islands of Boigu and Saibai just S of the PNG coast and the D'Entrecasteaux Archipelago of SE PNG. In Australia, the bird is resident and confined to the extreme NE of CYP N of Coen, where it is more abundant in

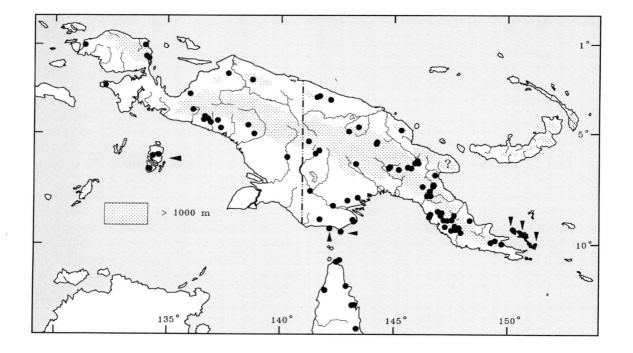

Trumpet Manucode *Manucodia keraudrenii* (Lesson and Garnot, 1826)

the E. Also occurs on Albany and Mai Is immediately adjacent to the northern tip of CYP (Frith 1994*a*). Where found in NG, the Trumpet Manucode coexists with at least one other manucode sp. (except at the highest sections of its elevational range).

Systematics, nomenclature, subspecies, weights, and measurements

SYSTEMATICS: morphologically the most divergent of the manucodes. Presumably this represents a sister lineage to the four other manucode spp. Has in earlier treatments been placed in its own genus, *Phonygammus*. A distinctive manucode with elongated lanceolate 'ear tufts' and 'hackles' and 9 subspp. of varying degrees of distinctiveness. Intraspecific variation deserves an in depth study as subspecific taxonomy has long been problematical. Cracraft (1992), applying the phylogenetic species concept, recently described two new 'species' which we here treat as subspp. It should be stressed that most of the 9 subspp. presently acknowledged have been described from limited material with inadequate samples of ad, subad, imm, and juvs of both sexes. Most subsp. characters relate to relative degrees and/or extent of iridescent feathering showing green, blue, purple, or violet colour and/or the length of lanceolate 'ear' tufts and neck feathering. There is clearly variation in some of these characters with age as ♀♀ and younger ♂♂, in at least some populations, exhibit more green in plumage than do ad ♂♂. Plumage wear must also be considered. This plus the fact that the collecting of birds to date involves numerous significant geographical gaps leaves the conclusions reached by Cracraft (1992) tentative at best. It is likely future research will show some of the following forms to be invalid. Hybridisation: unrecorded.

SPECIES NOMENCLATURE AND ETYMOLOGY

Manucodia keraudrenii (Lesson and Garnot)
Barita Keraudrenii Lesson and Garnot, 1826. *Bulletin des Sciences Naturelles et de Géologie Férussac* **8**, 110. Dorey, northwestern New Guinea. Type specimen MHN 5505.

Etymology: *keraudrenii* = after French Naval Inspector General of Medical Services, Pierre F. Keraudren.

SUBSPECIES, WEIGHTS, AND MEASUREMENTS

1. *M. k. keraudrenii* (Lesson and Garnot, 1826)

Range: westernmost NG, including the Vogelkop and Onin Penin and the Weyland Mts, at the westernmost terminus of the central ranges.

Measurements: wing, ad ♂ (*n* = 24) 144–166 (160), imm ♂ (*n* = 8) 148–165 (155), ad ♀ (*n* = 15) 139–158 (152), imm ♀ (*n* = 1) 145; tail, ad ♂ (*n* = 24) 110–133 (124), imm ♂ (*n* = 8) 122–129 (125), ad ♀ (*n* = 15) 108–125 (117), imm ♀ (*n* = 1) 117; bill, ad ♂ (*n* = 24) 31–36 (34), imm ♂ (*n* = 8) 32–35 (33), ad ♀ (*n* = 15) 29–34 (32), imm ♀ (*n* = 1) 33; tarsus, ad ♂ (*n* = 24) 30–37 (34), imm ♂ (*n* = 8) 33–37 (35), ad ♀ (*n* = 15) 31–35 (33), imm ♀ (*n* = 1) 31; ear tufts, ad ♂ (*n* = 24) 11–27 (19), imm ♂ (*n* = 8) 10–25 (17), ad ♀ (*n* = 15) 14–21 (18), imm ♀ (*n* = 1) 20. Weight: ad ♂ (*n* = 7) 130–175 (149), ad ♀ (*n* = 7) 130–140 (136), imm ♀ (*n* = 1) 126.

2. *M. k. aruensis* (Cracraft, 1992). *Cladistics* **8**, 10. Wanoem Bay, Kobror I., Aru Is., West Irian. Type specimen AMNH 677414.

Range: Aru Is, IJ.

Diagnosis: less green than the nominate form and differs from adjacent mainland NG *jamesii* in being generally darker and bluer, less green. Upperparts, particularly the back, suffused purple and lanceolate head feathering deeper, more purple, cobalt blue than in *jamesii*. Occipital tufts much shorter than *jamesii*, similar in length to nominate.

Measurements: wing, ad ♂ (*n* = 4) 166–170 (168), imm ♂ (*n* = 3) 154–162 (158); tail, ad ♂ (*n* = 4) 132–138 (136), imm ♂ (*n* = 3) 127–141 (134); bill, ad ♂ (*n* = 4) 33–34 (34), imm ♂ (*n* = 3) 33–34 (33); tarsus, ad ♂ (*n* = 4) 34–38 (36), imm ♂ (*n* = 3) 37–38 (37); ear tufts, ad ♂ (*n* = 4) 17–26 (22), imm ♂ (*n* = 3) 9–17 (13). Weight: none.

3. *M. k. jamesii* (Sharpe, 1877). *Catalogue of Birds in the British Museum* **3**, 181. Aleya, Hall Sound, British New Guinea. Type specimen BMNH 77.486.

Range: lowlands of southern NG, from the Mimika R., IJ, to the forests in the vicinity of Port Moresby, PNG (and further southeastward towards Milne Bay?).

Diagnosis: throat and breast dark metallic blue, washed greenish, lacking purple of nominate form. Occipital tufts longer than in nominate form. The type specimen was obtained near Yule I., SE NG by Dr James who was subsequently murdered by the people of that area.

Measurements: wing, ad ♂ (*n* = 26) 155–171 (164), imm ♂ (*n* = 1) 143, ad ♀ (*n* = 13) 148–166 (160), imm ♀ (*n* = 9) 142–159 (151); tail, ad ♂ (*n* = 26) 119–136 (128), imm ♂ (*n* = 1) 114, ad ♀ (*n* = 13) 119–135 (*n* = 13), imm ♀ (*n* = 9) 118–128 (123); bill, ad ♂ (*n* = 26) 30–36 (34), imm ♂ (*n* = 1) 32, ad ♀ (*n* = 13) 31–37 (33), imm ♀ (*n* = 9) 31–36 (33); tarsus, ad ♂ (*n* = 26) 33–39 (36), imm ♂ (*n* = 1) 37, ad ♀ (*n* = 13) 32–39 (36), imm ♀ (*n* = 9) 32–36 (34); ear tufts, ad ♂ (*n* = 25) 17–36 (27), ad ♀ (*n* = 13) 10–36 (24), imm ♀ (*n* = 9) 9–23 (17). Weight: ad ♂ (*n* = 2) 183–190 (187), ad ♀ (*n* = 2) 147–172 (160).

4. *M. k. neumanni* (Reichenow, 1918). *Journal für Ornithologie* **66**, 438. Lordberg. Type specimen ZMB 1.49.297.

Range: northern scarp of the central range, from the Lordberg to Schrader Ra and vicinity of Jimi and Baiyer R. watersheds (Sepik-Wahgi Divide). Identified by A. Engilis as occurring on the Bewani Mts (T. Pratt *in litt*.). A record of this sp. at Karawari near the Sepik R., PNG (Price and Nielsen 1991) and a sighting at Aiome at the E foot of the Schrader Mts at < 100 m (M. LeCroy *in litt*.) would involve this form and specimens are desirable.

Diagnosis: like similar-sized nominate form but lower back, rump, dorsal surface of tail, and wings dark blackish blue, washed iridescent deep violet purple rather than greenish. Breast and belly dark metallic blue. Occipital tufts on average shorter than all subsp. except for those of *adelberti*, which are comparable.

Measurements: wing, ad ♂ (*n* = 11) 150–162 (156), imm ♂ (*n* = 2) 142–148 (145), ad ♀ (*n* = 8) 142–152 (149); tail, ad ♂ (*n* = 11) 119–129 (124), imm ♂ (*n* = 2) 114–120 (117); ad ♀ (*n* = 8) 110–123 (117); bill, ad ♂ (*n* = 10) 30–36 (33), imm ♂ (*n* = 2) 32 (32), ad ♀ (*n* = 8) 30–32 (31); tarsus, ad ♂ (*n* = 11) 32–34 (33), imm ♂ (*n* = 2) 33–34 (34), ad ♀ (*n* = 8) 30–35 (33); ear tufts, ad ♂ (*n* = 9) 12–15 (13), imm ♂ (*n* = 2) 8–9 (9), ad ♀ (*n* = 8) 9–13 (11). Weight: ad ♂ (*n* = 5) 152–162 (157), imm ♂ (*n* = 1) 136, ad ♀ (*n* = 6) 133–154 (145).

5. *M. k. adelberti* (Gilliard and LeCroy, 1967). *Bulletin of the American Museum of Natural History* **138**, 72. Nawawu, Adelbert Mts. Type specimen AMNH 791016.

Range: inhabits the Adelbert Mts of N PNG.

Diagnosis: differs from nominate form by very short occipital tufts and greenish (not blue or purple) upperwings and tail.

Measurements: wing, ad ♂ (*n* = 10) 156–168 (163), ad ♀ (*n* = 1) 157, imm ♀ (*n* = 2) 146 (146); tail, ad ♂ (*n* = 10) 122–139 (130), ad ♀ (*n* = 1) 126, imm ♀ (*n* = 2) 120–122 (121); bill, ad ♂ (*n* = 9) 31–25 (33), ad ♀ (*n* = 1) 32, imm ♀ (*n* = 2) 29–32 (31); tarsus, ad ♂ (*n* = 9) 28–37 (34), ad ♀ (*n* = 1) 33, imm ♀ (*n* = 2) 34–35 (35); ear tufts, ad ♂ (*n* = 10) 11–17 (13), ad ♀ (*n* = 1) 14, imm ♀ (*n* = 2) 10 (10). Weight: ad ♂ (*n* = 6) 166–188 (175), ad ♀ (*n* = 1) 153, imm ♀ (*n* = 2) 128–135 (132).

6. *M. k. diamondi* (Cracraft, 1992). *Cladistics* **8**, 12. Awande, near Okapa, Eastern Highlands District, Papua New Guinea. Type specimen AMNH 809310.

Range: southern watershed of the Eastern Highlands, PNG, from near Okapa, possibly including portions of Kratke Ra, PNG, 1000–2000 m.

Diagnosis: similar to *purpureoviolacea* but defined as distinct from it in having back, upper wings and tail with strong violet purple sheens, the breast and belly dark metallic blue with little or no violet purple, and lanceolate head feathering bluish green as opposed to greenish blue washed violet-purple. Occipital tufts slightly longer than in *purpureoviolacea* and thus longer than all other forms.

Measurements: wing, ad ♂ ($n = 11$) 156–176 (171), imm ♂ ($n = 1$) 164, ad ♀ ($n = 4$) 159–167 (162), imm ♀ ($n = 1$) 172; tail, ad ♂ ($n = 11$) 121–134 (129), imm ♂ ($n = 1$) 127, ad ♀ ($n = 4$) 121–124 (123), imm ♀ ($n = 1$) 128; bill, ad ♂ ($n = 11$) 33–38 (36), imm ♂ ($n = 1$) 34, ad ♀ ($n = 4$) 31–35 (33), imm ♀ ($n = 1$) 34; tarsus, ad ♂ ($n = 9$) 34–40 (37), imm ♂ ($n = 1$) 38, ad ♀ ($n = 4$) 32–37 (35), imm ♀ ($n = 1$) 36; ear tufts, ad ♂ ($n = 11$) 26–53 (40), imm ♂ ($n = 1$) 25, ad ♀ ($n = 4$) 33–35 (34), imm ♀ ($n = 1$) 30. Weight: ad ♂ ($n = 3$) 168–190 (177), imm ♂ ($n = 1$) 174, ad ♀ ($n = 2$) 171–172 (172).

7. *M. k. purpureoviolacea* (Meyer, 1885). *Zeitschrift für die gesammte Ornithologie* **2**, 375, pl. **15**. Astrolabe Mountains. Type specimen SMT 8458.

Synonym: *M. k. mayri* (Greenway), 1942. *Proceedings New England Zoological Club* **19**, 51. Wau, Morobe District, NE New Guinea. Type specimen MCZ 168236.

Range: uplands of SE PNG, from the Kuper Ra and Wau SE through the Owen Stanley Ra. Details of distribution not well delineated.

Diagnosis: generally like nominate form but larger, with back, breast, and belly intensely-iridescent violet purplish, and occipital tufts far longer. The supposed evidence of a higher wing–tail index in *mayri* than in *purpureoviolacea* is meagre given limited specimen numbers for the former (Frith and Frith 1997*b*) and as no other characters appear to differentiate them, including occipital tuft length, we concur with Cracraft (1992) and combine the former into the latter form.

Measurements: wing, ad ♂ ($n = 42$) 157–175 (168), imm ♂ ($n = 5$) 159–174 (164), ad ♀ ($n = 34$) 148–180 (162), imm ♀ ($n = 6$) 151–161 (156); tail, ad ♂ ($n = 41$) 115–138 (128), imm ♂ ($n = 5$) 124–128 (126), ad ♀ ($n = 34$) 113–135 (123), imm ♀ ($n = 2$) 120–122 (121); bill, ad ♂ ($n = 40$) 31–38 (36), imm ♂ ($n = 5$) 32–36 (35), ad ♀ ($n = 34$) 32–37 (34), imm ♀ ($n = 2$) 29–32 (31); tarsus, ad ♂ ($n = 42$) 34–39 (37), imm ♂ ($n = 5$) 37–39 (38), ad ♀ ($n = 34$) 32–38 (36), imm ♀ ($n = 2$) 34–35 (35); ear tufts, ad ♂ ($n = 42$) 20–53 (38), imm ♂ ($n = 5$) 13–34 (26), ad ♀ ($n = 33$) 27–42 (34), imm ♀ ($n = 2$) 10. Weight: ad ♂ ($n = 20$) 159–240 (181), imm ♂ ($n = 1$) 182, ad ♀ ($n = 17$) 134–182 (156), imm ♀ ($n = 2$) 152–163 (158).

8. *M. k. hunsteini* (Sharpe, 1882). *Journal of the Linnean Society, London* **16**, 442. East Cape, New Guinea [in error for Normanby I., D'Entrecasteaux Archipelago]. Type specimen BMNH 1880.9.13.59.

Range: Goodenough, Fergusson, and Normanby Is of the D'Entrecasteaux Archipelago. Apparently inhabits only upland forests. On Fergusson I. seldom found below 450 m (Meek in Sharpe 1898).

Diagnosis: like nominate form but much larger (and larger than all preceding forms) with longer occipital tufts. Back, rump and dorsal surface of tail dark bluish purple (less green). Occipital tuft feathers less blue and more green than in nominate.

Measurements: wing, ad ♂ ($n = 13$) 179–194 (188), imm ♂ ($n = 3$) 174–181 (176), ad ♀ ($n = 8$) 179–186 (182), imm ♀ ($n = 4$) 169–175 (172); tail, ad ♂ ($n = 13$) 133–143 (139), imm ♂ ($n = 3$) 133–134 (133), ad ♀ ($n = 8$) 132–141 (135), imm ♀ ($n = 4$) 127–137 (131); bill, ad ♂ ($n = 13$) 35–38 (37), imm ♂ ($n = 3$) 35–36 (36), ad ♀ ($n = 8$) 33–37 (35), imm ♀ ($n = 4$) 33–34 (34); tarsus, ad ♂ ($n = 13$) 37–41 (39), imm ♂ ($n = 3$) 40–41 (40), ad ♀ ($n = 8$) 35–40 (38), imm ♀ ($n = 4$) 36–40 (38); ear tufts, ad ♂ ($n = 11$) 19–30 (26), imm ♂ ($n = 3$) 16–18 (17), ad ♀ ($n = 8$) 18–28 (22), imm ♀ ($n = 4$) 12–18 (16). Weight: none.

9. *M. k. gouldii* Gray, 1859. *Proceedings of the Zoological Society of London*, p. 158. Cape York. Type specimen BMNH 55.4.11.1.

Range: NE forests of CYP, Queensland, Australia.

Diagnosis: like nominate form but iridescence generally more green, less purple and particularly so on upperwing and tail. Occipital tuft feathers more narrowly pointed and much longer, and relative tail length much longer, than nominate

Measurements: wing, ad ♂ ($n = 35$) 155–177 (165), imm ♂ ($n = 1$) 155, ad ♀ ($n = 16$) 151–166 (158), imm ♀ ($n = 5$) 142–154 (147); tail, ad ♂ ($n = 35$) 124–142 (135), imm ♂ ($n = 1$) 137, ad ♀ ($n = 16$) 122–130 (127), imm ♀ ($n = 5$) 122–127 (124); bill, ad ♂ ($n = 32$) 32–36 (34), imm ♂ ($n = 1$) 35, ad ♀ ($n = 16$) 30–34 (32), imm ♀ ($n = 5$) 31–33 (32); tarsus, ad ♂ ($n = 35$) 34–39 (37), imm ♂ ($n = 1$) 38, ad ♀ ($n = 15$) 33–38 (35), imm ♀ ($n = 5$) 32–36 (34); ear tufts, ad ♂ ($n = 33$) 18–43 (34), imm ♂ ($n = 1$) 25, ad ♀ ($n = 16$) 22–36 (29), imm ♀ ($n = 5$) 16–24 (19). Weight: ad ♂ ($n = 6$) 150–184 (161), ad ♀ ($n = 1$) 138.

Habitat and habits Table 4.1

A near-obligate inhabitant of rainforest interior, rarely seen at forest-edge or in second-growth. At Brown R., Port Moresby, and about Kiunga, PNG, birds do, however, frequent forest and logged forest-edge (D. Bishop *in litt.*). These subspp., as currently designated, include lowland subspp. as well as ones inhabiting hill and midmontane forests. Occasionally frequents mangroves in Australia (Saenger *et al.* 1977). Single birds, pairs, and flocks, typically of 4–6, move steadily through forest canopy calling frequently; apparently spends less time in the lower strata of the forest than other manucodes (BB). The sp. responds actively to play-back of its calls by flying to perch close to the observer (D. Bishop *in litt.*).

Diet and foraging Table 4.2

Mainly frugivorous and apparently a fig specialist, but also consumes arthropods and small gastropods. Highly-aggressive; individuals on Mt. Missim (*M. k. purpureoviolacea*) displace all other birds of paradise (Blue Birds possibly excepted) and frugivorous passerines (T. Pratt *in litt.*). Insects, small crustaceans, and fruit were recorded in the diet by von Rosenberg (1875). Stomach contents of three birds from two locations contained only fruit remains (Schodde 1976). Bell (1983) observed birds gleaning arthropods from tree limbs and trunks but not amongst foliage. Hoogerwerf (1971) recorded small snails in the diet. Three faecal samples obtained at a nest contained snail shells, up to 8 mm dia (BB) (see Crested Bird). Other arthropods recorded in the diet include crickets, grasshoppers (Orthoptera), assassin bugs (Reduviidae), and ad and larvae of weevils (Curculionidae) and other beetles (Tmesisternine-Cerambycidae, Tenebrionidae, Staphylinidae, Elateridae), flies (including Tipulidae), and spiders (BB). A pair attending their CYP nest and eggs during Dec at times had conspicuous quantities of yellow flower pollen about the base of their bills indicating feeding at flowers but it remains to be learnt if pollen, nectar, or flower-frequenting insects are taken (Frith and Frith 1993*d*). The diet on Mt. Missim, PNG was predominantly (99%) fruit (22 spp. of plants—see Appendix 4), but birds were observed foraging for arthropods on limbs and in foliage (Beehler 1985*a*). Of 348 fruit-feeding bouts on 18 plants (see Appendix 4), 80% were on figs (*Ficus* spp.), 13% on other simple drupe or berry fruits, and 7% on capsule, or complex, fruits (Beehler 1983*a*; see Chapter 4). Most foods are sought in mid to upper forest levels, but birds mostly avoid outside of upper canopy, tending to keep to the lower canopy of larger trees. Foraging birds consist of a ♂/♀ pair or a group of up to seven (Gilliard and LeCroy 1967; Beehler 1983*a*, 1985*a*). Of 321 rainforest foraging records near Port Moresby, PNG, 93% were of birds in the forest canopy (63% in the lower canopy 25–30 m above ground and 30% in the upper canopy at 30–35 m), 4% in the sub-canopy (8–25 m), and 3% in the understorey at 2–8 m above ground (Bell 1982c). Of 58 complete foraging visits to fruit-bearing trees, average length of a visit was 5.7 min and average time spent feeding there was 5.3 min; birds spent 94% of time in the tree feeding; 97% of visits lasted 1–10 min, 2% 11–20 min, and 1% > 21 min (Pratt and Stiles 1983). Of 72 prey foraging records, involving seven birds, 83% were of feeding from 'debris' or dead foliage, 13% from tree branches, and 4% from tree trunks; all cases involved gleaning of prey (Bell 1984). Diamond (1987) reports that on lower mountain slopes, Trumpet Manucodes may generally accompany mixed-spp. foraging flocks including ♀-plumaged *Paradisaea* spp. and mixed flocks based upon Rufous Babblers and Rusty Pitohuis. Observations in the Lakekamu Basin, PNG, indicate that this sp. (*M. k. purpureoviolacea*) rarely joins mixed flocks in this lowland habitat (BB). May also be seen feeding with other spp. that happen to be seeking the same fruit (e.g. Marbled Honeyeater *Pycnopygius cinereus* and Brown Oriole *Oriolus szalayi* (BB)). Cooper (in Cooper and Forshaw 1977) saw a bird feeding on fruit of *Callicarpa pentandra*, also being visited by a Superb, Lesser, and Carola's Parotia.

Vocalisations and other sounds
Figure 9.14

Usually located by its varied vocalisations, this sp. has perhaps as many kinds as any bird of paradise. There is considerable evidence of an array of sex-specific calls, although details remain to be fully delineated.

♂♂ of the lowland populations of *jamesii* give a distinctive two-syllable song that from a distance is a high and clear *Whuu-oh* (Coates 1990; BB). At very close range this sounds quite harsh and would be transcribed as *kee-youwk!* (BB). Gregory (1995) notes this as a lovely, musical, whistled, fluky *ee-loo* from the Ok Tedi region, PNG. Populations of *diamondi* and *purpureoviolacea* have not been heard to give this song.

A courting male *adelberti* gave a *hrarrr* call followed by a pigeon-like, almost bell-like, musical, downslurred, slightly nasal, *oo* or *kyeu* (see below). Thane Pratt (*in litt.*) described calls of this subsp. as a forceful rasping *ha ...*, a clear and ringing *who!* and the display song a whistled far-carrying note like that of a peacock.

A commonly-heard call at the Wassi Kussa R. (*jamesii*), NG was 'a loud harsh squawk, somewhat recalling the call of [Yellow-faced Myna] *Mino dumonti*' (Rand 1938). Squawk calls of various forms are heard from both ♂♂ and ♀♀ of upland and lowland populations. Unless closely transcribed they are difficult to differentiate. Other transcriptions include 'a harsh gurgled *kyaakh*' (*jamesii*; Finch 1983); a 'commonly uttered short, loud belching bark, *chaw*, *chak* or *chow*, or a longer *cheow*, *cha-aw*' (*jamesii*; Coates 1990); *kyoup!*, *cawwp!*, and *geoaw!* (*purpureoviolacea*; BB); *scowlp*, *scowlk*, or *kwolk* note (*gouldii*; CF). This latter call is very similar in the Port Moresby (*purpureoviolacea*—Fig. 9.14(b)) and Cape York (*gouldii*—Fig. 9.14(d)) populations, is generally similar but more complex at Kiunga, PNG (*jamesii*—Fig. 9.14(a)) but is a much softer and briefer, somewhat parrot-sounding, note on Aru Is (Fig. 9.14(c)). Notes apparently given by ad ♂♂ tend to be low-frequency 'growling' and more resonant, drawn-out, and bugle-like sometimes with considerable vibrato (Fig. 9.14(e,f)). The single loud *kyowp!* is almost certainly that of a ♂ (BB). It is common to hear a pair of birds call back and forth for long periods while they forage in the forest. The putative ♂ will give the *kyowp!* call and the putative ♀ will respond with a harsh and grating note that ascends in pitch: *kraiiiggg* (BB).

The ♂ nuptial song of the upland population *purpureoviolacea* in PNG is a hollow, low, reverberating, and tremulous single note *wodldldldldldldl* of one pitch and 1–3 sec duration, that may be preceded by a *KAUP* note (Beehler 1978, and BB). This 'trumpet' song of the lowland Port Moresby, PNG and Silver Plains, Aust areas appears in Fig. 9.14 (e,f).

Calls heard in the Vanapa-Veimauri area PNG (form *jamesii*) included a sheep-like

236 Trumpet Manucode *Manucodia keraudrenii* (Lesson and Garnot, 1826)

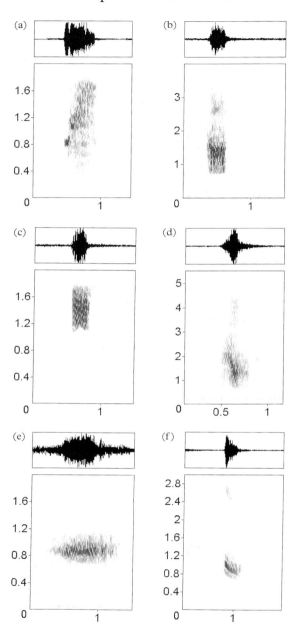

9.14 Sonograms from recordings of ad Trumpet Manucodes: (a) the *skowlp* note of the subsp. *jamesii* at Kiunga in Aug 1990; (b) *purpureoviolacea* at Brown R., near Port Moresby, PNG in July 1986; (c) *aruensis* of the Aru Is, Apr 1988, all by K. D. Bishop; (d) and of *gouldii* at Silver Plains, CYP in Aug 1996 by D. Stewart. Also shown are (e) the low-pitched 'trumpet' notes of *purpureoviolacea* and (f) *gouldii* from the above respective locations, dates, and recordists.

maaah (Finch 1983). A far-carrying hoarse *chauw, cher au-au-au-auw, chau-auw* was given by a displaying ♂ (Hoogerwerf 1971). Other calls by birds of unknown sex include a 'gloomy *chauw* or *chauw-chauw* often accompanied by a rolling *chooo-roooo* or a long drawn-out *krrrrooooo*' (Coates 1990). This may be the *krouw, rrauw* or *rrRRow* of c. 3 sec duration described by Coates (1990) for PNG as being delivered with the tail slightly spread and wings briefly held away from the body. Coates also heard a vibrating mammal-like *crrraaa*, a short, deep, harsh, nasal *chu* or *ch* note repeated 2–8 (usually 3–5) times in alarm and a short nasal, myna-like *agh* or *wha* that would appear to be the call described by Rand (1938).

Monotonous and repetitive calling, not eliciting any response from conspecifics, was given by a ♀ on Mt. Missim, PNG on 4 Apr, the bird calling 20 times in 3 min from a stationary position just below a ridgetop. After > 5 min of calling 2 ♂♂ answered nearby and she continued to call at the same rate for 10 min more. Here at 1457 hr on 16 Oct a ♀ calling continually in forest canopy called every 5–7 sec for 96 sec, opening her wings and extending the neck for each note (BB). Also here every morning for a week to 9 Nov a ♂ gave persistent pre-dawn *KAUP* calls every 8–10 sec, a ♀ eventually responding but only c. 1 per min. By dawn he was gone. Likewise, during 11–13 Dec a ♂ made monotonous predawn call notes (0510–0540 hr) every 5–10 sec (BB). Bell (1967a) reports calling all day near Balimo on the Fly R. estuary.

During observations at an Australian nest with eggs the incubating and foraging pair maintained contact by occasional call notes. When a bird was foraging or, far less often, when incubating it would give an infrequent, hollow, and far-carrying *skowlp, scowlk,* or *kwolk* note. This agrees well with the *kyowp!* of Beehler (1978). Another note that both sexes produced was a single *keow* that is presumably the 'musical, downslurred, slightly nasal *kyew* similar in quality to the call of [Helmeted Friarbird] *Philemon novaeguineae*' (Diamond 1982). As CF climbed an observation tree-platform near a nest tree, the birds produced a strong and brief 'creaking door' alarm call, rather cicada-like but deeper.

On 23 Sep, from 1100 hr, at S McIlwraith Ra, CYP a timed bird gave 32 regularly-repeated loud *skowlp* call notes during 5 min (1 per 9 sec). During 5 min from 1118 hr it gave 39 such notes, and during a timed 3 min from 1130 hr it gave 28 notes. These provide an average rate of one call note every 8 sec. This persistent calling at this time of year may have been mate attraction or nest site advertisement as an old nest of the previous season was immediately beneath the calling bird (CF).

Ever since the first ad ♂ was collected and skinned, the elongate and convoluted trachea of *Manucodia* has been a source of wonder (Lesson and Garnot 1826a). It has since been illustrated in numerous works on the birds of paradise, birds of NG, and more general ornithological works and, together with this feature in other manucodes, has been the subject of at least five papers (Pavesi 1874, 1876; Forbes 1882b; Beddard 1891; Clench 1978; Frith 1994b). As pointed out by Pavesi, only the ad ♂ has the fully-developed, convoluted, trachea (Fig. 1.5). However, ♀♀, like young ♂♂, have a larger than normal trachea for a passerine by having it simply looped (North 1901–4; Fig. 1.5). Frith (1994b) reviewed such trachea modification and pointed out that the common denominator in all birds concerned would be the need to communicate over greater distances than most birds. The Trumpet Manucode is, peculiarly (see Chapter 5), a monogamous, almost exclusively fruit-eating, yet non-territorial passerine (Beehler 1985a, 1989a). Such modified tracheas may, however, be the result of sexual selection by ♂♂ in vocal competition and/or by ♀♀ selecting for ♂♂ with the most impressive vocalisations. Another view is, however, that such tracheas 'acoustically exaggerate' the size of calling birds in non-visual intraspecific communication (T. Fitch *in litt.*).

Mating system

Monogamous and pair-bonding, and long known as such in Australia where early egg collectors reported pairs attending nests (Mathews 1925–7; Frith and Frith 1993d). Observations at one Australian nest showed that both sexes incubate (Frith and Frith 1993d). In NG, both sexes observed attending and feeding nestlings and some pairs remaining together for more than one season (Beehler 1985a). See Chapters 4 and 5 for discussions of the influence of diet upon the mating system in this manucode.

Courtship behaviour

Rand (1938) first noted that a larger, presumed ♂, (*jamesii*) bird chases a smaller, presumed, ♀ through the forest foliage and perches beside her to perform a frontal lunging display by raising the head and bill with erect head, neck, and throat feathering. A simultaneous harsh call note is given and the opened wings are raised and tilted forward to present the entire upper wing surface whilst the tail is widely fanned (Fig. 9.15). Hoogerwerf (1971) watched a similar display by the nominate form. Beehler (1983a) confirms this display description in broad terms (for *purpureoviolacea*), but says the ♂ 'drops his wings; cocks his tail, and gives a shivering display for 5–10 sec' before mounting the ♀.

Coates (1990) observed that the displaying ♂ (*jamesii*) 'stretches his neck, raises his head, and lunges forward while at the same time expanding his body plumage (including the occipital tufts and neck hackles), fully opening the wings and lifting them fan-like from behind, and fanning the tail. The display reaches its peak with the maximum volume of the call when the bird is at its most forward position, then quickly subsides as the call fades away'. Coates reports that calling indicative of display is mostly heard from late morning to late afternoon, mainly mid-to late-afternoon. The ♂ displays opportunistically wherever the ♀ happens to be. The display described is repeated at some 15 sec to several min intervals.

An apparent courtship was witnessed during Mar on the Adelbert Ra, PNG (G. Opit *in litt.*). A ♂ (*adelberti*) chased a smaller, open-mouthed, and panting ♀ with short tree-to-tree flights and branch-to-branch hopping until stopping to perch beside her on a horizontal branch *c.* 16 m above ground. He then suddenly started to perform very jerky head and neck movements as if looking about in a most excited way; then gave a *hrarrr* call as he spread his wings up and about him whilst slightly lowering his head and neck. This was followed by more jerky head and neck movements for *c.* 10 sec before a second raised-wings display, more head movements, and then a third but different display. The last involved a pigeon-like, almost bell-like, musical downslurred, slightly nasal, *oo* or *kyeu* as he stood tall and slightly stretched his neck while pointing his bill down towards his chest, and raised the head and upper neck feathers. This entire sequence of events was repeated several times with the first display type sometimes being given three times before the second type. After 5–6 min the ♀ flew off with ♂ in close pursuit.

Display of some kind in May at Ok Tedi, PNG involved a (*jamesii*) bird in flight giving a 'descending musical series as it dropped down, wings open, before flying back up to its perch. A harsh frog-like rattle was given by another perched bird with wings spread wide. At least three individuals seemed to be involved, two of them calling the frog-like rattle note with spread wings' (Gregory 1995).

Breeding Table 6.5

A nesting association between Trumpet Manucodes and Black Butcherbirds *Cracticus quoyi* in Australia (*M. k. gouldii*) was documented by Barnard (1911) and MacGillivray (1914) but subsequently overlooked or ignored other than being briefly alluded to by Cayley (1959) and noted in passing as occurring near Port Moresby, PNG (Watson *et al.* 1962). This association was then described and subsequently shown to CF by J. Young (Frith and Frith 1993*d*; D. Frith and C. Frith 1996). On CYP at least some Trumpet Manucode pairs apparently seek out and nest close to a nesting Black Butcherbird pair. Should the nest of the latter fail before the manucodes lay eggs they will desert their intended nest site or incomplete nest (Barnard 1911), strongly suggesting the aggressive and predatory butcherbirds are sought for protective purposes. Six active nests at Lockerbie, CYP were, on average, 133 (82–165) m from active Black Butcherbird nests. As an observer approached a manucode nest a butcherbird flew into a tree close to the manucode's nest and was attacked and driven off by a manucode (M'Lennan in MacGillivray 1914) leaving no doubt that this bird of paradise can dominate the butcherbird under such circumstances. At Wau, this manucode has been found nesting in the same tree with *Philemon, Dicrurus,* and *Cracticus* (BB).

9.15 Display calling posture of, what is assumed to be the ♂, Trumpet Manucode. After Coates (1990: 438), with kind permission.

NEST SITE: nest is suspended in the fork of a horizontally-forking tree branch. Numerous vine tendrils of the rim entwine the supporting branches of the nest which may be 9–22 m above ground, but far more often high than low down. Beruldsen (1990) observed a CYP nest 15 m high. Nest-building is by both sexes (MacGillivray 1914). Eight Australian nests were 6–27 m, averaging

15.2 m, above ground, two being in bloodwood trees *Eucalyptus terminalis* and another a 'mahogany' tree.

NEST: described by Shaw Mayer (BMNH clutch 1941.1.5.24) as an open and rather shallow cup, composed of strong, curly vine tendrils, woven together, lined with finer tendrils or (in Harrison and Frith 1970) as an open basin-shaped structure, composed of curly vine tendrils woven together, lined with fine creeping fern tendrils. The nest of this sp. is sparser than those of other manucodes and does not include larger leaves or pieces of dry wood (see Plate 3). Measurements of one nest were 152–203 mm in external dia by 102 mm total depth, and 82–102 mm internal dia by 38 mm internal (cup) depth (Campbell 1901). Another nest measured 152 mm in dia by 101 mm in depth, and 101 internal dia by 63 mm in depth (North 1901-4). Both of these authors published a photograph of a nest.

EGGS: one or two, pink-buff (6 and 219D but paler to washed-out or 121D but slightly darker and pinkish) spotted and blotched with browns, purples, and greys with these markings sometimes tending to longitudinal brush-like strokes more typical of other spp. of Paradisaeinae. Eggs are variable in size and colour (with subsp.). The identification of a number of clutches brought to ornithologists by local collectors in the field (notably some in the BMNH collection) are suspect. See Plate 13; 34 eggs average 35.9 (32.4–39.4) × 24.8 (22.9–26.8).

INCUBATION: by both sexes. Numerous changeovers at a nest with eggs were silent, with the sitting bird slipping over the nest rim to fly down and away from the nesting tree at the arrival of its mate. On one occasion, when the sitting bird was reluctant to depart, the relieving bird opened its bill and advanced towards the nest to displace its mate (Frith and Frith 1993d, 1996).

NESTLING CARE: Beehler observed parental activity at a nest containing two nestlings, at 1625 m on Mt Missim during 12–4 and 21 Dec 1978 for a total of 25.5 hr. During this time 87 nestling-feeding visits were made at an average of every 17 (range 1–38) min and at an average rate of 3.4 each hr (the average increasing from 2.9 to 4.3 an hr during each observation period as nestlings grew older). Both parents brooded and fed nestlings and usually, but not always, did so alternately. Parents maintained contact by calling when one or both were distant from the nest. All meals were regurgitated to nestlings and their faecal sacs swallowed immediately (BB). Nestling diet at one nest was at least 90% fig fruit pieces (Beehler 1985a). The collector W. Frost (in Rothschild 1930b) claimed that a Trumpet Manucode egg in the nest of a Greater Bird was laid there by a manucode behaving as an avian brood parasite. It would appear that Frost's idea of parasitic egg laying in *P. apoda* nests was encouraged by his observation (in Rothschild 1930b) that Trumpet Manucodes often appear to follow ♀-plumaged Greater Birds. We now know that this latter activity typically has to do with mixed-spp. foraging and not with nest parasitism. This notion of nest parasitism was rejected by Rand (1938) and Gilliard (1969) as erroneous, but Harrison and Walters (1973) again raised the possibility. They did point out, however, that the eggs and nests involved are more likely to be indicative of the manucode merely utilising newly-built nests for its own nesting activities. While Coates (1990) pointed out both possibilities would appear unlikely, egg dumping may be possible. As a final note on this subject, Ninga Kawa, while observing the nest of a ♀ *P. raggiana*, saw the nest visited by a ♀ *P. lawesii* in the absence of the owner. The ♀ *raggiana* returned to find the *Parotia* at her nest, and after driving the interloper away, she methodically dismantled the nest entirely (BB).

NESTLING DEVELOPMENT: unknown

Annual cycle   Table 6.5

DISPLAY: in the Port Moresby area a 'display flight' occurred in Aug (Eastwood 1989) and calls thought indicative of display were heard in all months (Coates 1990). A display and copulation were observed at 1122 hr on 14 Nov on Mt. Missim, PNG (BB). A display was noted at Ok Menga, PNG in May (Gregory 1995). BREEDING: at least May–Jan over the entire sp. range. While ♂♂ specimens have been collected during Feb–Apr and July–Dec with enlarged gonads, ♀♀ in such condition have been taken only in June (1) and Oct–Nov (4). EGG-LAYING: Oct–Jan and May. MOULT: 101 of 294 specimens were moulting, involving all months but predominantly during Jan–July (Appendix 3).

Status and conservation   Table 8.1

Von Rosenberg (1875) found the sp. very rare on the Aru Is, occurring there in woodland and on the coast and the interior forest only in some locations. During 28 Mar–12 Apr 1988 it was uncommon and patchily distributed in the Aru Is (J. Diamond and D. Bishop *in litt*). The mainland NG distribution is patchy as a result of restriction to an altitudinal belt on certain mountain ranges. In addition, some populations are isolated on smaller mountain ranges on the N coastal area of NG. Beehler twice visited the uplands of Goodenough I. in 1976 and 1980 on ornithological surveys and did not encounter this sp. John Gould (1865) suggested trans-Torres Strait migration between NG and Australia and a number of authors subsequently stated this to be the case as if fact. The Australian population has been noted as sparse in its fragmented and limited habitat by many authors. Indeed it would appear to have been this apparent sparseness that led Gould to consider the birds on CYP to be migrant or vagrant birds from NG (Frith 1994*a*). In fact no evidence exists for migration by this sp. Possibly as a result of the myth of migration, several sightings in Australia S of Coen on CYP have been reported. A review of these sightings resulted in them all being found unconvincing (Frith 1994*a*). It is more likely that the CYP population is a resident relict one. A home range of 200 ha or more has been calculated for birds on Mt. Missim, PNG (Beehler 1985*a*), with overlap by individuals of these non-territorial birds being very high. Aggregations of up to 10 birds occur at fruiting fig trees. Beehler noted the sp. to be common on Mt. Missim, encountering 7–10 birds in a fruiting tree a day. An estimate of density near Port Moresby, based on transect counts, was 5 birds (815 g of biomass) per 10 ha (Bell 1977).

Knowledge lacking and research priorities

This sp. offers a wealth of research topics, not least of which being the relationships of the various taxa presently constituting it. Geographical variation of the sp. needs to be reassessed, especially in light of an improved understanding of local vocal dialects as well as acknowledging the presence of upland and lowland forms (which implies incipient speciation). Locally, it would be valuable to detail ♂/♀ vocalisations and their significance to sp. behaviour and with regard to tracheal development (Frith 1994*b*). An understanding of the relationships of this bird to the foraging ecology of other manucode spp. is required. It will be important to attempt to confirm Beehler's suggestion that the sp. is a fig specialist.

Aviculture

Birds were in Taronga Park Zoo, Sydney during 1965–7 (CF). An egg in the BMNH (1941.1.2.97) was laid in captivity, apparently during Oct 1916 in Britain. Two live birds were delivered to London Zoo on 3 July and two more on 5 Oct 1908 by C. B. Horsbrugh and W. Stalker (Horsbrugh 1909). Also kept at the NYZS (Crandall 1935). An active and aggressive bird that does well in a large well-lit aviary with shady spots, perches for sunning, a pool, and access to rain/sprinkler for bathing (P. Shanahan *in litt*).

Genus *Paradigalla* Lesson, 1835

Paradigalla Lesson, 1835. *Histoire Naturelle des Oiseaux de Paradis et des Épimaques* p. 242. Type, by monotypy, *P. carunculata* Lesson.

The paradigallas include two **monotypic spp.** of medium-sized, stout, black-plumaged birds that form a superspecies in the western and central uplands of mainland NG (Frith and Frith 1997*a*). Preliminary studies of cranial osteology indicate that **skull** features are typical of the Paradisaeinae and, significantly, are most like those of *Astrapia* (Bock 1963: 100) with which paradigallas share several plumage characters. Nevertheless, Bock (1963: 122) placed this genus between his 'riflebird group' (*Ptiloris*, *Semioptera*, and *Seleucidis*) and the sicklebills (*Epimachus*) but did, however, note that the skull characters of his study were alone inadequate evidence. Anatomically and otherwise (tail form, dark ♀ plumage, nest) *Paradigalla* is most like *Astrapia*, but some of its other characters suggest close affiliation to *Manucodia* (plumage, egg, pointed tail feather tips), *Ptiloris*, and *Epimachus* (gapes, modified central tail feathers, plumage iridescence).

In overall **size** ♂♂ average 5% larger than ♀♀. Bill meets the skull high on the forehead, close to the fore-eye level, and the proximal culmen is sharply keeled. **Bill** straight, narrow, and strong, and considerably longer than head. Nostrils covered by feathers. **Tail** square with central rectrices shortest (ad) or slightly graduated (juv–imm) in *brevicauda*, to strongly graduated with central pair far longest in *carunculata*. In ♂ *carunculata* tail length increases with age but in *brevicauda* it becomes progressively shorter (Table 1.5). Tiny hair-like central points to tail feathers (as in *Macgregoria*, *Lycocorax*, and *Manucodia*). **Wing** long and rounded. P9 and P10 primaries pointed and mildly emarginated on inner vane. Secondaries may be as long as, or longer than, primaries. **Tarsus** length 28% of wing length, scutellation of the tarsus being prominent distally but indistinct proximally. Feet strong.

For both sexes, plumage deep black throughout, with brownish suffusion in places, and with fine, scale-like, iridescent, blue-green feathers on crown and ear coverts and decorated with unique bare fore-facial and lower-gape wattles but no ornate plumes (Plate 5). Juvs are similar but uniformly duller (unlike barred young of most Paradisaeinae) and nestlings hatch with fully-formed facial wattles. ♀♀ and young birds are unusual for Paradisaeinae in their uniformly-dark plumage—resembling the pattern of *Macgregoria*, *Lycocorax*, and *Manucodia*.

Gilliard (1969: fig 6.1) clearly considered *Paradigalla* polygynous, doubtless in the light of known **hybridisation** with three other genera (*Parotia*, *Lophorina*, *Epimachus*). This did not deter a number of subsequent authors from suggesting the paradigallas are monogamous, however, in view of their sexual monomorphism. While the systematic placement of *Paradigalla* has been problematical (see Fig. 3.3, Table 3.3) given the near-identical appearance of the sexes, the fact that at least *P. brevicauda* is now known to be **polygynous** significantly changes perceptions. The bulky open **nest** is typical of Paradisaeinae except those of *Manucodia*. **Eggs** (of a single *P. brevicauda* individual) are most like those of *Manucodia* (Plate 13).

Long-tailed Paradigalla *Paradigalla carunculata* Lesson, 1835

PLATE 5

Other names: Wattled Bird of Paradise, Carunculated Paradise Pie. *Happoa* in Arfak Mts, IJ (Beccari in Iredale 1950).

A little-known, large, jet black bird with conspicuous, colourful, facial wattles, and long and pointed graduated tail, confined to mid-montane areas of the extreme W NG mainland. Monotypic.

Description

ADULT ♂: 37 cm, tail strongly graduated. In folded wing, longest secondaries may extend to primary tips. Head, neck, and entire upperparts velvety jet black. Foreface decorated with bright yellow (157) to greenish yellow wattle with the look of melted plastic, originating at base of maxilla, and a small, swollen, sky blue (168B) wattle of similar consistency on the mandible base beneath, of which a small area of bare skin is orange-red. Entire head feathering iridescent oily bluish green, more apparent, and greenish blue, on scale-like crown and nape feathering. Velvety jet black of tertials, coverts, secondaries, mantle, back, rump, uppertail coverts, and dorsal surface of tail central pair with slight dark Purple (1) hue strongly overwashed with rich Olive-Green (47) iridescent sheen. Primaries and outer, shorter, tail feathers dark brown black (119) with slight Olive-Green iridescent sheen to outer edges. Tail feathers other than central pair terminally notched and with tiny hair-like central points. Upper breast to undertail coverts dark brownish black (19) with dull, deep coppery Purple, iridescent sheen in some lights. Underwing and ventral surface of tail as upperparts but glossy and slightly greyish. Bill shiny black, iris dark brown, legs and feet blackish, mouth colour unrecorded, but probably pigmented (pale blue or yellow?). According to Sharpe (1891–8), first two primaries pointed 'as in *Lophorina*' and notched at the ends.

ADULT ♀: 35 cm, as ad ♂ but generally slightly paler and duller and smaller (mean wing length 11% shorter).

IMMATURE–SUBADULT: as ad but slightly duller-plumaged with shorter tail.

NESTLING–JUVENILE: unknown but see Short-tailed Paradigalla.

Distribution

Endemic to the mountains of westernmost NG, including the Arfak Mts of the Vogelkop and (apparently) the Fakfak Mts of the Onin Penin (Gibbs 1994; Frith and Frith 1997a). Surprisingly, no known records from the Tamrau and Wandammen Mts of the Vogelkop or the Kumawa Mts of the Onin Penin; all three ranges might be expected to harbour this sp. Elevational range poorly defined, but probably c. 1400–2100 m.

Systematics, nomenclature, subspecies, weights, and measurements

SYSTEMATICS: the larger of the two paradigallas and of very different body proportions than its smaller and allopatric relative. The form *intermedia* is a junior synonym of *P. brevicauda* (Frith and Frith 1997a). Hybridisation: crosses

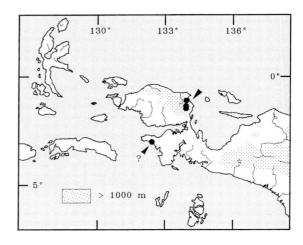

Long-tailed Paradigalla *Paradigalla carunculata* Lesson, 1835

documented with the Western Parotia, Superb Bird, and Black Sicklebill (Plate 14)—see Appendix 1.

SPECIES NOMENCLATURE AND ETYMOLOGY

Paradigalla carunculata Lesson, 1835. *Histoire naturelle des oiseaux de paradis et des épimaques* p. 242. Arfak Mts. Type specimen MHN 10324.

Etymology: *Paradigalla = Paradi* derived from *Paradisaea* + *galla* derived from *Gallus* jungle fowl, which like this genus, exhibits facial wattles; *carunculata* = modern L. *carunculatus*, carunculated, wattled (L. *caruncula* a small piece of flesh).

WEIGHTS AND MEASUREMENTS

Measurements: wing, ad ♂ ($n = 10$) 180–201 (186), imm ♂ ($n = 7$) 160–179 (171), ad ♀ ($n = 11$) 157–175 (165); tail, ad ♂ ($n = 10$) 122–137 (132), imm ♂ ($n = 7$) 115–130 (124), ad ♀ ($n = 9$) 118–131; central tail, ad ♂ ($n = 10$) 132–170 (160), imm ♂ ($n = 7$) 121–148 (133), ad ♀ ($n = 9$) 125–138 (132); bill, ad ♂ ($n = 8$) 38–45 (43), imm ♂ ($n = 7$) 42–45 (44), ad ♀ ($n = 10$) 39–44 (42); tarsus, ad ♂ ($n = 9$) 48–50 (49), imm ♂ ($n = 7$) 44–50 (48), ad ♀ ($n = 11$) 41–50 (46). Weight: ad ♀ ($n = 1$) 170.

Habitat and habits   Table 4.1
Midmontane forests. Beccari (in Sharpe 1898) found that it 'likes to sit on the tops of dead and leafless trees'. Bent Poulson (personal communication) reports seeing 'a pair' at 1725 m on the Vogelkop, IJ during July–Aug 1994, presumably an ad ♀ with offspring (see Short-tailed Paradigalla).

Diet and foraging   Table 4.2
Beccari (in Sharpe 1898) shot a bird as it fed upon small, fleshy, nettle (Urticaceae) fruits, possibly those of *Pipturus argenteus*.

Vocalisations and other sounds
Unrecorded.

Mating system
As it is known to hybridise with three other genera, and in view of what is known of the Short-tailed Paradigalla, it is highly likely that ♂♂ are promiscuous and ♀♀ alone attend the nest.

Courtship behaviour
Unknown.

Breeding   Table 6.5
Not recorded. Earlier record in literature was for *P. brevicauda*. NEST SITE, NEST, EGGS, INCUBATION, NESTLING CARE, AND DEVELOPMENT: unknown.

Annual cycle   Table 6.5
DISPLAY, BREEDING, AND EGG-LAYING: unknown. MOULT: 3 of 9 specimens were moulting, one in each of Jan, May, and July (Appendix 3).

Status and conservation   Table 8.1
The few records of this sp. come mainly as specimens from the Arfak Mts, and a sight record from the Fakfak Mts is presumably of this bird. More effort should be invested in documenting the distribution and abundance of this sp.

Knowledge lacking and research priorities
No bird of paradise is known from fewer records from fewer localities. This sp., being virtually unknown, merits attention. Efforts should be made to locate populations in the Tamrau, Wandammen, and Kumawa Mts. The provenance and taxonomic status of the populations in the Fakfak Mts require delineation (Gibbs 1994). Mating organisation, courtship, and diet require documentation.

Aviculture
Not known to have been kept in captivity.

Short-tailed Paradigalla *Paradigalla brevicauda* Rothschild and Hartert, 1911

PLATES 3,5; FIGURES 6.2, 9.16, 17

Other names: Short-tailed Wattled Bird, Blue-and-yellow Wattled Bird of Paradise. *Genat* of Western Dani (Ilaga) people, *Tenengel* of Uhunduni people, IJ (Ripley 1964); *Ibinimi* or *Ibidimi* of the Gimi people, PNG (Diamond 1972; D. Gillison *in litt.*); *War-cun-da-goona* at Mt. Hagen, PNG (Mayr and Gilliard 1954).

A stocky, all-black bird with a very short tail, long and narrow bill, and conspicuous yellow wattles on fore-face and a small blue wattle on the lower gape. Iridescent, scale-like, blue-green feathers adorn crown and nape. Encountered singly or as parent with offspring, usually in the middle and upper forest strata but at times within a few metres of the ground. Monotypic.

Description

ADULT ♂: 23 cm, head, neck, and entire upperparts velvety jet black. Foreface decorated with bright yellow (157) wattle with texture of melted plastic, originating at base of maxilla, and a small, swollen, sky blue (168B) wattle on the mandible base. Entire head has iridescent, oily, yellowish green feathering which is more apparent on scale-like crown and nape feathers. Jet black mantle and back with a slight deep Purple (1) hue overwashed with rich, silk-like, Olive-Green (47) iridescent sheen. Upperwings, rump, uppertail coverts, and shorter central pair of tail feathers with oily, Olive-Green, silk-like sheen particularly as broad leading edges to primaries and secondaries. Primaries and remaining tail blackish brown (119). Breast jet black and remaining underparts almost so but suffused with darkest brown and with slightest iridescent sheen of deep, dark, coppery purple (101) in some lights. Entire underwing and ventral surface of tail brownish black. Tail feathers not terminally notched as in Long-tailed Paradigalla. Bill black, iris dark brown, legs and feet purplish lead grey, mouth pale aqua blue. Mean central tail feather pair length 3 mm shorter than mean length of remaining rectrices.

ADULT ♀: 22 cm, as ad ♂ but smaller (*c.* 5%), tail longer, and plumage generally slightly duller, more brownish, the black and iridescence being more subdued than that of ad ♂. Mean central tail feather pair length 1 mm shorter than mean length of rest.

IMMATURE–SUBADULT: young of both sexes have a tail considerably longer than ad.

NESTLING–JUVENILE: 1-day-old nestling naked with purplish black skin, dull pale mustard yellow mandibular wattles, and tiny off-white egg tooth. At 17 days old; bill black, legs dark purplish grey, iris dark grey. A ♂ nestling close to fledging was much like ad in general appearance, with a tarsus length of 32 mm, wing 83 mm, and bill, from posterior of nostril, 1 6 mm (see Fig. 9.16). Body feathering soft and down-like, more so on underparts, sooty black with a tinge of deep brown throughout with a slight purple sheen on the back in certain lights. Crown and nape lack the glossy blue green plumage of ad, but uniform in colour with the rest of the plumage. Wattles the same shape and form as in ad ♀ and similar size proportionate to the bird; iris paler brown than ad. Egg tooth visible. All wing feathers approximately half grown, but tail not visible. Feathering on upper mandible and throat just beginning to grow. Juv generally similar to ad ♀ but entire plumage matte black, lacking iridescence or sheens. Facial wattles of a juv, gaining independence from its parent at 115 days old, were fully formed and ad sized but still pale whitish yellow (Ogilvie-Grant 1915*a*; Frith 1970; Frith and Frith 1992*b*). Note that the uniform, brownish black, nestling plumage of paradigallas is unlike the brown and barred nestling plumage of most

Short-tailed Paradigalla *Paradigalla brevicauda* Rothschild and Hartert, 1911

paradisaeinine birds of paradise but is like that of Macgregor's Bird and the manucodes.

Distribution
Western and central segments of the central ranges, from the Weyland Mts eastward including SE of Mt. Gigira, S Karius Ra, PNG (G. Clapp *in litt.*), to Mt. Karimui and the Bismarck Ra 1400–2580 m, mainly 1600–2400 m (Coates 1990). Not recorded E of the Watut-Tauri Gap, and eastern terminus of range is poorly delineated. Not recorded for the Kratke Ra, although it might be expected there.

Systematics, nomenclature, subspecies, weights, and measurements

SYSTEMATICS: the smaller of the two paradigallas and of very different body proportions than its larger and allopatric relative (Frith and Frith 1997*a*). Originally considered by some authors a short-tailed subsp. of the Long-tailed Paradigalla, with which it forms a superspecies (Mayr 1941*a*; Mayr and Gilliard 1954; Gyldenstolpe 1955*a*). In view of the marked differences in tail length and structure, in colour and structure of the mandibular wattle, and in overall size and proportions, and the lack of evidence of clinal size variation in *P. brevicauda*, the retention of two distinct spp. is appropriate (Frith and Frith 1997*a*). The population *intermedia* was defined on the basis of it supposedly having all facial wattles clear lemon yellow and the tail shorter than Arfak Mts birds. It has been recently demonstrated, however, that the specimens attributed to *intermedia* are, in fact, long-tailed imm individuals of the Short-tailed Paradigalla, and that *intermedia* is a junior synonym of *brevicauda* (Frith and Frith 1997*a*). Hybridisation: unrecorded but likely to occur in view of those involving the Long-tailed Paradigalla.

SPECIES NOMENCLATURE AND ETYMOLOGY
Paradigalla brevicauda Rothschild and Hartert, 1911. *Novitates Zoologicae* **18**, 159. Mt. Goliath, central Dutch New Guinea. Type specimen AMNH 678355.

Synonym: *P. carunculata intermedia* Ogilvie-Grant, 1913. *Bulletin British Ornithologists' Club* **31**, 105. Utakwa R., Nassau Ra, at 5500 ft. Type specimen BMNH 1916.5.30.1072.

Etymology: *brevicauda* = L. *brevis*, short; *cauda*, tail.

WEIGHTS AND MEASUREMENTS
Measurements: wing, ad ♂ ($n = 28$) 151–169 (158), imm ♂ ($n = 30$) 148–164 (159), ad ♀

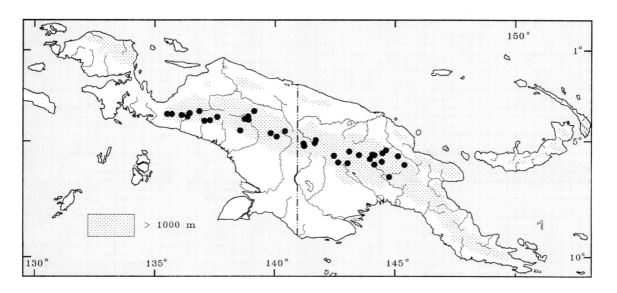

(n = 26) 144–154 (150), imm ♀ (n = 2) 152–153 (153); tail, ad ♂ (n = 25) 44–88 (53), imm ♂ (n = 29) 54–107 (78), ad ♀ (n = 26) 46–96 (68), imm ♀ (n = 2) 85–93 (89); central tail, ad ♂ (n = 24) 42–73 (51), imm ♂ (n = 26) 53–106 (78), ad ♀ (n = 23) 53–91 (67), imm ♀ (n = 2) 67–89 (78); bill, ad ♂ (n = 23) 40–49 (44), imm ♂ (n = 27) 40–48 (44), ad ♀ (n = 25) 42–48 (44), imm ♀ (n = 2) 44–45 (45); tarsus, ad ♂ (n = 25) 42–46 (44), imm ♂ (n = 30) 40–46 (43), ad ♀ (n = 26) 39–44 (42), imm ♀ (n = 2) 42–43 (42). Weight: ad ♂ (n = 3) 160–184 (173), imm ♂ (n = 4) 160–175 (169), ad ♀ (n = 2) 155–170 (163).

Habitat and habits   Table 4.1

Midmontane forests, including beech forest, forest-edge, secondary growth, and garden-edges. Usually seen in the crowns of the highest, isolated, old forest trees in which birds hop around in the moss layers of twigs and branches (Stein 1936). An unobtrusive bird, the habits of which are little known away from the nest. Rand (1942*b*) found it 'rather inactive in the lower substage'. Birds usually forage with rather slow movements but when seeking animal prey in epiphytic growth can be quite acrobatic in clinging to larger vertical tree trunks and boughs. A calling young bird at *c.* 1800 m in the Balim Valley, IJ on 16 Mar was persistently attacked by a Common Smoky Honeyeater, which dive bombed it more than 30 times, forcing it to drop below its perch limb to avoid being struck. In early Aug on the Ambua Ra, PNG an ad was seen fighting with this honeyeater several times and a ♀-plumaged Stephanie's Astrapia ejected an ad from a fruiting tree and followed it in flight chase (BB).

Diet and foraging   Table 4.2

One bird's stomach contained fruit only and another's fruit, insects, and a small frog (Rand 1942*b*). Birds appear preoccupied when foraging for fruits or epiphyte-dwelling arthropods and small vertebrates upon tree trunks, boughs, bark, and mosses. Seen feeding in same fruiting trees as Loria's Birds without interactions (D. Bishop *in litt.*). Diet was noted as 'chiefly vegetable matter, mainly seeds and small fruits' (Bulmer in Gilliard 1969) and mostly fruit but also insects and a small frog (Coates 1990) presumably derived from examination of stomach contents. A ♀ with a near-independent offspring fed upon fruits of *Planchonella* sp., *Sericolea pullei*, and *Perrotetia alpestris* (see also Appendix 4) and grubs from dead leaves and a hole in tree wood (Hicks and Hicks 1988*a*). Noted (on a specimen label) to feed on wild taro by Gilliard, who therefore suggested the wattles may keep the birds clean. Foods fed to a single nestling have been observed in some detail (Frith and Frith 1992*b*).

Vocalisations and other sounds

A melodic *hui* (Stein 1936). 'A high-pitched whistling call of very clear notes with a mournful tone; four ascending notes, the last being prolonged and rising markedly in pitch. While calling the bird opens wide its bill, and with the emission of each note the head is thrown back' (Cooper in Cooper and Forshaw 1977). At Wissel Lakes, IJ, a 'throaty croak' (King 1979). A melodious *hoo ee* and a *churr churr churr* similar to, but quieter than, that of the Superb Bird (Beehler *et al.* 1986). A near-independent imm was heard to beg at its mother with an often repeated, faint, high-pitched squeak whilst holding its head low with neck stretched towards the ad and its quivered wings held slightly away from the body (Hicks and Hicks 1988*b*). At Townsville, Ok Tedi area, PNG in Sept birds 'gave a quite loud rising, bell-like *zheee* call, reminiscent of the bell call of Eclectus Parrot [*Eclectus roratus*]; one would call and another reply' (Gregory 1995). At *c.* 1800 m in the Balim Valley, IJ on 16 Mar a juv gave a frog-or Astrapia-like questioning upslurred *kwee* incessantly every 2–4 sec, perhaps to attract its parent (BB). Wings in flight make an audible rattling or rustling noise much like that of the riflebirds, but softer.

Mating system

Assumed polygyny, in which ♂♂ are most probably promiscuous, dispersed, and solitary-displaying and ♀♀ uniparental in nest attendance (Frith and Frith 1992b). Intergeneric hybrids involving the closely-related Long-tailed Paradigalla support this assumption. Birds giving a rising bell-like *zheee* call appeared to be spaced at c. 150 m intervals in moss forest, suggestive of ♂♂ advertising from song posts (Gregory 1995).

Courtship behaviour
Unknown.

Breeding   Table 6.5

An individually-marked ♀ on the Ambua Ra, PNG, first seen nest-building on 5 Sept 1986, subsequently attended six different nests in an area of < 200 m² (*contra* < 100 m² of Frith and Frith 1992b) over 29 months. This same ♀ nested within this area during a further 4 years (J. Tano personal communication), CF seeing the bird at its nest in Jan 1993. Thus this ♀ nested in the same immediate area for at least 7 yr. Two nests, of successive seasons, were built in the same tree and that of a third in an adjacent one.

NEST SITE: all 7 nests examined by CF were placed in tree forks at an average height of 7 (range 5–11) m above ground. Four trees involved were well-foliaged saplings only a few m higher than the nest at the forest-edge or within gardens isolated from adjacent foliage and open to sky directly above. Two were in larger trees up to 15 m tall within the forest but close to a garden-edge.

NEST: the first known nest (BMNH N63.1), described and illustrated by Frith (1970), was decrepit and misshapen. A recently-used one described in the field was a substantial cup or bowl-shaped structure. The nest involved no sticks, the exterior being composed of considerable amounts of green mosses, fern fronds, and small-leaved epiphytic and climbing plants including orchids (some living), as decoration or camouflage. Within this looser exterior the nest consists of entwined woody vine tendrils some of which (> 450 mm long and up to 1.5 mm dia) entirely circle the nest rim and the outside of several supporting branches. The nest is well lined with dry leaves 75–125 mm long × 40–50 mm wide while the egg-cup lining, restricted to the small central bottom area of the egg-cup interior, is of very fine, dark, wire-like, tendrils c. 0.2–0.4 mm dia (Fig. 9.17). Mean external measurements of three nests: 168 (135–220) mm deep, 198 (190–215) mm dia; and internal nest cup depth 68 (65–70) mm and dia 120 (115–125) mm.

EGGS: only two known, both laid as single-egg clutches by the same ♀ in different seasons. The first egg measured 38.8 × 28.7 mm and weighed 17.4 g at c. 3 days old (= 9.7% of the ♀'s weight). It was smooth-surfaced and slightly glossy, pale pinkish buff (121D but much paler), finely spotted and blotched with light browns and purplish greys (see Plate 13). The second egg was unmeasured but was similar in shape, coloration, and markings (Frith and Frith 1992b).

INCUBATION: > 19 days. Activity at one nest was observed for > 30 hr. The individually-marked ♀ averaged 2 (1.7–2.3, n = 62) incubation bouts per hr: average bout length 17 min (range 13–26 min), average absence 13 min (6–24 min). An overall mean of 58% (range 44–86%) of total observation time was spent incubating (Frith and Frith 1992b).

NESTLING CARE: an egg was present in a nest at 1600 hr on 9 Jan but had hatched when next examined at 1445 hr the next day (= 1st day of life). Twenty nest-watches at this nest between 10 Jan and 3 Feb averaged 3 hr (90–360 min). The ♀ made an overall average of 2 nest visits per hr, of which 1.3 per hr were brooding bouts—these decreasing with increasing nestling age. Brooding bouts averaged almost 13 min. The average nest visit time was 10 min and mean absence 18 min.

248 Short-tailed Paradigalla *Paradigalla brevicauda* Rothschild and Hartert, 1911

Of 109 nestling meals fed by a single parent, on the Ambua Ra, PNG, 66 were identified. Of these, 65% were exclusively animal matter, 30% exclusively of fruit(s), and 5% were a mix. Of the 43 animal meals 34 were identifiable; 4 being of a complete (medium-sized) frog, 8 of frog pieces, 3 of complete skinks (Reptilia), 7 of white insect larvae, 4 of crickets, 2 of large beetles, 2 of earthworms and one each of a mantid, katydid, caterpillar, and large spider. All fruits regurgitated to the nestling were whole ones. Often an item of food was carried to the nest in the bill tip, particularly larger items, and the remainder of the meal subsequently regurgitated. The parent removed and ate all nestling faeces. This ♀ was seen feeding an imm of her size but with slightly duller plumage and less pure yellow facial wattles, which had left the nest on 28 June, on 14 Oct, or 108 days later and observations suggest she had ceased feeding this offspring a week later (Frith and Frith 1992b).

NESTLING DEVELOPMENT: the appearance of a nestling at 3 days old and of the same nestling, with its parent, at 24 days old is shown in Plate 3 and Fig. 9.16, respectively. At 6 days old the nestling's scapulars, primaries, secondaries, wing coverts, and lower back feathers were in pin and its eyes open. Scapulars, lower back, and lower abdomen feathers were out of pin at 10 days. At 13 days all primaries, save the first (innermost), and secondaries had burst from pin; crown feathers were in pin but the nape, breast, and abdomen feathers had burst. At 17 days old all feathers were well out of pin except on the throat, ear coverts, forecrown, and some malars, which were just bursting; and anterior lore feathers which were still in pin. Tail feathers remained in pin until 23 days old, when only 3 of the 16 mm central rectrices were out of pin. At 23 days the anterior feathers on the lores were just bursting, the chin and some malar feathers were in pin, and the iris colour was mid-grey. The nestling was in the nest at 0630 hr on 4 Feb but had left at 1530 hr that day, at 25 days old. Measurements of nestling weight gain and wing length development

9.16 Parent, presumed female, Short-tailed Paradigalla at nest with 24-day-old nestling. Ambua Lodge, Tari Valley, PNG, 3 Feb 1989. Photo: CF.

9.17 A nest of the Short-tailed Paradigalla, recently and successfully used. Ambua Lodge, Tari Valley, PNG. Feb 1989. Photo: CF.

appear in Fig. 6.2 but for more detail see Frith and Frith (1992b).

Annual cycle   Table 6.5
DISPLAY: unknown. BREEDING: nesting known in the one area during all months except Mar and Nov. EGG-LAYING: Jan, Apr–July, Sept, Oct and Dec. MOULT: 34 of 84 specimens were moulting involving all months collected (none in Apr and June) with least moult possibly being Aug–Nov (Appendix 3).

Status and conservation   Table 8.1
Widespread but patchily distributed. In some localities is common but is generally uncommon over most of its range. Commonly seen but not considered numerous about Sogolomik, Ok Tedi, PNG (Bell 1969). Not believed to be at risk.

Knowledge lacking and research priorities
Eastern extent of the species range requires better delineation. Mating organisation, courtship display, and diet require study.

Aviculture
Both the BRS and Nondugl Sanctuary have maintained captives, but these produced no offspring.

Genus *Astrapia* Vieillot, 1816

Astrapia Vieillot, 1816. *Analyse d'une Nouvelle Ornithologie Élémentaire* p. 36. Type, by monotypy, *Paradisea nigra* Gmelin.

    Synonyms: *Astrarchia* Meyer 1885 *Zeitschrift für die gesammte Ornithologie* **2**, 378. Type, by monotypy, *A. stephaniae* Finsch.

    Calastrapia Sharpe, 1898. *Monograph of the Paradiseidae, or Birds of Paradise, and Ptilonorhynchidae, or Bower-Birds*, Introd., p. 13. Type, by monotypy, *Astrapia splendidissima* Rothschild.

    Taeniaparadisea Kinghorn, 1939. *Australian Zoologist* **9**, 295. Type, by monotypy, *T. macnicolli* Kinghorn, *cf.* Mayr and Gilliard (1952) *American Museum Novitates* **1551**, 1–13 (= *A. mayeri*).

The astrapias are **five** smallish, compact-bodied birds with remarkably long tails, which inhabit only the uplands of mainland NG. They constitute a single, loosely-knit superspecies (*contra* Diamond 1986a), all being predominantly allopatric except for a **hybrid** zone where *A. mayeri* and *A. stephaniae* meet.

Other than the bill being more slender, the **skull** of *Astrapia* is almost identical in most respects to that of *Paradisaea*. The most significant cranial difference cited by Bock (1963) is that the ectethmoid plate is thicker than in *Paradisaea*, approaching the bulbous condition found in *Manucodia*. The lacrymal is lacking, the external naris is larger and more oval, with the anterior end being expanded towards the bill tip.

Astrapia appears to be the sister form to the higher paradisaeinine clade. It shares obvious characters with *Parotia*, *Ptiloris*, and *Lophorina*. The iridescent breast band of the ad ♂ *A. rothschildi* is formed by very elongate (up to 36 mm long) feathers tipped with iridescent copper—these, particularly to the sides of the breast, are much like the pectoral 'fan' plumes of sicklebills and the strongly-iridescent green-tipped feathers to the sides of their abdomen are most like those found in *Cicinnurus*. Intrageneric wild **hybrids** occur and the genus intergenerically hybridises with *Epimachus* (Plate 14 and Appendix 1).

♂♂ *c.* 10% larger than ♀♀. **Bills** *c.* 20% longer than the head and relatively small, fine and slightly decurved (Plate 6). They are similar in length in both sexes and are thus proportionately slightly larger in the smaller ♀♀. Nostrils partly covered by feathers. **Tails** markedly graduated with tiny hair-like central points on the feather tips. The central pair of rectrices are enormously elongate (6–16 times longer than shortest tail feathers) and either grossly broadened, narrowed into 'ribbons', or spatulate-tipped. In the ad ♂ *A. nigra*, tail feathers other than centrals curve to cross those on their opposite side, are V-shaped, and finely corrugated—all possibly to support the highly-modified central pair. The central pair of rectrices grow progressively longer with increasing ♂ age in all spp. except *A. splendidissima* in which they become shorter. In ♂ astrapias the rectrices other than the central pair become progressively shorter with age, thus increasing the contrast with the elongated central pair, except in *A. nigra* and *A. rothschildi* in which all rectrices grow longer with age (Table 1.5). **Wings** moderately long and pointed and outermost two primaries mildly emarginated on inner vanes. **Tarsus** length *c.* 25% that of wing. Scutellation of the tarsus is most prominent distally, and most indistinct proximally.

Ad ♂ plumage highly iridescent. ♀-plumaged birds also long tailed but overall drab, dull blackish brown on upperparts, head, and chest but rufous to blackish on the abdomen with fine blackish barring.

Polygynous, with promiscuous ♂♂, and the ♀ alone attending the nest. ♂♂ display at traditional arboreal sites, with at least some spp. doing so communally whilst others may be solitary. ♂ astrapias are atypical in being relatively minor vocalists. Astrapias construct a bulky, open, typical paradisaeinine **nest** involving numerous stems of epiphytic orchids, placed in a tree fork. Known **eggs** are typical of the Paradisaeinae, having clean, clear, elongate 'brush-stroke' markings as in many of the subfamily, particularly *Parotia*, *Ptiloris*, *Cicinnurus*, *Seleucidis*, and *Paradisaea*, but typically have fewer and shorter elongate markings (Plate 13).

Arfak Astrapia *Astrapia nigra* (Gmelin, 1788)

PLATE 6; FIGURE 9.18

Other names: Arfak Bird of Paradise, Black Astrapia, Arfak Astrapia Bird of Paradise, Long Tail, Incomparable Bird of Paradise.

A typical astrapia in body shape but ad ♂ exhibits a huge, blunt-tipped, markedly-broad, and long tail. Dorsal surface of tail and wings of ad ♂ are strongly iridescent deep purple. ♀-plumaged birds overall blackish brown with only the faintest and finest paler ventral barring. The only vaguely similar sp. in its range is the larger and longer-billed Black Sicklebill. Monotypic.

Description
Adult ♂: 60 (76) cm, head velvety jet black; crown and sides of face iridescent deep blue (70) to Purple (1) with coppery bronze washes particularly on larger, scale-like, nape feathers in suitable light. A conspicuous, central, nape-to-mantle 'cape' of large scale-like feathers, intensely-iridescent metallic yellowish green with purple blue to Mauve (75) sheens contrasts with elongate, plush, velvety jet black feathers, dully-iridescent deep blue to Magenta (2) sheens, to each side. Back to uppertail coverts sooty brownish black (119) with rich, dark, coppery Olive-Green (47) iridescent sheen. Exposed upperwing sooty black with deep, rich, violet purple (172) iridescent sheen that may appear deep blue and/or Magenta in some lights, concealed parts of flight feathers darkest brown-black. Dorsal surface of tail sooty black with intensely-iridescent sheens of rich deep violet purple and/or Magenta, far more so on central pair, in suitable light. In some lights tail feathers appear finely and uniformly barred owing to

Arfak Astrapia *Astrapia nigra* (Gmelin, 1788)

corrugations of the feather vanes. Malar area, chin, and throat feathers black with velvety iridescent deep blue with purple (172A) and/or Magenta washes in some lights. Upper breast 'cushion' of elongate, dense, and plush velvety jet black feathers with dull, coppery, violet purple, iridescent sheen in some lights, bordered below by a broad gorget of intensely-iridescent, bronzed coppery feathers, and/or Lime Green (159) in some lights, extending up either side of the breast and malar area to beneath the eyes. Underparts below the gorget to vent dully-iridescent deep Dark Green (162A) with large and broad scale-like feather tips highly-iridescent Cyan (164) and/or Sky Blue (66), in suitable light, down sides of lower breast and belly. Vent and ventral surface of tail coverts blackish brown (219). Underwing: coverts blackish, with slightest, deep blue, dull iridescent tipping, flight feathers brownish black (119). Undertail slightly glossy blackish with corrugations giving finely barred appearance in some lights. Bill shiny black, iris very dark brown, legs and feet brownish-black, mouth colour unrecorded.

ADULT ♀: 50 cm, entire head and nape black, slightly glossed, iridescent dark blue (74) in some lights. Mantle to uppertail coverts blackish brown with iridescent blue of nape grading onto upper mantle. Upperwing: coverts brownish black, remainder Fuscous (21) with Raw Umber (23) outer margins to feathers. Dorsal surface of tail dark blackish brown. Entire underparts dark matte blackish brown, slightly paler on belly to undertail coverts, extremely finely and uniformly barred pale Buff (124) on breast and becoming broader on lower parts. Underwing: coverts blackish brown barred Cinnamon (39). Flight feathers Olive-Brown (28) with broad Cinnamon trailing edges to all. Ventral surface of tail dark Brownish Olive (129).

SUBADULT ♂: as ad ♀ with few feathers of ad ♂ plumage intruding, to as ad ♂ with few feathers of ♀ plumage remaining.

IMMATURE ♂: as ad ♀, but tail on average longer, and after first year little to no paler barring on underparts except lower belly to undertail coverts.

NESTLING: unknown. JUVENILE: a single juv specimen (RMNH 10) has entire plumage far less blackish than imm and ad in ♀ plumage. Down on crown, mantle, throat, and breast blackish brown fading slightly (to 223) on belly and vent sparsely barred whitish buff. Upper primaries and secondaries narrowly outer-edged Cinnamon-Brown (33) and below these feathers have ochraceous (223C) trailing edges. Dorsal surface of tail as older ♀ plumaged birds.

Distribution

Mountains of the Vogelkop Penin. Virtually all records are from the Arfak Mts, with a single sight record from the Tamrau Mts (J. Mackinnon *in litt.*). 1700–2250 m.

Systematics, nomenclature, subspecies, weights, and measurements

SYSTEMATICS: a distinct sp., geographically isolated from its congeners and considered by Diamond (1986a) to form a superspecies with the Splendid Astrapia. First described from a trade skin and not subsequently discovered in

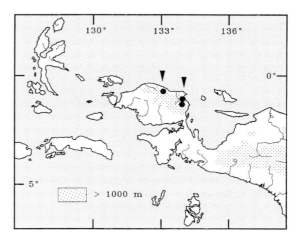

Arfak Astrapia *Astrapia nigra* (Gmelin, 1788)

its natural range until almost a century later by Odoardo Beccari, in 1872 (in Sharpe 1898). Hybridisation: known to occur only with the Black Sicklebill (Plate 14 and Appendix 1).

SPECIES NOMENCLATURE AND ETYMOLOGY
Astrapia nigra (Gmelin)
Paradisea nigra Gmelin, 1788. *Systema Naturae* **1** (1), 401. 'Oceanic Islands', restricted to the Arfak Mts, New Guinea. Type is an illustration.

Etymology: *Astrapia* = G. *Astrapios*, flashing or lightning *cf.* L. *astrapias*, a precious stone with gleams of light crossing its middle; *nigra* = *cf.* L. *niger*, black, shining black (*cf. ater*, matte or dead black).

WEIGHTS AND MEASUREMENTS
Measurements: wing, ad ♂ ($n = 25$) 179–193 (185), imm ♂ ($n = 6$) 173–180 (177), ad ♀ ($n = 17$) 157–182 (170); tail, ad ♂ ($n = 24$) 369–437 (399), imm ♂ ($n = 6$) 239–303 (266), ad ♀ ($n = 15$) 233–274 (258); central tail ad ♂ ($n = 23$) 518–756 (569), imm ♂ ($n = 6$) 299–377 (327), ad ♀ ($n = 14$) 290–332 (310); bill, ad ♂ ($n = 22$) 40–42 (41), imm ♂ ($n = 6$) 38–43 (41), ad ♀ ($n = 17$) 39–43 (41); tarsus, ad ♂ ($n = 25$) 39–43 (42), imm ♂ ($n = 6$) 41–43 (42), ad ♀ ($n = 17$) 37–43 (40). Weight: imm ♂ ($n = 1$) 190.

Habitat and habits   Table 4.1
Midmontane and upper montane forests, presumably to the summit of the highest peaks of the Arfak Mts. Habits virtually undocumented.

Diet and foraging   Table 4.2
The only information available is that of Beccari who noted that birds ate 'fruits of certain Pandanaceae, and especially those of the Freycinetiae' (climbing pandanus palms). The general diet is presumably similar to that of other astrapias.

Vocalisations and other sounds   Figure 9.18
Presumably the ♂♂ are relatively non-vocal. The only record is a simple, double, down-

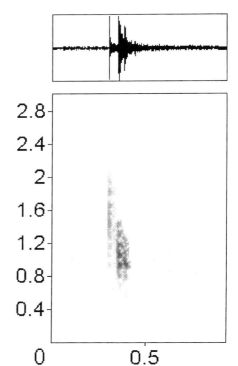

9.18 Sonogram from a recording of an Arfak Astrapia giving a simple, down-slurred, hollow-sounding, *clu-ck* double note from Mokwam, IJ in 1995 by D. Gibbs.

slurred, hollow *clu-ck* sound (Fig. 9.18) reminiscent of that produced with tongue and palate by people.

Mating system
Unknown but presumed polygyny with promiscuous ♂♂, and ♀♀ performing all nesting duties alone.

Courtship behaviour
Unknown

Breeding   Table 6.5
The only record appears to be that of a ♀-plumaged bird feeding a juv of similar appearance on 21 July on the Vogelkop, IJ and juvs were seen at Tanah Merah, Arfak Mts during July–Aug (B. Poulson personal communication).

Annual cycle   Table 6.5
DISPLAY: unknown. BREEDING: see above. MOULT: 5 of 11 specimens were moulting, involving the months Feb, Apr–June, and Aug (Appendix 3).

Status and conservation   Table 8.1
Beccari found the sp. 'common enough' but rarely encountered ad ♂♂. Amazingly, little is recorded since that 1872 observation. Common at higher elevations of Arfak Mts adjacent to Manokwari in late Aug 1995 (M. Carter *in litt.*). Because of its geographically-restricted range, it could be vulnerable.

Knowledge lacking and research priorities
Perhaps the least documented member of the family, notwithstanding claims made for that title for a number of other spp. It has been recently observed many times by people working in the Arfak Mts but no information is available in the literature. A thorough survey of this sp. is urgently required to assess its status, and would doubtless result in some basic knowledge of its biology. It is important to learn of the time ♂♂ take to acquire ad plumage and their longevity.

Aviculture
No recorded experience.

Splendid Astrapia *Astrapia splendidissima* Rothschild, 1895

PLATE 6; FIGURE 9.19

Other names: Splendid Astrapia Bird of Paradise, Splendid Bird of Paradise. *Erei* of Yabi people, Weyland Mts, IJ (Rothschild 1931). *Takena* of the Dani and *Degen* of the Uhunduni Peoples, IJ (Ripley 1964). *Dan* of the Hindenburg Ra, PNG (Gilliard and LeCroy 1961).

An astrapia distinctive for its short, blunt, white-based tail, the ♂'s prominently so. Inhabits highland forests of the western half of the central cordillera, potentially meeting the Ribbon-tailed Astrapia in western PNG. Polytypic; 2 subspp.

Description
ADULT ♂: 39 cm, entire crown, nape, and mantle brilliantly-iridescent metallic yellowish green; chin and throat metallic blue-green, with rich blue to purple highlights in suitable light. Back deep velvet black with rich Magenta (2) sheen. Rump and uppertail coverts matte black. Upperwing Dark Greyish Brown (20) with slight, deep purplish blue iridescence at leading edge above primary coverts. Variable extent of white bases and shafts to inner four pairs of otherwise blackish brown tail feathers with slight iridescent Magenta gloss to black, broad, spatulate tips of central pair. Tiny hair-like central points to tail feather tips. A broad, strongly-iridescent, metallic coppery red gorget narrows as it extends beyond the breast sides up the sides of the face to beneath the eyes. Beneath this, underparts are rich, silk-like, deep oily Dark Green (162A) with some larger plate-like feathers at breast sides with strong, paler, lime green (159) iridescence. Underwing variably Fuscous (21) with tiny, concealed, white bases to outermost primaries. Thighs, vent, and undertail coverts matte blackish brown. Undertail shiny Dusky Brown (19). Bill shiny black, iris dark brown to blackish brown, legs and feet fleshy lead-grey.

ADULT ♀: 37 cm, much duller than ♂. Entire head, nape, and throat brownish black with dull, bluish green iridescent gloss. Some individuals exhibit an area of rich Chestnut (32) feathers on nape that may extend onto crown and/or (rarely) behind ear coverts. Mantle, back, and upperwing variably Dark Greyish

Brown, slightly paler on wings, rump, and uppertail coverts. Upper breast Dark Greyish Brown and remainder of under parts quite variable in colour from a pale buff (223D) to rich buff-amber (123A) and conspicuously barred black throughout except the thighs which are Fuscous. Underwing pale fuscous to Olive-Brown (28) with tiny, concealed, white bases to outer primaries. Undertail Fuscous.

SUBADULT ♂: like ad ♀ with a few feathers of ad ♂ plumage intruding, to as ad ♂ with few feathers of ♀ plumage remaining. Individuals acquiring ad plumage do so initially with glossy-green feathering on the crown and throat (Gilliard 1969).

IMMATURE ♂: dark chestnut brown on crown and individually variable amount of chestnut-rufous in crown of some birds.

NESTLING–JUVENILE: like ad ♀ but plumage soft and fluffy particularly below. Less rufous in underparts, more grey in chin, throat, and upper chest than in ♀. Generally duller and less black above. Crown fluffier and may be suffused with deep chestnut. Tail feathers more pointed than in ad.

ABERRANT SPECIMENS: AMNH 342102 ssp. *helios* from Lake Habbema, IJ, is a partial albino in striking ♀ plumage; head, nape, throat, and upper chest white with a few black flecks. Upperwing (i.e. greater coverts) white as are the inner primaries and secondaries plus a tertial on one side. A few odd white feathers in breast, abdomen, and the vent. Legs and feet are conspicuously piebald with all claws pale. AMNH 302958 ssp *splendidissima* from Weyland Ra, IJ, is an undoubted ♀ (gonads drawn on label) but it has a single, iridescent Magenta, ad ♂-like feather in upper breast—also several larger lower back feathers have half the vane white.

Distribution

The Weyland Mts eastward through the central cordillera of IJ as far as the westernmost ranges of PNG (Star, Hindenburg, and

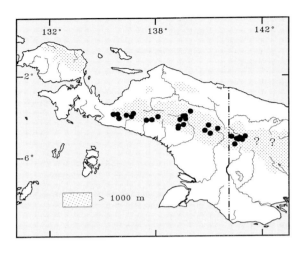

Victor Emanuel Ra). Probably contacts the Ribbon-tailed Astrapia in the vicinity of the Strickland gorge (see Coates 1990: 450). Eastern terminus of range has not been defined, and the relative distribution of *A. splendidissima* and *A. mayeri* on the central range N of the Ok Om/Lagaip catchment is unknown. Ranges from 1750 to 3450 m (mainly 2100–2700 m) (Coates 1990).

Systematics, nomenclature, subspecies, weights, and measurements

Systematics: the three cordilleran spp. of astrapia each exhibits distinct characters in a manner that is consonant with the hypothesis that the three are local isolates, the product of *in situ* fission of a widespread cordilleran parent form. Hybridisation: currently unrecorded, but *A. splendidissima* and *A. mayeri* may hybridise at points of contact in the Central Ra and Victor Emanuel/Muller Ra.

SPECIES NOMENCLATURE AND ETYMOLOGY

Astrapia splendidissima Rothschild, 1895. *Novitates Zoologicae* 2, 59, pl. 5. 'Probably Charles-Louis Mountains,' restricted to the Weyland Mts (Mayr 1941*a*). Type specimen AMNH 678036.

Etymology: *splendidissima* = L. *splendidissimus*, most brilliant, magnificent.

SUBSPECIES, WEIGHTS, AND MEASUREMENTS

1. *A. s. splendidissima* Rothschild, 1895

Range: the westernmost segment of the central cordillera, from the Paniai (Wissel) Lakes W to the Weyland and Charles Louis Ra.

Measurements: wing, ad ♂ (*n* = 16) 131–137 (134), imm ♂ (*n* = 7) 132–139 (135), ad ♀ (*n* = 21) 128–136 (132); tail, ad ♂ (*n* = 15) 148–175 (161), imm ♂ (*n* = 6) 162–192 (176), ad ♀ (*n* = 19), 166–204 (176); tail centrals, ad ♂ (*n* = 11) 193–213, (203), imm ♂ (*n* = 5) 181–233 (206), ad ♀ (*n* = 20) 189–232 (204); bill, ad ♂ (*n* = 15) 38–42 (40), imm ♂ (*n* = 7) 39–42 (40), ad ♀ (*n* = 21) 37–41 (39); tarsus, ad ♂ (*n* = 16) 35–39) (37), imm ♂ (*n* = 7) 35–38 (37), ad ♀ (*n* = 21) 33–39 (36). Weight: ad ♀ (*n* = 1) 120.

2. *A. s. helios* Mayr, 1936. *American Museum Novitates* **869**, 3. Mt. Goliath, Oranje Ra, Dutch New Guinea. Type specimen AMNH 448981.

Synonym: *A. s. elliottsmithi* Gilliard, 1961. *American Museum Novitates* **2031**, 3. Mount Ifal, 7200 ft [2200 m], Victor Emanuel Mountains, Mandated Territory of New Guinea. Type specimen AMNH 648726.

Range: the Central Ra of IJ from E of the Paniai (Wissel) Lakes to the Hindenburg and Victor Emanuel Ra of PNG. Eastern extent of distribution not delineated.

Diagnosis: like nominate form but crown, neck, and dorsal collar of ad ♂ more bluish and less golden-green. Spatulate tips of central tail feather pair broader than nominate. ♀-plumaged birds slightly darker above. Birds of both sexes and all age classes differ from nominate form in having extensive, unconcealed, white bases to underside of outer primaries except outermost two. These characters and the fact that individuals of *helios* are on average larger than those of the nominate form, more so in tail length, indicate *helios* is valid. We follow Cracraft (1992) in synonymising *elliottsmithi*. Cracraft expressed some reservation in doing so citing the slightly larger average size of *elliottsmithi* but this appears no more than the extreme of a W to E clinal size increase in the sp. The white primary bases of *elliottsmithi* are of the *helios* kind and are unlike that of the nominate form.

Measurements: wing, ad ♂ (*n* = 30) 133–145 (138), imm ♂ (*n* = 27) 133–147 (142), ad ♀ (*n* = 41) 130–145 (137); tail, ad ♂ (*n* = 23) 131–207 (182), imm ♂ (*n* = 27) 165–222 (200), ad ♀ (*n* = 39) 173–224 (192); tail centrals, ad ♂ (*n* = 25) 198–243 (223), imm ♂ (*n* = 22) 213–251 (230), ad ♀ (*n* = 34) 203–249 (216); bill, ad ♂ (*n* = 28) 38–43 (40), imm ♂ (*n* = 26) 36–42 (40), ad ♀ (*n* = 40) 36–43 (40); tarsus, ad ♂ (*n* = 30) 34–40 (38), imm ♂ (*n* = 27) 35–41 (38), ad ♀ (*n* = 41) 35–40 (37). Weight: ad ♂ (*n* = 12) 120–151 (138), imm ♂ (*n* = 10) 125–139 (132), ad ♀ (*n* = 16) 108–151 (125).

Habitat and habits   Table 4.1

Inhabits middle and upper montane forests to the treeline; also found at forest-edge and in associated secondary growth. Frequents the canopy down to undergrowth shrubbery within 3 m of the ground (D. Bishop *in litt.*). Generally solitary and rather secretive but pairs and threes recorded.

Diet and foraging   Table 4.2

Gilliard (1969) and D. Bishop (*in litt.*) each watched an ad ♂ methodically traversing heavily moss-covered tree trunks and limbs, constantly on the move as it worked along branches to peer at mossy surfaces apparently seeking small animal prey. Gilliard's ♂ was collected and had eaten numerous coral-coloured pandanus seeds averaging 11 mm long which Gilliard knew to be from a tree common above 1830 m that grows to *c.* 12 m tall. Of 15 Snow Mts, IJ, birds examined, 12 contained fruit remains, 1 a lizard, 1 a frog, and 2 the remains of insects (Rand 1942*b*). A ♀-plumaged and an ad ♂ near Wamena, IJ, at *c.* 1800 m were seen to feed quietly on fruits of epiphytic *Schefflera*, and an ad ♂ was seen to bark- and moss-glean for arthropods in open mid-canopy, on 16 Mar (BB). An ad ♂ also

Splendid Astrapia *Astrapia splendidissima* Rothschild, 1895

seen eating *Schefflera* fruits near Okbap, Star Mts, IJ (D. Bishop *in litt.*). Above Tembagapura, IJ, an ad ♂ consumed the tiny fruit of *Trema orientalis* on 10 Oct (see also Appendix 4). It has been tentatively estimated that the diet is 75% or more fruit and the remainder arthropods and other animal prey (Beehler and Pruett-Jones 1983).

Vocalisations and other sounds

Figure 9.19

♂, a nasal insect-like *to-ki*, the second note rising. A ♀-plumaged bird gave a similar call of *teek-teek* both on the same pitch (King 1979), this being described by D. Bishop (*in litt.*) as a dryish double note often given every 2–7 sec or less frequently. This appears to be the sharp, but softly-delivered and somewhat pathetic-sounding *jeet* note of birds of the Jayawijaya Mts, IJ (Fig. 9.19). A curious *tch tch tch* clicking call by ad ♂♂ was noted by Gregory (1995). An ad ♂ in dead twigs of an open tree top at Tembagapura, IJ, gave a frog-like dry *gree* call note at an average interval of 28 sec (*n* = 14) on 15 Oct (BB). A 'distinctive yelping call' was also heard on a few occasions by D. Bishop (*in litt.*). Ad ♂♂ were seen to produce a a 'mechanical/metallic whirring sound' vocalisation in PNG (A. Dennis *in litt.*). Wings do hiss in flight, but not much (BB).

Mating system

Presumed to be polygynous, with promiscuous ♂♂ displaying from arboreal courts. The classification of Beehler and Pruett-Jones (1983), extrapolating from limited available data, assumed ♂♂ were dispersed and non-territorial. However, at *c.* 1700 hr on 13 Nov at *c.* 2300 m in PNG (141° 20′ E, 5° 9′ S) four ad ♂♂ were found perched and calling equidistant (*c.* 40 m apart) around the forested edge of a natural meadow, perched on exposed branches near the tops of tall trees with some dead branches or completely-dead ones. Several other conspecifics were heard calling nearby but not seen. The ♂♂ called at *c.* 30 sec to 5 min intervals and regularly turned on their perches, often through 180° which emphasised their highly-iridescent plumage, for 30–40 min. After *c.* 10 min observation two ♀-plumaged birds were seen in the area but the ad ♂♂ paid them no particular attention other than vocalising in their direction now and then. The calling ♂♂ had a

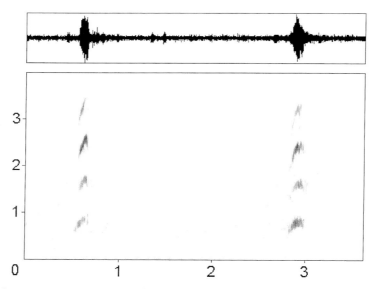

9.19 Sonogram from a recording of an ad ♂ Splendid Astrapia repeated *jeet* note at Jayawijaya Mts, IJ on 6 Apr 1993 by K. D. Bishop.

slightly fluffed-up appearance but no plumage was obviously extended or accentuated as if in display. The opposite side of the meadow was bordered with low bushes and bamboo and lacked astrapias (A. Dennis *in litt.*). These observations suggest the sp. might display in some form of lek aggregation.

Courtship behaviour
No information other than that which appears above.

Breeding   Table 6.5
Little known, other than nest-building seen in Mar, a juv in Aug, egg in Oct, and nestling in Nov. NEST SITE, NEST, EGGS, INCUBATION, NESTLING CARE, AND DEVELOPMENT: unknown.

Annual cycle   Table 6.5
DISPLAY: unknown. BREEDING: at least Mar (nest building) and July–Nov. EGG-LAYING: *c.* July–Oct. MOULT: 65 of 142 specimens were moulting, involving all months except June (Appendix 3).

Status and conservation   Table 8.1
In little demand for plumes in the Ilaga Valley, IJ, (Ripley 1964) and the Victor Emanuel Ra (Gilliard 1969). Common and widespread in the area of Okbap, Star Mts, IJ between *c.* 2600 and 3200 m but at lower elevation, to *c.* 2150 m, only ♀-plumaged individuals were seen (J. Diamond and D. Bishop *in litt.*). Not currently facing any significant threat, as most of its highland habitat is inaccessible to widespread development. A sighting at the 'Bailey bridge Tari Gap' (Tolhurst 1989) is presumed a misidentification (Frith and Frith 1992*a*, 1993*a*; Frith 1994*c*) and a report of this sp. briefly seen from a helicopter at the remarkably low altitude of 600–700 m near Tabubil in the Ok Tedi valley, PNG (Johnston and Richards 1994) requires confirmation.

Knowledge lacking and research priorities
It will be interesting to delineate points of range contact between this sp. and the Ribbontail, both in the Central Ra (to the N), and in the Victor Emanuel/Muller Ra area (to the S). Hybrids should prove readily identifiable. Determining the mating organisation of ♂♂ is also a priority, but if these birds are typical of astrapias, this will be a difficult task.

Aviculture
Has been kept at BRS, PNG.

Ribbon-tailed Astrapia *Astrapia mayeri* Stonor, 1939

PLATES 3, 5; FIGURE 9.20

Other names: Ribbon-tailed Bird of Paradise, Ribbontail Astrapia, Ribbon Tail, Shaw Mayer's Bird of Paradise. *Yaka Yan-gi* of the Fly, Yuat, Sepik watershed (Kinghorn 1939), *Kugo* or *Kugup* (W. Cooper personal communication) (M) and *Togi* or *Tooh-Ghee* (F) on Mt. Hagen (Gilliard 1969). *Elabe Yange* (Goldman 1981) or *Yange* of the Huli People (C. Ballard *in litt.*).

An astrapia with a very small cordilleran range and a very long tail. Ad ♂ is one of the most spectacular of the family, with two remarkably-long and narrow, white, central tail feathers, each of which is lax, black-tipped, and three times the length of the bird. ♀ may show some white in its smaller tail, which is narrower and longer than that of the ♀ Splendid Astrapia. Hybrids with Stephanie's Astrapia are common in some areas where the two spp. share the forest. Monotypic.

Description
ADULT ♂: 32 (125) cm, entire head, including naral tuft 'pom-pom' above base of maxilla, velvety jet black but intensely-iridescent,

Ribbon-tailed Astrapia *Astrapia mayeri* Stonor, 1939

metallic yellowish green and/or rich blue (68) and violet purple (172) and/or Magenta (2) in some lights, the latter predominantly so on ear coverts and elongate feathers at sides of head to nape. Mantle to uppertail coverts deepest velvety jet black with dull but rich iridescent sheen of bronzed Olive-Green (47) in suitable light. Upperwing: lesser coverts darkest matte brownish black (119) edged jet black, greater coverts and alula brownish black, flight feathers blackish brown (219). Dorsal surface of tail blackish brown but central pair grossly elongated (> 1 m) into unique cotton white 'ribbons' c. 2 cm broad, tipped (3–4 cm) blackish brown; the feather shaft is also blackish for the distal quarter of the tail. Tail feathers other than the central pair are short, form a graduated tail, and are narrow and sharply pointed (unlike those of ad ♂ Stephanie's Astrapia). Throat feathers, tending slightly to scale-like, of intense metallic yellowish green extend onto the upper breast to meet abruptly a broad area of deepest velvety jet black breast feathers with dull, iridescent, coppery bronze sheen in rare light. This dark area is bordered below by an extensive narrow gorget of highly-iridescent coppery red. Beneath this gorget, to the vent, feathers are dully-iridescent coppery red broken up by their exposed dully-iridescent deep green (146) bases. Vent and undertail coverts matte brownish black. Underwing: coverts velvety blackish with trace of dully-iridescent deep blue feather edging, flight feathers Dark Brownish Olive (129), paler at bases. Undertail Sepia (219). Bill blackish, iris very dark brown, legs and feet dark leaden blue grey, mouth pale greenish yellow.

ADULT ♀: 35 (53) cm, entire head and nape may appear jet black but is highly-iridescent, metallic bronzed yellowish green (more so in older individuals) and/or deep blue in some lights, duller and more matte blackish on the naral tuft above bill base. Blue iridescence more obvious on sides of face and lower throat. Mantle and back deep velvety blackish with dull, deep dark blue (74) and/or violet purple to Magenta in rare light. Rump and uppertail coverts browner, close to Sepia. Upperwing: coverts brownish black, flight feathers blackish brown and slightly paler on tertials and primary tips. Dorsal surface of tail as primaries, but longer central pair with white basal central shafts and with variable, longitudinal, white smudges down centre of each vane. Exceptionally the central tail feather pair may approach or be like that of ad ♂ (it is possible that it is older ♀♀ that attain this ad ♂ character). Breast as mantle but iridescence slightly green-influenced grading to remaining underparts of Cinnamon (123A) broadly barred blackish. Underwing: coverts variably Raw Umber (123) broadly barred to almost entirely blackish, flight feathers Dark Brownish Olive. Ventral surface of tail dark olive brown (121) with the longitudinal white smudges visible on central pair. Central pair of rectrices usually much shorter than those of ad ♂.

SUBADULT ♂: as ad ♀ with few feathers of ad ♂ plumage intruding, to as ad ♂ with few feathers of ♀ plumage remaining. Central pair or rectrices shorter than those of ad ♂ but remainder of tail significantly longer.

IMMATURE ♂: as ad ♀ but wing and tail, excluding central pair of feathers, on average longer. In first year like ad ♀ but crown and nape iridescence far duller, more blue than green. Brownish black mantle and back with dull iridescence less purple than ad ♀. Throat and breast less glossy than ad ♀, being matte black with slight dull bluish iridescence. Subsequently like ad ♀ but lower flanks, belly, and undertail coverts blackish with pale buffish brown barring; underwing coverts blackish brown; then upperparts become more iridescent, more pink-purple, sides of head and throat brighter oil to yellowish green, sides of neck and upper breast blackish strongly washed purplish blue. Lower breast and belly dusky with traces of barring, central rectrices variably (more so in older birds) broadly marked white (Gilliard 1969).

NESTLING–JUVENILE: hatchling naked and dark purplish grey to blackish, more pinkish dorsally with tiny white egg tooth. At 3 days old skin purplish black, mouth lime green. At 6 days old gape yellow, at 9 days legs bluish grey, and at 16 days old feathers of crown, nape, mantle, and breast matte black and abdomen barred matte blackish and brownish grey. Wings, lower back, and flanks dark sooty grey with egg tooth present but inconspicuous. At 23 days old plumage matte black with slight iridescence on mantle. Iris dark brown, bill black with no egg tooth (Frith and Frith 1993b). Juv like ad ♀ but duller and barred ventral plumage less brown, soft, and fluffy. Upperparts black with a slight deep bluish cast (less purple than ad ♀), underparts from chin to upper breast brownish black (dully glossed blue in older birds) and rest of breast and belly Ground Cinnamon (239) with broad brownish black barring.

Distribution

The central cordillera of western PNG, from Mt. Hagen and Mt. Giluwe NW through the Enga highlands and W to Doma Peaks, the Porgera/Mt. Liwaro highlands, and apparently westward to the Strickland R. (including the S Karius Ra according to informants to Frith and Frith 1993b and subsequently confirmed breeding there by G. Clapp *in litt.*). Also reported in Coates (1990) to occur W of the Strickland. Westernmost and NW limits of range require delineation, but the sp. presumably contacts the Splendid Astrapia on one or more ranges. Gregory (1995) notes that local Ok Tedi area informants indicated birds occur there at appropriate altitudes, but this is certainly in error.

Systematics, nomenclature, subspecies, weights, and measurements

Systematics: a distinctive sp. that forms the middle isolate of the three cordilleran *Astrapia* spp. The hybrid zone with Stephanie's is over a relatively-broad zone of sympatry at the lower altitudinal limits of the Ribbon-tail. The striking ad ♂ central rectrices, different bill proportions and shape, the conspicuous, elongate, naral tuft of feathers at the base of the maxilla and the oily-green iridescence of ♀-plumaged birds are, however, all diagnostic characters. Moreover the tail feathers, other than the central pair, are radically different, those of the Ribbon-tail being short, narrow, and sharply pointed and those of Stephanie's Astrapia broad and blunt-ended. Hybrids have outer tail feathers intermediate in shape and centrals pied in pigmentation. Hybridisation: known with only Stephanie's Astrapia, but possible with Splendid Astrapia (Plate 14 and Appendix 1).

SPECIES NOMENCLATURE AND ETYMOLOGY

Astrapia mayeri Stonor, 1939 (Feb) *Bulletin British Ornithologists' Club* **59**, 57. 'Eighty to a hundred miles west of Mt. Hagen' (Station) restricted to Mt. Hagen (Mayr in Mayr and Greenway 1962). Type specimen BMNH 1939.1.9.1.

Synonyms: *Taeniaparadisea macnicolli* Kinghorn, 1939 (Dec). *The Australian Zoologist*, **9**, 295. Type specimen AM O. 37058.

Astrapia recondita Kuroda, 1943. *Bulletin of the Biogeographical Society of Japan* **13**, 33. Type specimen Kuroda Collection 15457.

Etymology: *mayeri* = after Fred Shaw Mayer, naturalist, aviculturalist, and collector who

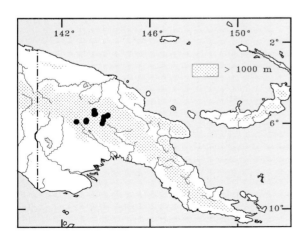

realised the significance of the sighting by J. G. Hides (see below).

Discovery: this is the latest species of bird of paradise to be described by scientists. First seen live by J. G. Hides (1936: 106) who, having watched two flying birds, instructed a member of his party to shoot an ad ♂ and preserve the tail feathers. Fred Shaw Mayer, who heard about and subsequently tried unsuccessfully to locate the (lost) Hides feathers, was given two central tail feathers and some other plumage of this sp. by a missionary in Aug 1938 who had obtained them from a Mt. Hagen man wearing them in his headdress. Shaw Mayer sent these feathers to C. R. Stonor as a donation to the BMNH, suggesting that they represented an unknown sp. of *Astrapia*. Stonor duly described a new sp. within a month on the basis of the pair of central and three outer tail feathers, two secondaries, and a greater wing covert of an ad ♂. Within a year J. R. Kinghorn of the Australian Museum, Sydney received three complete ad ♂ skins collected by J. L. Taylor and J. R. Black during their famous Hagen–Sepik patrol on behalf of the then Administrator of the Mandated Territory of New Guinea, Sir Walter Ramsay McNicoll. Kinghorn, unaware of the description by Stonor, quickly described the new sp. He illustrated it with a colour plate, noting that the new bird was most similar to the Splendid Astrapia, but he nonetheless erected a new genus for it, naming the bird *Taeniaparadisea macnicolli* (sic) after the Administrator. Kuroda (1943) described *Astrapia recondita* as a new sp., illustrated with a fine colour plate that shows what are identifiable as two imm ♂ Ribbon-tailed Astrapias.

WEIGHTS AND MEASUREMENTS
Measurements: wing, ad ♂ ($n = 30$) 173–185 (179), imm ♂ ($n = 10$) 156–175 (165), ad ♀ ($n = 27$) 150–163 (156); tail, ad ♂ ($n = 29$) 97–126 (111), imm ♂ ($n = 8$) 134–175 (162), ad ♀ ($n = 24$) 122–178 (152); tail centrals ad ♂ ($n = 26$) 657–1017 (893), imm ♂ ($n = 10$) 153–380 (300), ad ♀ ($n = 24$) 260–374 (308); bill, ad ♂ ($n = 25$) 29–35 (33), imm ♂ ($n = 10$) 31–35 (33), ad ♀ ($n = 25$) 31–35 (33); tarsus, ad ♂ ($n = 29$) 38–43 (41), imm ♂ ($n = 10$) 39–43 (41), ad ♀ ($n = 27$) 36–42 (39). Weight: ad ♂ ($n = 12$) 134–164 (147), imm ♂ ($n = 5$) 143–159 (148), ad ♀ ($n = 7$) 102–157 (132).

Habitat and habits   Table 4.1
Upper montane and subalpine moss forests. Not uncommonly foraging and nesting at the forest-edge, in isolated forest patches, and in selectively-logged and otherwise disturbed forest. Birds mostly frequent the upper third of the forest structure but may forage at all levels, including on the ground (Frith and Frith 1992a). Birds flying a distance, such as from one forest patch to another over subalpine grassland, do so with a shallow undulating flight consisting of 4–5 wing-beats followed by a brief and shallow downward glide with closed wings. In flight the horizontally-trailing streamers ripple in the air current.

Diet and foraging   Table 4.2
Birds forage at all levels of forest structure. Of 173 Mt. Hagen foraging records, 43% occurred at 0–15 m above ground, 31% at 15–25 m, and 26% at 25–35 m high (Kwapena 1985). Of 140 instances of birds seen foraging in Tari Gap during Sept–Jan, 1% was on the ground, 26% 0–5 m high, 31% 6–10 m, 34% 11–15 m, and 8% 16–20 m high. The fact that none was recorded above 20 m possibly reflects the difficulty of sighting upper canopy birds in the Tari Gap area but probably indicates a dislike of exposure on the upper canopy and sunshine.

Kwapena (1985) noted that of 181 observed foraging bouts on Mt. Hagen, 63% were upon tree branches, 17% upon fruits, 4% at leaves, and 4% at flowers. Of 145 observed fruit-feeding bouts 67% were on *Schefflera* spp., 13% a *Rhododendron* sp., 12% a *Fagraea* sp., and 3% or less on six other plant spp. Arthropod feeding was seen 52 times. Tari Gap birds were most often seen feeding upon epiphytic *Schefflera chimbuensis* (*fide* D. Frodin) fruits, a most significant food in the mid-storey.

Given that some observations of feeding birds involved foraging at more than one site, 49 involved foraging at *Schefflera* fruiting crowns; 33 at other fruits; 29 at mossy or other epiphyte-covered tree boughs; 9 at each of tree trunks, dead branches, and pandanus crown fronds; 5 at dead foliage; 3 at both live branches and on the forest floor; and once at both vine tangles and flowers (CF).

Arthropods are obtained by hopping along larger tree boughs or clinging to tree trunks and pecking into mosses, lichens, and other epiphytic growth. Birds also cling to dead and dying *Pandanus* fronds whilst picking and tearing at them in search of arthropods. Rotten wood may be deeply excavated by probing and tearing vigorously with the bill, often using the open-bill-probing technique. Birds may spend long periods foraging amongst foliage and/or epiphytes of a single tree.

Collected birds had fruit remains in their stomachs (Sims 1956; Gilliard 1969). Of faecal samples obtained from 13 birds mist-netted in Tari Gap during Sept–Oct, all contained exclusively fruit remains except two which each contained a small, 3-mm-long, beetle. Of the 13 birds, 12 had eaten *Schefflera* fruits, 5 *Acronychia kaindiensis* fruits, 3 *Timonius belensis*, 3 *Dimorphanthera alpinia* fruits, and one *Xanthomyrtus* sp. fruits (CF). See Appendix 4.

While the ad diet has been tentatively estimated to consist of 90% fruit (Beehler and Pruett-Jones 1983) repeated observations of birds methodically gleaning foliage and epiphyte-covered tree boughs and trunks suggest this is an overestimate of the fruit proportion (Kwapena 1985; CF). Of 140 Tari Gap foraging events, 56% were of birds eating fruit(s) and 41% of them apparently taking animal prey, one record involving the taking of both, and one being for feeding at flowers and another at flowers and flower buds. Of the arthropod feedings; 40 involved the bird(s) probing, 19 times into moss and other epiphytic growth on tree branches, 8 times on tree trunks, 7 times into dead branch wood, 4 times into the bases of pandanus crown fronds, once into live branches, and once into forest floor leaf litter. A bird was once seen to spiral around a tree trunk whilst foraging by probing and lifting bark and once a bird made a short sally flight to catch a flying insect. The latter instance, and once when a bird took a fruit to a perch to hold it by a foot to tear it apart, was the only time food was not eaten *in situ* at the foraging site. Open-billed-probing of dead branches and accumulated vegetable debris was observed. Of the 79 observed fruit feedings the plant could be identified in 69 cases; of these 50 were of *Schefflera chimbuensis*, 7 of *Fagraea salticola*, 4 of *Dimorphanthera alpinia*, 2 of *Pittosporum* sp., 2 of both *Timonius belensis* and *Syzygium* sp., and 1 of *Garcinia* sp. (CF). One bird was seen at *Schefflera chimbuensis* taking > 15 pieces of fruit in 350 sec (BB).

Of 140 Tari Gap foraging bouts, 86% involved a lone bird, 12% two birds, and one record each were of groups of three and five birds, respectively. Of the 17 instances of two birds feeding together, six involved an ad ♂ and a ♀-plumaged bird, six involved two ♀-plumaged birds, three involved a ♀-plumaged bird with a juv, one involved an ad ♂ with an imm ♂, and one involved two imm ♂♂. The group of three birds were all ♀-plumaged and the group of five consisted of two ad ♂♂ and three ♀-plumaged birds (CF). Single birds or groups of two or three may forage in the shrub layer and on the forest floor. Feeding assemblages of up to six and seven birds occur (M. Laska *in litt.*).

An ad ♂ once displaced and chased a Papuan Lorikeet *Charmosyna papou* from a fruiting *Schefflera* and another supplanted a ♀-plumaged Brown Sicklebill. A Belford's Melidectes *Melidectes belfordi* unsuccessfully attempted to deter an ad ♂ Ribbon-tail from approaching a food tree by flying out of that tree to dive-bomb it and on four other occasions this honeyeater sp. was seen to chase an ad ♂ Ribbon-tail off a *Schefflera*, one individual honeyeater chasing the same Ribbon-tail off four times. A Common Smoky Honeyeater supplanted another Ribbon-tail and chased it off (M. Laska *in litt.*).

Vocalisations and other sounds

The characteristically-common call at Tari Gap is a loud clear *waugh, or wock, whit-whit*, the first note being guttural and the immediately-repeated second note sharply whistled (Frith and Frith 1992*a*). An ad ♂ gave a *kenk!* and a scolding *skaw skaw* (BB). A short-tailed young ♂ gave typical, plaintive, astrapia-like, nasal and frog-like, single or double note *wreden, kuweep werwer, reaou reaou* or *wreden weep!* (Beehler *et al.* 1986). For display calls see below. Wings produce an audible rattling rustle and Cooper (in Cooper and Forshaw 1977) noted a sound produced by sidewise flicking of the tail feathers immediately prior to an ad ♂ flying from a perch.

Mating system

Polygyny with promiscuous ♂♂ but it is not yet clear if ♂♂, which do appear to have traditional display locations and perches, are solitary-displaying or do so in twos or larger numbers. In June in the Tari Gap several ad ♂♂ were displacing one another from tree perches when one grabbed another and they tumbled from the tree with claws locked (M. Laska *in litt.*). Uniparental nest attendance by the ♀ is confirmed (Frith and Frith 1993*b*).

Courtship behaviour

A captive ad ♂ performed a simple display to a ♀ in Taronga Zoo, Sydney. He perched 60 cm from the ♀, suddenly gave a raucous *grrrow, grr, grr* and then 'stood sideways as though looking at the ground with one eye, at the same time it let its wings droop, then held them over the back' and then suddenly flew forcefully from perch to perch (2 m apart) thumping the perches with its feet, 'then pausing to flick its tail like an angry cat' (Gilliard 1969). Cooper (in Cooper and Forshaw 1977) on 7 June watched an ad ♂ join a feeding ♀ in fruiting trees to chase her from branch to branch in a half-hearted way but once gave a series of *julak* notes whilst flicking his wings rapidly out from his sides.

Another display was seen on Mt. Giluwe at 1700 hr on 25 Aug. Two ad ♂♂ called apart then joined each other on tall *Schizomeria* sp. tree branches 35 m high, the longer-tailed bird slightly higher than the other, and they were then joined by three ♀-plumaged birds who perched on adjacent branches. The longer-plumed ♂ then jumped from branch to branch whilst giving a rough, harsh *hisss-sss-ssh* followed, by a loud, sharp *keaoo-ooo-ooo* call before raising his breast plumage and tail feathers to jump from branch to branch with much calling. As this display climaxed the ♂ flew between perches close together. He displayed for *c.* 10 min before flying off to join two of the three ♀-plumaged birds nearby. He subsequently displayed again and continued to do so until darkness at 1830 hr (Kwapena 1985).

Long-term observations at Tari Gap leave no doubt that ♂♂ typically court by jumping back and forth between two or more tree branches, sometimes rapidly but at times by spending up to 30–40 sec on each perch. Of a displaying group of three ad ♂♂ and four ♀-plumed birds watched for 1 hr the former repeatedly displaced the latter from perches in two adjacent trees (M. Laska *in litt.*).

An ad ♂ at Tari Gap was observed on 19 Sept to perform an extensive elaborate flight, unlike normal flight, presumably as a display. In this the wing-noise appeared significantly louder than usual and the 4–5 wing-beats were interspersed with a much deeper brief downward and then upward glide with closed wings before again wing-flapping. This enhanced the two white central tail feathers by waving them. An ad ♂ in the forest canopy, with a ♂ with half-grown white tail feathers nearby, displayed by hopping gently from side to side on-the-spot on his 18-m-high tree perch. This movement caused his central tail plumes to sway back and forth rhythmically through an expansive, maximised, arc. A second ♂ with its white tail feathers half grown displayed to a ♀-plumaged bird 6 m high on a *Schefflera* plant by jumping from side-to-side on his perch with the central tail plumes held stiffly apart and arched as they swung side-to-side behind him (CF).

Breeding Table 6.5

NEST SITE: of 14 forest nest sites described, seven were in disturbed forest with secondary growth and two were in small remnant forest patches within subalpine grassland. Ten were in conspicuously-isolated small trees or saplings with no immediately-adjacent tree branches or foliage, where the forest canopy was usually lacking directly above or at least there was a great deal of open space between the top of the nest tree and the forest canopy high above. Such nest sites may be limited in availability. That ♀♀ prefer such sites was clearly emphasised by the fact that several old nest structures were found by Kwapena (1985) and Frith and Frith (1993b) immediately adjacent to active ones, in the same tree, a situation also known in Loria's Bird and the Short-tailed Paradigalla (Frith and Frith 1992b, 1994b). This suggests that nesting in isolated trees may be a strategy to reduce nest access by tree-climbing predators. It is, however, also possible that this situation reflects the traditional use of a nest site close to a locally-significant resource, such as ♂♂ or food. Thirteen nests averaged 8 m high (range 3–18 m), placed in nest trees that averaged 11 m tall (range 3–22 m). All nests were built between near-vertical forking branches of trees or tree ferns except one 'on a tree stump 2 m high' (Kwapena 1985).

NEST: 13 active Tari Gap nests generally agree with the only two previous brief nest descriptions with respect to materials used: a frail to substantial deep cup of large leaves and pandanus palm frond pieces, sparsely to densely covered by fresh, green-leaved, epiphytic orchid stems (Frith and Frith 1993b). See Plate 3. So thin may the lower outer orchid covering be that a good deal of the inner leaf cup structure remains visible (see Fig. 9.20). The long green orchid stems are denser around the nest cup circumference and outer rim, which they encircle. The egg cup is lined with much finer, dead, straw-like, supple, leafless orchid stems. No nest had sticks incorporated into or beneath it but several had the odd piece of small fern frond 'decoration' externally. Average measurements (mm) of four are: total nest depth 131 (130–150), total dia 199 (180–215), interior cup dia 116 (110–125), exterior cup dia 167 (159–175), egg cup depth 64 (59–67), maximum dia of outer orchid stems 1.7 (1.5–1.9), and maximum length of them 865 (580–1310), maximum dia of lining orchid stems 1.1 (0.9–1.2), and maximum length of them 491 (398–610).

EGGS: a single-egg clutch. Boehm (1967) and Everitt (1973) described one of four single-egg clutches, laid in captivity in the USA, as 'medium brown streaked with darker brown heaviest at the thicker end' measuring 37 × 26 mm and being rather elongated in shape. An egg laid in captivity on 26 Oct 1961 at the Hallstrom Fauna Sanctuary, Nondugl, PNG, was elliptical-ovate, pale pinkish buff with large, slightly elongate, greyish purple blotches on and predominantly around the larger end and only a few about its middle and some larger russet blotches over other colours with one or two of these, particularly large, about the smaller end and two or three fine, but conspicuous, brown scribble lines about and on the smaller end. It was dully glossed, measuring 40.6 × 27.0 mm (Bishop and Frith 1979). An egg collected on Mt. Giluwe on 20 June was 'creamy white with reddish brown spots' and measured 40 × 25 mm (Kwapena 1985). Four fresh Tari Gap eggs averaged 38.6 (38.3–39.0) × 27.8 (27.3–28.5) mm and

14.9 (14.8–14.9) g in weight. In ground colour they were very pale washed-out buffish pink to Beige (219D) marked with blotches and elongate broad streaks of purplish greys and browns to russet mostly at and around the larger end. See Plate 13.

INCUBATION: exclusively by the ♀ parent. The only recorded incubation period was 21 days—for an egg in captivity (Boehm 1967). During 19 hr of time-budgeted activity of a ♀ incubating at Tari Gap during 31 Dec–6 Jan (CF) the ♀ averaged 2 (2–4) nest visits an hr, spending an average of 66 (61–84)% of time incubating. Incubation bouts averaged 16 min (13–23 min, $n = 47$) and 43 absences from the nest 9 min (5–11 min, $n = 43$) (Frith and Frith 1993b).

NESTLING CARE: during 15 hr at a Tari Gap nest (16–29 Oct) the ♀ parent averaged 3 visits an hr (2–4, $n = 44$) to feed and/or brood, spending an average of 58 (47–78)% of time at the nest of which 55 (43–77)% was spent brooding. Average brooding period was 11 min (8–23 min, $n = 44$). An average of 2 (1–3, $n = 27$) feeding visits an hr were made to the young and average time absent was 9 (7–10) min. Parental activity was observed for 51 hr at another Tari Gap nest at which the otherwise normal looking ♀ in attendance had a fully-developed pair of white central tail feathers like those of ad ♂♂ and could thereby be identified (Plate 3). She brooded for 52% of observation time when the nestling was 1 day old but this decreased to 3% by the time it was 17 days old, after which it was not brooded. Brooding bouts varied in average length from 4 to 12 min. The average length of parent absences varied from 6 to 15 min, being generally longer with increased nestling age, the overall mean absence being 9 min. The percentage of time the ♀ was at the nest varied from 5 to 66%, averaging 24% overall. The average number of her feeding visits was 3–7 an hr, average 5 overall. The ♀ ate the faeces of the nestling until its last few days in the nest, when they accumulated about the nest rim. All nestling meals were regurgitated to the young by the ♀. Nestling diet has been studied at one Tari Gap nest. Of 147 identified meals, 63% were exclusively of fruit, 31% exclusively animal, and 6% consisted of some fruit and animal components combined. A single meal of *Schefflera* fruit may consist of up to 23 tiny pieces, regurgitated and each fed to the young one at a time in quick succession. Of 93 exclusively fruit meals, 82 were identifiable to plant taxa and of these 60% consisted of *Schefflera* fruit pieces (mean count = 9.9, range 2–23), 28% of several pieces of *Dimorphanthera alpina* (mean count = 4.7, range 2–11), 5% of *Alpinia tephrochalmys*, and 1% of a *Pittosporum* sp. (see Appendix 4). Of 18 nestling faecal samples, 14 contained remains of *Schefflera* sp., 12 *Fagraea salticola*, 11 *Dimorphanthera alpina*, 11 *Alpinia tephrochalmys*, 4 *Timonius belensis*, 3 *Tetrastigma* cf. *dichotoma*, 1 *Coprosma novoguineensis* and 1 *Cissus* sp. fruit remains (Frith and Frith 1993b).

The majority of 46 animal nestling meals could be identified only as arthropod(s) or winged insect(s) but on six occasions a cricket was identified, once a spider, a dragonfly, and part of a frog. Of 18 nestling faecal samples, 12 contained remains of small to medium-sized beetles (Frith and Frith 1993b).

NESTLING DEVELOPMENT: weight gain and wing length development of a single nestling brood, the second mentioned above, is summarised in Fig. 6.2 where age is expressed as days of life. For more detail of growth see Frith and Frith (1993b). Nestling's eyes were closed at 3 days old but open at 6 days when pin feathers were present except on the naked head. At 9 days old primary and secondary feathers were in pin and breast feathers were still mostly in pin; abdomen, back, and scapular feathers had just burst from pin. At 16 days its throat and ear coverts were still in pin, there were no lore feathers but the rest of its contour and head feathers were well out of

9.20 A nest of the Ribbon-tailed Astrapia recently and successfully used. Tari Gap, PNG, Nov 1986. Photo: CF.

pin. At 23 days the lores were feathered but the 'cere' was still bare. At 0915 hr on 4 Feb the nestling hopped onto and perched upon an adjacent old Ribbon-tail nest and was still there at 1145 hr on 7 Feb. At 0720 hr the next day it was perched beside the old nest but had gone at 1130 hr that day. It is most likely that prolonged late morning and afternoon rains during 4–7 Feb delayed this nestling's departure from the nest, and it is probably correct to consider that it fledged on 4 Feb, at 25 days old, rather than on 8 Feb at 29 days old (Frith and Frith 1993b). Boehm (1967) stated his captive nestling remained in the nest for 22–26 days but Everitt (1973), Boehm's aviculturalist, gave 24 days for the same nestling.

Annual cycle Table 6.5
DISPLAY: June, Aug, Dec. BREEDING: recorded during May–Mar and therefore clearly possible during most of the year. EGG-LAYING: at least May, June, and Sept–Feb inclusive and therefore possibly all months. MOULT: 28 of 83 specimens were moulting, involving all months but Oct (Appendix 3). While one bird showed body and tail moult in Feb all other moult was July–Nov. Mayr and Gilliard (1952) found that most birds had completed moult shortly prior to July, but one ad ♂ was in full moult with most flight and body feathers wholly or partly in pin on 15 July. Of 12 individual birds examined for moult in Tari Gap, Southern Highlands, PNG during Sept, Oct, Nov, and Jan, 40% of the five caught in Sept, 83% of the six in Oct, and the only bird caught in Nov were moulting (Frith and Frith 1993a). A bird of unknown age taken to Sydney grew black central tail feathers for its first two moults, one short black one and the other longer with much white in its third moult, both long and white with black edges and tips in its fourth moult, and both white with only a little black remaining in the fifth moult. Thus this individual would have taken at least 6 years to obtain fully-developed ad ♂ tail plumage (McGill 1951: 250).

Status and conservation Table 8.1
Common to abundant in some places, ♀-plumaged birds being far more numerous than ad ♂♂ but even the latter not uncommon in the Tari Gap. Unlikely to be threatened in the near future because much of its range is uninhabited and inaccessible. Earlier concerns that this sp. might be genetically swamped by Stephanie's by hybridisation appear unfounded.

Knowledge lacking and research priorities
The western and northern boundary of the geographical range of this PNG endemic is poorly known and a survey of the full extent of its distribution, plus geographical extent of hybridisation with Stephanie's Astrapia, is required. Contact zone and possible hybridisation with Splendid Astrapia also merit study. Details of courtship and mating organisation require attention as well.

Aviculture
At least two birds hatched in aviaries in New Guinea (Hallstrom 1962). Has been kept for long periods in captivity, with one successful breeding in an outside aviary in the USA (Boehm 1967).

Stephanie's Astrapia *Astrapia stephaniae* (Finsch and Meyer, 1885)

PLATE 6

Other names: Princess Stephanie's or Stephanie Bird of Paradise, Princess Stephanie Astrapia, Stephanie Astapia. *Mek* on Mt. Hagen (Cooper in Cooper and Forshaw 1977); *Tawanta* (♂) and *Okai* (♀-plumaged) of the Fore people, *Melo* of the Gimi and *Kwekwe* of the Daribi people (Diamond 1972); also *Meulo* or uncommonly *Lulumane* of the Gimi (D. Gillison *in litt.*); *Ksks* (♂) and *Bdon* (♀) of the Kalam people (Majnep and Bulmer 1977). *Kabale* of Huli people, Tari Valley, PNG; *Kis-a-kis*, *Kus-a-kus* or *Kos-A-kos* (♂ only—and doubtless derived from the wing-noise) and *Bon-dong* (♀) of the Schrader Mts region (Gilliard and LeCroy 1968). *Manglupuni* of Murray Pass, PNG (W. Cooper personal communication). *Ga-din* on Mt. Wilhelm, *Tong Lee* or *Omgi-nil* in Mung R., Kubor Mts area (specimen labels).

The cordilleran astrapia distinguished by its long black tail. ♂ is unmistakeable with the huge tail, short bill, and greenish head; ♀ is less striking but tail large and all-black unlike the other cordilleran astrapias; similar but long-billed Black and Brown Sicklebill ♀♀ can be mistaken for this sp. if poorly seen. The local abundance of Stephanie's × Ribbon-tailed Astrapia hybrids in west-central PNG complicates identification there. Polytypic; 2 subspp.

Description

ADULT ♂: 37 (84) cm, head similar to Ribbon-tail but lacking 'pom-pom' above bill base and intensely-iridescent crown feathers less yellowish green, much more Cobalt Blue (168) and/or violet purple (172) and Magenta (2), and more extensively onto nape and thence to sides of face. Upperparts generally like the Ribbon-tail but iridescence stronger and paler, more bronzed Lime Green (159). Upperwing darker than Ribbon-tail, being jet black throughout with stronger and richer iridescent violet purple and/or Magenta sheens on greater coverts, outer edges of flight feathers, and on tertials. Dorsal surface of tail: central pair velvety jet black with intensely-iridescent sheens of rich deep violet-purple and/or Magenta and conspicuous, white, central feather shafts for basal third of their length, the remaining tail blackish brown (219) with slightest coppery sheen. Uppertail coverts are not 'lost' (*contra* Goodfellow 1926*a*). Underparts generally like Ribbon-tail but chin to upper breast feathers finer, more silky and not tending to scale-like, and slightly more bluish green. Underwing: coverts black with trace of deepest blue iridescence, flight feathers blackish. Ventral surface of tail brownish black (119) but central pair glossy black. Bill shiny black, iris very dark brown, legs and feet purplish leaden grey, mouth pale or lime green. Note that rectrices other than the central pair much larger, broader and square-ended than in the Ribbon-tail.

ADULT ♀: 53 cm, generally similar to ad ♀ Ribbon-tail but with some significant differences. Iridescence on the dark brownish black head not intense and more deep blue than yellowish green. Mantle to uppertail coverts somewhat variably blackish brown and, to less extent, brownish black. Upperwing brownish black with indistinct, blackish brown, outer edging to coverts and flight feathers, primaries slightly paler. Dorsal surface of tail brownish black with white, basal, central feather shafts for a sixth of the length of the central pair. Chin to breast dark brownish black with only slightest trace of green-blue iridescent sheen grading briefly to Raw Umber (23) and then Cinnamon (123A) throughout with blackish barring (narrower than in ♀ Ribbon-tail and thus giving a paler appearance). Underwing as Ribbon-tail. Ventral surface of tail brownish black.

Stephanie's Astrapia *Astrapia stephaniae* (Finsch and Meyer, 1885)

SUBADULT ♂: as ad ♀ with few feathers of ad ♂ plumage intruding, to as ad ♂ with few feathers of ♀ plumage remaining.

IMMATURE ♂: as ad ♀ in first year but in some specimens with some rich rufous remaining in nape feathering. Birds then with iridescence as ad ♂ initially on forehead, crown, and chin and subsequently with entire head, throat, and upper breast with ad ♂ iridescence and thereafter more of ad ♂ plumage acquired. Writing of changes from ♀-type to ad ♂ plumage, Majnep (in Majnep and Bulmer 1977) states that first the head plumage changes, then the breast is replaced by the dark greenish blue breast feathers and lastly the long black tail grows; the complete change taking 2 years as only a partially ♂-coloured tail is acquired in the first year of mature ♂ plumage acquisition.

NESTLING: unknown.

JUVENILE: plumage soft and fluffy, black above with faintest blue cast with variable amount of deep rich rufous nape feathering that may extend onto rear crown. Underparts buff with blackish barring. Iris brown-grey.

Distribution

Easternmost of the cordilleran astrapias, ranging from the Owen Stanley Ra of SE PNG northwestward through the Eastern and Western Highlands of PNG into Enga and Southern Highlands Provinces (including Bismarck, Schrader, Kubor, Hagen, Tondon, and Doma Peaks Ra). Western terminus of range has not been well delineated, but in the N extends at least to the Central Ra between Wabag and Kompiam, and in the S to the mountains of the Tari Basin (Coates 1990). Hybridises extensively with Ribbon-tailed Astrapia from the Hagen/Giluwe Ra westward. Ranges from 1280 to 3500 m (mainly 1500–2800 m), with generally a lower mean elevational amplitude than that for the Ribbon-tail.

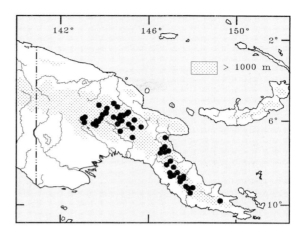

Systematics, nomenclature, subspecies, weights, and measurements

SYSTEMATICS: one of the three cordilleran astrapias, which we presume constitute members of a superspecies. Although there are superficial similarities between Stephanie's and Huon Astrapias, we do not consider that these two are more closely related than the three cordilleran spp. (*contra* Diamond 1986a). Poorly-defined subspeciation has occurred in Stephanie's Astrapia, with the slightly larger nominate form *A. s. stephaniae* occupying the Owen Stanley Ra of the SE and the smaller *A. s. feminina* the ranges to the NW. Forshaw (in Cooper and Forshaw 1977), Diamond (1972), and Cracraft (1992) questioned the validity of *A. s. ducalis*, Cracraft (1992) accepted *A. s. feminina* (into which he subsumed *A. s. ducalis*) and Coates (1990) doubted the justification of any subspp. We tentatively retain *feminina* only, given its disjunct range and morphological/biometrical distinctiveness. See remarks under Ribbon-tailed Astrapia. Hybridisation: known only with Ribbon-tailed Astrapia (see Plate 14 and Appendix 1).

SPECIES NOMENCLATURE AND ETYMOLOGY

Astrapia stephaniae (Finsch and Meyer)
Astrarchia stephaniae Finsch and Meyer, 1885.

Stephanie's Astrapia *Astrapia stephaniae* (Finsch and Meyer, 1885)

Zeitschrift für die gesammte Ornithologie 2, 378. Hufeisengebirge [Horseshoe Mt. = Mt. Maguli], southeastern New Guinea. Type specimen SMT C 8156 (8463), (lost during WW II). The sp. was discovered in the Owen Stanley Mts in 1884 by Carl Hunstein, who later drowned off New Britain.

Etymology: *stephaniae* = after Princess Stephanie of Belgium, wife of Crown Prince Rudolf (see Blue Bird) of Austria-Hungary.

SUBSPECIES, WEIGHTS, AND MEASUREMENTS

1. *A. s. stephaniae* (Finsch and Meyer, 1885).

Synonym: *A. s. ducalis* Mayr, 1931. *Mitteilungen aus dem Zoologischen Museum, Berlin* 17, 711. Dawong, Herzog Mts. Type specimen ZMB 30.1991. We follow the comments of Gilliard and LeCroy (1968) in subsuming *ducalis* into the nominate form. This also accords with the assessment of Cracraft (1992) that *ducalis* does not merit any independent taxonomic status.

Range: from the southeasternmost highlands of the Owen Stanley Ra (Mt. Dayman) northwestward to the central and southern ranges of the central cordillera as far as Mt. Giluwe, Mt. Hagen, Nipa, and Doma Peaks. Distribution includes Mt. Ialibu, Kubor Ra, Mt. Karimui, Mt. Michael, and the Kratke and Herzog Ra (Diamond 1972).

Measurements: wing, ad ♂ (n = 32) 160–182 (171), imm ♂ (n = 19) 153–179 (162), ad ♀ (n = 34) 145–165 (156); tail, ad ♂ (n = 27) 132–182 (154), imm ♂ (n = 18) 146–247 (203), ad ♀ (n = 30) 170–230 (201), tail centrals, ad ♂ (n = 20) 574–693 (640), imm ♂ (n = 16) 316–380 (352), ad ♀ (n = 30) 279–358 (328); bill, ad ♂ (n = 26) 34–41 (39), imm ♂ (n = 18) 36–42 (39), ad ♀ (n = 34) 36–42 (38); tarsus, ad ♂ (n = 30) 38–45 (42), imm ♂ (n = 19) 40–44 (42), ad ♀ (n = 34) 36–43 (40). Weight: ad ♂ (n = 4) 144–169 (160), imm ♂ (n = 1) 164, ad ♀ (n = 4) 139–159 (148).

2. *A. s. feminina* Neumann, 1922. *Verhandelungen der Ornithologischen Gesellschaft in Bayern* 15, 236. Schraderberg. Type specimen ZMB 7, 50.370.

Range: Schrader Ra, Bismarck Ra, and Sepik-Wahgi Divide, N of the range of the nominate subsp. (Diamond 1972; Coates 1990).

Diagnosis: like nominate form but less contrast between colour of crown/nape and the back and the crown and nape more bluish black in ad ♂♂. Ad ♀ wing/tail ratio lower than that of the nominate form, the wing being shorter, the tail longer.

Measurements: wing, ad ♂ (n = 26) 156–173 (166), imm ♂ (n = 25) 152–165 (159), ad ♀ (n = 39) 144–165 (152); tail, ad ♂ (n = 22) 107–185 (131), imm ♂ (n = 23) 148–225 (191), ad ♀ (n = 33) 151–217 (186); tail centrals, ad ♂ (n = 21) 565–727 (633), imm ♂ (n = 20) 246–467 (339), ad ♀ (n = 33) 268–362 (317); bill, ad ♂ (n = 26) 34–40 (37), imm ♂ (n = 25) 33–39 (36), ad ♀ (n = 37) 34–40 (37); tarsus, ad ♂ (n = 26) 38–44 (42), imm ♂ (n = 25) 38–47 (41), ad ♀ (n = 38) 37–43 (40). Weight: ad ♂ (n = 5) 145–156 (150), imm ♂ (n = 10) 130–164 (149), ad ♀ (n = 15) 123–148 (134).

Habitat and habits Table 4.1

Montane and subalpine forests, including forest-edge, selectively-logged forest, and forest disturbed in other ways mixed with secondary growth. Mostly frequents upper vegetation levels, less frequently the middle strata. Goodfellow (1926a) noted that birds tended to keep to the ridges and avoid valleys. Similar in flight to the Ribbon-tail—a number of rapid wing-beats followed by a downward glide resulting in a series of shallow undulations. Foraging birds are unobtrusive but are not particularly shy or wary when encountered. Both western ornithologists and PNG informants have noted that whilst ♀-plumaged birds quite often frequent forest-edges and gardens ad ♂♂ usually remain deeper in forests. Birds bathe at forest pools, where they may be killed from hides by local hunters (Majnep and Bulmer 1977).

Diet and foraging

Table 4.2

Consumes fruit, arthropods, and presumably small vertebrate prey (frogs, skinks). Fruit probably predominates in diet. Most commonly observed foraging for animal prey by searching limbs and moss in the middle and upper strata of montane forest.

Of five stomachs from Mt. Hagen specimens, one contained seeds, two fruits and seeds, one seeds and vegetable matter, and one only insect remains (Bulmer in Gilliard 1969). Shaw Mayer watched a bird foraging for arthropods in mosses upon tree limbs and saw it take what was probably a spider (in Sims 1956). An ad ♂ was observed moving about for 20 min on the crown limbs of a forest tree heavily covered with epiphytic plant growth, hopping agilely about larger horizontal or gently-sloping limbs, probing and excavating mosses with its bill. It also once flew to a nearby fruiting tree to eat small green fruits, 'crawling almost cat-like on tiny outer branches bearing the fruit' (Gilliard in Mayr and Gilliard 1954). Fruits of a *Pittosporum* sp. tree and *Schefflera chimbuensis* were consumed at Mt. Gahavisuka Park, PNG, 21 Aug 1984 (BB). A ♂ seen feeding upon *Schefflera* fruits 17 m above ground was displaced by a Great Cuckoo-Dove *Reinwardtoena reinwardtii* (G. Opit *in litt.*). A ♀-plumaged bird seen to chase an ad Short-tailed Paradigalla from a small fruiting tree and pursue it in flight on 9 Aug 1991 at Ambua Lodge, PNG (BB).

Stevens (in Greenway 1935) recorded 'pebbles, earth and seeds' in several birds' stomachs. In another five stomachs were fruits of 3–14 mm dia, in two fruit and a large hard insect and in another a 20-mm-long insect (Diamond 1972). Three birds from two locations contained 70–100% fruit and 0–30% arthropods (Chilopoda, Chrysomelidae, Clubionidae, Curculionidae, Muscidae, and larvae of Lepidoptera; Schodde 1976). Diet estimated to be 85% fruit and the remainder arthropods and other animal prey (Beehler and Pruett-Jones 1983). A captive ate a moth by holding it to a perch by a foot to tear its wings off with the bill (CF).

Of 475 instances of birds observed foraging on Mt. Giluwe, 298 (63%) were recorded on tree branches, 74 (16%) at fruits, 28 in foliage, 26 at twigs or buds, 23 at flowers, 17 on dead wood, 2 on the ground, and 7 undefined. Of 446 records of birds observed eating identifiable food, 76% were taking fruits and 24% animal prey (Kwapena 1985). Of the at least 34 different food plants, *Schefflera* spp. represented 45% of the 341 fruit-feeding records, indicating the great significance of these epiphytic plants to the diet (Kwapena 1985). Only a single instance of feeding upon the fruit of pandanus palms was observed by Kwapena, a resource much used by sympatric sicklebills (but see below). See Appendix 4 for food plants utilised.

Usually forages singly or in loosely-associated small groups of two or three, sometimes more, individuals. Simpson (1942) watched one and up to six birds feeding in a single pandanus tree, presumably upon the fruit. Birds may join mixed-spp. foraging flocks with Brown Sicklebills and several were once seen in company with five or six Crested Birds which foraged directly below the astrapias (Coates 1973a). A ♀-plumaged bird was insect foraging with a ♀-plumaged Brown Sicklebill on the same branch 3 m apart, the birds following each other as they moved about on 9 Mar 1975, at 2150 m on Mt. Kaindi, PNG (BB). Uses 'open-billed probing' in foraging (Harrison 1964). Two individuals were seen to fly at and displace an ad ♂ Ribbon-tailed Astrapia (M. Laska *in litt*). Laska also observed individuals displace two ♀-plumaged Ribbon-tails, an ad ♂ Brown Sicklebill, and an ad ♂ King of Saxony.

Of 403 sightings of birds taken to assess avain preference of forest strata, 36% were of birds 0–15 m above the ground, 36% 15–25 m, and 28% 25–35 m (Kwapena 1985).

Vocalisations and other sounds

Atypical of the polygynous court-displaying birds of paradise, the ♂ is not known for loud and far-carrying advertisement vocalisations. This (and other astrapias) are among the least vocal of the birds of paradise. Few calls noted. A captive ♂ gave a shrill *quee quee* and a ♀ an occasional weak cat-like *meow* (Gilliard 1969). A cat- or frog-like, upwardly-inflected, *whenh* once every 3–5 sec (Beehler *et al.* 1986, and BB) would seem to be the call alluded to by Gilliard (1969) and Diamond (1972). Calls of wild ♂♂ notated as *nge, nge, nge* and, more softly, *ss, ss, ssw, ssw* and of a ♀ a squeaking sound (Majnep and Bulmer 1977). Also rapid, drawn-out, harsh notes sometimes followed by *Melidectes* honeyeater-like *hoo-hee-hoo-hee* ..., given by several birds at once. Flight of ad ♂♂ produces a distinctive and clearly-audible wing rustle. The Kalam and Schrader Mts peoples' name for ad ♂♂ refer to the wing-flight noise.

Mating system
Polygyny, with promiscuous ♂♂ performing communal displays at leks of 2–5 birds, and ♀♀ performing nesting duties alone. During a year in the Bismarck Mts, PNG, Healey (1978*b*) located four lek sites, nearest neighbour distance *c*. 1.5–2.0 km, out of visual and auditory contact, on the crest or side of steep-sided ridges. A display site consisted of 4–7 trees *c*. 25 m tall with their bare main limbs forking from the trunks at *c*. 17–18 m above ground, one or two of which served as the main display tree(s). Display trees typically have long, straight, bare branches lacking foliage and epiphytes beneath the canopy and sloping up at a fairly steep angle from the trunk (Majnep and Bulmer 1977; CF, BB). If 5–6 ad ♂♂ arrive two or three of them perch at each end of a display limb to display and then change places from one end of the perch to the other (Majnep and Bulmer 1977).

Courtship behaviour
Healey (1978*b*) studied two leks, one of which stood in a partly-cleared and regenerating area and the other in (uncleared) forest interior. The former site had been known as a traditional display site for two decades and ad ♂♂ had been hunted for their plumes there throughout the period. Birds started to gather at a lek shortly after dawn (0600 hr), the number of ♂♂ averaging 3 (1–5, $n = 8$) during July–Sept. During the eight occasions when at least one ♂ was present at the lek an average of 1 (1–2) presumed ♀♀ were also present on five occasions, none joining the ♂♂ on three occasions. Having arrived at the lek, individual ♂♂ remained for *c*. 2 hr, during which they would leave only briefly in order to feed. Whilst present, ♂♂ perched quietly in separate trees, each occasionally calling and hopping through the branches of his tree or to an adjacent tree, or preening plumage. The display of one ♂ would stimulate an immediately-subsequent display by another, usually in the same tree, whilst other excited birds would call and hop about the perches of a tree or fly from tree to tree.

Two distinct levels of ♂ display intensity were definable. During both of these, modified feathering of the ♂♂s' head and breast was raised. In the case of a single high-intensity display, a ♂ raised a distinct naral tuft of feathers at each side of the upper mandible base, like that visible at all times on the ad ♂ Ribbon-tailed Astrapia. Low-intensity display was when ♂♂ hopped between two perches of *c*. 35 mm dia and on the same level, an intervening perch sometimes present and being touched by the bird's feet only momentarily in passing, but a pause occurred at each of the two main perches. If the displaying birds turned to lift the tail across the perch, before hopping in the other direction or not, was not noted. In this display the body was held near vertically and, at the end of each traverse between perches, the tail was swung forward in an inverted V under the perch. During the latter movements a flicking sound was audible, possibly made by rapidly and partially opening and closing wings or by the tail striking the intervening perch. This form of activity lasted up to 2 min and no other sounds were produced during it. Low-intensity display was observed six times, during Aug.

In high-intensity courtship a ♂, having flown to a display tree from its perching tree, hopped 9–10 times very rapidly between two

Stephanie's Astrapia *Astrapia stephaniae* (Finsch and Meyer, 1885)

perches c. 1.5 m apart during less than c. 4 sec, its tail streaming behind it rather than swinging below the perch as in low-intensity display. The body was held horizontally, with the shoulders hunched and the bill cocked slightly upward. This high-intensity display was seen only three times, during July.

High-intensity courtship occurred in the one tree, this being visited in turn by different ♂♂ from their individual traditional perching trees. Only once were two ad ♂♂ seen perching in the same tree. ♂♂ displayed both in the presence and absence of ♀-plumaged birds and the presence of the latter did not always result in display by ♂♂. Only what Healey described as the low-intensity display was performed in the presence of ♀-plumaged birds.

At 1020 hr on 20 Nov on Mt. Giluwe two mature ♂♂ and three ♀-plumaged birds were perched in a tree near a garden. 'The big mature ♂ was first seen jumping from branch to branch with its breast plate raised and the wings were spread slightly. The ♂ then chased one ♀ from branch to branch while the other ♂ and two ♀♀ were looking on. After 5 min of display, the other ♂ was chased away by the displaying ♂. Two other ♀♀ flew away. The displaying ♂ and the remaining ♀ were left alone. These two flew away later and sat on a nearby tree 15 m high. While perching on the same branch, a [the] ♂ repeated its display for another 10 min. This time the ♂ reached the courtship climax and mounted the ♀ for a couple of sec before dismounting. There was no fighting when copulation occurred' (Kwapena 1985).

Prior to mating, a pair captive in the USA were seen to fight viciously by grappling one another with their powerful feet by the thighs and breast. This could well be an artefact of captivity in view of subsequent knowledge (Boehm 1967; Everitt 1973). Boehm noted that after such a fight the ♂ 'will crest his hood and shield in a display' immediately prior to copulation. The 'hood' presumably refers to the modified ear coverts of ad ♂♂, first noted by Sharpe (1898) and by Goodfellow (1926a) as 'only seen, however, when the bird displays, and are then one of its most striking features'. These feathers are not obvious when birds are not performing.

Kwapena (1985) was informed that a Mt. Giluwe man found a pair of birds 'mating and that they had fallen down from a tree while still grasping each other by the thighs. He collected the pair in a string bag and brought them home'. The birds were possibly fighting and not mating. At 1636 hr on 10 Aug at a Tari Gap, PNG roadside two ad ♂♂ and three ♀-plumaged birds were involved in much aggression and flight chases, one ad ♂ flying a long distance to displace a ♀-plumaged bird near another ♀-plumaged individual by locking claws and the two tumbling 20 m to near the ground (BB).

A Mt. Hagen ♂ finished feeding and then flew to a horizontal limb 15 m high and beneath the forest canopy. It then perched across the limb 'with its tail hanging in a wide inverted V' to display by 'lifting the wings in a most peculiar manner as though stretching, so that the primaries were held up at right angles to the body. The head was pulled down in a crook. This position was held for at least three seconds'. The solitary silent bird then flew off (Gilliard 1969).

Breeding Table 6.5

NEST SITE: Kwapena (1985) records an active nest 10 m high in the forked branch of a young *Nothofagus* tree. An active nest on the Ambua Ra was c. 10 m high in a forked tree branch, supported and partly concealed by the tendrils and foliage of associated vines (CF). A nest near Wau was 3.8 m above ground in climbing bamboo in mature forest (T. Pratt *in litt.*).

NEST: a 'thick shallow structure of large leaves and creeper, lined with root fibres' (Bulmer in Gilliard 1969). External dimensions 191 mm dia and 89 mm deep. Internal dimensions 115 mm dia and 44 mm deep. Described as being an open structure like that of Lawes' Parotia and the Superb Bird (Majnep and Bulmer 1977). The nest collected by Kwapena was oval, measuring 160 mm × 100 mm and

40 mm deep. The inside cup was lined with rootlets, small leaves, and bark debris. The outside was of orchid and fern stems and leaves (like the Ribbon-tail nest). An active nest on 20 Nov 1973 at the Bulldog Road, Wau, PNG was large and bulky, and constructed predominantly of epiphytes (T. Pratt *in litt.*).

EGGS: a single egg in the BMNH, laid by a captive *A. s. stephaniae* in the UK, measures 42.9 × 28.5. A single egg collected in the Owen Stanley Mts by A. S. Anthony on an unknown date is smooth-surfaced and glossy, Pale Pinkish Buff (121D) with deep brown longitudinal dashes from the thick end downwards, a few spots on the thick end, and paler brownish red spots; measuring 36.5 × 25.4 mm. A single-egg clutch near Wau, PNG on 20 Nov 1973 measured 39 × 27 mm and weighed 15.4 g (T. Pratt *in litt.*). Two single-egg clutches thought to be of this sp. (Bishop and Frith 1979) were apparently laid by a captive *c.* 6 month-old ♀ resulting from hybridisation between a ♂ Ribbon-tail and a ♀ of another sp. (Shaw Mayer and Peckover 1991), presumably a Stephanie's Astrapia. These eggs measured 36.1 × 25.9 and 36.1 × 26.1 mm. Two single-egg Tari Valley clutches measured 36.1 × 25.9 and 36.1 × 26.1 mm (CF). All recorded clutches have been of a single egg.

INCUBATION: observation of captive breedings and limited evidence from wild nestings make it clear that the ♀ alone incubates, and will defend the nest and egg from a ♂ in the same aviary (Goodfellow 1926a; Yealland 1969; Boehm 1967; Everitt 1973). In captivity incubation for one nesting was 22 days.

NESTLING CARE: by ♀ alone, who regurgitates all nestling meals.

NESTLING DEVELOPMENT: a nestling resulting from a USA captive breeding left the nest at 26 days whilst one in a UK aviary did so at 27 days old (Yealland 1969; Everitt 1973).

Annual cycle Table 6.5

DISPLAY: ♂♂ display at all times of the year but dry season is the peak of courtship, although less so during the moult of *c.* June–Aug. Kwapena (1985) saw displays on Mt. Giluwe during Nov–Jan. BREEDING: at least May–Dec. Said by Healey's (1978b) informants near the Jimi R., PNG to be during Nov–Feb. Ad ♂♂ have been collected with moderately to much-enlarged gonads during each month of the year, but more so during Mar–Oct. ♀ gonadal activity has been noted in June–Aug and Oct–Dec. EGG-LAYING: June and Oct–Dec, the Mt. Giluwe nest with young that fledged *c.* 25 July (Kwapena 1985) indicating a laying date of *c.* 9 June, and the Ambua Ra fledging date of 18 Dec a laying date of *c.* 1 Nov (CF). MOULT: 72 of 174 specimens were moulting, involving all months but Dec, concentrated during Mar–Nov and particularly Mar–June (Appendix 3). According to informants of Healey (1978b) moult of ad ♂ tail feathers on the S Bismarck Mts takes place *c.* June–Aug, when display activity wanes. An ad ♂ well below Tari Gap on the road into the Tari Valley was in newly-acquired fresh plumage including a fully-grown tail on 12 Feb 1997 (CF).

Status and conservation Table 8.1

Common in some areas but sparse in others of apparently suitable habitat. Bulmer (in Majnep and Bulmer 1977) noted a decline in this sp. in the upper Kaironk Valley, PNG between the early 1960s and the 1970s which he thought may have reflected the pressure of hunting for the plume trade. Given the large range and elevational distribution of the sp., there is no reason to believe it is at risk in the near future.

Knowledge lacking and research priorities

Relatively well known. The westernmost extent of its range requires definition, especially in the N and S extremities of the central cordillera. Lek behaviour and mating require study.

Aviculture

Walter Goodfellow delivered seven live birds to E. J. Brook of Scotland in 1909 and three to London Zoo in the mid 1920s, the longest-living of the former shipment being 8 yr in captivity (Goodfellow 1926*a*). Goodfellow found the birds remarkably tame and confiding, taking insects from his fingers within half an hr of being captured. Several young reported to have been hatched in NG aviaries but subsequently eaten by other captive bird of paradise spp. (Hallstrom 1962). Bred in the UK and USA (see above).

Other

During 13 Aug–12 Oct 1973 a bird in ad-♂ plumage was seen several times (T. Pratt *in litt.*) to collect pieces of moss, some epiphytic orchids, and berries and to add them to a nest structure containing egg shell fragments. The significance of this peculiar observation is obscure. Was the bird a ♂, or a ♀ in aberrant ♂-like plumage (*cf.* Ribbon-tailed Astrapia account)?

Huon Astrapia *Astrapia rothschildi* Foerster, 1906

PLATE 6; FIGURE 9.21

Other names: Rothschild's Astrapia, Huon Bird of Paradise, Huon Astrapia Bird of Paradise, Rothschild's Long Tail, Lord Rothschild's Bird of Paradise. *Samun* (♂) and *Sunumaak, Samun kwawat* or *Kwawat* (♀) of the Nokopo people, Finisterre Ra, PNG (Schmid 1993).

An inconspicuous, little-known, stout-bodied astrapia with a very long, broad, and blunt-tipped tail, most similar to the Arfak Astrapia. The only astrapia on the Huon Penin; sympatric with Wahnes' Parotia, being generally similar but shorter-tailed. ♀-plumaged birds almost entirely sooty black save fine, inconspicuous, pale barring on abdomen and undertail coverts. Monotypic.

Description

ADULT ♂: 69 cm, entire head and nape black with strong green-blue iridescence, more so on fine, scale-like, crown feathers. Enlarged, plate-like, feathers of hindneck and mantle form a distinctive cowl, rich iridescent Dark Green (262), the exposed bases broadly-tipped rich iridescent Magenta (2), more coppery red against nape. Back blackish with rich Olive-Green (47) sheen, rump blackish. Upperwing black, with deep blue purple (172A) sheen in some lights. Dorsal surface of tail velvet-black with strong rich bluish purple to pinkish purple (1) in some lights. Breast like head and bordered below by strongly-iridescent, distinct, fine, coppery orange gorget. Lower breast and belly rich, silk-like, deep oily dark green (162A) with some larger plate-like feathers down breast sides strongly-iridescent paler, lime, green (159). Underwing blackish, with deep blue gloss on coverts. Thighs, vent, and undertail coverts matte brownish black. Ventral surface of tail glossy black. Bill shiny black, iris dark brown, legs and feet fleshy lead-grey.

ADULT ♀: 47 cm, unlike ad ♂ and with wing and tail significantly shorter. Entire upperparts Dark Greyish Brown (20), head and breast a darker blackish brown, often with paler broad nape band of finely tan (119B) tipped feathers, with slightest of dull, deep blue iridescent sheen. Dorsal surface of tail slightly paler, more brownish, than upper body. Upperwing like back, with paler, more brownish (119A) leading edges to coverts, primaries, and secondaries. Remaining underparts blackish (119) with fine off-whitish barring on lower breast to undertail coverts. Slightest of dull, deep blue, iridescent sheen to head and upper breast.

Odd individuals may have buff barring on the nape. Bare parts and iris as ♂.

SUBADULT ♂: as ad ♀ with few feathers of ad ♂ plumage intruding, to as ad ♂ with few feathers of ♀ plumage remaining. As birds age head becomes increasingly intensely iridescent and barring on underparts becomes obscured by increasing blackish plumage.

IMMATURE ♂: like ad ♀ but slightly larger, lacking pale nape barring and abdomen barring much reduced. Blue iridescence on head increasing and abdomen barring decreasing with age.

NESTLING–JUVENILE: unknown.

Distribution
Finisterre, Saruwaged, Rawlinson, and Cromwell Mts of the Huon Penin, 1460–3010 m. Most records from the eastern section of range, and only one sight record from the Finisterre Mts.

Systematics, nomenclature, subspecies, weights, and measurements
Systematics: On geographical grounds, it can be surmised that the predecessor population to this sp. 'budded' from the predecessor of what is now *A. stephaniae*. The plumage similarities between the two non-cordilleran isolates *A. nigra* and *A. rothschildi* are curious. One explanation is that this plumage may represent the ancestral state that has been lost in the cordilleran vicariants. Hybridisation: unrecorded.

SPECIES NOMENCLATURE AND ETYMOLOGY
Astrapia rothschildi Foerster, in Foerster and Rothschild, 1906. *Two New Birds of Paradise*, p. 2. Rawlinson Mts, northeastern New Guinea. Type specimen AMNH 678080.

Synonym: *Astrapia alboundata* Reichenow, 1918 *Journal für Ornithologie* **66**, 244. 'Probably the eastern part of Kaiser Wilhelmsland'. Type specimen not located.

Etymology: *rothschildi* = after Lord Lionel Walter Rothschild. Collected by Carl Wahnes in the Rawlinson Mts and sent to Lord Walter Rothschild's private museum at Tring, Hertfordshire, U.K. by Professor F. Foerster who named the sp. in an obscure pamphlet issued by the Tring Museum 1 Oct 1906.

WEIGHTS AND MEASUREMENTS
Measurements: wing, ad ♂ ($n = 25$) 182–194 (188), imm ♂ ($n = 4$) 158–176 (169), ad ♀ ($n = 21$) 152–180 (164); tail, ad ♂ ($n = 20$) 271–387 (352), imm ♂ ($n = 4$) 202–241 (225), ad ♀ ($n = 19$) 208–233 (220); tail centrals, ad ♂ ($n = 16$) 367–486 (443), imm ♂ ($n = 4$) 231–283 (255), ad ♀ ($n = 16$) 240–283 (256); bill, ad ♂ ($n = 24$) 37–41 (39), imm ♂ ($n = 4$) 38–40 (39), ad ♀ ($n = 21$) 37–41 (39); tarsus, ad ♂ ($n = 25$) 38–44 (42), imm ♂ ($n = 5$) 39–43 (41), ad ♀ ($n = 20$) 39–44 (40). Weight: ad ♂ ($n = 5$) 186–225 (207), imm ♂ ($n = 2$) 195–199 (197), ad ♀ ($n = 9$) 143–200 (159).

Habitat and habits   Table 4.1
Midmontane and upper montane forests. Flight is an undulating series of easy flaps followed by a glide with closed wings (T. Pratt *in litt.*).

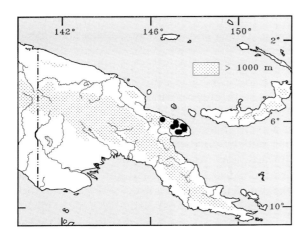

Diet and foraging
Table 4.2
Recorded eating arillate seeds of a *Pittosporum* sp. and fruits of *Schefflera* sp. once and the fruits of climbing pandanus *Freycinetia* sp. several times (e.g. M. LeCroy *in litt.*) including the use of a foot to hold fruit to the perch. See Appendix 4 for other food plants. Four or five ♀-plumaged birds once seen feeding together. Birds bite and tear at mosses on upper branch surfaces and twigs as small as 12 mm dia, and inspect and probe tree knot holes (T. Pratt *in litt.*). A bird was seen feeding upon berries of the introduced Canadian Elder shrub *Sambucus canadensis*, at 1600 m about Ogeranang village, PNG, during 28 June–2 July (Lambley 1990). In this general area, astrapias generally kept high in the canopy but also came low to feed (Draffan 1977).

Vocalisations and other sounds
Captives call little, giving only an occasional thin American Jay-like *kak, kak, kak, kak!* apparently as a call or an alarm note (Crandall 1932). Calls of wild birds said to resemble those of Stephanie's as is a muffled scolding *jj, jj* or *jiw, jiw*. One ad ♂ gave a 5-note call 'more melodious than a parotia scream but not the yelp of Stephanie's Astrapia' (T. Pratt *in litt.*). Ad ♂♂ make a clearly-audible, continuous, loud, sharp, dry, hissing, rattle sound in flight much like Stephanie's. An ad ♂ diving to chase a ♀-plumaged bird caused a loud whistling noise (T. Pratt *in litt.*).

Mating system
Unknown but, in view of what sister species do, presumably polygyny with solitary-displaying (but possibly lekking?) promiscuous ♂♂ and uniparental nesting ♀♀.

Courtship behaviour
Captives at NYZS performed two simple displays: an ad ♂ 'stood erect on the perch, his tail pushed forward beyond the perpendicular and slightly spread. The dark blue gorget or breast plate was widely spread and flattened, its fiery golden margin glowing conspicuously. The green feathers of the breast were also expanded laterally. The bird remained rigidly in this position for about 10 sec, making no sound.' A few days later, the same posture was seen, except that on this occasion the bird rapidly opened and closed the lateral tail feathers. This display was seen almost daily during 17 May–26 Aug after which a more complex display replaced it: 'The bird turned backward under the perch, at right angles to it, the body nearly horizontal, but the anterior portion slightly lower than the posterior. The head and neck were turned upward at one extremity, the tail at the other, so that the bird formed an approximate semicircle. The gorget and the feathers of the abdomen were widely spread, as before, the gold margin of the former being very conspicuous. The ear coverts were spread upward around the head, joining the elevated ruff on the nape. The wings were pressed tightly against the body, and the feathers of the back were expanded so as to partly cover them. The tail was held upright and widely expanded at first, than the lateral feathers were rapidly opened and closed, the middle pair remaining stationary. The display was enacted four times, at intervals of from 4–5 min, each period lasting from 10 to 15 sec' (Crandall 1932). At no time was any sound heard. Thane Pratt (*in litt.*) observed wild birds performing the latter display between 0650 and 0950 hr during 31 Jan–7 Feb. This inverted display was performed on various perches within different kinds of display saplings and trees and was interspersed by each solitary-displaying ♂ hopping back and forth between adjacent display tree perches. Display trees were located on the broad crest of a ridge and often at the edge of a windfall clearing. In all cases the ♂ dropped backward to adopt the inverted display posture, after some shaking and expanding of plumage. Once inverted, ♂♂ typically hold the head up with closed bill pointed skyward and the tail cocked to curve upward. The throat and upper breast feathers are fanned and/or flattened and appear to be expanded to join the raised, elongated, nape feathers to form a disk. Ad ♂♂ were seen to perform the typical inverted display up to six

times. On a couple of occasions the same ♂ fanned his upward-pointing tail during his inverted display (Fig. 9.21) and once rapidly and repeatedly opened and closed it (cf. Crandall 1932). One inverted ad ♂ swayed himself gently from side to side, and took one step to his left while in display (T. Pratt *in litt.*).

Breeding Table 6.5

NEST: two collected by missionary Christian Keysser during 1908–11 are similar except in size (mm), the smaller being 154 in external dia and 82 in depth and 95 internal dia and 40 deep (BMNH N 193.255), and the larger 204 in external dia and 74 in depth (BMNH N193.254). These nests are a firm shallow cup mostly of vines, rootlets, and creepers built onto a conspicuous foundation of large, strong, broad leaves and leaf pieces, leaf skeletons, and pieces of moss. There are also additional odd pieces of moss on the outside of the structure, mostly on the rim. The larger nest also had some fine hair-like rootlets as egg-cup lining (Frith 1971). Nests reported to be made of stems of a *Bulbophyllum* orchid known by Nokopo people as 'house of the Huon Astrapias' (Schmid 1993).

EGGS: three single-egg BMNH clutches average 35.3 (33.8–36.7) × 27.4 (26.3–28.1) mm with a ground colour of Pale Pinkish Buff (121D) or paler and more pinkish or with faintest purple grey wash. These are marked with blotches and (mostly) broad elongate 'brush-stroke-like' markings, predominantly at and around the larger end, of lavender grey overlaid with same of dark chestnut-browns.

INCUBATION, NESTLING CARE, AND DEVELOPMENT: unknown.

Annual cycle Table 6.5

DISPLAY: unknown other than during 31 Jan–7 Feb. BREEDING: a ♀ collected in each of Mar and Aug had enlarged gonads and in each of Oct and Nov were nesting with a clutch. Two ♂♂ had moderately-enlarged gonads in Mar and one in Oct. Nesting known during Oct–Nov. EGG-LAYING: Oct–Nov. MOULT: 22 of 44 specimens, involving all months except May and Sept, were moulting during Jan–July (Appendix 3). In NYZS a captive ad ♂ started moult on *c.* 1 Jan and finished *c.* 1 May and a second bird moulted during *c.* 15 Jan–30 May (Crandall 1932).

Status and conservation Table 8.1

Common in forest at Ogeranang, PNG at 1600 m (Draffan 1977). Skins and single tail feathers of ad ♂♂ remain treasured items for personal ornament by Nokopo people (Schmid 1993). Given the intact status of most of the montane forests in the Huon Penin, we assume this sp. is under no threat.

Knowledge lacking and research priorities

This sp. is no better known than the Arfak Astrapia and thus urgently merits a thorough survey of range, habitat, and status. Details of mating organisation are lacking. Does this sp. display in a lek or do ♂♂ attend solitary courts? Diet merits study, to determine the degree to which it differs from the cordilleran spp. A molecular systematic study of the five astrapias might prove enlightening.

Aviculture

Birds have been maintained in captivity at BRS, PNG, and the NYZS but have not bred.

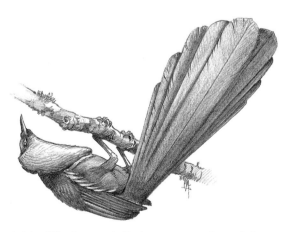

9.21 The inverted display posture of an ad ♂ Huon Astrapia: tail fanned and breast-shield and ear covert erected. After field sketch by T. K. Pratt (*in litt*).

Genus *Parotia* Vieillot, 1816

Parotia Vieillot, 1816. *Analyse d'une Nouvelle Ornithologie Élémentaire* p. 35. Type, by monotypy, 'Sifilet, Buffon' = *P. sefilata* Pennant in Forster.

The parotias are **four species** that inhabit the midmontane forests of mainland NG. They form a closely-allied group, and might be considered a superspecies, but for the considerable plumage divergence of *P. carolae*. *Parotia* has skull and bill characters most like those of *Cicinnurus* and *Pteridophora* (Bock 1963). Some characteristic plumage and courtship display traits indicate a sister relationship to *Pteridophora*. Intergeneric **hybridisation** known with *Paradigalla, Lophorina*, and *Paradisaea*.

♂♂ c. 10% larger than ♀♀. **Bill** half to slightly more than half the head length and relatively fine and short (c. 23% of wing length). **Tail** grossly (*wahnesi, sefilata*), slightly (*lawesii*) to barely (*carolae*) graduated; in former two spp. central tail feather pair longest and in latter two spp. these are as long as (in ad ♂) or shorter than (♀-plumaged) the rest. **Wing** moderately long and rounded. Two outermost primaries greatly emarginated on inner vanes and P9 sharply attenuated. Mean **tarsus** length is longer proportionate to mean wing length (32%) than in all other Paradisaeidae except the cnemophilines and *Macgregoria*. Distal portion of tarsus scuted, less strongly proximally.

Ad ♂♂ generally velvet-black with a narrow **nuchal bar** of scale-like iridescent feathers and three grossly-elongated, wire-like, **occipital plume** shafts, terminally tipped with small ovate 'flags' of black vane, arising from above and behind each eye (Bleiweiss 1987). Also complex silver- or golden-bronze tipped **naral tufts** and/or **frontal crests** of erectile feathers and a compact **breast-shield** of large, scale-like, highly-iridescent feathers. Females cryptically plumed in drab browns above with blackish head and paler underparts regularly barred blackish. Iris in both sexes blue and yellow except Carola's in which it is yellow in both sexes (Plate 7).

Polygynous, with promiscuous terrestrial court-display and solitary to exploded **lekking** ♂♂; the ♀♀ alone conduct nesting duties. **Nest** is a bulky open cup placed in a tree fork and typical of the Paradisaeinae as are the **eggs** (Plate 13).

Western Parotia *Parotia sefilata* (Pennant, 1781)

PLATE 7; FIGURES 9.22, 23

Other names: Arfak Six-wired Bird of Paradise, Arfak Six-wired Parotia, Arfak Parotia, Arfak Six-plumed Bird of Paradise, Greater Six-plumed Bird of Paradise. *Kourangen* or *Koorangan* of Biak plume traders and *Amde* of Karoon people (Mayr and Meyer de Schauensee 1939a) but Bergman (in Gyldenstolpe 1955b) gives *Kourang* as Arfak Mts name and D'Albertis (1880) gives *Niedda* or *Coran-a* of Hatam, Arfak Mts. *Na Day* on W. Tamrau Mts, IJ. (specimen label).

Endemic to the western peninsular region of NG, the Western Parotia is typical of the genus but for the relatively-elongated tail of the ♂ and the olive-brown back, wings, and tail of the ♀. Monotypic.

Description

ADULT ♂: 33 cm, entirely velvety jet black. Head decorated by an erectile, triangular, frontal crest of elongate, finely-pointed, silver feathers atop the forecrown and a broad nuchal bar of intensely-iridescent scale-like feathers rich blue anteriorly and rich deep violet purple (172) to red purple (1) with Magenta (2) hues posteriorly in suitable light. Crown between these head adornments and feathers in front of the silver frontal crest and on sides of face, with rich, dark coppery bronze, iridescent sheen in some lights. From behind each eye grows, amid an 'ear tuft' of elongate narrowly-pointed feathers, three long, erectile, wire-like, bare, occipital plumes tipped with near-circular 'flags' of normal feather vanes which, when the bird is perched at rest, extend posteriorly down over the tertials below the lower mantle margin. Mantle to dorsal surface of tail, including entire exposed wing, with dull, but rich, coppery bronze, iridescent sheen in some lights. Inner, concealed, vanes of wing and tail flight feathers jet black. Velvety blackish chin and throat with slightest iridescent sheen of coppery bronze to rich purple in some lights, this grading into an otherwise discrete breast-shield of large scale-like feathers intensely iridescent as in Wahnes' Parotia. Remaining underparts, including ventral surface of wing and tail, as in Wahnes'. Bill shiny black, iris Cobalt Blue (168) with pale yellow outer ring, legs and feet purplish lead grey, mouth greenish yellow.

ADULT ♀: 30 cm, head blackish brown (19) with slight lustre of deep chestnut, with broad pale greyish malar stripe flecked blackish brown and the slightest of a superciliary stripe of a line of tiny, pale greyish buff, central feather streaks. Mantle to uppertail coverts rich reddish brown (223A), upperwing and tail slightly darker and browner Raw Umber (223) with indistinct, narrow, reddish brown, outer edges. Chin and submalar area dusky Fuscous (21), throat pale greyish flecked Fuscous grading into remaining underparts of pale greyish buff to warm pale buff (223D) uniformly barred dark grey-brown (119A). Underwing: coverts as underparts, flight feathers dark brown (121) with pale buff trailing edges to all but outermost two primaries. Ventral surface of tail Raw Umber. Iris blue with yellow outer ring (Hartert 1930).

SUBADULT ♂: unknown but presumably as other parotias.

IMMATURE ♂: as ad ♀. The shape of the outer primaries in the imm ♂ is almost normal but in a bird in ♀-plumage save for one iridescent nape feather, a few white streaks in the forehead, and velvety feathers on the occiput, the outer two primaries are shaped as in ad ♂ (Mayr and Meyer de Schauensee 1939a).

NESTLING: unknown. JUVENILE: young 'in first plumage' have rufous red outer and inner edging to the 'quills and upper wing-coverts' (Hartert 1930).

Distribution
The midmontane forests of the Vogelkop and Wandammen Penin (Tamrau, Arfak, and Wondiwoi Mts), 1100–1900 m (B. Poulson personal communication).

Systematics, nomenclature, subspecies, weights, and measurements
SYSTEMATICS: the westernmost geographical representative of *Parotia*, and certainly a sister form to the other all-black parotias, although its relationship to *P. carolae* is uncertain (*contra* Diamond 1986a). Hybridisation: known with Long-tailed Paradigalla and Superb Bird (Plate 14, Appendix 1).

SPECIES NOMENCLATURE AND ETYMOLOGY
Parotia sefilata (Pennant)

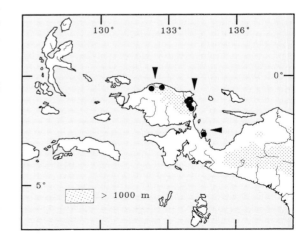

Western Parotia *Parotia sefilata* (Pennant, 1781)

Paradisea sefilata Pennant, 1781. *Specimen Faunulae Indicae*, in Forster's *Zoologia Indica selecta*, p. 40. New Guinea, restricted to Arfak Mts. Type is an illustration.

Synonym: *Paradisea sexipennis* Boddaert, 1783. *Table Planches Enluminées* p. 38.

Etymology: *Parotia* = Gr. *parotis*, a lock or curl of hair by the ear; *sefilata* = L. *sex fili*, six threads or filaments; *atus*, possessing.

WEIGHTS AND MEASUREMENTS

Measurements: wing, ad ♂ (n = 32) 161–170 (166), imm ♂ (n = 17) 154–167 (162), ad ♀ (n = 26) 141–163 (152); tail, ad ♂ (n = 29) 125–137 (130), imm ♂ (n = 17) 126–147 (136), ad ♀ (n = 24) 122–134 (129); bill, ad ♂ (n = 31) 30–40 (35), imm ♂ (n = 16) 35–38 (37), ad ♀ (n = 25) 30–37 (35); tarsus, ad ♂ (n = 31) 50–57 (53), imm ♂ (n = 17) 50–56 (53), ad ♀ (n = 24) 44–54 (48). Weight: ad ♂ (n = 12) 175–205 (192), imm ♂ (n = 5) 160–200 (183), ad ♀ (n = 14) 140–185 (172). Central pair of rectrices average 2 mm longer than mean remaining tail length.

Habitat and habits   Table 4.1

Midmontane forests with a possible preference for well-established secondary forest with an abundance of small saplings and a fairly open canopy (R. Kirby *in litt.*). Two ad ♂♂ and three ♀-plumaged birds reported joining a mixed-spp. foraging flock including the Black-fronted White-eye *Zosterops atrifrons*, a scrub-wren *Sericornis* sp., Black Fantail *Rhipidura atra*, Vogelkop Whistler *Pachycephala meyeri*, Dwarf Whistler *Pachycare flavogrisea*, Mountain Drongo *Chaetorhynchus papuensis*, and Western Smoky Honeyeater *Melipotes gymnops*. Birds were seen to probe bark and epiphytes on the underside of branches, probably in search of arthropods (B. Poulson personal communication).

Diet and foraging   Table 4.2

D'Albertis found fruits of a nutmeg *Myristica* sp. and a fig *Ficus* sp. in the stomach of one ♂. Stresemann (1931) reports a ♂ holding food to a perch to take pieces from it. Cinnamon Ground-Doves *Gallicolumba rufigula* and Pheasant Pigeons *Otidiphaps nobilis* have been seen repeatedly visiting courts to feed upon fruit seeds voided by the parotias.

Vocalisations and other sounds   Figure 9.22

A very harsh *gned gned* (D'Albertis 1880) or *gnaad gnaad* and cockatoo-like notes (Beehler *et al.* 1986). ♂♂ at their courts give loud squawking notes but ♀-plumaged birds only very quiet, high-pitched, mewing sounds (R. Kirby *in litt.*). Of captives in the USA, only ♂♂ produced a harsh single squawk that may be repeated so rapidly that, except in quality, it somewhat resembled the full call of a *Paradisaea* (Crandall 1932). Bergman (1957*b,c* and in Gyldenstolpe 1955*b*) described the harsh call as 'cawing' very similar to that of the European Jay *Garrulus glandarius* (Corvidae) which is probably the harsh nasal screech *wengh* shown in Fig. 9.22.

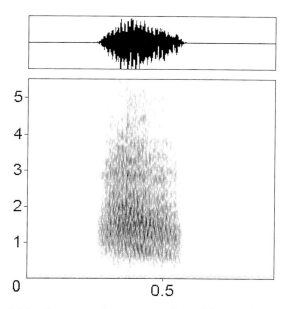

9.22 Sonogram from a recording of the nasal screech *waugh* or *wengh* court advertisement note of an ad Western Parotia at Arfak Mts in Oct 1994 by R. Kirby for the BBC.

Western Parotia *Parotia sefilata* (Pennant, 1781)

Mating system

Clearly, as other parotias, polygyny with terrestrial court-maintaining ♂♂ in all probability being promiscuous and ♀♀ uniparental in nest attendance (see Lawes' Parotia).

Courtship behaviour

Ripley (in Mayr and Meyer de Schauensee 1939*a*) described several courts in the Tamrau Mts at 1524 m as consisting of clearings on the forest floor 1.5 m in dia that were 'apparently jealously guarded and frequently cleared of falling twings or leaves'. Branches directly above these courts were so well used as to be worn and stunted. Bergman (1958) noted courts were 'generally less than one square metre in diameter [*sic*] and of unspecific shape' in dense forest and usually overshadowed by large trees. He noted that an essential prerequisite for a court site was 'one or two horizontal branches' over the cleared court. Other courts have been noted as *c.* 2 m dia and to extend upward to include vines and saplings directly above the cleared ground to a height of *c.* 3–4 m. These are stripped of their leaves to provide a range of suitable clean perches on which birds interact (R. Kirby *in litt.*).

Wild birds were recently filmed and observed by Richard Kirby (*in litt.* and in Attenborough 1996) during Sept–Oct. The ad ♂ at the most-observed court frequently tolerated 3–4 ♂ conspecifics on his court perches at once, but none of them on the court floor. All of these perched birds might be in ♀ plumage (and be ♀♀?) or some of them in subad ♂ plumage. When the ad ♂ flew up to his perches these visiting birds quickly made way for him except for a soliciting ♀. ♂♂ displayed at their court between 0630 and 0810 and again between 1300 and 1500 hr. The focal ♂ spent the first hour of daylight flying about the forest canopy, rarely far from his court, advertising with loud squawking calls and appearing to 'shepherd' several ♀-plumaged birds towards his court by chases progressively lower through the canopy and sub-canopy. During this activity the ♀-plumaged birds produced only soft mewing notes. The impression was gained that once several ♀-plumaged birds had been attracted the ♂ starts to display. A summary outline of his behaviour follows.

Courtship display is started by the ♂ on a horizontal perch *c.* 60 cm above the edge of the court. Here he hops sideways back and forth along the perch while he repeatedly flicks his wings half open and fans his tail (in doublets). He stops now and then to wipe his beak on the perch. He then hops down onto the court and continues flicking his wings, fanning his tail, and begins pecking at the ground in such a way as to flash his silver-white naral tuft prior to a little hop forward. The behaviour looks like ritualised pseudo-court-clearing and lasts *c.* 20–30 sec and is followed by a brief wing preen, a squawk, and often with departure from the court unless followed by a hopping display. The latter involves the ♂ adopting a stiff horizontal posture and hopping in a bouncing gait directly across the entire court, usually in a pattern of two or three long hops followed by four short ones, and back again. During this repeated traversing of the court the closed wings are sometimes flicked half out with each hop. Occasionally he adopts a brief, rigid *Upright Sleeked Pose* (Fig. 9.23(a)). He may hop back and forth a dozen times, passing beneath the ♀♀ perched above, before stopping in the centre of the court to preen briefly before starting the 'ballerina' dance.

Again adopting a stiff erect posture he faces the ♀ or ♀♀ and directs a slow deep *Initial Display Bow* (Fig. 9.23 (b)) to his audience. He then rises up fully with tail cocked to one side, elongate flank plumes erected to form a 'skirt' or 'tutu' about him and the six occipital plumes raised and brought forward to near-horizontal in front while shaking the head rapidly side-to-side. This is the *Ballerina Pose* (Fig. 9.23(c)). He then moves alternately sideways, often in a shallow semi-circle, side-to-side in front of and below the perched audience by side-stepping in a mincing gait, increasing the tempo and semi-circular nature of the dance after the first few seconds. Positioning himself directly beneath and just in front of the ♀ he then briefly stands still with feet wide apart as the occipital plumes are

Western Parotia *Parotia sefilata* (Pennant, 1781)

raised to *c.* 45° and widely separated. He then slightly lowers his skirt (as in closing an umbrella) and slowly raises it to near-horizontal with the ground before suddenly snapping himself downward to crouch on horizontal tarsi and into intense and extensive head and neck waggling (Fig. 9.23(d)) while raising himself up onto fully-extended legs. He then makes a couple of large hops to each side during which his head, with now continuously forward-pointing bill, is pumped repeatedly and sharply down into his neck/clavicle—this action repeatedly flicking his breast-shield up and down. He then flies directly up and onto the ♀ to mate for 2–3 sec before both birds fly off.

At any time during this display the ♂ may stop dancing to attempt copulation with a ♀ by approaching her with wing flicking or hopping directly onto her rump. A ♀ will solicit by squatting low on the perch with beak agape and gentle wing fluttering. During a copulation of 2–3 sec he holds her neck feathers in his bill and then squawks loudly before flying off as all other birds present fly off. Filmed display sequences

9.23 Some courtship display postures of ad ♂ Western Parotia: (a) initial *Upright Sleeked Pose*, (b) *Initial Display Bow*, (c) *Ballerina Pose*, (d) head and neck waggling. See text.

were always the same although a ♂ may omit one or more elements. Kirby once saw three copulations at one court in one morning. In addition, young ♂♂ may appear to attempt to copulate with a perch after a dancing display.

Italian explorer and naturalist D'Albertis first witnessed an adult ♂ in display in the Arfak Mts in Sept 1872. German ornithologists E. Stresemann (1931) and W. Johnas (1932) observed captive ♂♂ displaying at London and Berlin Zoo, respectively. All behaviour included in the descriptions and illustrations of these three men are described and illustrated here.

From a hide in the Arfak Mts, Swedish ornithologist Sten Bergman (in Gyldenstolpe 1955*b*) watched an ad ♂ display at a court on 8 Aug and witnessed some of the above behaviour. He noted that he never saw more than one ♂ in full ad plumage, but had sometimes seen ♀-plumaged birds, and that he had the impression each court was owned by a single ad ♂ who kept a more or less continuous watch on the place all day. He noted that most display took place during morning and late afternoon. Bergman subsequently (1958) wrote much about the displays as a result of maintaining six captive ♂♂, adding significantly to his observations of displays in the wild but nothing not documented in Kirby's films. Occasionally a bird would go directly into the *Ballerina Pose* for a brief display, omitting the *Upright Sleeked Pose* and *Initial Display Bow*. Birds also displayed on perches, by performing the *Upright Sleeked Pose* and the *Initial Display Bow* and performing any dance steps along the perch. In such displays birds tended, however, to remain static on the perch but performed all other body movements of the display. He did note that his birds nearly always gave a harsh call at the end of a display.

The choreography of ♂ display is less complex than that of Lawes' Parotia and does not involve the back-forward stepping movements across the court performed by the latter sp.

Breeding   Table 6.5
NEST SITE, NEST, EGGS, INCUBATION, NESTLING CARE AND DEVELOPMENT: **unknown**.

Annual cycle   Table 6.5
DISPLAY: ♂♂ seen displaying in the wild in early Aug–late Oct (D'Albertis 1880; Bergman in Gyldenstolpe 1955*b*; Kirby *in litt.*). Bergman's (1958) captives in Sweden displayed all year save during their 4-month moult. BREEDING: ♂ gonad activity indicates breeding during *c.* July–Jan but data are few. MOULT: 11 of 64 specimens, involving all months but Oct–Dec, were moulting during Jan and Apr–July but most predominantly during the latter period (Appendix 3). Six ad ♂♂ collected during 15 July–20 Aug in the Arfak Mts were in fresh plumage (Gyldenstolpe 1955*b*). An ad ♂ moulted during *c.* 4 months, 5 May–6 Sept, in NYZS (Crandall 1932). Bergman (1958) also noted a 4-month (summer) moult in his six captive ♂♂ in Sweden.

Status and conservation   Table 8.1
Common in the S Arfak Mts, at Ditschi and Siwi, and more so at Wondiwoi on the Wandammen Peninsula (Gyldenstolpe 1955*b*; Mayr in Hartert 1930). Common in Arfak Mts during mid Aug 1995 (M. Carter *in litt*). In spite of its restricted distribution, there is no reason to suspect this sp. is currently vulnerable, as there is ample upland forest habitat available.

Knowledge lacking and research priorities
Data on nesting biology, court dispersion, and diet are needed, although these might be expected to approximate closely those for Lawes' Parotia.

Aviculture
Birds have been maintained in Europe (UK and Sweden) and USA.

Other
D'Albertis (1880) ate the flesh of four birds in one meal without remarking on their palatability, good or bad, which is noteworthy in view of other birds of paradise proving inedibly bitter (see end of Chapter 3).

Lawes' Parotia *Parotia lawesii* Ramsay, 1885

PLATE 7; FIGURES 1.4, 4.4, 5.2, 9.24–27

Other names: Lawes' Six-wired Bird of Paradise, Lawes' Six-wired Parotia, Lawes's Six-plumed Bird of Paradise, Helena's Parotia *Parotia helenae*. *Kiara* of the Fore and Gimi people and *papa* of the Daribi people (Diamond 1972). *Keaga* (♂) and *Agule* (♀) of the Gimi people of Crater Mt., PNG (D. Gillison *in litt.*). *Kabay wog-dep* or *Kabay pok* of the Kalam people (Majnep and Bulmer 1977). *Kandi* (Goldman 1981), *Kandi Ayu* or *Guru Kendi* of the Huli people (C. Bollard *in litt*). *Manisau* of the Waria Valley and *Pabuga* of Kubor Ra, PNG (W. Cooper *in litt.*).

The parotia of E NG's central cordillera; the ♂ is all-black, chunky, broad-winged, and short-tailed, foraging in the forest canopy but dancing at a terrestrial display court. ♀♀ have the typical paradisaeinine plumage pattern but with a distinctive black hood and contrasting paler malar streak. Polytypic; 2 subspp.

Description

ADULT ♂: 27 cm, entirely velvety jet black. Head decorated by a silver, erectile, naral tuft atop the basal half of the maxilla, and immediately behind this a frontal crest of deep coppery brown feathers that can be raised between the silver naral tuft feathering, and a narrow nuchal bar of intensely-iridescent, deep blue to pink-purple, scale-like feathers. From behind each eye, amid an 'ear tuft' of elongate, narrowly-pointed feathers, grow three long, erectile, wire-like, black, occipital plumes with roughly circular 'flag' tips. Mantle and back feathering with dull and silk-like sheen that may appear coppery-bronze and/or deep green (46) in some lights. Remaining upperparts velvet jet black with deep coppery-bronze sheen in some lights; primaries paler, more brownish black (19) but exposed leading edges dull iridescent green (146) in some lights. Chin and throat velvety jet black with iridescent sheen of rich Purple (1) in some lights, this grading into an otherwise discrete breast-shield of large scale-like feathers intensely-iridescent, bronzed metallic emerald green to greenish yellow and, in some lights, Purple, Magenta (2) to Bluish Violet (172B) with jet black feather bases visible at lower sides of the shield. Remaining underparts, including elongate flank plumes, jet black with deepest coppery sheen in some lights. Underwing: coverts brownish black (119), flight feathers and ventral surface of tail blackish brown (219). Bill shiny black, iris Cobalt Blue (168) with narrow, cream-yellow, outer ring. Legs and feet purplish lead grey, mouth lime yellow to lime green. Most remarkable about the iris is that the bird apparently can alter the inner and outer colour, shifting the eye from mostly blue to mostly yellow (BB).

ADULT ♀: 25 cm, head brownish black, more brownish on nape, with a short indistinct malar stripe of dirty buff spotting. Some ♀♀ with an indistinct patch of deep chestnut on the nape (as with some astrapias). Mantle, back, rump, and uppertail coverts rich chestnut brown (223A) and of the upperwing the lesser coverts and tertials just a little darker while the remainder darker still (219), but the majority of greater coverts and flight feathers with rich chestnut brown leading edges. Dorsal surface of tail as wing flight feathers. Chin and throat dirty buff (223D) grading to rich brown cinnamon (121C) on remaining underparts, darker on thighs and undertail coverts, all strongly and uniformly barred brownish black. Underwing: coverts brown cinnamon, flight feathers and undertail blackish brown (219). Bill brownish black, eye, legs, and feet as ad ♂.

SUBADULT ♂: as ad ♀ with few ad ♂ feathers intruding, to as ad ♂ with few feathers of ♀ plumage remaining. Ad head plumage is attained first, wings and tail are similar to those of ad and breast-shield has a coppery sheen rather than the clearer green of fully-plumaged ♂♂ (Mayr and Rand 1937).

IMMATURE ♂: as ad ♀ but iris said to be duller, more greyish to brownish than blue, in younger birds. Imm ♀♀ lack indentation on the second primary, but tail feathers not pointed (Greenway 1935).

NESTLING–JUVENILE: 5–6-day-old nestling naked with bluish black skin and some sparse down feathers (CF)

Distribution

The eastern third of NG's central cordillera, from the far SE Owen Stanley Ra northwestward to the Bismarck, Schrader, Hagen, and Giluwe Ra, and to Nipa and Tari. The western extent of distribution is poorly delimited, but there is a Bishop Museum specimen from Oksapmin, just W of the Strickland R. Shares the westernmost section of its range with Carola's Parotia (Mt. Giluwe, Sepik–Waghi Divide, Kubor Ra, Crater Mt.). Recorded from 500 to 2300 m (mainly 1200–1900 m) (Coates 1990).

Systematics, nomenclature, subspecies, weights, and measurements

SYSTEMATICS: this sp., *P. sefilata*, and *P. wahnesi* form a rather compact superspecies. The population *helenae* of far SE PNG is treated by some authorities as a full sp. (Schodde and McKean 1972, 1973). Hybridisation: known with Blue Bird of Paradise only (Christidis and Schodde 1993; Frith and Frith 1996a). It is surprising that no hybrids have been found between *lawesii* and *carolae*, the only two sympatric parotia spp. See Plate 14 and Appendix 1.

SPECIES NOMENCLATURE AND ETYMOLOGY

Parotia lawesii Ramsay, 1885. *Proceedings Linnaean Society New South Wales* **10**, 243. Astrolabe Mts—but subsequently considered to be the Aruma Apa-Maguli Ra of the Owen Stanley Ra (Schodde and McKean 1973).

Exploration: Carl Hunstein was the first western collector to obtain scientific specimens, of both sexes, from Mt. Maguli inland of Port Moresby.

Etymology: *lawesii* = after the British missionary Rev. William G. Lawes who apparently assisted in the acquisition of trade skins of this sp.

SUBSPECIES, WEIGHTS, AND MEASUREMENTS

1. *P. l. lawesii* Ramsay, 1885. Type specimen AM B6386.

Synonyms: *P. l. exhibita* Iredale, 1948. *Australian Zoologist* **11**, 162. Hoiyeria, Mt. Hagen district. Syntypes AM O.38563-64. (Diamond 1972; Schodde and McKean 1973).

P. l. fuscior Greenway, 1934. *Proceedings of the New England Zoological Club* **14**, 2. Mt Missim, Morobe district. Type specimen MCZ 167,002. (Schodde and Mason 1974; Cracraft 1992).

Range: western and southern highlands of PNG southeastward into peninsular PNG and including all but the range of *helenae*.

Measurements: wing, ad ♂ (n = 45) 148–163 (155), imm ♂ (n = 20) 144–157 (152), ad ♀ (n = 35) 141–159 (147); tail, ad ♂ (n = 44) 73–84 (80), imm ♂ (n = 21) 91–104 (99), ad ♀ (n = 34) 92–103 (98); bill, ad ♂ (n = 43) 29–37 (33), imm ♂ (n = 19) 27–35 (33), ad ♀ (n = 35) 26–36 (32); tarsus, ad ♂ (n = 42) 46–53 (50), imm ♂ (n = 19) 45–54 (50), ad ♀ (n = 35) 40–53 (46).

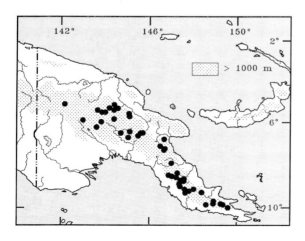

Weight: ad ♂ (*n* = 48) 153–195 (167), imm ♂ (*n* = 18) 151–180 (166), ad ♀ (*n* = 60) 122–180 (144).

2. *P. l. helenae* De Vis, 1897. *Ibis* p. 390. Neneba, upper Mambare River, north of Mt. Scratchley. Syntypes QM O. 19684–86, O. 20698.

Range: N watershed of peninsular PNG from the Waria SE to Milne Bay and perhaps continuing around onto the S watershed in the far SE (specimen data lacking).

Diagnosis: frontal crest of ad ♂ is bronzed brown, not silvery white, and ♀ dorsal plumage slightly less russet, but all measurements of both sexes near identical to nominate form. Originally described as a sp. and then subsequently long considered a subsp. of *P. lawesii* until Schodde and McKean (1973), Cooper and Forshaw (1977), and Cracraft (1992) resurrected it as a sp. We concur with Gilliard (1969), Diamond (1972), Beehler and Finch (1985), Beehler *et al.* (1986), and Coates (1990), however, in treating the taxon as a subsp. of *P. lawesii*.

Measurements: wing, ad ♂ (*n* = 23) 148–160 (154), imm ♂ (*n* = 9) 147–154 (150), ad ♀ (*n* = 14) 143–156 (148); tail, ad ♂ (*n* = 23) 78–84 (80), imm ♂ (*n* = 9) 92–103 (98), ad ♀ (*n* = 14) 14) 95–107 (99); bill, ad ♂ (*n* = 22) 29–39 (34), imm ♂ (*n* = 9) 30–36 (34), ad ♀ (*n* = 13) 34–36 (35); tarsus, ad ♂ (*n* = 22) 46–53 (49), imm ♂ (*n* = 8) 47–51 (50). Weight: ad ♂ (*n* = 5) 162–177 (170), imm ♂ (*n* = 2) 160–173 (167), ad ♀ (*n* = 5) 135–158 (149). All described subspp. are consistently similar in all measurements.

Habitat and habits Table 4.1

Midmontane forests including primary mixed oak forests, disturbed forest, secondary growth, and large to small remnant forest patches within areas of extensive village gardens; not common at forest-edges. Flight swift and buoyant, consisting of *c*. 4 wingbeats followed by a short glide before the next wing-beat. Rounded wings and short tail give a distinctive silhouette.

Diet and foraging Table 4.2

Diet mainly fruits with small quantities of arthropods. Smyth (1970) often saw birds feeding in fruit-bearing trees near gardens in timbered areas about Tari, PNG. Stomach contents of eight birds from two locations suggested a diet of 95–100% fruit, the arthropod groups represented being Chilopoda, Curculionidae and Dermaptera, and there was also the remains of a skink (Schodde 1976). Analysis of > 150 faecal samples collected at courts over a year indicated the diet to be 95% fruit (Beehler and Pruett-Jones 1983). Of 129 fruit-foraging bouts (Beehler 1983*b*), 36% were of *Schefflera* sp., 19% of *Ficus* spp., and 17% of *Gastonia* sp. Beehler (1989*a*) lists 10 plants at which 249 observations of this sp. taking fruits were made, the most important being *Gastonia spectabilis* (50%) and *Schefflera pachystyla* (17%); 57% of the bouts were on plants producing simple drupes or berries, 26% capsular fruits, and 17% figs (see Chapter 4). Birds were seen near Tari, PNG eating *Omalanthus* sp. fruits (T. Pratt *in litt.*).

Reported eating fruits of *Trema orientalis* with Superb and Blue Birds at 1800 m near Wau, PNG (Hopkins 1988). Cooper (in Cooper and Forshaw 1977) reported a ♂ holding a nutmeg fruit *Myristica* sp. to its perch with a foot in order to extract the aril. A ♀-plumaged bird was seen feeding on *Myristica subaluata* fruits on 17 July and an ad ♂ on ripe epiphytic *Schefflera pachystyla* fruit on 3 Aug. A bird ate fruit of *Myristica longipes* on 14 Aug on Mt. Missim, PNG (BB). For other plant foods see Appendix 4. A captive was seen to pull green leaves through its aviary wire and eat them (Frith and Frith 1981).

Of 68 complete foraging visits to fruit-bearing plants, average duration per visit was 5.5 min and average time spent actively feeding therein 5.1 min. Thus 95% of time in a feed tree was spent eating and 90% of foraging visits lasted 1–10 min, 9% 11–20 min, and only 1% longer than 21 min (Pratt and Stiles 1985).

Observed to strip epiphytes from tree limbs in search of arthropods (A. Mack *in litt.*). Said to break open and eat large snails on the forest

Lawes' Parotia *Parotia lawesii* Ramsay, 1885

floor (Majnep and Bulmer 1977), but this may originate from ♂♂ being seen with snail shells at or en route to their court (i.e. Healey 1976, 1980). Three individuals seen foraging for arthropods at Mt. Missim at 20–30 m above ground on dry tree limbs and in foliage. One ♀-plumaged bird foraged for arthropods in mixed-spp. flock including a Raggiana Bird on 2 May (BB).

Vocalisations and other sounds

Figure 9.24

Very loud, raucous, growling, metallic and rasping or grating anguished, often questioning, *graannh-graannh* or unpleasant *kraanh! kraanh!* (Fig. 9.24), the note sometimes being given up to six times, mostly early morning and late afternoon. A call note at Mt. Missim, PNG on 20 Apr was a Magnificent Bird-like *whennh* and above Kanga, Mambare Valley, PNG (subsp. *helenae*) at 1000 m, a double harsh *kschack-kschack!* heard on 8 July (BB). ♂♂ give a nasal twittering about the court. Prior to display ad ♂♂ will utter a faint short *sip* note about the court (Coates 1990). ♂♂ in captivity make soft corvid-like guttural cawings, high-pitched screetchy *waagh-waagh-waagh* calls, a hoarse guttural *kack-kack-kack*, and a softer screechy *ked-ek, ked-ek, ked-ek* quickly repeated (Frith and Frith 1981).

Mating system

Polygyny with promiscuous ♂♂ at their courts dispersed into exploded leks (Fig. 9.25). Only ♀♀ nest-build, incubate, and raise young. Some ♀♀ show fidelity to particular ♂♂, by returning to the same individual to mate year after year (Pruett-Jones and Pruett-Jones 1990). The court of ♂♂ is an area of forest floor, 0.5–20 m^2, meticulously cleaned of all leaves, litter, and debris to leave an area of exposed soil contrasting with surrounding litter and live plants. The court must have several perches above it upon which ♀♀ perch to watch courtship displays. The owning ♂ 'decorates' his court with various items—see below (Pruett-Jones and Pruett-Jones 1990). Mean nearest neighbour distance between the courts of 25 ♂♂ was 77 (5–350, SD = 106.3) m. Clearly, most courts were near to those of neighbours but those of five ♂♂, or 20% of the study population, were solitary. Each court was maintained by a single ad ♂ and in three cases an individual bird main-

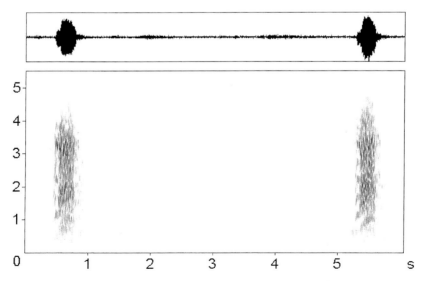

9.24 Sonogram from a recording of two ad ♂ Lawes' Parotia court advertisement song notes at Tari Valley, PNG in Oct 1991 by K. D. Bishop.

tained two adjacent courts (mean distance 81, range 69–100 m between them). ♂♂ with courts < 15 m apart were in some instances in visual contact and frequently disrupted each other's courtship displays or engaged in cooperative ones. Owners of courts 20–70 m apart were in auditory contact only and seldom disrupted each other or co-operated in display. Of 72 mating interactions, however, 42% were disrupted to some degree by ♂♂ from other courts but these did not affect mate choice by ♀♀. The five solitary ♂♂ could apparently hear only the louder advertisement vocalisations of other ♂♂, rarely interacting with them in any way. Interactions between ♀♀ and ♂♂ were observed only at the courts of ♂♂ (Beehler and Pruett-Jones 1983; Pruett-Jones 1985). ♂♂ advertisement-sing from court perches or forest canopy above, but remain mostly silent when interacting with ♀♀. ♀♀ alone or in groups of up to eight visited courts, which they approached directly or from the canopy in company with the court-owning ♂. Of 84 independent court visits by ♀♀, 39% resulted in mating, 69% involved a lone ♀ and 31% a group of ♀♀. Of the 26 group visits by ♀♀, 39% involved more than one ♀ mating. All ♀♀ that solicited a mating (63% of total matings) were not seen to mate subsequently with any other than the ♂ originally solicited. Individual ♀♀ visited courts for up to 12 weeks during a display season, up to 6 weeks before and 6 weeks after an observed mating. On average, ♀♀ had the courts of 17 individual ♂♂ within their home range, each ♀ visiting most or all of them and the behaviour of ♀♀ suggested that they were well aware of court locations. During 3 consecutive years, 44, 30, and 41% of ♂♂ (mean sample = 21, range 18–23), respectively, obtained matings; the nine successful ♂♂ mated an average of 0.1–1.2 times per 10 hr of observation. There was no evidence of an age-related increase in mating success. The most statistically significant of numerous variables analysed was the probability that a ♂ would display to a visiting ♀. No displays were specific to ♂-♂ interactions (Pruett-Jones and Pruett-Jones 1990).

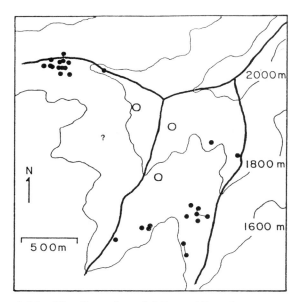

9.25 The dispersion of ♂ Lawes' Parotias on Mt. Missim, PNG: solid circles indicate display courts of 25 ♂♂, those connected by a line were maintained by the same male, open circles indicate areas where ♂♂ were heard calling but no courts found. The area indicated by a question mark was not thoroughly searched. The bold line traces four main ridges. After Beehler and Pruett-Jones (1983), © Springer-Verlag with kind permission.

Courtship behaviour

The forest floor court of ♂♂ was first described by A. P. Goodwin (1890), as a 'playground' cleared of debris to expose the soil where 'from six to eight' birds may occur together.

♂♂ bring objects such as sloughed snake skin, mammal dung, chalk pieces, fur, feathers, and bone (in order of apparent preference) to place on their cleared court floor and\or smear on the court perches. ♂♂ do not use objects in display (as do ♂ bowerbirds at their bowers), but do steal them from each other's courts. ♀♀, however, remove these objects quickly (usually within 24 hr) after their placement by ♂♂ throughout the nesting season. It was suggested that snake skin might be used as nest-lining (actually a nest 'decoration' as it would likely be placed about the rim

288 Lawes' Parotia *Parotia lawesii* Ramsay, 1885

and not in the egg cup) and the chalk eaten as a mineral supplement (to aid egg-formation). Moreover, it was thought mammal scats placed on or near nests may deter mammals (Pruett-Jones and Pruett-Jones 1990). ♀♀ were assumed to benefit from court objects, but there was no correlation between ♂♂ with more court objects and their relative mating success. ♂♂ with objects on courts did, however, receive longer visits by ♀♀, which may provide them with an indirect advantage (e.g. a ♀'s presence attracts additional ♀♀) and thereby provide selection for the intriguing habit of collecting court objects.

Frith and Frith (1981) observed 60 displays consisting of two or more distinct elements by an ad ♂ at Rotterdam Zoo and another at Baiyer River Sanctuary, PNG, on courts they had cleared. A typical complete sequence of display consisted of a ritualised set of danced steps and movements accompanied by simultaneous intricate feather movements as complex as any bird display. The sequence is started by a ♂ clearing debris from his court floor at an increasingly-rapid rate as his movements become more rigid. They become ritualised to the extent that he stoops and lowers his bill to the ground but does not actually pick up or remove anything. As this activity quickens he occasionally flicks his wings and ruffles his flank plumes. He may perform an *Upright Sleeked Pose* like that of the Western Parotia (see Fig. 9.23(a)) during court clearing but not immediately prior to further display.

Suddenly the ♂ briefly adopts an *Initial Display Bow* which is a 'frozen' court-clearing pose like that of the Western Parotia (see Fig. 9.23(b)) This pose is held for < 1 sec as the black frontal crest feathers are brought forward and above the silvery white naral tufts. From the *Initial Display Bow* the ♂ flicks his wings sharply and slightly out from the body to free his flank plumes. He then stretches his body upward fully on extended legs as his 'skirt' is raised and spread and his six occipital plumes raised, separated, and brought around and above either side of his head to project forward almost parallel to the ground whilst his head is shaken vigorously side-to-side. This signals initiation of the *Ballerina Pose* (Fig. 9.26(a)). Continuing to shake his head from side-to-side, the ♂ starts his dance display from the point on the court he adopted the *Ballerina Pose* (Point 1). He moves sideways with minute mincing steps in a semi-circle while rotating himself and bowing, or curtsying, slightly to face point 1 as he dances backward to stop c. 0.5 m away, at Point 2, and now facing Point 1. Here he pauses momentarily but continuously shakes his head to wave the occipital plumes about vigorously, their spatulate tips thrashing about in front of him like hovering flies. He then dances directly backwards to start a *Back-forward Dance* phase

(a)

(b)

9.26 Some display postures of ad ♂ Lawes' Parotia: (a) *Ballerina Pose*, (b) the *Agonistic Pose* after Coates (1990, 483), with kind permission. See text.

(Fig. 9.27(g–m)) consisting of stepping backward from point 2 to a point, < 0.5 m directly behind (Point 3 in Fig. 9.27), and then immediately forward to Point 2 again and repeats this back–forward dance twice more in rapid succession. Once again, during the backward-stepping movement of this dance phase, the front of the 'skirt' is lowered to make the bird appear to bow, or curtsey, and is then raised to parallel to the ground during the forward steps—the occipital plumes remaining fully forward throughout (Fig. 9.27 (g–m)). Seven displays timed from their start took an average of 12 sec to reach the completion of this *Back–forward Dance* phase.

Again at Point 2, the ♂ follows his *Back–forward Dance* with a *Stationary Phase* by crouching slightly, to bring his skirt closer to the ground and remain momentarily motionless (Fig. 9.27 (g)) while bringing his widely-separated occipital plumes upward and backward to nearly or (rarely) just beyond vertical. The occipital plumes are then immediately brought forward and downward to an angle of 45° as the bird draws himself stiffly erect and freezes in this posture (Fig. 9.27(h)). This *Stationary Phase* with feather manipulation is infrequently performed on reaching Point 2 for the first time (i.e. before the *Back–forward Dance*).

Suddenly the ♂ crouches again, snapping his head directly downwards several times as the breast-shield is flicked outward and upward (Fig. 9.27(i)) resulting in a flash of iridescent colour. Remaining crouched, he then bobs his head rapidly in a reptilian manner, first to one side and then to the other, three or four times, while retaining the flag plumes at 45°. Throughout this peculiar movement the bill points forward and at the low point of each bob the head is central and the breast-shield flicks out and upwards. As suddenly as it started, this bobbing stops and, after a brief pause, the bird again draws itself erect into a momentary frozen posture and brings his gathered occipital plumes around and backward to a position above the back (Fig. 9.27(j)) and then immediately forward again to just forward of vertical and widely separated (Fig. 9.27(k)).

The ♂ then crouches again and pumps his head and neck vigorously up and down two to four times with his bill pointing forward. During this, the flags remain almost vertical and separated and the frontal crest is conspicuously projected forward over the naral tufts and bill. As his head reaches its lowest point, the breast-shield is conspicuously but rapidly flicked upward and outward at its lower edge and adjusted to bring either side of the upper edge slightly forward (Fig. 9.27(l)). This rapid movement of the breast-shield, sometimes almost at right angles to the bird's breast, results in dramatic flashing of changing iridescent colours. So vigorous is the last series of vertical movements that the bird's feet, though remaining the same distance apart throughout, sharply hop a short distance to one side and back to the original spot with each pump of the head.

The complete performance always ends, after final head-pumping, by the bird suddenly lowering all feathers to their normal positions (Fig. 9.27(m)) and flying to a low perch above or adjacent to his court immediately or after a brief bout of feather preening or court clearing. Components performed out of sequence or alone by ♂♂ reflect idiosyncrasies between the two individual ♂♂. When performing the entire *High-intensity Display* both birds, however, follow the same sequence.

♂♂ displaying to a ♀ in the wild conclude the above high-intensity display by *Lunging* across the court with sleeked plumage, and head and bill thrust forward and held horizontal to the ground, before hopping up onto a court perch there to give the posture described above as the *Initial Display Bow* (called a pre-copulation *Bill-point* by Pruett-Jones and Pruett-Jones 1990) before copulation. This *Lunge* preceded 69% of all (70) matings and the *Bill-point* immediately preceded 73% of all copulations in the wild (Pruett-Jones and Pruett-Jones 1990). It appears to be the same behaviour as the initial ritualised hopping across the court performed by Western

Parotias. A behaviour performed by a ♂ on his court perch(es) as a ♀ approaches from the canopy is a *Bounce* display in which he 'bounces on the display perch, reversing direction, while flicking his wings' (Pruett-Jones and Pruett-Jones 1990). This would seem analogous to the *Hopping-on-the-spot* display seen performed by a captive ad ♂ Carola's Parotia (Frith 1968).

When a captive ad ♂ directed a display at another ♂ the *Initial Display Bow* was presented either side-on (in profile) or facing directly away from the other ♂. From this posture the displaying bird suddenly turns around to assume the *Ballerina Pose* and, 'with head and neck arched slightly forward, immediately walk in the direction of the neighbour's court' (in an adjacent aviary). It would some-

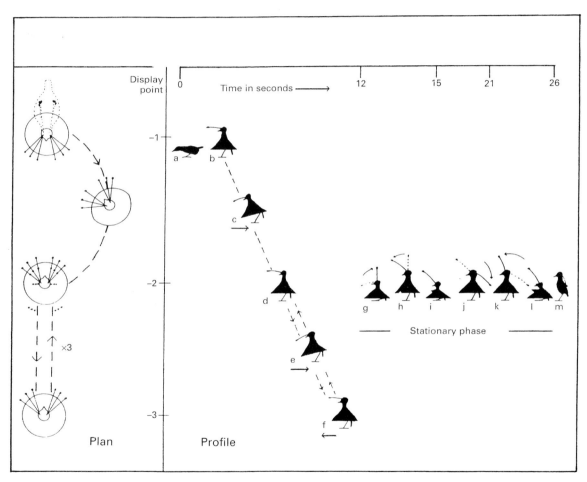

9.27 High intensity display of Lawes' Parotia. **Plan** (left) = positions of displaying ♂ from *Initial display bow* (dotted figure) at Point 1 through the semicircular dance to Point 2 and the back-forward dance between Points 2 and 3 to remain at Point 2 for the *Stationary Phase* of display. Various positions of the occipital plumes in the *Stationary Phase* of display at Point 2 are indicated. **Profile** (right) = silhouette of ♂ in display. Horizontal arrows indicate direction of movement. Movements of occipital plumes are indicated by dotted line representing position from which these are moved and the solid line to where they are moved. Note crouched posture of *g*, *i*, and *l* and breast-shield position in *i* and *l*. From Frith and Frith (1981), with kind permission.

times then turn around, at the edge of its court and dance away again, but often it would simply gradually bend forward, head and neck arching horizontally and occipital plumes pointing towards the ground ahead to present the *Agonistic Pose*—Fig. 9.26(b) (Coates 1990).

Crandall (1931) described the dancing display of a NYZS captive ad ♂ in general terms that agree with the above. While no detailed account of courtship by wild birds is published the above appears typical of them (Pruett-Jones and Pruett-Jones 1990).

Breeding Table 6.5
NEST SITE: high up in a large tree (Shaw Mayer in Harrison and Frith 1970), 11 m above ground in outer canopy foliage of a large tree (CF), 12 m above ground in a forest tree (Coates et al. 1970) and 5–10 m high in a thick *Cissus aristatus* vine tangle at the top of a *Macaranga* tree (T. Pratt *in litt.*).

NEST: a nest collected by A. S. Anthony was described by Hartert (1910) and quoted by Gilliard (1969) and Cooper and Forshaw (1977) as such, but the original description is somewhat reminiscent of a manucode nest. Fred Shaw Mayer described an undoubted Lawes' Parotia nest, at which he collected the attending ♀ and single-egg clutch, as an 'open and rather shallow structure [38 mm deep inside], composed of fern tendrils and creeping fern stems woven together; lined with very fine tendrils. On the outside of nest a few green fern fronds'. A nest seen from the ground 11 m below was large and bulky but shallow and built atop a few sticks that were probably already in the tree fork (CF). An active parotia's nest, very probably of this sp., was reported to have pig dung on its rim (Majnep in Majnep and Bulmer 1977). This is most significant in view of the finding that ♂♂ decorate their courts with mammal dung and that ♀♀ visiting courts remove it (Pruett-Jones and Pruett-Jones 1988*b*).

EGGS: clutch said to be of one egg (Gilliard 1969; Pruett-Jones and Pruett-Jones 1990).

The egg described by Hartert (1910) is small, measuring 33 × 24 mm and, while not manucode-like in appearance (as is its nest description), it lacks the longitudinal markings found on most typical bird of paradise eggs. Thus there is some doubt about this clutch which appears to be missing from the BMNH collection or has been re-identified (M. Walters personal communication). A single-egg clutch from Boneno, Mt. Mura, *c.* 50 km NW of Mt. Simpson, SE PNG measured 38.4 × 27.8 mm and was a glossy light buff (39 but paler, more creamy) with broad longitudinal streaks of brown, grey, and purple-grey with some small spots and blotches (Harrison and Frith 1970). See Plate 13.

INCUBATION: a ♀ was collected with her nest and egg. Probably exclusively by ♀, as in other polygynous birds of paradise.

NESTLING CARE AND DEVELOPMENT: unknown.

Annual cycle Table 6.5
DISPLAY: on Mt. Missim, courts may be maintained for up to 10 months (May–Mar) each year, but all copulations observed occurred from mid-Sept to mid-Feb (Pruett-Jones and Pruett-Jones 1990). BREEDING: June–Jan. ♂♂ have been collected with moderately-enlarged gonads during Jan–July and much enlarged ones during Aug–Dec, and ♀♀ with moderately-enlarged oocytes during Aug–Dec. EGG-LAYING: the Boneno clutch was found on 8 Dec and the nest found with a 5–6 day-old nestling on 26 Nov must have been laid in at the end of Oct or 1 Nov. MOULT: 48 of 140 specimens were moulting during all months but mostly Jan–June (Appendix 3).

Status and conservation Table 8.1
Said to be fairly common in the Owen Stanley Mts and is common in the Tari Valley. In an area of *c.* 4 km^2 on Mt. Missim 28 courts of 25 ♂♂ were found and three other ♂♂ called regularly there but apparently lacked a court

(Fig. 9.25). Inhabits forest at an elevation popular for human settlement, but the sp. is apparently under no current threat.

Knowledge lacking and research priorities

One of the better-studied sp. Still, nagging questions remain. Do ♀♀, in fact, add objects to their nest that were obtained from the courts of ♂♂ and if so to what purpose? Nesting biology remains unknown. ♀-plumaged and subad ♂♂ appear in the Crater Mt. Biological Research Station area during Mar–Sept, probably migrants from higher altitudes. This parotia overlaps extensively with Carola's in this area and at higher elevations here both can be observed on the same display courts (Mack and Wright 1996; M. LeCroy personal communication). Hybridisation between the two here would seem a possibility given that both have hybridised with other bird of paradise genera (see Plate 14 and Appendix 1). Interactions between the two spp. (including hybridisation) would be of great interest.

Aviculture

Not uncommonly kept in PNG, Europe, and USA in the past but not reported to have bred. An active bird that adapts well to a well-lit aviary with accessible dark areas, a pool with shallows and deeper (10 cm) water, and sunning perches. Somewhat aggressive to smaller spp. and ♀♀ sometimes likewise to conspecific ♂♂ (P. Shanahan *in litt.*).

Other

An ad ♀ caught and banded by T. Pratt on Mt. Missim 28 Oct 1978 was recaptured at the same place by S. Pruett-Jones 7 Dec 1986, > 8 yr 1 month later (Anon 1989). The odd observation has been made of an ad ♂ Blue Bird apparently associating in some way with an ad ♂ Lawes' Parotia at its immediate court area. One parotia associated with in this way was repeatedly seen sitting on a court perch watching the Blue Bird display (D. Gillison *in litt.*).

Wahnes' Parotia *Parotia wahnesi* Rothschild, 1906

PLATE 7; FIGURES 9.28,29

Other names: Wahnes' Six-wired Bird of Paradise, Wahnes' Six-wired Parotia, Wahnes's Six-plumed Bird of Paradise, Huon Parotia. *Nyat nyat* and *Dakwalova* of the Nokopo people of the Finisterre Ra, PNG (Schmid 1993). *Agoten mornia* near Mt. Mengam, Adelbert Ra (G. Opit *in litt.*).

Endemic to the mountains of the Huon Penin and Adelbert Ra, PNG. Large wedge-shaped tail and a prominent bronzy 'pom-pom' above the base of the bill. Within its limited range, could be confused only with the longer-tailed Huon Astrapia. Monotypic.

Description

ADULT ♂: 43 cm, entirely velvety jet black. Head decorated by a large, erectile, naral tuft of elongate coppery bronzed feathers above the nostrils and culmen and a narrow nuchal bar of intensely-iridescent, deep blue to pink-purple, scale-like feathers. From behind each iris grows, amid an 'ear tuft' of elongate, narrowly-pointed feathers, three long, erectile, wire-like, bare, occipital plume shafts tipped with spatulate 'flags' of normal black feather vane. Mantle to dorsal surface of tail with a slight plum-purple sheen in some lights. Velvety black chin and throat with slight iridescent sheen of rich Purple (1) in some lights, this grading into an otherwise discrete breast-shield of large scale-like feathers, intensely iridescent, bronzed emerald green to green-yellow and, in some lights, Purple,

Magenta (2) to Bluish Violet (172B) with jet black feather bases visible to lower sides of the shield. Remaining underparts, including elongate flank plumes, jet black with deepest plum-purple sheen in certain light. Underwing: coverts jet black, flight feathers brownish black (119). Bill shiny black, iris Cobalt Blue (168) with greenish cream outer ring, legs and feet purplish lead-grey, mouth colour unrecorded.

ADULT ♀: 36 cm, head matte brownish black, more brownish on nape which may be spotted with rich chestnut feathers (as in astrapias), with an off-white, narrow, superciliary stripe originating from behind the eye and a broad malar stripe flecked with blackish grey. Brownish black of nape breaks up into barring, then to broad and finally to fine streaking on the rich Raw Sienna (136) mantle, back, wing coverts, rump, and uppertail coverts but not onto this pure colour on the dorsal surface of tail. Upperwing also Raw Sienna except the normally-concealed parts of the flight feathers which are blackish brown (121). Chin and throat whitish grey flecked and barred brownish black grading to remaining underparts of brown cinnamon (121C) uniformly barred brownish black. Underwing: coverts brown cinnamon barred brownish black, flight feathers largely this brown cinnamon with extensive dark brown (119A) tips. Iris deep cobalt-blue with pale creamy-blue outer ring (Schodde and Mason 1974).

SUBADULT ♂: unknown but presumably as other parotias.

IMMATURE ♂: iris as ad ♂. First-year plumage recognised by pointed tail feathers and lack of indentation of P9 (Mayr 1931: 649).

NESTLING–JUVENILE: unknown.

Distribution
Endemic to a small section of the N coastal ranges of PNG, Cromwell, Rawlinson,

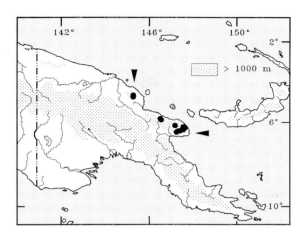

Saruwaged, and Finisterre Mts of the Huon Penin, and the Adelbert Ra 1100–1700 m.

Systematics, nomenclature, subspecies, weights, and measurements
SYSTEMATICS: the long tailed, geographically restricted, isolate considered the 'primitive' parotia by Gilliard (1969). Hybridisation: unrecorded.

SPECIES NOMENCLATURE AND ETYMOLOGY
Parotia wahnesi Rothschild, in Foerster and Rothschild, 1906. *Two New Birds of Paradise*, p. 2. Rawlinson Mts. Type specimen AMNH 678233. Discovered by Carl Wahnes in 1906.

Etymology: *wahnesi* = after the German collector Carl Wahnes.

WEIGHTS AND MEASUREMENTS
Measurements: wing, ad ♂ ($n = 21$) 157–166 (161), imm ♂ ($n = 8$) 152–161 (157), ad ♀ ($n = 20$) 142–157 (149); tail, ad ♂ ($n = 19$) 159–198 (187), imm ♂ ($n = 7$) 140–172 (161), ad ♀ ($n = 19$) 140–164 (153); tail centrals, ad ♂ ($n = 16$) 201–225 (213), imm ♂ ($n = 7$) 165–184 (175), ad ♀ ($n = 17$) 161–175 (166); bill, ad ♂ ($n = 18$) 29–33 (31), imm ♂ ($n = 4$) 31–33 (32), ad ♀ ($n = 19$) 28–33 (31);

294 Wahnes' Parotia *Parotia wahnesi* Rothschild, 1906

tarsus, ad ♂ (n = 21) 49–53 (51), imm ♂ (n = 8) 43–52 (49), ad ♀ (n = 20) 44–52 (46). Weight: ad ♂ (n = 3) 170–172 (171), imm ♂ (n = 1) 176, ad ♀ (n = 7) 144–154 (146).

Habitat and habits   Table 4.1

Midmontane forests. An active and noisy sp. that makes jerky wing and tail flicking movements and often cocks the conspicuously-long tail slightly upward or to the side.

Diet and foraging   Table 4.2

Foraging ecology appears similar to that of Lawes' Parotia. A ♀-plumaged bird in the Adelbert Ra, PNG, during Mar foraged for arthropods before hanging head down on an epiphytic ginger to pluck and swallow some 15 of the small yellow clustered fruits (G. Opit *in litt.*). A wild ♀-plumaged bird was seen to hold an insect to its perch while it bit it into small pieces. This bird was subsequently attacked and pursued in flight by a ♀-plumaged Superb Bird (T. Pratt *in litt.*). In the same area, T. Pratt (*in litt.*) watched two ♀-plumaged birds foraging below the tree tops by probing into moss and epiphytes for arthropods, reminding him of astrapias. A Baiyer River Sanctuary captive ad ♂ and a ♀ were both seen to rip pieces of leaves, or pluck an entire leaf, and hold them to the perch by a foot to tear them into smaller bits and eat them (Frith and Frith 1979). See Appendix 4 for food plants.

Vocalisations and   Figure 9.28
other sounds

Notes given by ad ♂ close to display courts include a harsh double roar *khh kaakkk*, with a cockatoo-like quality, possibly court-advertisement. Notes of similar harsh screetching quality notated as a sharp and dry *wetch* or *snatch* are given in twos (Fig. 9.28(a)) or may form part of longer and more complex song based on them (Fig. 9.28 (b)) in court advertisement. Birds also gave a single loud nasal *Garr* note (G. Opit *in litt.*). The scream of ad ♂ ♂ is described as essentially the same

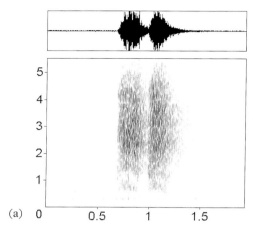

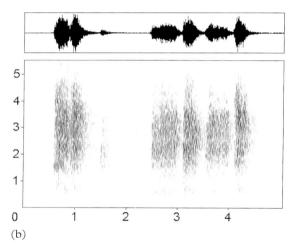

9.28 Sonogram from a recording of the harsh, dry, sharp, double *wetch* screech note given singly (a) or in series (b) by ad ♂ Wahnes' Parotia in court advertisement song in Jan 1975 by T. K. Pratt, at Bambu near Wasu, Huon Penin, PNG.

as that of Lawes' Parotia but slightly higher-pitched, quieter, and given more often (T. Pratt *in litt.*). Thin and soft cheeping notes are given as the ♂ is about to descend to the court floor from low perches above it. Also a low nasal twittering is given in flight or immediately after landing, or when active on perches. Ad ♂ ♂ in flight typically produce an audible rustling sound, but appear able to reduce or eliminate this.

Mating system

Doubtless polygynous, in view of court and display characters, with promiscuous ♂♂, and ♀-only nest attendance, as in Lawes' Parotia. Courts are patches of forest floor cleared of litter situated beneath several horizontal or near-horizontal perches as in other parotias. Two courts were found *c.* 5 m apart atop a broadly-rounded ridge at 1590 m, in thick shrubbery beneath a small gap in the canopy of mixed primary montane forest, dominated by an overstorey of *Nothofagus, Dryadodaphne, Lithocarpus,* and *Podocarpus.* Both were flattish areas that had been scrupulously cleared of all debris to expose the first layer of interwoven fine rootlets, and incorporated several well-spaced small saplings from which most leaves had been removed. One court was circular and 1.5 m in dia and the other oblong and 1.5 × 1 m. At least one ad and one imm ♂ attended these courts (Schodde and Mason 1974). Courts tend to be adjacent to or within windfall 'clearings' and thus directly beneath a hole in the forest canopy (T. Pratt *in litt.*).

Courtship behaviour

The basic courtship display pattern is similar to that of Lawes' Parotia with some variation (T. Pratt *in litt.*). Displays of a NYZS captive ♂ are described by Crandall (1940): 'When about to display on the ground, the bird stands with its body in a horizontal position, with the wings held closed but high, in order to clear the slightly loosened flank feathers. The tail is turned sideways [Fig. 9.29(a)], usually to the left. The bird feigns picking at the ground, then suddenly throws the body forward and downward with the head turned under the breast, so that its crown is nearly parallel with the ground [Fig. 9.29(b)]. The tail now becomes the center of attention, for it is thrown straight upward, behind the crouching body. While all else remains immobile, the graduated lateral feathers are now rapidly opened and closed, the central pair remaining fixed. After this fan-like effect has been obtained perhaps five or six times, the bird suddenly drops the tail and throws its body into an upright position. Simultaneously, the long feathers of the back and flanks are raised to form the 'umbrella' well known in other forms of *Parotia*, and the head plumes are thrown far forward, three on each side of the crown [Fig. 9.29(c)]. The brilliant breast plate is noticeably flat and lies between a slight extension of the feathers of the upper neck and the erected flank and breast plumes. With the head now extended to its greatest possible height, the bird turns it rapidly from side to side, causing the tabs on the plumes to rotate in the typical manner. Several mincing, short steps are now taken along the ground, usually to the right. After moving perhaps a foot in this manner, the bird suddenly becomes immobile except for the neck, which is rapidly moved from side to side, between the head and breast plate. The display usually ends with this manoeuvre. During the upright form of the display, the tail takes a negligible part and may turn to either side or even drag on the ground' [Fig. 9.29(d)].

The display differs from that of Lawes' Parotia mainly in that it is preceded by little court clearing, naral tufts are opened but there is no crest to extend forward to cover them, the occipital plumes are thrown forwards before the bird stands erect, and this species uses the tail and wings when crouched (T. Pratt *in litt.*).

In preparing to display off the ground the bird moves about the perches, body stiffly horizontal and tail held to one side. When the chosen spot is reached, he throws himself directly into the upright form, the tail-fanning being omitted. The display is then identical to that performed on the ground, including movement along the perch and 'wobbling' of the neck. The tail fanning display is also performed on perches (T. Pratt *in litt.*). Gary Opit (*in litt.*) observed birds on the Adelbert Ra,

296 Wahnes' Parotia *Parotia wahnesi* Rothschild, 1906

PNG between 29 Feb and 22 Mar. He was attracted by loud nasal calls to a young ♂ in ♀-plumage 'except for a bumpy, glossy-feathered head and six flag-tipped wires coming from the head above the eyes'. This subad ♂ was chasing and displaying to a ♀-plumaged bird. He faced the presumed ♀ and 'spread his wings open slowly while calling a very loud harsh, nasal, parrot-like *Garrr* then twisted back and forth on his perch, very like a fantail [*Rhipidura* spp.] will do, causing his head wires to whip back and forth. In another tree a hundred yards off were three imm ♂♂, without head wires, all twisting back and forth and leaping from branch to branch calling noisily. One came down to within ten feet of the ground, twisting back and forth. Others could be heard calling in various directions a couple of hundred yards off'.

A wild ad ♂ court owner tolerated ♀-plumaged and ad plumaged ♂♂ on his court but was also once seen to attack another ad ♂ and knock it off a court perch (T. Pratt *in litt.*).

Coates (1990) wrote of a captive ♂ at Baiyer River Sanctuary: 'While on the court and on nearby ground the ♂ hops about with a bouncing motion, body slightly elevated showing the tibia, and tail cocked and sometimes angled to one side. Sometimes he hops about at speed, and sometimes, probably when in the mood to display, he hops swiftly across the court and about its vicinity with a peculiar mammal-like appearance, hunched, like a hopping rodent. A

9.29 Some courtship display postures of ad ♂ Wahnes' Parotia: (a) pre-display pose, (b) *Initial Display Bow*, (c) *Ballerina Pose*, (d) head and neck waggling. See text.

display would commence when, after moving about the court and nearby, tending it, he was at the edge of the court.

Suddenly he adopts the *Initial Display Bow*, a peculiar crouching pose in which the body and tail are in a fairly straight line, the head and neck are extended, head towards the ground, tail slightly elevated, and the golden tuft over the forehead extended forward. This pose is held for perhaps 2 sec. He then stands upright, at the same time raising and expanding the elongated body plumage to form a wide, circular skirt, giving the *Ballerina Pose*. When fully upright, in an instant just before he begins to dance, he flicks his head briefly at right angles to one side, giving a fleeting glimpse of a bright jewel-like flash (the eye or nape). Then the occipital wires are brought forward and, leaning forward and with the tail touching the ground, he begins the dance by walking forwards and to one side, head bobbing from side to side, presenting a sideways view. Reaching a certain point he stands still, but the sideways head-bobbing increases in tempo causing the flag-tipped wires to wave about wildly, and he may tilt his body slightly to one side then the other; the tail, which is not featured, hangs slightly to one side. This phase of the display continues in silence for many sec (perhaps a min or more) then he suddenly stops, shuffles his plumage and resumes normal behaviour'. Thane Pratt (*in litt.*) observed wild birds perform the above display on the court. He also noted both an imm and ad ♂ end a display by darting onto a court perch and pseudocopulate on it.

Breeding   Table 6.5
NEST SITE AND NEST: unknown.

EGGS: two eggs laid by a Baiyer River Sanctuary captive on consecutive days were dully glossed and 'pale cream with heavy streaking at the large end becoming sparser half way along the egg and practically absent at the small end. The marks vary from small dots to elongate broad streaks and are grey or tan in about equal proportions, with the tan marks in some places overlaying the grey'. The measurements of this clutch were given as 39.6 × 26.7 and 40.2 × 25.3 mm by Mackay (1990) who noted that the latter egg was badly cracked. This egg measured 38.6 × 25.3 mm prior to damage (CF).

INCUBATION AND NESTLING CARE: unknown but presumably by ♀ only.

NESTLING DEVELOPMENT: unknown.

Annual cycle   Table 6.5
DISPLAY: Coates suggested that the display period is centred on the dry season. Thane Pratt observed waning court attendance in Jan. BREEDING: At least c. Nov–Feb but data meagre. Nesting recorded in captivity, in early Jan. EGG-LAYING: the BRS clutch was laid on 1–2 Jan. MOULT: 16 of 34 specimens, involving all months but Aug, were moulting during Oct–May (Appendix 3). Crandall (1940) reported a 4-month moult (20 Oct to end of Feb) by an ad ♂ captive in NYZS. Basing his conclusion on observations of captives at BRS, Coates (1990) suggests the moult occurs during the wet season.

Status and conservation   Table 8.1
Locally common on the Cromwell Ra (Lindgren in Coates 1990). Uncommon to rare on the Adelbert Ra, PNG during Mar (G. Opit *in litt.*).

Knowledge lacking and research priorities
A comprehesive survey of this bird's range and abundance is required. It is a little-known sp. that would be a worthy and accessible subject of in-depth studies.

Aviculture
Fred Shaw Mayer provided the London Zoo with a ♂ in 1931 and more birds in 1939 from which one ad ♂ was sent to the NYZS where it arrived on 17 Aug 1939.

Carola's Parotia *Parotia carolae* Meyer, 1894

PLATE 7; FIGURE 9.30

Other names: Queen Carola's Six-wired Bird of Paradise, Queen Carola's Parotia, Queen Carola of Saxony's Six-plumed Bird of Paradise, Queen of Saxony's Bird of Paradise, Carol Six-plumed. *Dul* of the Telefomin people (Gilliard and LeCroy 1961), *Kiawoi* of the people of the Jimi Valley (Healey 1976) and the Bismarck Ra (W. Cooper *in litt.*). *Kabay pok* of the Kalam people (Majnep and Bulmer 1977). *Kulele* (♂) and *Agule* (♀) of the Gimi people, Crater Mt., PNG (D. Gillison *in litt.*). *Kesa* of the Yabi people, Snow Mts (specimen label).

The aberrant parotia, with a distinctive array of vocalisations, unique erectile head, throat, and flank plumes, and instances of sympatry with Lawes' Parotia. Polytypic; 6 subspp.

Description

ADULT ♂: 26 cm, head velvety jet black with dull coppery bronze sheen and highly-ornate decorations: a short, erectile, blackish bronzed, frontal crest tipped silver white sits between jet black, elongate, and vertically-raised lore and foreface feathering, tipped silver white, from nostril to above the eye. Eyes broadly encircled by rich, iridescent, coppery-gold feathering. Feathering atop the slightly concave crown as around the eye and immediately behind this a narrow nuchal bar of highly-iridescent scale-like feathers that appear blue-green to Purple (1) and/or Magenta (2) in some lights. From behind each eye, and amid an 'ear tuft' of elongate and narrowly-pointed feathers, grow three long, erectile, wire-like, occipital plumes with relatively small spatulate 'flag' tips which usually sit above the lower mantle edge. Velvety jet black mantle to dorsal surface of tail with dull coppery bronze sheen in some lights, including the upperwings except primaries and their coverts which are brownish black (19). Chin dusky Olive-Brown (28) smudged blackish with paler tips to elongate 'whiskers' surrounded by malar area and throat of Buff (124) with dull golden sheen. Lower central throat more whitish, flecked Cinnamon (123A), with fine long 'whiskers' to either side, grading into an otherwise discrete breast-shield of large scale-like feathers, intensely-iridescent bronzed Yellow-Green (58) and/or Magenta to pink (9) in suitable light with jet black feather bases visible on lower sides of the shield. Remaining central underparts, to undertail coverts, deep rich blackish Sepia (219) with iridescent, lustrous, deep coppery sheen becoming browner to dark reddish brown (223A) adjacent to an extensive patch of cotton white, elongate, and inwardly-curving flank plumes. Underwing: coverts variably red brown to dark Buff (24) and blackish, flight feathers dark greyish brown (219) and ventral surface of tail more blackish (119). Bill black, iris Sulphur Yellow (157), legs and feet blackish grey, mouth colour apparently pale green (BB).

ADULT ♀: 25 cm, head Olive-Brown, darker to Dark Brownish Olive (129) on ear coverts, above, behind, and finely around the eyes and on the rear crown. Lores, forehead, and nape Buff flecked Olive-Brown. A broad supercilium and another broad line above the Olive-Brown malar area dirty white, flecked Olive-Brown, with some paler flecks of malar extending onto anterior ear coverts. Mantle to uppertail coverts Olive-Brown, slightly darker and browner (129) on lesser wing coverts and dorsal surface of tail. Greater wing coverts, alula, and flight feathers Dark Brownish Olive with broad rich Raw Sienna (136) outer edges. Concealed flight feather vanes blackish brown (219). Chin and adjoining broad malar stripe dusky Olive-Brown, flecked and faintly barred greyish brown. Throat pale greyish white, flecked greyish brown to become barring of blackish brown (219) on upper chest and throughout remaining underparts of pale cinnamon (223D), darker on flanks. Underwing: coverts Ground Cinnamon (239) barred brownish grey, flight feathers Dark Brownish

Olive with broad Ground Cinnamon trailing edges to all but outer two primaries. Ventral surface of tail Olive-Brown with paler central feather shafts. Tail longer than that of ad ♂. Iris pale grey, cream, or yellow (possibly varies with age) but requires confirmation.

SUBADULT ♂: as ad ♂ with few feathers of ad ♂ plumage intruding, initially about head, to as ad ♂ with a few feathers of ♀ plumage remaining.

IMMATURE ♂: as ad ♀ but iris pale grey.

NESTLING–JUVENILE: unknown.

Distribution
Western two-thirds of NG's central cordillera, from the Weyland Ra eastward to the Giluwe and Hagen Ra, the Sepik–Wahgi Divide, Schrader and Bismarck Ra, and Crater Mt. (Eastern Highlands). Easternmost distribution in the southern watershed of the cordillera appears to be much less continuous than it is in the N. Also recorded from Mt. Bosavi. Ranges from 1100 to 2000 m, mainly 1450–1800 m (Coates 1990). For details of overlap with *P. lawesii* see account for that sp. The two spp. have been seen attending the same court in the Crater Mt. Reserve (Mack and Wright 1996; M. LeCroy personal communication). A ♀-plumaged individual was observed by D. Bishop (*in litt.*) foraging with five Lawes' Parotias at the SE edge of the Wahgi Valley where it meets the N Kubor Ra.

Systematics, nomenclature, subspecies, weights, and measurements

SYSTEMATICS: morphologically the most distinctive parotia and one showing considerable and inadequately-understood and rather poorly-defined variation. The three 'black' parotias (*wahnesi*, *lawesii*, *sefilata*) are unambiguous sister forms that constitute a superspecies. Their relationship with *P. carolae* is less certain, and a mode of differentiation that would produce such a geographical pattern (with *carolae* sandwiched between *sefilata* and *lawesii*) is unclear. Given the limited sympatry between *lawesii* and *carolae* it is possible that *carolae* is a full member of this superspecies, in spite of the apparent character evolution. The ♀♀ of this sp. and those of the Superb Bird exhibit what is apparently parallel variation. An explanation of this is that the smaller Superb has evolved to mimic the larger and more aggressive ♀ Carola's Parotia (Diamond 1982). Hybridisation: known with Superb Bird (Plate 14, Appendix 1).

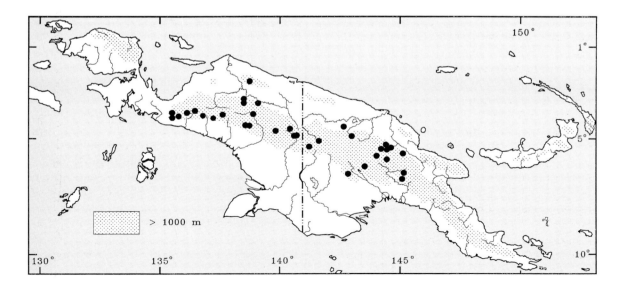

Healey (1976) and LeCroy (personal communication) have seen both this sp. and Lawes' Parotia at the same court, indicating the potential for the two to hybridise.

SPECIES NOMENCLATURE AND ETYMOLOGY

Parotia carolae A. B. Meyer, 1894. *Bulletin of the British Ornithologists' Club* **4**, 6. 'Amberno River' [= Mamberamo R.], but apparently from Weyland Mts (Mayr 1941).

Etymology: *carolae* = after Queen Carola of Saxony, wife of King Albert of Saxony.

SUBSPECIES, WEIGHTS, AND MEASUREMENTS

1. *P. c. carolae* A. B. Meyer, 1894. Type specimen MT 14319.

Range: Weyland Mts E to Paniai (Wissel) Lakes region of westernmost central cordillera of IJ.

Measurements: wing, ad ♂ ($n = 12$) 152–156 (154), ad ♀ ($n = 13$) 133–146 (143); tail, ad ♂ ($n = 12$) 73–82 (75), ad ♀ ($n = 13$) 87–96 (91); bill, ad ♂ ($n = 10$) 31–38 (35), ad ♀ ($n = 12$) 34–39 (37); tarsus, ad ♂ ($n = 12$) 48–53 (50), ad ♀ ($n = 13$) 43–47 (45). Weight: none.

2. *P. c. meeki* Rothschild, 1910. *Bulletin British Ornithologists' Club* **27**, 35. Setekwa R., Dutch New Guinea. Type specimen AMNH 678161.

Range: Snow Mts of IJ E of Paniai Lakes and S of (but not including) Doorman Mts (northern scarp facing Mamberamo R.), E to range of *P. c. clelandiae*.

Diagnosis: like nominate form in size (bill slightly larger) and appearance, but chin and sides of throat blackish.

Measurements: wing, ad ♂ ($n = 4$) 151–154 (153), imm ♂ ($n = 2$) 147–148 (148), ad ♀ ($n = 12$) 136–146 (141); tail, ad ♂ ($n = 4$) 73–83 (76), imm ♂ ($n = 2$) 87–94 (91), ad ♀ ($n = 12$) 86–96 (91); bill, ad ♂ ($n = 2$) 37–39 (38), ad ♀ ($n = 7$) 35–39; tarsus, ad ♂ ($n = 4$) 50–52 (51), imm ♂ ($n = 2$) 47–50 (48), ad ♀ ($n = 11$) 45–48 (46). Weight: none

3. *P. c. chalcothorax* Stresemann, 1934. *Ornithologische Monatsberichte* **42**, 145. Doormanpaad, upper Mamberamo (Idenburg R.). Type specimen ZMB 30.32890.

Range: Doorman Mts, an isolated range on the northern scarp of the central cordillera of IJ, just S of the lower Idenburg (Taritatu) R.

Diagnosis: like nominate form but significantly-longer occipital plumes (of *clelandiae* length) and upperparts with bright coppery sheen, underparts more coppery, and long loral feathering less intense black, being slightly brownish.

Measurements: wing, ad ♂ ($n = 2$) 158–160 (159), ad ♀ ($n = 5$) 145–148 (146); tail, ad ♂ ($n = 2$) 75–79 (77), ad ♀ ($n = 4$) 90–95 (93); bill, ad ♂ ($n = 1$) 34, ad ♀ ($n = 4$) 33–36 (35); tarsus, ad ♂ ($n = 2$) 51–52 (52), ad ♀ ($n = 5$) 43–47 (45). Weight: none.

4. *P. c. berlepschi* Kleinschmidt, 1897. *Ornithologische Monatsberichte* **5**, 46. New Guinea; the Van Rees Mts, IJ have been suggested as the unknown home of this distinct subsp. (Mayr in Mayr and Greenway 1962) as have the Foya Mts, IJ (Diamond 1985). Type specimen not located, possibly destroyed in ZMB during WW II.

Range: Diamond (1985) located a population of this sp. in the Foya Mts that he hypothesises may be this 'lost' subsp., whose original specimens were collected from an unknown locality. Since Diamond only encountered ♀♀ and the subsp. is described from the ♂ plumage, confirmation of this range will depend upon further fieldwork. While Diamond (1985) considered this form only as a subsp., he argued it is a more distinctive one than is *helenae* of Lawes' Parotia which Schodde and McKean (1972, 1973) elevated to sp. status.

Diagnosis: like nominate form in size (bill slightly shorter) and appearance but occipital plumes slightly longer and upper neck, nape, and mantle heavily bronzed and with throat black. Spatulate tips to occipital plumes relatively small. Only four specimens known.

Measurements: wing, ad ♂ ($n = 2$) 156 (156); tail, ad ♂ ($n = 2$) 77 (77); bill, ad ♂ ($n = 2$) 33–36 (34); tarsus, ad ♂ ($n = 2$) 47 (47). Weight: none.

5. *P. c. clelandiae* Gilliard, 1961. *American Museum Novitates* **2031**, 5. Telefolmin [= Telefomin] 5000 ft. [1524 m], Victor Emanuel Mts, Mandated Territory of New Guinea. Type specimen AMNH 708171.

Range: From PNG border southeastward probably to the southern watershed of the Eastern Highlands at least as far as Crater Mt. (but subspp. designation of southeasternmost populations has not been carried out). The range of *chrysenia* lies entirely to N of *clelandiae*, but details of the boundary of these two ranges has not been adequately delineated.

Diagnosis: like nominate form but upperparts darker, more jet black, less brown and on average larger, bill slightly shorter, and with near exclusively-longer occipital plumes. Disposition of the two subspecies in PNG poorly delineated and subject to disagreement in the literature. Specimens for analysis are too few to differentiate *clelandiae* and *chrysenia* adequately in PNG.

Measurements: wing, ad ♂ ($n = 8$) 149–161 (157), imm ♂ ($n = 3$) 151–152 (152), ad ♀ ($n = 5$) 141–153 (147); tail, ad ♂ ($n = 8$) 77–86 (80), imm ♂ ($n = 3$) 94–102 (98), ad ♀ ($n = 5$) 93–97 (94); bill, ad ♂ ($n = 6$) 31–33 (32), imm ♂ ($n = 3$) 37 (37), ad ♀ ($n = 5$) 35–39 (36); tarsus, ad ♂ ($n = 7$) 49–53 (50), imm ♂ ($n = 3$) 50–54 (52), ad ♀ ($n = 5$) 46–52 (49). Weight: ad ♂ ($n = 1$) 205, ad ♀ ($n = 1$) 163.

6. *P. c. chrysenia* Stresemann, 1934. *Ornithologische Monatsberichte* **42**, 147. Lordberg, Sepik Mts. Type specimen ZMB 33.1106.

Range: N scarp of the Central Range of PNG, including Lordberg, Hunstein Ra, and probably Schrader and N scarp of the Bismarck Ra. Apparently localised to the ranges that drain northward.

Diagnosis: traditionally said to differ from nominate form by having the long, black, loral feathering with a coppery sheen (like eye ring but darker) but several specimens lack this character, their lores being pure black. Tail longer than all other subspp., however, and occipital plumes longer than all except *chalcothorax*.

Measurements: wing, ad ♂ ($n = 2$) 158–159 (159), imm ♂ ($n = 2$) 151–154 (153), ad ♀ ($n = 9$) 138–154 (147); tail, ad ♂ ($n = 2$) 81–82 (82), imm ♂ ($n = 2$) 98–100 (99), ad ♀ ($n = 9$) 91–97 (95); bill, ad ♂ ($n = 1$) 37, imm ♂ ($n = 2$) 33–35 (34), ad ♀ ($n = 6$) 29–37 (34); tarsus, ad ♂ ($n = 2$) 52–53 (53), imm ♂ ($n = 2$) 48–51 (48), ad ♀ ($n = 9$) 45–48 (47). Weight: imm ♂ ($n = 1$) 210, ad ♀ ($n = 3$) 110–152 (130).

Habitat and habits   Table 4.1

Primary and secondary midmontane forests; also observed in areas of regrowth and of abandoned gardens; ad ♂♂ tend to remain in forest whilst ♀-plumaged birds more commonly range into latter habitats. Quite sociable, especially ♂♂ at and about courts (BB). Groups of ♂♂ and ♀-plumaged individuals forage in fruiting trees and elsewhere, 'scurrying along moss-covered horizontal branches, clambering among vines, and flying short distances between perches of tree crowns of the forest upper midstorey' (D. Bishop *in litt.*).

Diet and foraging   Table 4.2

Eats fruits and frequently forages for arthropods by stripping epiphytic mosses off large tree limbs and tree trunks to branches c. 35 mm dia (A. Mack *in litt.*; T. Pratt *in litt.*). One bird seen to snatch a grasshopper from cushion moss and then hold it to perch by a foot and tear at it with its bill. Another bird was flushed from a mound of earth on the ground (T. Pratt *in litt.*). Cooper (in Cooper and Forshaw 1977) recorded two birds eating the small red fruits of the tree *Callicarpa pentandra* in the regrowth of old gardens also frequented by a Trumpet Manucode, Superb, and a Lesser Bird (and quite close to human habitation). See Appendix 4 for other food plants. Forages predominantly in the upper and middle stories of the forest. A captive ♂ was regularly seen to eat green leaves (Coates 1990).

Vocalisations and   Figure 9.30
other sounds

The ♂♂ possess a significant vocal repertoire that includes several loud harsh notes, a musical 'wolf whistle', and a contact note. On the Utakwa R., Snow Mts, IJ, birds often a gave a *prat, prat* call in the early morning (Ogilvie-Grant 1915a). A UK captive ♂ gave soft twitterings and cluckings when close to a ♀ followed by an extremely loud and resounding *Kuck Kurrk* repeated 2–3 times (Frith 1968). The latter appears to be the loud, high-pitched and frantic-sounding, initially ascending and then descending, two-note shriek (the second note being almost whistled) recorded at Tabubil, PNG (Fig. 9.30). Characteristic of the ♂ is a conspicuously-upward inflected *kwoi* note given singly or up to four, typically two, times that terminates abruptly (Cooper in Cooper and Forshaw 1977). Healey (1980) records a loud advertisement song as *scree scree scree, oo-wit, oo-wi-oo*, the first three notes being short and grating and the latter two phrases powerfully whistled and low-pitched with the first, of two notes, rising and the second, of three notes, being a repetition of the first and reminiscent of a human 'wolf-whistle'. He also records a grating *chack* cry, a 'bell-like squeaking' call and a loud, grating *cor cor cor*. In approaching his court an ad ♂ gave 'metallic *shre* noises and little pleading notes' (Beehler in Coates 1990). A musical and breezy *whee o weet*, sometimes slurred into a brief *kwoieet* is characteristic of ♂♂—perhaps a group contact vocalisation (Beehler *et al.* 1986). One call note is described as like that of a large parrot, loud and raspy, and audible at great distance (King 1979). A two or three syllabled rather raspy, *Melidectes* honeyeater-like, *wrenh* was the usual call in the Ok Tedi area (Gregory 1995). At Mt. Pugent, 1800 m asl, PNG in late June duetting by ♂♂ was common, a ♂ giving a *KWEER* and another responding with a loud and ventriloquial *Kweer*. A four-note, ventriloquial whistle was also distinctive. A variety of notes, including a bleating whistle, a metallic note suggestive of Lawes' Parotia, and some quiet musical notes also heard. Ad ♂♂ produce a clacking with their wings but younger ♂♂ and ♀♀ appear not to have this capacity (BB).

Mating system

That ♂♂ display at a terrestrial court leaves no doubt the sp. is polygynous like Lawes' Parotia. Ad ♂♂ possibly maintain exploded leks, are promiscuous, and each ♀ attends to nesting duties without assistance.

Courtship behaviour

Observations of a UK captive ad ♂ and presumed ♀ showed that, whilst the ♂ mostly displayed by dancing across an area of aviary floor he had meticulously cleared of debris, he also did so on perches. Before displaying, this ♂ gave soft twittering and clucking calls near the ♀ which then were interspersed by a loud *Kuck Kurrk* call repeated 2–3 times. The first display performed consisted of the ♂, on a perch level with that of the ♀, turning away from her to crouch and raise and expand his white flank plumes over his primaries for *c.* 15 sec. On the same perch, the ♂ then repeatedly performed a 'sideways facing' display by hopping-on-the-spot very rapidly to face in opposite directions with each hop, in order to present his entire alternate profile to the ♀ (Frith 1968: fig. 3).

The ♂ then perched on a branch below and in front of the perched ♀, with his back to her, and watched her intently from over his shoulder. He then suddenly pointed his bill, head, and neck rigidly upward and swayed his entire body rhythmically side-to-side whilst fluttering his wings slightly and rapidly and alternately

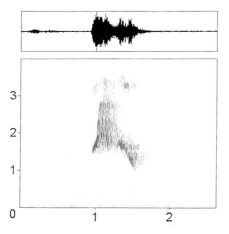

9.30 Sonogram from a recording of an ad ♂ Carola's Parotia loud, frantic-sounding, two-note, near-continuous shriek, the second being almost whistled, Aug 1993 by K. D. Bishop at Tabubil, PNG.

rapidly parting and closing the silver-tipped foreface feathers to reveal and erect his frontal crest. The latter complex rapid action presented an alternately flashing silver and gold focal point to this display. An elaboration of this display followed, in which the ♂, still below the ♀ and with his back to her in the same posture, with fluttering wings and crest movements, repeatedly raised and lowered his body from a crouched position on bent legs to a high point on fully-stretched straight legs at which point the wings were snapped opened fully either side of his body on the same plane as his back (Frith 1968: figs 4 and 5).

These, increasingly complex, initial displays culminated, 10 min after the last, in the ♂ suddenly performing an *Initial Display Bow*. This consisted of a rigid jerk of the body to point the bill downward as the tail was flicked and held briefly upward and the wings flicked slightly but sharply. The latter action permitted the long flank plumes to be released and raised, and the body then brought suddenly upright with a simultaneous spreading of the elongate breast and flank feathers to form a complete ballerina-like tutu that met over the central back. Thus far the behaviour is similar to that of Lawes' Parotia (see Figs 9.26(a), 27). With the attainment of this posture the crest feathering was again brought into constant play and the six 'flagged' occipital plumes separated and thrown forward in front of the ♂'s face and the 'whisker-like' throat feathers 'bristled' by continuous movement. This display lasted *c.* 90 sec, during which the only movement was that of the crest and throat feathering, before the bird suddenly and completely adopted a normal posture and plumage position. Another such display involved the ♂ simultaneously 'dancing' along perches with short, sharp, side-ways, mincing steps. The ♀ often somersaulted backward from her perch to another perch below, or to the ground, after the ♂ displayed. The significance of this odd, possibly aberrant, behaviour is unclear (Frith 1968; Frith and Coles 1976).

A court found in July on the S slopes of the Bismarck Mts consisted of a cleared patch of forest floor 1.5 × 1 m, with five or more thin branches 30–75 cm above it, on a 20–25° slope and partly concealed by light undergrowth but with a forest canopy gap above permitting sunlight to reach it at times (Cooper in Cooper and Forshaw 1977).

On 22 Oct *c.* 10 km N of the Jimi Valley, PNG, Healey (1980) observed an ad ♂ perform a display on his court similar to the court of a Wahnes' Parotia (Schodde and Mason 1974). Shortly after 0834 hr, the ♂ hopped onto a half-buried log at the centre of his court in an upright posture and fanned out his flank plumes to form the 'ballerina tutu' and stretched his neck fully upward for a few sec before rapidly bobbing his head side to side 3–4 times and then bobbing the body up and down several times on flexed legs. He then shuffled on his perch to turn 90° and extended his 'skirt' further forward to an angle to the vertical of *c.* 45° while raising the skirt sides even further laterally to almost horizontal, so that the two areas of white flank plumes almost met over his back. He then bobbed up and down again, with the now expanded breast-shield pulsed in and out so that at its maximum extension it was raised to be almost at a right angle to the breast. The skirt plumes were then returned to the normal position but the breast-shield remained expanded and was pulsed several times more when, still in his upright posture, the ♂ waggled his elongate throat feathers 'like a beard' several times before returning all plumes to normal and flying to a low tree above the court to call loudly before flying off. The entire display lasted *c.* 1.5–2 min and was silent. The six occipital plumes were not raised in this display but the silver and golden-bronze crest feathering was noticed and was therefore probably manipulated (Healey 1980).

A court watched for 3 days in the second half of June on Mt. Pugent, PNG, was attended by 3–5 birds, including an ad ♂ additional to the ad ♂ owner and subad ♂♂, with all birds occasionally joining in excited calling together. The ad ♂ 'court-owner' was seen to evict other birds and to perform several high-intensity ballerina-like dance displays to a ♀-plumaged bird and one of these bouts continued for half an hour, with a few intervals of *c.* 15–60 sec during which the ♀ rebuffed attempted mountings (Beehler in Coates 1990). To BB, the most interesting aspect of his observations was the social nature of the 'convergence' of birds at the court. A group of excitedly-calling ♂♂ could be heard approaching from a distance, and the various ♂♂ descended, calling, into the vicinity of the court. This may have been related to ♂♂ guiding a ♀ down to a display site.

Breeding Table 6.5
NEST SITE, NEST, EGGS, INCUBATION, NESTLING CARE, AND DEVELOPMENT: unknown.

Annual cycle Table 6.5
DISPLAY: courts maintained in the Jimi R., PNG area in June, July, and Oct, or the dry season. In the Herowana area, PNG ad ♂♂ actively court and mate during Oct–Dec (D. Gillison *in litt.*). BREEDING: gonadal activity in ♀♀ indicates at least Sept–Oct but data are few. MOULT: 22 of 100 specimens were moulting, representing all months except Apr, May, July, and Nov with most activity during Sept and Nov–Jan (Appendix 3).

Status and conservation Table 8.1
On a 1-km walk three, possibly four, ♂♂ were heard calling in an area on the S slopes of the Bismarck Mts, PNG, where the sp. was considered reasonably common (Cooper in Cooper and Forshaw 1977). Ad ♂♂ thought to be outnumbered by ♀-plumaged birds by a 3 : 1 ratio in the Ok Tedi area (Gregory 1995). This widespread sp. is under no current threat.

Knowledge lacking and research priorities
Details of distribution of the subspp. are poorly delineated and merit research, especially for mysterious *berlepschi*. Nidification and nesting biology require study as do interactions at courts with Lawes' Parotia. Any examples of hybridisation would be significant. Mechanics of court behaviour of this sp.—especially how it differs from *P. lawesii*—would be valuable.

Aviculture
Kept briefly in NG. An ad ♂ sent to Honolulu Zoo in 1954 lived and displayed there until dying in Nov 1968 (Gilliard 1969). Birds were maintained and performed displays at Chester Zoo, UK (Frith 1968; Frith and Coles 1976).

Genus *Pteridophora* Meyer, 1894

Pteridophora Meyer, 1894. *Bulletin of the British Ornithologists' Club* **4**, 11. Type, by monotypy, *P. alberti* Meyer.

The King of Saxony Bird of Paradise, invariably treated as a **monotypic genus** (Diamond 1972 excepted), is endemic to the highlands of NG's central cordillera. **Skull** generally like that of *Cicinnurus* and *Parotia* (Bock 1963). Crown markedly dished behind and above the orbit to accommodate large muscles required to manipulate the occipital plumes of ad ♂♂. External naris oval, the nasal bone broad and *Cicinnurus*-like in shape. Lacrymal medium-sized with slightly expanded foot resting on the jugal bar, similar to *Cicinnurus*. Palate of lighter construction but similar in all respects to *Cicinnurus*. The **systematic position** of *Pteridophora* is not well defined, but recent observations of courtship display are concordant with anatomical and plumage characters that link it to *Parotia* (Frith and Frith 1997c).

Bill straight and fractionally shorter than head length, and the proximal culmen quite sharply keeled. **Tail** squared-off, length 71% of wing length, the central pair of rectrices not longer than the rest. Two outermost primaries slightly emarginated on inner vanes. **Tarsus**

length 26% that of wing length and therefore proportionately far more like that of *Cicinnurus* than longer-legged *Parotia*. Lower tarsus scuted, more faintly proximally.

Highly sexually dimorphic. Gilliard (1969) considered this genus closest to *Parotia*, mainly because of the erectile cape, elongate and uniquely highly-modified occipital plumes, and metallic-like iridescent breast-shield (Plates 7 and 10). Unlike ad parotias, however, ad King of Saxony Birds do not have a pale iris and their ♀♀ lack the pale superciliary eye stripe and/or malar stripe of ♀ parotias. Moreover, the yellow underparts and extensive ochraceous wing patch of ad ♂♂ and the generally grey ♀ and imm ♂ plumage, scalloped and not barred below, are unique within the Paradisaeinae.

The promiscuous ad ♂ sings and postures from canopy perches before descending to a suspended vine, within a few metres of the ground, to perform a bouncing courtship display. Typical ♂♂ appear to be solitary and territorial, but in some areas may aggregate into exploded leks (Gilliard 1969). The single known nest site, **nest** structure, and **egg** (Plate 13) are typical of members of the Paradisaeinae, particularly *Astrapia, Parotia, Cicinnurus,* and *Paradisaea*.

King of Saxony Bird of Paradise *Pteridophora alberti* Meyer, 1894

PLATE 10; FIGURE 9.31,32

Other names: King of Saxony's Bird of Paradise, Enamelled Bird of Paradise, The Enamelled Bird. *Kis-ba* in Bismarck Mts behind Nondugl, *Kis-a-ba, Kis-a-baa* or *Kissaba* in Kubor Mts behind Kup (Mayr and Gilliard 1954; Gilliard 1969). *Wigelo* of the Dani people and *tat* by those of Uhunduni, Ilaga Valley, IJ (Ripley 1964). *Kongbuk* (ad ♂) and *Doongbi* (♀-plumaged) in the Central Highlands, PNG (Sims 1956). *Nopd kolman* (ad ♂) and *Nopd neb* (♀-plumage) of the Kalam people (Majnep and Bulmer 1977). *Gangade* (Goldman 1981) or *Tima Gangade* of the Huli people (C. Ballard *in litt.*), *Wárale* of the Fore people (Diamond 1972) and *Leme* of the Ialibu area, PNG (Kwapena 1985). *Petre* of Yabi people, Weyland Mts IJ (Rothschild 1931). *Oretha* of the Gimi people, Crater Mt., PNG (D. Gillison *in litt.*). *Inem*, Hindenburg Ra, PNG (skin label data).

A small bird of paradise of the highlands of the NG central cordillera, invariably heard more often than seen. Ad ♂ black with pale yellow underparts, ochre wing patch, and two remarkably long, erectile, head plumes. ♀♀ and imm ♂♂ grey-brown above and off-white below with greyish scalloping. Monotypic.

Description

ADULT ♂: 22 (50) cm, entire head, mantle, and back deepest velvet black with dull sheen of bronzed green, particularly on elongate 'cape' feathers of mantle, in some lights. From behind each eye, amid an 'ear tuft' of elongate feathers grows a uniquely-modified occipital plume up to 50 cm of bare central feather shaft with 40–50 plastic-looking 'flags' decorating its outer side only. The upper surface of each 'flag' is an enamel-looking glossy Sky (168C) to True Blue (168A) and the underside Sepia (219). Rump, uppertail coverts, and dorsal surface of tail matte black washed very dark brown as upperwings except for exposed bases and broad leading edges of secondaries and all but outermost two primaries which are Cinnamon (123A). Chin and throat as upper head but large and sparse scale-like feathers at centre and lower border of black throat narrowly-tipped, strongly iridescent, green-blue to purple in some lights (being a sub-obsolete breast-shield). Remaining underparts dark egg-yolk yellow (18), brighter on breast, paler and duller on vent and undertail coverts, with creamy feathering on flanks. Underwing: coverts and bases to most flight feathers Tawny Olive (223D), with blackish tipping to the former, and the extensive flight feather tips and ventral surface of tail blackish brown (19). Bill black, eye dark brown, legs and feet dark brown-grey, mouth pale to rich aqua-green.

ADULT ♀: 20 cm, entire upperparts grey to sooty grey, darker and browner (219) on lower back. Warmer, ashy, grey (79) on head and neck with faint trace of blackish tipping giving finely-scalloped appearance. An elongate, narrowly-pointed feather or two often present behind each eye. Upperwing blackish brown (219) with narrow paler, buffy, outer edges to coverts and narrow off-whitish outer edges to flight feathers. Chin, throat, and neck buffy grey with broad dark brownish grey barring giving scalloped appearance. Breast to vent whitish, heavily marked with open, shallow, blackish brown chevrons. Underwing: coverts pale Buff (124) broadly barred blackish brown, flight feathers pale brown-grey (121) with narrow pale Buff trailing edges. Thighs pale fawn (27) and undertail coverts Buff (24) with blackish brown open chevrons. Ventral surface of tail pale brown-grey. Bill black, iris dark brown, legs and feet bluish grey to greyish brown.

SUBADULT ♂: as ad ♀ with few feathers of ad ♂ plumage intruding, to as ad ♂ with few ♀ plumage feathers remaining. First sign of ad ♂ plumage is black nasal tuft feathering and primaries darker with orange (not grey) concealed bases followed by more black head plumage and some yellow in breast.

IMMATURE ♂: as ad ♀ but upperparts said to be paler, more uniform brownish grey to grey, less scalloped. Underparts whiter, owing to less dark barring and spotting (Gilliard 1969). Iris red-brown.

NESTLING: A hatchling with wings 10 mm long, head + bill length of 20 mm and tarsus of 11 mm was naked with dark greyish purple skin, paler on abdomen, with a mid grey gape, off-white claws, and tiny white egg tooth. Three days later this nestling's skin was blackish purple, gape and claws dull off-white (Frith and Frith 1990c). JUVENILE: undescribed.

Distribution

Western and central two-thirds of the central cordillera of NG, from the Weyland Ra of IJ eastward possibly to the Kratke Ra of central PNG, including the Snow and Star Mts, the Victor Emanuel, Bismarck, S Karius, Ambua, Giluwe, Hagen, Schrader, and Kubor Ra, and Mt. Bosavi (Coates 1990; M. Hopkins via

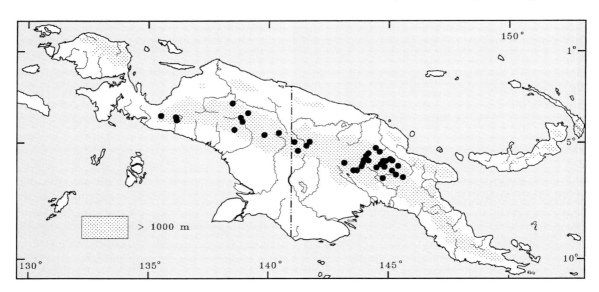

T. Pratt *in litt.*; G. Clapp *in litt.*). Said to be as far E as Kaintiba and Menyamya (informants to T. Pratt *in litt.*). Ranges from 1400 to 2850 m, mainly 1800–2500 m.

Systematics, nomenclature, subspecies, weights, and measurements

SYSTEMATICS: both the ♂ and the ♀ plumages are distinct from all other paradisaeid plumages, making placement within the phylogeny difficult. Gilliard (1969) considered *Pteridophora* closely related to *Parotia* and *Lophorina* citing the expandable dorsal cape, complex occipital plumes, and the sub-obsolete metallic breast-shield as shared characters. The fact that all three of these occur in both *Pteridophora* and *Parotia* together with shared courtship display components adds weight to their proposed close relationship (Frith and Frith 1997*c*). We concur with comments of Gilliard and LeCroy (1968), Diamond (1972), Coates (1990), and Cracraft (1992) in treating *P. alberti* as monotypic. We do note, however, that birds of what has been considered the form *P. a. hallstromi* have on average ($n = 67$) slightly longer wings (6%) and tail (4%) than the two other populations, which are almost identical in size ($n = 77$) but whose overlap in ranges of measurements is considerable (Frith and Frith 1997*b*). Hybridisation: unrecorded.

SPECIES NOMENCLATURE AND ETYMOLOGY

Pteridophora alberti Meyer, 1894. *Bulletin of the British Ornithologists' Club* **4**, 11. Mts on the Ambernoh R. [= Mamberamo R.], but restricted to the Weyland Mts by Mayr (1941*a*). Type specimen not located.

Synonyms: *P. a. hallstromi* Mayr and Gilliard, 1951. *American Museum Novitates* **1524**, 12. Forests above Tomba, south slope of Mt. Hagen, Central Highlands, Mandated Territory of New Guinea. Type specimen AMNH 348210.

P. a. bürgersi Rothschild, 1931. *Novitates Zoologicae* **36**, 253. Schraderberg, Sepik Mts. Type specimen AMNH 678693.

Etymology: *Pteridophora* = Gr. *pteris, pteridos*, a feathery fern; *phoros* = carrying (with reference to the structure of the two head plumes of ad ♂♂); *alberti* = after Albert or Albrecht, King of Saxony.

WEIGHTS AND MEASUREMENTS
Measurements: wing, ad ♂ ($n = 53$) 119–133 (126), imm ♂ ($n = 40$) 112–129 (122), ad ♀ (n 45) 108–128 (115); tail, ad ♂ ($n = 54$) 79–97 (86), imm ♂ ($n = 40$) 78–95 (88), ad ♀ ($n = 45$) 77–91 (84); bill, ad ♂ ($n = 52$) 22–27 (24), imm ♂ ($n = 39$) 21–27 (24), ad ♀ ($n = 44$) 22–26 (24); tarsus, ad ♂ ($n = 54$) 30–35 (32), imm ♂ ($n = 37$) 28–34 (32), ad ♀ ($n = 45$) 28–34 (31). Weight: ad ♂ ($n = 13$) 80–95 (87), imm ♂ ($n = 12$) 71–93 (81), ad ♀ ($n = 12$) 68–88 (77).

Habitat and habits Table 4.1

Mid- to upper montane forest and forest-edge and in lightly-disturbed areas of forest about hunting lodges and tracks. In Tari Gap, PNG, ad ♂♂ tend to be seen in the forest canopy at their advertisement perches whereas ♀-plumaged birds are frequently seen at all levels of the vegetation structure down to shrubbery near the ground. While this apparently is reflected in relative numbers of birds caught in mist-nets from ground to 3 m high there during Sept–Feb, 3 ad ♂♂ and 14 ♀-plumaged birds, it could also reflect relative densities of the different plumage types. A ♀-plumaged Tari Gap bird caught and ringed on 12 Sept 1986 was retrapped on the same site on 15 Oct 1987 and again on 21 Jan 1989. David Bishop (*in litt.*) saw an ad ♂ calling from the same perch as a ♂ Loria's Bird which displaced it briefly.

Diet and foraging Table 4.2

Forages primarily for fruit in the canopy and sub-canopy of montane forest; with an apparent predilection for green fruit. Fruits only noted in stomachs of birds collected in Weyland Mts (Rothschild 1931). Two Central Highlands, PNG, birds and Mt. Kominjim birds contained small green berries (Sims 1956; Gilliard and

LeCroy 1968). Three individuals from two locations in PNG contained 75–90% fruit and 10–25% arthropods, animal groups represented being Araneida, Chrysomelidae, Dermaptera, and larvae of Lepidoptera (Schodde 1976).

Of 59 sightings of foraging birds on Mt. Hagen, 53% were feeding from tree branches, 24% at fruits, 9% at flowers, 9% from leaves, 4% at dead wood, and 1% at twigs or buds. Of 51 records of locations of birds within the forest when first seen, 35% were at 0–15 m above ground, 47% at 15–25 m, and 18% at 25–35 m high. Fruits were seen to be eaten 73 times and arthropods 11 times. Of the 73 records of birds taking fruits, 27% were of a *Fagraea* sp., 19% *Medinilla markgrafii*, 10% *Eurya tigang*, 7% *Perotettia alpestris*, and 5–1% at least 15 other plants (Kwapena 1985). For other food plants see Appendix 4.

Of 24 sightings of feeding birds at Tari Gap, PNG during Sept–Oct, 3 occurred 3–5 m above ground, 5 at 6–10 m, 11 at 11–15 m, and 5 at 16–18 m high. All birds were lone foragers except for one group of 3 ♀-plumaged birds and one ♀-plumaged bird with a begging juv; 20 foraged at fruits, 2 on dead branches, and 2 on tree limbs covered with mosses and other epiphytic growth, 1 at foliage, and 1 in dead foliage; 2 birds foraged at 2 sites. Of the 27 feeding individuals, 85% ate fruit and 19% arthropods, one bird feeding upon both. All birds ate the food at the foraging site. Of the 20 fruit feedings for which plants could be identified, 65% were of *Timonius belensis*, 30% of *Fagraea salticola*, and 5% of *Dimorphanthera alpina*. Birds feeding upon arthropods tore and probed at mosses and lichens beneath relatively-small branches and, once, a bird took a dead fern frond to a perch to hold it by one foot as it tore it apart with the bill. Another was seen to do likewise with a piece of dry moss torn from beneath a branch (CF).

Of 17 faecal samples collected from mist-netted birds at Tari Gap during Sept–Oct, 15 contained *Timonius belensis* fruit remains, 12 *Fagraea salticola*, and one the remains of each of *Garcinia* sp., *Acronychia* sp., *Pittosporum* spp, and *Riedelia* sp. Four of the 17 samples contained insect remains, the largest being a weevil 6 mm long and a beetle 3 mm long (CF). It has been tentatively suggested that the diet consists of 80% fruit (Beehler and Pruett-Jones 1983).

An ad ♂ displaced a Common Smoky Honeyeater while another was supplanted by a ♀-plumaged Brown Sicklebill and another by a White-breasted Fruit-Dove *Ptilinopus rivoli* (M. Laska *in litt.*).

Vocalisations and other sounds

Figure 9.31

The ad ♂ gives a unique advertisement 'song', whereas birds in ♀ plumage give a distinctive scolding note. The advertisement song of ad ♂♂ is notoriously difficult to describe. It has been likened to the sound of 'the squeaking of rusty iron' (Shaw Mayer in Sims 1956) and said to start 'as a drawn-out hissing note (sounding like escaping steam) and terminate in an explosive rasp that can be heard for almost a mile over the crown of the cloud forest' (Gilliard 1969). Diamond (1972) described the song well, as lasting *c*. 3 sec, gradually increasing in volume and of a weird unbird-like quality 'a very dry rattling, a spitted jumble of insect-like notes poured out at a machine-gun pace and suggestive of bad static on the radio, which it briefly turns into a twittering at the climax of the crescendo'. See Fig. 9.31(a). The singing of an ad ♂ perched on his bare advertisement perch at the apex of an emergent tree in Tari Gap, PNG was timed during 0920–1020 hr on 15 Nov. Each song lasted 4–5 sec and 48 were given at an average duration of 1 per 71 sec. Songs were more frequent during 0920–0950 hr (every 56 sec) than during 0950–1020 hr (every 84 sec) (CF). On 6 Nov 1987 a courtship display was observed by D. Frith who, having heard a bird giving a continuous hissing or whirring sound in the forest, suddenly sighted the ad ♂ perched on a horizontal branch of a small sapling 2 m above ground. The call is basically a subsong of hissing sound within which can be heard soft high-pitched clucks, chatterings,

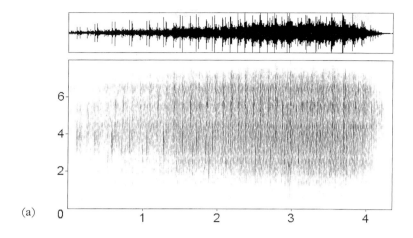

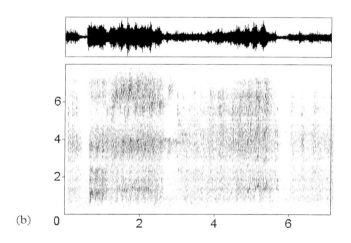

9.31 Sonograms from a recording of an ad ♂ King of Saxony Bird giving (a) the complex loud canopy advertisement song and (b) the substage courtship display subsong in Dec 1995 by P. Hurrel and M. Potts for the BBC at Tari Gap, PNG.

mewings, and squeakings, some like a soft form of the advertisement song, the whole somewhat reminiscent of Common Starling *Sturnus vulgaris* subsong, but not audible at any great distance (Fig. 9.31(b)). It is possible some vocal avian mimicry is incorporated, especially of scrubwrens *Sericornis* spp. (CF) but this requires confirmation as such is unknown in wild birds of paradise. Young ♂♂ monotonously repeat a jeering call reminiscent of that of the Hooded Cuckoo-shrike *Coracina longicauda*: four or five notes, the series rising slightly in pitch, each note harsh, rolling, and down-slurred *chweer chweer chweer chweer* (Beehler *et al.* 1986). ♀-plumaged ♂♂ are noisier than ad ♂♂ (Frith and Frith 1992*a*).

Mating system
Polygynous, with the promiscuous ♂ advertisement-singing solitarily from a number of traditional, exposed, emergent perches, and displaying to ♀♀ from a slim vine in the forest substage. Two ad ♂♂ sung with each calling every *c.* 45 sec, within earshot of each other 100 m apart at Mt. Gahavisuka Park, PNG at 0700–0709 hr on 22 Aug (BB). What has been described as the territory of an ad ♂ has been approximated to be 137 m in dia on Mt. Hagen in mid-April, the bird calling from emergent trees 30–37 m tall about this area and occasionally flying down to feed upon the fruits of vines and trees at 15–21 m above the ground. This ♂ remained active in this way throughout 2 days of observation, but his calling was most frequent during the mid-afternoon (Beach 1975). Certainly our observations and those of others do suggest that solitary ♂♂ occupy and attend a traditional limited area including a number of exposed calling perches in emergent trees, but it remains to be demonstrated that ♂♂ defend an all-purpose territory. An ad ♂ was observed during July–Aug of each year throughout 1985–95 to occupy the same perch and to call during late afternoons. While it is not known if the same individual was involved, this does demonstrate traditional use of a canopy calling site (D. Bishop *in litt.*). Gilliard (1969) wrote of clans of ♂♂ dancing 'in isolated ill-defined groups' and in 'isolated areas in which the ♂♂ gather to display' by which he was referring to relatively-large tracts of habitat. In the Kubor Mts he encountered three groups of 3–7 ♂♂ at inter-group distances of *c.* 8–16 km apart. The ♂♂ of two groups each displayed on individual perches. He noted ♂♂ to be spaced some 366 m or more apart and tentatively suggested that in this sp. ad ♂♂ form exploded leks, but evidence for this is absent and confirmation is required. Song perches of ♂♂ appear to be regularly dispersed through the extensive Tari Gap forests (CF). In noting that 'the ♀♀ outnumber the ♂♂ by a ratio of about three to one' Gilliard and LeCroy (1961) were alluding to the more numerous ♀-plumaged birds (which include imm ♂♂) and not known ♀♀. The ♀ alone attends the nest (Frith and Frith 1990*c*).

Courtship behaviour
Forest canopy advertisement singing and display: ♂♂ sing and perform advertisement displays from perches beneath some foliage of emergent trees. Here ♂♂ perform relatively simple displays by gaping as they sing and by moving their occipital flag plumes about. The latter may be brought well forward, to be held almost parallel in front and with the mantle cape and breast-shield erected (Fig. 9.32(a)). Alternatively the occipital plumes may be brought around to be horizontal and at right angles to the head and twisted so that the flags are below the shaft, and the mantle cape and head plumage raised, as the mouth is opened wide to sing (Fig. 9.32(b)). On numerous occasions the occipital plumes of the singing ♂ may be moved synchronously and/or independently about the head at all possible angles (Fig. 9.32(a,b,c)).

The ♂ descends from his traditional canopy perch into the understorey to court a ♀ on vines *c.* 2–15 m above ground (most often closer to the former height; Rothschild 1931; Gilliard 1953; 1969; Beach 1975; Frith and Frith 1997*c*).

Understorey bouncing display: ♂♂ descend from high advertisement perches to understorey vines to court and mate, but it remains to be determined if ♀♀ are led, chased, or make their own way there. The ♂ perches up to 50 cm below the ♀ on a vine giving a continuous hissing or whirring sound with soft high-pitched chatterings and squeaking, not audible at any great distance. He may gape, thus exposing his vivid aqua-green mouth interior. The vocalising ♂ clings tightly to his perch and extensively, vigorously, and repeatedly pumps or flexes his legs from the tibiotarsal joint to bounce up and down in rhythmic unison, while retaining his grip on the bouncing vine with erect mantle cape and breast-shield, and occipital plumes held over his back (Fig. 9.32(d-e)). A ♀ clinging to the vertical stem above the ♂ (her perch being at right angles to his) is bounced about by the vine undulations generated by the ♂.

The occipital plumes are mostly held out behind the ♂ and near horizontal to the ground (Fig. 9.32(e)), so that their distal sections sway up and down in the vertical plane as the bird bounces (Fig. 9.32(d-e)). Now and then, however, he might repeatedly raise his occipital plumes to bring them widely apart and slightly forward of his head and then return them to their normal position over the back. Both birds rise and fall in unison with the vine movement, the ♀ watching the ♂ but without active participation. She may turn away from the ♂ and hop a little up her perch. This causes the ♂ to stop bouncing instantly, but not his hissing, and to hold his closed wings only the slightest away from his body to shiver or tremble them rapidly in a juvenile fashion while holding his head low with bill directed at the ♀ and occipital plumes just above his back. Apparently in response to this change in ♂ display the ♀ usually turns and approaches him and he again starts to bounce vigorously. At this point the occipital plumes might, however, be repeatedly brought round wide apart then to be brought together out in front of the ♂ (Fig. 9.32(f)). This wing-shivering is usually given when the ♀ shows signs of disinterest or departure, the bouncing display being continued as soon as her interest is renewed. If the ♀ remains perched on the vibrating perch the ♂ slows and weakens his bouncing while bringing his occipital plumes widely around his head to unite them together in front of him and direct them at the ♀ more frequently. Throughout the bouncing and wing-shivering displays the ♂ gives the hissing subsong continuously, his mandibles held slightly apart.

At this point in the courtship the ♂ starts stiffly and, initially slowly, to rotate his upper body and simultaneously waggle his increasingly erectly-held head and neck side-to-side in a peculiar motion. His mantle cape, breast-shield, and head feathering are erected to their maximum and occipital plumes raised to $c.$ 45° and held widely apart (Fig. 9.32(g)) then to be brought around either side of the head finally to project forward of and above him. He hops, initially slowly but progressively more quickly, up the vertical perch in this stiffly upper-body-rotating, head-waggling, and occipital-plume-swirling display to mount and mate the ♀ for $c.$ 3 sec. During copulation the ♂'s bill tip is pointed down into the ♀'s nape feathers, his fully-open wings are flapped to maintain balance, and his much-fanned tail is brought around and beneath the left side of the ♀'s tail and his occipital plumes held vertical above his head and apart at $c.$ 45°. Immediately after the filmed copulation the ♂ fluttered down to perch lower down on the display perch as the ♀ flew to an adjacent one and the ♂ then directed a renewed juvenile wing-shivering display at her until she departed the area.

The above complete courtship display agrees well with partial descriptions (F. Shaw Mayer in Rothschild 1931; Gilliard 1953, 1969; Beach 1975; Healey 1975). The people of Tsuwenkai Village, Bismarck Ra informed Healey (1975), however, that both sexes perch on a vertically-hanging vine and that the ♂ jerks back and forwards, parallel to the ground rather than bouncing in a vertical plane.

Breeding Table 6.5

Gilliard (1953) wrote that after a brief courtship and mating the ♀ leaves to carry out nesting duties alone, but this was clearly an assumption based on his observations of ♂ behaviour, as at that time—and indeed not until 1988 (Frith and Frith 1990c)—no nest was known.

NEST SITE: the only known nest (Frith and Frith 1990c) was found being constructed by a

312 King of Saxony Bird of Paradise *Pteridophora alberti* Meyer, 1894

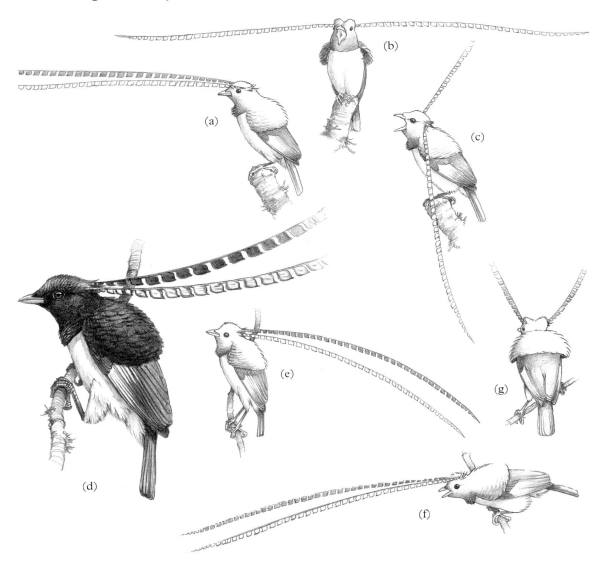

9.32 Some courtship display postures of ad ♂ King of Saxony Bird: (a–c) *Forest canopy displays*, (d–e) *Understorey bouncing display*, (f) female-enticement display, (g) pre-copulation perch accent with head waggle. See text.

♀-plumaged bird on 24 Dec, 11 m above ground in a large, trifurcate, upright branch fork of a *Timonius belensis* tree on a gentle slope at 2665 m in the Tari Gap, PNG. Three other *T. belensis* trees were in the immediate area, which consisted of disturbed mixed beech forest with few large trees other than *Pandanus tectorius*. The nest tree was within 25 m of a disused pandanus-thatch hunting hut and associated forest disturbance and 30 m from where the forest-edge met extensive subalpine grassland.

NEST: a shallow open cup of *c.* 170 mm outside dia, and internal egg-cup depth *c.* 55 mm, consisting of a fairly-loose accumulation of numerous and varied fine epiphytic orchid stems, probably of *Glossorhyncha* spp., including some *Bulbophyllum*-like orchids, and very fresh, green, 'comb-tooth', fern fronds up to 250 mm long and probably of *Blechnum* or *Doodia* spp; no sticks used. Egg-cup interior sparsely lined with fine epiphytic orchid stems or tendrils and a few of other kinds of plant tendrils.

EGGS: smooth-surfaced with a slight lustre or gloss and pale Buff (124) in ground colour, with numerous longitudinal blotches of purplish greys, purple, grey, and russet, underlaid with spots and flecks of these colours. The longitudinal blotches form a band around the larger end (Plate 13). Both ends of egg rather bluntly rounded; egg oblong oval or elliptical, measuring 33.6 × 23.5 mm and weighing 9.7 g when *c*. 10 days old.

INCUBATION: a presumed ♀ was watched nest-building during 24–28 Dec but during 29–31 Dec was watched but not seen to add to the structure. On 2 Jan and thereafter the ♀ was obviously incubating. Incubation activity, exclusively by ♀, was observed for 19.5 hr over 15 watches of 1–2 hr (mean = 1 hr 18 min) during 3–20 Jan (at egg age of 3–20 days assuming it was laid the morning of 1 Jan). The ♀ made an average of 4 (2–6) nest visits to incubate per hr, spending a mean of 8.5 min (5–18 min) incubating per visit, and 58% of total observation time at the nest. Absences from nest averaged 4 (2.5–6.4) per hr, each averaging 6.5 min (4–12 min). Incubation period > 22 days. During incubation ♀ seen to chase away small passerines from within 1–2 m of nest on three occasions.

NESTLING CARE: observations of a ♀'s care of her single nestling were made for 17.5 hr during six watches of 1–4 hr during 24–30 June, at a nestling age of 1–7 days. The ♀ made an average of 4 (3–5) feeding and/or brooding visits to the nest per hr, the brooding bouts averaging 6.5 min (6–8 min). Mean time the ♀ was at the nest was 7 min, including time spent perched on the nest rim (= 4.4% of total observation time). An average of 3.5 (0–5) feeding visits per hr were made to the nestling, averaging 45 (6–55) sec long before the ♀ subsequently brooded or left the nest. On four occasions she fed the nestling then departed without brooding it, and on seven occasions she brooded the nestling without first feeding it. She brooded the nestling for 41% of total observation time. Her absences from the nest and nestling averaged 4 per hr, with a mean duration of 9 min (5–11 min). All nestling meals, which could not be identified, were regurgitated to the young by the ♀ (Frith and Frith 1990c).

NESTLING DEVELOPMENT: unknown.

Annual cycle Table 6.5
DISPLAY: at least Sept–Apr. On 11 September 1986 CF observed an ad ♂ apparently chasing a ♀ to a display vine *c*. 6 m above ground. On 6 Nov 1987 a courtship display was observed by D. Frith (personal communication). BREEDING: would appear possible at any month but presumably peaks *c*. July–Feb (Table 6.5). EGG-LAYING: known only for Jan (Tari Gap). MOULT: 53 of 139 specimens were moulting, with moulting birds from all months (Appendix 3). Of nine birds caught in the Tari Gap during Sept, 44% were moulting, of eight Oct birds, 63% were moulting, and the single Jan bird caught was in moult (Frith and Frith 1993a).

Status and conservation Table 8.1
Some areas of extensive and apparently-suitable habitat appear to lack ad ♂♂ whilst in others a number of ♂♂ can be heard singing from their advertisement perches where they are dispersed within auditory contact (Gilliard 1969). Gilliard (1953) found three ♂♂ calling and displaying at trees within 1.6 km of each other. There appeared to be *c*. 1 bird to each 4.5 ha at Tsuwenkai in the Bismarck Ra, PNG (Healey 1986). Common in the Tari Gap region, PNG.

Uncommon on Mt. Hagen; but common at 2745 m, and moderately so at 2105–2560 m in the Okbap area, Star Mts, IJ (J. Diamond and D. Bishop *in litt.*). ♀-plumaged birds are obtained by Kalam people by setting perches above forest pools and from a bush hide shooting birds as they come to bathe (Fig. 7.3) or drink (Majnep and Bulmer 1977). The suggestion that this widespread and common bird be classified as seriously endangered (Peckover 1990) is unjustified.

Knowledge lacking and research priorities
Appears predominantly frugivorous, specialising in fruits of the 'false fig' *Timonius belensis* but confirmation required. Whether ♂♂ aggregate into exploded leks merits study. Finally, it would be useful to investigate the nestling diet (which may be high in fruit content).

Aviculture
Gilliard (1953) noted that only one bird had been taken alive to a zoo, but that it died almost immediately.

Other
A recent novel discovery concerns this bird and the sympatric Archbold's Bowerbird *Archboldia papuensis*. This large bowerbird is restricted to upper montane moss forest in the central ranges of NG (Frith *et al.* 1995). ♂♂ do not construct a typical bower but accumulate live and dead fern fronds on the forest floor to form a thick 'mat' directly beneath several perches which they drape with numerous orchid stem decorations. In addition to decorating the 'mat' with piles of snail shells, beetle wing cases, pieces of fungus, charcoal and tree resin, fruits, and other objects, ad ♂ owners of larger and finer bowers also lay the occipital plumes of ad ♂ King of Saxony Birds on their central mat area. Of 24 bowers examined the six largest were so decorated with an average of three and a maximum of six King of Saxony occipital plumes. It is believed that this use of King of Saxony plumes by ♂ Archbold's Bowerbirds to decorate their bower is an example of promiscuous ♂ bowerbirds using objects that are rare in the birds' environment in order to impress potential mates with their ability to find and retain such uncommon decorations (Frith and Frith 1990b, 1991; Diamond 1991; Frith *et al.* 1996).

Genus *Ptiloris* Swainson, 1825

Ptiloris Swainson, 1825. *The Zoological Journal* **1**, 479. Type, by monotypy, *Ptiloris paradiseus* Swainson.

Synonyms: *Craspedophora* Gray, 1840. List of the Genera of Birds. ed. 1, add. and err. p. 1. Type, by original designation, *Falcinellus magnificus* Cuvier (not preoccupied by *Craspedophorus* Hope 1838 The Coleopterist's Manual **2**, 165).

Mathewsiella Iredale, 1922. *Bulletin of the British Ornithologists' Club* **43**, 39. Type, by original designation, *Craspedophora magnifica claudia* Mathews. To replace *Craspedophora* Gray (1840), not *Craspedophorus* Hope (1838).

The riflebirds are **three species** (or possibly four *cf.* Beehler and Swaby 1991) of medium to large, stout birds that include one or two (*P. paradiseus*/*P. victoriae* and *P. magnificus*/ *P.* [*magnificus*] *intercedens*) superspecies (Plate 8). The largest sp., *P. magnificus*, occupies much of the lowland forest of mainland NG and the extreme N tip of CYP, Australia, whereas the two smaller spp. inhabit the wetter coastal areas of E Australia.

Skull generally like that of *Paradisaea* but with an ectethmoid, while much as in

Paradisaea, with a much larger lacrymal reaching to the frontal dorsally and to the jugal bar ventrally. Bones of the palate more elongate than in *Paradisaea* but otherwise similar. On this evidence Bock (1963) tentatively grouped *Ptiloris* with *Semioptera* and *Seleucidis*. Our analysis supports placing *Ptiloris* in a clade with *Lophorina*, which in turn, is paired with a clade including *Parotia* and *Pteridophora*. Given the recent nature of the radiation, it is not surprising that other assessments place *Ptiloris* closer to *Seleucidis*—the rapid evolution of characters has produced an array of reversals, and relative placement of these taxa depends on what characters have been employed in assessment (see Chapter 3).

Sexually dimorphic in **size** (10–20% in wing length) and slightly to moderately in bill length and decurvature. **Bill** twice the head length, laterally compressed to narrow, and slightly decurved. Nostril partly to near entirely covered by feathers. **Tail** is not graduated (*P. magnificus* and *P. victoriae*) or is rounded, barely graduated, (*P. paradiseus*) with central pair of rectrices shortest and highly-iridescent metallic green or blue in ad ♂♂. **Wings** moderately long and highly modified in ad ♂, being much rounded, with square ends to secondaries and most primaries which are almost equal in length. Outer three primaries inwardly curving and pointed. Outermost, or outermost two (*P. magnificus*), primaries mildly emarginated on inner vane. **Tarsus** length c. 22% of wing length. Lower tarsus scuted but more faintly above. Strongly sexually dimorphic in plumage. Ad ♂♂ basically deep velvet black with a moderate to large throat/breast-shield of highly-iridescent metallic and scale-like feathers, which also adorn the crown. Feathering about throat and surrounding breast-shield and that of the nape soft, thick, and plush. Flank plumes in *P. magnificus* long and filamentous like *Paradisaea* and suggestive of *Seleucidis*-like wires at terminal tips of some. Abdominal feathers of other two *Ptiloris* spp. are grossly broadened and decorated with dull iridescent green (Plate 8).

Polygynous, ad ♂ solitary and promiscuous with traditional display perch; ♀ alone conducts nesting duties. ♂ courtship displays show some *Paradisaea* characteristics (Frith and Cooper 1996) but also many display characters that are distinct. Nests placed in dense epiphytic vegetation or in frond bases of pandanus tree crowns. **Nests** of all three spp. characteristically involve numerous *Polypodium* fern stems and fronds as outer material, and sloughed snake skin added to nest rim by the two smaller spp. **Eggs** much like those of other typical Paradisaeinae (Plate 13).

Magnificent Riflebird *Ptiloris magnificus* (Vieillot, 1819)

Plate 8; Figures 5.2, 9.33–36

Other names: Magnificent Rifle Bird, Albert Rifle-Bird, Prince Albert's Rifle-bird, Prince Albert's Rifle Bird, Scale-breasted Paradise Bird. *Yagoonya* or *Yogoonga* of Cape York Aboriginal People (Macgillivray in Elliot 1873) and *Dowde* of Darnley I. (Elliot 1873). *Kan'buda* (M) *Tsido* (♀-plumaged) at Lake Kutubu, PNG (Schodde and Hitchcock 1968). *Aror* in the Kemp Welch River, E of Port Moresby, PNG (Loria, in Sharpe 1891). *To-owa* of the Daribi people (Diamond 1972) and *Urumun* of Wanuma people of the Adelbert Mts (T. Pratt *in litt.*). *Iwe'* of the Fair language at Bomakia and *Huwo* of the Kombai language at Uni, Irian Jaya (Rumbiak 1994).

Stout, round-winged, and short-tailed, with powerful legs and feet, a long and decurved bill that is laterally compressed and powerful. Often clings to dead wood which it vigorously excavates for boring insects. Promiscuous ♂♂ give far-carrying, whistled or growled notes from the forest canopy. Ad ♂♂ elusive and difficult to see. Polytypic; 3 subspp.

Magnificent Riflebird *Ptiloris magnificus* (Vieillot, 1819)

Description

ADULT ♂: 34 cm, head velvety jet black decorated from lores to nape with scale-like feathers intensely-iridescent metallic green-blue, washed purple and Magenta (2) in some lights. Sides of face, chin, and throat dully-iridescent violet-purple (172) and/or rarely Olive-Green (46) sheen visible. Mantle to uppertail coverts, lesser wing coverts, and tertials velvety jet black with strong iridescent sheen of rich violet-purple washed Magenta and, rarely, deep blues and/or Olive-Green sheen visible. Greater coverts, alula, and exposed wing flight feathers iridescent deep dark blue (74), rarely seen to be washed Magenta. Dorsal surface of tail velvety jet black, basal outer edges of vanes with iridescent deep blue (74) sheen, but shorter central pair intensely-iridescent metallic dark blue-green to green-blue with violet-purple and/or Magenta sheens in some lights. Narrow central throat broadening to upper breast decorated with extensive delta-shaped breast-shield of scale-like, intensely-iridescent, metallic greenish blue feathers with narrow jet black centres exposed at lower centre of the shield that in some lights is washed with rich violet-purple and/or Magenta sheens. The lower shield edge bordered by a narrow breast band of velvety jet black feathers with violet-purple iridescent sheen in some lights and immediately beneath this an even narrower band of highly-iridescent bronzed yellow-green (58). Remaining underparts to vent, including elongate filamental flank plumes, matte brownish black overwashed with dully-iridescent deep rich Maroon (31) to Carmine (8) and/or, rarely, Olive-Green (47) particularly against the breast band. Vent and undertail coverts matte blackish, as are tips of longer flank plumes. Underwing brownish black but broad tips to secondaries that become narrower on all but the outermost two or three primaries reflect iridescent blue to white in appropriate light (see Fig. 9.36(b)). Ventral surface of tail brownish black. Bill black, gape pale yellow and visible at all times, iris dark brown, legs and feet dark leaden-grey to blackish, mouth pale lemon yellow to lime green.

ADULT ♀: 28 cm, bill and tarsus slightly shorter and wing length significantly shorter than that of ♂ ♂ of all ages. Entire crown, from nostril to nape, Raw Umber (23) finely flecked and streaked with pale buff central feather shafts. Lores, ear coverts, and sides of face darker, more Fuscous (21), and similarly streaked. A broad supercilium and malar stripe dirty white, finely flecked, variably buff to brown. Entire upperparts, including wings and tail, Raw Umber with Amber (36) wash to outer edge of exposed primaries and secondaries, their concealed parts being darker Raw Umber (223). Chin whitish to greyish white, throat flecked and spotted blackish grey, these marks grading to fine barring on breast and then to broader and slightly-paler barring on remaining greyish white underparts. Underwing: coverts as breast, flight feathers Olive-Brown (28) with broad Ground Cinnamon (239) trailing edges to outer primaries which occupies entire trailing vane of inner primaries and the secondaries. Ventral surface of tail Cinnamon-Brown (33) with paler central feather shafts.

SUBADULT ♂: as ad ♀ with few feathers of ad ♂ plumage intruding, to as ad ♂ with few feathers of ♀ plumage remaining.

IMMATURE ♂: as ad ♀

NESTLING: at 1–2 days old naked, dark blackish purple above and paler below.

JUVENILE ♂: Ogilvie-Grant (1915*a*) writes of a juv 'A quite young ♀ still retains the nestling-plumage on the underparts; the feathers are much softer and more downy than those of the subsequent plumage, and the dark bars on the breast and belly are sooty-brown and much wider apart; the upperparts including the crown are paler and more rufous.'

OTHER: an aberrant ad ♂ specimen (RMNH; un-numbered) and apparently shipped from Goenong Tobi, NW NG lacks almost all iridescent colour, there being a mere trace of iridescent purple-blue in the lores only. It is a washed-out, matte, pale, smoky brownish grey above (reminiscent of *Semioptera*), the wings and tail being paler. There is a single spot of iridescence the size and shape of a tear-drop on one central tail feather 12 mm above its tip. The crown feathering is scale-like, but is matte, smoky, brownish grey and is soft. That of the breast-shield is normal in shape and structural stiffness, but is all slightly-glossy brownish grey with no iridescence. The tract of feathers between the sides of the breast-shield and mantle are near velvet black, being dark brownish grey. Remaining underparts are blackish brown (closest to 121) with no sign of the usual claret red or green sheens. Quill shafts of primaries, secondaries, and tail feathers are white clearly emphasising this abnormal plumage. A publication in Dutch noted the bird had certainly not been preserved in fluid (Büttikofer 1895). An ad ♂ specimen in the MHN has the rear chin and throat feathers of the otherwise typical, iridescent, breast-shield glossy blackish each with a pale, off-whitish, centre and broad outer edging (C. Frith 1998).

Distribution

Lowland and hill forests of NG and CYP of NE Australia, from sea level to 1450 m (mainly up to 700 m only). Widespread in NG but absent from the forests of the Trans-Fly, the savanna areas of the Fly–Digul R. region, and the fringing islands off the coast of NG. In Australia, ranges from Cape York S to the Rocky R. area of the McIlwraith Ra on the E coast and the Weipa area on the W. Recorded on Albany I. off Cape York (Blakers et al. 1984).

Systematics, nomenclature, subspecies, weights, and measurements

SYSTEMATICS: is the sister form to the two Australian endemic riflebird spp. The eastern NG population (*intercedens*) may prove to be specifically distinct (Beehler and Swaby 1991; Cracraft 1992). The first riflebird sp. described, originally known from trade skins and first described by Levaillant (1807) and Cuvier (1817); it has a complex pre-Linnaean taxonomy that we do not present here; in the past the sp. has been attributed to the genera *Epimachus*, *Craspedophora*, and *Paradisaea*. The first naturalist to find the sp. in the wild was A. R. Wallace (1869) who collected specimens near Dorey (= Manokwari, Vogelkop, IJ). Hybridisation: known with the Superb, Twelve-wired, and Lesser Birds. See Plate 15 and Appendix 1.

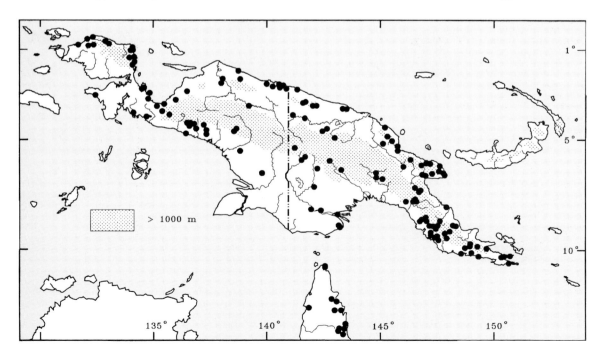

Magnificent Riflebird *Ptiloris magnificus* (Vieillot, 1819)

SPECIES NOMENCLATURE AND ETYMOLOGY

Ptiloris magnificus (Vieillot)
Falcinellus magnificus Vieillot, 1819. *Nouveau Dictionnaire d'Histoire Naturelle, Nouvelle Édition* 28, 167. La Nouvelle-Guinée, restricted to Dorey, Vogelkop, New Guinea (Mayr 1941a). Type none designated.

Etymology: *Ptiloris* = Gr. *ptilon*, a feather; *rhis* the nose (alludes to feathered nostrils or bill base); *magnificus* = L. grand, splendid, or magnificent.

SUBSPECIES, WEIGHTS, AND MEASUREMENTS

1. *P. m. magnificus* (Vieillot, 1819)

Range: western and central NG, from the Vogelkop eastward to PNG, in the N ranging as far as the Wewak area, and in the S as far as the Purari R. A bird giving the distinct vocalisation of this subsp. has, however, been recorded from Yalumet, on the N watershed of the Huon Penin (considerably E of the Sepik). Note also that the Lake Kutubu record for the *intercedens* song-type (Schodde and Hitchcock 1968) is an error (R. Schodde *in litt.*). In the Wabo area of Gulf Province, PNG, the vocal dialect of the nominate form is known to occur on the E bank of the Purari (A. Mack personal communication).

Measurements: wing, ad ♂ ($n = 31$) 182–200 (192), imm ♂ ($n = 20$) 169–183 (177), ad ♀ ($n = 25$) 147–165 (157); tail, ad ♂ ($n = 31$) 92–110 (102), imm ♂ ($n = 20$) 102–118 (110), ad ♀ ($n = 25$) 91–113 (99); tail centrals, ad ♂ ($n = 29$) 88–108 (97); bill, ad ♂ ($n = 31$) 54–64 (60), imm ♂ ($n = 20$) 51–63 (59), ad ♀ ($n = 25$) 45–57 (53); tarsus, ad ♂ ($n = 29$) 37–42 (40), imm ♂ ($n = 20$) 37–41 (40), ad ♀ ($n = 24$) 33–37 (36). Weight: ad ♂ ($n = 20$) 180–230 (207), imm ♂ ($n = 9$) 168–190 (180), ad ♀ ($n = 8$) 120–185 (142).

2. *P. m. intercedens* Sharpe, 1882. *Journal of the Linnean Society London, Zoology* **16**, 444. Milne Bay and East Cape, southeastern New Guinea. Type specimen BMNH 80.9.13.49 (a ♂, not ♀ as labelled).

Range: central and eastern PNG, E of the range of the nominate subsp. (see above); its western limit being about the Wanuma, Adelbert Mts area. Because of the distinct vocalisations, it is relatively simple to delineate the range boundaries between the two. Songs at the Crater Mt. Biological Research Station area, near Haia, Purari Basin, PNG are almost invariably of the *P. m. magnificus* type (see below) but that of *P. m. intercedens* was once recorded (H. Gomez de Silva in Mack and Wright 1996) making this the second area where two subspp. have been known to occur in sympatry.

Diagnosis: flank plumes equal to or shorter than the tail. In almost all mean measurements of both sexes and all age classes all but identical to nominate form except for bill length which averages a mere *c.* 4 mm longer in the present form. Bill straighter than in *alberti*. The distinctive song of this SE NG population and the fact that only a small portion of its culmen base is unfeathered (whereas that of nominate *magnificus* is mostly unfeathered) may be indicative of a previously-unrecognised sp. (Beehler and Swaby 1991). Because the Australian population exhibits considerable geographical variation in vocalisation (but not *intercedens*-like) within its small range (MacGillivray 1918 and CF) we here, however, treat *intercedens* as a subsp. pending further study.

Measurements: wing, ad ♂ ($n = 25$) 188–199 (193), imm ♂ ($n = 20$) 162–187 (175), ad ♀ ($n = 27$) 145–162 (152); tail, ad ♂ ($n = 25$) 96–106 (101), imm ♂ ($n = 20$) 103–116 (108), ad ♀ ($n = 27$) 92–103 (98); tail centrals, ad ♂ ($n = 25$) 90–105 (96); bill, ad ♂ ($n = 24$) 52–62 (56), imm ♂ ($n = 19$) 51–58 (55), ad ♀ ($n = 27$) 45–51 (48); tarsus, ad ♂ ($n = 23$) 37–43 (40), imm ♂ ($n = 19$) 38–43 (41), ad ♀ ($n = 27$) 33–38 (35). Weight: ad ♂ ($n = 12$) 155–214 (184), imm ♂ ($n = 7$) 165–189 (176), ad ♀ ($n = 15$) 102–149 (126).

3. *P. m. alberti* Elliot, 1871. *Proceedings of the Zoological Society of London*, p. 583. 'Cape York', Australia. Type specimen not located.

Synonyms: *Craspedophora magnifica claudia* Mathews, 1917. *The Austral Avian Record* **3**,

72. Claudie R., North Queensland. Type specimen AMNH 677613.

Craspedophora magnifica yorki Mathews, 1922. *The Austral Avian Record* **5**, 8. Cape York, North Queensland. Proposed as a new name for *Ptiloris alberti* Elliot, which proved not to be preoccupied.

Range: Cape York, S to the Rocky R. area of the McIlwraith Ra on the E coast and the Weipa area on the W. Albany I. off Cape York (Blakers *et al.* 1984).

Diagnosis: mean wing length *c*. 10 mm shorter, tarsus and tail fractionally shorter, and bill narrower and conspicuously more decurved than in other two subspp. Flank plumes longer than tail, as in nominate form. Extent of feathered culmen base intermediate between that of nominate *magnificus* and *intercedens*. Central pair of rectrices of ad ♂ average 8 mm shorter than rest of tail and thus are relativelly shorter than in other two subspp. (both 5 mm shorter than remainder of tail).

Measurements: wing, ad ♂ (*n* = 40) 174–187 (181), imm ♂ (*n* = 6) 151–169 (161), ad ♀ (*n* = 24) 139–165 (149); tail, ad ♂ (*n* = 38) 94–105 (99), imm ♂ (*n* = 35) 97–104 (101), ad ♀ (*n* = 21) 88–108 (96); tail centrals, ad ♂ (*n* = 35) 83–98 (91); bill, ad ♂ (*n* = 34) 51–57 (55), imm ♂ (*n* = 6) 51–56 (54), ad ♀ (*n* = 23) 46–56 (51); tarsus, ad ♂ (*n* = 39) 36–40 (38), imm ♂ (*n* = 5) 38–39 (38), ad ♀ (*n* = 24) 31–44 (35). Weight: ad ♂ (*n* = 4) 143–171 (160), imm ♂ (*n* = 1) 131, ad ♀ (*n* = 3) 94–112 (104).

Because of differences in ♂ song, culmen feathering, and length of flank plumes, it has been suggested that *P. m. intercedens* might represent a distinct sibling sp. A specimen (CSIRO 4112) that may represent a hybrid *P. m. intercedens* × *P. m. magnificus* was collected at Putei, just E of the Purari R. The Purari has been mentioned as the range boundary between the two populations. Treating *magnificus* and *intercedens* as two species does not, however, resolve the status of *alberti* as clearly as has been suggested (Beehler and Swaby 1991; Cracraft 1992).

Habitat and habits Table 4.1

Lowland, hill, and lower montane rainforests; also known from monsoon, swamp, and gallery forest and forest-edge. Also occasionally visits mangroves (Saenger *et al.* 1977) and timber plantations. Will forage in monoculture teak plantations, on tree trunks and branches for arthropods (Peckover 1990). Ad ♂♂, although commonly heard, more difficult to see than ♀-plumaged birds. Commonly joins mixed-spp. foraging flocks.

Birds, at least those in ♀-plumage, commonly use forest pools to bathe in both NG and Australia (Coates 1990; CF). Ad ♂, particularly in SE NG, will occasionally respond to play-back of their calls by flying close to or past the observer (D. Bishop *in litt.*).

Diet and foraging Table 4.2

Most frequently forages alone but occasionally small groups of loosely-associated birds are encountered, usually at a fruiting tree, and birds may join a mixed-spp. foraging flock led by the rufous-backed (see below) Rusty Pitohui or Rufous Babbler, ♀-plumaged birds possibly more so than ad ♂♂. Also seen in association with the Crested Pitohui *Pitohui cristatus* foraging on the ground (Coates 1990). One bird was seen associating with a flock of Lesser Birds (Hoogerwerf 1971).

On 20 July at Puwani R., PNG, a ♀-plumaged bird was part of a mixed-spp. foraging flock consisting of a Jobi Manucode, a King Bird, 5 Rufous Babblers, 3+ Rusty Pitohuis, Yellow-breasted Boatbill *Machaerirhynchus flaviventer*, Frilled Monarch, Rufous Monarch *Monarcha rubiensis*, Fairy Gerygone *Gerygone palpebrosa*, Yellow-bellied Gerygone *Gerygone chrysogaster*, Black Berrypecker *Melanocharis nigra*, Spangled Drongo, Black Cuckoo-shrike, and Rufous-backed Fantail *Rhipidura rufidorsa* (BB).

Another mixed-spp. flock with an ad ♂ riflebird as part of it on 10 Aug at Krissa, PNG, included an ad ♂ Twelve-wired and King Bird, Variable Pitohui *Pitohui kirhocephalus*, Rufous Babbler, Frilled Monarch, Spangled Drongo, Black Cuckoo-shrike, and Rufous-backed

Fantail. The riflebird ate *Cyrtostachys* palm fruits. Another such flock seen 2 days previously included an ad ♂ and a ♀-plumaged Magnificent Riflebird and had all of the latter spp. with the addition of Rusty Pitohui and an ad ♂ Lesser Bird (Beehler and Beehler 1986; BB).

Of 18 individuals observed in various mixed-spp. foraging flocks seeking insects, 30% did so from leaves, 27% from tree trunks, 23% tree limbs, and 10% from both tree branches and dead tree trunks or limbs (Bell 1977). Foraging for arthropods on dead and living wood is animated, acrobatic, and vigorous, and falling wood and debris often attracts the observer.

Stomachs of five birds from two E NG locations contained up to 100% fruits or arthropods. The following invertebrate groups were represented: Araneida, Blattidae, Chelisochidae, Chrysomelidae, Curculionidae, Formicidae, Lepidoptera (ad and larvae), Stenopelmatidae, Tenebrionidae, and Tettigoniidae (Schodde 1976). A centipede (myriapod) and scarabeoid (BB) are also recorded. Circumstantial evidence suggested a ♀-plumaged bird killed a smaller bird held in a mist net (A. Mack *in litt.*).

Of 259 records of birds foraging in lowland rainforest near Port Moresby, PNG 2% occurred in the upper canopy (30–35 m above ground), 35% the lower canopy (25–30 m), 29% the sub-canopy (8–25 m), and 34% in the understorey (0–8 m high), (Bell 1982c). For fruit foraging, however, 92% of records were in the main canopy and 8% in the lower understorey. Of 150 recorded foraging events (involving 30 sightings of birds) 24% were from each of the site categories of debris, tree trunks, and aerial, 20% from tree branches, and 8% from rotten wood (Bell 1984). All observations, other than the 36 instances of aeriel foraging, involved birds gleaning (i.e. both birds and prey perched).

Gary Opit (*in litt.*) reported that in Oct on Mt. Missim, PNG at 1500 m, 'birds are often associated with pandanus palms and I have often seen them hunting for insects amongst them. Now that the pandanus is fruiting they are especially attracted to them and would examine every pandanus they found as they flew through the forest. One that I closely observed ate seven orange pandanus fruits.'

It has been suggested that the diet consists of equal proportions of fruit and animal matter but it is likely that either will dominate at different times of year and that, perhaps, the overall greater proportion is of animal foods. Bell (1977) thought birds mostly eat arthropods, obtained by clinging to limbs and branches and probing wood, predominantly of the lower canopy and sub-canopy. This is certainly consistent with other observations of feeding birds and with the predominant foraging and diet of the two Australian riflebirds (CF).

At Varirata National Park, PNG on 28 Apr the sp. was seen feeding on fruiting *Elmerrillia* with a Raggiana Bird, Brown Orioles, and Fawn-breasted Bowerbirds *Chlamydera cerviniventris* (BB).

Of 34 observations of fruit foraging at Mt. Missim, PNG, 47% were upon fruits of *Chisocheton cf. weinlandii*, 24% upon *Omalanthus novoguineensis*, and 21% upon *Gastonia spectabilis*; the other three plant spp. representing only 3–6% of the fruit diet. Of the 34 records, 71% were upon capsular fruits, 8% upon drupes/berries, and 6% upon figs (BB; see also Appendix 4 and Chapter 4).

Australian birds fed on fruits (including those of *Ficus* spp. and the introduced custard apple *Annona muricata*), flowers, and seeds, and on ad and larvae of insects including ants and *Chalcopterus* beetles of the Tenebrionidae (Barker and Vestjens 1990). The above is consistent with foods found in stomachs of birds by early collectors (Macgillivray 1852; D'Albertis 1880; Greenway 1935; Jardine *in litt.* to North 1901; Simpson 1942; Hoogerwerf 1971).

Vocalisations and other sounds Figures 9.33, 34

♂♂ of each subsp. produce distinct advertisement vocalisations. In *P. m. magnificus* of W and central NG (Fig. 9.33 (a)) and *P. m. alberti*

Magnificent Riflebird *Ptiloris magnificus* (Vieillot, 1819)

of CYP the call is a powerful, clear, upslurred, but variable (Fig. 9.34) 'reversed wolf-whistle' *woiiieet-woit*. Notes of the nominate population in the Kumawa Mts, IJ are lower, upward-inflected, clear, hollow, delivered as more gibbon-like (*Hylobates* spp.) 'hoots' (Fig. 9.33 (b)). In *intercedens*, of SE NG, the very different call (first noted by Hunstein in Sharpe 1891) is variously described as (1) '*hrraah-hraoou* the first syllable rising the last part of the second falling' (Finch 1983), (2) a deep guttural growled *CRRRAIY-CRRROW* (Coates 1990) or (3) *uRAUow-urauow* (Beehler and Swaby 1991) (Fig. 9.33(c)). Above Wasu, Huon Penin, PNG the call of *intercedens* is again different, being the typical double growl-like notes but delivered with a harsher, rasping quality and the second note being upward or downward inflected (T. Pratt *in litt.*). This appears to be the only bird of paradise sp. additional to the Trumpet Manucode in which geographical races are known to have a strongly-distinctive song. The call of the Australian form *alberti* is variable, however. At Cape York at the northernmost tip of CYP it consists of two identical, upward-inflected, clear, whistled notes (Fig. 9.34 (a)). The first of these may sometimes be delivered flatter and softer, and the two sometimes repeated as a series in song (Fig. 9.34(b)). At the Claudie R., Iron Ra it 'differs remarkably from its note as heard at Cape York' (MacGillivray 1918; CF) in that the first note is lower and briefer and is continuous with the second in being fluidly connected by a downward inflection prior to the sharper last part (see Fig. 9.34(c)). This sounds just like a person whistling to attract another. At Rocky R., Silver Plains, at the southernmost point of the sp. range in

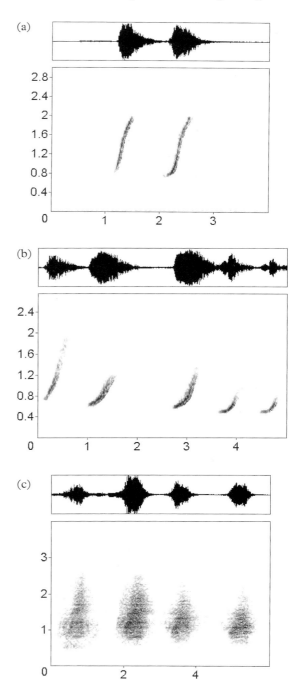

9.33 Sonograms from New Guinea recordings of ad ♂ Magnificent Riflebirds showing geographically-variable advertisement songs: (a) the whistled clear notes of the most widely-known call of the nominate subsp., recorded at Kiunga, PNG in Aug 1990 by K. D. Bishop; (b) the distinct call, almost gibbon-like (*Hylobates* spp.) in delivery, of the nominate form recorded in the Kumawa Mts, Bomberai Penin, IJ in Sept 1987 by J. Diamond; (c) the stikingly different gutturally-growled notes of *P. m. intercedens* at Varirata National Park, PNG in Aug 1987 by K. D. Bishop.

322 Magnificent Riflebird *Ptiloris magnificus* (Vieillot, 1819)

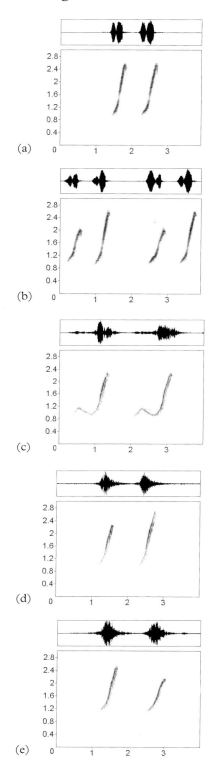

Australia, the call (Fig. 9.34(d,e)) is again like that of the same subsp. at Cape York (Fig. 9.34(a.b)) and the nominate subsp. of the trunk of mainland New Guinea (Fig. 9.33(a)). Thus of the *P. m. alberti* populations, the calls at Iron Ra are distinct from birds to both the N and S within Australia (see Fig. 9.34). In the Cyclops Mts, N IJ, at *c.* 400 m, and at Nawen Hill, near Wewak, PNG at 100 m the nominate song was a *KOIT-KOIT KOIT KOIT* (BB). At Setakwa R., S IJ the call was described as a long drawn-out *oo*û ending in two sharp and loud notes *wah wah* by Ogilvie-Grant (1915*a*) who noted of NG birds 'They will often call during the night if disturbed by the falling of a tree or other cause.' Reported by Andrew Mack (*in litt.*) to be less vocal during Nov–March at Ubaigubi, PNG. ♀-plumaged birds said to call, but not so loudly and well. A call of nominate *magnificus* in the Adelbert Mts was noted to be less loud than the call of *intercedens* near Wau and to be two, sometimes three, frog-like croaks (T. Pratt *in litt.*). Coates (1990) has noted that ♂♂ may give peculiar croaking sounds from their perches and that their advertisement singing not infrequently attracts ♀-plumaged Raggiana Birds. In flight, the wing feathers produce a loud rustling *wish wish* sound (Oglivie-Grant 1915*a*) and the wings of an ad ♂ in display a *rush-rush-rush-rush rushrushrushrush* sound audible from some distance (BB). The wing-noise of flying ad ♂♂ is described as a whistling sound that lacks the hissing quality of the Twelve-wired Bird's flight (Beehler in Finch 1983). The 'deep blowing' briefly alluded to by Coates (1990) appears to us to be the sound produced by the wings in its courtship display.

9.34 The geographically-variable song of the Australian population of the Magnificent Riflebird *P. m. alberti* is demonstrated by recordings made at: (a,b) Cape York proper in mid-Oct 1987 by K. Uhlenhut; (c) Iron Ra by R. Swaby July 1985; and (d,e) at Rocky R., Silver Plains by M. McGuire in Sept 1993.

Mating system

J. A. Thorpe (in North 1901) proposed that the sp. is polygynous, having observed ad ♂♂ displaying to several apparent ♀♀. Barnard (1911), who examined more than 50 CYP nests stated 'The ♂ bird is never seen near the nest' and this has been confirmed by all subsequent observers. That only ♀-plumaged birds attend nests (Frith and Frith 1993*d*) and that ad ♂♂ attend traditional singing/display perches to which numerous ♀♀ are attracted indicates a polygynous and promiscuous ♂ mating system. Each ♂ displays solitarily from a traditional arboreal perch(es), and ♂♂ are regularly arrayed through the forest (see Fig. 9.35). Whether the ♂♂ are territorial remains to be determined. Along the Ok Ma road, Ok Tedi area, PNG, ♂♂ called c. 500 m apart with up to 12 being heard in an afternoon (Gregory 1995).

Courtship behaviour

Brief courtship displays of captive and wild ♂♂ of various subspp. have been described (Barnard in Campbell 1901; Jardine in North 1901; Selous 1927; Crandall and Leister 1937; Crandall 1938; Diamond 1972; Hooper 1972; Coates 1973*b*, 1990; Anon 1974; Opit 1975*a*; Bell 1977; R. Mackay 1990).

We recently viewed video tape of several wild and solitary ad ♂ *alberti* displaying to ♀-plumaged birds on CYP but not culminating in copulation. The displaying ♂ statically perches upon a low horizontal vine, branch, or large bough and upon sighting a ♀ becomes stiffly sleeked and agitated. Stretching his head and bill in the direction of the ♀ he makes repeated, small, sharp hops side-to-side on the spot, sharply flicking his wings slightly, and now and then quickly runs his bill down the underside of a wing as 'displacement' preening. As the ♀ approaches more closely, the ♂ adopts a ritualised, sleeked, upright, bill-up posture, breast towards the ♀, while now and then sharply jerking his entire body and/or wings and continuously pulsing his breast-shield feathers which results in the conspicuous play of light upon them (Fig. 9.36(a)). Instantly his wings are suddenly fully outstretched either side of his body, with an accompanying sharp, loud, rustling sound, as his upwardly-pointed neck and bill are rhythmically swung back and forth between the leading edges of his wings with increasing speed (Fig. 9.36(b)). During this he synchronously raises and lowers himself on his legs, at times with a side-to-side rocking motion, and raises and lowers his wings with a rustling sound audible from > 50 m and produced in a rising tempo. Coates (1990) noted that when the ♂ *intercedens* is at maximum height one leg may be stretched more than the other depending on which wing the head is lying along. This uneven stretching of the legs causes the body to rotate slightly about a vertical axis. Brief non-vocal gaping may be performed immediately prior to and initially after the wings are opened (*contra* Frith and Cooper 1996). The tail is cocked up to about the horizontal.

In this posture the ♂ may hop on flexed tarsi along the display perch for up to 4–5 m in a light but deeply-bouncing action towards, and sometimes directly backward away from, the ♀ now on his perch. Opit (1975*a*) noted that in this posture the (*intercedens*) bird's wings at times almost touched each other over the head forming an 'almost perfect disc with a central hole around the birds shoulders and head.'

The ♂'s (*alberti*) fully-extended wings are held out from either side of his body, with tail raised to the horizontal behind, as he performs

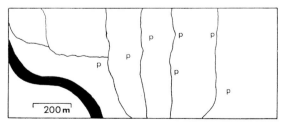

9.35 Display dispersion of ♂ Magnificent Riflebirds at Sii R., Kakoro, PNG: P's indicate the locations of the song perches of ♂♂ and lines the transects. After Beehler and Pruett-Jones (1983), © Springer-Verlag with kind permission.

a bouncing dance along his display bough by hopping from a crouched position, concealing his lowered legs (Fig. 9.36(c)), up and forward some 15 cm to crouch again, and so on. With each upward hop, on extended legs and feet off the perch, the stiffly-pointed neck, head, and closed bill are rapidly swung from along the leading edge of one wing through an extensive arc to that of the other (Fig. 9.36(d)). If the end of his display bough is reached he instantly swings around to dance in the opposite direction. As well as performing whilst facing his audience along his perch (i.e. wings outstretched at right angles to the perch) he also hops along sideways with wings in alignment with the perch (as in Fig. 9.36 (c,d)).

Of a long and regularly-executed captive *intercedens* display of this kind, Crandall and Leister (1937) wrote 'the head was moved from side to side 35 times, punctuated, in perfect rhythm, by the rustling snap of the extended wings. No two displays are exactly alike, for they are varied by changing the regular, rhythmic beat of these co-ordinated movements. Action may be slowed down for a few sec, then speeded up to the original beat ... Voice seems to play no part in the display and, unlike Selous, we have never heard a vocal sound during dozens of closely-observed displays.'

Of 17 courtship displays near Port Moresby, PNG, 76% were in the lower understorey (0–4 m above ground), 6% in the upper understorey (4–8 m), and 18% in the sub-canopy (*c.* 8–25 m) but none in the main canopy 25–40 m high (Bell 1982*c*).

Diamond (1972) observed display between two ♀-plumaged (*magnificus*) birds, which he considered to be probably an imm ♂ and a ♀, on 7 Aug 1964 the two birds 'alighted next to each other on a branch 4.6 m above the ground at the top of a bush at the edge of the forest. One bird faced the other, reared up on its legs, spread out its wings and bent them backwards, threw back its head, and remained in this uncomfortable posture for some time. The second bird then assumed this posture, and finally both assumed it simultaneously before flying off. No calls had been given. On the second occasion (12 Sept 1965) two birds perched facing each other on a branch 9.1 m above the ground in forest. One tilted its body up until it was vertical, held the tail back at right angles to the body so that it was horizontal, opened the wings and bent them at the shoulder so that they nearly met behind the back and the breast was arched towards the partner, and then rose up and down on its legs like a horseman posting on a trotting horse. The pair flew off to a branch of another tree and repeated the display.' In view of this behaviour it would seem more likely that both birds were imm ♂ (see Victoria's Riflebird).

On 22 April Coates (1973*b*) witnessed another *intercedens* display near Ower's Corner, PNG. He saw two ad ♂♂ fly through low shrubbery to a near-horizontal, twisted, woody vine 3 m above ground to face each other less than 1 m apart. 'Then one bird, which was perched higher and facing me, pointed up its bill and stretched its neck and expanded the gleaming blue breast and throat shield. The wings were suddenly fully spread in the classic pose and the head, with neck still stretched was bent to one side to touch the outstretched wing. The bird then elevated itself to full height at the same time emitting what sounded to me like a deep blowing sound gradually increasing in volume until full height was reached at which it immediately subsided as the bird lowered itself again, stopping when the initial height was reached. Once maximum height was reached the bird lowered itself again. Just as it was beginning to come down the head was turned rapidly to the other outstretched wing. The complete motion lasted perhaps less than two sec. This was repeated five and more times in one go—the effect being a slow motion up and down bobbing action accompanied by a rising and falling blowing noise with the head being moved in swift action from one wing to the other with each motion.' Coates then saw the second ♂ perform an identical display and, as he viewed this from behind the bird, he could see the tail was cocked up. He subsequently considered this display to have been agonistic in context.

Magnificent Riflebird *Ptiloris magnificus* (Vieillot, 1819)

9.36 Some courtship display postures of ad ♂ Magnificent Riflebird: (a) pre-display pose, (b) full display posture, (c–d) the bouncing dance display. See text.

Interestingly, Roy Mackay (1990) twice saw a Baiyer River Sanctuary captive ♂ (*magnificus*) conclude display by turning to the ♀ with his wings still extended to clap her between his open wings several times, as is typical of Victoria's Riflebird courtship behaviour. A BRS ad ♂ captive once displayed for 9 sec to a wild ♀-plumaged Lesser Bird perched above his aviary (CF).

According to Coates (1990) ad ♂♂ display from time to time during mid to late morning, at midday, and during the afternoon, the season for display being the late wet and throughout the dry season but not during the early wet when birds are moulting. In the Lakekamu Basin in Aug 1994, an ad ♂, under observation at his forest interior display post on a fallen tree trunk ca. 2.5 m above the ground, displayed regularly early each morning and also in the late afternoon, less than 100 m from the display post of a Twelve-wired Bird (BB).

Breeding   Table 6.5

NEST SITE: during 2–12 Nov Frith and Frith (1993*d*) found five new, empty, *alberti* nests in monsoonal vine forest at Iron Ra, CYP, built between the bases of living fronds of pandanus tree crowns and atop 'basket fern' (*Drynaria* spp.) epiphytes on tree trunks at a mean height of 4 (range 2–5) m. Another nest was 6 m high atop a broken-off tree trunk with shoot regrowth and contained two eggs on 2 Nov. These are the favoured sites in Australia (Barnard in Sharpe 1891; Barnard 1911; MacGillivray 1918; CF). In addition to more typical CYP nesting habitat of dense monsoon vine-forest on creek lines, birds also nested in the more open adjacent forest with pandanus palms, numerous epiphytic ferns, and dominated by melaleuca trees. Another 16 Australian nests were 0.5–16 m, averaging 4.2 m, above ground. Gould and Sharpe (1875) stated that Dr Beccari's NG nest (*magnificus*) was found by one of Mr Bruijn's hunters in the branches of a *Calophyllum inophyllum* tree.

NEST: according to Barnard (1911) 'The nest is composed of large dead leaves and vine tendrils very loosely put together. Unlike the two southern spp., the Albert Rifle-bird [*P. m. alberti*] does not decorate its nest with snake skins. I examined about 50 nests, and did not find snake skins in a single instance. ... If a nest was found containing one egg, and left untouched in order to secure the full clutch, on returning next day the egg was sure to have disappeared; but if a single egg were taken, and the nest visited on the following day, the second egg would be found in the nest. I had the same experience during my former visit to the locality, in 1896.' MacGillivray (1918) described an apparently new but empty *alberti* nest as constructed of 'broad leaves and twigs wound round with a parasitic climbing plant pulled in its green condition. The lining was of fine midribs of leaves and fibres.' A black and white photograph of a CYP nest in North (1901) shows no sticks.

EGGS: clutch usually two but sometimes one; 68 eggs, mostly of *P. m. alberti* average 33.1 (30.5–38.8) × 23.6 (22–27.6) mm (CF). Ground colour Pale Horn (92) with faintest pinkish wash, or whiter, to Pale Pinkish Buff (121D) but paler marked with broad longitudinal 'brush' strokes of browns and purplish greys. Eggs relatively small for the bird compared with those of its congeners. See Plate 13.

INCUBATION: only the ♀ incubates. No details known.

NESTLING CARE: a recently-hatched young collected by MacGillivray (1918) contained 'insects, grasshoppers and beetle remains'. A CYP nest containing two eggs on 12 Nov contained two young on 14 Nov and was subsequently observed over 15 days. The ♀ that attended was extremely wary, and she regurgitated all meals (Frith and Frith 1993d). A presumed ♀ was seen attending a dependent young on CYP on 29 Sept (RAOU NRS).

NESTLING DEVELOPMENT: unknown

Annual cycle Table 6.5
DISPLAY: Bell (1982b) reported frequently hearing and occasionally seeing courtship display from Apr to Sept near Port Moresby, PNG. Display reported from late wet to throughout dry season (Coates 1990). BREEDING: at least June–Feb over the sp. range as a whole. ♀♀ collected with enlarged oocytes during Feb, June, July, and Oct–Nov. Some ♂♂ collected with moderately to much-enlarged gonads during all months, but smallest in Feb–Mar. EGG-LAYING: predominantly early Sept to mid Feb but a nest with eggs in early June at Moroka area, PNG (Ogilvie-Grant 1912) indicates a longer season. MOULT: 63 of 213 specimens were moulting involving all months but the vast majority being during Dec–Mar and, to lesser degree, Oct–Nov and Apr–May (Appendix 3). Said to occur during early part of wet season by Coates (1990).

Status and conservation Table 8.1
Mostly common where found. Tolerant of selectively-logged forest. Bell (1982a) calculated a population near Port Moresby to be 6 birds per 10 ha. In a 100 ha study area near Kakoro, PNG, Beehler and Pruett-Jones (1983) found a minimum of seven apparently resident ad ♂♂, dispersed at a mean nearest neighbour distance of 175 (range 140–275) m, which they

considered to be non-territorial (Fig. 9.35). In the area of Kiunga, PNG calling ad ♂♂ were common, and *c.* 1–200 m apart, in slightly-rolling topography in both low swampy and rather poor more elevated forest. Calls could be heard up to *c.* 1 km distant and birds were shy (D. Bishop *in litt.*). The sp. was uncommon and calling ♂♂ sparse on the S Cyclops Mts, IJ between 300 and 1200 m in July 1990 (J. Diamond and D. Bishop *in litt.*). Jardine (*in litt.* to North 1901) thought the population of ♀-plumaged birds to be 'perhaps 15 or 20 to one [ad] ♂'. Thorpe (in North 1901) shot 106 ad ♂♂ and 80 ♀-plumaged birds during his 17 months near Cape York, these figures possibly reflecting the preference of a professional collector for more spectacularly-plumaged ad ♂♂ and/or the tendency to be attracted to them by their vocalisations. The sp. is under no threat.

Knowledge lacking and research priorities
Incubation and nestling periods unknown. Further work on the taxonomic status of the NG forms is required. It would be especially enlightening to carry out transect surveys of singing ♂♂ across the two contact zones (in the N and S of central PNG). Are there sharp breaks in the two vocal forms or do they show range overlap? Are there 'hybrid' vocal types?

Aviculture
Much of what is known of display has been learnt from captives. A ♀ *intercedens* was delivered to London Zoo 5 Oct 1908 by C. B. Horsbrugh and W. Stalker (Horsbrugh 1909) and a further three pairs arrived there about this time (Seth-Smith 1923*a*). This form has been kept at NYZS (Crandall 1935). A ♀ escaped from the aviaries of Mrs Johnstone at Groombridge, UK, and lived in the woods for several months. A single egg of *P. m. intercedens* in the BMNH collection was laid in the Hoddam Castle aviaries, Dumfries, Scotland, in 1913. Numerous birds have been held in European aviaries. A lone male has been maintained at the Vogelpark Walsrode, Germany, throughout the 1990s to time of writing. An active, non-agressive, and rather solitary sp. that does well in large aviaries with some hiding places, adequate flight area, pool, and access to rain/sprinkler to bathe (P. Shanahan *in litt.*).

Other
It is hypothesised that the rufous dorsal coloration of the ♀ plumage of the two NG subspp. is an adaptation to mixed-spp. flocking, this being mimicry of the dorsal coloration of spp. typically leading such flocks (which are characteristic of the NG avifauna—see Diamond 1987; Coates 1990; Beehler and Swaby 1991). The ♀ plumage of the Australian population has less rufous upperparts and mixed-spp flocks are not typical of its rainforest habitat.

Paradise Riflebird *Ptiloris paradiseus* Swainson, 1825

PLATE 8; FIGURES 9.37,38

Other names: Paradise Rifle-bird, Rifle-bird, Riflebird, Rifle bird, Velvet Bird, New South Wales Rifle Bird of Paradise. Aboriginal name at Richmond and Clarence R. *Yass* or *Jass* (Strange, in Gould 1865), after the call note but also *Bong Bong* (Goodwin, in Sharpe 1891–8), and in the Big Scrub *Bung-bung* meaning shining or silky (Campbell 1901).

A compact, strong-footed, and strong-billed, sexually-dimorphic sp. of the forests of eastern Australia. Bill long, decurved, and laterally compressed. Ad ♂ jet black with iridescent blue-green breast-shield, crown, and central tail feathers. ♀ cryptically-coloured brown and rufous above and buff barred blackish below. ♂ larger than ♀, but with shorter bill. ♀ broadly similar to but larger than *P. victoriae*. Monotypic.

Description
ADULT ♂: 30 cm, head and entire upperparts except crown and central pair of tail feathers deepest velvet jet black with rich, silk-like, iridescent sheens of Purple (1) and/or Magenta (2). Mantle may show deep green sheens in some lights. Entire crown of scale-like feathers intensely-iridescent, metallic, greenish blue to bluish green, washed rich Purple, particularly on nape, in some lights. Central pair of tail feathers short and intensely-metallic deep bluish green to greenish blue. Chin to lower breast velvet jet black but central throat to upper breast adorned with an extensive, roughly-triangular, breast-shield of scale-like feathers, intensely-metallic, iridescent, greenish blue, washed rich purple in some lights. Broad feathering below breast-shield to undertail coverts with dull, silk-like, iridescent, deep oil-green (146) broken by extensive jet black feather bases conspicuously visible at upper and lower border of this area. Underwing and tail glossy black. Bill shiny black, gape pale yellow, iris dark brown, legs and feet black, mouth bright yellow or lime yellow.

ADULT ♀: 29 cm, bill slightly larger than ♂'s, but wing significantly shorter than ♂'s of all ages. Upperparts predominantly Dark Brownish Olive (129) but crown, lores, and sides of face darker (21) with feather shafts finely-streaked pale buff contrasted by a broad, whitish, superciliary stripe. Dorsal surface of wing and tail strongly washed Raw Umber (23), particularly on leading edges to the generally more blackish primaries and secondaries. Malar area, chin, and throat whitish, becoming Buff (124) on upper breast to a pale cinnamon (223D) on remaining underparts, strongly marked with blackish brown deep chevrons that deteriorate to barring on the flanks and undertail coverts. Underwing and tail pale Olive-Brown (28) with broad pale cinnamon (223C) inner edges to all but outermost primaries, underwing coverts Buff. Bare parts as ad ♂.

SUBADULT ♂: as ad ♀ with few feathers of ad ♂ plumage intruding, to as ad ♂ with few feathers of ♀ plumage remaining.

IMMATURE ♂: as ad ♀.

NESTLING-JUVENILE: hatchling naked and black-skinned (Jackson in Campbell 1901). Jackson (1907) inexplicably noted that hatchlings of a brood of two had 'large horny cones on their bills, just above the nostrils and something similar to that found on the upper mandible of the Friar Bird or Leather Head [= *Philemon corniculatus*: Meliphagidae]'. Perhaps these were some form of 'egg tooth'. Nestlings of Victoria's Riflebird show no such character.

Distribution
The most southerly bird of paradise, its entire range being S of the tropics. E coastal Australia, from immediately S of Rockhampton (Calliope Ra), Queensland to just N of Newcastle, NSW on the Great Dividing Ra, plus some areas of the lowlands to the E. Norris (1964) did not find the sp. below an altitude of 840 m at the Tooloom Scrub whereas G. Holmes (*in litt.*) seldom found it below 200 m except in winter when seen down to sea level. Most numerous above 500 m and now rarely encountered below 200 m (G. Holmes *in litt.*). Jackson (1907) stated the southern limit of the sp. to be the Manning R. district of NE NSW but the southern limit is now known to be the northern Hunter Valley near Barrington Tops (Morris *et al.* 1981).

Systematics, nomenclature, subspecies, weights, and measurements
SYSTEMATICS: forms a superspecies pair with Victoria's Riflebird. This superspecies, in turn, is the sister-lineage to the Magnificent Riflebird. Has been treated by some authorities as conspecific with Victoria's Riflebird

Paradise Riflebird *Ptiloris paradiseus* Swainson, 1825

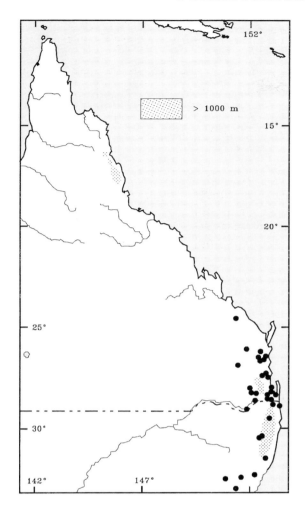

northern New South Wales. Type specimen MHN 10351.

Synonym: *Ptiloris paradisea queenslandica* Mathews, 1923. *The Austral. Avian Record* 5, 42. Mathews (1923) described the subsp. *P. p. queenslandica* from the Blackall Ra but later (1930) invalidated it. The first specimen is said to have been shot by a convict named Wilson in 1823 (Jackson 1907).

Etymology: *paradiseus* = L. *paradisus*, paradise.

WEIGHTS AND MEASUREMENTS

Measurements: wing, ad ♂ (n = 37) 153–165 (160), imm ♂ (n = 5) 139–154 (148), ad ♀ (n = 26) 137–156 (144); tail, ad ♂ (n = 37) 92–103 (98), imm ♂ (n = 4) 89–98 (94), ad ♀ (n = 26) 86–99 (91), tail centrals, ad ♂ (n = 37) 84–95 (89); bill, ad ♂ (n = 34) 49–62 (54), imm ♂ (n = 5) 52–61 (55), ad ♀ (n = 25) 51–66 (59); tarsus, ad ♂ (n = 35) 30–36 (33), imm ♂ (n = 5) 31–35 (33), ad ♀ (n = 24) 30–36 (32). Weight: ad ♂ (n = 3) 134–155 (141), imm ♂ (n = 2) 113–115 (114), ad ♀ (n = 6) 86–112 (104).

Habitat and habits Table 4.1

Subtropical and temperate rainforests, but also wet sclerophyll forest adjacent to rainforest and dry sclerophyll forest to > 1 km from rainforest, particularly during winter months. Now occurs predominantly in the hill forests of the Great Dividing Ra because of the extensive clearing of suitable lowland habitat in its original natural range. When not displaying on their traditional perches ♂♂ spend much time there in advertisement singing and preening.

Diet and foraging Table 4.2

Usually encountered as lone birds, but aggregations of up to six or seven may occur in fruiting trees. Depending on time of year, possibly even time of day, dominant part of diet is arthropods, mostly insects dwelling in wood and dead foliage, or fruits of forest trees and vines. An ad ♂ was once seen feeding upon fruits of a *Polyscias elegans* tree in which a ♂ and two ♀-plumaged Regent Bowerbirds

(e.g. Storr 1984) but in view of the morphological differences, particularly the differences between them in ♀ ventral plumage, body size, and sexual dimorphism in bill length and shape, we concur with most recent authors (Gilliard 1969; Cooper and Forshaw 1977; Blakers *et al.* 1984; Beehler and Swaby 1991; Christidis and Boles 1994; Donaghey in Strahan 1996; C. Frith 1997) that they are better treated as distinct spp.

SPECIES NOMENCLATURE AND ETYMOLOGY

Ptiloris paradiseus Swainson, 1825. *Zoological Journal* **1**, 481. No locality but accepted as

Sericulus chrysocephalus, two Green Catbirds *Ailuroedus crassirostris*, and two ♀-plumaged Satin Bowerbirds *Ptilonorhynchus violaceus* were feeding, near Brisbane in June (CF). Such gatherings with sympatric bowerbirds at fruiting trees are quite common but do not represent any special form of association.

Mostly remains high in the forest. When feeding on arthropods, clambers about tree trunks and limbs much like Australian tree creepers *Climacteris* spp. Similar in both foraging techniques and diet to Victoria's Riflebird although its relatively-longer bill suggests that it excavates rotten wood and probes epiphytic vegetation and litter to a greater depth and extent. Other than lifting bark pieces and probing dead wood, presumably for beetles and their larvae and spiders, often probes into and tears at the bases of larger *Asplenium* and *Platycerium* spp. epiphytic ferns and accumulated litter therein. Ramsay (1919) clearly observed a bird to pick at wood and then 'lay her ear against it listening for any movements of insects within.'

Recorded foods include beetles and their larvae and the larvae of butterflies and moths (Lea and Gray 1936), a phasmid (G. Holmes *in litt.*), fruits of *Synoum glandulosum* (Ramsay 1919), and native tamarind (*Diploglottis* sp.). For other food plants see Appendix 4—but note that it does not include the following food plants recorded by Church (1997): *Acacia maidenii, Anthocarapa nitidula, Cupaniopsis flagelliformis* var *australis, C. foveolata, Dendrocnide excelsa, Diploglottis australis, Dysoxylum fraserianum, Elaeocarpus obovatus, Elattostachys xylocarpa, Guioa semiglauca, Jagera pseudorhus, Mischocarpus anodontus, Sarcopteryx stipata.* Stomachs contained plant material and seeds, crickets, spiders, ants, a cockroach, a caterpillar, and a wasp (Barker and Vestjens 1990). Everitt (1962) noted that captive birds would often take flying insects by sallying after them. The powerful, sharply-clawed, and highly-manipulative feet are often used to hold prey and fruits to a perch for dissection, such as one holding a *Ficus coronata* fruit by a foot to tear off and swallow small pieces (G. Holmes *in litt.*). A bird was once observed to cling to the underside of a branch for more than 40 min whilst eating insects (Ramsay 1919). Grant (in North 1901–4) collected a ♀ that was feeding on the ground. Often drinks water from tree cavities.

The 'reversed' sexual dimorphism in bill size, in which the smaller ♀ has a bill 8% larger than the ♂, is a rare phenomenon that probably reflects intraspecific character displacement produced by the relatively-limited resource of wood-dwelling invertebrates (Jamieson and Spencer 1996; Moorhouse 1996; Frith 1997).

Vocalisations and other sounds Figure 9.37

Advertisement song of ♂ is an explosive, once repeated *yaassss* lasting < 1 sec (Fig. 9.37), very similar in structure to that of Victoria's Riflebird (Fig. 9.39). Campbell (1901) suggests that younger (♀-plumaged) ♂♂ give a single *yass* note, older ♂♂ give the note twice somewhat hurriedly, and still older ♂♂ in ad plumage leave a measured interval between the two notes of the call. He reports that the call can be heard from a distance of at least 800 m. Grant (in North 1901–4) states that seasonal calling by ♂♂ starts in Aug. One ♂ observed 'often extending double note into series by adding closely-spaced less harsh notes' (G. Holmes *in litt.*). The latter type of calling is often produced by Victoria Riflebird ♂♂ in ♀ plumage and possibly involves relatively-young/inexperienced individuals (CF). A deep mellow whistle ending with an upward inflection is recorded; as is a short whistle, low chatter, and a Noisy Friarbird- (*Philemon corniculatus*) like loud *cluck* note (G. Holmes *in litt.*). During high intensity courtship display the alternately-opened wings produce a loud *woof-woof* sound (Ramsay 1919). Regular monthly surveys of bird call frequencies at the Binna Burra section of Lamington National Park by G. Holmes (unpubl. data) during Feb 1990–Feb 1993 showed the period of least calling to be Mar–Aug with a significant increase in Sept and by far most vocal activity during Oct–Feb.

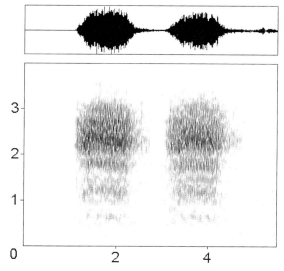

9.37 Sonogram from a recording of the typically-double advertisement song screech notes of an ad ♂ Paradise Riflebird at Lamington National Park by D. Stewart, and after Stewart (1996).

Mating system

Given the distribution of this bird, near major human populations, it remains remarkably little known. ♂♂ are clearly promiscuous, solitary-displaying birds that seasonally attend a display territory containing several regularly-used, exposed, bare, canopy or sub-canopy limbs or branches. It remains to be determined the degree to which the sp. defends a territory. It has been postulated that ♂♂ are non-territorial (Beehler and Pruett-Jones 1983) but, in view of what is known of other riflebird spp, it is more likely ♂♂ are territorial to at least the extent of the area of their display sites. It was long ago noted that during July–Aug ♂♂ move about but during Sept–Nov they remain in one place, and if a ♂ is shot during the latter period another will take its place (Goodwin in Sharpe 1891). This implies defence of display perches. Foster (in Campbell 1901) claims to have observed 'a pair of birds, i.e. both ♂ and ♀, building' on c. 10 Nov; the ♂ involved appeared immature in not being in entirely black plumage. It can only be assumed that this ♂ was merely following the nest-building ♀ and did not actually take part as implied. While not yet proved, ♀♀ are no doubt uniparental at the nest. Pizzey (1980) states that the Common Koel *Eudynamys scolopacea* (Cuculidae) is a brood parasite of the Paradise Riflebird, but no such record was mentioned in a review of Australian cuckoo hosts (Brooker and Brooker 1989).

Courtship behaviour

The courtship display of ad ♂♂ has often been described in general terms, but no detailed study of the postures and movements involved in a typical progression of courtship to copulation is available. Jackson (1907) described an ad ♂ displaying on a thick horizontal limb of a Red Cedar tree *Toona australis*. This bird would 'open his wings to their fullest extent, then suddenly bring them over in front till both ends touched the limb, and with his head well thrown back would walk majestically up and down the limb a distance of about three feet, bobbing up and down, and causing the wings to make an extraordinary noise resembling the rustling of a piece of new silk; then he would suddenly turn round and round, and every few seconds make quite an unusual sound, resembling the faint croaking of a frog.'

A ♂ may attend his traditional display limbs at all times of day, but particularly so early in the morning. There he sings to advertise his location and preen frequently. At the first sign of a ♀ in his immediate area he commences display posturing directed at the ♀ throughout. The ♂ instantly transforms himself by fully opening and extending his conspicuously-rounded wings and holding them horizontal to the ground, to either side and in front of himself, whilst cocking the tail upward and progressively fluffing the abdomen and flank feathers (Fig. 9.38). Should the ♀ be below he will peer downward, with lowered head and bill, or if above him will look upward bill uppermost. During this initial phase he may occasionally give his loud call note or may simply gape his mouth widely open to reveal his bright yellow mouth interior to the ♀.

As the display intensifies his wings are twisted back to become more (but not) vertical

332 Paradise Riflebird *Ptiloris paradiseus* Swainson, 1825

9.38 The high-intensity courtship display posture of ad ♂ Paradise Riflebirds. After Frith and Cooper (1996), with kind permission.

to the ground and their leading edges held at about 45° and the tail lowered. As the ♀ approaches he brings his wing tips forward to touch (or nearly touch) one another as the outer primaries are held in an unnatural contorted position. The ♂ has by this time raised his head and bill vertically upward, sleeked his head and neck plumage, and started to sway his head rhythmically from side to side with increasing tempo. As his head reaches its maximum arc to his right his wings are swayed to his left and vice versa. Should the ♀ be seriously interested in mating she will by this time join the ♂ by perching immediately in front of him.

The ♂'s wings now all but encircle or embrace the ♀'s head. He then vigorously throws his head from side to side as the ♀ peers into the hole framed by his wings, occasionally appearing to pick or peck at his throat or head. The displaying ♂ now leans backward and supports himself by propping upon his lowered tail whilst ever increasing the tempo of his movements until suddenly lowering his wings and mounting the soliciting ♀ to copulate. Immediately after copulation the ♀ vigorously flutters her wings and departs, if not chased off by the ♂ (Frith and Cooper 1996).

A displaying ad ♂ at Mt. Glorious, Brisbane was reported to hang completely inverted from a broken bare display bough for 2–3 min and to remain still, save silently gaping the mouth widely (A. Hiller *in litt.*).

Breeding   Table 6.5

NEST SITE: one nest was 35 m up in large mistletoe in the top of a eucalypt in woodland almost 800 m from rainforest, another in a mistletoe 39 m up in a eucalypt at the rainforest edge (Ramsay 1919), and another 19 m high at the very apex of a Sassafras *Doryphora sassafras* tree where it was enveloped by vines (M. Goddard in G. Holmes *in litt.*). The most frequently-used site appears to be a dense tangle of vines, sometimes lawyer vines *Calamus* spp., 5–40 m above ground. Eight nests were 5–15 m high, averaging 8.6 m. Jackson (1907) records a nesting in a Scrub Cherry *Cassine australis* covered with *Calamus* and *Flagellaria* vines and another in the same tree sp. apparently lacking vines; and one in dense vines at the top of a Booyong tree *Argyrodendron actinophyllum*. Bailey (in Campbell 1901) found an active nest with one egg, adjacent to two old nests in the same tree. Sidney Jackson (1907) reported 'Two nests found were built on top of old ones of the same kind, and another was placed eighteen inches away from the new one; this proves that they certainly build year after year in the same tree' (see Chapter 6).

NEST: the first nest (and egg) was found by I. Foster at Rous, Richmond R. district, NSW, in the second week of Nov 1897 and described by Campbell (1897). Nest larger and bulkier than that of Victoria's and is a shallow bowl shape of vine stems lined with finer vines, fibres, and rootlets. The outer rim is, in many cases, decorated with lengths of sloughed snake skin, often of pythons, smaller pieces sometimes accumulating within the egg cup. Fresh green fronds of climbing ferns of the genus *Polypodium*, including the sp. *P. confluens*, typically decorate the rim of the structure. Some large dead leaves are incorporated into the nest base. Campbell (1901) gives the overall nest dia as 203–228 mm and the total depth as 102 mm, and the internal cup as 102 mm dia and 51 mm deep. Jackson (1907) noted that, between the time of nest completion and egg-laying (1–2 weeks), one nest was filled with leaves (not of trees of the immediate area), and he was convinced the bird did this. A nest to which Jackson (1907) watched the presumed ♀ owner bring and add five pieces of snake skin had a clutch of two eggs hatch in it 6 days later (see Victoria's Riflebird and Lawes' Parotia for discussion of similar behaviour) and he considered the use of sloughed snake skin to be to deter nest-robbing birds and mammals. He recorded skins of the Black Snake *Pseudechis porphyricus*, Carpet Python *Morelia spilota*, and Death Adder *Acanthophis antarcticus* to be used.

EGGS: usually two, sometimes one. Ground colour darker and richer, and longitudinal markings smaller, more numerous, and evenly distributed, than in Victoria's eggs. First collected by W. McEnery and S. W. Jackson at Booyong Scrubs, c. 25 km NE of Lismore in the Richmond R., NSW on 2 Nov 1899 (Jackson 1907). 16 eggs (7 two-egg clutches, 2 one-egg clutches) average 32.9 (29.7–34.7) × 23.8 (23–24.8) mm (CF).

INCUBATION: the incubating ♀ is said to leave the nest twice each day to feed, usually at about mid-morning and in the late afternoon (Cooper and Forshaw 1977), but this is unlikely to be the case as other incubating birds of paradise feed far more frequently. Incubation period stated to be 15–16 days (Schodde and Tidemann 1988), but no evidence indicates this relatively-short period is correct. That of Victoria's Riflebird suggests 18–19 days is more likely (Frith and Frith 1995*b*, 1998).

NESTLING CARE: nestling period c. 4 weeks according to Schodde and Tidemann (1988). A brood of two 'newly born-looking' young on 6 Oct, thought to be 'almost fully fledged on 27 Oct, had left the nest by 5 Nov having been in it at least 21 days (RAOU NRS). Note that Victoria's Riflebird's nestling period is 14 days (Frith and Frith 1995*b*, 1998*a*).

NESTLING DEVELOPMENT: unknown. A presumed ♀ was seen feeding two fledglings at Lamington National Park on 7 Mar (RAOU NRS). Another was accompanied by a downy juv at Binna Burra on 15 Feb 1990 and another fed a juv at Styx R. State Forest on 16 Mar 1979 (G. Holmes *in litt.*).

Annual cycle Table 6.5

DISPLAY: at least Aug–Dec. BREEDING: certainly Sept–Jan and probably at least Aug–Feb. Two nests were under construction during the first half of Nov (Ramsay 1919). EGG-LAYING: Sept–Jan. MOULT: 12 of 56 specimens were moulting but as monthly samples are small we can only observe that most moult occurs Jan–Apr with some in Aug–Dec (Appendix 3).

Status and conservation Table 8.1

Sharpe (1891) wrote that large numbers of ad ♂ skins were sent to London for the millinery trade and that the 'area of country inhabited by the Rifle-bird cannot be considered very extensive, and if the present mode of destruction continues to be carried on, there is no doubt that the bird will soon become extinct.' While the sp. is still with us it has been extirpated over much of its original range. Other than the isolated N population on the Calliope Ra, the sp. has a continuous distribution down the Great Dividing Ra. It is still common within the upland forests of the northern part of its range but would appear to be less abundant in remaining habitat to the S, where it has suffered loss of lowland habitat over extensive areas. While occasional sightings were made in the *Araucaria* plantation with remnant notophyll vine forest strips of East Nanango Forestry area, c. 8 km E of Nanango, SE Queensland in the 1940s and early 1950s, the only recent record from there was of a single ♀ netted in Sept 1988 (Templeton 1992). The only measure of abundance is that an average of two birds was seen on each of at least six of 20 3.5 km walks in the Tooloom Scrub (Norris 1964). Strange (in Iredale 1950) noted that ♂♂

frequent three or four *Callitris macleayana* cypresses about 180 m apart. Lucas (in North 1901–4) seldom found more than one ad ♂ to every 400 m of scrub, each ♂ being sedentary. The sp. was 'rare or absent in rainforest patches 2.5 ha or smaller' in NE NSW (Howe 1986). Holmes (1973) recorded six birds in 112 ha of subtropical rainforest and 13 in 102 ha of wet sclerophyll forest near Dorrigo; i.e. one bird per 10–20 ha. Holmes (*in litt.*) recorded extirpation from remnants even as large as 60 ha in the Big Scrub. May become a pest of soft fruit crops, such as pawpaw (papaya) *Carica papaya*, and as a result are killed by farmers. A permit may be issued to kill any birds that damage fruit crops in Queensland. In the case of this sp., which has lost so much ground to habitat destruction, this ability to obtain permits to destroy birds is most questionabe.

Knowledge lacking and research priorities

The morphology of birds of the recently-reported and apparently-isolated population on the Calliope Ra near Rockhampton requires investigation. Given the ease of access to and availability of populations of this most southerly bird of paradise we would entreat biologists to perform intensive studies of its foraging ecology, courtship displays, mating system, and nesting biology.

Aviculture

London Zoo obtained a bird on 4 Apr 1882 (Seth Smith 1923*a*) and Goodwin (in Sharpe 1893) took at least one bird there, where it lived for several years. Talbot (in North 1901–4) kept a bird he had shot and broken a wing of for 3 months, finding it preferred to eat centipedes and remained lively in its cage. The sp. has been kept at Taronga Zoo, Sydney.

Victoria's Riflebird *Ptiloris victoriae* Gould, 1850

PLATE 8; FIGURES 9.39–43

Other names: Queen Victoria's Rifle-Bird, Victoria Riflebird, Victorian Rifle Bird, Victoria Rifle Bird of Paradise, Barnard Island Riflebird, Lesser Rifle Bird. *Eur-a-lum* of Aboriginal people of Tinaroo Scrub, Atherton Tableland (Jackson 1909).

A small riflebird of tropical N Queensland, generally similar to the Paradise Riflebird, but smaller, shorter-billed, and ♀ with rich cinnamon underparts. Monotypic.

Description

ADULT ♂: 25 cm, much like Paradise Riflebird but breast-shield narrower and smaller and the oil-green underparts more bronze-yellow and uniform (no exposed black feather bases) at its upper (anterior) edge. Bare parts as for Paradise Riflebird.

ADULT ♀: 23 cm, bill slightly longer, and tarsus on average shorter, than ♂♂ of all ages, and wing considerably shorter than that of ad ♂. Generally like ♀ Paradise Riflebird on upperparts but less contrastingly dark on crown and sides of face, slightly more greyish on mantle and back, and slightly less washed Raw Umber (23) on wings and tail. Only the malar area and chin whitish, plumage becoming pale Cinnamon (39) on throat to rich dark Cinnamon (123A) on breast and remaining underparts marked throughout with small blackish brown chevrons, so compact on some birds as almost to form spotting, that becomes barring on the flanks. Undertail coverts almost unmarked dark Cinnamon. Underwing: coverts pale Cinnamon and the remainder, as tail, like Paradise. Bare parts as ad ♂.

SUBADULT ♂: as Paradise Riflebird.

IMMATURE ♂: as ad ♀, with individuals apparently gaining increasingly stronger and broader blackish flank barring until the first sign of ad ♂ plumage attained.

NESTLING–JUVENILE: generally as ad ♀/imm ♂ but underparts indistinctly barred, not spotted, brownish. Bill pale bluish grey smudged dark grey, legs and feet pale bluish grey.

Distribution

The Atherton Region of tropical NE Queensland, Australia, including all suitable habitat within the Wet Tropics World Heritage Area and some off-shore islands. From Big Tableland (S of Cooktown) southward to Mt. Elliot (c. 30 km S of Townsville), from sea level to 1200 m (Williams *et al.* 1993).

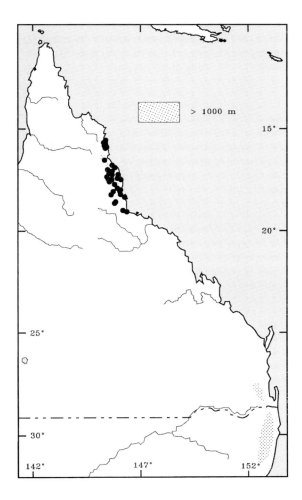

Systematics, nomenclature, subspecies, weights, and measurements

SYSTEMATICS: see previous sp. account. ♀-plumaged birds at extreme N of sp. range, Bloomfield R. to Big Tableland, are, on average, far more uniform (less marked with blackish) on their underparts than are birds elsewhere (museum specimens; CF).

SPECIES NOMENCLATURE AND ETYMOLOGY

Ptiloris victoriae Gould
Ptiloris Victoriae Gould, 1850. (between Jan and June). *Proceedings of the Zoological Society of London* 1849, p. 111, pl. 12. Barnard I., North Queensland. Type specimen BMNH 1850.7.20.135.

Synonym: *P. paradisea dyotti* Mathews, 1915. *The Austral Avian Record* 2, 133. Cairns, North Queensland. Type specimen not located. Mathews described *P. p. dyotti*, from Cairns, but subsequently (1930) declared it a junior homonym.

Exploration: discovered by John Macgillivray during the survey of the N Queensland coast by officers aboard the H.M.S. Rattlesnake in 1848; the first of many birds of paradise to be named after a monarch.

Etymology: *victoriae* = after Queen Victoria of England.

WEIGHTS AND MEASUREMENTS

Measurements: wing, ad ♂ (n = 51) 133–148 (139), imm ♂ (n = 25) 121–134 (129), ad ♀ (n = 34) 116–136 (125); tail, ad ♂ (n = 51) 77–85 (81), imm ♂ (n = 23) 75–85 (81), ad ♀ (n = 33) 72–85 (78); bill, ad ♂ (n = 51) 40–47 (43), imm ♂ (n = 24) 40–46 (43), ad ♀ (n = 33) 40–49 (45); tarsus, ad ♂ (n = 51) 30–35 (32), imm ♂ (n = 24) 30–35 (32), ad ♀ (n = 31) 28–33 (30). Weight: ad ♂ (n = 22) 91–119 (105), imm ♂ (n = 22) 82–104 (93), ad ♀ (n = 25) 77–96 (86). Mean lengths of central pair of rectrices average 2 (♀) and 4 (♂) mm shorter than rest.

Habitat and habits   Table 4.1

Mainly lowland to hill rainforests, but also adjacent eucalypt- and melaleuca-dominated wet sclerophyll woodland, swamp woodland, and the landward edge of mangrove communities. Most commonly seen agilely foraging for arthropods on tree limbs and trunks, often treecreeper-like, or at fruiting trees and vines. In winter, may form temporary feeding aggregations, and will move out of rainforest into adjacent wet sclerophyll woodlands, typically frequenting the ecotone to the W of the rainforests of the Great Dividing Ra. Between completion of annual moult and start of singing and courtship activity (i.e. c. May–July) territoriality of ad ♂♂ breaks down and up to four or more individuals may aggregate at a single food resource where fighting occurs only rarely (CF). Birds in both ad ♂ and ♀ plumage have been seen to displace most other passerine spp. at feeding trees within the rainforest, including the much heavier Spotted Catbird *Ailuroedus melanotis*. Occasionally seen to cling to and climb a vine or *Calamus* stem by using the bill parrot-like to assist as a 'third foot' (the Superb Bird is also recorded doing so). Flight is undulating and laboured.

Diet and foraging   Table 4.2

Of 65 feeding records made at Paluma, at c. 1000 m, near Townsville, Queensland, 75% occurred in forest canopy (10–25 m), 14% in sub-canopy (5–10 m), 8% in understorey (1–5 m), and remainder on emergent trees > 25 m above ground (Frith 1984). Birds do, however, feed close to and on forest floor, where they may forage by turning over leaves, twigs, and small stones in search of arthropods (CF).

Of the above 65 feeding records, 65% were of foraging for arthropods, 32% upon fruits, and 3% upon flowers (Frith 1984). Thus diet is predominantly wood- and dead foliage-dwelling arthropods, and fruits taken from upper stages of forest (Frith and Frith 1995b).

Arthropods are often sought in dead curled leaves, birds pulling them off twigs with the bill and then taking them to a perch to hold them down by a foot (or both feet) whilst tearing them apart (CF). Fruits may dominate the diet at some times of year when these are particularly abundant.

Beetles and their larvae, centipedes, grasshoppers, and spiders have been found in stomachs, as well as fruits of the palm *Archontophoenix alexandrae*, of *Alphitonia petrei* and of the introduced custard apple *Annona muricata* and orange *Citrus sinensis* (Barker and Vestjens 1990). Observed to come to within 1 m of the ground to take *Hibbertia scandens* fruits to an adjacent perch and there hold them to the perch by a foot to feed (Robinson in Cooper and Forshaw 1977). When in flower (Austral spring) the succulent internal bases of the bright red bracts of climbing pandans *Freycinetia* spp. are a popular consumable throughout the rainforest as are the subsequent fruits (CF). Birds were once observed feeding at flowers of the cauliferous *Syzygium cormiflorum* (R. Keller and D. Keller *in litt.*). Birds regularly visited a bird feeder to eat 'canary seed' (put out for estrildid finches) after which they tended to drink much (A. Griffin *in litt.*). Fruit seeds frequently regurgitated and dropped from perches and some indigestible arthropod matter may presumably also be discarded in this way. See Appendix 4 for additional food plants. A ♀-plumaged bird pulled apart the nest of a Red-browed Finch *Neochmia temporalis* in an isolated garden tree near the rainforest edge, took something in the bill and flew off with it (A. Dennis personal communication). The 'reversed' sexual dimorphism in bill size, in which the smaller ♀ has a bill 4% larger than that of the ♂, is a rare phenomenon—see previous species account. Why the smaller ♀ possesses the larger bill remains unexplained.

Vocalisations and   Figure 9.39
other sounds

The typical vocalisation is the advertisement song of ♂♂ which is an explosive loud *sssssshh*

Victoria's Riflebird *Ptiloris victoriae* Gould, 1850

or *yaaaass* that in the S of the sp. range is typically immediately repeated (as in Paradise Riflebird), but on the Atherton Tableland is usually delivered as a single note (Fig. 9.39). In the latter area it has been notated as *shhhhh* and is very similar to advertisement call notes of ♂ Superb Birds (BB). An aggressive note given to conspecifics at a disputed food resource is a sharp repeated *kek* delivered with fluffed mantle and flank feathers. In feeding small blind nestlings a ♀ gave a soft *kruk* note repeatedly. In an apparent attempt to call a large nestling off the nest a parent ♀ gave a soft, low, repeated *kuk* on the nest rim followed by a more musical 'bubbling' form of such notes as she flew from the nest and then called from nearby perches. When a fledgling only 30 min out of the nest was approached by several people, the parent ♀ hopped about within several metres and gave sharp agitated *kek* or *kruk* notes (Frith and Frith 1995b). Wing-noise produced by ad-plumaged ♂♂ in flight is a sharp, dry, rattling rustle that is so loud and characteristic that it probably functions in courtship and/or territorial contexts as a signal to potential mates, to rival ♂♂, or to all conspecifics.

Mating system

♂♂ sedentary, promiscuous, and court-displaying during breeding season and territorial at least to the extent of defending a number of exclusive display posts and perches, but it remains to be determined if (as appears likely) a more extensive foraging territory is seasonally maintained. Vast majority of display sites of ad-plumaged ♂♂ consist of the top of a broken-off vertical tree or tree fern trunk stump of c. 10–20 cm dia with the apex at several to many metres above ground. High and larger, bare, living or dead tree boughs may, only occasionally, be used. ♂♂ immediately reply to advertisement song of others, even from rainforest separated by 0.75 km of intervening pasture (CF). ♀♀ construct and attend nests alone.

Courtship behaviour

Typical courtship consists of three discrete components (Frith and Cooper 1996). First is the single or double *yaaass* call note given by ad ♂ on or close to his display stump to advertise his presence. Calling birds in ♀ plumage, subsequently seen to display and therefore assumed to be ♂, call with less power. This calling is a display component because it involves widely gaping the mandibles to expose the yellow mouth directed at ♀♀ (Fig. 9.40(a)).

Once a ♀ is close to an ad ♂ he performs a *Circular wings and gape* display. This is directed at the ♀, the ♂ turning slowly or spinning quickly in order to continuously face a ♀ moving around him. At high intensity this display involves the wings meeting, if not overlapping, above, or above and in front of, the ♂'s head and held rigidly still as his mouth is

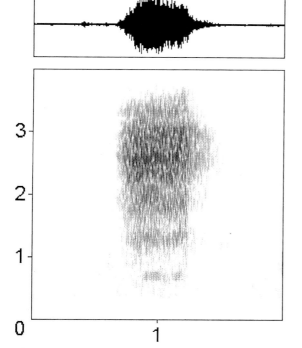

9.39 Sonogram from a recording of the typically-single advertisement song screech note of an ad ♂ Victoria's Riflebird near Malanda, Atherton Tableland by M. McGuire in Sept 1993.

widely gaped. His tail is cocked up to or just above horizontal and his abdomen feathering raised (Fig. 9.40(b)). The only movement now involves him repeatedly, slowly and rhythmically, raising and lowering himself on his legs at a rate of *c.* once every 1.5–3 secs, or turning to follow a moving ♀. Figure 9.40(b,c) shows the relative positions of the wings, tail, bill, head and neck, abdomen plumage, and legs at the highest and the lowest point of this body raising and lowering in display. No call is given, the bill remaining open to expose its yellow interior to the ♀ as the ♂ watches her intently. If the ♀ is in the canopy above the ♂ his head and neck are thrown well back beyond his vertical wings and his tail held further down. Under the extreme circumstances of this upwardly-directed static display the abdomen feathers are much raised to become slightly inverted (Fig. 9.40(d)).

The movement of a ♀ away from an ad ♂ in the intense *Circular wings and gape* display may cause him to relax his posture to a greater or lesser degree, presumably in accord with his perception of the ♀'s level of receptivity. This he does by progressively closing, or lowering, the primaries and then the secondaries of his raised wings and/or closing his bill, thus significantly changing the shape of the silhouette that his velvet-black plumage presents. The wings are closed by an action similar to that involved in closing a fan; each primary and secondary, from the innermost, sliding behind the next until the wing is progressively closed. The opening and closing of wings produces a dry rustling sound, like that produced by rubbing two rough, thin, wood shingles across each other.

The arrival of a ♀ upon the apex of the display stump is initially and briefly greeted by the ♂ leaning back and away from her in *Circular wings* display with his head and closed bill hidden behind the leading edge of one wing, possibly concealing most or all of his throat-shield. He may sway his body slowly towards one side (Fig. 9.41(a)). His wings are, however, now held forward of vertical, towards the ♀, their concavity forming a shallow parabolic dish into which she peers. If the ♀ appears nervous and inclined to depart the ♂ retains this posture rigidly; but if not he commences the *Alternate wings clap* display. He sharply raises and fully extends one wing, let us say the right, and brings it around towards the ♀ as he slightly closes and lowers the left wing and hides his upward-pointing bill behind the leading edge of the raised wing. During this rapid action he extends his right

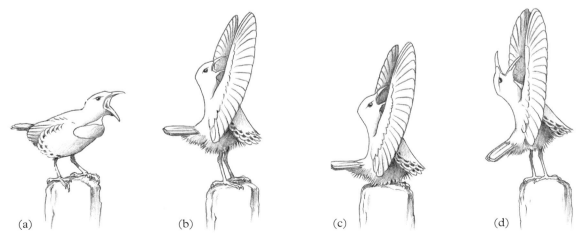

9.40 Some initial courtship display postures of ad ♂ Victoria's Riflebird: (a) advertisement singing, (b–d) *Circular wings and gape* display. After Frith and Cooper (1996), with kind permission. See text.

leg to raise that side of his body whilst lowering his left side. He then suddenly raises his left wing to hit his right above his head with an audible dull thud and instantaneously swings his rigidly upright-held head and bill across to hide behind its leading edge. Simultaneously he half closes and lowers his right wing and extends his right leg while lowering his left (Fig. 9.41(b,c)). The ♂'s bill is usually closed during the *Alternate wings clap* display but it may sometimes start with the bill open then to be closed as the tempo increases. This rigidly-stiff and rapid behaviour is then repeated numerous times (initially at a rate of one wing-lift per sec), the raised head and bill swinging like an inverted pendulum between alternately-raised wings as the body sways side to side and twists slightly from the leg/pelvis joint.

As the display progresses it intensifies as the tempo of the ♂'s wing and head movements and associated body swaying and twisting increases until the movements are so sharp and quick as to appear impossibly rapid. The bill remains closed and the iridescent throat-shield appears as a bright shining green, sometimes blue, vertical line shot with electric-white highlights as it is now repeatedly jerked, rather than swung, from one side to the other.

Young ♂♂ in ♀ plumage performing this display appear clumsy and un-coordinated in their movements and postures compared with an ad, often 'tripping over themselves'. They lack the square-ended primaries and secondaries of ad plumage and their wings thus do not form a smooth-edged circle but a rather irregular tattered one.

If at any point in the *Alternate wings clap* display the ♀ loses interest, the ♂ freezes his display at the point his head and bill are concealed behind one fully-raised wing. He remains motionless (Fig. 9.41(d)) until such time as the ♀ appears settled before continuing his display. As this display increases rapidly in tempo the ♂ moves and leans progressively towards the ♀ and twists each wing more and more with each alternate body sway so that they become increasingly less vertical and more to the horizontal. Thus his swinging head must also move progressively towards the horizontal, and the ♀, from its initially near-vertical orientation (Fig. 9.41(e)).

♀♀ often leave the display stump as the ♂'s swaying and body twists become vigorously rapid (when his movements may cause the upper display stump to sway or rotate violently). If a ♀ remains, however, the ♂ advances and adjusts his now violently-rapid *Alternate wings clap* display so that the cupped, or concave, primaries of his extended wings all but beat the ♀ who is, thus, alternately 'embraced' by each wing (Fig. 9.41(e,f)). So extremely rapidly does each wing now alternately partly encircle her (*c.* twice per sec) that she becomes effectively continuously embraced by the ♂'s wings. A suitably receptive ♀ rapidly flutters her closed wings and the ♂ then makes his last sway and body twist to one side more extensive in order to hop onto the ♀'s rump to copulate. Each of two copulations lasted an average of 4.3 sec. after which the ♂ hopped back onto his stump to recommence the *Alternate wings clap* display until the ♀ departed. He then reverted to the *Circular wings* display posture which then slowly subsided into a normal perching posture.

The ♀-plumaged, presumed ♀, birds we have seen remain on the display stump throughout ad ♂ display and mounting have initially appeared to be uninterested. They look about, peck at the stump and/or preen during the earlier part of display, but as the tempo of the *Alternate wings clap* display increases they peer intently at the centre of the ♂'s alternately-raised wings to watch his rigidly-upright throat and bill swing rapidly from one wing to the other. At this point in display it often became apparent that the watching ♀-plumaged bird may in fact have been a ♂ because it would suddenly start to perform simultaneously the *Alternate wings clap* display. This occurred quite often on film we viewed and as the ad ♂ continued to display while the presumably younger ♂ departed we imagine they were filmed early in the display season when the ♂'s tolerance was high. We

340 Victoria's Riflebird *Ptiloris victoriae* Gould, 1850

have gained the impression that ad ♂♂ are less tolerant during the peak mating season.

Some variation in the *Circular wings and gape, Circular wings*, or *Alternate wings clap* displays occurs if the relative positions of the ♂ and ♀ differ from face to face atop the display stump. When the ♀ arrives by clinging to the side of the stump some distance below the ♂ he will move to the edge of the stump apex and direct a *Circular wings and gape* display downward at her by leaning forward (Fig. 9.41(g)). If, on the other hand, the ♀ is atop the stump and the ♂ is clinging to its side he will direct his *Circular wings and gape* or *Alternate wings clap* display upward at her whilst clinging firmly to the vertical trunk and holding himself out horizontally (Fig. 9.41(h)), then walk upward towards her in this

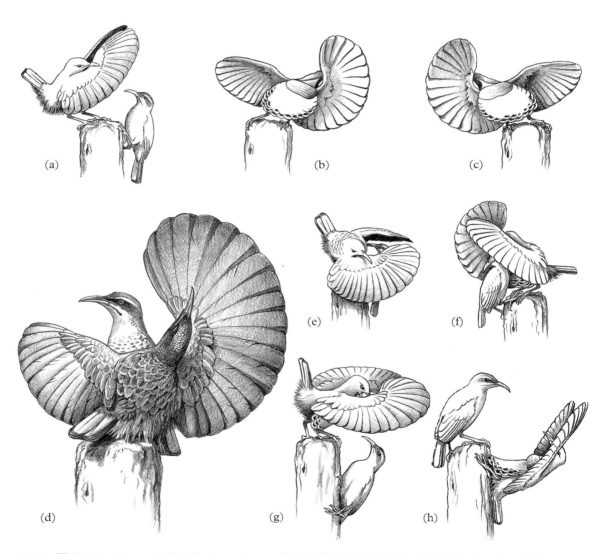

9.41 High-intensity courtship display postures of ad ♂ Victoria's Riflebird: (a) *Circular wings* display, (b–d) slow tempo *Alternate wings clap*, (e–f) fast tempo *Alternate wings clap*, (g) *Circular wings and gape* to ♀ below, (h) *Circular wings* to ♀ above. After Frith and Cooper (1996), with kind permission. See text.

posture until atop the stump where he can then commence or continue the *Alternate wings clap* display. Figures 9.40(b-d) and 41(a-h) show the relative positions of wings, tail, head and bill, legs, and plumage during different parts of typical *Circular wings and gape*, *Circular wings* and *Alternate wings clap* display.

We have occasionally observed a ♀-plumaged, presumed ♂, perform the *Circular wings* display to a passing bird of another sp.; to a Musky Rat-kangaroo *Hypsiprymnodon moschatus* and once to a moth as it flew slowly past (Frith and Cooper 1996; CF).

A gaping posture is sometimes directed downward involving the ♂ leaning forward and down with raised tail (Fig. 9.42(a)). An apparently aggressive posture involved an ad ♂ gaping downward while presenting his crown to a conspecific and raising his mantle feathers as he made short hops towards his opponent and repeated a soft sharp note. An ad ♂ was seen to fly to and cling onto a vertical tree trunk to be followed by a ♀-plumaged bird that clung to the trunk 1 m above him. He then repeatedly and sharply flicked his tail sideways through an arc (Fig. 9.42(b)) for 1 min or more before flying off to be followed by the conspecific (Frith and Cooper 1996).

A ♂ was once seen hanging upside down, tail fanned, wings spread, from his perch (Breeden and Breeden 1970: 113). Two birds, one ad ♂ and one in ♀-plumage or two ♀-plumaged birds, have not infrequently been seen to perch breast to breast atop a tree stump and display to one another (Bourke and Austin 1947; CF). In these cases the ♀-plumaged birds are most likely imm ♂♂. It has been implied that cases of such 'mutual' displays involve both sexes (Schodde and Tidemann 1988). In view of what is known about other birds of paradise and the present sp. in particular this interpretation is unlikely to prove correct and it is more likely that both birds in such circumstances are ♂. Two to four or more presumed young ♂♂ may in fact at times gather to direct displays at each other. Four ♀-plumaged birds did so on exposed

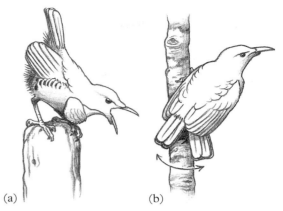

9.42 Two displays of ad ♂ Victoria's Riflebird: (a) a downward-directed gaping display, probably courtship advertisement to a ♀ below the display perch, (b) a sideways tail flicking, possibly ♀-enticement, display. After Frith and Cooper (1996), with kind permission. See text.

telegraph poles and adjacent wires c. 50 m from the rainforest edge at Paluma, near Townsville, one displaying to the others in turn (CF). The above account includes all previously-described displays (e.g. Bourke and Austin 1947; Sharland 1977).

Breeding Table 6.5

Apparently one record of cuckoo *Eudynamys scolopacea* parasitism (Brooker and Brooker 1989)

NEST SITE: 1.5–20 m above ground, 17 nests averaging 8 m, typically so well concealed as to be invisible from below. Often atop a broken-off tree trunk with prolific new foliage shooting from it, in a vine-covered tree, or in the centre of a *Cyathea* tree fern crown. Most frequent sites in lowlands are pandanus, fan palms *Licuala ramsayi*, and cordylines, the nest being placed upon frond bases immediately adjacent to the trunk but in the uplands dense vine tangles on trees or atop a broken-off living tree stump with dense new foliage shooting are favoured (Barnard in Campbell 1901; Hislop in North 1901; Jackson 1907; CF). Meek (1913) found that nests at Cedar Bay were 'usually built in the heads of the umbrella palms and were woven out of old vines'. A nest containing an egg was found in the crown of a banana tree at Mission Beach on 15 Nov (S. Breeden personal communication). Indicative of traditional nesting is that Hislop (in North 1901) wrote 'On one Screw Pine [*Pandanus* sp.], I found the nests of three successive seasons' and another was being built there. A ♀ nest-building 4.5 m high in an ornamental garden plant immediately beside rainforest on 20 Sept was, having almost completed the nest, vigorously attacked by another ♀-plumaged conspecific on the nest site, the two grappling each other to fall to the ground pecking at each other. This fight brought the nest-building to a halt until 16 Nov when another was started in the same place (Frith and Frith 1995*b*).

NEST: a nest collected 3 days after a nestling left it was circular and 93 mm in total external depth; 185 mm in external dia (excluding protruding material); and 100 mm in internal egg cup dia which was 44 mm deep (Fig. 9.43(a)); average upper nest wall thickness 43 mm. Outer structure consisted of 80 *Pyrrosia dielsii* fern rhizomes, of fresh to dry fronds 140–400 mm long, some with associated moss growth, and 20 additional bare stems averaging *c*. 300 mm long that were also predominatly of *P. dielsii*. These plants encircled the entire outer nest rim and walls with only a couple of them supporting the structure directly beneath. Thus, the base of the discrete, leafy, internal nest cup (Fig. 9.43(b)) was clearly exposed beneath. Within this outer structure was a substantial cup-shaped structure of 43 dead brown leaves (Fig. 9.43(b)), 20 (47%) from the vine *Melodinus bacellianus* and the rest from trees: 6 of *Beilschmiedia tooram*, 5 of *Litsea leefeana* (both Lauraceae), 3 of *Mischocarpus macrocarpus* (Sapindaceae), 2 of *Neolitsea dealbata*, 1 of *Endiandra sankeyana* (both Lauraceae), and 1 of *Cupaniopsis flagelliformis* (Sapindaceae); 5 unidentifiable. Within this leaf cup was the egg-cup lining, a neatly-constructed accumulation of *c*. 240 supple, fine, woody, tapering,

flower inflorescence stems of the vine *Maesa dependens* (Myrsinaceae) curved into place to form a springy, dry, shallow cup (Fig. 9.43(b)). These stems reached to the top of the nest rim and measured 0.6–1.5 mm in dia and 85–320 mm in length. The nest was constructed within numerous near-vertical, outwardly-curved, large leaves of an exotic *Dracaena* plant, from which it was easily lifted out entire after use. Nests of four subsequent nestings in adjacent sites and probably by the same ♀ were similar in materials and relative numbers

(a)

(b)

9.43 Victoria's Riflebird nest photographed 3 days after nestling departure: (a) entire and viewed from above and (b) the constituent structural parts of external *Pyrrosia dielsii* fern plants (left), internal cup-shaped structure of dead leaves (right) and nest-cup lining of rainforest vine inflorescence stems (below). See text. Photo: CF.

of constituent parts (Frith and Frith 1998a). Meek (1913) wrote of finding many nests that always had sloughed snake skin as part of the materials, and skins of Black Snake and Carpet Python have been identified.

EGGS: clutch usually two but clutches of one (and one of three, B. Gray personal communication) are known. 48 eggs averaged 32.2 (29.1–34.9) × 23 (22.1–24.6) mm (CF). At 18–19 days old the eggs of a two-egg clutch weighed 8.6 and 8.2 g (Frith and Frith 1995b). Ground colour Pale Pinkish Buff (121D) but slightly paler/pinker to between 121D and Pale Horn (92) boldly marked with broad longitudinal 'brush' strokes of browns and purplish greys. Plate 13. An egg collector with much experience of this sp. believed two broods are sometimes raised in a season and he obtained three clutches of two eggs from the one ♀ by taking each clutch as it was completed (Hislop in Iredale 1950).

INCUBATION: exclusively by ♀. One ♀ with a two-egg clutch, studied for 128.5 hr over the last 17 days (mean = 7.5 hr per day) of the incubation period, of 18–19 days, spent an overall average of 71% of time incubating, incubation bouts averaging 44 min, at 1 bout per hr. ♀ absences averaged 18 min (Frith and Frith 1995b). A ♀, quite possibly the same one, nesting within c. 6 m of the nests from the last 3 years, later and similarly incubated her clutch of two for 18–19 days to their hatching (Frith and Frith 1998a).

NESTLING CARE: nestling care is exclusively by the ♀ who carries larger-bodied insects to the nest in the bill tip, but also brings some fruit and regurgitates additional food items or may produce a meal of softer items exclusively by regurgitation. In 6.5 hr of observation at a nest over 3 days (during 9–13 Oct) just before the brood of two left it only a ♀-plumaged bird attended. Food was regurgitated to the young, with the first item of each meal often being carried in the parent's mandible tips. Of 10 items seen in the parent's bill, four were insects (including a small beetle, winged termite, and a large cricket), two were of fruit, and four were unidentified. The ♀ was present for a total of 7.5% of the time, each presence averaging 1.5 min and each absence c. 18 min, with an average of 3.4 visits per hr. During 149 hr of observation at a nest (nest 2) that contained two nestlings for 2 days and only one nestling for the subsequent 13 days, the ♀ brooded for 21% of the time overall but did not brood after the nestling was 8 days old. On three occasions she did, however, subsequently shelter the nestling from rain by standing in the nest with her wings open over it. She averaged 0.7 brooding bouts per hr, each of 18.5 min, 1.8 nest visits per hr and 1.7 nestling meals per hr with 2 min perched on the nest rim in feeding, nest sanitation, and nestling/nest maintenance. Absences averaged 24 min (Frith and Frith 1995b). Three years later a ♀, probably the same individual, was studied for 140 hr successfully raising her brood of two (nest 3). The ♀ brooded for 24% of the time overall and again did not do so after a nestling age of 8 days save once to shelter the 10-day-old brood from torrential rain. She averaged 0.7 brooding bouts per hr each of 21 min, 3 nest visits per hr, and 3 meals per hr with 1.7 min perched on the nest rim in feeding, nest sanitation, and nestling/nest maintenance. These figures clearly demonstrate the greater demand placed upon the ♀ by a brood of two compared with one (Frith and Frith 1998a). A total of 256 nestling meals at nest 2 involved 280 items. Of these 11% were too small to be identifiable but were, therefore, animal matter (being too small to be fruits), 39% were medium-sized to large orthopterans, 14% unidentifiable animal/insect foods, 10% unidentifiable insects, 6% spiders, 2% each of cockroaches, beetles, and insect larvae (probably of beetles), and 1% each of centipedes, cicadas, and wood lice; 7% were unidentified fruits (except that at least those of *Ardisia brevipedata* and *Pothos longipes* were included), and 4% were of the nutmeg *Myristica insipida* fruit aril. Thus a minimum of 78% and a potential maximum of 89% of items fed to the nestling were animal and 11% were fruit. Less comprehensive records of nestling meals at nest 3 show a generally similar nestling diet, but add skinks (Frith and Frith 1998*a*).

NESTLING DEVELOPMENT: Schodde and Tidemann (1988) state young leave the nest in 'about four weeks' without further comment. The single nestling, from an original brood of two, left nest 2 at 15 days old. Its growth and development are summarised in Fig. 6.2. The brood of two departed nest 3 at 13–14 days old (Frith and Frith 1995*b*, 1998*a*) and the ♀ was seen feeding a juv begging with soft vocalisations and slightly-quivering wing tips 74 days after it left the nest (W. Cooper personal communication). At Lake Eacham, Atherton Tableland on 23 Nov a presumed ♀ led two young unable to fly, with well-developed primaries but rectrices only just starting to sprout, into dense foliage by hopping in front of them (RAOU NRS). At Mountain Creek, Paluma Ra on 2 Feb a ♀ fed a full-grown fledgling a black beetle, and then both birds preened for 5 min and the ♀ occasionally plucked down feathers from her young. This imm was like the ♀ parent but its breast was more beige, less rufous, and it was more speckled about the throat. It regurgitated a small seed which the ♀ took and dropped. On 20 Mar at Paluma a ♀ fed a young of her own appearance with *Omalanthus novoguineensis* fruit pieces as it begged with drooped quivering wings and soft pleading vocalisation (CF). A ♀ was feeding a mobile, well-feathered, and flying imm at Mt. Lewis on 6 Jan 1990 (D. Whitford personal communication).

Annual cycle   Table 6.5

DISPLAY: July–Dec. Ad and imm ♂♂, but particularly young ♂♂ in ♀-plumage, commonly perform post-moult displays during Mar–Apr (CF). BREEDING: at least Aug–Feb, ♂♂ with moderate to much-enlarged gonads having been collected during Jan, Apr, July, and Sept–Oct, and ♀♀ in Oct–Dec. EGG LAYING: late Aug–early Jan. MOULT: only 3 of 73 specimens were moulting, involving one ♂ in body and tail moult in Apr and two ♀♀ in general moult in Dec (Appendix 3). Ad ♂♂ complete their moult by late Feb to mid Mar (CF).

Status and conservation   Table 8.1
Once heavily collected for the millinery trade and scientific collections. Campbell (1901) writes of shooting 10–17 birds a day on the Barnard Is. Common throughout its remaining habitat, having lost much rainforest on the Atherton Tableland to clear-felling for dairy and other farming. Bourke and Austin (1947) estimated that each ad ♂ occupied and defended an area of about 2 ha or more. Reported to be more common on the Bloomfield R. during July–Jan than at other times of year, but this may well merely reflect relative vocalisations and conspicuousness. May be a locally serious pest to soft fruit orchards, such as pawpaw, and birds are both legally (permits issued if commercial crops are damaged) and illegally killed at the time of writing this, sometimes in large numbers.

Knowledge lacking and research priorities
A study of the taxonomic status of birds at the extreme N of the sp. range may prove rewarding (see Systematics above).

Aviculture
Small numbers have been kept but no documentation is known to us.

Other
An ad ♂ and a ♀-plumaged bird were observed mobbing a Black Butcherbird *Cracticus quoyi* near the nest of the presumed ♀ (Le Souëf in Campbell 1901: 75). One was predated by a Lesser Sooty Owl *Tyto tenebricosa* (Schodde and Mason 1980). One bird taken by a Grey Goshawk *Accipiter novaehollandiae* 5 Sept 1992 (CF). A ♂ caught and banded at Paluma in ♀ plumage on 6 July 1982 was in subad ♂ (partly black) plumage during Jan and Dec 1984. This bird had almost completed moult-

ing into ad plumage on 27 Jan 1986 and had done so by Mar of that year. Thus this ♂ took a minimum of, and doubtless much longer than, 3.3 yr to acquire ad plumage and lived at least 15 yr before being killed by a domestic cat in Dec 1996 (CF).

Genus *Lophorina* Vieillot, 1816

Lophorina Vieillot, 1816. *Analyse d'une Nouvelle Ornithologie Élémentaire* p. 35. Type, by monotypy, 'Le Superbe, Buffon,' = *Paradisea Superba* Pennant.

The Superb Bird of Paradise, the sole species in the genus *Lophorina*, is distributed extensively through the uplands of mainland NG. It has **skull** and bill characters showing similarities to *Cicinnurus* and *Parotia* and was tentatively grouped with *Parotia*, *Pteridophora*, and *Cicinnurus* by Bock (1963). Intergeneric **hybrids** with *Paradigalla*, *Parotia*, *Ptiloris*, and *Cicinnurus* are known. The slightly-decurved **bill** is just slightly longer than the head. **Tail** mildly graduated, central pair longest; older ♂♂ have shortest tails. **Wing** relatively long and pointed. Outermost two primaries emarginated on inner vane. **Tarsus** length 25% of wing length. Distal section of tarsus scuted, fading to smooth proximally.

The most striking features of *Lophorina* are the mass of velvety black erectile cape feathers and a winged pectoral shield of modified and highly-iridescent, scale-like feathers of the ad ♂ (Plate 10). Plumage patterns of the sexes are remarkably *Ptiloris* and *Parotia*-like.

Polygynous, with ad ♂ territorial, solitary-displaying, and promiscuous, and ♀ conducting nesting duties alone. Courtship display somewhat mid-way between that of riflebirds and parotias in several respects and the three genera would also appear to be close in plumage and other characters. **Nest** site and structure generally much like *Ptiloris* spp., particularly in use of living *Polypodium* fern stems and fronds, but other constituent plant genera insufficiently known. **Eggs** very like those of typical Paradisaeinae (Plate 13).

Superb Bird of Paradise *Lophorina superba* (Pennant, 1781)

PLATE 10; FIGURES 1.4, 5.2, 9.44–46

Other names: Lesser Superb Bird of Paradise. *Niedda* (D'Albertis 1880). *Kera* of Yabi people, Weyland Mts, IJ (Rothschild 1931), *Neni* (M) and *Piyo* (♀ and imm ♂) of Fore people. *Nine-oba* or *Nene Koba* (♂) and *Araro* (♀ and imm ♂) of Gimi people (Diamond 1972; D. Gillison *in litt.*). *Kabay bl, Kabay kl, Kabay bl-bad, Kabay yb* or *Kabay cgaij* of Kalam people (Majnep and Bulmer 1977). *Kongerai* of Kubor Mts people (Mayr and Gilliard 1954) and also *Trang, Gua-nam* there; *Isaap* of Biak I. plume traders (Mayr and Meyer de Schauensee 1939a). *Teten* of the Ilaga Valley, *Terel ha* of the Baliem Valley, Dani people, and *Yel* of Uhunduni people (Ripley 1964). *Nyat nyat* of Nokopo people of the Finisterre Ra, PNG (Schmid 1993). *Kei Galinch* at Mt. Hagen. *Yagama* (M) and *Ponge* (F) of Huli people, Tari Valley (Ballard *in litt.*). *Kandeyaka* at Kompiam, PNG (R. Whiteside *in litt.*). *To-gompa* or *To-go-pa* (Mt. Hagen, PNG).

A compact bird of paradise of NG montane forests. Common and widespread, and ecologically more tolerant of disturbed and fragmented habitat than most members of the family. The jet black ad ♂, with iridescent blue-green breast-shield and vast cape of elongate nape

Superb Bird of Paradise *Lophorina superba* (Pennant, 1781)

plumes, is unmistakable. ♀♀ and imm ♂♂ are extremely similar in appearance to ♀♀ of some parotias and the Magnificent Riflebird. Polytypic; 6 subspp.

Description

ADULT ♂: 26 cm, entire head velvety jet black with dully-iridescent, dark coppery green sheen, except the crown of scale-like, highly-iridescent, metallic green-blue feathers with purple to magenta sheens in some lights. A tuft of elongate, erectile, velvety jet black feathers above and behind each nostril. Feathers at the base of the latter and on the lores and elongate forward-pointing chin feathers with dull purple to magenta iridescent sheen. Nape and rear neck feathers grossly- and uniquely-elongated and modified to form a vast erectile nape 'cape', the outer feathers of which being spatulate to fan-shaped (producing a symmetrical shape when raised—see Fig. 9.46(e)), these velvety jet black feathers dully-iridescent, rich, dark Olive-Green (46) like those of the mantle beneath. Back and rump black, uppertail coverts as mantle. Upperwing: coverts and tertials velvety matte black, flight feathers blackish brown (19) with narrow, dull, olive-brown leading edges to primaries. Dorsal surface of tail as wing flight feathers except longest central pair which is velvet jet black with rich iridescent sheen of violet-purple (172). Throat velvet-black with dark Olive-Green grading to rich Purple (1) iridescent sheens with Magenta (2) washes in some lights. Scale-like feathers, greatly elongated laterally, form a shallow, winged, delta-shaped breast-shield of intensely-iridescent metallic greenish blue with deep bluish green to violet purple sheens in suitable light. Remaining underparts slightly glossy black with trace of dull, dark, Olive-Green and Purple sheens on belly in rare light. Underwing: coverts jet black, flight feathers Dark Greyish Brown (20). Ventral surface of tail blackish brown. Bill black, iris very dark brown, legs and feet blackish, mouth lemon-yellow to lime-green.

ADULT ♀: 25 cm, wing of ♀ shorter, with virtually no overlap with ♂♂. Entire head and nape except chin and throat blackish brown with a short line of tiny whitish spots as a post-ocular, sub-obsolete stripe and a similar malar stripe immediately beneath the gape. Mantle to uppertail coverts dark reddish brown (223A). Upperwing: lesser coverts as mantle, greater coverts, flight feathers, and tertials Fuscous (21) with rich Chestnut (32) outer edges. Dorsal surface of tail Fuscous with pale chestnut outer edges to feathers. Chin and throat whitish-grey grading to pale Buff (124) on upper breast to darker buff (223D) on flanks, thighs, and undertail coverts all uniformly narrowly barred brownish black except on chin and upper throat where this breaks down to fine flecking. Underwing: coverts dark buff (223C) barred dark brown, flight feathers, and undertail dark olive brown (119A).

SUBADULT ♂: as ad ♀ with few feathers of ad ♂ plumage intruding, initially the outer primaries becoming black (BB), to as ad ♂ with few feathers of ♀ plumage remaining.

IMMATURE ♂: as ad ♀ but wing and tail average longer.

NESTLING: the first nestling hatched at Chester Zoo, UK, was killed by the ♂ at 5 days old when it was naked with blackish flesh skin, paler below; bill greyish flesh, gape white, mouth flesh, and feet greyish. A Mt. Missim chick (19 g) was naked, black skinned, with a greenish gape and no down (T. Pratt *in litt.*).

JUVENILE: like ad ♀ but dark crown, rear, and sides of neck barred ochraceous. Plumage generally soft and fluffy, particularly on crown and underparts.

Distribution

Virtually all upland forests of NG: the central cordillera, the mountains of the Vogelkop and Huon Penin, and Adelbert Mts ranging from 1000 to 2300 m (mainly 1650–1900 m). Also Mt. Bosavi (T. Pratt *in litt.*)

Superb Bird of Paradise *Lophorina superba* (Pennant, 1781)

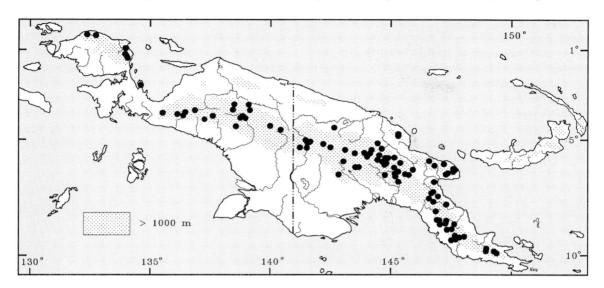

Systematics, nomenclature, subspecies, weights, and measurements

SYSTEMATICS: *Lophorina superba* is a sp. representing a monotypic genus with characters that link it to a number of the core paradisaeinine genera (as evidenced by the many generic recombinations argued by Diamond [1972]). The sp. exhibits complex infraspecific variation, which has been analysed in the literature. We recognise six subspp. As the taxa accepted by Cracraft (1992) are in broad accord with those delineated by, and with most doubts expressed by, Gilliard (1969), Diamond (1972), and Coates (1990), we accept these six tentatively, until a comprehensive re-evaluation of the systematics of the sp. provides a more authoritative reassessment (Frith and Frith 1997*b*). Hybridisation: recorded to have crossed with more bird of paradise spp. than any other—with Long-tailed Paradigalla, Western and Carola's Parotias, Magnificent Riflebird, Black Sicklebill, and Magnificent Bird (Fuller 1995; Frith and Frith 1996*b*). See Plates 14, 15 and Appendix 1.

SPECIES NOMENCLATURE AND ETYMOLOGY

Lophorina Superba (Pennant)
Paradisea superba Pennant, 1781. *Specimen Faunulae Indicae*, in Forster's *Zoologica Indica Selecta*, p. 40 (based on Daubenton 1774, *Table des Planches Enluminées d'Histoire Naturuelle*, pl. 632). New Guinea, restricted to Arfak Mts. Type is an illustration in Daubenton, drawn by F. N. Martinet.

Etymology: *Lophorina* = Gr. *lophos*, a crest; Gr. *rhis, rhinos*, nose, beak; *superba* = L. excellent, splendid.

SUBSPECIES, WEIGHTS, AND MEASUREMENTS

Owing to the complexity of minor variation in the species, its widespread montane distribution, and the inadequate material available for analysis for many local populations, the following subspecific analysis must be considered provisional. In particular, populations from the Adelbert Mts, the Tari region, the Hunstein Ra, and other outlying populations, require further diagnosis of subspecific affinity.

1. *L. s. superba* (Pennant, 1781)
Range: Arfak and Tamrau Mts of the Vogelkop Penin, IJ.
Diagnosis: head and side of throat of ♀ entirely blackish; underparts tinged faintly with buff and barred with blackish brown (Gilliard 1969: 167); mean ad ♂ tail length equals 73% of mean wing length.

Superb Bird of Paradise *Lophorina superba* (Pennant, 1781)

Measurements: wing, ad ♂ ($n = 24$) 131–149 (140), imm ♂ ($n = 7$) 129–135 (134), ad ♀ ($n = 16$) 122–133 (126); tail, ad ♂ ($n = 20$) 97–109 (102), imm ♂ ($n = 7$) 101–107 (104), ad ♀ ($n = 16$) 93–110 (98); bill, ad ♂ ($n = 23$) 28–33 (30), imm ♂ ($n = 7$) 29–30 (30), ad ♀ (n 16) 26–32 (30); tarsus, ad ♂ ($n = 23$) 28–33 (31), imm ♂ ($n = 7$) 31–32 (31), ad ♀ ($n = 15$) 28–32 (30). Weight: ad ♂ ($n = 8$) 81–95 (90), ad ♀ ($n = 5$) 68–85 (74).

2. *L. s. niedda* Mayr, 1930. *Ornithologische Monatsberichte* **38**, 179. Wondiwoi, Wandammen Mts. Type specimen ZMB 1418 (destroyed in WW II?).

Range: Mt. Wondiwoi of Wandammen Penin, IJ.

Diagnosis: female plumage is distinguishable from nominate form by darker and more ochraceous underparts (Gilliard 1969).

Measurements: wing, ad ♂ ($n = 4$) 137–142 (139), ad ♀ ($n = 2$) 123–124 (124); tail, ad ♂ ($n = 4$) 98–105 (101), ad ♀ ($n = 2$) 99–105 (102); tarsus, ad ♂ ($n = 4$) 28–32 (30), ad ♀ ($n = 2$) 30–31 (30); bill, ad ♂ ($n = 4$) 31 (31), ad ♀ ($n = 2$) 30–31 (31). Weight: ad ♂, ($n = 2$) 100–105 (103), ad ♀ ($n = 2$) 70–71 (71).

3. *L. s. feminina* Ogilvie-Grant, 1915. *Ibis, Jubilee Supplement* **2**, 27. Utakwa R., Nassau Ra. Type specimen BMNH 1916.5.30.1043.

Range: mountains of the central cordillera of IJ, from the Weyland Ra eastward to the Victor Emanuel and Hindenburg Mts of westernmost PNG. Boundary of range of this and the following subsp. has not been delineated.

Diagnosis: ♀ differs from nominate form (to the W) and the similar *latipennis* (to the E) by the broader superciliary stripe, which extends narrowly across nape to other eye (see Plate 10). Ad ♂ wing length as in nominate form but tail/wing proportion shorter (63%), and ad ♀ smaller in all measurements. In general and proportionate measurements most like *latipennis*.

Measurements: wing, ad ♂ ($n = 27$) 134–146 (139), imm ♂ ($n = 11$) 120–134 (131), ad ♀ ($n = 18$) 115–127 (122); tail, ad ♂ ($n = 26$) 83–99 (88), imm ♂ ($n = 10$) 77–92 (84), ad ♀ ($n = 18$) 73–88 (81); bill, ad ♂ ($n = 28$) 29–33 (31), imm ♂ ($n = 11$) 29–31, ad ♀ ($n = 18$) 29–33 (31); tarsus, ad ♂ ($n = 26$) 30–34 (32), imm ♂ ($n = 11$) 30–34 (32), ad ♀ ($n = 18$) 29–32 (30). Weight: ad ♂ ($n = 5$) 60–95 (81), ad ♀ ($n = 2$) 54–71 (63).

4. *L. s. latipennis* Rothschild, 1907. *Bulletin of the British Ornithologists' Club* **19**, 92. Rawlinson Mts, Huon Penin. Type specimen AMNH 678271.

Synonyms: *L. s. connectens* Mayr, 1930. *Ornithologische Monatsberichte* **38**, 180. Dawong, Herzog Mts. Type specimen ZMB 30.1961.

L. s. addenda Iredale, 1948. *Australian Zoologist* **11**, 162. Mt. Hagen district. Type specimen AM O.37689.

Range: Central and Eastern Highlands of PNG, eastward to the Herzog, Kuper, and ?Ekuti Ra where, presumably somewhere SE of Wau, the subsp. meets *L. s. minor*. Also mountains of the Huon Penin and perhaps the Adelbert Ra.

Diagnosis: head of ♀ dark brown, chin and throat whitish, with broad whitish superciliary stripe; white streaking on forehead, crown, and nape. Upperparts variably olive brown. Overall smaller and lighter than nominate form, mean tail length as a proportion of mean wing length smaller (68%) and ad of both sexes slightly larger than *minor*.

Measurements: wing, ad ♂ ($n = 59$) 127–145 (136), imm ♂ ($n = 25$) 119–135 (128), ad ♀ ($n = 55$) 110–131 (119); tail, ad ♂ ($n = 55$) 82–102 (92), imm ♂ ($n = 24$) 82–93 (87), ad ♀ ($n = 54$) 75–88 (82); bill, ad ♂ ($n = 59$) 28–33 (30), imm ♂ ($n = 24$) 28–32 (29), ad ♀ ($n = 55$) 28–32 (29); tarsus, ad ♂ ($n = 59$) 29–35 (32), imm ♂ ($n = 25$) 29–35 (32), ad ♀ ($n = 54$) 28–37 (30). Weight: ad ♂ ($n = 39$) 78–99 (87), imm ♂ ($n = 13$) 64–86 (77), ad ♀ ($n = 38$) 54–78 (67).

5. *L. s. minor* Ramsay, 1885. *Proceedings of the Linnean Society of New South Wales* **10**, 242. Astrolabe Mts. Type specimen AMNH 032935.

Synonyms: *L. s. lehunti* Rothschild, 1932. *Annals and Magazine of Natural History*. ser.

10, **10**, 126. Mekeo, SE New Guinea. Type specimen is a ♀ said to be in BMNH but not located.

Range: southeastern PNG, including the entire Papuan Penin, NW through the entirety of the Owen Stanley Ra.

Diagnosis: ♀ blacker-plumaged than *latipennis*, with head and throat blackish brown and upperparts rich dark chestnut. Superciliary stripe largely absent, appearing as a small sub-obsolete post-ocular streak. No or little pale nape marking. Wing shorter than all other subspp. and tail shorter than all but *feminina*.

Measurements: wing, ad ♂ (n = 33) 126–140 (132), imm ♂ (n = 14) 117–131 (122), ad ♀ (n = 24) 110–126 (116); tail, ad ♂ (n = 32) 82–94 (88), imm ♂ (n = 13) 77–92 (87), ad ♀ (n = 21) 73–91 (81); bill, ad ♂ (n = 31) 28–34 (30), imm ♂ (n = 13) 29–31 (30), ad ♀ (n = 24) 28–31 (29); tarsus, ad ♂ (n = 32) 28–33 (32), imm ♂ (n = 14) 29–33 (31), ad ♀ (n = 24) 27–32 (29). Weight: ad ♂ (n = 8) 77–93 (83), imm ♂ (n = 4) 67–84 (72), ad ♀ (n = 5) 56–68 (63).

6. *L. s. sphinx* Neumann, 1932. *Ornithologische Monatsberichte* **40**, 121. Locality unknown. Type specimen MCZ 153639.

Range: unknown, but hypothesised by Mayr (in Mayr and Greenway 1962) to come possibly from the mountains of far SE PNG.

Diagnosis: 'More reddish brown on upperparts than *minor*, and eyestripe less extensive and forehead and neck without white flecks' (Gilliard 1969: 168). We disagree with Cracraft (1992), who synonymised *sphinx* with *minor*, because the unique ♀ type specimen is considerably larger than *minor* (Frith and Frith 1997b).

Measurements: single imm ♂ specimen; wing, 137; tail, 91; tail centrals 96; bill, 28.2; tarsus, 31.7.

Note: '*L. s. pseudoparotia*' Stresemann, 1934. *Ornithologische Monatsberichte* **42**, 144. Hunsteinspitze, middle Sepik. Type specimen ZMB 3.49.296 is a hybrid cross of *Lophorina superba* × *Parotia carolae* from the Hunstein Mts (Frith and Frith 1996b). No other specimen of (pure) *L. s. superba* is known from this mountain range, although A. Mack reports (personal communication) recording this sp. in a 1989 field survey of the Hunstein Mts. The subspecific designation of this Hunstein population of *L. superba* remains to be determined.

Habitat and habits   Table 4.1

Mid- to upper montane forests, disturbed forests, and forest patches among gardens and other cleared areas. In the Tari and Wahgi Valleys, PNG, ♂♂ have been seen perched atop thatched village houses, presumably foraging for insects (D. Bishop *in litt.*). Solitary and slow-moving but joins mixed-spp. foraging flocks. Ad ♂ sings high in the crowns of leafy trees. In disturbed areas, display trees are often at the forest-edge or in partly-felled areas. ♀-plumaged birds spend twice as much time in the understorey as ad ♂♂, the latter never seen in company with the former (Diamond 1972) but LeCroy (1981: 3–4) allowed the possibility of a brief pair bond. May use the bill parrot-like, to assist in climbing about and typically uses a foot to hold food against the perch while feeding (D. Frith and C. Frith 1988; T. Pratt *in litt.*). Frequents forest pools to drink and bathe. Both sexes of a pair in NYZS allopreened the head and neck feathers of the other (Crandall 1932). Two captives in PNG also allopreened but the ad ♂ usually (n = 19) stimulated the ♀-plumaged to preen his crown by presenting it in a posture similar to that of *Initial display activity* (see below) and therefore possibly as a display (Fig. 9.46(b)) but in a more relaxed manner and with drooped wings. If ignored by the ♀-plumaged bird he held the pose for up to 20 sec. Only once did the ♀-plumaged bird try to solicit the ad ♂ to allopreen its crown (D. Frith and C. Frith 1988). More surprising was the behaviour of a confirmed ad ♂ and ♀ at Chester Zoo, UK involving the ♂ 'feeding the hen and displaying to her' (Timmis 1968, 1970). This ♂ also offered the ♀ mealworms as she moved about the aviary during the period she was incubating her clutch. These instances of allopreening and feeding between the sexes are possibly artefacts of captivity and are unlikely to occur in the wild. This said, ♂ and ♀ of this sp. seem more tolerant of each other than all other birds of paradise spp. maintained in captivity. A ♀-plumaged bird attacked and chased off in flight a ♀-plumaged Wahnes' Parotia (T. Pratt *in litt.*).

Diet and foraging Table 4.2

Thane Pratt (*in litt.*) made numerous sightings of foraging birds in the Adelbert Ra, PNG during 29 Feb–22 Mar 1975. Birds were typically 4–9 m above ground, and all but one fed upon arthropods while hopping about small trees and tree limbs covered with moss and other epiphytic growth, which they probed and tore into. Birds commonly bite, tear, and probe curled dead leaves, bark, rotten wood, mosses, and other epiphytic vegetation on tree limbs and trunks in the creeper-like manner of a riflebird. Individuals may forage in this way for up to 6 min on trunks and branches 2–25 cm dia and at 45–60° from horizontal. Of 41 sightings of feeding birds on Mt. Missim 25 were at branches, 9 at fruits, 4 at leaves, 2 at flowers, and 1 at dead vegetation (BB). ♀-plumaged birds were seen feeding on 'yellow figs' in the middle and upper limbs of original forest and in clumps of secondary growth in gardens hundreds of metres from the forest-edge Gilliard (1969). A ♀-plumaged bird was watched in old garden secondary growth feeding acrobatically on fruits of two *Callicarpa pentandra* trees in company with Trumpet Manucodes, Carola's Parotia, and Lesser Birds (Cooper in Cooper and Forshaw 1977). Individuals actively participated in a mixed-spp. foraging flocks including Trumpet Manucode, Magnificent, and Raggiana Birds at Mt. Missim, PNG on 24 Oct. A bird foraged with seven other spp., including Trumpet Manucode (BB). Uses a foot to hold items.

Stomach contents of 14 birds, from four locations, varied from all-arthropods to all-fruit. Diet based on stomach samples, faecal samples, and random encounters with foraging birds was assessed to be 24% fruit and 76% arthropods (Beehler and Pruett-Jones 1983). Arthropod groups were: Cerambycidae, Chilopoda, Chrysomelidae, Curculionidae, Dermaptera, Elateridae, Gryllacrididae, Lepidoptera (caterpillars), Formicidae, and Tettigoniidae (Schodde 1976). Grasshoppers are also eaten. Goodfellow (1908) considered the sp. 'chiefly insectivorous', a view we assume he gained as a result of observing foraging birds and stomach contents of specimens his expedition collected ($n = 17$).

In an analysis that contrasts earlier studies, Beehler (1989a) found 79% of 319 foraging and dietary samples were for fruit and 21% for arthropods. The majority of fruits eaten are of the highly nutritious 'capsular' type (see Chapter 4). Healey (1986) observed 14 plant spp. to be utilised.

Of 65 feeding bouts by wild birds upon fruits observed by Beehler (1983a) at 1430 m on Mt. Missim, PNG during 249 days of field observation over July 1978–Nov 1980, 35% were upon those of *Chisocheton cf. weinlandii* and 26% those of *Omalanthus novoguineensis*. Of all fruit types eaten 69% were capsule, 19% drupe/berry, and 12% figs (see Chapter 4).

The observation of 106 complete visits (arrival to departure) to fruit-bearing plants enabled Pratt and Stiles (1983) to calculate an average length of visit of 4.8 min to a feeding tree and of 4.3 min, or 90% of time in it, spent feeding in the plant; 94% of visits lasted 1–10 min.

Birds on Mt. Giluwe, PNG fed upon arthropods 12 times and upon fruit 71 times involving 25 plants of which the vine *Medinilla markgrafii* (15%) and the epiphyte *Schefflera lasiophaera* (14%) were most important. Mt. Bosavi, PNG, birds ate epiphytic wild pepper *Piper* sp. fruits by sallying to snatch a pendant bunch and then taking it to a perch, and consuming it while holding it down with a foot (T. Pratt *in litt.*). See Appendix 4 for other food plant spp.

Of 38 recorded heights at which birds were sighted, 37% were at 0–15 m, 45% at 15–25 m, and 18% at 25–35 m above ground (Kwapena 1985).

Vocalisations and Figure 9.44
other sounds
The characteristic advertisement song is a loud, metallic, and nasal series of 4–7 or more

(sometimes many) *shre* or *scheee* notes (Beehler 1978; Beehler *et al.* 1986; Coates 1990), the initial notes being softer and more slowly delivered, each of < 1 sec, and becoming slightly louder and faster, and then more slowly spaced towards the end, lasting *c.* 7–10 sec (Fig. 9.44). This is heard throughout the day, but particularly during the morning. A bird repeats the series from a song perch at *c.* 10-min intervals (Diamond 1972). The notes are much like those of the two small Australian riflebirds. A call of a captive ad ♂ was *ka-a, ka-a, ka-a* (equivalent to the call cited above) and an alarm of both sexes was a series of rather sharp thin notes, rapidly repeated. The latter call was also given in the evening by the birds going to roost when both sexes joined in a chorus. The notes of the ♀ were then higher and thinner than that of the ♂ (Crandall 1932). Local informants state that wild birds can be sexed by calls (Healey 1986), but the sexing of ♀-plumaged individuals is questionable. Song described on the Kubor Ra, PNG as a harsh, grating screetch resembling *au-aa-aa-aah* starting softly and building in volume, and each note, while fading at the end, is longer than the preceding one. The last note is approximately twice as long as the first (Cooper in Cooper and Forshaw 1977). Continuously-repeated call notes of Vogelkop ♂♂ were reported, however, to 'resemble the syllables *mjat-mjat*' (Bergman in Gyldenstolpe 1955*b*). Display is occasionally preceded by a clockwork-like hiss (R. Whiteside *in litt.*). During courtship display the ad ♂ produces a loud, sharp, single, or continuously-repeated double click *tick-tick* or *tick* by a sudden movement of a wing(s) and/or primaries. Typically 1–3 double clicks are followed by a sequence of single ticks in display. In describing this sound by wild displaying ad ♂♂ one observer refers to both bill and wing snapping (T. Pratt *in litt.*). A sharp ticking of this kind may also be produced in flight (as by the Magnificent Bird). A captive ad ♂ gave a soft mew-like sound *pe-er, pe-er, pe-er* on three occasions as he approached a ♀ and the ♀ gave this call once. The ♂ also gave a quiet nasal snort now and then (D. Frith and C. Frith 1988). A

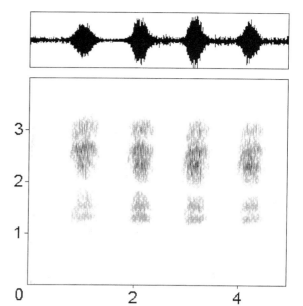

9.44 Sonogram from a recording of a typical ad ♂ Superb Bird advertisement song of repeated, sharp, screech notes at Mt. Hagen, PNG by K. D. Bishop. This song was started by a single softer note not shown in the figure.

rasping nasal note virtually indistinguishable from that of a ♀ Blue Bird near her nest was given as an alarm or scold note by a subad ♂ and a ♀-plumaged bird (R. Whiteside *in litt.*). A fully-developed juv gave an audible peeping when begging from its parent (T. Pratt *in litt.*). Ad ♂♂ in flight produce a rustling sound like that of ad ♂♂ riflebirds but not as loud.

Mating system
Polygynous. Promiscuous ♂♂ solitary, territorial (to other ♂♂—see Fig. 9.45) and sedentary with each ♂ using a number of high forest tree perches to call loudly and frequently within auditory contact of other ♂♂. A 14-month study of 6 ha of forest on Mt. Missim, PNG resulted in the location of nine resident ad ♂♂ (Fig. 9.45). Territories of three ad ♂♂ were non-overlapping and abutting and each on average 1.5 ha, at an average distance of 140 m between birds that were at all times in vocal or visual contact (Beehler and Pruett-Jones 1983). At Tsuwenkai, Bismarck Ra,

PNG density of ♂♂ was estimated to be 1 per 2.7 ha of primary forest and 1 per 1.8 ha at the forest-edge and 1 per 2.3 ha overall (Healey 1986). A general impression of abundance at Mt. Missim, PNG, at 1430 m is indicated by 2–4 birds being observed at a fruiting tree per day (Beehler 1983a). Observation of a single, vocal ad ♂ on Mt. Mengam, Adelbert Ra, PNG for 11 hr over 26–27 Oct indicated a small area over which he actively ranged (D. Frith and C. Frith 1988). On Mt. Bosavi, PNG, Thane Pratt (*in litt.*) estimated an individually-marked (banded) ad ♂ territory to be 150 m in dia, with other ♂♂ audible, while near Kompiam R. Whiteside (*in litt.*) thought territory to be significantly smaller. Pratt found territories centred on ridge crests (unlike sicklebills on ridge flanks), with birds preferring steeply-sloping spur ridges, on the flanks of large ridges, for singing perches (like sicklebills). Uniparental nest attendance by ♀. It has been suggested ♀♀ nest within the territory of the ♂ they mate with (Beehler and Pruett-Jones 1983). Some display postures may be performed on high perches but, contrary to the assumption that ♂♂ typically display upon high perches, it now appears Crandall (1931) was correct in stating display takes place close to the forest floor. Stein (1936) reported that a hunter claimed he shot an ad ♂ as it danced in display in a low tree. Crandall (1931) stated that ♂♂ dance near the ground where they are easily trapped by local people. That ad ♂♂ do display close to and on the forest floor in the wild was confirmed when an ad ♂ was seen to court a ♀ several times on a fallen log only 1 m above ground in the Adelbert Mts, PNG before mating (D. Frith and C. Frith 1988). Moreover R. Whiteside (Whiteside and Feignan 1998; and *in litt.*) repeatedly watched several solitary individual ad ♂♂ direct a high intensity display to a ♀-plumaged bird on a low perch or flat piece of ground in the forest and another on a not-quite-horizontal fallen tree trunk *c.* 1 m high between 0600 and 1130 hr while making loud clicking noises. On Mt. Missim T. Pratt (*in litt.*) twice recorded an ad ♂ displaying on logs at ground level.

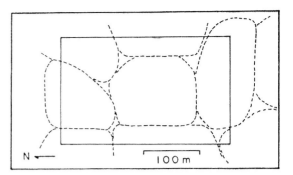

9.45 Display dispersion of ♂ Superb Birds on Mt. Missim, PNG: dashed lines within the rectangle indicate territory boundaries for eight adjacent ♂♂. From Beehler and Pruett-Jones (1983), © Springer-Verlag with kind permission.

Courtship behaviour

Displays by both captive and wild ad ♂ birds have been of two basic kinds: *Initial display activity* (IDA) and *High intensity display* (HID). In the former the ad ♂ slightly crouches with breast-shield sleeked tightly against himself, cape held back and down against the back, wings and tail held normally, head and bill pointed upward with eyes fixed on the ♀, and the naral tufts projecting conspicuously forward and bifurcated (Fig. 9.46 (a)). This pose is followed by a repeated, sudden, upward and outward extension of the breast-shield, with head and bill still pointing at the ♀ and the naral tufts unchanged. Interspersed with this 'breast-shield flashing' is a 'cape-flicking' action in which the nape cape is flicked well forwards above and over the head (but not fanned, or spread) by an exaggerated downward movement of the bill and head that presents the iridescent crown to the ♀ (Fig. 9.46(b)). Breast-shield flashing and cape-flicking increase in tempo as IDA progresses. Bouts of breast-shield flashing and cape-flicking alternate during this rapid performance, while wings and tail are held in their normal positions. Forecrown feathering is raised to the maximum and reflects a pure white light. If the ♀ moves about during IDA the ♂ takes several short steps to continue facing her but after each such movement he briefly reverts to the slightly crouched, sleeked, posture before recommencing IDA. The ♂ may occasionally open his mouth widely to expose its pale interior (Fig. 9.46(c)), sometimes call and/or give a clockwork-like hiss, elevate and depress the tail, and produce a loud *click* sound by a sudden wing movement (Morrison Scott 1936; D. Frith and C. Frith 1988).

In the HID, the breast-shield is thrust forward and expanded fully, the naral tufts erected and bifurcated, and the nape cape flicked forwards and spread laterally to form a complete concave semi-circle over the head and down either side to below the breast-shield. Below the lower edge of the breast-shield the black nape cape feathers extend round and inwards to meet the body feathers, thus forming a complete if irregular ellipse of black relieved only by the iridescent breast-shield (Fig. 9.46(d-e)). Below the central cape, above eyes and bill, a conspicuous, iridescent, white 'light' appears, which is striking and may appear pale bluish green when suitably lit. The closed bill is not discernible unless raised to bisect the white patch into two spots. The bird's eyes can only just be seen below the lower edge of each of the white spots. The white patch is formed by light refracted by the much-elevated iridescent forecrown feathers, the elevated matte black naral tufts and bill bisecting this to form the two spots. The tail is cocked up at *c.* 35–45° to the horizontal and the mouth kept closed. In this posture the ♂ proceeds in short, sharp, but deeply bouncing, hops to dance round the ♀ (D. Frith and C. Frith 1988; Attenborough 1996), each step or hop accompanied by a simultaneous sharp double flick or flutter of the briefly partly-opened wings as the tail is sharply raised. These wing movements cause the sharp *tick* or *click* sounds. Near Kompiam, PNG, in Jan, an ad ♂ was observed to direct HID to a ♀ on a flat piece of ground in forest, possibly after having performed some pre-display on a fallen log 10 m away. Another ad ♂ which came repeatedly to display on a not-quite-horizontal fallen tree trunk within 20 m of this perch, including a rotting log half underground, a horizontal tree trunk lying on the ground, and an arched sapling *c.* 3 m above ground. A moulting imm ♂ puffed out his chest feathers and made the clicking sound associated with HID while fluttering its wings and hopping on a small muddy mound (R. Whiteside *in litt.*).

On 26 Oct in primary rainforest at *c.* 1500 m on Mt. Mengam, PNG an ad ♂ was singing about his territory as he approached a fallen tree trunk 0.5 m in dia sloping from 0.5 to 1 m above ground, known to be an oft-visited display perch. At 1135 hr the ♂ flew onto the log and immediately performed the IDA for 15–20 sec, directing it upward at a ♀ perched near and above him. She then flew to 30–50 cm below the ♂ on the log and he performed HID at her for 35 sec. She approached him in a submissive crouched posture with slightly-drooped wings. The ♂ remained stationary in the HID posture until the approaching ♀ reached him and then proceeded to dance around her moving in quick, short, bouncing side steps and hops. She exaggerated her submissive pose. He danced past her in semi-circles, twice down the log and back up. During the dance their bills almost touched and she side-stepped to rotate her body so as always to face him. On his final return the ♂ lowered his plumage and mounted for < 5 sec to mate, the ♀ then flying up into saplings. Copulation was also seen in the wild by R. Whiteside (*in litt.*) after a much shorter initial display and HID on a low, directly sunlit, perch.

In a fallow garden with sparse tree cover, at Kikita village near Tari, PNG, an ad ♂ perched in the dead branches of an emergent tree and sung and preened frequently. As a Yellow-browed Melidectes *Melidectes rufocrissalis* hopped up to within 1 m of him he directed a brief HID at the honeyeater (CF).

Breeding Table 6.5

Apparently possible at any month over the sp. range as a whole. Goodfellow (1908) claimed

354 Superb Bird of Paradise *Lophorina superba* (Pennant, 1781)

9.46 Some courtship display postures of ad ♂ Superb Bird: (a–c) *Initial display activity*, and the *High intensity display* dancing (d) and static (e) postures. After Frith and Frith (1988) and Attenborough (1996). See Text.

to have seen 'a pair' of birds 'carrying nesting material' in Apr but presumably only the ♀ was in fact involved. A ♀-plumaged bird was seen carrying nest material to and 'testing' a nest site by crouching down onto it, with another ♀-plumaged bird watching from < 1 m away on 17 Nov at Mt. Missim, PNG (BB). The extra bird may have been a young of the preceding year.

NEST SITE: 'high up' and 'thirty feet' in a forest tree (Harrison and Frith 1970) are contrary to the average of only 2.1 (range 1.5–3.5; $n = 8$) m above ground on Mt. Missim (T. Pratt *in litt.*). One nest in the centre of the crown fronds of a tree fern (Pruett-Jones and Pruett-Jones 1986) and presumably therefore < 4–5 m high. Of eight active Mt. Missim nests, five were in a *Pandanus* crown, three in a *Calamus*, and one in a *Calyptrocalyx* palm (T. Pratt *in litt.*). This nesting in the crown of palm or structurally palm-like plants is also found in the riflebirds and, to a lesser extent, the sicklebills.

NEST: A. S. Anthony first collected a nest at *c*. 1525 m in the Owen Stanley Mts during 1895–6. It consisted of 'dry and partially decayed leaves, dry twigs and rootlets' (Hartert 1910). A nest (BMNH 1907.7.9.56) collected with an egg at Kagi, Owen Stanley Ra, PNG measured 124 mm in dia and 85 mm deep. The nest was a very loosely-made rough cup, mostly a mesh of dark springy rootlets and fibres with a few large dead leaves and small *polypodium*-type fern fronds incorporated, together with some strands of a *selaginella*-like creeping plant, *c*. 70 mm deep and 110 mm external dia, and 50 mm deep and *c*. 75 mm dia internally (Harrison and Frith 1970). A colour photograph of a ♀ at her nest shows the structure to be a deep substantial cup of leaves with vine-like tendrils, quite possibly epiphytic orchid stems, and some living *Pyrrosia* fern stems with fronds on the outside and the rim (Pruett-Jones and Pruett-Jones 1986). Only the ♀ of a captive breeding pair constructed the nest (Timmis 1968, 1970).

EGGS: the two eggs of the clutch collected by A. S. Anthony measure 32 × 22.4 and 31.8 × 20.6 mm. One is brownish buff, longitudinally-splashed and marked with pale brownish grey and some apparently deeper-lying pale brown markings. The other is more cream in colour with rufous spots and longitudinal markings and underlying pale bluish grey longitudinal splashes and spots, with some brown points and dots. A partially-incubated single-egg clutch on 5 Jan measured 31.8 × 22.3 mm. Its ground colour was a pale creamy-buff with some very fine speckling all over and with heavy longitudinal streaking of dull brown and lavender (Harrison 1971). An undated single egg clutch collected by Captain J. R. Barton at Kagi, at *c*. 1830 m, Owen Stanley Ra, PNG measured 31.3 × 21.3 mm. All eight nests found on Mt. Missim by T. Pratt (*in litt.*) contained a single egg or nestling.

INCUBATION: only the ♀ of a captive breeding pair incubated, incubation lasting 18 and 19 days in two consecutive nesting attempts (Timmis 1970).

NESTLING CARE: exclusively by the ♀ of a captive breeding pair (Timmis 1970) and confirmed true of wild birds in PNG by T. Pratt (*in litt.*), who saw the female feed her nestling a long-antennaed orthopteran. A ♀ with a fully-developed but begging juv was seen on Mt. Bosavi, PNG on 28 Feb–16 Mar, the ♀ feeding it by regurgitation. A ♀ with a chick was once observed scolding a parotia and another ♀ attacked a parotia (T. Pratt *in litt.*).

NESTLING DEVELOPMENT: a nestling opened its eyes at 6 days old and at 8 days had feather quills down the centre of its back and those on the wings had begun to burst. At 10 days old the head, back, and wings were feathered. This nestling first left the nest at 18 days old (Timmis 1970). A Mt. Missim chick examined on 21 Sept weighed 19 g and had just opened its eyes. It was infested by bot fly larvae and died 2 days later (T. Pratt *in litt.*).

Annual cycle   Table 6.5

DISPLAY: late Sept–Dec in the wild (D. Frith and C. Frith 1988; R. Whiteside *in litt.*), main season Aug–Jan (mid dry to early wet season) according to Coates (1990). BREEDING: recorded throughout yr over sp. range. EGG-LAYING: eggs found in the wild during Jan, Mar, June, and Nov. MOULT: 58 of 304 specimens examined show some sign of moult, involving all months of the year but very predominantly Jan–Mar with some during Apr–May but very little during the other months (Appendix 3). Crandall (1932) found an ad ♂ and a ♂ just acquiring ad plumage moulting heavily in Oct at Deva Deva, PNG, when all other individuals of this sp. examined showed no moult. In captivity, in the USA, Crandall noted that ♂♂ took *c*. 5 months to complete a moult.

Status and conservation　　Table 8.1
Noted as common in riverine forests bordering the Wahgi (1524 m) and Kubor foothills up to 2226 m 'which is apparently the ceiling for the sp.' (Mayr and Gilliard 1954). This remained the case in 1989 (D. Bishop *in litt.*). Not uncommon in the edges of tall midmontane forest but apparently not found above *c*. 1800 m in the Victor Emanuel and Hindenburg Mts, PNG (Gilliard and LeCroy 1961). More abundant at the lower elevations of its range, where individuals present were predominantly in ♀ plumage, during the non-breeding season (Pruett-Jones and Pruett-Jones 1986). The commonest bird of paradise within its altitudinal range of *c*. 1067–2134 m in the Eastern Highlands, PNG, representing 3% of the avifauna of Mt. Karimui at 1646–1829 m (Diamond 1972). A marked change in the composition of the population was noted with altitude: at 1220–1280 m only a single imm ♂; at 1341–1448 m two ♀ plumaged birds; at 1448–1643 m ♀-plumaged birds and an ad ♂ calling at 1560 m; at 1643–1817 m ad ♂♂ were common and ♀♀ were noted only up to 1738 m. Above this altitude the few birds seen were all ad ♂♂. At lower forests on Mt. Michael, few birds were found (Diamond 1972). Very common in forests of the Adelbert Ra, PNG during 29 Feb–22 Mar in the middle and upper canopy. At 1500–1600 m ad ♂♂ predominated in the population while at 1200 m ♀♀ were much more the commoner sex (Pratt 1982*b*).

Knowledge lacking and research priorities
The monotypic genus represented by this bird appears to be closely related to the riflebirds but its other affiliations remain puzzling to some extent. The results of molecular phylogenetic studies are eagerly awaited. The territorial nature of the ♂♂ merits additional study—how are these all-purpose territories established and maintained? Interactions between the sexes in the wild require clarification.

Aviculture
A single ♀ kept in Sydney and 'mated [presumably socially bonded only] to a ♂ Eastern Whipbird *Psophodes olivaceus*' built a nest and laid an egg on 6 Dec 1919 only to destroy it next day (Coles 1920). Frequently kept in zoos about the world but only rarely bred until recent and repeated success at the Honolulu Zoo, Hawaii. A restless non-aggressive sp. that does well in a large flight aviary (which need not be particularly tall) but must have adequate bathing facilities and access to rain/sprinklers as birds otherwise decline. Can be kept with other smaller spp. (P. Shanahan *in litt.*).

Other
The breast-shield of the ad ♂ plumage is the most typical, almost invariable, centre piece of the ceremonial wig of the Huli Wigmen of the Tari Valley and adjacent areas (Fig. 2.1). The Kalam people refer to the cape of ad ♂♂ as their *kd-wad* or 'net-bag at the back' and their breast-shield as *bet-wad* or 'the net-bag at the front' because Kalam women 'with many things to carry, carry their net-bags in these ways' (Majnep and Bulmer 1977).

Genus *Epimachus* Cuvier, 1817

Epimachus Cuvier, 1817. *Le Règne Animal* 1, 407 (Dec 1816). Type, by monotypy, *Upupa magna* Gmelin = *Promerops fastuosus* Hermann (1783).

The *Epimachus* 'sabretailed' sicklebills are **two species** of large, extremely long-tailed birds with narrow sickle-shaped bills and strong sexual dimorphism in size and plumage (Plate 9). They are confined to the mountains of NG, including the central cordillera, the N Coastal ranges of PNG, and the mountains of the Vogelkop. According to Bock (1963) the **skull** of both the *Epimachus* and *Drepanornis* sicklebills share common characters, supporting the hypothesis that the four spp. in the two genera form a natural clade. It is for this reason and others that the two genera are, in some treatments, combined into *Epimachus*. The lacrymal in all is large and fills most of the space between the ectethmoid and nasal bones, an adaptation typical of the longer, slender-billed birds of paradise (*Ptiloris, Semioptera, Seleucidis*). In at least *E. fastuosus*, the combined ectethmoid–lacrymal complex thus forms a bulbous mass very similar to that found in *Ptiloris magnificus*. In most other respects the skull is *Ptiloris*- and *Paradisaea*-like. This said, the combining of *Drepanornis* with *Epimachus* is probably the least convincing lumping action proposed by Diamond (1972). The two *Drepanornis* sicklebills share a number of significant characters lacking in the two *Epimachus* spp. (see *Drepanornis* account).

E. fastuosus has been recorded to **hybridise** with *Paradigalla, Astrapia,* and *Lophorina*. ♂♂ are c. 15% larger than ♀♀. **Bill** narrow, decurved, and sickle-shaped, but powerful and more than two times longer than the head. Nostril only proximally verged by feathering. Bill meets the skull in a conspicuous 'hinge', presumably as an adaptation to vigorous arthropod foraging that permits some 'give' at the skull/bill junction. **Tail** graduated, c. 133–166% of wing length. The tail feathers other than the central pair are shallowly V-shaped (as in *Astrapia nigra*) and curved to cross those on the opposite side of the tail—probably to support the large central pair. Subad ♂♂ with black in wing and tail feathers lack the specialised morphology of the ad ♂ tail feathers. **Wings** long and somewhat rounded, outermost two primaries mildly emarginated on inner vanes. **Tarsus** length c. 27% of wing length. Distal section of tarsus scuted, becoming fainter to smooth proximally.

Huge 'axe-head' shaped pectoral display plumes in ad ♂ plumage. Filamental flank plumes in ad ♂♂ are of the *Ptiloris–Paradisaea* type in structure and texture.

Polygynous, with ad ♂♂ territorial and solitary, displaying at traditional arboreal perches with auditory contact, and ♀♀ alone carrying out nesting duties. Known ♂ courtship displays are solitary and relatively static. **Nest** undescribed but known to be placed at the base of pandanus tree crown fronds. **Eggs** are typical of Paradisaeinae but ground colour browner rather than pinkish (Plate 13).

Black Sicklebill *Epimachus fastuosus* (Hermann, 1783)

PLATE 9; FIGURES 9.47–8

Other names: Black Sickle-billed Bird of Paradise, *Epimachus fastosus*, Greater Sicklebill, Black Saber-tailed Bird of Paradise. *Amkurmat* of the Karoon people, Tamrau Mts, Vogelkop, IJ (Mayr and Meyer de Schauensee 1939a), *Deawa* of the Yabi people, Weyland Mts, IJ (Rothschild 1931) and *Blak Blak* of people of the Telefomin region (Gilliard and LeCroy 1961). Local Wissel Lakes, IJ, name is *Oewawa* (Junge 1953). Mt. Hagen, PNG, name is *Karae-tumbo* (Mayr and Gilliard 1954). Ad ♂ known as *gwlgw, kalaj* or *yng-mlek* and ♀ plumaged birds as *galknen* by the Kalam people (Majnep and Bulmer 1977). *Gula, Gula Gula* or *Kula kula* of the Huli people, Tari Valley, PNG (C. Ballard and D. Bishop *in litt.*). *Kumpi* of the Fore, *Kurikuri* of the Gimi and *padukwa* of the Daribi people, E Highlands, PNG (Diamond 1972). Also interchangeably *Fureda, Buk Bk,* and *Sino* of the Gimi people, Crater Mt., PNG (D. Gillison *in litt.*). *Boo-ALI* of the Yali people of the Balim Gorge, IJ (BB), and *Bok-bok* of the Ketengban people, Star Mts, IJ (J. Diamond and D. Bishop *in litt.*).

A striking, long-tailed, and sickle-billed inhabitant of the canopy of Papuan montane forest. Ad ♂ is a spectacular black bird with a huge sabre-shaped tail; ♀ is smaller, plumed

Black Sicklebill *Epimachus fastuosus* (Hermann, 1783)

in browns, with a dark face and profuse, dark, ventral barring. Occurs mainly at an elevation lower than that of the pale-eyed Brown Sicklebill; can best be distinguished by its remarkable voice and red iris colour. Polytypic; 3 subspp.

Description

ADULT ♂: 63 (110) cm, entire head black but in suitable light scale-like feathers show intense iridescence of metallic green-blues with purple (172) and/or Magenta (2) washes, the crown, malar area, and ear coverts particularly so, but chin 'beard' tuft and throat blackish with only slightest Magenta sheen in rare light. Mantle to uppertail coverts velvety black with intense deep violet-purple and/or Magenta iridescence but highly-modified, large, scale-like, central back feathers with highly-metallic blue-green iridescence. Upperwing velvet-black with variable rich deep blue to violet-purple and/or Magenta iridescent sheens, primaries blackish-brown (19). Uppertail blackish brown (19) with rich, deep, iridescent purple except central pair which is intensely-iridescent, metallic, deep blue-purple (72) to violet-purple and/or rich Magenta in suitable light but jet black under other conditions. Upper breast brownish black (119) grading to the slightly paler and browner (19) lower breast and even more so, to Sepia (219), on elongate filamental flank plumes, vent, and undertail coverts. All underparts may show slightest deep plumpurple (1) sheen in some lights, but appear jet black in others. Greatly-enlarged, axe-headshaped pectoral plumes jet black with metallic, dark Magenta, iridescent gloss and shorter ones overlaying these jet black and broadlytipped, intensely-iridescent, metallic blue (270), Purple (1), and/or violet in suitable light. Elongate modified feathers to each side of the belly and vent with highly-iridescent, metallic, deep purple-blue to violet-purple broad tips, tapering distally, which may appear more blue-green in some lights. Underwing blackish brown (19) and ventral surface of tail this colour to black and slightly glossy. Bill shiny black, iris bright blood red, legs and feet blackish-grey, mouth bright yellow.

ADULT ♀: 55 cm, forehead, crown, and nape Chestnut (32). Lores, sides of face, chin, and throat Sepia. Entire upperparts Olive-Brown (28), browner on upper mantle and upper tail coverts. Dorsal surface of wing: lesser coverts Olive-Brown, greater coverts and flight feathers the same but with broad Raw Sienna (136) outer edges. Dorsal surface of tail Raw Umber (23), slightly washed olive on central pair. Upper breast Sepia with finest of pale Buff (124) barring becoming increasingly broad down onto brownish black lower breast and belly, grading to a slightly paler colour with even broader buff barring on flanks, vent, and undertail coverts. Underwing: coverts Pale Pinkish Buff (121D) barred blackish, flight feathers Olive-Brown with broad Ground Cinnamon (239) trailing edges except on outermost two primaries. Undertail Raw Umber (123) washed pale olive, central feather shafts paler. Mean wing length 15% and weight 30% less than those for ad ♂.

SUBADULT ♂: ranges from identical to ad ♀ with a few ad ♂ feathers intruding, to much like the ad ♂ with a few feathers of ♀ plumage remaining. Individuals may appear in striking patchy plumage (see Plate 9). It is notable that museum specimens suggest the iridescent central tail feathers do not replace ♀ plumage ones but intermediate feathers are worn, for at least one season, these being ♀-type feathers with an elongated 'smudge' of the rich iridescence down the centre of each vane.

IMMATURE ♂: as ad ♀ but tail averages longer and primaries are said to be more tapering and pointed (Mayr and Meyer de Shauensee 1939*a*).

NESTLING: unknown.

Black Sicklebill *Epimachus fastuosus* (Hermann, 1783)

JUVENILE: like ad ♀ but plumage is soft and downy below, more brownish than blackish on chin and throat. Crown and upperparts are more rust-red than the brown of ad. Underparts are more buff than the off-white of ad.

Distribution

Mountains of western and central NG, including those of the Vogelkop and Wandammen Penins, the central cordillera from the Weyland Mts eastward to the Kratke Ra, Mt. Bosavi, and the Bewani and Torricelli Ra of the N coast of PNG (Mts Menawa and Somoro), from 1280 to 2550 m (mainly 1800–2150 m).

Systematics, nomenclature, subspecies, weights, and measurements

SYSTEMATICS: a sister form to *Epimachus meyeri*, but this species-pair has evolved to the extent that the two spp. have established extensive geographical sympatry along the central cordillera. Microgeographically, the two spp. sort out elevationally, with *fastuosus* inhabiting a narrow elevational zone immediately below that of *meyeri* (Diamond 1972). The question remains as to how the two spp., in areas of sympatry, maintain this elevational displacement. First discovered and described from plume trade skins of unknown origin. The early synonymy of this, first known, sicklebill is complex (see Salvadori 1881: 541–2) with both sexes being described as a distinct sp. and the remarkable bill causing ornithologists to consider various groups, including hoopoes (Upupidae) and wood-hoopoes (Phoeniculidae), as the closest relatives of this form. Hybridisation: known to cross with Arfak Astrapia, Long-tailed Paradigalla, and Superb Bird. See Plate 14 and Appendix 1.

SPECIES NOMENCLATURE AND ETYMOLOGY

Epimachus fastuosus (Hermann)
Promerops fastuosus Hermann, 1783. *Tabula Affinitatum Animalium* (Argentorati), p. 194 (based on *Planches Enluminées*, pls 638–639). New Guinea, restricted to the Arfak Mts, Vogelkop by Hartert (1930) *Novitates Zoologicae* **36**, 33. Type specimen none designated.

Synonyms: *Falcinellus striatus* or *speciosus* auctorum. *Epimachus magnus* Cuvier, 1817. *Règne Animal* 1, p. 400.

Etymology: *Epimachus* = Gr. *epi*, over; and Gr. *machaira*, bent sword, referring to the

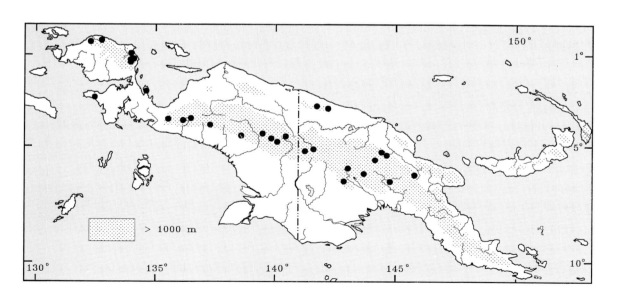

decurved bill; or = Gr. *epimakhos*, a fighter (alludes to scimitar, sword-like bill); *fastuosus* = L. *fastuosus*, proud, haughty (*fastus* = pride).

SUBSPECIES, WEIGHTS, AND MEASUREMENTS

1. *E. f. fastuosus* (Hermann, 1783).

Range: Tamrau and Arfak Mts of the Vogelkop. By geography, the population from the Wandammen Penin (Mt. Wondiwoi) might be expected to be intermediate between this and *atratus*; Mayr (1941a) assigned it to the latter subsp.

Measurements: wing, ad ♂ ($n = 24$) 188–205 (197), imm ♂ ($n = 8$) 167–182 (177), ad ♀ ($n = 12$) 152–187 (166); tail, ad ♂ ($n = 23$) 340–539 (392), imm ♂ ($n = 8$) 226–257 (242), ad ♀ ($n = 11$) 204–249 (229); tail centrals, ad ♂ ($n = 23$) 593–791 (710), imm ♂ ($n = 7$) 283–321 (304), ad ♀ ($n = 12$) 270–321 (291); bill, ad ♂ ($n = 22$) 73–80 (77), imm ♂ ($n = 6$) 68–81 (75), ad ♀ ($n = 10$) 66–77 (72); tarsus, ad ♂ ($n = 23$) 47–54 (50), imm ♂ ($n = 8$) 49–53 (51), ad ♀ ($n = 12$) 43–52 (47). Weight: ad ♂ ($n = 7$) 250–280 (266), imm ♂ ($n = 7$) 205–25 (222), ad ♀ ($n = 7$) 160–235 (191).

2. *E. f. atratus* (Rothschild and Hartert, 1911). *Novitates Zoologicae* **18**, 160. Mt. Goliath, Oranje Mts, Dutch New Guinea. Type specimen AMNH 677957.

Synonym: *E. f. stresemanni* Hartert, 1930. *Novitates Zoologicae* **36**, 34. Schraderberg, Sepik Mts, type specimen AMNH 6777964, is synonymised with *atratus* based on the comments of Gilliard and LeCroy (1961), Cracraft (1992), and Diamond (1969) concerning clinal nature of the supposedly distinguishing characters.

Range: mountains of Wandammen Penin and the central cordillera eastward to the Kratke Ra.

Diagnosis: *E. f. atratus* is thinly differentiated from the nominate subsp. by the darker ventral plumage of the ad ♂. It differs from the N coastal population (*ultimus*) by the longer bill. Quite variable in size.

Measurements: wing, ad ♂ ($n = 35$) 188–232 (205), imm ♂ ($n = 14$) 164–205 (191), ad ♀ ($n = 36$) 154–193 (174); tail, ad ♂ ($n = 31$) 342–537 (413), imm ♂ ($n = 14$) 207–306 (261), ad ♀ ($n = 34$) 171–280 (222); tail centrals, ad ♂ ($n = 27$) 640–946 (757), imm ♂ ($n = 13$) 273–424 (338), ad ♀ ($n = 27$) 214–387 (294); bill, ad ♂ ($n = 31$) 71–86 (80), imm ♂ ($n = 14$) 65–83 (78), ad ♀ ($n = 35$) 72–89 (77); tarsus, ad ♂ ($n = 35$) 48–55 (51), imm ♂ ($n = 14$) 46–57 (52), ad ♀ ($n = 36$) 43–55 (48). Weight: ad ♂ ($n = 8$) 255–315 (281), imm ♂ ($n = 2$) 275–297 (286), ad ♀ ($n = 5$) 160–255 (206).

3. *E. f. ultimus* Diamond, 1969. *American Museum Novitates* **2362**, 31. Summit of Mt. Menawa, Bewani Mts, Sepik, District, Mandated Territory of New Guinea. Type specimen AMNH 789765.

Range: known from Mt. Menawa of the Bewani Ra and Mt. Somoro of the Torricelli Ra, both in PNG's N Coastal ranges. Perhaps to be expected from a few other of the highest peaks of the Bewani Ra, but only above 1350 m.

Diagnosis: bill is shorter than that for all other populations, and the tail of ♀ is longer than in *fastuosus* and most *atratus*. Differs from the nominate form, but like *atratus*, in being more black, less brown, on underparts of ad ♂ and more olive, less rufous, on dorsal surface of tail of ad ♀.

Measurements: wing, ad ♂ ($n = 4$) 193–198 (196), imm ♂ ($n = 6$) 172–186 (183), ad ♀ ($n = 10$) 161–168 (165); tail, ad ♂ ($n = 4$) 374–391 (380), imm ♂ ($n = 6$) 254–282 (264), ad ♀ ($n = 9$) 225–248 (235); tail centrals, ad ♂ ($n = 3$) 665–761 (725), imm ♂ ($n = 6$) 320–382 (347), ad ♀ ($n = 8$) 307–341 (320); bill, ad ♂ ($n = 4$) 70–73 (72), imm ♂ ($n = 6$) 66–74 (70), ad ♀ ($n = 10$) 70–73 (71); tarsus; ad ♂ ($n = 4$) 49–52 (50), imm ♂ ($n = 6$) 49–52 (51), ad ($n = 10$) 44–47 (46). Weight: ad ♂ ($n = 2$) 275–280 (278), imm ♂ ($n = 6$) 223–252 (239), ad ♀ ($n = 6$) 184–207 (190).

Habitat and habits   Table 4.1

Inhabits a narrow elevational zone in midmontane forest, occasionally at the forest-edge, and more rarely in adjacent second-growth and garden-edges, where ad ♂♂ are at risk from plume hunters (Coates 1990). Given the array of birds of paradise that inhabit this zone, it is uncertain whether the sp. elevational range is a product of diffuse competition or particular preferences for certain forest types. The latter explanation is unlikely, given that this zone is particularly variable in terms of floristics, structure, degree of mossing, and epiphyte luxuriance. Ad ♂♂ behave cryptically and are difficult to follow and observe (T. Pratt *in litt.*), although ♂♂ make their presence known by regularly calling from arboreal perches on steep hillsides below ridgecrests (Diamond 1972). Flight of ad ♂♂ is like that of other sicklebills in being slow, undulating, and silent. Ad ♂♂ fly short distances with slow leisurely flight, a few initial wing-beats followed by a glide. Foraging birds cover much distance on foot—by bounding from perch to perch (T. Pratt *in litt.*). Occasionally forages within audible range and sight of Brown Sicklebill (D. Gillison *in litt.*). Birds occasionally respond dramatically to tape play-back of their calls by flying past or perching close to the observer (D. Bishop *in litt.*).

Diet and foraging   Table 4.2

Consumes a mix of fruit and arthropods (and probably small vertebrates as well) obtained by foraging primarily in middle and upper stages of forest interior. Said to feed on arthropods obtained by clinging to tree trunks and boughs and tearing and probing at bark and epiphytic vegetation. Beccari (in Salvadori 1881: 547) wrote that the sp. lives 'on the fruits of certain Pandanaceae, and especially on those of *Freycinetia*, which are epiphytous on the trunks of trees.' Captive birds have been observed to regurgitate fruit seeds, expelling each well clear of the long bill with a sharp head movement (CF). Birds on Mt. Bosavi were seen to eat *Freycinetia* fruits by T. Pratt (*in litt.*), who also saw an ad ♂ foraging on the ground in leaf litter. Birds were filmed on the Vogelkop, IJ, repeatedly visiting pandanus trees to feed on their ripe fruit. An ad ♂ tended to be intolerant of a conspecific in the pandanus crown but several ♀-plumaged birds were seen to congregate at a single large fruit without conflict. Of 24 foraging records made by Kwapena (1985) on Mt. Giluwe, 14 involved foraging on branches, 5 on dead vegetation, 1 on twigs or buds, 2 in foliage, 1 at flowers, 1 on fruits. His foraging records were made of individuals in the lower forest (0–15 m high) 11 times, the middle (15–25 m) three times, and the upper (25–35 m high) forest nine times. Shaw Mayer (in Rothschild 1931) found small pandanus fruits and a large grasshopper in the stomach of a juv ♂, and cockroaches and other large insects in the gut of two ♀♀. Of 48 records of feeding upon fruit made by Kwapena (1985: 113), 11 were upon *Pandanus brosimos* fruits (see Appendix 4 for a list of known food plants). He also recorded 16 instances of arthropod foraging. The proportion of fruit and animal foods in the diet has been estimated at 50 : 50 (Beehler and Pruett-Jones 1983).

Vocalisations and   Figure 9.47
other sounds

The advertisement call is regionally variable, but the best known rendering is a pair of musical, ringing, upslurred whip-like notes *QWINK QWINK* which may be repeated (Fig. 9.47(a)) or broken into a series of single notes (Fig. 9.47(b)). They have also been annotated as: *whik whik*, or *kwik kwik, bwink bwink, blick blick, blit blit, buk buk*, or *bek kek* (Mayr and Meyer de Schauensee 1939a; Ripley 1957; Gilliard and LeCroy 1961; Diamond 1972; Beehler *et al.* 1986; T. Pratt *in litt.*; CF transcription of BBC tapes). The above calls are repeated at unpredictable intervals of c. 1 min to > 40 min (D. Bishop *in litt.*). The sp. also uncommonly gives a range of other notes and series, including guttural notes and honking. Figure 9.47c shows one such, being a growled deep *grr-grrk grr K-WICK!* or *guck-er-ruk bl-whit!* the latter two-syllable note sometimes repeated. The degree to which there is significant geographical variation in advertisement

Black Sicklebill *Epimachus fastuosus* (Hermann, 1783)

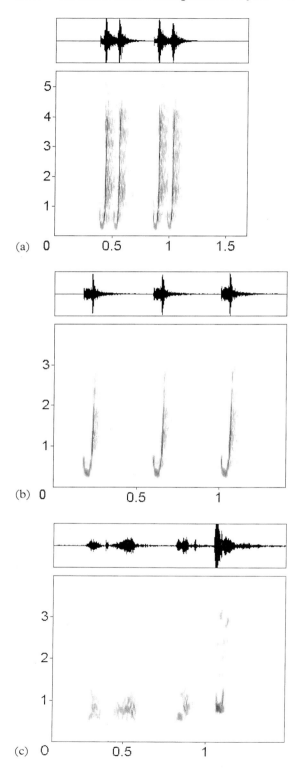

vocalisation is unclear, as the various renderings may simply be a product of the notational style of the fieldworkers. Gilliard (1969) recorded it as 'powerful, liquid, whip-like notes which carried about half a mile and were by all odds the most penetrating sounds to be heard in the forests at that altitude'. When singing, the ♂'s head is bent down (as if to regurgitate) then, as the first note is produced, it is jerked up as the bill is opened and the note thrown out as the whole body is shaken (T. Pratt *in litt.*). Pratt also heard an ad ♂ give a 'peculiar set of frog or animal-like guttural noises, interspersed with a soft honking note and, more rarely, a song'. In giving a knocking call the bird assumes the same posture as for the song but the neck appears swollen, perhaps inflated. The latter may be true of song but the movements are too quick to see if the neck is inflated. Ad ♂♂ on Mt. Bosavi started daily song at *c*. 0630–0700 hr and sang *c*. three times a min. Songs sounding like those of the Brown Sicklebill were possibly produced by imm ♂♂ (T. Pratt *in litt.*). Vocalisation of the isolated population from Mt. Menawa (*E. f. ultimus*) was variable; some gave the typical two-note call, and others gave a very rapid staccato, machine-gun-like burst of notes ending with a single liquid note apparently intermediate between the above calls of this sp. and the typical 'machine-gun' call of the Brown Sicklebill (Diamond 1969). These latter were perhaps produced by an imm ♂, as such are known to give garbled or unusual renditions of advertisement calls. A very loud *du dug … du dug* advertisement song was given from the forest canopy and display perch and a

9.47 Sonograms from recordings of ad ♂ Black Sicklebills giving: (a) typical clear, explosive advertisement song notes of the nominate form above Manokwari, Arfak Mts, IJ in Oct 1994 by R. Kirby for the BBC; (b) very similar notes of what is probably *E. f. atratus* at Bonaria Road, Tari Valley on 23 Aug 1990 and (c) a more gutteral song (see text) of same by K. D. Bishop.

'quiet rattle, sounding exactly like a distant machine gun' in horizontal display posture (R. Kirby *in litt.*) was heard on video tape to be a series of very low-pitched, hollow, dry, knocking notes accompanied by peacock display-like plumage rustling (CF). Ripley (1957) noted a loud penetrating whistle, sounding like the syllable *whick*, given by an ad ♂ in full display posture prior to a display flight (see below). The wings do not appear to produce a particular sound in flight as is common in other birds of paradise.

Mating system

Polygynous, with promiscuous ♂♂, although mating has not yet been observed. Ad ♂♂ are solitary and sedentary, each frequenting a home range containing one or more traditional courtship display perches reasonably high in the forest structure (Gilliard 1969; Majnep and Bulmer 1977). One ad ♂ observed to display on a fallen log on the ground at Ubaigubi, PNG (B. Whitney *in litt.*).

Courtship behaviour

Remarkably little known. A courtship display was described by Ripley (1957) at 1585 m on Mt. Kourangen, Tamrau Ra, Vokelkop, IJ in Mar 1938. An ad ♂ was perched on a bare branch, just above a ♀-plumaged bird, high in a huge *Agathis* sp. tree. His pectoral plumes were 'spread out and upwards like two raised arms. The tail was partially spread showing the shorter, outer feathers.' The ♂ suddenly called (see above) and then, too rapidly to be seen, retracted the pectoral plumes to turn and dive downward approximately 30 m to within metres of the forest floor shrubbery before turning upward with spread wings to sail up to the original perch—all so quickly that it took Ripley an instant to realise it had taken place.

Bell (1969) briefly saw an ad ♂ in mature beech forest above Lake Sogolomik, near the headwaters of the Ok Tedi R., PNG, in Apr display by 'quivering the pectoral plumes round its head, almost forming an arch, while it stood very erect and croaked'. Film and observations of an ad ♂ at his 6 m high display perch were made at 2100 m in IJ during Sept–Oct 1994 (R. Kirby *in litt.*; Attenborough 1996). This bird was on a tall, vertical, dead tree snag in the mid-canopy surrounded by forest on three sides but open to the side where the ridge fell steeply to the valley below. This ♂ was active between 0520 and 0530 hr each morning, first moving about the forest canopy for a few min, stopping only to call. He then flew to the top of his display perch, or to a metre or so below it, then immediately hopped to its apex. He always faced to the downslope forest clearing when singing and displaying. There was always at least one ♀-plumaged bird close by when he displayed. The ♂ always left his display perch area by 0545 hr each morning and no afternoon displays were observed.

The ♂ begins display with a loud *du dug … du dug* advertisement song and then jerks his pectoral plumes backwards and forwards finally bringing them right over his head to meet. He stretches forwards and assumes an almost horizontal position on the perch and sways from side to side while giving a quiet rattle call like the sound of soft knocking on an empty box or of distant machine gun firing (Fig. 9.48). This lasts only a few sec before he assumes the more upright posture again and gives the *du dug … du dug* song. This performance may be repeated up to five times. In the fully-horizontal posture a conspicuous and striking, continuous, shining, 'electric' blue line is formed down the entire length of the elongated bird's shape by the broad iridescent tipping to the pectoral tufts, the triangular iridescent tipping to the flank plumes, and the highly-iridescent tail feathers (Fig. 9.48). This display is almost certainly followed by a precopulatory display if and when a ♀ arrives on the display perch (not seen). The longest the ♂ spent on his display perch was 10 min. On the two mornings that R. Kirby (*in litt.*) did not see the ad ♂ appear 'juv [imm] ♂♂ used the vacant perch to practice their display'.

Observations at Ubamurai, near Crater Mt., PNG, in mid July confirm that above display

Black Sicklebill *Epimachus fastuosus* (Hermann, 1783)

9.48 The horizontal courtship display posture of the ad ♂ Black Sicklebill. After Attenborough (1996). See text.

postures are performed by an ad ♂ atop short vertical tree stumps and that once two ad ♂♂ displayed facing and displacing each other (D. Gillison *in litt.*).

An ad ♂ perched atop his vertical display stump first responded to a ♀-plumaged bird flying into an adjacent tree by instantly adopting the horizontal display pose. Raising his pectoral plumes several sec later he remained static until the ♀-plumaged bird flew off. This ♂ also performed a head-down crouched pose with tail cocked at > 45° and his mouth widely gaped at a ♀-plumaged conspecific below in nearby trees. This display appears homologous to that performed by Victoria's Riflebirds in the same circumstances (Frith and Cooper 1996; see Fig. 9.42(a)).

Breeding   Table 6.5
NEST SITE: said by people of the Telefomin area, PNG, to nest in the middle storey of the forest in pandanus trees during Feb–Apr (Gilliard and LeCroy 1961). A nest found by Kwapena (see below) was placed on a forked branch of a tree 10 m above ground.

NEST: a nest found at 2030 m on Mt. Giluwe on 1 Nov 1978 was 'composed wholly of orchid stems with vines and moss on the base' (Kwapena 1985).

EGGS, INCUBATION, NESTLING CARE, AND DEVELOPMENT: unknown.

Annual cycle   Table 6.5
DISPLAY: observed in Sept, Oct, and Feb–Apr. Peak display activity at Ubamurai, near Crater Mt., PNG is late Sept (D. Gillison *in litt.*). BREEDING: at least *c.* Nov–Feb. ♂♂ have been collected with enlarged gonads during Apr–Sept. A. S. Meek collected a fledgling (AMNH 677963) on 17 Feb that was out of the nest only a few weeks, on Mt. Goliath, Oranje Mts, IJ. Kwapena (1985) found a Mt. Giluwe, PNG nest on 1 Nov and observed a juv there in late Jan; he suggested that the mean breeding season is the wet season of Nov–Apr inclusive. In the Ok Tedi area, PNG ♂♂ were heard singing in Sept–Dec, Jan, and June (Gregory 1995). EGG-LAYING: Ralph Bulmer dissected a ♀ with large eggs in the oviduct on 5 Feb in

the Baiyer Valley, PNG. MOULT: 47 of 150 specimens were moulting, involving all months except Mar and the majority during June–Feb (Appendix 3). Of two ♂♂ and a ♀ taken on the Hindenberg Mts during 10–11 Apr the ♂♂ were in fresh plumage with traces of moult in the tail only and the ♀ had traces of moult on the neck and chest only (Gilliard and LeCroy 1961).

Status and conservation Table 8.1

Sometimes locally common, such as above 1800 m on Mt. Bosavi, PNG, but for the most part uncommon or rare. Its favoured elevational zone is heavily settled by highland people, thus many potentially-suitable areas have been cleared for human habitation and agriculture. Kwapena (1985) found this sp. uncommon on Mt. Giluwe where its mid-montane habitat had been degraded by logging and agriculture. He observed that it does not successfully colonise secondary forest, a point noted by Diamond (1972) and Peckover and Filewood (1976). Apparently heavily hunted with firearms by the labour force during initial work on the Ok Tedi copper mining development, resulting in a decline in numbers so dramatic that it caused investigators to suggest they had possibly approached a 'minimum survival level' (Coates and Lindgren 1978). As recently as 1994 birds were being hunted with shotguns in the Ok Tedi area. At that time a fine specimen was offered for sale at 250 PNG Kina, a high price indeed when specimens of other birds of paradise of the area brought only 5–20 Kina (Gregory 1995). Prior to Crater Mt., PNG, becoming a Wildlife Management Area the Gimi people of Ubaigubi considered this species virtually extinct there, apparently because of plume trading and ritual gift giving. After a hunting ban (several men were fined for killing sicklebills during 1996) birds appear to have made a strong recovery (D. Gillison *in litt.*). Common around Gunung Ngribou, Arfak Mts, in mid Aug 1995 (M. Carter *in litt.*). This sp. is at risk in all populated areas of its range. As highland human populations increase, it may face severe threats in all but the most remote regions. The combination of plume-hunting and forest conversion give reason for concern about its future. The ban on commercial trade of its plumes should be strictly enforced.

Knowledge lacking and research priorities

This poorly-studied sp. offers a range of interesting research priorities. Full display and mating have not been described; nor has the spatial distribution of displaying ♂♂ through the forest; it is unknown whether ♂♂ are territorial. The relationship between Black and Brown Sicklebills requires study. In particular, it is important to discern how the spp. interact where they make contact: is there interspecific territoriality or do their ranges overlap? Do these spp. ever forage together? Finally, how does the relative morphology of the bills of the two spp. correlate with their respective foraging and diets of fruit and animal prey?

Aviculture

Known to have been kept in the USA at the aviaries of Edward Marshall Boehm.

Other

Alfred Russel Wallace was assured by different people of NG on a number of occasions that this bird makes its nest in a hole underground, or beneath rocks, always choosing a place with two apertures, so that it may enter at one and leave by the other. He pointed out that this seemed unlikely, but also noted that such local information invariably turned out to be correct. In this instance it is clear, however, that Wallace's informants were reciting to him the same sort of story people believed to be true of the Buff-tailed Sicklebill—that this bird lays the eggs that the monotreme Long-beaked Echidna *Zaglossus bruijnii* hatches from! It may be that the echidna actually dens in a two-entrance burrow, but even this is not currently known to mammalogists (Flannery 1995*a*).

Brown Sicklebill *Epimachus meyeri* Finsch, 1885

PLATE 9; FIGURES 9.49–51

Other names: Brown Sickle-billed Bird of Paradise, Meyer's Sickle-billed Bird of Paradise, Long-tailed Sicklebill, Meyer's Sickle Bill, Long-tailed Bird of Paradise, Gray Saber-tailed Bird of Paradise. *Deawa* of Yabi people, Weyland Mts, IJ (Rothschild 1931). *Jbjel* (ad ♀) and *Galknen* (♀-plumage) of the Kalam people, PNG (Majnep and Bulmer 1977). *Gula* or *Gula Gula* of Tari Valley, PNG (C. Ballard *in litt.*). The local name is *Tomba* on Mt. Hagen and *Ti-tumba* on Mt. Kubor for ♂♂ and *Pan-da-biam* for the F. *Bebu* of Waria Valley, *Queer* of Murray Pass, *Sumba* (♂) and *Agubugu* (♀) of Tomba peoples, PNG (W. Cooper *in litt.*). *Ken-ti-co* or *Ken-ti-low* for the ♀ on Mt. Wilhelm (Mayr and Gilliard 1954). *Quen-quen* of the Victor Emanuel Mts (Gilliard and LeCroy 1961), *Garag-nan* of the Schrader Mts region (Gilliard and LeCroy 1968). *Feta* or *Dafefetai* of the Gimi people, Crater Mt., PNG (D. Gillison *in litt.*). The Grand Valley, IJ, Dani name is *Malabecok* (Ripley 1964).

A montane sp. distinctive for its very pale iris, sickle-shaped bill, 'jack-hammer' call, and long graduated tail with sabre-shaped central rectrices. The ad ♂ is blackish with a huge tail and dusky-brown breast and belly, whereas the ♀ is brown dorsally, barred below, with a tail half the length of that of the ♂ (the ♀ Black Sicklebill is nearly identical but for iris colour). Polytypic; 3 subspp.

Description

ADULT ♂: 49 (96) cm, entire head may appear black but in suitable light the scale-like feathers of crown and face show intense iridescence of metallic green-blues with purple (172) and/or Magenta (2) washes. Chin, throat, and entire neck black with rich Magenta iridescent feather tipping. Mantle and back velvet-black with green-blue and/or Magenta iridescent sheens, with highly-modified, large, scale-like, central back feathers highly-iridescent, metallic blue-green. Rump velvet-black with purple or plum gloss. Upperwing velvet-black with variable rich blue-green, blue to purple or plum gloss or sheen. Uppertail brownish black with rich, deep blue, iridescent sheen to outer vanes and the grossly-elongated central pair intensely-iridescent, metallic green-blue and/or rich Magenta in suitable light, but jet black in others. Breast rich, warm, dark brown (121) increasingly overwashed with deep plum-Purple (1) to sides. Greatly-enlarged, axe-head-shaped, pectoral plumes jet black with metallic, dark Magenta, iridescent gloss and shorter ones overlaying these jet black, broadly-tipped, intensely-iridescent, metallic blue (270), Purple, and/or Violet in suitable light. Elongate modified feathers to each side of breast, belly, and vent with highly-iridescent, metallic Purple and/or Magenta, broad but tapering tips which may appear blue-green in suitable light. Sparse, filamental, flank plumes variable fawn brown (25) with paler, straw-coloured, central shafts. Underwing and thighs blackish-brown (119). Vent and undertail coverts Olive-Brown (28). Undertail glossy blackish-brown, almost black. Bill black, eye very pale chalk-blue, legs and feet dark greyish to blackish, mouth bright yellow.

ADULT ♀: 52 cm, forehead, crown, and nape rich dark reddish Amber (36). Lores and sides of face dark blackish-brown and chin to throat likewise but finely flecked dirty buff. Mantle to uppertail coverts Olive-Brown washed dark reddish Amber, more so on upper mantle and uppertail coverts. Dorsal surface of wing: coverts Olive-Brown, greater coverts and flight feathers dark Brownish Olive (129) with indistinct broad Raw Umber (23) outer edges. Dorsal surface of tail Olive-Brown slightly washed Raw Umber. Upper breast becoming paler than throat to a dirty buff (223D), regularly barred blackish-brown throughout underparts. Underwing coverts barred as lower underparts, remainder dark grey-brown.

Ventral surface of tail Dark Drab (119B) with paler, straw-coloured, feather shafts. Mean wing length and weight 14 and 29% smaller than ad ♂, respectively.

SUBADULT ♂: variable, from much like ad ♀ but with a few feathers of ad ♂ plumage intruding, to as ad ♂ with few feathers of ♀ plumage remaining. Young first attain darker crown than ad ♀ and blackish feathering about eyes, lores, base of bill, chin, and upper throat. This is followed by a predominantly to entire ad ♂ head plumage being acquired and with subsequent moults an increasing proportion of plumage becoming that of ad ♂, with one of the last ♀ plumage characters lost being barred belly feathering. A captive received as a juv 13 Sept 1978 had finally assumed ad ♂ plumage except that his 'long tail feathers' were still 'brown with black patches' on 1 Apr 1984 when 'at least six years old' (M. Mackay 1984). See Aviculture below.

IMMATURE ♂: as ad ♀ but tail longer.

NESTLING–JUVENILE: like ad ♀ but mantle and crown brighter and more rust-coloured with general plumage soft and fluffy especially on the abdomen (Mayr and Gilliard 1954; Gilliard 1969). A bird with 'nestling plumage on the upper back' was brown, the feathers tipped with rufous; the lower back and rump dark earthy brown. The throat feathers were sooty black, those of the breast and abdomen similar to those of the imm ♂ plumage in colour, except for being slightly tinged buffy, and easily distinguished by their fluffy texture. Upperwing coverts distinctly margined with rufous (Mayr and Rand 1937). Crandall (1932) and Aruah and Yaga (1992) noted that ♂♂ wear subad plumage for at least 2 years before acquiring full ad appearance.

Distribution

The central ranges of NG from the Weyland Mts of IJ to extreme southeastern Owen Stanley Mts, PNG, from 1500 to 3200 m (mainly 1900–2900 m; Coates 1990). Where it abuts the Black Sicklebill range the Brown Sicklebill occurs at higher elevations, but above Ubaigubi, PNG Peckover (1990) observed that 'their ranges do overlap on a 2120 m ridge' and that here both 'have display stations almost within throwing distance of each other'. The sp. was not recorded in an elevational transect of the northern flanks of English Peaks (above the Mambare R. valley) and may be locally absent (BB).

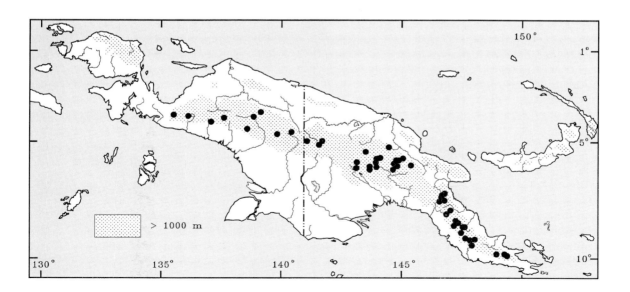

Systematics, nomenclature, subspecies, weights, and measurements

SYSTEMATICS: a sister form to *Epimachus fastuosus*, but the two do not constitute a superspecies, having established a considerable degree of sympatry. By current distribution of this species-pair and other analogous paradisaeid spp. pairings (*Paradigalla, Parotia, Astrapia*), it appears likely that the parental form to this pair inhabited the mountain chain from Vogelkop southeastward. Subsequent speciation was then produced by a vicariance event that isolated what would become *fastuosus* on the mountains of the Vogelkop, and the predecessor to *meyeri* on the central ranges. Presumably, morphological evolution and differentiation in elevation distribution permitted subsequent establishment of sympatry. Hybridisation: unrecorded in the wild but has been crossed with Ribbon-tailed Astrapia in captivity (Bishop and Frith 1979). Sympatric with Buff-tailed and Black Sicklebills in limited areas. Hybrids between Brown and Black Sicklebills have not been reported.

SPECIES NOMENCLATURE AND ETYMOLOGY

Epimachus meyeri Finsch

Epimachus Meyeri Finsch, 1885. *Zeitschrift für die gesammte Ornithologie* 2, 380. Hufeisengebirge, southern New Guinea (= Mount Maguli, Owen Stanley Mts, PNG). Discovered by Carl Hunstein, a German professional collector, in 1884. Type specimen SMT C 8140 (8450) ♀ (destroyed in World War II).

Etymology: *meyeri* = after Finsch's associate Adolf Bernhard Meyer, ornithologist at the Dresden Museum.

SUBSPECIES, WEIGHTS, AND MEASUREMENTS

1. *E. m. meyeri* Finsch, 1885.

Range: mountains of southeastern NG, northwestward to the Ekuti Ra and Mt. Missim, E of the Watut/Tauri Gap, which presumably serves as the topographic discontinuity separating the nominate subsp. from *E. m. bloodi*.

Measurements: wing, ad ♂ ($n = 30$) 177–206 (185), imm ♂ ($n = 19$) 150–184 (173), ad ♀ ($n = 24$) 147–185 (163); tail, ad ♂ ($n = 29$) 235–282 (261), imm ♂ ($n = 18$) 180–260 (217), ad ♀ ($n = 24$) 178–250 (208); tail centrals, ad ♂ ($n = 23$) 631–822 (748), imm ♂ ($n = 18$) 265–372 (335), ad ♀ ($n = 21$) 257–359 (306); bill, ad ♂ ($n = 25$) 78–90 (84), imm ♂ ($n = 17$) 77–92 (84), ad ♀ ($n = 22$) 66–91 (81); tarsus, ad ♂ ($n = 30$) 48–54 (51), imm ♂ ($n = 19$) 45–54 (51), ad ♀ ($n = 24$) 42–53 (47). Weight: ad ♂ ($n = 7$) 253–310 (281), imm ♂ ($n = 4$) 186–260 (236), ad ♀ ($n = 6$) 162–188 (173).

2. *E. m. bloodi* Mayr and Gilliard, 1951. *American Museum Novitates* **1524**, 10. Mt. Hagen, Central Highlands, Mandated Territory of New Guinea, 8300 ft. Type specimen AMNH 348211.

Range: the central segment of the central ranges, presumably from the Kratke Ra just W of the Watut/Tauri Gap, northwestward to Mt. Giluwe, Mt. Hagen, and probably to the Doma Peaks/Tari region (westernmost extent yet to be determined).

Diagnosis: like the nominate form but considerably smaller (but ad showing nearly identical bill lengths and therefore with a proportionately longer bill). Ad ♂ flank plumes are paler, more dirty-whitish, than dirty pale-brownish of *meyeri*.

Measurements: wing, ad ♂ ($n = 16$) 171–181 (177), imm ♂ ($n = 9$) 151–169 (162), ad ♀ ($n = 23$) 144–165 (153); tail, ad ♂ ($n = 16$) 218–264 (237), imm ♂ ($n = 9$) 194–235 (219), ad ♀ ($n = 20$) 171–226 (197); tail centrals, ad ♂ ($n = 16$) 595–766 (670), imm ♂ ($n = 7$) 306–348 (324), ad ♀ ($n = 18$) 262–341 (288); bill, ad ♂ ($n = 13$) 82–89 (84), imm ♂ ($n = 8$) 68–87 (80), ad ♀ ($n = 23$) 78–89 (82); tarsus, ad ♂ ($n = 16$) 46–51 (49), imm ♂ ($n = 9$) 46–50 (48), ad ♀ ($n = 23$) 41–50 (45). Weight: ad ♂ ($n = 6$) 144–230 (187), imm ♂ ($n = 2$) 190–200 (195), ad ♀ ($n = 4$) 140–202 (163).

3. *E. m. albicans* (van Oort, 1915). *Zoologische Mededelingen*, Leiden **1**, 228. Treubbivak

(2366 m), Treub Mts, central New Guinea. Type series = LM 497 and LM 630.

Synonym: *E. m. megarhynchus* Mayr and Gilliard, 1951. *American Museum Novitates* **1524**, 10. Gebroeders Mts, Weyland Ra, Dutch New Guinea, 6000–7000 ft. Type specimen AMNH 677998. The subsp. *megarhynchus* was erected on the basis of a larger bill than the others, but bill lengths of the only three specimens known (♂ 82, ♀♀ 79, 83 mm) fall within the range of all subspp. See also Cracraft (1992).

Range: the western segment of the central ranges, from the Weyland Mts, IJ, southeastward to the Hindenburg and Victor Emanuel Mts, PNG. The exact eastern boundary of this subsp. requires definition in relation to that of *E. m. bloodi*.

Diagnosis: similar to the nominate form but slightly smaller and ad ♂ flank plumes are whitish unlike those of *meyeri* and *bloodi*.

Measurements: wing, ad ♂ ($n = 5$) 171–188 (177), imm ♂ ($n = 7$) 162–171 168), ad ♀ ($n = 18$) 142–170 (154); tail, ad ♂ ($n = 5$) 234–265 (247), imm ♂ ($n = 7$) 171–232 (207), ad ♀ ($n = 17$) 151–248 (196); tail centrals; ad ♂ ($n = 3$) 522–656 (574), imm ♂ ($n = 6$) 250–370 (315), ad ♀ ($n = 11$) 213–349 (280); bill, ad ♂ ($n = 4$) 82–87 (85), imm ♂ ($n = 6$) 77–90 (85), ad ♀ ($n = 17$) 69–88 (83); tarsus, ad ♂ ($n = 4$) 48–50 (50), imm ♂ ($n = 7$) 38–52 (48), ad ♀ ($n = 18$) 39–47 (45). Weight: ad ♂ ($n = 1$) 189, ad ♀ ($n = 4$) 145–175 (160).

Etymology: *albicans* = L. *albus*, white; L. *canens*, becoming grey.

Habitat and habits Table 4.1

Original and disturbed mid- and upper montane forests, secondary growth, and forest-edge. This is the sicklebill of NG's moss forest. The trees *Nothofagus grandis*, *Phyllocladus hypophyllus*, and *Pandanus tectorius* are abundant in the Tari Gap where the sp. is common. Found in disturbed forest dominated by *Nothofagus* spp. and podocarps on Mt. Hagen by Cooper (in Cooper and Forshaw 1977). Said to favour large central tree limbs (of *Castanopsis acuminatissima* and other oaks), usually moss-covered, in the higher mid-montane forests by Gilliard (in Gilliard and LeCroy 1961). Ad ♂♂ are retiring and not easy to observe except at a display perch. ♀-plumaged birds are more common and are considerably easier to observe, especially when they are busy foraging, when they can be quite oblivious to human observers. At least during parts of the year, if not throughout, birds vigorously search through the significant accumulated debris at the base of live and dead pandanus crown fronds. This activity involves much acrobatic clambering about the vegetation and the casting off of debris which causes much noise and often first attracts the attention of the observer. Birds involved in this foraging may fly from one pandanus to another in a preoccupied fashion. Bill-wiping and the regurgitation of fruit seeds are frequent activities. A captive ad ♂ seen drinking from a pool on several occasions did so by turning the head almost upside-down, to lower the then upward-curving bill into the water then to lift the uprighted bill only slightly in order to swallow. A wild bird seen feeding on the ground covered some 50 m during about 5 min, and a captive was noted often to use its bill 'as a lever to flick over stones and bark with a sideways motion' (M. Mackay 1984). Flight is undulating, with a brief bout of flapping followed by a glide in which the wings are folded tightly against the body, producing a dip in trajectory before the next bout of wing-beats that brings the bird upward again. Said to roost in tall pandanus crowns (Majnep and Bulmer 1977). Occasionally forages within audible range and sight of Black Sicklebill (D. Gillison *in litt.*)

Diet and foraging Table 4.2

Consumes fruit, arthropods, and small vertebrates. Of 117 dietary records collected by a variety of methods, 52% included fruit and 48% included animal matter (Rothschild 1931; Simpson 1942; Sims 1956; Gilliard and LeCroy 1968; Diamond 1972; Beehler and Pruett-Jones 1983; Kwapena 1985; D. Frith personal communication; CF; BB). Forages widely in the forest vegetation, from the ground to the canopy, but most commonly in the middle stages of the forest interior. Most conspicuous when foraging for animal prey in pandanus crowns or on large moss-covered limbs. Often solitary, but not uncommonly forages in company of conspecifics or other spp. of birds of paradise (especially *Astrapia*).

Animal prey include Orthoptera, '40 mm grubs', and a small tree frog. Fruit taken include *Elaeocarpus* sp., *Timonius belensis*, *Fagraea salticola*, *Garcinia* sp., and *Omalanthus* sp. (CF); *Perottetia alpestris*, *Freycinetia inermis*, *Maesa edulis*, *Aralia* sp., *Pandanus brosimos*, *Cucurbita* sp., and Urticaceae sp. (Kwapena 1985); see also Appendix 4.

The following assortment of foraging observations gives a feel for the behaviour of the sp. Of 12 foraging bouts observed by Kwapena (1985) on Mt. Giluwe, PNG, 8 were of birds on tree branches, 3 in foliage, and 1 in dead wood; 8 were 25–35 m above ground and 4 15–25 m above ground. As his observations included 17 records of fruit feeding and 9 of birds feeding upon arthropods individual birds were obviously seen feeding upon > 1 item.

A ♀ captive in England was twice seen to catch a half-grown frog and then wedge it into a wood cavity before puncturing and tearing it to eat. The same bird also caught a house mouse and killed it by beating it vigorously against a perch before wedging it into a crevice to eat it piece by piece (Timmis 1972).

Birds seen feeding in the Tari Gap, PNG, foraged at all levels in the forest except on the top of the upper canopy, mostly by probing into and tearing at epiphytic plant growth and, mostly, using the bill to seek animal prey within accumulated debris between the bases of pandanus palm fronds (CF). Two ♀-plumaged birds in the Tari Gap in Nov foraged on the ground at road-side forest-edge for 4 min, probing into mud and small plants with the bill (D. Frith personal communication). As many as four birds may forage in the same tree. An ad ♂ was displaced and chased off from a fruiting *Schefflera* by a Belford's Melidectes while a ♀-plumaged individual of the former supplanted one of the latter and a Common Smoky Honeyeater. Another of the Melidectes flew at and grabbed a ♀-plumaged sicklebill, dislodging feathers, before the latter flew off (CF). An ad ♂ sicklebill was observed twice to supplant an Archbold's Bowerbird (M. Laska *in litt.*).

♀-plumaged birds are often seen in loose feeding aggregations of up to some half a dozen birds. Occasionally joins mixed-spp. foraging flocks that sometimes also include astrapias.

Of 41 sightings of feeding birds at Tari Gap, PNG, during Sept–Jan, 32 involved a lone bird and 9 two birds. Of the latter, 5 instances involved an ad ♂ in company with a ♀-plumaged bird. Of the 41 sightings, 6 were of birds 1–5 m high, 13 were 6–10 m high, 17 were 11–15 m high, and 5 were 17–21 m high (CF).

Film taken by H. Sielmann shows a ♀-plumaged bird in the wild plucking and eating one after another small cauliferous figs. Majnep (in Majnep and Bulmer 1977) states fruit of many trees and vines are consumed, including some found on the forest floor, such as those of taro *Alocasia* sp. and wild raspberries *Rubus* sp. It is notable that this sicklebill is not known to consume the fruit of *Schefflera chimbuensis* so important to the Ribbon-tailed Astrapia in the central highlands of PNG.

All but one of the 22 records of feeding upon arthropods involved the bird using the long decurved bill to probe vigorously into debris at the base of pandanus fronds or into epiphytic growth or dead tree wood (CF).

Ad ♂♂ appear to forage within their home range, exclusive of other ad ♂♂, but they permit ♀-plumaged birds to feed there. That this sp. and the Black Sicklebill competitively exclude each other where they occur on the same mountain ranges is suggested by the fact that in southeastern NG where the Black Sicklebill is absent, the Brown Sicklebill ranges down to elevations considerably lower than where the two spp. co-exist.

Vocalisations and other sounds

Figure 9.49

The far-carrying advertisement song of the ad ♂ is one of the most characteristic sounds of

Brown Sicklebill *Epimachus meyeri* Finsch, 1885

moss forests of the central ranges. It sounds like the burst of a pneumatic jack-hammer—percussive and entirely unmusical. Slight variations of the call apparently exist; birds of the SE of NG give a *Tat-at, tat-at, tat-at* whilst those of the Central Highlands produce a *Tat-at-at-at, Tat-at-at-at-at-at* (B. Whitney in Beehler *et al.* 1986). In the Eastern Highlands of PNG the call consists of three double-notes, lasts 2 sec, and is repeated at intervals of about 2 min. (Diamond 1972). This sound, which can be heard by a human from a distance of 2 km, is accompanied by exertions by the ♂ that include throwing the head sharply back and vigorously pumping the throat as the chest and wings are sharply jerked with each note. Sonograms of this song (Fig. 9.49(a,b)) show it to be structurally a much speeded up delivery of Black Sicklebill-like notes (see Fig. 9.47(a,b)). Well-dispersed ♂♂ can clearly hear the song of immediate neighbours, and it is probable that each mature ♂ can hear the calls of at least three immediate rivals (Fig. 9.50). A minor foraging or contact note is a nasal *nreh!* or *wahn?*, similar to the call note of the Blue Bird (Beehler *et al.* 1986; BB). A note given by alarmed captives of both sexes was a deep guttural grunt delivered with rapid jerking of the wings and body. Call notes given by ♀-plumaged birds every 5–20 sec are described as 'like a chicken', and a barking *ugh!* (M. Laska *in litt.*). Other vocalisations and mechanical sounds clearly remain to be adequately described. Gilliard (1953) reports a ♂ following the jack-hammer call by drumming 'like a grouse, beating its wings against its sides and making loud, cracking reports which quite mystified us. It seemed impossible that wings alone could make such a fuss.' Gilliard (in Gilliard and LeCroy 1961) heard a ♂ over several successive days during May in the Victor Emanuel Mts make very loud, snapping or cracking noises. These sounds were repeated with alarming suddenness about every half hour during the mornings and the afternoons. A second ♂, *c.* 1.6 km away from the first, also gave these calls. Excited birds give a gurgling series of notes, rather like an

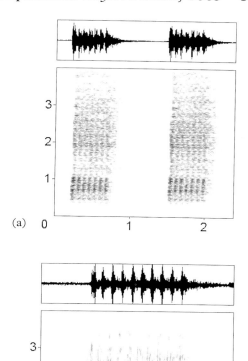

9.49 Sonograms from recordings of an ad ♂ Brown Sicklebill explosive advertisement song at (a) Jayawijaya, IJ on 6 Apr 1993 by K. D. Bishop and (b) at Tari Gap, PNG during Mar 1993 by A. Griffin.

underwater version of the jack-hammer vocalisation (BB). A recently-fledged juv uttered a 'quiet honking *ur, ur*' (T. Pratt *in litt.*).

Mating system

Polygynous, with promiscuous ♂♂ maintaining large territories that appear to be defended against neighbouring ♂♂. In some locations vocal ♂♂ are regularly dispersed along ridges (Fig. 9.50). Counter-singing at territorial boundaries may maintain ♂ dispersion but most singing is performed at display sites

Brown Sicklebill *Epimachus meyeri* Finsch, 1885

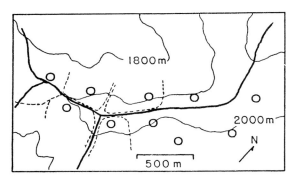

9.50 Display dispersion of ♂ Brown Sicklebills on Mt. Missim, PNG: open circles indicate regular calling sites of 10 ♂♂; dashed lines delimit territory boundaries of 5 ♂♂. No other ♂♂ were known in the area indicated. The bold solid line follows the main ridge crests. From Beehler and Pruett-Jones (1983), © Springer-Verlag with kind permission.

consisting of several high tree display perches nearer the centre of the territory.

Courtship behaviour
Display perches of ad ♂♂ at traditional sites are usually near-horizontal or gently-sloping branches in the forest canopy, the upper midstage of the forest or in the substage (Coates 1990). Wild display and courtship are unrecorded but the apparently typical behaviour has been observed in captive birds.

Three forms of display by two captive ♂♂ were defined as *Pumping, Leaning,* and *Upright*. In *Pumping* the bird was 'in an upright position, with the breast feathers spread. The tail was jerked wide open, then tightly shut, the alternation being very rapid. The wings, which were closed against the body, were moved up and down along its sides, the upward movements coinciding with the opening of the tail. The display was continued for a min or more …' (Crandall 1932).

In *Leaning* the bird leans backward to about 45° from the vertical (Fig. 9.51(a)). The perched ♂ 'expanded the feathers of the breast, taking some time to arrange the short decorative flank plumes, which extended outward, forming a fringe around the sides. He then gave his rattling call and turned the breast upward, his feet retaining their original position. The breast feathers were now spread to their fullest extent, the bird's body appearing flattened. The short feathers of the upper breast turned upward about the head, circling the throat so closely that the iridescent black of face and throat became very conspicuous. The wings were closely folded and the tail was slightly spread, though not vibrating or moving. The beak was closed. The long pectoral shields were folded beneath the plumage, so that they were entirely invisible (they take no part in this display). As in the imm stage, this pose was held rigidly for about 10 sec, when the bird returned to normal position' (Crandall 1932). When performed by a ♂ in ♀ plumage this display followed the 'jack-hammer' song. The bird then leaned to one side to rotate his body so that his breast was directed upward, the breast feathers being spread as widely as possible, to conceal the tightly-closed wings and the tail partly fanned. This posture, held for *c.* 10 sec, was followed by the bird moving rapidly about the aviary to return to his perch to perform a *Leaning* display again.

In the *Upright* display a normally-perched bird starts preening his extended breast feathering and pectoral fan plumes. 'Suddenly, without calling, the body is drawn erect, with tail very slightly opened, wings tightly closed. The breast feathers, encircled by the decorative flank plumes, are widely spread. The pectoral shields are thrown straight upward, so that they extend far above the head, wrapping it closely. At the upper extremity, the shields are narrow and compressed; at their bases, they broaden gradually, to pick up the line of the spread breast feathers. The beak is widely opened, to show the bright yellow lining of the mouth (Fig. 9.51(b)). This position is usually held rigidly for about 5 sec, when the bird resumes his alternate preening and displaying.

'On rare occasions, usually late in the afternoon, a further development of this display

Brown Sicklebill *Epimachus meyeri* Finsch, 1885

has been seen. Stiffly maintaining the upright position just described, and with the feet firmly grasping the perch, the bird rotates his body in a series of short jerks, pausing for several sec at the end of each, until it is at right angles with the axis of the perch. He than jerks slowly in the opposite direction, until he has again come to a right angle with the perch but is facing the other way. This movement may be continued for 2–5 min. Throughout, the bird is obviously exerting himself to the utmost to maintain his tense attitude. There is no movement of tail, wings or plumes, and no sound, once the position has been struck' (Crandall 1932).

A second captive ♂ also regularly performed the *Upright*, or high-intensity, display. In so doing he did not leave a small space directly above his head between the pectoral fan plumes raised from each side of the body, but brought them to meet to form a continuous vertical oval above the head. He simultaneously, rapidly and repeatedly, fanned and closed all outer tail feathers whilst maintaining

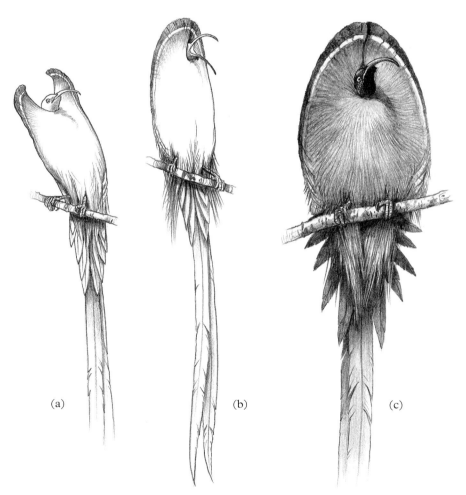

9.51 Three forms of display of ad ♂ Brown Sicklebills: *Leaning* (a) and *Upright* (b,c). After Crandall (1932, 1946*a*). See text.

this posture (Fig. 9.51(c)). Much was made of this difference between the two birds at the time but in all probability the less-intense *Upright* display of the first bird merely indicates a less motivated or experienced individual (Crandall 1946(*a*,*b*)). The ♂ of a pair at Chester Zoo, UK was seen to perform all of the above displays (Timmis 1972).

In Apr 1977 an ad ♂ in Rotterdam Zoo performed the *Leaning* and *Upright* displays to an ad ♂ Magnificent Riflebird, which would often display in return. Prior to the *Leaning* display the ♂ often bill-wiped and picked at his perch, apparently as displacement activity. This was followed by the raising of the pectoral fan plumes outward and upward to either side of the neck and head repeatedly. In this display the ♂ would occasionally hop towards the Magnificent Riflebird and lean backwards as do displaying ♂ riflebirds. The high-intensity, or *Upright*, display often followed this, but was also performed without preamble (CF).

In the *Upright* display the ♂'s eyes were concealed behind the raised pectoral fan plumes with the bill protruding through the plumage. In this posture he vibrated his entire body very slightly and slowly rotated it side-to-side. Where the upper pectoral fan plumes meet above the head the plumes from one side overlap those from the other rather than simply abutting. A completely-different brief display involved the ♂ bowing, his bill lowered almost to the perch between his feet with the pectoral fan plumes semi-erect (CF).

Breeding Table 6.5

NEST SITE: a nest in the Kubor Mts was 3.7 m above ground in the slender crotch limbs of a thickly-leaved small tree, which grew in the forest substage on top of a flat ridge beneath tall open trees (Gilliard 1969). A ♀ was observed repeatedly bringing food to a nestling in a nest (not visible to the observers) 8 m high in the crown of a pandanus (Cooper and Forshaw 1977). A nest found by Kwapena (1985) was 7 m high in the crown of a pandanus near a garden, and another thought to be of this sp. was 3 m high in the centre of a tree fern crown. A presumed ♀ was seen repeatedly to visit a nest with food for young 12 m high in the frond bases of the crown of a pandanus in an area of disturbed secondary forest adjacent to disused hunting huts in Tari Gap, PNG (Frith and Frith 1992*a*).

NEST: two were described by Gilliard (in Mayr and Gilliard 1954; Gilliard 1969) with conflicting measurements (below). The Mt. Kubor nest was a cup of stringy living mosses and slender vines attached to slender, pencil-sized, crotch limbs of a small tree and lined with slender, brownish rootlets and many skeletonized leaf fragments measuring 175×150 (or 100) mm deep externally and 95×45 (or 63 mm) internally. The Mt. Hagen nest with nestling was attached to a simple crotch and consisted of much stringy 'furry' mosses, leaves, and grasses, lined with slender rootlets, dried fern tips, and some small leaves; measuring 150×100 outside and 80×45 mm inside. A nest found by Kwapena (1985), presumed to be of this sp., was oval-shaped, $200 \times 100 \times 57$ mm deep with inside walls of orchid and fern stems and the outer wall constructed of moss and fern fronds. Of a confirmed pair in an aviary only the ♀ carried out nest construction, building two complete and three incomplete nests (Timmis 1972).

EGGS: the first egg described was laid by a NYZS captive on 28 Feb 1941. In ground colour it is Pale Pinkish Buff (121D) heavily blotched and streaked at the large end, the marking decreasing towards the almost clear smaller end. The under markings are greyish white; those on the surface are reddish-brown. It measured 44.3×28.5 mm and weighed 18.9 g at laying (Crandall 1941). The long-incubated egg from the Kubor Mts is Pale Pinkish Buff (121D) with the faintest pinkish wash with bold brown, grey, and reddish-brown longitudinal streaks (mostly on larger end with a sprinkling of small brown (221B) and watery lavender spots. It measures 39.3×26.7 mm (CF). An egg in the BMNH measuring 36.5×25.5 mm (CF) said to be of this sp.

(Schönwetter 1944) was considered by Grant (BMNH clutch card) to be that of the Raggiana Bird. As it is less pink than many eggs of the latter but is like those of the riflebirds, except darker and duller, and not unlike the egg laid in the NYZS, its original identification may well be correct (CF). See Plate 13.

INCUBATION: before collecting the Kubor Mts nest and egg, Gilliard's assistant watched a ♀-plumaged bird visit the nest three times.

NESTLING CARE: casual observation at two nests suggests this is exclusively by the ♀. Cooper and Forshaw (1977) noted a presumed ♀ regularly carrying small red fruits in the bill to a nestling. This parent was also seen to chase and attack an ad ♂ Ribbon-tailed Astrapia that approached the nest tree, the two grappling birds flapping almost to the ground before separating. A presumed ♀ was seen repeatedly to visit a nest with unseen but audible young and once carried a large mole-cricket in the bill tip to this nest (Frith and Frith 1992a).

NESTLING DEVELOPMENT: unknown. A recently-fledged juv had a very short tail and bill shorter than ad (T. Pratt *in litt.*).

Annual cycle Table 6.5
DISPLAY: on Mt. Hagen ♂♂ sing at least during May–Aug (Mayr and Gilliard 1954; BB). Little known, but certainly Sept–Dec in the Tari Gap, PNG. A nest with an egg collected in early Aug on Mt. Giluwe indicates courtship in July, and Kwapena (1985) here witnessed display in mid-Dec and Feb and copulation on 26 Jan. BREEDING: Apr–Jan, and probably possible during all months. ♂♂ have been collected with much enlarged gonads during July–Dec and moderately-enlarged ones in Jan–Feb, Apr, and June. A ♀ with much enlarged oocytes was collected in Jan and with moderately-enlarged ones in Mar, July, and Dec. On 19 Dec 1974 at Bulldog Road, Wau a recently-fledged juv was seen in company of an ad (T. Pratt *in litt.*). On 12 Mar 1975, on Mt. Kaindi, Morobe, PNG, a juv was seen to beg from a ♀-plumaged bird (BB). EGG-LAYING: from scanty available data, at least Apr–Oct. MOULT: 60 of 146 specimens were moulting, involving all months but least activity during June–Sept and most during Oct–May (Appendix 3). Two birds collected live at Deva Deva, PNG in Oct were in moult when they arrived in New York on 29 Mar. Both birds were dropping feathers in mid-Mar of the following year, and completed moult by mid-June. In the next year both birds started to moult on 5 Apr and the ♀ had finished moulting by 10 July but the ♂ was acquiring ad plumage until 10 Sep, over *c.* 5 months. In the next year the ♂ started moult on 10 May and ended by 3 Oct, another 5 month moult period.

Status and conservation Table 8.1
May be uncommon or absent in some apparently suitable habitats, but common to abundant over much, if not most, of its range. At least 10 ad ♂♂ recorded in a 350 ha study area on Mt. Missim, PNG, with a mean nearest neighbour distance of 332 m (range = 270–440 m) (Fig. 9.50). Ripley (1964) found ad ♂♂ rare in the Baliem and Ilaga Valley areas of IJ and considered this a direct result of traditional hunting for head-dresses. Owing to its higher altitudinal range, above most clearing for human habitation and agriculture, this sicklebill is in no immediate danger, unlike its close relative the Black Sicklebill. Birds in the Kaironk Valley, Schrader Ra, PNG, were 'much less in evidence' in the 1970s than in the 1960s—possibly reflecting hunting for trade (Majnep and Bulmer 1977).

Knowledge lacking and research priorities
Display and mating in the wild remain undocumented. For purposes of interspecific comparisons, it would be particularly useful to record the full display/mating sequence in the field. Black and Brown Sicklebills' contact

Aviculture

Kept in PNG (Peckover and Filewood 1976; Aruah and Yaga 1992), the USA (Crandall 1935; Everitt 1973), the UK (Timmis 1972), and Europe (CF; BB). First in captivity was a pair of *E. m. meyeri* captured by Walter Goodfellow in 1909 and taken to Hoddam Castle in Scotland for the private collection of E. J. Brook. NYZS record receiving an imm pair in 1929 that were collected in Oct 1928 (Crandall 1935), the ♀ dying of an accident in 1935, and the ♂ living until Sept 1936. A wild-caught young ♂ in ♀ plumage obtained by BRS, PNG on 13 Sept 1978, from Wabag, first gave the 'jack-hammer' call on 29 July 1982 (Aruah and Yaga 1992). On 1 Aug 1983 some ad ♂ plumage was first noticed and a close examination on 25 Aug revealed black wing coverts, upper leg, under chin, and nape feathers, all other plumage being typical of ♀. By Nov 1983 the wings were iridescent, dark purplish black, head and chin were black, forehead iridescent light blue, back spangled iridescent blue, underwing body feathers soft lavender, side breast plumes blue-black, flank plumes fawn, the outer tail feathers black and the inner pair brown, and the iris pale blue. This plumage pattern was no different on 26 July 1984 but the full ad plumage was attained by 13 May 1985. Aruah and Yaga considered this bird to be 1 year old when acquired, the transformation from ♀ plumage to that of a full ad ♂ having taken at least 7.5 yr, possibly 8.5 yr if small ♂ plumage characters acquired with the 1982 moult were overlooked. A ♂ lived for 7.5 yr in the in NYZS. Goodfellow (1926b) found this sp. remarkably tame, individuals taking insects from his fingers within a half hour of capture and being confiding in other respects. This sp. is active, vocal, and does well in large aviaries with adequate flight area and a pool with stepping stones and shallows. Will probe and prise everywhere (and often on the ground) and thus prone to damage their fine, long bill in captivity. Prone to infestations of internal parasites (P. Shanahan *in litt.*).

Genus *Drepanornis* Sclater 1873

Drepanornis Sclater, 1873, *Nature* **8**, 192. *Proceedings of the Zoological Society of London*, p. 560. New name for *Drepanephorus* Sclater, 1873. *nec* Egerton, 1872 (Pisces). Type, by original designation, *Drepanornis albertisi* Sclater.

Synonym: *Drepananax* Sharpe, 1894. *Bulletin of the British Ornithologists' Club* **4**, 15. Type, by original designation, *Drepanornis Bruijnii* Oustalet.

Our resurrection of the genus *Drepanornis* (post Diamond 1972) is prompted by a character analysis of the epimachine lineage. *Drepanornis* is distinct from *Epimachus* in the following **characters**: different relative body proportions, negligible sexual dimorphism in overall size (Tables 1.3 and 1.4), tail morphology, lack of sexual dimorphism of mantle plumage, dull ♂ iris colour, presence of extensive facial apterium, and musically-whistled calls.

The *Drepanornis* sicklebills are **two species** of medium-sized forest-dwellers with medium-length rounded tails, narrow, strongly-decurved bills, and relatively-limited sexual dimorphism (Plate 9). The two spp. of *Drepanornis* share a number of significant characters lacking in *Epimachus*. In marked contrast to their larger relatives, they show little sexual dimorphism in size, tail length, and plumage, have different overall proportions, and structurally quite different tails, lacking grossly-elongated central rectrices. They also have extensive **bare facial skin**, structurally-different bills, dark irides in both sexes, a different type of ventral barring, and distinct head feather structure and coloration. A number of these features are more *Cicinnurus*-like than *Epimachus*-like.

♂♂ are larger than ♀♀, but sexual dimorphism in **size** (wing length) is a mere 3%. As with *Epimachus*, **bill** decurved, narrow, and sickle-shaped and more than two times longer than head. Nostrils marginally verged by feathers, proximally, in ♂♂ but ♀♀ lack feathering near nostrils. Bill meets the skull in a conspicuous 'hinge', presumably as an adaptation to vigorous arthropod foraging that permits some 'give' at the skull/bill junction. **Tail** rounded by mild graduation, *c.* 71–84% of wing length.

Wings long and somewhat rounded, outermost primary not modified. **Tarsus** length *c.* 22% of wing length. Lower tarsus scuted, becoming fainter to smooth proximally.

Pectoral display plumes are *Cicinnurus*-like. Iris in both sexes dark brown.

Polygynous, with ad ♂♂ territorial and solitary, displaying at traditional perches, and ♀♀ alone conducting nesting duties. Known ♂ courtship displays are solitary and relatively static. The single inadequately-described display appears typical of the Paradisaeinae. **Eggs** typical of Paradisaeinae but with darker, more brown, and less pink ground colour (Plate 13).

Buff-tailed Sicklebill *Drepanornis albertisi* (Sclater, 1873)

PLATE 9; FIGURES 9.52–5

Other names: Black-billed Sicklebill, Black-billed Sicklebill Bird of Paradise, Short-tailed Sicklebill, D'Albertis's Bird of Paradise, *Epimachus albertisi*, *Drepanornis albertisii*. *Quarna* by people of the Mt. Arfak area (D'Albertis), *Sagroja* in the Arfak Mts, IJ, *Vaun* on the Huon Penin (Sharpe 1892), and *Sodadi* by the Daribi people of the Karimui Basin, PNG (Diamond 1972). *Habanaoesa* of the Gimi people, Crater Mt., PNG (D. Gillison *in litt.*). *Zabalong* of Cromwell Mts people (skin label).

A drably-plumed, secretive, medium-sized bird of midmontane forest, with a remarkable black sickle-shaped bill, and a conspicuous and rounded buff-coloured tail. Uncommon and local; joins mixed-spp. foraging flocks and probes bark and dead wood for arthropods. Vocalisations of the ♂ are the best means of discovery. The ad ♂ has head, pectoral, and flank plumes, whereas the ♀ has distinctively barred underparts. The similar Pale-billed Sicklebill inhabits lower elevations and the Brown and Black Sicklebills, whose ♀ plumages are similar, have a longer, pointed tail and a rufous-chestnut crown. Polytypic; 2 subspp.

Description
ADULT ♂: 35 cm, head plumage variably Cinnamon-Brown (33), more Raw Umber (23) on crown where feathers dully-tipped, almost redundantly, iridescent coppery purple. Lores, chin, and throat of dense plush feathers strongly-iridescent deep Leaf Green (146). Elongate, dark, forecrown feathers form a 'horn' above each fore-eye of which some shorter feathers are tipped strongly-iridescent purple-blue and the longer ones Magenta (2) in suitable light. Bare facial skin dark maroonish grey. Upperparts and upperwing Cinnamon-Brown, paler and more Cinnamon (123A) on rump and uppertail coverts. Tertials and primaries with broad Cinnamon leading edges. Upper surface of tail pale Cinnamon to almost Clay (123B) of central feather shafts. Upper breast Olive-Brown (28) and separated from lower breast and flank plumes of same colour by a breast band of feathers broadly-tipped, deep, iridescent violet-purple (172A) with a

Buff-tailed Sicklebill *Drepanornis albertisi* (Sclater, 1873)

Magenta wash. Loosely-structured pectoral plumes extensively, and richly-iridescent bronze with Magenta wash in suitable light. Similarly-structured, elongate, flank plumes broadly-tipped, deep, rich Royal Purple (172A). Underwing: coverts variable paler browns and buff, flight feathers pale Olive-Brown with extensive broad Tawny Olive (223D) trailing edges. Central belly, vent, and undertail coverts white. Under surface of tail pale Sayal Brown (223C). Bill shiny black, iris dark brown, mouth pale green or pale yellow.

ADULT ♀: 33 cm, bill on average slightly longer than ad ♂'s. No iridescent feathering, but otherwise entire upperparts similar to ad ♂ but slightly darker and more Amber (36) on dorsal surface of tail. Chin and throat Cinnamon-Brown with fine, pale, buffy, central feather shafts. Underwing: coverts Tawny Olive, barred dark brown; flight feathers as ad ♂. Remaining underparts variably Cinnamon (39), to Tawny Olive on vent and undertail coverts, uniformly barred dark brown (121) throughout except on lower breast where barring of each feather forms chevrons. Under surface of tail Sayal Brown.

SUBADULT ♂: varies from similar to ad ♀ with a few feathers of ad ♂ plumage intruding, to much like ad ♂ but with a few feathers of ♀ plumage remaining. The ad plumage is exhibited patchily for at least one breeding season, the over-eye 'horns' usually being acquired with the iridescent green throat, iridescent Royal Purple-tipped flank plumes, and extensive central breast patch of Olive Brown, with barred ♀ plumage remaining above and below the latter.

IMMATURE ♂: as ad ♀ but with a longer tail than ad of both sexes.

NESTLING–JUVENILE: unknown.

Distribution

Uncommon and patchily distributed in all or virtually all mountain ranges of New Guinea, including the central cordillera, the mountains of the Huon Penin and Vogelkop, as well as the Foya, Fakfak, and Kumawa Mts of IJ. Apparently not recorded (and presumably absent) from much of the central cordillera of central and eastern IJ and westernmost PNG (from Paniai/Wissel Lakes eastward to the Lordberg, PNG). Also not recorded from the Adelbert and N Coastal Ra of PNG. Ranges from 600 to 2250 m, but mostly 1100–1900 m.

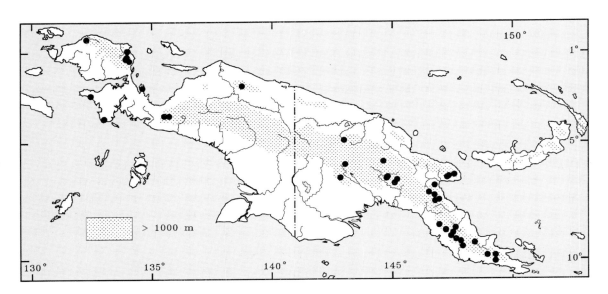

Systematics, nomenclature, subspecies, and measurements

SYSTEMATICS: earlier authorities (see Diamond 1986a) have considered *D. albertisi* and *D. bruijnii* a superspecies. We believe these two may have evolved beyond superspecies status. This is suggested by their non-complementary distributions, differences in habitat use, bill morphology, and vocalisations. Hybridisation: unrecorded.

SPECIES NOMENCLATURE AND ETYMOLOGY

Drepanornis albertisi (Sclater)
Drepanephorus albertisi Sclater, 1873 (June). *Nature* **8**, 125. *Proceedings of the Zoological Society of London* (June 17, 1873), p. 558, pl. 47—Mount Arfak [restricted to Hatam by Mayr 1941a]. Type specimen SMT 14067, C. 11199.

Etymology: *Drepanornis* = Gr. *drepanon*, a sickle; *ornis*, a bird; *albertisi* = after Luigi Maria d'Albertis, Italian naturalist and explorer, the first westerner to encounter and collect this sp., on Mt. Arfak in 1872. D'Albertis immediately realised it represented a new genus and sp. of bird of paradise. This sp. was also encountered in the same year and on the same mountain by A. B. Meyer (Iredale 1950).

SUBSPECIES, WEIGHTS, AND MEASUREMENTS

1. *D. a. albertisi* (Sclater, 1873).

Synonym: *Drepanornis geisleri* Meyer, 1893. *Abhandlungen und Berichte des Königl. Zoologischen und Anthropologisch-Ethnographischen Museums zu Dresden* **4** (3), 15. Sattelberg, Huon Peninsula.

Range: the non-cordilleran ranges of NG: Mts of Vogelkop and Huon Penin (and presumably also the Foya, Fakfak, and Kumawa Mts of IJ). This fragmentary and dispersed range is peculiar, in the extreme, and we suggest it may be an artefact. A future reviser may determine that *D. albertisi* is monotypic, with minor regional variation.

Diagnosis: tail darker and richer chestnut that that of *cervinicauda* (*cf.* Gilliard 1969). *Geisleri* is not sufficiently distinct to warrant subsp. status and Diamond (1972) has argued that for morphological reasons, *geisleri* should be subsumed into the nominate form. Although this is distributionally odd, it links all of the non-central-cordilleran forms into a single taxon.

Measurements: wing, ad ♂ ($n = 22$) 149–161 (155), imm ♂ ($n = 10$) 144–158 (151), ad ♀ ($n = 9$) 145–156 (150); tail, ad ♂ ($n = 20$) 128–141 (135), imm ♂ ($n = 10$) 120–137 (132), ad ♀ ($n = 7$) 125–133 (130); bill, ad ♀ ($n = 18$) 72–83 (78), imm ♂ ($n = 9$) 66–85 (77), ad ♀ ($n = 7$) 79–91 (84); tarsus, ad ♂ ($n = 21$) 33–36 (34), imm ♂ ($n = 10$) 33–37 (35), ad ♀ ($n = 9$) 33–37 (34). Weight: ad ♂ ($n = 5$) 105–125 (112), imm ♂ ($n = 2$) 105–116 (111), ad ♀ ($n = 2$) 105–138 (118).

2. *D. a. cervinicauda* Sclater, 1883. *Proceedings of the Zoological Society of London* p. 578. Vicinity of Port Moresby. Type specimen BMNH 1883.12.17.3.

Synonym: *Drepanornis a. inversa* Rothschild, 1936. *Mitteilungen aus dem Zoologischen Museum Berlin* **21**, 188. Mt. Kunupi, Weyland Mts. Type specimen AMNH 302362.

Range: central cordillera of NG, but unsampled or actually absent from the Nassau Ra/Snow Mts of IJ, eastward to the Lordberg and to Tari of W PNG.

Diagnosis: uppertail coverts and tail paler chestnut than in nominate form. Underparts of ♀ plumaged birds pale buff, barred strongly with paler brown than nominate. Smaller than nominate and tail proportionately shorter.

Measurements: wing, ad ♂ ($n = 29$) 145–159 (152), imm ♂ ($n = 1$) 150, ad ♀ ($n = 25$) 143–153 (147); tail, ad ♂ ($n = 25$) 118–130 (125), imm ♂ ($n = 1$) 129, ad ♀ ($n = 21$) 115–132 (122); bill, ad ♂ ($n = 25$) 69–80 (75), imm ♂ ($n = 1$) 77, ad ♀ ($n = 24$) 68–84 (78); tarsus, ad ♂ ($n = 28$) 32–36 (35), imm ♂ ($n = 1$) 34, ad ♀ ($n = 25$) 32–35 (34). Weight: ad ♂ ($n = 7$) 103–120 (112), imm ♂ ($n = 1$) 103, ad ♀ ($n = 7$) 92–128 (108).

Habitat and habits Table 4.1

Interior of lower and midmontane forests; less commonly in logged areas and at the forest-edge. ♀-plumaged birds have been observed in mixed-spp. foraging flocks but ad ♂♂ have not. Diamond (1987) sighted this sp. in five of nine mixed flocks he encountered in the range of the sp. Skulking and difficult to observe. Forages over epiphyte-covered branches of the lower canopy of tall trees (D. Bishop *in litt.*).

Diet and foraging Table 4.2

Primarily insectivorous, but also consumes a range of fruits in small quantities. Beehler and Pruett-Jones (1983) estimated the diet to be 94% arthropod and 6% fruit. Beehler (1987a) noted 89% of feeding records were of arthropods. The bird uses its very specialised bill to glean, prise, and probe bark, dead wood, limb surfaces, mosses, dead leafy debris, and knot holes for arthropod prey. Larger prey are held to a perch by a foot while torn apart with the bill (T. Pratt *in litt.*). Spends much of each day foraging in the middle and upper stages of the forest interior for arthropods (Beehler 1987a). Most foraging observations have been made of the ♂ because it is vocal and more easily located than the ♀. An early account noted birds foraging on dead trees and fallen trunks (in Rand and Gilliard 1967). Beccari (in Sharpe 1891–8) found only insects, predominantly ants, and the larvae of a moth in stomachs. The stomach of a ♂ collected by Gilliard (in Gilliard and LeCroy 1970) contained small beetles, three 'worms', two long orange insect casings, and many small hard insects. Stomachs of four individuals contained only arthropods, of the following groups: Orthoptera (4 families), Dermaptera, Coleoptera, and Lepidoptera (larvae only) (Schodde 1976). Beehler (1987a) found arthropod prey to include Araneida (2), Chilopoda (2, one 7.5 cm long), Orthoptera (1 Gryllacrididae, 1 Gryllidae, 1 Stenopelmatidae, 1 Tettigoniidae, 1 unidentified), Coleoptera (1 white 'grub' larva, 6 unidentified beetles), Lepidoptera (3 larvae, 1 a hairy caterpillar), and 5 unidentified arthropods. Gary Opit (*in litt.*) watched an ad ♂ feeding for 30 min in the middle storey of a *Castanopsis* forest by probing and picking into mosses on branches for arthropods. Of seven records of fruit feeding on Mt. Missim five were upon the capsule-fruited *Chisocheton* cf. *weinlandii*, one upon the capsular *Elmerrillia papuana* (= *tsiampaca*), and one a fig *Ficus* sp. (Beehler 1987a). At least four plant spp. have been recorded as used (see Appendix 4), but fruits appear to represent a small proportion of the ad diet. Goodfellow (1908) reported the sp. to frequent 'lower forest growth' and saw it foraging 'a short distance above the ground' Diamond (1985) reports the bird to forage mainly in the lower canopy (records from 3 m up to the canopy). Beehler (1987a) observed 45 insect-foraging events, all of individuals feeding 8–28 m above ground involving gleaning or probing into tree bark and knot holes of branches 1.5–15 cm diameter (mean 6.1 cm, $n = 20$). The use of the sicklebill is worth noting. In most cases it is simply used as forceps to pick up prey or to remove fruit from capsular husks but the entire bill length may be inserted into tree holes. In some instances the bill is opened widely and either the maxilla or mandible is used to probe into cavities for prey (T. Pratt *in litt.*; BB).

Vocalisations and other sounds Figure 9.52

There are three primary vocalisations: a single contact note, an advertisement song, and a display song. The contact note is a plaintive *wrenh?*—reminiscent of the call notes of other paradisaeinine spp. The advertisement song is a high, musical, powerfully-whistled series of notes, each note: *dyu dyu dyu dyu dyu dyu dyu* ..., increasing in speed and rising (sometimes falling) in pitch. Each song comprises an average of 18 notes ($n = 24$, range 13–27, SD 4) and lasts an average of 3.2 sec ($n = 24$, range 2.1–4.2, SD 0.5). The series (Fig. 9.52) can be heard from as far away as 450 m. The display song is quite varied, but is typically an elaboration of the advertisement song: higher, more insistent, faster, and terminating with a series of sibilant notes: *tish-tish-tish-tish-tish-*

Buff-tailed Sicklebill *Drepanornis albertisi* (Sclater, 1873)

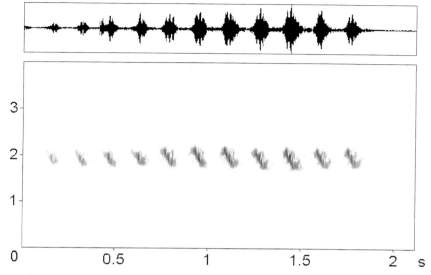

9.52 Sonogram from a recording of an ad ♂ Buff-tailed Sicklebill advertisement song at Mt. Missim, PNG on 30 Oct 1980 by B. M. Beehler, 1976.

tish-tish-tish (G. Opit *in litt.*), or *ki-ki-ki-ki-ki-ki-ki* (BB). The advertisement song is given in very early morning and late afternoon from a traditional song post high in the canopy or from a perch that stands above the canopy. It is also given regularly throughout the ♂'s range while he is foraging (least frequently during the middle of the day). The display song is typically given in the lower part of the forest, sometimes within 1 m of the ground. The contact note is given throughout the ♂'s habitat, whilst foraging. No vocalisations by the ♀ have yet been recorded. Gary Opit (*in litt.*) also recorded an occasional loud *Yapp* note, and a soft down-slurred rasp *ksp* is known (Diamond 1985). Vocalisation of the recently-discovered Fakfak Mts population is distinct from that from the Arfak Mts. Gibbs (1994) described the Arfak song as a Whimbrel-like *Numenius phaeopus* call of 3–4 sec duration, rising in volume and accelerating slightly. Fakfak birds, however, give a series of down-slurred whistles slightly decelerating and much less rapidly delivered—rather reminiscent of a slowed-down yaffle of a Green Woodpecker *Picus viridis* (Gibbs 1994). The begging of a juv was a 'soft barely audible piping call' (T. Pratt *in litt.*).

Mating system

A 28-month study in 200 ha of forest on Mt. Missim, PNG, found five ♂♂ singing from dispersed posts in the forest—what appeared to be exclusive year-round all-purpose territories. The sp. is exceptional in combining court-displaying polygyny with near-total insectivory. Adjacent ♂♂ were in auditory contact. The mean nearest neighbour distance between display sites of five ♂♂ was 450 m (range = 410–520 m) (Beehler and Pruett-Jones 1983), see Fig. 9.53. Male A (Fig. 9.53) was colour-ringed and his territory was *c.* 14 ha. His traditional display site, used for at least 4 successive years, consisted of a small area in old growth forest near the centre of his home range (all of which was in the forest interior). A radio-tagged ♀ was silent as she moved about a 43 ha range over the 8 days she was tracked.

Courtship behaviour

The solitary ad ♂ sings in the early morning from a high song post, descends to a sapling to display to visiting ♀♀, and also conducts display calling and display from thickets near the ground. Mating has not been observed.

Buff-tailed Sicklebill *Drepanornis albertisi* (Sclater, 1873)

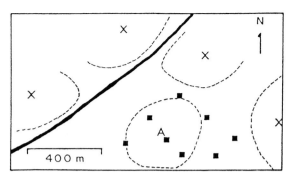

9.53 Display dispersion of ♂ Buff-tailed Sicklebills on Mt. Missim, PNG: dashed lines enclose territories of 5 adjacent ♂♂. Approximate centres of territories, and known or suspected display sites, are indicated by an × or A. Territory A was of a colour-ringed ♂ studied in detail. A permanent stream is indicated by the bold black line. From Beehler and Pruett-Jones (1983), © Springer-Verlag with kind permission.

Opit (1975b and *in litt.*) heard a ♂ calling from the canopy of tall trees and then saw him fly down to perch upon a slightly-sloping branch *c.* 5 cm in dia, some 15–18 m above ground, and below the forest canopy amid foliage. This perch and all branches within the bird's reach had been cleared of leaves. The ♂ whistled *to to to to to to to to* followed by a harsh but softer note and began to display by slightly raising his wings, calling loudly, and dropping forward to hang upside-down. The flank fan plumes were fully erected, then the pectoral fan plumes, and with a convulsive jerk both were made to form a complete perfect feather disk around the body. The upper pectoral fan plumes were held behind the head and then sloped down across to the front of the body to become contiguous with the flank fan plumes (Fig. 9.54). This posture was held for *c.* 30 sec as the bird was backlit, the sunlight showing through the feather disk but for a dark patch to each side of the head.

The displaying ♂ then resumed normal perching to pull at and remove leaves near his display perch. He repeated the above inverted display and followed it with foliage removal several times, always calling immediately before adopting the inverted posture. This display bout lasted *c.* 10 min.

Beehler (1987a) found all courtship occurred at a fixed site consisting of at least three perches or perching areas. An advertisement song perch attended daily by an ad ♂ over 6 months of the year was in dead branches in the open crown of a 40-m-tall emergent tree (*Toona sureni*, Meliaceae). A display perch was situated in a sapling just below the song perch. The ♂ also displayed at a site near the ground in a thicket adjacent to a terrestrial court belonging to a Magnificent Bird of Paradise, some 45 m from this advertisement perch. He once sang and displayed for > 8 min in another low thicket about 125 m from his advertisement perch. Based on numerous observations of parts of one ad ♂ display and, in particular, longer displays recorded over 3 days in Sept–Oct, Beehler (1987a) compiled a composite description of display activities: the ♂ flew to his open-canopy singing perch at 0534 hr to give 69 advertisement songs during 30 min. At 0603 he gave the first display song and started to intersperse this between the advertisement singing. At 0607 he dropped to his display sapling in the leafy middle story and three ♀-plumaged birds moved warily close to hop about surrounding branches. The ♂ chased the presumed ♀♀ out of sight and returned to the sapling area to perch 12 m high to continue alternating advertisement with display singing. Suddenly he leaned backward to about 50° from upright, flared his flank plumes, opened his bill, and held his pose for about 25 sec. Returning to upright he continued singing for several min. He then dropped back until entirely inverted and flared all his plumes to hold his pose (Fig. 9.54) while vibrating his body for 30 sec during which he gave one display song. He then turned upright and began stripping leaves from branches closest to his perch. At 0620 he returned to his canopy song perch to sing from it and adjacent perches until 0645, when he dropped into the low display thicket to perch within 50 cm of the ground to repeat display song for 10 min. He then left to forage and sing in the forest

Buff-tailed Sicklebill *Drepanornis albertisi* (Sclater, 1873)

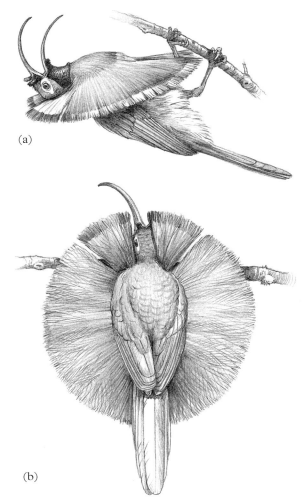

9.54 An inverted courtship display posture of ad ♂ Buff-tailed Sicklebill seen in profile (a) and as viewed from behind the bird (b). After Beehler (1987*a*), with kind permission. See text.

canopy before returning to his thicket site to sing from 0720 to 0750. Mary LeCroy (*in litt.*) saw an ad ♂ perform an odd display for *c*. 15 sec at the edge of a parotia court she had under observation at Heroana, at *c*. 1525 m, Eastern Highlands, PNG. While clinging head uppermost to a vertical sapling trunk about 1 m above ground, with his body in line with the sapling, he 'reared backward to *c*. 20° from vertical, [he] spread [his]' plumes around until they formed a horizontal vibrating half-circle in front of the body, with the iridescent pectoral plumes prominent at the sides'. The bird was silent, apparently alone, and used only his legs to support his odd posture in what is most reminiscent of a sickletail (*Cicinnurus* spp.) display.

Breeding Table 6.5

Little known. The observations of a calling and displaying ♂ on Mt. Missim (BB) suggests mating occurs mostly during and either side of Oct–Nov in central PNG. A begging juv was seen following its mother on 28 Nov 1974 (T. Pratt *in litt.*).

NEST SITE: between a horizontal bough in a fork of a thin branch.

NEST: a thin, rather flat structure with a slight depression of about 25 mm deep; network of wire rootlets stretched across a tree fork provided a foundation for the nest proper which consisted of wiry grasses of a light reddish-brown colour, the platform being of black wiry roots (Ramsay in Sharpe 1895). It is possible the 'wiry grasses' were in fact orchid stems or other plant matter, as grasses are an unlikely nest material in view of what is known of other spp.

EGGS: clutch is one or two. Pale dull cream in ground colour, with a reddish tinge, spotted all over with oblong dashes of reddish brown and light purplish grey that are denser on the larger end (Plate 13); 3 eggs average 31.4 (30.6–32.1) × 22.9 (22.0–24.1) mm.

INCUBATION, NESTLING CARE, AND DEVELOPMENT: unknown.

Annual cycle Table 6.5
DISPLAY: One ad ♂ on Mt. Missim was vocal at his traditional canopy song perch for the 7-month period May–Nov each year, and virtually silent the intervening months (Fig. 9.55). BREEDING: at least Sept or Dec–March. EGG-LAYING: C. Keysser collected an undated two-egg clutch on the Huon Penin and, according to Gilliard (1969), he collected bird of paradise specimens there during Sept and Dec–Mar. The eggs may have been collected at another time. MOULT: 26 of 73 specimens were moulting, involving all months except July and Oct–Nov and most moult occurring during Dec–Mar (Appendix 3). An imm ♂ collected on the Vogelkop in early Oct had his forehead, sides of face, throat, and pectoral plume feathers in pin (Gyldenstolpe 1955b: 301).

Status and conservation Table 8.1
Nowhere abundant, and may be absent from apparently suitable areas of habitat (especially eastern IJ and westernmost PNG).

Knowledge lacking and research priorities
The sp. was not encountered by J. Diamond and D. Bishop (*in litt.*) in the Star Mts, N slope of the Snow Mts, the Van Rees, and Cyclops Mts. The apparent absence of this sp. in central and eastern IJ and westernmost PNG needs confirmation. Data on nesting behaviour and the diet fed to nestlings would

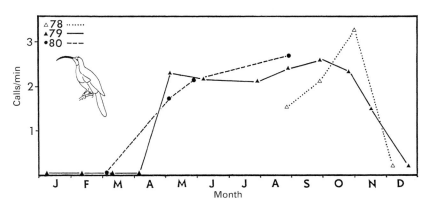

9.55 Seasonality of morning song of a ♂ Buff-tailed Sicklebill over 29 months (1978–80) on Mt. Missim, PNG, based on a mean of samples (each of 5 min) recorded 30 min after first light, from near the ♂'s display post. From Beehler (1987a), with kind permission.

be valuable. Intraspecific variation requires further study.

Aviculture
D. a. cervinicauda was apparently held in the Brook collection in Scotland, and one of the population previously known as *D. a. geisleri* was collected for the London Zoo by F. Shaw Mayer (Crandall 1935). Goodfellow (1908) reported a bird taken to the UK where it became very tame.

Other
At 1620 hr on 25 Aug on Mt. Missim, PNG an ad ♂ was seen rain-bathing by sitting 10 m up in a sapling and preening for 8 min in a heavy downpour (BB). Solitary ♂♂ were seen several times bathing in small pools (T. Pratt *in litt.*). The displaying ad ♂ observed by Opit described above was first discovered by Thane Pratt, who subsequently found (in 1974) the remains of the bird, consisting only of the bill and feathers, apparently taken by a goshawk *Accipiter* sp. Within days this dead ad ♂'s site was occupied by a singing ♀-plumaged ♂ and in 1977 the ♂ resident there was in ad plumage (T. Pratt *in litt.*). A. E. Pratt (1906) found a bird inedible (see end of Chapter 4). The people of the Arfak Mts, IJ are said to believe that the monotreme Long-beaked Echidna *Zaglossus bruijnii* hatches from an egg of this bird of paradise (but see also Black Sicklebill).

Pale-billed Sicklebill *Drepanornis bruijnii* Oustalet, 1880

PLATE 9; FIGURES 9.56–7

Other names: White-billed Bird of Paradise, White-billed Sicklebill Bird of Paradise, Lowland Sicklebill, Bruijn's Bird of Paradise, *Epimachus bruijnii*, *Deatai* of people of Biri within the Meervlakte area (Diamond 1981), *Yokwa* at Jayapura and *Kurawai* on the upper Tor R., IJ (Ripley 1964).

A peculiar, dull-plumaged canopy-dweller of lowland rainforests of northwestern NG—perhaps the most cryptic and difficult-to-observe bird of paradise. Its only distinctive features are the pale sickle-shaped bill and prominent russet tail. But for the weird and distinctively-whistled vocalisations, the sp. would be invariably overlooked. The only lowland sicklebill. Monotypic.

Description
ADULT ♂: 35 cm, bill on average longer than that of ad ♀. Crown and nape variable Fuscous (21) with lores to up and over eye, where elongate feathers form erectile 'horns', iridescent rich blue-purple (172A) and/or red-purple (1) in suitable light. Chin, throat, and ear coverts darker, velvety, fuscous, almost blackish, with deep Leaf Green (146) iridescence. Extensive bare facial skin dull lead grey with slight purple hue. A small roughly-circular patch of fine scale-like feathers below the eye, over lower mandible base, strongly-iridescent deep blue (70) to Purple (1) in suitable light. Mantle and upper back Olive-Brown (28) becoming Cinnamon-Brown (33) on lower back, rump, and uppertail coverts. Upperwing as adjacent mantle and back with paler leading edges to coverts and tertials and the primaries darker, more Fuscous. Uppertail Cinnamon-Brown. Upper breast dark Olive-Brown with extensive tips of longer feathers strongly-iridescent Olive-Green (46) including elongate feathers covering bases of pectoral plumes. Pectoral plumes Dark Greyish Brown (20), the shorter row being broadly-tipped, strong, iridescent coppery red and these overlaying a row of longer ones finely-tipped, strong, iridescent Purple and/or deep blue in suitable light. Grey feathers of sides of lower breast, usually

concealed beneath wings, decorated with dull, Leaf Green, iridescent, broad tipping and beyond these a line of jet black feathers finely tipped with strong, iridescent, Purple and/or deep blue in suitable light. Underwing: coverts greyish brown to Tawny Olive (223D), remainder greyish brown with fine buff trailing edges to basal primaries. Remaining breast and belly dark warm grey with a dark lavender wash, the thighs, vent, and undertail coverts paler and more greyish brown. Ventral surface of tail Ground Cinnamon (239) with off-white central feather shafts. Bill ivory whitish, iris very dark brown, legs and feet purplish brown.

ADULT ♀: 34 cm, head paler Fuscous than ad ♂ and bare facial skin similar but slightly duller. A finely-brownish-flecked, off-whitish moustachial stripe, chin, and throat are separated by an obvious broad, fuscous, malar stripe. All upperparts like ad ♂ but slightly paler, more buff, less brown. Underparts different: chin, throat, and upper breast dirty Buff (124) to Warm Buff (118) on lower breast to pale and dilute Cinnamon (39) on belly, vent, and undertail coverts entirely and regularly barred, except on chin and throat where finely flecked, blackish brown. Underwing: coverts Tawny Olive barred dark brown, remainder, including ventral surface of tail, as ad ♂.

SUBADULT ♂: variable, some resembling ad ♀ with a few feathers of ad ♂ plumage intruding, to as ad ♂ with few feathers of ♀ plumage remaining. The dark throat feathers and pectoral plumes of ad ♂ apparently attained first with partial warm grey of underparts only, while much barring of the imm (♀) plumage retained. More warm grey dorsal feathering acquired as barring diminishes with subsequent moults.

IMMATURE ♂: resembles the ad ♀ but tail longer than ad of both sexes.

NESTLING–JUVENILE: unknown.

Distribution
Restricted to the lowlands of northern IJ and northwesternmost PNG. Inhabits the N watershed of NG from the E side of Geelvink Bay eastward through the Meervlakte (Lakeplain) and presumably the Idenburg (Taritatu) Basin and N coastal lowlands to the Vanimo area of PNG and northwestern reaches of the Sepik R. drainage (Utai). Eastern limits of the range require determination. Ranges from sea level to 180 m.

Systematics, nomenclature, weights, and measurements
SYSTEMATICS: postulated to comprise a superspecies with the Buff-tailed Sicklebill (Diamond 1972: 309), although there are sufficient morphological differences between them that this requires further assessment; in addition, the relative geographical and elevational distribution of the two weigh against superspecific status. Nonetheless, these two spp. are unambiguously sister forms, and the *Drepanornis* lineage is sufficiently divergent from that of *Epimachus* to warrant generic status (*contra* Diamond 1972). Hybridisation: not recorded.

SPECIES NOMENCLATURE AND ETYMOLOGY
Drepanornis bruijnii Oustalet
Drepanornis Bruijnii Oustalet, 1880. *Annales des Sciences Naturelles*, ser. 6, **9**, art. 5, p. 1 also cited

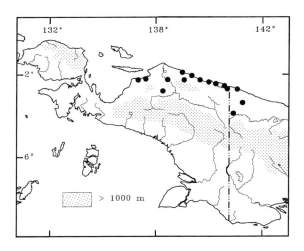

in: 1880 *Bulletin Hebdomadaire de L'Associátion Scientifique des France*, sér. 2, I, no. 11, p. 172. Coast of Geelvink Bay between 136° 30′ and 137° of longitude. Type specimen MHN 12 (No. 50), M and MHN 1033a/154, F.

Etymology: *bruijnii* = after the Dutch plume trade merchant Anton August Bruijn (Bruijn is Dutch for 'Brown' and is pronounced 'braon').

Exploration: collected by L. Laglaize hunting for the plume trader A. A. Bruijn in 1880, 4 years after Bruijn had first become aware of the sp.

WEIGHTS AND MEASUREMENTS
Wing, ad ♂ (*n* = 28) 153–165 (160), imm ♂ (*n* = 12) 146–164 (157), ad ♀ (*n* = 14) 148–162 (155); tail, ad ♂ (*n* = 26) 103–116 (110), imm ♂ (*n* = 12) 109–128 (115), ad ♀ (*n* = 14) 107–116 (111); bill, ad ♂ (*n* = 26) 74–83 (79), imm ♂ (*n* = 12) 62–81 (75), ad ♀ (*n* = 12) 68–79 (75); tarsus, ad ♂ (*n* = 26) 31–35 (33), imm ♂ (*n* = 12) 30–34 (32), ad ♀ (*n* = 14) 30–34 (32). Weight: ad ♂ (*n* = 4) 160–164 (161), ad ♀ (*n* = 3) 144–149 (146). Mean central tail feather pair length average 1 (♀) and 3 (♂) mm shorter than the rest.

Habitat and habits　　　　　　Table 4.1
Interior of original and disturbed lowland rainforests. Always inconspicuous and difficult to observe. Reputed to have a predilection for lowland riverine forest, and can be found within a km or two of the coast. Also occurs in original and selectively-logged forest in limestone hills, where more abundant than in a nearby site in lowland alluvial forest (Beehler and Beehler 1986). Ripley (1964) specifically noted birds in 'very tall *Agathis* trees.' Spends most of its time in the canopy, but descends to the substage to join mixed-spp. foraging flocks or to display. Both sexes look very peculiar from some angles because of the large facial apteria and the narrow strip of dark, short, crown feathering that gives the appearance of a 'mohawk haircut'.

Diet and foraging　　　　　　Table 4.2
Forages for fruit and arthropods in the forest canopy and both ad ♂♂ and ♀-plumaged birds forage in the forest substage when in the company of a mixed-spp. flock. Apparently more highly frugivorous than the Buff-tailed Sicklebill; its heavier, broader bill accords with this. In eight observations of arthropod-foraging (Beehler and Beehler 1986; Whitney 1987) birds clambered about on canopy limbs peering under branches, poking the bill into cracks, knot holes, lichen, and holes in dead wood, searching dead leaves, and inspecting vertical or sloping bark or limb surfaces. Beehler and Beehler (1986) observed birds at Puwani and Krissa camps, W Sepik Province, PNG, foraging on the following food plants: a canopy tree with a large dehiscent woody capsule (*Sloanea* sp.), an understorey palm (*Cyrtostachys* sp.), and an unidentified fruit with a single smooth spheroid seed (33 of which were regurgitated by an ad ♂ that was loafing in a large *Sloanea* tree). Diamond (1981) observed a ♀-plumaged bird *c.* 5 m above the ground, foraging as a member of a mixed-spp. flock in the Meervlakte of IJ. The flock included Variable and Rusty Pitohuis and the Rufous Babbler. Beehler and Beehler (1986) observed this sicklebill in mixed flocks, usually composed of several other spp. of birds of paradise (Jobi Manucode, Magnificent Riflebird, Twelve-wired, King, and Lesser Birds of Paradise), the Spangled Drongo, and the same pitohuis and babblers as were in Diamond's flock.

Vocalisations and　　　　Figure 9.56
other sounds
The ♂ gives an advertisement song, a display bill rattle, and minor contact notes; the ♀ is not known to give vocalisations. The advertisement song is sufficiently distinct that it is not readily recognisable as that of a sicklebill; components of this song are reminiscent of songs of the Lesser Bird (Whitney 1987) and the western population of the Magnificent Riflebird (Ripley 1964). It is a string of hoarse or hollow, whistled, musical notes, sometimes preceded or followed by one or several low musical gurgles or coughs, for example *wik-kew KWEER KWEER kwer? kor kor kor*, and it

varies considerably within and between individuals (Beehler and Beehler 1986). It is moderately high-pitched, carries a considerable distance, and has a plaintive tone. The main body of the song is a series of 4–6 notes, each of which either rises or falls in pitch; to this main segment is appended soft or gurgled notes, usually at a lower pitch. Length averaged 5.2 sec (n = 27, range = 3.7–7 sec.); median number of call notes was 8 (n = 27, range = 5–10 sec). Beehler and Beehler (1986) reported that on four clear mornings the first songs of a ♂ were given between 0631 and 0636 hr. Calling frequency was highest in the first half-hour of the day's calling, when the ♂ gave, on average, 1.2 songs per min (n = 81). After 0900 the rate of singing dropped considerably (but did not end entirely), and picked up again late in the afternoon. Ripley (1964) recorded the advertisement song of an ad ♂ in July as 'a loud, not unmusical series of descending whistles reminiscent of the [western NG Magnificent] rifle bird. The first note starts at the tone and pitch of the second note of the rifle bird's call. From this start came a series of descending whistles, repeated over again, several at a time. In between these extremely loud calls, this ♂ as it moved about in the tree made several gruff, churring notes rather like a typical *Paradisaea*.' Whitney (1987) described the call of birds near Vanimo, PNG as differing from those heard by Ripley in Irian: 'Rather, there was usually a pause of one to three min between songs, which consisted of one loud, musical descending series of four to ten whistles that sometimes ended with several lower, harsher whistles or "yuree" phrases. A typical song of about seven descending whistles lasts about five sec.' See Fig. 9.56. Whitney found ♂♂ had one or two favoured singing trees from which they called every 1–3 min, mostly during mornings before 0930 hr but also between 1630 and 1730 hr. From one spot he could hear four ♂♂, dispersed linearly along a road at about 200-m intervals, singing. Thus the ♂♂ were within auditory contact at their favoured singing sites. Their response to play-back of taped calls and imitations of calls left Whitney with the clear impression ♂♂ are territorial. The call note, given on occasion while foraging, is a quiet interrogative note *whehn?* very similar to that given by a number

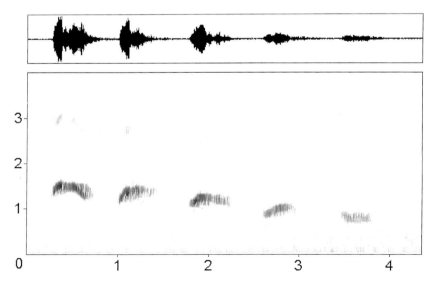

9.56 Sonogram from a recording of an ad ♂ Pale-billed Sicklebill advertisement song—a series of whistled hollow notes delivered as a somewhat gibbon-like (*Hylobates* spp.) sound—on 21 July 1984 at Puwani R., south of Vanimo, PNG by B. M. Beehler.

of other typical birds of paradise. In the height of display the ♂ rattles his bill rapidly but gives no vocalisation (Beehler and Beehler 1986).

Mating system
Typical court display polygyny, with ♂♂ exclusively occupying (and defending) a large display/foraging territory, not unlike Buff-tailed Sicklebill (but apparently with some sharing of foraging range as evidenced by the observation of a pair of ad ♂♂ in a mixed-spp. flock). ♂ has an activity pattern similar to that of Buff-tailed Sicklebill as well as to that of Superb Bird (Beehler and Pruett-Jones 1983). ♂ patrols his territory daily, singing at different points and counter-singing and counter-displaying to neighbouring rival ♂♂ (Beehler and Beehler 1986). This last is taken as evidence of territorial defence.

Courtship behaviour
♂ displays to a visiting ♀ in the lower middle story of the forest interior (c. 7–10 m above ground) while perched on a small horizontal branch. Display is preceded by advertisement song. On five occasions Beehler and Beehler (1986) observed a single stereotyped posturing display in which the ♂ sits erect on the perch, his short upper pectoral plumes erected vertically, the longer lower feathers erected into a wide skirt, and the tail fanned (Fig. 9.57). The ♂ holds this position with little movement for as long as 10 sec, and rattles the bill conspicuously. Mating has not been observed, and we are certain our description is not a complete account of the ♂'s courtship displays.

Breeding   Table 6.5
No details of nesting are available.

Annual cycle   Table 6.5
DISPLAY: Aug (Beehler and Beehler 1986). BREEDING: 'testes of ♂♂ in some cases greatly enlarged. Therefore, probably, the nesting season will be Nov.' [for Holtekong, IJ] (Hartert 1930). MOULT: 13 of 30 specimens were moulting, involving only the months Mar,

9.57 A display posture, probably an initial one, of ad ♂ Pale-billed Sicklebill during which the bill is rattled. After Beehler and Beehler (1986), with kind permission. See text.

Aug, and Oct–Nov with most occurring in Aug and Oct (Appendix 3). A series of birds obtained by Ernst Mayr at Hol, Humboldt Bay in Aug and Oct were 'mostly still in moult' (Hartert 1930). Of three ♂♂ and three ♀♀ collected by Ripley (1964) during late July–early Aug 1960, one ♀ was moulting her primary wing coverts on 1 Aug.

Status and conservation   Table 8.1
Within its rather restricted geographical distribution the sp. may be widespread and is locally quite common, even in selectively-logged forest (Whitney 1987). At one roadside locality, Whitney could hear four ♂♂ singing from a single vantage point. Beehler and Beehler (1986) documented that the range of a focal ♂ near the Puwani R. over 7 days was c. 15 ha, and no other vocal ♂♂ were located. At the Krissa camp, Beehler and Beehler found two neighbouring vocal ♂♂.

Knowledge lacking and research priorities
Details of displays and breeding biology unknown. Diamond (1981) posed the questions: Is there an altitudinal gap between *E. bruijnii* and *E. albertisi*? Is *E. bruijnii* widespread in the Meervlakte? How widespread is the sp. distribution in the Sepik region?

Aviculture
Undocumented.

Genus *Cicinnurus* Vieillot, 1816

Cicinnurus Vieillot, 1816. *Analyse d'une Nouvelle Ornithologie Élémentaire* p. 35 Type, by monotypy, 'Manucode, Buffon' = *Paradisea regia* Linné.
  Synomyns: *Diphyllodes* Lesson, 1834 *Histoire Naturelle des Oiseaux dev Paradis et des Epimaques*.

The Sickletails are **three small species** with a recurved central pair of tail wires. Two spp. were formerly placed in the genus *Diphyllodes*. Two spp. are found extensively over mainland NG at differing altitudinal zones and one (*C. respublica*) is confined to the Western Papuan Is. According to Bock (1963) the **skull** is predominantly *Paradisaea*-like but shows some *Astrapia*-like characters and others that suggest affinity with the *Parotia, Pteridophora, Lophorina* group. Bock observed that part of the gap in skull characters between the cnemophilines and paradisaeines is bridged by characters in *Cicinnurus*. This notwithstanding, a now considerable body of morphological and life history evidence indicates the sickletails are most closely related to the *Paradisaea* cluster. Intrageneric **hybridisation** is relatively common between *C. regius* and *C. magnificus* and intergeneric crosses are known with *Lophorina* and *Paradisaea*.

Bill straight and slightly longer than head. Culmen sharply keeled to knife-edge in *C. regius* and *C. respublica* (as in *Semioptera*). **Tail** not graduated, central pair of ad ♂ tail feathers highly modified into slightly-to tightly-coiled 'wires' vaned only on their outer sides and highly iridescent. **Wings** not markedly short or rounded, secondaries in ad ♂ *C. regius* highly modified and as long as primaries. Indentation on inner vane of P2 of ad ♂ *C. regius* but absent on ♀-plumaged birds and in all *C. magnificus* and *C. respublica*. **Tarsus** length c. 27% of wing length. Tarsus scuted distally, fading to smooth proximally.

Central pair of narrow to wire-like rectrices in ad ♂♂ show progression from open curled (*C. magnificus*) to tightly-curled (*C. respublica*) to highly-modified, tightly-curled disks of iridescent outer feather vanes (*C. regius*). **Facial apterium** slightly developed in both sexes of *C. magnificus* and highly developed in *C. respublica*. All three spp. have bright blue legs and feet.

All spp. extremely sexually dimorphic in **plumage**, with ad ♂♂ exhibiting some intensely-brilliant iridescent dorsal plumage, including erectile capes, and extensive to much reduced breast-shield and modified flank feathers (*C. regius* only). ♀♀ are drab and cryptically feathered (Plate 10).

Nest of *C. magnificus* is typical of Paradisaeinae and built in tree branches but, uniquely in the family, the nest of *C. regius* is placed in a tree crevice (nest of *C. respublica* unknown). **Eggs** are typical of the polygynous spp. of the subfamily and are most like those of *Ptiloris* and *Paradisaea* (Plate 13).

Magnificent Bird of Paradise *Cicinnurus magnificus* (Pennant, 1781)

PLATE 10; FIGURES 1.4, 5.1, 9.58–60

Other names: *Sabelo* on Salawati I. and *Arung-arung* on Misool I. (Gould and Sharpe 1875–88). *Firak-kait* or *Firak* on upper Fly R. (trade skin), *Kellelo* at Dinawa, NE of Yule I., S PNG (Pratt 1906). A Malay name is *Balarottan* (Mayr and Meyer de Schauensee 1939a). *Kombok* of the Kubor Mts (Mayr and Gilliard 1954). *Timonsen* of the Telefomin region (Gilliard and LeCroy 1961). *Iyamanu* (ad ♂), *Tsido* (♀ plumage) at Lake Kutubu (Schodde and Hitchcock 1968). *Tarotaro* (♂) of the Fore; *yaro* or *Daro* of the Gimi people, Crater Mt., PNG (Diamond 1972; D. Gillison *in litt.*) and *mobudali* of the Daribi peoples of the E Highlands, PNG (Diamond 1972). *Kabay asdal* of the Kalam people (Majnep and Bulmer 1977). *Pengaluo* of the Maring people, Jimi Valley, PNG (Healey 1986). *Pamanane* of the Fair language at Bomakia and *Hanggo* of the Kombai language at Uni, IJ (Rumbiak 1994). *Sau* of the Wanuma people, Adelbert Ra (T. Pratt *in litt.*). *Pena* of Papuans and *Blarotang* of Malays near Sorong (Mayr specimen label), *Gasava* of Waria Valley and *Biengmai* of Jimi R., PNG (W. Cooper *in litt.*), *Die Yon* in the Tamrau Mts, IJ (specimen label).

A small and, unless calling, easily-overlooked sp. of hill and lower montane forest interior. ♂ appears compact and oddly patterned, as the colours of the bright display plumes are generally not clearly discernible in the forest interior, although tail wires can be obvious. ♀ is drab-coloured and patterned typical of the core paradisaeinine ♀♀ (brown above, finely barred below). More often heard than seen. Polytypic; 3 subspp.

Description

ADULT ♂: 19 (26) cm, plumage complex: upper head and nape feathers short, stubbly, and variably brown, from Olive-Brown (28) on central crown, where darker feather tipping gives scalloped appearance, to Raw Sienna (136) to sides and on ear coverts. A small spot of lore feathering immediately in front of eye intensely-iridescent dark green (146) and a semicircle of feathers above the eye jet black. Grossly-elongate, square-ended hind neck feathers form a discrete Sulphur Yellow (157) nape cape with intense iridescence of white light giving spun glass appearance. Several elongate feathers to sides of this yellow cape variably Olive-Brown to Raw Sienna tipped blackish. Back feathers form a semicircle, emerging from beneath the yellow cape, of dark Carmine (8) bordered jet black with spun glass intense iridescence. Lower back and rump variably Raw Sienna and Dark Brownish Olive (129) with spun glass iridescence on former only. Uppertail coverts Dark Brownish Olive washed Raw Sienna with dull dark green (146) iridescence in some lights. Upperwing: lesser coverts Olive-Brown washed Raw Sienna, greater coverts, alula, and flight feathers Dark Brownish Olive with fine rich Buff (24) leading edges to all but outermost primaries and much broader such edges to secondaries. Tertials modified into extensive and frill-edged, rich Buff washed Orange Yellow (18), feathers with spun glass, white-light, iridescence. Dorsal surface of tail Dark Brownish Olive but central pair modified into fine 'sickles' which cross near the normal tail tip and have narrow vanes on their outer edge only, highly-iridescent, deep, metallic, green-blue above and Olive-Brown below. Elongate and plush chin and upper throat feathering brownish black with Dark Green (162A) iridescent sheen. Smooth lower throat to lower breast feathers form an extensive breast-shield of highly-glossy Dark Green, the elongate lower feathers of which extend onto the belly. A line of broadly-elongated, scale-like feathers extending along the ventral mid-line, from central throat to upper breast intensely-iridescent Cyan (164) interspersed with dark green barring. Feather tips of the lower and outer border of the breast-shield are also narrowly-tipped iridescent Cyan to Cobalt

(168) blue in some lights. Entire breast-shield may in some lights produce rich Purple (1) and/or Magenta (2) iridescence in some light, and in other may appear jet black. Belly to undertail coverts blackish brown (19) with slightest wash of deep violet purple (172) in certain lights. Underwing: coverts Pale Pinkish Buff (121D) tipped Olive-Brown, flight feathers Olive-Brown with broad Pale Pinkish Buff trailing edges except outermost primary. Ventral surface of tail dark olive brown (119A). Bill chalky pale grey blue, iris dark brown, legs and feet deep blue, mouth very pale green.

ADULT ♀: 19 cm, tail considerably longer than that of ad ♂ (excluding central pair of rectrices). Upper head Olive-Brown strongly washed Raw Sienna, the area directly above the eye finely spotted and barred grey white, and central crown feathers with darker tips giving scalloped appearance. Mantle Brownish Olive (29) and back to uppertail coverts this colour washed Raw Sienna. Upperwing Olive-Brown with golden Buff, fine outer edges to greater coverts and innermost primaries, and broader ones on the secondaries and tertials. Dorsal surface of tail Olive-Brown, strongly-washed Raw Sienna. Chin Olive-Brown, malar and throat dirty white, flecked Dark Brownish Olive, and remaining underparts pale Buff (124) dilutely washed Cinnamon (39), more strongly on flanks and undertail coverts, uniformly barred Dark Brownish Olive throughout. Underwing: coverts Tawny Olive (223D) barred olive brown, flight feathers Olive-Brown with broad Tawny Olive trailing edges to all but P9 and P10. Bare parts as ad ♂ but bill slightly darker/duller.

SUBADULT ♂: as ad ♀ with few feathers of ad ♂ plumage intruding, initially about head, to as ad ♂ with few feathers of ♀ plumage remaining. A wild-caught bird in ♀ plumage in Aug 1969 was first noted to begin acquiring ad ♂ plumage on 20 Sept 1975, at an age of at least 6 yr (Aruah and Yaga 1992). Another individual that hatched in an aviary on 19 June was, however, thought to be showing signs of attaining central tail wires of the ad ♂ plumage on 21 Aug, 3 yr later. Initial 'wires' of the central tail feather pair can be spatulate-tipped, as in *Paradisaea* spp.

IMMATURE ♂: as ad ♀ but bill blackish brown and older birds with orange on secondaries and wing coverts. Younger birds ($c. < 2$ yr) have an orange mouth. Older ♀ plumaged birds develop narrow, pointed, and increasingly longer central rectrices prior to acquiring the wire-like 'sickles' of adults.

NESTLING–JUVENILE: hatchling naked with bright pink skin and yellow gape, the skin darkening on the second day until dark bluish-slate. Mouth colour orange.

OTHER: a partial albino ad ♂ specimen (11289, III, 174) in the Dresden Museum is white throughout except for mostly iridescent green underparts (not on chin and throat), a few brownish back and wing feathers, and one or two mustard yellow tertials. A similar specimen is in the AMNH together with at least one showing scattered white feathers through its plumage (Frith 1988).

Distribution

Hill and lower montane forest throughout mainland NG, plus Yapen I. and Salawati I., IJ. Mees (1965) has cogently argued that the sp. does not occur on Misool I., IJ, in spite of early reports. Ranges from near sea level (where hilly) to as high as 1780 m (but mainly up to about 1400 m).

Systematics, nomenclature, subspecies, weights, and measurements

SYSTEMATICS: a distinctive sp., which together with Wilson's Bird constitutes the subgenus *Diphyllodes*. We follow Diamond (1972) in combining this into *Cicinnurus*. While four subspp. have been long and widely accepted we consider *intermedius* invalid (see below). Hybridisation: known with the King and the Lesser Bird. See Plate 15 and Appendix 1.

Magnificent Bird of Paradise *Cicinnurus magnificus* (Pennant, 1781)

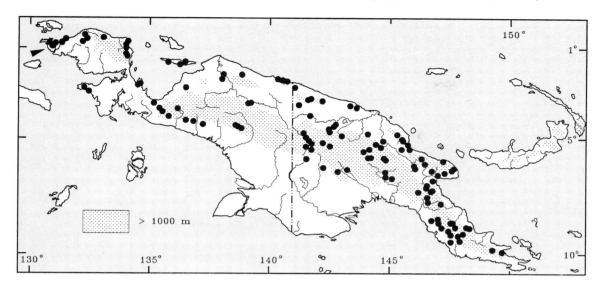

SPECIES NOMENCLATURE AND ETYMOLOGY

Cicinnurus magnificus (Pennant)
Paradisea Magnifica Pennant, 1781. *Specimen Faunulae Indicae*, in Forster's *Zoologia Indica Selecta*, p. 40 (based on Daubenton 1774, *Planches Enluminées* pl. 631 drawn by F. N. Mertinet). New Guinea, restricted to Arfak Mts (Mayr 1941a).

Etymology: *Cicinnurus* = Gr. *kikinnos*, a curled lock of hair; *-ouros*, tailed; *magnificus* = L. grand, splendid

SUBSPECIES, WEIGHTS, AND MEASUREMENTS

1. *C. m. magnificus* (Pennant, 1781)
Synonym: *Diphyllodes rothschildi* Ogilvie-Grant, 1915. *Ibis, Jubilee Supplement* **2**, 24. Salawati I. Type specimen BMNH 1881.5.1.1737.

Range: westernmost IJ, including Salawati I., Vogelkop, Wandammen, and Onin Penin.

Measurements: wing, ad ♂ ($n = 26$) 111–121 (116), imm ♂ ($n = 6$) 114–124 (117), ad ♀ ($n = 16$) 108–117 (113); tail, ad ♂ ($n = 25$) 34–43 (39), imm ♂ ($n = 6$) 60–67 (64), ad ♀ ($n = 14$) 57–69 (63); tail centrals, ad ♂ ($n = 21$) 213–319 (269); bill, ad ♂ ($n = 25$) 26–31 (29), imm ♂ ($n = 6$) 29–30 (29), ad ♀ ($n = 16$) 28–33 (29); tarsus, ad ♂ ($n = 25$) 29–34 (32), imm ♂ ($n = 6$) 30–34 (31), ad ♀ ($n = 16$) 28–33 (30). Weight: ad ♂ ($n = 13$) 85–110 (97), imm ♂ ($n = 2$) 82–94 (88), ad ♀ ($n = 7$) 73–88 (79).

2. *C. m. chrysopterus* (Elliot, 1873). *Monograph of the Birds of Paradise*. p. 13. Jobi I. (= Yapen I., IJ). Type specimen BMNH 1881.5.1.1733.

Synonym: *C. m. intermedius* (Hartert, 1930) *Novitates Zoologicae* **36**, 36. Snow Mts (= Upper Setakwa R., IJ). Type specimen AMNH 678407.

Range: western and central NG, including the central cordillera, from the Weyland Mts E to western PNG and along the N scarp of the cordillera to the mountains of the Jimi R. and Sepik–Wahgi Divide, as well as the Adelbert Ra.

Diagnosis: in view of comments by Cracraft (1992) and all measurements of both sexes of *intermedius* being extremely similar to those of *chrysopterus* (Frith and Frith 1997b) we synonymise the former. In all measurements of both sexes *chrysopterus* is almost identical to the nominate form but differs from it in secondary coverts and outer edges of flight feathers being more orange, less yellow, and with the crown darker.

Measurements: wing, ad ♂ (*n* = 40) 112–120 (116), imm ♂ (*n* = 37) 111–120 (116), ad ♀ (*n* = 47) 104–125 (110); tail, ad ♂ (*n* = 38) 33–34 (39), imm ♂ (*n* = 36) 57–72 (63), ad ♀ (*n* = 40) 53–69 (60); tail centrals, ad ♂ (*n* = 29) 229–321 (282); bill, ad ♂ (*n* = 37) 27–35 (30), imm ♂ (*n* = 37) 29–32 (30), ad ♀ (*n* = 46) 26–33 (30); tarsus, ad ♂ (*n* = 39) 29–33 (32), imm ♂ (*n* = 37) 30–33 (32), ad ♀ (*n* = 47) 28–33 (30). Weight: ad ♂ (*n* = 36) 92–119 (102), imm ♂ (*n* = 26) 74–108 (98), ad ♀ (*n* = 23) 81–113 (91).

3. *C. m. hunsteini* (Meyer, 1885). *Zeitschrift für die gesammte Ornithologie* **2**, 389, pl 21. Hufeisengebirge, southeastern New Guinea (= near Astrolabe Mts). Type specimen SMT C 8146 (8466).

Synonym: *Diphyllodes magnificus extra* Iredale, 1950 [not 1948 as Iredale claimed in his work]. *Birds of Paradise and Bowerbirds*, p. 111. Mt. Hagen district. Type specimen none designated.

Range: eastern PNG, westward to the Huon Penin, the Wahgi region, and along the S scarp of the central cordillera to the upper Fly R. region.

Diagnosis: like nominate form but paler on head and back (Gilliard 1969) and orange on secondary coverts and outer edge of flight feathers a brighter, deep orange. All measurements of both sexes average fractionally smaller than other forms.

Measurements: wing, ad ♂ (*n* = 38) 105–117 (113), imm ♂ (*n* = 33) 105–117 (112), ad ♀ (*n* = 45) 104–115 (108); tail, ad ♂ (*n* = 38) 35–43 (39), imm ♂ (*n* = 32) 59–71 (33), ad ♀ (*n* = 31) 53–66 (60); tail centrals, ad ♂ (*n* = 32) 225–303 (272); bill, ad ♂ (*n* = 37) 29–33 (31), imm ♂ (*n* = 33) 29–32 (30), ad ♀ (*n* = 45) 29–32 (30); tarsus, ad ♂ (*n* = 35) 29–33 (31), imm ♂ (*n* = 33) 30–34 (32), ad ♀ (*n* = 45) 27–32 (30). Weight: ad ♂ (*n* = 33) 75–105 (91), imm ♂ (*n* = 27) 72–108 (87), ad ♀ (*n* = 37) 62–94 (77).

Habitat and habits Table 4.1

Hill to midmontane forests, rarely in lowland rainforest. ♀♀ and imm ♂♂ are described as solitary, individuals moving about inconspicuously in the middle and lower strata of forests. Ad ♂♂, which spend much time feeding in the understorey, are far more often heard than seen and do not appear to associate with ♀-plumaged birds except at their display courts. Rand (1940*a*) found the sp. active and spending much time high in the tree tops and substage and on the slopes of the Idenburg R., IJ occasionally encountered parties of 4–6 ♀-plumaged birds moving through the forest. Commonly joins mixed-spp. foraging flocks. At Mt. Missim, PNG on 7 Nov a bird seen foraging with seven spp. of small insect-eaters, including the Mountain Drongo (BB).

Diet and foraging Table 4.2

Claude H. B. Grant (in Ogilvie-Grant 1915*a*) shot several birds feeding apparently upon flowers in large trees, in company with sunbirds and honeyeaters. Diamond (1972) found only fruit in 10 birds' stomachs and he observed that birds may perch acrobatically to obtain them. Schodde (1976) found that in nine individuals from three locations the stomachs contained from 90 to 100% fruit and 0–10% arthropods, the latter including the Tettigoniidae. Animal foods identified in faecal samples include ants, spiders, beetles, cockroaches, crickets, lepidopteran larvae, and an assassin bug (BB). Pratt and Stiles (1983) observed 16 complete feeding visits to fruit-bearing plants by birds: average length of a visit was 4 min, birds fed throughout the period; 93% of all visits lasted 1–10 min, 7% 11–20 min. Fruits of *Trema orientalis*, *Psychotria*, *Myristica*, *Pipturus argenteus*, *Sloanea*, *Harretia* and *Piper* spp. recorded eaten in PNG (Healey 1986; Hicks 1988*a*; A. Mack *in litt.*). Birds on Mt. Bosavi, PNG, in late Feb clung to and hung upside down from the complex Magnolia-type fruit of *Elmerrillia papuana* (= *tsiampaca*) to eat its tiny arillate fruit (T. Pratt *in litt.*). The diet was considered to be 70% fruit by Beehler and Pruett-Jones (1983). Of 261 wild fruit feeding instances observed on Mt. Missim by Beehler (1983*a*), 40% were upon *Omalanthus* sp. (a capsule), 22% upon *Gastonia* sp. (a modified capsular fruit), and of 19 other plants

Magnificent Bird of Paradise *Cicinnurus magnificus* (Pennant, 1781)

recorded eaten 8 were capsule (47 records), 6 drupe/berries (28 records), and 5 figs *Ficus* spp. (24 records). Thus, overall fruit diet consists of 58% capsule, 33% drupe/berry, and 9% figs (see Chapter 4 for discussion of this). See Appendix 4 for many food plants. Coates (1990) observed the use of the foot by a bird to hold a fruit to a perch in order to tear it apart, and that ♂♂ on their court occasionally regurgitated fruit seeds and dropped them. Courts are often visited by pigeon species such as the Cinnamon Ground-Dove and Thick-billed Ground Pigeon *Trugon terrestris* to forage for seeds dropped by the ♂.

Vocalisations and other sounds

The ♂ produces several vocalisations, all commonly heard on or near the display court. First is a series of plaintive, downslurred *churrs*, growing louder and more insistent (Fig. 9.58). Second is a single, metallic, querulous, downslurred *kyong* note. Third is a rolling, trilled series of repeated *churr* notes, each downslurred. Fourth is a loud, sharp, slightly musical *kyerng!* And, fifth, given only at the court, is a scolding *ksss-kss-ks-ks-ks-kss* (BB). The advertisement song of ♂♂ at the court has been notated as 'a violent or strident *ca cru cru cru*; a loud clear *car* or *cre* repeated a number of times; a hoarse or squealing *caaat ca ca ca*'. Ad ♂♂ disturbed at their court produced 'low spitting and clucking notes and a scolding *char*'. In low-intensity display or in following a ♀ about the court a ♂ would give 'low enticing, questioning calls of *eek* or *eee*' (Rand 1940b). An ad ♂ gave a *cheeeung* call prior to displaying on his court (Thair and Thair 1977). The call of ad ♂♂ is described as loud, harsh, buzzy, downslurred, low-pitched notes on one pitch but progressively increasing in volume, given from high concealed tree perches. That of ♀♀ and imm ♂♂ 1–3 harsh, downslurred, quiet *chew* notes, or a shortened soft form of the ad ♂ call (Diamond 1972). The former was described by Coates (1990) as 4–6 loud, throaty, downslurred *chaws*. He described the above *kyerng* note as a single, loud, sharp, high-pitched, and penetrating *kyeng*, used in advertisement, which may start

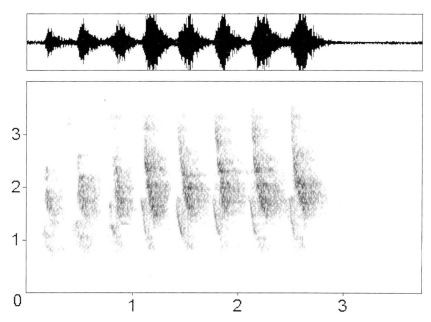

9.58 Sonogram from a recording of an ad ♂ Magnificent Bird's court advertisement song at Weri, IJ, in July 1990 by K. D. Bishop.

Magnificent Bird of Paradise *Cicinnurus magnificus* (Pennant, 1781)

at a frequency of 8 notes in 10 sec to slow to 3 in 10 sec before ending. The *kyerng* call may be given every *c.* 30 sec (BB). A call reported only by Coates (1990) accompanies the *Dancing display* of the courting ad ♂. This is a 'peculiar, low, rhythmic, hard, clicking, buzzing song with higher and lower notes'. Ad ♂♂ produce an audible clacking or rattling sound 'like rapidly striking two pebbles together' in flight (Beehler *et al.* 1986) but Coates notes this appears to be controlled by birds as he observed ad ♂♂ flying about courts silently. He also noted that ♀-plumaged birds may produce the flight noise.

Mating system

Polygyny, with the ♂ promiscuous and the ♀ performing all nesting duties. The ad ♂ meticulously clears a traditional terrestrial court and keeps it clean of all litter. The area is several to many square metres in extent (Rand 1940b; Cooper and Forshaw 1977; Coates 1990). Within this cleared area stand the vertical stems of several sapling trees (Fig. 5.1). The court-owning ♂ spends much time on these vertical perches, and he removes leaves from them and their stems become worn clean and smooth by the bird's feet. Both T. Pratt and A. Mack (*in litt.*) note that courts occur on steep slopes in treefall or landslip areas, which is to say beneath a gap in the forest canopy. Beehler (1983a) found that some courts were maintained for at least 3 consecutive years. He found six courts of five ad ♂♂ dispersed (Fig. 9.59) at a mean nearest neighbour distance of 209 (170–280) m with the owners being non-territorial. One ♂ maintained two courts simultaneously, 60 m apart. At Baiyer R., PNG, another two courts were found 70 m apart in early June that appeared to be maintained and attended by a single ad ♂ (BB). Whilst apparently out of auditory contact from court to court (BB could hear ♂♂ at no more than 175 m distant), ♂♂ maintain auditory contact by calling as they move about their home ranges, which overlap considerably with those of one or more other ad ♂♂. ♀♀ also had home ranges overlapping extensively.

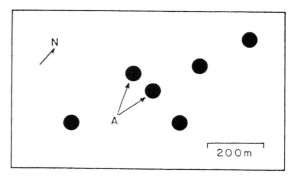

9.59 Display dispersion of 5 ♂ Magnificent Bird display courts, indicated by solid circles. The two labelled 'A' were both maintained by one ♂. From Beehler and Pruett-Jones (1983), © Springler-Verlag with kind permission.

The display court has often been described as a 'bower' which is a term used for a structure constructed or accumulated by members of the bowerbird family Ptilonorhynchidae. 'Court' is a more appropriate term. Tweddle *et al.* (1998) argue that the use of the term 'bower' should be applied only to certain bowerbird structures.

Courtship behaviour

Important observations at several courts above the Idenburg R., IJ over 13 days during 2 Mar–5 Apr 1939 were made by Rand (1940b). When alone, a ♂ spent much time simply sitting on favoured horizontal court perches 0.6–4.6 m high for periods of up to 30–40 min, often preening. Three phases of display were defined, almost always performed within half a metre of the ground upon the vertical stems of saplings growing within the cleared court. The first and most common simply involved breast-pulsing (*Breast display*) in which the modified breast-shield plumage appears to 'inflate' and 'deflate' as it is raised and lowered. The upper margins of the breast-shield rise either side of the head (Fig. 9.60(a)) and undulations of the breast feathers send iridescent shimmers across it. Iridescent spots in front of the eyes are at times conspicuous. This display is given from perches within 30 cm or so of the court floor without preamble, both by a ♂ alone and at the approach of a ♀.

Far more stereotyped behaviour is the horizontal display (*Cape display*) given by a ♂ clinging to the side of a vertical sapling stem *c*. 30 cm above ground (Fig. 9.60(b,c)) alone or in the presence of a ♀. When given by a ♂ alone this display was always preceded by a period of court floor cleaning. The third display Rand witnessed, which he called the pecking display (*Dancing display*), followed copulation. It involved a ♂ perching on a vertical sapling stem without his nuptial plumage erect save the tail and tail wires held at right angles to the body, or vertically behind it, and the bill held open to present the mouth colouration to the ♀ (Fig. 9.60(d)).

At 1440 hr on 28 Mar Rand witnessed a ♂ displaying to and mating with a ♀ on his court. The ♂ was on the floor clearing his court of debris when the ♀ flew into the area, at which point the ♂ immediately flew to perch 30 cm high on a vertical sapling stem within the court and the ♀ flew to perch 1 m above him on the same sapling. The ♂ performed the *Breast display* which he directed at the ♀ as she moved from one court sapling to another at *c*. 1.2 m above ground. Once he followed her to another sapling and hopped up it with his breast parallel and close to the trunk whilst giving *Breast display*. If the ♀ made as if to leave, the ♂ immediately turned away and hopped onto the court floor to clear or as if to clear it, thus enticing the ♀ to remain on or descend the sapling. Finally, as the ♂ again perched below the ♀ performing *Breast display* the ♀ hopped down towards him at which he pressed his breast closer to the sapling whilst continuing *Breast display* and gave a repeated low eager call note. The ♀ paused *c*. 30 cm above the ♂ and he performed the *Cape display* with breast-shield lengthened and flattened (as in Fig. 9.60(b,c)) with the iridescent central line of large 'scale-like' feathers conspicuous and the cape held at right angles to the back. His tail was held in line with the body and vibrated but he was otherwise motionless. After *c*. 30 sec the ♀ hopped down towards the ♂ a little and he then hopped up to her with deliberate movements and mounted her to copulate. Immediately after mating the ♂ hopped onto the sapling beneath the ♀ and performed the pecking display or *Dancing display* (Fig. 9.60d) as he vigorously pecked the nape of the ♀, drawing back after each peck with gaping mouth. A few moments later the ♀ flew off.

On 1 Apr Rand saw the ♂ perform the *Breast display* and *Cape display* to a visiting bird. He also watched an ad ♂ perform the *Breast display* at what he thought was an imm ♂ before rushing at it. The ad ♂ then performed the *Dancing display* at which point the presumed young ♂ rushed the ad, possibly striking it, before flying off. Rand thought it likely imm ♂♂ visit courts in the absence of their ad owners in order to display and perhaps to mate should the opportunity present itself.

Brian Coates (1990) improved understanding of the complex courtship display by intensive observation at a court on the Sogeri Plateau, PNG. The ad ♂ appeared at his court shortly after first light and remained at or near it until the sun was well up, but departed during 0820–0920 hr before sunlight fell on the court. He was to be occasionally heard about the area but did not return until early afternoon.

A ♂ would occasionally fly onto a vertical sapling to perform a *Cape display* after court clearing activity. Only two sapling trunks were used for display and Coates gained the impression their central court location was significant to the bird(s). Whilst the ♂ would quickly fly to a vertical sapling to perform only the *Breast display* when another Magnificent was about, Coates could discern a clear sequence of display postures as follows: *Back display*, a previously unrecorded silent display initially given by the ♂ upon landing on his vertical display perch in which he inverts with tail uppermost and head lifted slightly upward (Fig. 9.60(e)). This posture is held briefly before he turns head uppermost but if held longer may be enhanced by jerking his body side to side to hold the position to each side briefly. The function of this elaborate *Back display* appears to be to present the bright back coloration to a wider audience area.

The *Back display* is followed by *Breast display* (Fig. 9.60(a)), or breast-pulsing, accompanied by low 'conversational' pleading notes. *Breast display* is not necessarily preceded by *Back display*. A more intense form of the *Breast display* with more urgent calling, involves the ♂, always in the presence of a ♀, suddenly changing body position to a rigid near-horizontal one with head slightly uptilted and all plumage sleeked tightly against the body except the breast plumage which is expanded and pulsed (the *Horizontal display* of Coates 1990).

The silent *Cape display* is performed if a visiting ♀ moves closer to the ♂. This involves the ♂ leaning, to as far as horizontal (Fig. 9.60(c)), with extended neck and bill directed at the ♀, the cape erect and fully fanned and the breast-shield dramatically flattened and elongated whilst the tail wires rest in their normal position of repose. The *Cape display* is often held by the slightly trembling ♂ for many sec as the ♀ hops closer down the sapling. When she is within c. 30 cm he may suddenly perform the *Dancing display*. He pulls his head and neck back into his body, puffs and expands most plumage including the pulsated breast-shield, lowers the fanned cape and cocks his tail over his back to present the iridescent upper surface of the central tail wires to the ♀ (Fig. 9.60(d)). He jerks his head from side to side with bill opening and closing to flash his brilliant mouth coloration. The tail is quivered from side to side as the ♂ dances forwards up and then backwards down the sapling with short and jerky movements of body and head that emphasise the fanned bright yellow cape. During this *Dancing display* the ♂ emits a 'peculiar low, rhythmic, hard, clicking, buzzing song with higher and lower notes' (Coates 1990). This display was occasionally performed by a lone ♂ and sometimes without a prior *Cape display*.

Whilst Coates did not witness copulation his observations suggest the *Dancing display* may constitute part of pre-copulation display, notwithstanding the fact that Rand (1940b) saw a copulation preceded only by *Breast display* and *Cape display*. In this case, however, it is clear from Rand's description and illustration that the *Dancing display* posture was used as a post-copulation display.

The *Cape display* is frequently performed, often repeatedly so, by a lone ♂ on his court, mostly during the early morning but also mid to late morning and late afternoon. During these lone *Cape displays* the posture may be held for a few to many sec and the ♂ sometimes opens and closes his bill during the last sec.

At one court a ♂ was observed in Aug to perform the *Breast display* to four ♀-plumaged birds with another ad ♂ present in his court area, without signs of aggression (Thair and Thair 1977). As the ♂ subsequently was performing the *Cape display* on the ground to a ♀ 15 cm above him on a vertical sapling trunk, all birds were disturbed and flew off.

The above display account includes all four postures described by Seth-Smith (1923b).

Breeding Table 6.5

NEST SITE: the few recorded were as follows: (1) 3 m above ground in a pandanus tree crown on the slope of a ridge (Campbell 1977); (2) 4.3 m above ground in dense foliage of a parasitic plant in secondary growth; (3) 76 cm high in a low bush in kunai grass; and (4) 3.6 m high in a pandanus tree crown at the edge of a garden (Bulmer in Gilliard 1969).

NEST: one nest constructed of green mosses, a little mammal fur, and lined with fibres and rootlets, measured 165 mm dia and 114 mm deep externally and 90 mm in dia and 44 mm deep internally. A second nest was of mosses, dry leaves, and weeds and lined with fine rootlets and plant-like fibres, measuring 152 mm in dia and 102 mm deep externally and 76 mm dia and 57 mm deep internally (Bulmer in Gilliard 1969). A nest built in a USA aviary, from which two nestlings successfully fledged, was constructed entirely by the ♀ (Everitt 1965, 1973).

EGGS: clutch usually two, sometimes one. 13 eggs average 31.2 (29–32.9) × 22.6 (21.2–23.8) mm (CF). Ground colour Pale Horn

Magnificent Bird of Paradise *Cicinnurus magnificus* (Pennant, 1781)

9.60 Some courtship dispay postures of ad ♂ Magnificent Bird: (a) *Breast display*, (b–c) *Cape display*, (d) *Dancing display*, (e) *Back display*. After Coates (1990), with kind permission. See text.

(92) but with slightest pinkish to pink-tan wash. Boldly marked with broad, elongate, brush-stroke-like streaks and blotches of purplish greys overlaid with tan to dark brown ones, from around larger end down most of egg length, much like eggs of other higher Paradisaeinae. Plate 13.

INCUBATION: exclusively by ♀. Commencing with first egg of clutch and lasting 18–19 days in captivity (Everitt 1965, 1973). Bulmer (in Gilliard 1969) found three active wild nests all attended by a presumed ♀ only, on 24 and 30 Sept and 24 Dec on the N watershed of Mt. Hagen, PNG.

NESTLING CARE: exclusively by the ♀ of a captive pair, the ♂ being removed from the aviary because of his potentially damaging behaviour, which is clearly typical of the sp. (Everitt 1965, 1973). A wild ♀ regurgitated large red and numerous small black berries to its nestlings (Campbell 1977).

NESTLING DEVELOPMENT: by 5 days old feather quills are visible under the skin of the wings (Everitt 1965). A captive hatchling weighed 6.6 g on the day it hatched. Eyes opened at 6 days and at 7 days quills were visible down the centre of the back with those on the wings beginning to burst. At 11 days head, back, and wings were feathered. One young fledged a day before the other in a captive brood of two, their nestling periods being 17 and 18 days (Everitt 1965, *contra* 16 days of Everitt 1973). They became independent of the ♀ parent at 38 days old (Everitt 1965).

Annual cycle Table 6.5
DISPLAY: possible at any month but predominantly July–Feb. In Feb on Mt. Bosavi, PNG, some display occurred but was much on the wane (T. Pratt *in litt.*). BREEDING: at least July–Dec but probably at all months over sp. range as a whole. ♂♂ with much-enlarged gonads have been collected during Mar–Nov inclusive and ♀♀ in breeding condition during Feb–Mar and May–Nov over the sp. range as a whole. Rand (1940a) noted that ♂♂ in ♀ plumage were not infrequently collected in breeding condition, with enlarged gonads. A ♀ netted on Mt. Missim, PNG, on 6 Nov had an egg in her oviduct (BB). EGG-LAYING: known during Mar and Aug–Dec inclusive but ♀ gonad activity suggests any month over sp. range as a whole. A captive ♀ laid the two eggs of two different clutches on consecutive days (Everitt 1965, 1973). MOULT: 84 of 285 specimens were moulting, involving all months but the least activity being during Feb–Sept and the greater during Oct–Jan (Appendix 3).

Status and conservation Table 8.1
Widespread and common and apparently under no current or long-term threat. Gilliard (1969) found the display courts of five ad ♂♂ in an elongate area of forest *c.* 2.4 km long near Kup, Wahgi Valley at 1524 m. He found two more in the Adelbert Mts, PNG, *c.* 800 m apart. Diamond (1972) found this one of the most abundant birds in hill forest of the E Highlands of PNG. In the Mt. Karimui area he found it represented 1.7% of the local avifauna at 610 m where imm ♂ birds were twice as numerous as ♀♀, and ad ♂♂ were few; 2.7% of the avifauna at 1113 m where ♀♀ were slightly more common than imm ♂♂, and ad ♂♂ were few; 2.4% of avifauna at 1220–1280 m; 2.3% at 1341–1448 m where ad ♂♂ were increasingly common; and 1.3% of avifauna at 1448–1549 m where most ad ♂♂ occurred. He observed, as have others, that the altitudinal range of the sp. clearly correlates with that of the Superb Bird, the latter sp. replacing the former at higher altitude. Beehler (1983a) calculated seeing some 3–4 birds feeding at a fruiting tree during an average day on Mt. Missim, PNG. During a 28-month study within 50 ha of forest on Mt. Missim, six courts of five ad ♂♂ were located and plotted (Fig. 9.59). During July 1990 found commonly, often in mixed-spp foraging flocks, in primary lowland, swampy, and hill forest at *c.* 45–1070 m on N slope of Snow Mts and moderately common on S slopes of Cyclops Mts, IJ from *c.* 760 to 1700 m. Common from *c.* 300 to 1340 m on W Van Rees Mts., IJ in Sept 1994 (J. Diamond and D. Bishop *in litt.*) Gilliard (1969) noted quite frequently seeing men of the Wahgi Valley wearing the central tail sickle-wires of the ad ♂ in their hair.

Knowledge lacking and research priorities
A well-known sp. and an excellent subject for detailed studies of ♂ court behaviour and ♂ investment in advertisement activities. Easily netted and common, and thus a good subject for mark-and-recapture studies.

Aviculture

Successfully bred at BRS, PNG in Dec 1983–Jan 1984 (Anon 1984). As recently as 1994 it bred in the Honolulu Zoo, where the single young was raised by hand. Does well in a large but not necessarily tall flight area if supplied adequate animal foods and access to a pool and rain/sprinkler for bathing. Can be maintained together with Superb Birds but few other birds of paradise spp. (P. Shanahan *in litt.*).

Wilson's Bird of Paradise *Cicinnurus respublica* (Bonaparte, 1850)

PLATE 10; FIGURES 9.61–2

Other names: Waigeu or Waigeo Bird of Paradise, Bare-headed Little King Bird of Paradise. Malay name *Kapala kruis* [= cross head] (Ripley 1950).

A very small forest-dwelling sp. endemic to Waigeo and Batanta, two islands off the western tip of NG. An inconspicuous inhabitant of hill forest interior, best located by its distinctive vocalisations. Monotypic.

Description

ADULT ♂: 16 (21), cm crown, and nape largely without feathering—the exposed skin bright Cerulean Blue (67) with fine feathering, velvety jet black with coppery bronze iridescent sheens forming fine compact lines across the naked crown. A discrete semicircular nape cape of brilliant yellow (55) contrasts with a mantle of deep Crimson (108) bordered by a broad jet black line. Upperwing; lesser coverts blackish brown, broadly-edged paler Raw Umber (123), greater coverts and alula dark Raw Umber (223) with fine, dark, crimson edging. Tertials dark Crimson (100) tipped jet black, secondaries brownish black with dark crimson leading edges not extending to tips and the primaries dark brown (121) with finest of deep crimson leading edges. Rump, upper-tail coverts, and dorsal surface of tail blackish brown (219) but modified central tail 'sickles' richly-iridescent violet-purple (172) with Magenta sheens in suitable light. Chin and upper throat velvety jet black with coppery bronze to Purple (1) gloss, Mid-throat to belly occupied by an extensive breast-shield of smooth, oily, and dull-glossy emerald green plumage that appears partly to entirely deep blue purple (172) to red purple (1) in some lights. Tiny central spots of several central throat feathers strongly-iridescent Turquoise Blue (65), and the broadest outermost feathers of the lower breast-shield broadly-tipped, highly-iridescent Turquoise Green (64), in some lights (when not concealed beneath wing). Remainder of underparts sepia brown (219) with deep violet to Purple sheen in some lights. Underwing: coverts dark brown, flight feathers dark Olive-Brown (28) with broad cinnamon (26) trailing edges. Ventral surface of tail dark Olive-Brown. Bill blackish, gape pale yellow but probably concealed beneath feathers, iris very dark brown, legs and feet rich blue (69), mouth bright creamy yellow to yellow-green.

ADULT ♀: 16 cm, same size as ad ♂ except for central rectrices. Upper head plumage brownish black (119), more brownish on nape contrasting with rich blue naked crown and nape. Entire remaining upperparts dark olive-brown (119A) but wings, particularly secondary leading edges, and tail strongly-washed reddish amber (36). Chin, malar area, and lower ear coverts cream to pale buff, finely-flecked greyish, and remainder of underparts dilute, or washed-out, Buff (24) finely- and uniformly-barred brown-black. Underwing: coverts Cinnamon (123A) indistinctly-barred blackish brown, flight feathers Olive-Brown with broad pale cinnamon trailing edges. Ventral surface of tail Olive-Brown. Legs and feet slightly darker than ad ♂.

SUBADULT ♂: as ad ♀ but few feathers of ad ♂ plumage intruding, initially about head and breast, to as ad ♂ but with few feathers of ♀ plumage remaining.

IMMATURE ♂: tail, other than central feather pair, markedly longer (as in ad ♀) than ad ♂ and rectrices pointed at their tips.

NESTLING–JUVENILE: unknown.

OTHER: a unique ad ♂ specimen (SNHM 15522) has all underparts deep rich fawn colour (27) as are its upperparts, save that the yellow cape and the red of the central back (bordered posteriorly by a black line) and wing patch of normal plumage are present, but the back and secondary coverts are less red, more orange, than usual. Bare head skin is not black as in all other specimens but is brown, suggesting it was not the usual blue colour in life. The usually dense and plush black head and throat feathers are silvery-buff in this bird. The central tail 'wires' are missing (Frith 1988).

Distribution
Known only from Waigeo and Batanta Is, in the Rajah Ampat group off the W coast of the Vogelkop, IJ; ranges in the hills (300 m) to perhaps the highest summits of these two islands (c. 1000–1200 m). Details of its distribution on these islands are lacking.

Systematics, nomenclature, subspecies, weights, and measurements

SYSTEMATICS: although most ornithologists have acknowledged the close relationship of Wilson's and the Magnificent Bird to the King, it was not until recently that the logical step of placing all three into a single genus (*Cicinnurus*, by priority) was suggested (Diamond 1972) and widely accepted. We suggest that the two sister forms (*respublica*, *magnificus*) be considered constituting a subgenus (*Diphyllodes*). Gilliard (1969) suggested that the bare blue head of this bird may have evolved as an isolating mechanism resulting from sympatry with its close relative the King Bird on Waigeo I. Mees (1982) has pointed out, however, that the latter sp. has not in fact been confirmed as present on that island (see King Bird account). Hybridisation: unrecorded. Wilson's Bird is sympatric only with the Red Bird, and as hybridisation is known between Magnificent and Lesser Birds the crossing of the former two spp. must be considered theoretically possible.

SPECIES NOMENCLATURE AND ETYMOLOGY
Cicinnurus respublica (Bonaparte)
Lophorina respublica Bonaparte, 1850 (Feb). *Comptes Rendus des Séances de l'Academie des Sciences*, Paris **30**, 131. 'New Guinea', restricted to Waigeo I. Type specimen ANSP No. 3152. In the very brief footnote that serves as the original description, Prince Bonaparte compares this new form to the Superb Bird, and he placed it in that genus *Lophorina*. He corrected this mistake a month later (*Comptes Rendus des Séances de l'Academie des Sciences*, Paris, **30**, 291) by more accurately using *Diphyllodes respublica*.

Synonym: *Paradisea Wilsonii* Cassin, 1850 (Aug). *Proceedings of the Academy of Natural Sciences Philadelphia* p. 67; based on specimen seen by Bonaparte. See Sclater, 1857. *Pro-*

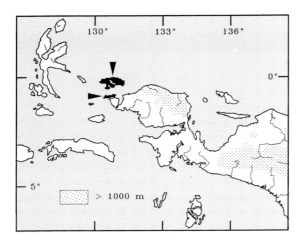

ceedings of the Zoological Society of London p. 6. *Schlegelia calva* Bernstein, 1864. *Natuurkunde Tijdschrift Nederlandische Indië* **27**, 79. *Schlegelia wilsonii* Sharpe, 1877. *Catalogue of Birds in the British Museum* **3**, 175.

Etymology: *respublica* = L. *respublica*, the republic.

For an historically interesting account of the controversial naming of this sp. from a single, badly-damaged, native trade skin see Iredale (1950: 113–5) and Gilliard (1969: 211–2). The sp. was first discovered in the wild by H. A. Bernstein on Waigeo I. in 1863 whilst collecting for the Leiden Museum.

WEIGHTS AND MEASUREMENTS
Measurements: wing, ad ♂ (n = 28) 94–101 (98), imm ♂ (n = 5) 99–102 (101), ad ♀ (n = 16) 93–101 (97); tail, ad ♂ (n = 28) 35–41 (39), imm ♂ (n = 5) 51–54 (52), ad ♀ (n = 16) 49–59 (52); tail centrals, ad ♂ (n = 22) 121–150 (140); bill, ad ♂ (n = 27) 23–28 (25), imm ♂ (n = 5) 25–27 (26), ad ♀ (n = 14) 23–28 (25); tarsus, ad ♂ (n = 27) 26–28 (27), imm ♂ (n = 5) 27–30 (28), ad ♀ (n = 15) 25–29 (27). Weight: ad ♂ (n = 6) 53–67 (61), imm ♂ (n = 3) 61–65 (63), ad ♀ (n = 6) 52–60 (56).

Habitat and habits Table 4.1
Primarily hill forest but more rarely also lowland rainforest and midmontane forest of the hilly interior of Waigeo and Batanta Is. Moderately vocal throughout much of the day but difficult to observe (J. Diamond and D. Bishop *in litt.*). On Mt. Besar, Batanta I., common and vocal at 460 m but almost silent at twice that altitude (Gilliard via M. LeCroy *in litt.*).

Diet and foraging Table 4.2
Only two wild records; F. Day collected a bird 'feeding fairly high up with a wave of other birds', apparently a mixed-spp foraging flock (M. LeCroy *in litt.*), and a ♀-plumaged bird was seen taking small orange fruits from the lower crown of a tree (D. Bishop *in litt.*). In captivity, birds regurgitate various matter to re-swallow or drop from the bill tip. Grasshoppers are taken in the bill to a perch and there held by one or both feet and dismembered before eating. During this activity an ad ♂ closed his yellow cape and held it gathered down the central mantle, presumably to prevent soiling (Frith 1974). The Pheasant Pigeon *Otidiphaps nobilis* and Cinnamon Ground-dove *Gallicolumba rufigula* have been repeatedly seen to visit courts where they seek seeds Wilson's Birds have voided.

Vocalisations and Figure 9.61
other sounds
The wild ad ♂ will start calling as early as 0610 hr, from perches as high as *c.* 9 m high near his court (Bergman in Gyldenstolpe 1955b). Vocalisations recorded from a captive ad ♂ are: a sharp explosive single *keeetch* < 1 sec in duration, and audible at *c.* 15 m, and more common was a pleasant soft, high-pitched, whistle, *teel*, repeated 5–6 times. A stronger and louder call similar to the last was a *too-too-too-too-too-too-wit*, the last note being sharply raised in pitch to give a whip-crack-like ending. A louder and more penetrating form of the last call was *too-too-too-too-too-too-zeet* given with the neck stretched and head thrust upward with each note. Given the loudness of the latter it is possible it is given as court-advertisement. A ♀ housed with the calling ♂ did not vocalise (Frith 1974). Recorded by the BBC, an ad ♂ singing at his court gave a clear powerful, whistled, *cheew, chau, chow* or *twou* note repeated up to 12 times in *c.* 5 sec, the first couple being softer and twanging in quality (Fig. 9.61(a)). This call may at times be given as a series of sharper notes of metallic, frog-like, wooden clicks, and the second sonogram (Fig. 9.61(b)) shows this followed by three clearer and sharper click notes. During display ad ♂♂ produce complex squeeky twitterings in subsong punctuated with bowerbird-like, guttural, burring notes (Fig. 9.61(c)) some of which are much like display notes produced by ♂ Red Birds in display (CF). Scold notes are harsh and churring ones

Wilson's Bird of Paradise *Cicinnurus respublica* (Bonaparte, 1850)

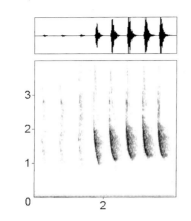

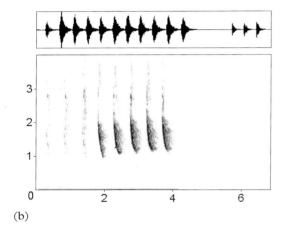

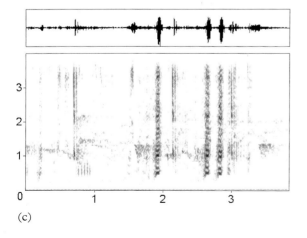

typical of the family. Ad ♂♂ in flight about their court produce a rapidly-repeated, loud, sharp, clear, dry *tick* much like some male Neotropical manakins (Pipridae) in the same context. It is not known if this sound is produced when the ♂ is foraging in the forest.

Mating system
The court-based behaviour of ad ♂♂ (see below) clearly indicates that the sp. is polygynous, and the ♂♂ dispersed and promiscuous, as is the case in the two other *Cicinnurus* spp.

Courtship behaviour
Ad ♂♂ kept in captivity in Europe, America, and Asia have long been observed to clear and maintain a ground court and associated vertical sapling trunks as do Magnificent Birds. Bergman (in Gyldenstolpe 1955*b*) observed this in the wild and wrote that 'These display-grounds are usually of about a half up to one meter in circumference [more likely the diameter] and invariably situated in dense forests, preferably in a small clearing around a fallen tree or at the close vicinity of it. That clearing is always surrounded on all sides by a very dense and tangled vegetation which almost completely conceals it from view except at very close quarters. The display-ground reminds of the stage of a theatre where the brilliant rays of the sun form the bright roof-illumination. The actual display-arena is cleansed with utmost care by its owner who has removed all such leaves and other vegetable matters that might have dropped down on it since the owner's last visit. It was so carefully cleansed that the bare soil was always visible. However, small bushes and some slender, low trees are always left intact although the bird had deprived them from [of] their foliage.'

9.61 Sonogram from recording of an ad ♂ Wilson's Bird's typical court advertisement song (a); another in which the notes are sharper and are followed by clear sharp clicking (b) and a segment of display song (c) in Feb 1986, Batanta I., IJ by K. D. Bishop.

Wilson's Bird of Paradise *Cicinnurus respublica* (Bonaparte, 1850)

Gilliard (in Greenway 1966) wrote of a court 'in a bramble of saplings and vines in an open area of the forest near a landslide. It is on a 45° slope and is approximately 7 ft by 10 ft [2 × 3 m] in size. The key display stage seems to be a limbless sapling 1 ft 4 ins [clearly erroneous—probably 1.4 ins or *c*. 3 cm] thick at ground level and about 8 ft [2.4 m] tall. A fallen tree marks the upper edge of the ground clearing and a loose splay of vines averaging one half inch [13 mm] thick are in and over part of the cleared area. These may be used in display. All leaves have been removed from the tree and the vines. Looking straight up from the court one sees only the open sky across a breadth of 10 ft [3 m], then on one side is the feathery frond of a tree fern. Around the sides of the court are many vines and small saplings.'

Ad ♂ clears the court floor by flicking litter aside with the bill while perched head-down on the base of saplings or by standing directly on it. Leaves, even twice the size of the bird, are easily removed and dropped from live saplings by grasping and sharply twisting the base of the petiole.

Film made by the BBC (some appearing in Attenborough 1996) shows an ad ♂ initially responding to a ♀-plumaged individual arriving near his court with a 'frozen' posture on the base of a vertical sapling (Fig. 9.62(a)). As the ♀ then perches above him he directs his head and bill at the ♀ with all his plumage sleeked and neck and tail held normally. He sharply flicks his head and neck only the shortest of distance to either side, with an odd clockwork-like motion (Fig. 9.62(b)) while producing a softly-whistled 'whisper' song of ticking and buzz-like sounds. As the ♀ approaches him he suddenly pulls head and neck back into his body while continuing to look vertically up the perch and to flick out and expand his flattened breast-shield, and cock his tail up to 90° (Fig. 9.62(c)). William Timmis (personal communication) had seen a captive ♂ give this display whilst moving up the vertical stem until almost

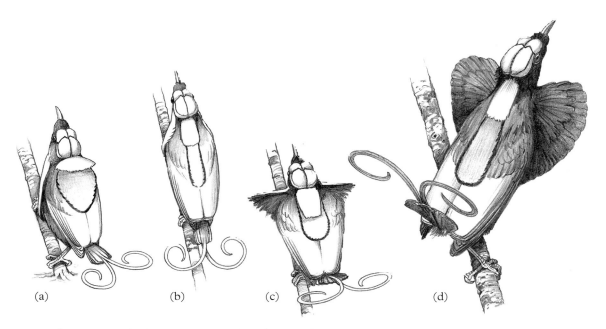

9.62 Some courtship display postures of ad ♂ Wilson's Bird: (a) initial 'frozen' posture, (b) sleeked upward-stare with extended head and neck sharply flicked finely to either side, (c) head pulled in as breast-shield and tail elevated in full display, and (d) sleeked and extended 'stare' in full display. After Attenborough (1996). See text.

touching the ♀ and then turn and maintain the same posture whilst moving down the stem. This display form may be interspersed with the first posture, the male giving a complex soft subsong of mechanical twitterings punctuated with burring and ticking notes (reminiscent of Red Bird display vocalisations) the whole while.

At this point the ♀ always flew to a nearby court sapling. At this the stationary ♂, sometimes giving a harsh, scolding, growl note, again sleeks himself and stretches his neck, head, and bill towards her with tail still cocked but with his breast-shield position modified (Fig. 9.62(d)). At a Batanta I. court the court-owning ♂ and a ♀-plumaged imm ♂, less than 1 m apart, simultaneously performed this posture, together redirecting their 'pointing' bill at a ♀ as she moved from perch to perch around the court. So simultaneously did the stiffly-pointing ♂♂ swing to one side or the other to follow the ♀ that they moved as two compass needles, remaining stiffly parallel to each other (T. Schultze-Westrum film *in litt.*).

If the ♀ then returns to perch above him on the sapling the ♂ reverts to the 'head and neck in' pose (Fig. 9.62(c)). This behaviour is generally similar to initial courtship of the Magnificent Bird. As no behaviour subsequent to the last posture (c) was filmed with a ♀-plumaged bird remaining above the ♂ on his perch it is quite possible that ♂♂ might follow this with a dance phase of display like that of the Magnificent Bird but this requires confirmation.

Two London Zoo ♂♂ and a Chester Zoo ♂ performed three phases of 'frontal' display in agreement with the above except that the bill was occasionally gaped open during postures (c) and (d). A much less elaborate *Back display* was also observed, however, in which the bird 'bends down the body to a position parallel with the perch and depresses the head, standing broadside to the observer. The head is then slowly moved from side to side. It suddenly turns and resumes the frontal display' (Winterbottom 1928a: Frith 1974). A horizontal display, like that of the Magnificent Bird, was sometimes performed after the upright one (Fig. 9.62(b)) by the Chester Zoo captive ♂ (Timmis in Frith 1974).

Breeding   Table 6.5
No information save that ♂♂ have been collected with enlarged gonads during May–June and Oct and ♀♀ with moderately-enlarged oocytes in June.

NEST SITE, NEST, EGGS, INCUBATION, NESTLING CARE, AND DEVELOPMENT: unknown.

Annual cycle   Table 6.5
DISPLAY AND BREEDING: unknown (see Breeding, above). MOULT: 8 of 34 specimens were moulting during May–June, but with no samples for Jan and Nov (Appendix 3).

Status and conservation   Table 8.1
During Sept–Oct 1948 Bergman (in Gyldenstolpe 1955b) found the sp. 'in fair numbers' on Waigeo I. and 'fairly common' on Batanta I. where he noted it was 'much more abundant than on Waigeo'. The sp. was moderately common in tall, mature secondary forest, selectively logged and primary lowland and hill forest c. 60–700 m on the N slopes of Batanta I. in Feb 1986. Much lowland habitat on Batanta has been lost to swidden agriculture or has been severely changed or degraded (J. Diamond and D. Bishop *in litt.*).

Knowledge lacking and research priorities
A thorough survey of the status and biology of this insular endemic sp. is needed in order to assess the extent of suitable habitat and any potential threat to it. It would be interesting to know the behavioural significance of the uniquely-coloured bare skin of the crown found in both sexes of this sp.

Aviculture
F. Shaw Mayer, W. Frost, and S. Bergman all took live birds to Europe and birds have been maintained at the NYZS. Birds may be seen in the odd aviary, such as in Singapore, Jakarta/Bogor, and on Bali, at the time of this writing.

King Bird of Paradise *Cicinnurus regius* (Linnaeus, 1758)

PLATE 10; FIGURES 9.63–6

Other names: Little King Bird of Paradise. *Burong rajah* of Malay language and *Goby-goby* of the Aru Is. (Wallace 1869). *Saya* (Lesson 1834–5). *Keping Keping* of Malay people of W NG (Ripley in Mayr and Meyer de Schaunsee 1939a). *Wowi-wowi* on Aru and Sapaloo Is. (Stresemann 1954). *Kabini buna* (ad ♂), *Kuguru* (♀-plumaged) at Lake Kutubu (Schodde and Hitchcock 1968). *Naburo* of the Daribi people (Diamond 1972) and *Kerego* of the Gimi people, Crater Mt., PNG (D. Gillison *in litt.*). *Wahyo* of the Fair language at Bomakia and *Kyalo* of the Kombai language at Uni, IJ (Rumbiak 1994).

This, the smallest bird of paradise, is an inconspicuous but vocal inhabitant of the interior lowland and hill forest. The ♂ habitually attends a vine rich mid-storey song post, and ♀-plumaged birds are commonly seen in the company of mixed-spp. foraging flocks. The ♂ appears dark except in good light, and the ♀ exhibits the typical paradisaeinine plumage pattern: brown above, finely barred below. Polytypic; 2 subspp.

Description

ADULT ♂: 16 (31) cm, culmen sharply keeled (as in *Semioptera*). Head Crimson (108), forecrown to tuft of elongate plush feathering over culmen base more orange (16); a discrete spot of jet black feathering above the central eye which in certain lights shows iridescent dark green (146). Elongate mantle 'cape' feathers, back, rump, tertials, and some wing coverts deep Crimson to Carmine (8) with intense, glossy, white highlights giving a spun glass look in some light. Uppertail coverts slightly duller, more orange, less glossy crimson. Upperwing predominantly glossy crimson but, when visible, some coverts and the flight feathers variably Raw Umber (23) with brown orange (140) outer edges. Dorsal surface of tail Dark Brownish Olive (129) with brown orange outer edges. The elongated central pair reduced to fine, bare, red brown, central shafts with spiral terminal disks formed of inner feather vane, highly-iridescent metallic Dark Green (262) with bronzed yellow sheen. The lowermost crimson throat feathers are very finely-tipped pale Buff (124) where they meet a narrow breast-shield of intensely-iridescent Dark Green feathers that may appear jet black to burnished green yellow in some lights. To either side of the green breast-shield grow several elongate, erectile, fan-shaped, Olive-Brown (28) pectoral plumes that become briefly Pale Pinkish Buff (121D) immediately prior to a broad tip of intensely-iridescent, metallic, bright green (62). Remaining underparts cotton white. Underwing: coverts cotton-white, flight feathers variably Chestnut (32) with paler, more Hazel (35), broad trailing edges to all but outer three primaries. Ventral surface of tail Medium Neutral Gray (84) with Chestnut outer edges to feathers and underside of terminal central rectrices 'disks' dark burnished Cinnamon-Brown (33). Bill ivory yellow, iris pale brown to dark brown, legs and feet violet blue (170), mouth very pale aqua-green.

ADULT ♀: 19 cm, head mousy Olive-Brown very finely streaked above and behind eye and on lores, ear coverts, malar area, chin, and (more densely) throat with pale buff. Entire upperparts, including wing and tail, Olive-Brown but greater wing coverts and all flight feathers with rich Amber (36) outer edging and tertials with a more olive (29) wash. Underparts variably buff-grey (119D) to Tawny Olive (223D) but washed very pale Cinnamon (123A) on breast, flanks, and lower belly and finely uniformly-barred Olive-Brown. Underwing; coverts salmon pink (6) barred Olive-Brown, flight feathers brownish grey with broad salmon pink trailing edges to all but outer three primaries. Ventral surface of tail brownish grey. Legs slightly duller than those of ad ♂; iris and mouth as in ♂.

408 King Bird of Paradise *Cicinnurus regius* (Linnaeus, 1758)

SUBADULT ♂: as ad ♀ with few feathers of ad ♂ plumage intruding, to as ad ♂ with few feathers of ♀ plumage remaining.

IMMATURE ♂: as ad ♀. Younger (darker billed) individuals have much orange rufous on wing coverts and outer edges of flight feathers but steadily lose this as they attain an increasingly-paler bill before attaining the red of ad ♂ plumage. Younger birds also have an orange rufous wash on the upper breast and, more so, to sides of it. Gyldenstolpe (1955*b*) notes 'All imm ♂♂ have—as usual—the tail feathers distinctly pointed at their tips. The presence of such pointed tail feathers, is a good characteristic for imm ♂♂, even when they are still in ♀ garb.' Figure 9.63 shows the form ♂ tail development takes with age.

NESTLING–JUVENILE: hatchling naked and skin dark red and nestling to 6 days old still naked with vivid yellow gape. The appearance of the young at 25 days old was 'beak dark horn-coloured, lighter at the base. Iris grey-brown. Upperparts of the head brown tinged with russet. The back grey-brown. The wings grey-brown like the tail; the wing-feathers decidedly darker brown. The greater wing-coverts and outer webs of the primaries red-brown. The underside light grey with dark cross-bars, curved on either half of the underside. Closer dark cross-bars on the undertail coverts. A light bar immediately above the eyes, and over this a dark spot. The chin grey-brownish yellow with small descending streaks and points. The gape vivid yellow. The inner anterior part of the lower mandible and its borders yellowish-green, as also the upper mandible. The legs and feet were blue, paler than the ♀'s.' At nearly 2 months old the young changed all feathers on the sides and tops of their heads and the entire back became a pure grey colour. They had also replaced their ventral feathering with stronger cross-bars, especially on the flanks. These bars were broader in the young birds, on a colour less brown than in the ad ♀, and their greater wing coverts, primary coverts, and the outer webs of their primaries were vivid red-brown (Bergman 1957*d*).

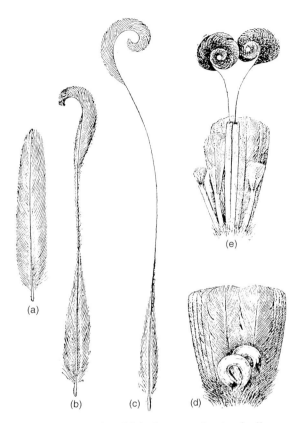

9.63 The way in which the central pair of tail feathers develops with age in male King Birds to become the ornately-tipped 'wires' of adult plumage: (a) first and second year plumage, (b–c) third year plumage, (d–e) fourth year plumage. From Ogilvie-Grant (1913), with kind permission.

Distribution
Widespread through the lowland forests of NG, Aru Is, Misool and Salawati Is (of the Rajah Ampat group), and Yapen I. (Geelvink Bay), IJ. Ranges from sea level to 950 m (mainly no higher than 300–400 m). In many hill forests it is found with the Magnificent Bird.

Systematics, nomenclature, subspecies, weights, and measurements
SYSTEMATICS: a distinctive sp. long treated as the sole member of *Cicinnurus*, but now accepted as sharing the genus with the two other sickletails. The subspecific division of

King Bird of Paradise *Cicinnurus regius* (Linnaeus, 1758)

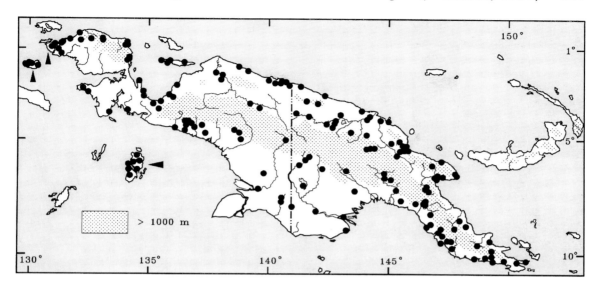

this sp. has been the subject of debate. Six subspp. had been long established (Mayr 1941a; Mayr and Greenway 1962), when it was suggested that only two, *C. r. regius* and *C. c. coccineifrons*, be retained and that the roundish green eye-spot of the former and the more elongate shape eye-spot in the latter be a 'tolerably constant character' Mees (1964, 1965). Gilliard (1969) acknowledged six subspp. but pointed out that they are all very similar. Mees (1982) then enlarged upon his view for reducing the subspp. to two having again reviewed a considerable series of specimens. We follow Mees, particularly in view of the comments of Gilliard (1969), Diamond (1972), Forshaw (in Cooper and Forshaw 1977), and Cracraft (1992) with respect to the weak morphological differentiation of most populations, until additional evidence suggests otherwise. We find Cracraft's criteria for acknowledging four subspp. (sp. in his view) inconsistent (e.g. he synonymises *gymnorhynchus* into *regius* stating it is 'nearly 100% diagnosably distinct' but retains three taxa of the *coccineifrons* group because he views them as 'nearly 100% diagnosable'). Hybridisation: known with the Magnificent Bird from many ad ♂ specimens. See Plate 15 and Appendix 1. Hybrids in ♀ plumage are not recorded but given the ♀♀ of the parental forms are so similar, hybrids between them might easily be overlooked in both field and laboratory. Of course many more ad ♂♂ were collected during the period of trade plume hunting than the unwanted ♀ plumage. *C. lyrogyrus* Currie, 1900. *Proceedings of the United States National Museum* **22**, 497 from NG; *C. goodfellowi* Ogilvie-Grant, 1902. *Bulletin of the British Ornithologists' Club* **19**, 39. Cyclops Mts; and *Diphyllodes gulielmitertii* A. B. Meyer, 1875. *Proceedings of the Zoological Society of London*, p. 31, 'Waigeu' (in error) are all *C. regius* × *C. magnificus* hybrids.

SPECIES NOMENCLATURE AND ETYMOLOGY
Cicinnurus regius (Linnaeus)
Paradisea regia Linnaeus, 1758. *Systema Naturae* edn. 10, p. 110. East Indies, restricted to Aru Is. (Berlepsch, 1911. *Abhandlungen der Senckenbergischen Naturforschenden Gesellschaft* **34**, 59.). Type specimen none designated.

Etymology: *regius* = L. royal, magnificent (*regis*, a king).

SUBSPECIES, WEIGHTS, AND MEASUREMENTS
1. *C. r. regius* (Linnaeus, 1758).
Synonyms: *C. r. rex* (Scopoli, 1786). *Deliciae Florae et Faunae Insubricae* (Picini) pt. 2, p. 88

King Bird of Paradise *Cicinnurus regius* (Linnaeus, 1758)

(based on Sonnerat 1776, *Voyage a la Nouvelle Guinea*, p. 156, pl. 95).

C. spinturnix Lesson, 1835, *Histoire Naturelle des Oiseaux des Paradis et dev Epimaques*, p. 182. Dorei = Manokwari, IJ.

C. r. claudii Ogilvie-Grant, 1915, *Ibis*, Jubilee Supplement **2**, 16. Mimika R., southwesten NG.

C. r. gymnorhynchus Stresemann, 1922. *Journal für Ornithologie* **70**, 405. Type specimen ZSM 11.606.

Range: Aru Is, Misool and Salawati Is, the Vogelkop and all of S NG, and the N watershed of SE PNG from Huon Gulf to Milne Bay. Repeatedly reported to occur on Batanta I. but this is probably erroneous (Mees 1982).

Diagnosis: birds on the Aru Is (formerly *regius*) are on average slightly larger than birds of southern NG (formerly *rex*) but variation is great and the slight differences in plumage morphology are inadequate to justify retention of the latter subsp. (Mees 1964; Cracraft 1992).

Measurements: wing, ad ♂ ($n = 81$) 93–110 (101), imm ♂ ($n = 31$) 96–110 (103), ad ♀ ($n = 48$) 94–107 (100); tail, ad ♂ ($n = 79$) 28–38 (32), imm ♂ ($n = 26$) 52–64 (59), ad ♀ ($n = 42$) 51–63 (58); tail centrals, ad ♂ ($n = 70$) 134–182 (162); bill, ad ♂ ($n = 79$) 25–31 (27), imm ♂ ($n = 29$) 25–28 (27), ad ♀ ($n = 48$) 25–30 (27); tarsus, ad ♂ ($n = 79$) 22–29 (27), imm ♂ ($n = 31$) 23–28 (27), ad ♀ ($n = 48$) 25–28 (26). Weight: ad ♂ ($n = 32$) 43–65 (53), imm ♂ ($n = 4$) 42–53 (50), ad ♀ ($n = 10$) 38–58 (49).

2. *C. r. coccineifrons* Rothschild, 1896. *Novitates Zoologicae* **3**, 10. Jobi I. (= Yapen I.) Type specimen AMNH 678666.

Synonyms: *C. r. similis* Stresemann, 1922. *Journal für Ornithologie* **70**, 405. Stephansort, Astrolabe Bay. Type specimen not located.

C. r. cryptorhynchus Stresemann, 1922. *Journal für Ornithologie* **70**, 405. Taua, lower Mamberamo R. Type specimen ZMB (field No. 260; destroyed in World War II).

Diagnosis: we follow Mees (1964) in combining these taxa. We do not find Cracraft's (1992) tentative plumage characters in support of retaining *cryptorhynchus* as a distinct taxon convincing and all measurements of that population are entirely compatible with those of Yapen I. (*coccineifrons*) and northen NG (*similis*) populations. In fact almost all measurements of the three populations previously known as *coccineifrons*, *cryptorhynchus*, and *similis* are clinal from larger to slightly smaller, W to E. The even more easterly population of *gymnorhynchus* does not fit this pattern, however, having a markedly-larger bill and being therefore more like birds of southern NG (*C. r. regius*). The central pair of rectrices and bill length as a proportion of mean wing length in ad ♂♂ of *C. r. coccineifrons* as here defined are on average 5% and 2% longer than in the nominate form, respectively.

Range: N watershed of the main body of NG, from the E coast of Geelvink Bay eastward to the Ramu R. Birds from the N coast of the Huon have not been assigned to subspecies.

Measurements: wing, ad ♂ ($n = 60$) 93–105 (99), imm ♂ ($n = 26$) 96–106 (100), ad ♀ ($n = 31$) 93–103 (98); tail, ad ♂ ($n = 59$) 28–37 (32), imm ♂ ($n = 24$) 56–63 (59), ad ♀ ($n = 28$) 52–63 (58); central tail, ad ♂ ($n = 51$) 136–170 (153); bill, ad ♂ ($n = 56$) 23–28 (25), imm ♂ ($n = 24$) 23–26 (25), ad ♀ ($n = 31$) 23–28 (25); tarsus, ad ♂ ($n = 59$) 24–28 (26), imm ♂ ($n = 26$) 25–27 (26), ad ♀ ($n = 30$) 25–28 (26). Weight: ad ♂ ($n = 47$) 46–64 (53), imm ♂ ($n = 14$) 50–60 (54), ad ♀ ($n = 5$) 49–57 (52).

Habitat and habits Table 4.1

Common but inconspicuous bird of lowland rainforest, monsoon forest, gallery forest, and forest-edge, including disturbed areas and tall secondary forest. Ad ♂♂ usually forage alone but frequently join mixed-spp. foraging flocks as these pass by. Once a ♂'s display area is located, the ♂ can be readily located and is not particularly shy there. ♀-plumaged birds can be solitary or active participants of mixed-spp. foraging flocks that may include Magnificent Riflebirds and *Paradisaea* spp. These active birds move quickly and may be highly acrobatic, but ♂♂ may perch motionless for long periods in their display trees. In dry weather,

Coates (1990) found that both ad ♂ and ♀-plumaged birds regularly visited forest pools but that whilst the former would only drink, the latter would also bathe.

Diet and foraging   Table 4.2

Diet includes fruit and arthropods, and the sp. forages in a fashion typical of most paradisaeinine birds of paradise. Of 223 sightings of birds in forest near Port Moresby, 20% occurred in the lower canopy of the forest 25–30 m above ground, 60% in the sub-canopy 8–25 m high, and the remainder 0–8 m above ground (Bell 1983). Limited sightings of the sp. elsewhere were of birds 3–18 m above ground (Diamond 1972). At the Puwani R., PNG at 200 m, on 20 July a ♀-plumaged bird seen vine gleaning for insects in mixed-spp. foraging flock of Rufous Babblers with a Lesser Bird (BB). When joining mixed-spp. foraging flocks birds typically forage low in the forest interior. Bell saw King Birds present in 40 such flocks at an average of 1.0 SD 0.2 birds per flock. When in mixed-spp. flocks, King Birds fed predominantly upon insects, and Bell (1983) considered the general diet of this sp. to be more arthropod than fruit. Of 144 foraging bouts performed by 65 individuals foraging in mixed flocks, 27% were at fruit, 25% involved foraging from logs/stumps, 22% from leaves, 13% from branches, 7% upon tree trunks, and 6% upon vines (Bell 1984). Lesson (1834–5) wrote of the bird's liking for the small fruits of teak-trees (presumably *Tectonia grandis*) at Dorey Harbour. Simpson (1942) reported the crop of one bird to contain wild banana *Musa* sp. pulp and seeds. Stomachs of nine Vogelkop, IJ birds contained fruit remains only (Hoogerwerf 1971). The fruits of *Pipturus argenteus* and *Dysoxylum* sp. were eaten at c. 200 m, c. 70 km NW of Port Moresby, PNG (Hicks 1988a). Near Lae, PNG on 20 Jan a ♀-plumaged and ad ♂ were observed feeding heavily on fruit of *Pipturus argenteus* near the ♂'s display perch; fruits of a *Chisocheton* sp. also seen eaten at several locations, on 11 July and 9 Sept (BB). ♂♂ on their display perches often regurgitate and drop fruit seeds. The sp. is clearly omnivorous but conflicting assumptions concerning the relative proportions of animal/fruit foods taken by ad birds (Bell 1982c, 1983; Diamond 1986a; Beehler and Pruett-Jones 1987; Beehler 1989a; Coates 1990) require clarification.

Vocalisations and   Figure 9.64
other sounds

The ♂ on his song post is an inveterate caller, perhaps more persistent than any other bird of paradise. The most unusual aspect about ♂ calling is that it is carried out throughout the day, broken by departures for foraging. The following are typical: a nasal and plaintive series of *ca*, *wa*, or *wau* notes (Fig. 9.64), the series rising slightly in pitch; a lower-pitched, trilled, querulous, slowly-delivered *rahn rahn rahn rahn*; a pair of notes, high, plaintive, and nasal *Ki-kyer*, often repeated several times a min; finally a high but insistent *Ki kyer kyer kyer kyer kyer kyer kyer kyer kyer* rapidly delivered and dropping in pitch and speed towards the end of the series. Some vocalisations are strongly reminiscent of those of typical *Paradisaea* calls, whereas others closely approximate calls of the Magnificent Bird. A series of song notes often starts with several muffled and pathetic slower-delivered ones to be followed by the increasingly-loud and rapid typical song notes. Some songs are more rapid, higher-pitched, and urgent and last 9–10 sec or more. When undisturbed an ad ♂ frequently gave a harsh somewhat drawn-out and plaintive *caaar* from his display tree but when someone climbed into it the bird flew about giving a hissing *chee* alarm note (Rand 1938). Gilliard (1969) notes the following calls as given by an ad ♂ from his display tree: '*waa-waa-waa-waa-waa*, much like [the Lesser Bird] *P. minor* but with a higher pitch; it also called *kii-kii-kii-kii-kii-kii-kii* much in the manner of *P. minor*, but again higher in pitch. Other calls heard while the ♂ was on the display perches were *quaa-quaa-quaa-qa-qa*, rather deep and raspy, with the wings slightly open and the head bent downward; a single *kee* every 30 to 50 sec while squatting on the perch; a kingfisher-like *kreea, kreea*; and a drawn-out *kaa, kaa*. Also heard were insect-like buzzing and harsh, sharp *quaa* notes.' ♂♂

King Bird of Paradise *Cicinnurus regius* (Linnaeus, 1758)

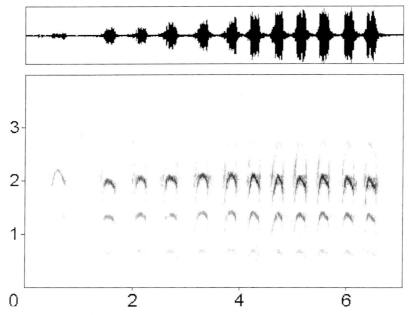

9.64 Sonogram from a recording of an ad ♂ King Bird's advertisement song at Brown R., Port Moresby on 31 July 1986 by K. D. Bishop. Note that vocalisations are structurally most *Paradisaea*-like.

call frequently from their display site(s) and occasionally two or several ♂♂ have been observed to associate and call much together (BB). The 'feeding call given by all birds' is a plaintive *weeo-weeo* with each note downslurred and descending, and fairly loud for the size of the bird (Finch 1983). Of Vogelkop birds Hoogerwerf (1971) described the 'fairly inconspicuous call' as a high-pitched unmelodius *kie-kwew-kwew-kwew* or *kwew-kwew-kwew*. An ad ♂ at Karawari R., PNG gave a repeated soft *wauk* (with a soft k) note 9–10 times (CF). Coates (1990) found the most common call near Port Moresby PNG to be reminiscent of ♂ Raggiana calls but higher-pitched and much less loud, the most frequent being a pair of subdued *whei-wha* or *hiia-haa* notes in which the first is higher. Sometimes the second note alone is given, or the *wha* may be repeated 2–3 times. A descending *whei-wher-wha* and a ringing and falling *wher-whei-wha* vigorously-given 7 to c. 15 notes represent characteristic advertising calls, with much variation in tone and volume. A descending *Queu-queu-queu* and a deep *Kraaa-Kraaa-Kraaa* is also given in advertisement, as is a rising series of c. 7 loud, throaty, *cho-chow-chaw-chaw-chaw-chaw-chai* notes similar to those of the Magnificent Bird. Other calls are single notes and a series of *caaw-caaw-caaw-caaw* that are cat-like in quality. Also a harsh, rasping, buzzy threat or alarm call note. Ad ♂♂ give a *Paradisaea*-like scold (BB). During courtship displays the ad ♂ produces specifically-different vocalisations, continuous for up to 1 min, that are subdued chittering or rhythmic twittering, churring, buzzing, and grating notes. Nestlings produce a 'suppressed, feeble peeping sound, rather like small mice' (Bergman 1957*d*). The wings of an ad ♂ noted to make a flipping sound in flight (BB).

Mating system

The promiscuous ♂ attends a court in thick sub-canopy vines in the shaded forest interior. Dispersion of ♂♂ is variable, from solitary to clustered, but the modal dispersion appears to be paired ♂♂, with mean distance between individuals of a pair 70 m. Earlier observation suggested ad ♂♂ occupy traditional display

King Bird of Paradise *Cicinnurus regius* (Linnaeus, 1758)

sites of one or two adjacent, tall trees with thick, dark, leafy canopies and abundant, thin, hanging vines (Rand 1938; Gilliard 1969). ♂♂ defoliate favoured perches and their immediate area. Berggy (1978) noted a 'clan' of displaying ♂♂ within earshot of one another in flooded forest, mostly of tall secondary growth with remnant emergent trees supporting heavy growth of figs and other climbers providing the display trees, near the Naru and Gogol R., PNG. Cooper (in Cooper and Forshaw 1977) noted that display trees of two adjacent ad ♂♂ were c. 100 m apart. Recent systematic studies show that ad ♂♂ appear to be dispersed predominantly in pairs c. 45–90 m apart although the odd group of > 2 may aggregate and form effective leks (Fig. 9.65). The exploded leks of pairs of ♂♂ were found between 150 to > 530 m apart and these appear to be dispersed into a ♂ mating system that is intermediate between a solitary one and an exploded lek in which ♂♂ are clumped into larger groups. Ad ♂♂ invest significantly more time attending their display perches than do ♂ Raggiana Birds at the lek (BB). An ad ♂ will tolerate another briefly in and about his display tree without overt aggression but he will follow the visitor about closely and may adopt a threat posture by leaning the front of his body forward and downward as he moves about whilst giving vocalisations indicative of his mood. Bishop (in Coates 1990) once saw two ad ♂♂ fall to the ground locked in combat. At Veimauri, PNG on 8 Aug three ad ♂♂ called 50 m apart. Territoriality of an ad ♂ was demonstrated at Laloki R., PNG on 10 Aug by it responding three times to play-back of a call by flying > 70 m to cross the river and perch above the point of play-back (BB).

Courtship behaviour

One of the first descriptions of display involves a behaviour not subsequently seen. Goodfellow (in Ogilvie-Grant 1915a) watched an ad ♂ 'soar into the air like a Sky-lark' [*Alauda arvensis*] from the canopy of river-edge rainforest to rise c. 10 m upward before seeming to collapse and drop back into a tree canopy as if shot. Reference to the Skylark probably implies the rising ♂ was singing, as a song of the sp. has been likened to that of the Skylark (Bergman 1957d). Gilliard (1969) noted that an ad ♂ 'suddenly dropped from its display arena like a rock, falling four to ten or more feet before it opened its wings.'

A ♂ will hop up vertical branches or vines of the display tree by 'reversing its body at each hop' (Rand 1938) zigzagging from right to left (Bergman 1968) or 'switching its tail and body from side to side' (Gilliard 1969). This action is performed by Raggiana and Red Birds and congenerics under similar circumstances (Frith 1976, 1981).

Most components of displays described for captives by Bergman (1956, 1957d) were reconfirmed by Coates (1990), who noted two basic display types by wild birds, one emphasising the pectoral fans and tail wires and the other the opened and vibrating wings. He provides a description of seven phases of display

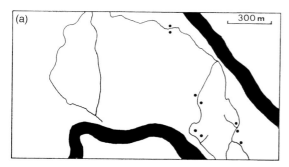
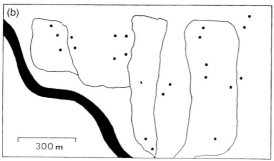

9.65 Display dispersion of ♂ King Birds at (a) the Nomad R. and (b) Kakoro areas. River lines are indicated by black areas. Solid circles are ♂ display sites and the lines are transect routes. From Beehler and Pruett-Jones (1983), © Springer-Verlag with kind permission.

King Bird of Paradise *Cicinnurus regius* (Linnaeus, 1758)

given in sequence which we here quote: 'Alone on his perch, the ♂ begins such a display sequence with (i) an initial brief partial spreading (cupping) and vibrating of both wings while more or less perched upright; this is the *Wing-cupping Phase* [Fig. 9.66(a)]. This is followed by (ii) the *Dancing Display*, given with the body more or less parallel to the perch. In this display he fluffs out his plumage, squats low to the perch, spreads the pectoral fans high, and cocks the tail so steeply that the wires are tilted forwards reaching above and beyond his head [Fig. 9.66(b,c)]. He then dances by vibrating his body and shaking the tail wires about, all the while singing his weak song. If a ♀ is present this display is given facing away from her (see below). The pectoral fan may be flexed, sometimes extending higher than the head, and he may make little steps along the branch and repeatedly lunge at the branch ahead, striking it with the bill. At times he pauses to look about, then continues. This display continues for very many sec, sometimes mins, then suddenly he turns completely about and, while in the same pose and still singing, performs for several sec (iii) the *Tail Swinging Phase* in which the tail is switched from side to side causing the wires to sweep from one side to the other in a wide arc. That usually concludes the display sequence but he may continue, to give (iv) the *Horizontal Open Wings Display* by spreading both wings out and forwards (as viewed from below) with the body parallel to the branch. With wings vibrating he rocks his body several times from side to side. This display is given in silence and lasts several sec then, suddenly, holding the same pose, he flips under the branch to repeat the display whilst inverted giving (v) the *Inverted Phase* of the *Open Wings Display*. This time the bill is held open and the head is turned from side to side [Fig. 9.66(d)]. This display usually lasts for only a few sec when, still under the branch, he closes his wings and

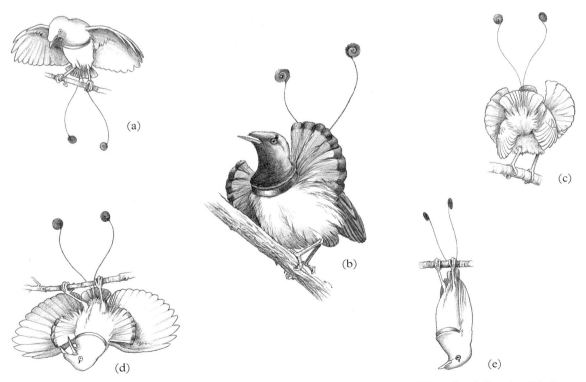

9.66 Some courtship display postures of ad ♂ King Bird: (a) *Wing-cupping Phase*, (b–c) *Dancing Display*, (d) *Inverted Open Wings Display*, (e) *Pendulum Display*. After Coates (1990), with kind permission. See text.

hangs suspended like a brightly-coloured fruit (or *Loriculus* hanging parrot). Then, looking about, he swings several times like a pendulum, from side to side to give (vi) the *Pendulum Display* [Fig. 9.66(e)]. This is brief, lasts only a few sec, and is also given in silence. He may then right himself on his perch or may, as a finale, drop like a stone from the branch into the clear space below to fly to another perch.'

Coates (1990) witnessed a partial display sequence and mating near Port Moresby at 1330 hr on 17 Oct. Both birds were perched on a horizontal branch just beneath the canopy and the ♂ was performing the *Dancing Display* while facing away from the ♀: 'The ♀, perching more or less across the branch, was very close to the ♂. The ♂ did not move along the branch and continued to display in this way for well over a min. Then, in the same pose and still singing, he began to quiver his wings which he held slightly open. The ♀ responded by slightly opening and quivering her wings. The ♂ then swiftly turned about and, still in the same display posture but with the bill partly open, rocked back and forth several times, each time striking the ♀ about the head with its bill. The ♀ adopted a passive, hunched posture and was turned slightly away from the ♂. He then briefly mounted, copulated, then flew forwards, downwards and away to perch in another part of the arena. The ♀ remained a few moments longer on the same perch before flying off.' On another occasion Coates watched an ad ♂ perform similar displays to an imm ♂ at only 6 m above the ground. Coates pointed out that his observations of displays agree, except for minor variations, with those recorded by Ingram (1907) and Bergman (above) and summarised the differences as follows: 'Bergman noted that a captive ♂ of the race *rex* [= *C. r. regius*] from western IJ, in the presence of a ♀, first raised his tail (and wires) at right angles to the body before performing a sequence comprising the *Wing-cupping, Horizontal Open Wings, Inverted Open Wings* and *Pendulum Display*. Also in a variation on the *Tail Swinging Phase* of the *Dancing Display*, Bergman noted that his bird usually, but not always, had the fans drawn in under the wings and that the body remained stationary.'

At Kakoro, PNG, at 70 m on 8 Sept a display court with an ad ♂ in full display was found (BB). He performed a 'wing display', a lot like a Raggiana Bird, on a sloping branch. His tail and head arched and bowed down, and wings out and fanned downward, tail wires below him, he performed a 'rocking display' consisting of violent rocking from side to side on the limb. This was followed by a 'wings inverted display' the bird dropping quickly upside down with wings out as before and performing a stiff, completely-inverted display with wings folded tightly, legs straightened, mouth wide open, and rocking from side to side with tail pointed skyward. All performed silently. The bird preferred to perch on vertical or near-vertical vines.

Breeding Table 6.5

NEST SITE: the only nest found in the wild (Aru Is) was in a tree hole 2.1 m above ground with an entrance aperture hole of *c.* 38 mm dia and a cavity *c.* 46 cm deep (Frost 1930).

NEST: the interior of the above tree cavity or hole was 'filled to within a few inches of the lip with palm fibres'. Frost's (1930) find was met with so much scepticism, as no other bird of paradise had been (or is) known to nest in a tree hole, that he attempted to breed the sp. in captivity in Java. He provided a pair with a vertical nesting log with internal cavity measurements of 128 mm dia and 153 mm depth. The ♀ did nest in this but her efforts went unobserved, during Frost's absence, until two nestlings were found dead in the log. The same ♀ again went to nest in the same log, moved to another aviary, and laid a single-egg clutch but did not incubate it. Bergman (1956, 1957*d*, 1968) subsequently bred the sp. in his aviaries in Sweden. The ♀ was given the choice of three nest boxes and she used the one with the largest entrance dia (85 mm).

EGGS: the two eggs of a clutch collected at Kobroor, Aru Is, IJ in late Mar were creamy-pink and marked with dark brown streaks on the larger end. Frost (1930) remarked that they were similar, except in their size (both 27.5 × 21 mm), to the eggs of the Greater Bird. A two-egg clutch laid in captivity in Sweden was 'cream-coloured, with numerous chocolate-brown streaks on the large end' (Bergman 1956). Bergman's captive ♀ laid the first of her two eggs 'either in the early morning or forenoon' of 11 June and circumstantial evidence suggested to him that the second egg was laid the following day. A single BMNH egg measures 27.2 × 20.4 (CF).

INCUBATION: started with the first egg and was entirely by the ♀ in Bergman's aviaries. Incubation lasted 17 days and bouts varied from a few to 43 min (Bergman 1968).

NESTLING CARE: Bergman noted that his captive ♀ ate egg shell pieces several times a day subsequent to her eggs hatching, although she had not eaten this while incubating. The ♀ fed her nestlings exclusively by regurgitating the meals, which included grasshoppers that she first dismembered. All nestling faecal sacs were eaten by the ♀ until nestlings reached 14 days old when she removed some in her bill. Bergman noted that, like several other birds of paradise, the ♀ parent ate predominantly fruits whilst giving her nestlings as much animal food as possible. In contrast to this the young were fed fruits in addition to grasshoppers once they left the nest, but not whilst within it. Bergman's captive ♀, despite being accustomed to him, performed a distraction display as he examined her all-but-fledged young by creeping 'with expanded wings and tail along the ground and among the branches and leaves' and at one point landed on his head to peck him and beat him with her wings. Coates (1990) records two ♀♀ seen with newly-fledged young in May. Finch and Lowry (1980) saw a ♀ with two young 'only just able to fly' near Port Moresby, PNG on 4 May.

NESTLING DEVELOPMENT: eyes opened at 5 days old with only wing flight feathers in pin. At 12 days nearly all their feathers were still in pin, just one or two wing feathers having just burst. At 14 days all feathers had sprouted and the young were being fed almost exclusively grasshoppers. The nestlings fledged at 14 days old, possibly a little earlier than would be the case in the wild because Bergman disturbed them and/or because of the availability of optimal food/ water and shelter. At this point the young birds' tail feathers were c. 30 mm long and they were fed fruits every day in addition to numerous grasshoppers by their mother. At 1 month old the young birds continued to beg with fluttering wings and feeble cheeps and their tail feathers were nearly as long as the ♀'s and their underparts were much paler than hers.

Annual cycle   Table 6.5

DISPLAY: observed in Port Moresby area during Oct–Jan (Coates 1990). BREEDING: at least Mar–Oct and clearly all months over the sp. range as a whole (Table 6.5). ♂♂ have been collected with much-enlarged gonads during Jan–Nov with the greater proportion of them in Apr–Oct. A ♀ with enlarged oocytes collected in Aug. Coates (1990) witnessed a copulation in the Port Moresby area on 17 Oct. Coates (1990) records a ♀ carrying nest material in mid-June. EGG-LAYING: late Mar is the only known wild laying. MOULT: 77 of 294 specimens were moulting involving at least one for each month but conspicuously the least moult during June–Sept and most during Oct–Jan with moderate activity during Feb–May (Appendix 3). Of a captive ad ♂ in Sweden, Bergman (1956, 1957d) noted that its moult started with the two tail wires, these being dropped on consecutive days, and the entire moult taking just over 3 months.

Status and conservation   Table 8.1

An estimation of 6 individuals per 10 ha has been made for forest near Port Moresby, PNG (Bell 1982a). Twenty-two ad ♂♂ were found during 11 days in 100 ha of mature forest at Kakoro, PNG, but only 9 ad ♂♂ in 150 ha of similar habitat during 6 days at Nomad R., PNG (Fig. 9.65). Diamond (1972) found the sp. to be fairly common and to represent c. 1.3% of the local avifauna at Soliabeda at

2000 m, near the Purari R., PNG. The sp. is under no threat.

Knowledge lacking and research priorities
The relationship established between paired or grouped ♂♂ in exploded leks merits examination. How much interaction occurs between ♂♂ in vocal but not visual contact? Is this sp. an obligate hole-nester? Diet requires clarification.

Aviculture
Numerous birds have been kept in NG, Europe, Asia, and the USA but all too few have been bred successfully, even though the sp. is relatively easy to breed (Frost 1930; Bergman 1956, 1957d, 1968).

Genus *Semioptera* Gray, 1859

Semioptera Gray, 1859. *Proceedings of the Zoological Society of London*, p. 130. Type, by monotypy, *Paradisea wallacii* Gray, 1859.

The Standardwing Bird of Paradise represents a **monotypic genus** geographically isolated together with *Lycocorax* on northern Moluccan Is. **Skull** characters typical of the Paradisaeinae, but particularly similar to those of *Ptiloris*, *Seleucidis*, and *Paradisaea* (Bock 1963). Recent studies of courtship displays and the finding of a nest and egg support the suggested close affinity with these genera and closely-related *Cicinnurus* (Frith 1992; Frith and Frith 1995b; Frith and Cooper 1996; Frith and Beehler 1997). No hybridisation known.

The ad ♂ c. 10% larger than the ♀. **Bill** about one-and-a-half-times longer than head, laterally compressed, decurved, and powerful. Culmen sharply keeled (as in *Cicinnurus*). **Tail** not graduated, central pair of rectrices markedly shortest in ad ♂♂ (as in *Ptiloris*). **Wing** long and rounded. Outermost primary mildly emarginated on inner vane. **Tarsus** 27% of wing length. Tarsus scuted distally, grading to smooth proximally.

Ad ♂♂ have two uniquely-elongated wing coverts on the bend of each wing that form the 'standards' used in display (Stephan 1967; Jeikowski and Stephan 1972), plus a large, winged breast-shield of iridescent green feathers. Note relict elongate flank plumes (*cf. Ptiloris magnificus* or *Paradisaea* spp.).

Sexual dimorphism considerable but plumage of both sexes far drabber than other dimorphic spp. (Plate 12). **Polygynous**, with promiscuous ad ♂♂ **lekking** to perform *Paradisaea*-like displays on perches plus a unique aerial 'parachute' display flight. Courtship display characters and unbarred plumage of ♀♀ are *Paradisaea*-like (Frith 1992). ♀♀ presumed to be uniparental in nesting duties. **Nest**, only recently described in general terms, is an open bowl-shaped structure typical of the Paradisaeinae (see below).

Standardwing Bird of Paradise *Semioptera wallacii* (Gray, 1859)

PLATE 12; FIGURES 9.67–8

Other names: Wallace's Standardwing, Wallace's Standard Wing Bird of Paradise, Wallace's Bird of Paradise, Standard Wing, Standard-winged Bird of Paradise. *Burong Plet* (or *Pleti*) of Halmahera I. (Goodfellow 1927). Bahasa Indonesian name is *Bidadari* and on Halmahera it is *Weka Weka*.

An aberrant form, the only ornamented bird of paradise inhabiting the Moluccan Is. Distinctive for its washed-out plumage and weird bill/head profile (both sexes), and, in the

Standardwing Bird of Paradise *Semioptera wallacii* (Gray, 1859)

lekking ad ♂♂, the striking white wing standards and the emerald green breast-shield. Polytypic; 2 subspp.

Description

ADULT ♂: 26 (28) cm, head and dorsal coloration subdued but complex. Culmen sharply keeled. A conspicuous tuft atop the upper mandible base and lores Buff (124) grading onto the somewhat flattened crown of dully-iridescent Lavender (77) grey with Lilac (76) sheen in some lights and this washing out to pale Beige (219D) sides of face with slightest of Lilac sheen. Mantle and back rich olive-brown (119A) becoming paler on rump and uppertail coverts to Dark Drab (119B) to even paler Light Drab (119C) on dorsal surface of tail with whitish central feather shafts. Central pair of tail feathers shortest. Upperwing coverts, tertials, and secondaries Dark to Light Drab. Alula and primary bases Drab-Grey (119D) fading to off-white at their tips. Two grossly-elongated, white lesser coverts form standards as long as, or longer than, the wing. Malar area, chin, and throat dark buffy (223D) with richly-iridescent bronzed yellow green feathering increasingly dense to form an extensive breast-shield of rich intensely-iridescent emerald green with bronzed green yellow sheen that in some lights may appear deep green-blue. Unlike breast-shields of riflebirds, the Superb Bird, and the parotias, that of the Standardwing is not of scale-like feathers. It has no clearly defined upper and lower central margin, the iridescent emerald green feathers thus encroaching onto the olive-brown lower breast and belly. Here they become progressively ill-defined until no more than a yellow-green sheen which does not reach the undertail coverts. Ventral surface of tail Drab-Grey. Underwing: coverts olive-brown, flight feathers Drab-Gray. Bill pale horn colour (92), iris dark brown, legs and feet yellow-orange, mouth colour unrecorded.

ADULT ♀: 23 cm, wing significantly shorter but tail longer than that of ad ♂ and central tail feathers considerably so. Generally uniform olive-brown, being darker above (28) and paler (119B) below. The Buff tuft above the nostril smaller and sparser than in ad ♂ and crown lacking the lavender-grey and with only slightest of discernible Lilac sheen. Upperwing as in ad ♂ but generally darker and the primaries more so, being pale cinnamon (223D). Dorsal surface of tail Dark Drab with whitish central feathers shafts. Lores, ear coverts, malar area, chin, and throat pale fawn (119D) and remaining underparts Dark Drab. Underwing and ventral surface of tail similar to, but darker than, that of ad ♂.

SUBADULT ♂: varies from as ad ♀ with a few feathers of ad ♂ plumage intruding, to as ad ♂ with a few feathers of ♀ plumage remaining; wing and tail on average intermediate between those of ad and imm ♂.

IMMATURE ♂: as ad ♀ but with longer tail. Wallace noted that ♀-plumaged birds far outnumbered fully-plumaged ♂♂ and suggested it is therefore possible that ♂♂ do not acquire ad plumage until third or fourth year of life. What we now know about other sexually-dimorphic birds of paradise suggests this may be an underestimate.

NESTLING–JUVENILE: unknown.

Distribution

Endemic to N Moluccas Is of Halmahera, Bacan, and Kasiruta. Ranges from low hills to 1100 m (on Bacan). Apparently patchily distributed on Halmahera.

Systematics, nomenclature, subspecies, weights, and measurements

SYSTEMATICS: its remarkable appearance, with the unique wing 'standards' of the ad ♂, the long pale bill, the generally faded or washed-out plumage in both sexes, as well as its geographical isolation in the Moluccas, has ensured this bird's placement in a monotypic genus. Recent observations of courtship displays and plumages indicate a closer relation-

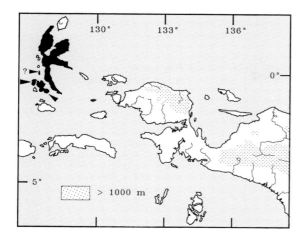

ship to *Cicinnurus/Paradisaea* than previously perceived (Frith 1992). This is one of several forms whose systematic position might be readily clarified by the use of molecular techniques. Hybridisation: unrecorded. Sympatric only with Paradise Crow.

SPECIES NOMENCLATURE AND ETYMOLOGY

Semioptera wallacii (Gray)
Paradisea (Semioptera) wallacii Gray, 1859. *Proceedings of the Zoological Society of London*, p. 130. Type specimen BMNH 73.5.12.141, near Labuha Village, Batchian I. [Bacan I.].

Recent debate on the correct spelling of both the generic and specific name of this bird (McAlpine 1979; LeCroy 1983, 1988; Mlikovsky 1989; LeCroy and Bock 1989) is the product of very confused and secondary accounts of a verbal description of the bird by Gray at a London scientific meeting. Gray's verbal account was based on a letter and accompanying sketch received from Wallace, prior to receipt of the specimens from the field. A submission to International Commission on Zoological Nomenclature resulted in conservation of *Semioptera wallacii* (Anon 1990).

The name of the bird's discoverer, Alfred Russel Wallace, has long been incorporated into the common name of this fascinating sp. The name Wallace's Standardwing continued this tradition without the wordy suffix 'Bird of Paradise,' but provided no clue as to what the animal is (another 'standardwing' is the Standard-winged Nightjar *Macrodipteryx longipennis* of Africa), as does unadorned 'Standardwing' (Johnsgard 1994). For this reason and because 'Standard Wing' was the original vernacular (Gould 1869; Wallace 1869) we adopt Standardwing Bird of Paradise rather than Wallace's Bird of Paradise. Moreover Standardwing (or Standard Wing) Bird of Paradise has been as widely used recently (Iredale 1950; Gruson 1976; White and Bruce 1986; Lambert and Young 1989; Bishop 1992; Frith 1992; Graves 1995) as the alternative names combined. Two long acknowledged subspp.

Etymology: *Semioptera* = Gr. *semeion* = sign, mark, star, standard or flag; and *ptera*, of *pteron*, = wing; *wallacii* = in honour of Alfred Russel Wallace (1823–1913), author of *The Malay Archipelago* and other books, who discovered the sp. on Bacan I. in 1858.

SUBSPECIES, WEIGHTS, AND MEASUREMENTS

1. *S. w. wallacii* (Gray, 1859)

Range: Bacan I., from low hills to 1150 m (Lambert 1994). The population recently reported on Kasiruta (Lambert 1994) may also prove to be referable to this subsp.

Measurements: wing, ad ♂ ($n = 29$) 152–162 (156), imm ♂ ($n = 11$), 141–156 (150), ad ♀ ($n = 26$) 135–154 (145); tail, ad ♂ ($n = 28$) 71–90 (78), imm ♂ ($n = 11$) 85–90 (88), ad ♀ ($n = 25$) 78–94 (84); tail centrals, ad ♂ ($n = 27$) 60–81 (66); bill, ad ♂ ($n = 26$) 41–48 (45), imm ♂ ($n = 11$) 44–47 (45), ad ♀ ($n = 24$), 42–47 (44); tarsus, ad ♂ ($n = 28$) 39–44 (42), imm ♂ ($n = 11$) 36–44 (41), ad ♀ ($n = 25$) 37–43 (40). Weight: ad ♂ ($n = 3$) 152–174 (163), imm ♂ ($n = 2$) 150–155 (153), ad ♀ ($n = 3$) 126–143 (135).

2. *S. w. halmaherae* Salvadori, 1881. *Ornitologia della Papuasia e della Molucche* **2**, p. 573. Halmahera. Type specimen MCG holds 8 syntypes = C.E. 25675–77, 22510–12 and 22514.

Range: Halmahera I., forests of low hills up to c. 1000 m.

Diagnosis: similar to nominate form but crown and nape (and in some individuals also mantle) with rich, pinkish, coppery purple iridescence. Mean wing length equivalent to nominate form; tail longer; other measurements fractionally smaller.

Measurements: wing, ad ♂ ($n = 32$) 153–162 (157), imm ♂ ($n = 8$) 147–157, (154), ad ♀ ($n = 25$) 138–151 (144); tail, ad ♂ ($n = 31$) 79–93 (85), imm ♂ ($n = 8$) 84–94 (90), ad ♀ ($n = 25$) 81–90 (86); tail centrals, ad ♂ ($n = 30$) 67–93 (74); bill, ad ♂ ($n = 31$) 40–48 (44), imm ♂ ($n = 7$) 42–47 (44), ad ♀ ($n = 24$) 41–46 (43); tarsus, ad ♂ ($n = 30$) 39–44 (42), imm ♂ ($n = 8$) 40–43 (42), ad ♀ ($n = 25$) 36–40 (39). Weight: none.

Habitat and habits Table 4.1

Lowlands to > 1000 m in forests on Halmahera and to > 1150 m on Bacan. On Halmahera, considered to occur mainly above 250 m, in forest characterised by tree genera including *Canarium*, *Eugenia*, *Diospyros*, *Vitex*, and *Agathis* with a diverse understorey flora dominated by palms with *Calamus* thickets. Believed to be absent from the forests of the flat lowlands and to be patchily distributed on the relatively steep, hilly terrain, particularly in limestone-based hill-forest. Rarely, in mature secondary woodland (Coates and Bishop 1997). Birds frequent the lower forest canopy and sub-canopy, and are inconspicuous, shy, and difficult to observe away from leks, where they usually forage in densely-foliaged forest canopy (Coates and Bishop 1997). In 1926, on Halmahera, ♂♂ at their lek were found to be curious and bold, birds coming down to within 5 m to investigate Goodfellow (1927). Birds are infrequently seen away from leks, but are then usually alone or in twos, but occasionally groups of 3–4 may be seen. Sometimes they join mixed-spp. foraging flocks, when they are difficult to observe well. From the behaviour, vocalisations, and numbers of birds at the lek after sunset and prior to sunrise Bishop (1992) concluded that ♂♂ roosted at the lek overnight.

Diet and foraging Table 4.2

Wallace (1869) noted birds foraging, apparently for arthropods, by clinging to tree trunks 'almost as easily as' (not 'like' as is oft-quoted) a woodpecker. Goodfellow (1927) did not observe this behaviour but Bishop (1984, 1992) did and noted it is perhaps more appropriate to liken this foraging to that of other birds of paradise, such as the sicklebills *Epimachus* spp. than to the unrelated woodpeckers. Birds have been seen foraging in groups of 3–4, including ad ♂♂, and feeding upon small fruits of < 10 mm dia. A bird collected on 28 Oct at Patani, SE Halmahera, had two nutmeg fruits, *Myristica fragrans*, in its gizzard (Bishop 1992). Wallace thought these birds primarily frugivorous, with the diet supplemented with arthropods, but Goodfellow's experiences with them led him to consider them 'very insectivorous' birds that preferred 'green-coloured insects, with a pronounced partiality for the larger soft-bodied grasshoppers [? = katydids or phasmids], nearly 5 or 6 inches long, which live on the branches of the coconut palms'. This is apparently based on his observation of dietary preferences of birds he held captive for export to England. He wrote of them 'dissecting these insects' and large tropical cockroaches with great skill, by which he probably referred to them holding such prey to a perch with one foot whilst removing their legs and wings as is performed by many typical birds of paradise. The proportion of fruit in the diet has been tentatively approximated to be 50–75% (Beehler and Pruett-Jones 1983), but this is likely to prove an over-rather than an under-estimate. Food found in a nest cup with a single nestling was fruit remains (D. Bagali *in litt.*). A 7 Apr 1995 Halmahera sighting involved two individuals possibly participating in a mixed-spp. foraging flock of 10 Golden Bulbuls *Ixos affinis*, 4 Spectacled Monarchs *Monarcha trivirgatus*, 1 Slaty Monarch *Myiagra galeata*, 1 Golden Whistler *Pachycephala pectoralis* and 2 Spangled Drongos (M. Poulsen *in litt.*).

Standardwing Bird of Paradise *Semioptera wallacii* (Gray, 1859)

Vocalisations and other sounds

Mostly produces sounds similar to those of the typical birds of paradise. Four vocalisations are apparent: (1) an advertisement call that was either a single or as many as 6–7 upslurred, loud, nasal, strident *wark* (barking of Bishop 1992) notes; (2) an aerial display call of far louder, raucous, and nasal upslurred barks; (3) musical chatter and twittering associated with high intensity display; and (4) contact bark notes given away from the lek. Fig. 9.67(a) shows call (1) of six clear notes followed by five churring notes of lower amplitude. The most commonly-heard vocalisation is (4), a series of loud, upslurred, nasal *wark* notes given far less intensely than in advertisement ((1)) by foraging ad ♂♂ or by birds moving about the forest canopy away from a lek, which would appear most likely to be a contact call in view of the context in which it is given. These may be heard by an observer from as far as *c*. 300 m (Coates and Bishop 1997). Two basic kinds of vocalisations given by lekking ♂♂ are described (Bishop 1992):

Advertisement Calls—The first calls prior to dawn were single, loud, nasal, upslurred 'bark' notes somewhat like the notes given away from the lek in daylight. Within 15 min calls increased in volume, intensity, and complexity; the main call included a series of 6–7 loud, raucous, coarse, nasal upslurred barks increasingly loud towards their conclusion. These calls are distinctly pulsed, each note being well structured to form a complex sound (Fig. 9.67 (a)). Such calls were often answered by other ♂♂. ♂♂ were noted to give an individually-distinct call by Bishop (1992), possibly reflecting age/experience (e.g. Fig. 9.67(b)). Another call recorded by the BBC (Fig. 9.67(c)) consisted of *c*. 9 notes as the last

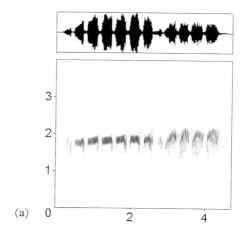

(a)

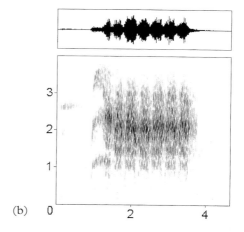

(b)

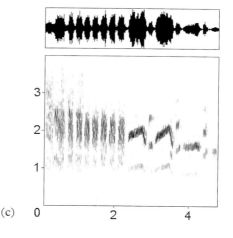

(c)

9.67 Sonograms from recordings of ad ♂ Standardwings' typical (a) and less typical (b,c) lek advertisement song at *c*. 20 km inland of Sidangoli, Halmahera I. in June 1987 by K. D. Bishop.

described but faster-delivered and followed by three distinctive friarbird (*Philemon* spp.)-like *waa-kuck* notes (each initially upward-but ending with downward-inflection) descending as the call fades out.

Aerial Display Call—Immediately prior to performing the *Aerial Display* (see Courtship Behaviour) the calling of ♂♂ increases in intensity, becoming almost cacophonous, as ♂♂ leap upwards from their perches. During this display ♂♂ call by giving loud, raucous, nasal, upslurred barks whilst others give a more musical chatter and/or twitter lasting 5–10 sec. These vocalisations attracted ♀-plumaged birds to the lek area, and usually indicated a ♀ had arrived at the lek.

Coates (in Frith 1992) considered 'a peculiar mildly explosive cacophony of harsh buzzy notes lasting *c.* 2–5 sec given in synchrony by at least two ♂♂' to be lek advertisement calls. ♂♂ vigorously hopping about their lek gave variable, harsh, loud, churring *waughh* notes repeated several to many times and sometimes also a clearer *wau-wau-wau* ... note more like that of the Twelve-wired Bird and plumed *Paradisaea* spp. Vocalisations have struck several ornithologists with a knowledge of several living bird of paradise spp. as much like those of *Cicinurus* and *Paradisaea* spp. David Bishop (1992) observed that as ♂♂ landed on their display perch after the *Aerial Display* they gave a series of high-pitched, sweet, rapidly-repeated twittering notes quite distinct from their other calls. The twittering call was given when ad ♂♂ were displaying on their perch with breast-shield and wings extended and standards erect. It was also given prior to the *Aerial Display* but not with the same intensity. This call would appear to be that described by Goodfellow (1927) as that 'reminding one of our English Starling [*Sturnus vulgaris*] on the chimney stacks, bubbling, gurgling, and producing occasional explosive sounds which shake their whole body'. Friedmann (1934) described another (the same?) call as 'a series of low, guttural, almost whisper-like sounds like that of faint cracking of twigs. The throat dilates slightly as the successive notes are given.' Daily vocal activity peaks close to 0800 hr when chances of observing birds are four times greater than 1.5 hr before that time (Anon 1995; M. Poulsen *in litt.*). Flying birds produce an audible and distinctive wing-rustling but this is not as loud as that produced by the Magnificent Riflebird or the Twelve-wired Bird. Individuals have been heard clicking their mandibles in unison at the lek (see Courtship Behaviour).

Mating system

Walter Goodfellow was the first to provide any clue to the mating system. In 1926 he was led to an area near Patani, Halmahera I. where he saw at least 30 birds, all but two in ad ♂ plumage, in comparatively low trees. The birds were constantly animated, flying back and forth 'from tree to tree with a great fluttering of their wings, and at times hanging in all sorts of positions from the slender branches, some turning round and round like a cartwheel, and all the time making a variety of squawks and calls.' He felt sure that 'these low trees formed a dancing place' and that what he saw the birds perform was the display, as it was clearly not an area the birds fed in. He observed that the birds left the area, or lek, at midday (Goodfellow 1927). Recent observations of communal displays by ♂♂ aggregating in a few adjacent trees to form a lek with traditional (worn) perches over at least 4 yr leave no doubt the sp. is polygynous and ♂♂ are promiscuous. ♀♀ almost certainly nest alone and unaided. The lek visited by Goodfellow was in a 'stunted kind of jungle' with enormous trees here and there but the majority of trees straight saplings 9–12 m high with a few quite short branches and very little undergrowth and with none at all in places. Birds frequented 'comparatively low trees', too slender to take the weight of an adult person, and were flying back and forth from tree to tree. Wallace also noted that the bird frequents 'the lower trees of the forests and is in constant motion'. ♂♂ vigorously hopping about the lek tree(s) cause foliage to move about constantly in a way that is conspicuous from a distance. The best-known lek, located and studied by Bishop (1992) on Halmahera, was situated at the ridgecrest of a range of low hills cloaked in tropical rainforest with little to distinguish the site from any other. The lek was on a slight knoll on a N-facing slope *c*. 50 m above and from a large, ephemerally-flooded, boulder-strewn creek. The surrounding moist limestone hill-forest had been selectively logged, mostly for *Agathis*, but the overall forest canopy remained intact. Display perches were mostly thin, near-vertical branches or twigs, 25–30 m above ground in the lower crown of tall forest canopy trees. Typically 5–7 ad ♂♂, but up to 12 birds including ♀-plumaged ones, commenced lek activity at *c*. 0535 hr, approximately 20 min before sunrise, with a series of intermittent, tentative calls. Vocalisations became progressively more complex and intense, with increasing individual participation which reached a crescendo at *c*. 0615 hr. Although too dark to observe until *c*. 0610, individual birds could be heard rustling, flapping, and 'cracking' their wings and clicking their bills suggesting that display was underway. Bouts of intense calling combined with progressively-intense display postures continued until *c*. 0730 hr. During this period ♀-plumaged birds and one or two imm ♂♂ appeared at the lek. Copulation occasionally occurred during this period and was observed on 4 of 27 observation days. Activity continued thereafter but with less zest after 0800 hr, until *c*. 0830 hr and occasionally until 1030 hr if dull or overcast—the birds perhaps adverse to displaying in full sun. ♀♀ and imm ♂♂ invariably departed the lek before ad ♂♂, birds leaving singly or in groups of 3–4, and birds rarely, if ever, remained at the lek over the middle of the day (Bishop 1992). At this location in June 1990 Coates (*in litt.*) observed one ♂ displaying low in the understory *c*. 1 m above ground on a hill slope. Standardwings remove foliage from lek perches by plucking or tearing off leaves, as do plumed *Paradisaea* birds and other spp. displaying at traditional perches or a court. One lek at Tanah Batu Putih, Halmahera I. was said to have up to 100 birds at it (Anon 1995) but while three other visitors to it estimated 80–100 birds to be in the general area they saw no more than 30–40 birds in the lek tree(s) at once (M. Poulsen *in litt.*).

Courtship behaviour

Bishop (1984, 1992) found the display of the ♂ Standardwing to be complex and to include several components. He discovered that ad ♂♂ perform a hitherto unsuspected and spectacular communal *Aerial Display*. Interestingly, the major constituent phases of the display, different in character though some of them are, can be categorised into the main components of display performed by ♂♂ of most *Paradisaea* spp.

Convergence Display: initially a ♂, or several ♂♂, charges along or about branches of several adjacent sapling crowns of the lek whilst frequently beating partly-opened wings 4–5 times with a rowing-like action and giving rapidly-repeated sharp call notes. During this *Charging* the breast-shield is not utilised and the wing standards are only slightly raised (Fig. 9.68 (a)) as birds flutter from perch to perch, travelling horizontally and/or vertically and often pursuing a ♀-plumaged bird. This display sometimes develops into a *Wing Pose* posture (Fig. 9.68 (b)) or a *Wing Standard Display* (Fig. 9.68 (c–f)).

Aerial Display: is performed after an initial period of excited display as above. Fully-plumaged ad ♂♂ perform this one after another and individual birds may perform it at least twice in immediate succession. It is started by a ♂ perching upright on his steeply-angled perch with wings closed, standards and breast-shield relaxed. As he leans forward and begins to call he stretches his head and neck upward and opens his breast-shield slightly. Duetting and counter-singing with and against other ♂♂ continues for 2–4 min with increasing speed and intensity until he leaps vertically upward from his perch by explosively, fully, extending his legs, and instantaneously following this with the first deep down beat of fully-open wings. Wing-beats vary in number from 4 to 23 (mean = 12) during the vertical or nearly vertical ascent flight and are extremely powerful, deep, and rapid; so rapid (mean = 9/sec, range = 2–15/sec) that they can only be clearly observed during slowed playback of film of display.

At the apex of the ascent flight, which averages 7 m (range = 4–11 m) and 1.3 sec (range = 1–2.5 sec) in duration, the bird stops flapping and holds the wings fully extended and open, turns to point bill downward–and holds his fully-open wings in slightly-decurved profile just below horizontal with the back (see Fig. 9.68 (g)). In this posture the bird 'floats' downward whilst rapidly vibrating the conspicuously-white primaries vertically through a small arc so that they appear as a pale blur at

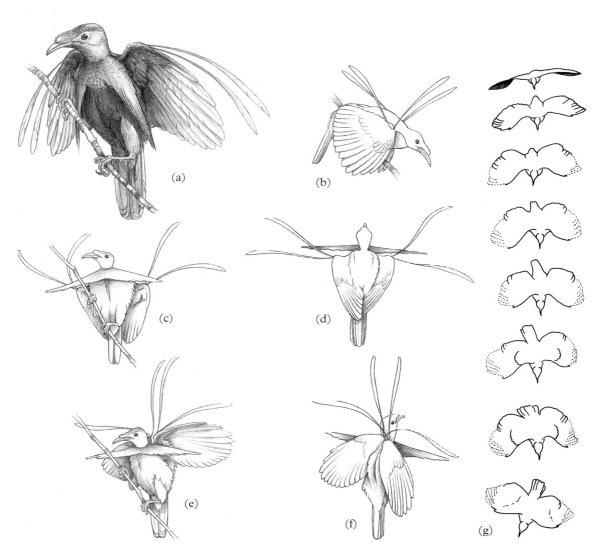

9.68 Some courtship display postures of ad ♂ Standardwing: (a) *Charging*, (b) *Wing Pose*, (c–f) *Wing Standard Display*—with carpal joints apart (c–d) and together (e–f), and (g) *Aerial Display*. After Frith (1992), with kind permission. See text.

either wing tip. In most descents, which last an average of 2.2 sec (range = 1–3 sec), the tail is closed but occasionally it is fanned for the latter part of the descent, presumably as a brake for landing. In 14 complete *Aerial Displays* observed, the ♂ returned to the perch he took off from or to an immediately-adjacent one 10 times, and in only four instances did he alight on another perch close by. Owing to the speed of events, it is difficult to be sure of fine feather detail but it would appear that the wing standards are held vertically upright above the upperwing surface and these are blown backward by the airflow.

A captive ♂ was apparently unable to perform the *Aerial Display* owing to lack of space but did perform the fully-opened wing posture of this display at the appropriate point in his display sequence while perched (Friedmann 1934: fig. 5). This *Aerial Display*, unique to the sp., may have originated, or evolved, from initial charging about the lek from perch to perch and/or from inverted open-winged displays such as are performed by ♂♂ of most *Paradisaea* spp. as a *Static Display* (see Chapter 5) with parallel evolution of the wing standards.

Static Display: it is not clear from the only copulation reported (Bishop 1992) where this event fitted into the context of the complete display sequence or, if indeed, this particular mating was typical of the sp. with respect to ♂ display. It would seem probable, however, that copulation usually follows the perched *Static Display* postures that typically immediately follow *Aerial Display*. This consists of the *Wing Standard Display* and it is usually performed on vertical or near-vertical lek perches, rarely on horizontal ones. The ♂'s body is held rigidly upright and the bill usually slightly raised as the bird repeatedly changes the position of his vibrating wings from only partly opened and slightly lifted out and away from the body with the carpal joints held conspicuously wide apart (Fig. 9.68 (c,d)) to half or nearly fully open with carpal points held together (Fig. 9.68 (e,f)). The sudden alternating of these two positions causes striking gyrations of the four raised white carpal standards. The carpal joint movements appear analogous to *Paradisaea* behaviour. A crack sound (the 'wing crack' of Bishop 1992) like the faint cracking of twigs is produced, which Friedmann (1934) clearly saw to be a vocalisation, and the 'rowing' of vibrating primaries is made conspicuous by their contrasting pale whitish coloration. During such perch displaying the crown feathering is raised and lowered, enhancing its iridescent lilac-purple colours, and the upper breast-shield meets behind the neck. During this display the bird may also sharply click his mandibles.

Copulatory sequence: at 0645 hr on 10 May one of two ♀-plumaged birds approached seven ad ♂♂ as they performed the *Wing Standard Display* and flew to the most frequently-occupied display perch. The ♂ perching there immediately increased his vocalisations and bounced from perch to perch with horizontally-extended, quivering wings as he snapped his breast-shield open and held his standards vertically and then horizontally. All other ♂♂ simultaneously performed frenzied vocalisations and displays. The ♀ leant forward on a horizontal branch attached to the display perch of the ad ♂ and he mounted her to copulate. While copulating, his wings were extended horizontally but somewhat backwards, his standards vertical and his breast-shield erect. As copulation took place the other ♂♂ on the lek performed *Aerial Display* (Bishop 1992).

The above account includes all display postures described by Friedmann (1934, 1935) and Crandall (1936). ♂♂ at the lek give the impression of usually competing with each other in a non-physical way, by display, but on one occasion two neighbouring ad ♂♂ did grapple with one another to tumble to the forest floor and fought there until caught and disentangled (Bishop 1992).

Breeding   Table 6.5

An observation of copulation at a lek at 0645 hr on 10 May 1985 was the only indication of breeding (Bishop 1992) until a Halmahera nest containing a single egg was discovered on 9

May 1995 by Demianus Bagali (Frith and Poulsen in press).

NEST SITE: only one described, 10 m above ground between the foliage of a large epiphytic orchid and the trunk of an unidentified palm tree where the frond bases meet immediately beneath its canopy (D. Bagali in Frith and Poulsen in press).

NEST: the only one known (Anon 1995) is described as a open cup structure 25 × 15 cm and 7 cm deep (total depth or inside cup?) with the bottom constructed of dried leaves (D. Bagali in Frith and Poulsen in press).

EGGS: the only clutch described consisted of a single 'oval' egg (D. Bagali in Frith and Poulsen in press).

INCUBATION, NESTLING CARE, AND DEVELOPMENT: unknown.

Annual cycle   Table 6.5

DISPLAY: low display intensity and freshly-plumaged birds noted by Bishop (1992) during Apr suggested display started early in the dry season, *c.* Apr, and finished with the onset of the wet season *c.* Dec but this requires confirmation. BREEDING: at least May–Sept. ♂♂ have been collected with much-enlarged gonads during Apr–Oct and a single ♀ in Sept. EGG-LAYING: the nest with one egg found 9 May had a nestling on 9 June which left the nest in that month. MOULT: 20 of 105 specimens were moulting, fewer during July–Oct and most predominantly during Nov–June (Appendix 3).

Status and conservation   Table 8.1

Moderately common on Bacan I. in both primary and logged forest between 70 and 1150 m and rather scarce in flat areas of logged forest close to Ra R. (Lambert 1994). Also observed on Kasiruta I. (Lambert 1994). Bishop (1992) failed to locate birds in the remaining lowland forests around Sidangoli and N along the coast to the town of Gilolo and at the foot of Gunong Gamkanora on the W coast of the northern peninsula of Halmahera I. where Ripley (1959) also failed to find it. Apparently rare in the area of Tobelo. Also absent from fine-looking coastal forest on the E shores of the E penin (D. Bishop *in litt.*). Generally common on Halmahera and Bacan but may be locally so on the former (Coates and Bishop 1997). In 1976–8 the area of forest remaining on Halmahera and Bacan was approximated to be 19 000 km^2 and less than 10% of this was on limestone, which Bishop (1992) thought Standardwings prefer. Subsequent logging, agriculture, and clearance for transmigration settlements has resulted in the near complete destruction of the lowland forests and considerable damage to the hill forest, which Bishop (1992) considered a serious threat to the survival of this bird. An expedition to Halmahera performed systematic (variable circular plot) bird surveys in July–Sept 1994 resulting in the average figures of 0.2–0.3 Standardwings per ha on limestone and non-calcareous sedimentary rock while few were seen on igneous rock and in montane areas (Mackinnon *et al.* 1995). BirdLife International ornithologists performed systematic (variable circular plot) survey counts during 1994–5 to calculate relative abundance in primary lowland forest on Halmahera. Results were averages of 0.4–0.8 birds per ha with the higher number on volcanic substrate (Frith and Poulsen in press). Further work is required to explain these differences.

Knowledge lacking and research priorities

Both the conservation status and subspecific identification of the population on Kasiruta I. are a priority. Resolution of the relationship of this genus to the remainder of the family might be possible with a molecular systematic study. Nesting biology awaits study as does ad and nestling diet and foraging ecology. It would be most valuable to confirm the existence of the 'mega-leks' with as many as 100 ♂♂—which seems inconceivable given lek size in other birds of paradise. It would be particularly interesting to determine inter-lek distance for such giant leks, should they be located.

Aviculture
Other than the birds collected live by Wallace (1869), the paper by Friedmann (1934) based on the ♂ in the zoo in Washington, D.C., and the captive photographed at NYZS (Crandall 1936) there appears to be no further published records of this sp. in captivity. Wallace found it a most hardy bird even in the most trying of circumstances.

Genus *Seleucidis* Lesson, 1835

Seleucidis Lesson, 1835. *Histoire Naturelle des Oiseaux de Paradis et des Épimaques*, Synopsis p. 28, pl. 35. Type *Seleucidis acanthilis* Lesson, *ibid* = *Paradisea melanoleuca* Daudin.

The Twelve-wired Bird of Paradise, the sole sp. of **monotypic** *Seleucidis*, is distributed extensively through the lowlands of mainland NG and Salawati I. According to Bock (1963) **skull** characters are little different from those of *Paradisaea*, but bill long, thin, and slightly laterally compressed, much like *Ptiloris*. Lacrymal large with an expanded base resting on the jugal bar as in *Ptiloris*. We link the genus to the *Semioptera/Cicinnurus/Paradisaea* complex, based on characters of its courtship display (Frith and Beehler 1997), plumage, pale leg and eye colour, and vocalisation, but other authorities argue that the sister form to *Seleucidis* is *Ptiloris*. We acknowledge that there are characters that link *Seleucidis* to both lineages. Known to **hybridise** intergenerically with *Ptiloris* and *Paradisaea*.

Adult ♂♂ c. 5% larger than ♀♀. **Bill** long, straight, and laterally compressed to very narrow, like that of *Ptiloris* but less decurved, and over twice the head length. Nostril slightly covered by feathers. **Tail** short and square-ended, central pair of feathers not longest. **Wing** moderately long; primaries P9 and P10 inwardly curving and pointed (adult ♂) but not emarginated. **Tarsus** length 23% of wing length. Tarsus is scuted distally, but this grades to only faintly scuted proximally. Feet extremely powerful and claws particularly large and curved, and grey, contrasting the coral-pink feet.

Dissection of a single adult ♂ documented a unique modification of the **trachea**, involving eight tracheal rings that are slightly dilated and flattened antero-posteriorly and ossified at the middle whilst remaining cartilaginous at their borders, which are flared. The intervals between these eight modified rings were deeper than elsewhere and consisted of delicate membrane that made this section of the trachea highly elastic. Presumably this unusual tracheal modification influences ♂ vocalisation (Forbes 1882a); examination of additional material is desirable. **Tongue** very long and extensible, but flat and a little fibrous distally, much like that of *Paradisaea* (Wallace 1869). **Thighs** in adult ♂♂ unfeathered, brightly pigmented as are the legs, and are featured in courtship display (Frith and Beehler 1997).

Strikingly **sexually dimorphic**, the ♂ and ♀ sharing virtually no plumage characters. Six more heavily-shafted plumes on each flank of the adult ♂ are grossly elongated and sharply recurved to rise dorsally and flare outward to each side of the bird (Plate 8). If straightened (as they are in flight) these wires extend to the length equivalent to the flank plumes of typical *Paradisaea* spp., and we believe they share a common origin. Adults of both sexes red-eyed.

Polygynous, with ad ♂♂ territorial and solitary, displaying at traditional perches, the ♀ managing all nesting duties. **Nest** typical of Paradisaeinae and situated in the base of the leafing head of a pandanus. **Egg** as in typical Paradisaeinae and most like those of *Ptiloris* and *Paradisaea* (Plate 13).

Twelve-wired Bird of Paradise *Seleucidis melanoleuca* (Daudin, 1800)

PLATE 8; FIGURES 4.5, 9.69–71

Other names: Twelve-wired Paradise Bird; *Seleucides ignotus, Seleucides ignota*. *Isop* at Etna Bay, IJ (van Oort 1909). *Palengo* on Biak I. (Mayr and Meyer de Schauensee 1939*a*) and near Sorong (Bergman 1957a). *Man* or *Karagambo* of the middle Sepik region (Gilliard and LeCroy 1966). *Iwwe* of the Fair language at Bomakia and *Hanggo'* of the Kombai language at Uni, IJ (Rumbiak 1994). *Laba* of the Paiwaian and Gimi people of Crater Mt, PNG (D. Gillison *in litt.*).

A retiring but robust, broad-winged, and short-tailed inhabitant of lowland forest canopy. Both sexes have a long bill, very powerful legs, and large-clawed feet. The bright pink tarsi and lack of pale supercilium and malar markings distinguish ♀-plumaged birds from all other birds of paradise. Best means of discovery is the powerful and resonant advertisement calls of the ♂. Polytypic; 2 subspp.

Description

ADULT ♂: 33 cm, entire head velvety jet black with dull, iridescent, coppery Olive-Green (47) sheen throughout in some lights but a rich, deep, Purple (1) iridescence on entire crown also apparent. Mantle to uppertail coverts, including lesser wing coverts, velvety jet black with iridescent oily sheen of rich coppery Olive-Green but with Purple washes in certain lights. Greater wing coverts, alula, tertials, secondaries, and dorsal surface of tail highly-iridescent, deep violet purple (172) and/or Magenta (2) in suitable light. Primaries virtually jet black. Chin, throat, and entire breast velvety jet black with slight, coppery yellowish green, iridescent sheen in rare light. Large feathers bordering lower breast broadly-tipped, highly-iridescent, rich emerald green (62) with, only slightly exposed, rich violet purple adjacent bases, to form a gorget that extends up each side of the breast. Breast feathers elongate and dense, forming an extensive breast 'cushion.' Remaining underparts, including grossly-elongated, filamentous, and inwardly-curving flank feathers, brilliant yellow (55). The white central shaft of six flank plumes on each side of the bird are grossly elongated into black 'wires' beyond their vanes that after recurving upward are again white, and the finely-tapered tips spread about the lower bird like the spokes of a wheel. Underwing entirely brownish black (119), ventral surface of tail even blacker. Bill shiny black, iris bright blood red, legs (including bare thighs) and feet deep pink (108B), claws pale mid grey or grey brown, mouth aqua-green.

ADULT ♀: 35 cm, generally smaller than ad ♂ but longer in the tail. Upper head and upper mantle sooty black with dull, iridescent, deep Purple sheen in some lights. Remaining mantle and entire upperparts rich chestnut brown (223A) slightly paler, more Raw Sienna (136) on exposed primaries. Inner, concealed, vanes of wing flight feathers blackish. Malar area, throat onto upper breast greyish white, flecked, spotted, and then barred, respectively, blackish. Remaining pale buff (123D) underparts, washed Cinnamon (39) on breast, flanks, vent, and undertail coverts, uniformly barred blackish. Underwing: coverts Ground Cinnamon (239) with broad darker greyish brown barring, flight feathers Dark Neutral Gray (83) with broad Ground Cinnamon trailing edges except outer two primaries. Ventral surface of tail Cinnamon-Brown (33). Legs and feet slightly duller pink than in ♂.

SUBADULT ♂: as ad ♀ but with a few feathers of ad ♂ plumage intruding, to as ad ♂ with a few remaining feathers of ♀ plumage. Individuals can be striking—with the plumage part-♀ and part-ad ♂, in most instances with ad ♂ head and breast plumage with part-♀, part-ad ♂ wings, tail, and lower underparts (see Plate 8). Iris varies from yellow to almost red with age.

Twelve-wired Bird of Paradise *Seleucidis melanoleuca* (Daudin, 1800)

IMMATURE ♂: like ad ♀, but in some specimens obviously paler and plumage washed with a sandy orange yellow. Iris pale brown, turning to yellow with age. Some individuals have a brown base to the mandibles. Central tail feather pair become increasingly shorter and blunter with age. Tail much longer than that of ad ♂.

NESTLING–JUVENILE: nestling initially naked and pale-skinned. A nestling's iris colour was dirty yellow but when 9 months old it was mid grey-blue (Khin May Nyunt personal communication).

Distribution

Lowland forests of all of NG except NE PNG and the PNG penin E of Port Moresby, where there is little or no coastal plain (from the Ramu R. SE to Milne Bay). Also Salawati I., IJ. Mainly near sea level, but has been recorded as high as 180 m.

Systematics, nomenclature, subspecies, weights, and measurements

SYSTEMATICS: Lesson erected the monotypic genus *Seleucidis* in 1835 for this bird and it has remained as such almost ever since. The position of the genus within the family has, however, been far from settled (see Table 3.3 and Fig. 3.3). Ad ♂♂ show characters of more bird of paradise genera than any other: plumes of *Paradisaea*, pectoral 'fan-like' plume feathers of *Epimachus*, bill form and plush chest 'cushion' feathers like *Ptiloris* and *Paradisaea*, iridescent purple and green of *Astrapia* and *Ptiloris*, a buff tail and ventral barring like *Epimachus* and *Drepanornis* (in imm), black head in ♀ plumage as in *Parotia* and *Lophorina*, and a red iris as in *Epimachus*. If this sp. were today known from only one or two ad ♂ skins it would probably be considered a hybrid! Hybridisation: known with Magnificent Riflebird and Lesser Bird (Plate 15 and Appendix 1).

SPECIES NOMENCLATURE AND ETYMOLOGY

Seleucidis melanoleuca (Daudin)
Paradisea melanoleuca Daudin, 1800. *Traité d'Ornithologie* (Lesson) **2**, p. 278. Waigiou, in error for Salawati or the Vogelkop. Type specimen none designated.

Synonyms: *Paradisea ignota* Forster, 1781. Zoologica Indica, pp. 31, 36, No. 6 (ex Valentyn).

Etymology: *Seleucidis* = L. migrant birds sent by the gods to destroy locusts (i.e. birds from paradise). Also: the Seleucidans were people

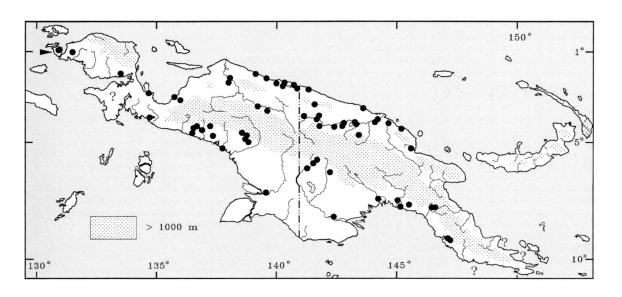

inhabiting Seleucia of the Middle East. *Seleucus Nicator* was a general of Alexander the Great's army; *melanoleuca* = Gr. *melas*, black; *leukos*, white.

SUBSPECIES, WEIGHTS, AND MEASUREMENTS

1. *S. m. melanoleuca* (Daudin, 1800)

Range: Salawati I. and W and S NG (from the Vogelkop eastward in the N watershed to the Mamberamo R., IJ; E in the southern watershed to Milne Bay).

Measurements: wing, ad ♂ (n = 34) 167–185 (177), imm ♂ (n = 22) 163–184 (177), ad ♀ (n = 21) 158–171 (165); tail, ad ♂ (n = 34) 62–86 (71), imm ♂ (n = 21) 102–120 (110), ad ♀ (n = 19) 102–113 (108); bill, ad ♂ (n = 32) 65–75 (71), imm ♂ (n = 22) 64–73 (70), ad ♀ (n = 21) 57–69 (65); tarsus, ad ♂ (n = 33) 31–44 (41), imm ♂ (n = 22) 38–43 (41), ad ♀ (n = 20) 37–40 (38). Weight: ad ♂ (n = 4) 190–217 (205), imm ♂ (n = 3) 182–220 (202), ad ♀ (n = 2) 186–188 (187).

2. *S. m. auripennis* Schlüter, 1911. *Falco* 7, 2. Dallmannshafen (= Wewak), German New Guinea. Type specimen AMNH 677823.

Range: northern NG, from the Mamberamo R., IJ, E to the Ramu R., PNG.

Diagnosis: similar to but on average generally smaller than the nominate form (*c.* 6%), the difference most prominent in bill length. Underparts of ♀ darker, more brownish, and more heavily barred than nominate.

Measurements: wing, ad ♂ (n = 25) 163–173 (168), imm ♂ (n = 15) 161–169 (166), ad ♀ (n = 19) 152–169 (158); tail, ad ♂ (n = 25) 61–70 (65), imm ♂ (n = 15) 95–112 (100), ad ♀ (n = 19) 93–111 (99); bill, ad ♂ (n = 25) 63–68 (65), imm ♂ (n = 14) 60–69 (64), ad ♀ (n = 18) 56–64 (61); tarsus, ad ♂ (n = 25) 37–42 (39), imm ♂ (n = 15) 37–40 (39), ad ♀ (n = 19) 34–40 (37). Weight: ad ♂ (n = 12) 170–202 (189), imm ♂ (n = 5) 158–188 (175), ad ♀ (n = 3) 160–163 (161).

Habitat and habits   Table 4.1

Flat lowland rainforest of various types, but apparently most common in swamp forest that is permanently or seasonally flooded and that supports pandanus and sago palms. Said to favour the seaward, river-, and lake-edge of forests, but this may well reflect the bias of water-bourne natural history collectors and ornithologists, where travel and visibility are easier, and not a real preference by the birds. Considerably more widespread than originally believed; often overlooked because of its retiring habits. Ad ♂♂ have been found by all observers to be shy and retiring, being heard far more often than seen, other than at display perches. Common on uplifted karst hill forest at 180 m in the West Sepik Province of PNG (BB). The 12 wire-like flank plumes normally radiate about the bird (see Plate 8) rather than as usually depicted by artists. Grant (in Ogilvie-Grant 1915a) described the flight as swift and graceful, and CF noted it as direct and consisting of shallow-dipping, brief, gliding undulations interspersed with 3–4 wing flaps accompanied by a swishing wing-noise. In flight the 12 pectoral plume wires are somewhat straightened by air flow to more than two-thirds the length of the body. Multiple ♂♂ occasionally seen together. At Krissa, West Sepik, PNG, on 9 Aug 3 ♂♂ of varying ages were together: an ad, a subad with chestnut wings, and a ♀-plumaged bird except for black throat and breast and tan flanks (no barring). Oldest and youngest birds touched bills several times then played follow-the-leader through the canopy (BB). Observations of local peoples' hunting techniques made by Wallace (1869), Guillemard (1886), and Bergman (1968) clearly indicate that birds use traditional night roost perches in low bushy trees, at *c.* 3–4 m above ground.

Diet and foraging   Table 4.2

Diet is fruit and arthropods, the relative proportions being tentatively thought to be *c.* 50 : 50 (Beehler and Pruett-Jones 1983). May specialise on pandanus fruit, but otherwise a typical paradisaeinine forager, in the canopy of fruit trees, scansorially hunting for bark arthropods, and also joining mixed foraging flocks. D'Albertis (1880) wrote 'I may say the *Seleucidis* is common here and is not difficult to get at, if we discover a Pandanus bearing fruit'. Guillemard (1886) noted that this bird is 'exceedingly fond of the scarlet fruit of the Pandanus'. During boat travel on the upper Karawari R., PNG, in Jan, birds were often seen clinging to and feeding upon the large fruits of *Pandanus lauterbachi*, a plant obviously of considerable significance to the ecology of this bird when flowering and fruiting (CF). Flower nectar is also taken, as Wallace (1869) watched birds frequenting flowering trees and 'sucking the flowers' of pandanus and sago palms and whose field assistant found that birds' stomachs contained nothing but a 'brown sweet liquid, probably the nectar of the flowers on which they had been feeding'. May be acrobatic, hanging upside down, clambering, etc. to probe holes in branches, first just with maxilla, then with whole beak, then just with mandible (BB). D'Albertis, who claimed to have necropsied more than 50 specimens, found fruit and no animal matter in their stomachs with the single exception of one small lizard and noted that Dr Beccari told him that he had found fruit only and one frog (D'Albertis 1880). Meyer (in Sharpe 1891) observed that it 'searches for insects under the bark of trees, but also eats fruit'. Three Sepik R. specimens examined by Gilliard (in Gilliard and LeCroy 1966) contained only fruit remains, but local informants told him the bird captures insects from between the bases of fronds of pandanus and sago. One was seen to probe between pandanus frond bases (Layton 1971). Two ♀-plumaged birds near Port Moresby, PNG, in July were inserting their bills into large gourd-like fruits and eating the white contents. An ad ♂ at the Vanapa R. Oxbow, PNG, was seen on 21 Aug eating fruits of 15-m-high epiphytic aroid *Epipremnum* sp., for 4.5 min and taking > 20 fruit. Also seen to eat fruits of *Glochidion* sp. and an unidentified small palm (BB). Rand and Gilliard (1967) noted the use of the feet in holding food in order to reduce it in size by tearing it with the bill. Joins mixed-spp. foraging flocks. One ad ♂ seen at Krissa, West Sepik, PNG, at midday on 5 Aug took some small *Cyrtostachys* palm fruits while in a flock containing an ad ♂ Pale-billed Sicklebill, a ♀-plumaged Lesser Bird, an ad ♂ Magnificent Riflebird, Variable Pitohui, Golden Monarch *Monarcha chrysomela*, Rufous-backed Fantail, Spangled Drongo, and Olive Flycatcher *Microeca flavovirescens* (Beehler and Beehler 1986).

Vocalisations and other sounds   Figure 9.69

The trachea of ad ♂♂ is modified in a unique fashion, in a manner considerably different from that of the manucodes. We presume this modification permits the far-carrying properties of vocalisations (Forbes 1882*a*; Frith 1994*b*). See genus account above. At least three distinct vocalisations are given by ♂♂ on or in the vicinity of display perches, two apparently as advertisement and one in the context of display. The first consists of single throaty, slightly nasal, resonant, mournful, and down-slurred *harnh* or *hahn* notes produced every few to *c*. 30 sec. The second is a series of 3–8 but usually 4–5 notes: *hahr—haw haw haw*, the first note highest-pitched and followed by a brief pause and the final 3–7 notes in rapid succession (Fig. 9.69(a)). Published transcriptions of vocalisations from elsewhere in the bird's range might superficially indicate regional variation. BB's experience with the bird's in Central, Gulf, E Sepik, and W Sepik provinces of PNG, and in two sites in IJ indicate, however, little or no variation in vocalisation. When display is imminent, usually because a ♀ has arrived near the display site, the ♂ gives a *Paradisaea*-like convergence call, distinct in intensity and tone. This is a higher-pitched and more insistent series of notes with an almost whining quality and with a nasal twang—*hahng-hahng-hahng-hanhg-hahng*, the series falling in pitch. This is often given as a ♂ flies from an adjacent perch(es) to his display perch. Also given is a series of upslurred notes *koi—koi koi koi koi*. The *koi* call is most similar to some calls of the Raggiana Bird, which are less deep and more raucous (Beehler *et al.* 1986). Wallace described the ♂'s advertisement song as *Ca'h, ca'h* repeated 5–6 times in a descending scale, but the 'long drawn *ooû* ending in two sharp and loud notes, *wah wah'* of Grant (in Ogilivie-Grant 1915a) is more typical. Meyer (in Sharpe 1891) considered the call a loud *wau-wau*, in a high key in the throat. Finch (1983) described the song as 'a single *ouw* dropping in tone after a series of single notes there may be a series of notes together each successive one lower *oww .. oww .. ouw-ou-ou-ou'*. Coates (1990) transcribed advertisement calls of the Port Moresby area as 'a rather lethargic sounding, loud *hauw* (reminiscent of Raggiana Bird but lower in tone and not as resounding), given singly or repeated at intervals of a few sec [Fig. 9.69(b)]. The latter single note may be given and followed by a soft descending distinctly whistle-like *twi* note [Fig. 9.69(c)]. When stimulated by the presence of other birds a five-note call may be given; this consists of a note similar to the advertising call, followed immediately by a descending series of four similar but shorter, slightly upslurred notes: *houw-wah-wah-wah-wah* (3 sec).' On the Karawari R., PNG, the advertisement song was a *Paradisaea*-like rapidly-repeated *Wah wah wah wah wah* or, at times, *Wauk wauk wauk* (CF). Near Madang, PNG, on 25 Oct a ♂ first called at 0604 hr, calling being sparse but regular between 0604 and 0620 hr, then quiet, then at 0638, 0650, 0653, 0654, 0701, 0704, 0706, 0713 hr. The ♂ sat stationary in the same spot at the edge of old grassland, the typical song being nine rapidly-delivered *qua* notes. The ♂ sang from *c.* 12 m up hidden in the middle storey. A ♀ made quiet purring sounds at her nest (BB). The hissing of wings in flight is described as *tss tss tss tss* (Beehler *et al.* 1986) and has been likened to the song of the Black Sunbird *Nectarinia aspasia* (Finch 1983). It is produced by the power stroke, the glide being silent (BB). Some ♀-plumaged birds, possibly

(a)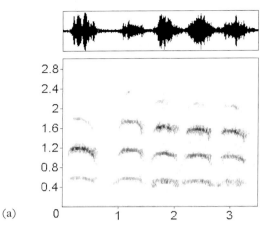

(b)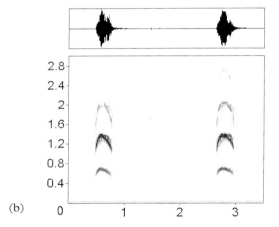

(c)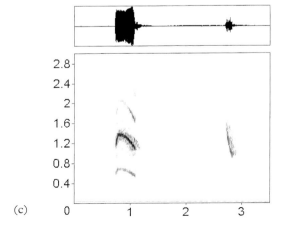

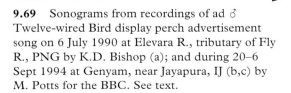

9.69 Sonograms from recordings of ad ♂ Twelve-wired Bird display perch advertisement song on 6 July 1990 at Elevara R., tributary of Fly R., PNG by K.D. Bishop (a); and during 20–6 Sept 1994 at Genyam, near Jayapura, IJ (b,c) by M. Potts for the BBC. See text.

♀♀ only, lack the wing-noise in flight, whereas ad ♂♂ and some other ♀-plumaged birds, possibly imm ♂♂, produce it (BB).

Mating system
Undoubtedly a polygynous sp. in which dispersed, solitary ad ♂♂ are promiscuous. Six vocal ♂♂, three mapped at each of Nomad R. and Kakoro, PNG (Fig. 9.70), had a mean nearest neighbour distance of 730 m (range 420–940 m) (Beehler and Pruett-Jones 1983). At least some display sites are traditional, one on the Elevara R., PNG being in use during July–Aug throughout 1986–96 (D. Bishop *in litt.*). Only the ♀ builds and attends the nest.

Courtship behaviour
The ad ♂ frequents one or several adjacent near-vertical or vertical, leafless, and typically dead tree 'snags', stumps, or spires that protrude conspicuously from the surrounding canopy vegetation. No other bird of paradise

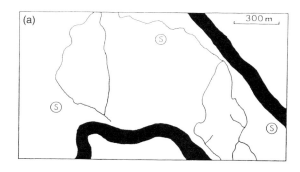

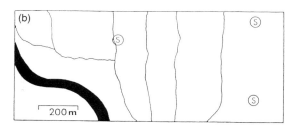

9.70 Display dispersion of ♂ Twelve-wired Birds at (a) the Nomad R. and (b) Kakoro area. River lines are indicated by black areas. Open circles with an S within indicate sites of calling ♂♂ and the lines are transect routes. From Beehler and Pruett-Jones (1983), © Springer-Verlag with kind permission.

displays from such an exposed perch. During the first light of morning over many months of the year the ♂ vocalises loudly to advertise its presence.

Upon arrival each morning a ♂ sits atop his perch to preen and vocalise for as little as 30 min to a maximum of 1 hr or so, starting before first light, far more briefly than other birds of paradise except perhaps the Black Sicklebill. Although ♂ vocalisation is an important part of advertisement and display, on occasion (invariably subsequent to the initial ♂-♀ interaction of the morning) no vocalisations are given prior to, during, or after display and mating.

Singing is often accompanied by holding the wings half-opened (rarely almost fully opened) above the back at *c.* 45°, an action enhancing visibility of the yellow flank plumes (Fig. 9.71(a)). Two variations of this posture involve the bird (1) leaning far forward and downward with bill and head below the feet while perched atop the stump and (2) performing this posture while perched horizontally across the vertical stump below its apex. In addition, the bird repeatedly rotates itself around the circumference of the stump apex or immediately below it while calling and may also raise the wings while doing so. The extremely large powerful feet enable the ♂ to rotate himself around a vertical perch in this way, and by so doing, he may enhance his advertisement by broadcasting vocalisations through 360° (Frith and Beehler 1997).

♀♀ mate at the display site shortly after first light. Upon arrival of a ♀ at the display site the ♂ immediately directs his bill at the ♀ with breast-shield erected, wings held tightly into the body and yellow plumes expanded upward towards the wings, the outermost ones spread upward and curving over the leading edges of the wings to conceal their leading edges and tips. This instantly changes his appearance from predominantly black to predominantly yellow. The raising of the yellow plumes towards the wings tightens or sleekens his lower central abdomen and belly feathers against his body, which fully exposes the bare and colourfully-pigmented thighs. He then starts pecking or probing at the ♀'s bill tip. If she is on the apex, this usually causes her to hop or flutter down over the ♂ to perch beneath him. He then turns bill-downward to parry at the ♀'s upward-pointing bill and pursues her in a ritualised, jerky, hopping gait and repeatedly sharply flicks his head and bill downward between parrying as she moves backward and downward away from him. During this pursuit the sleeked underpart feathering and the way the legs are used and held greatly enhance visibility of the bare thighs. If she backs downward he follows, head first, but if she spirals down the perch he spirals in pursuit—the birds remaining mostly bill-to-bill throughout.

A point in this pursuit is reached (after a *c.* 30–90 cm descent) at which the ♀ breaks away to hop and/or fly up past/over the ♂ onto the perch apex. The ♂ immediately rotates himself 180° to face bill-uppermost and hops vertically up towards the ♀ in his ritualised jerky gait until he can recommence bill 'fencing' with her. His breast-shield feathering is often rhythmically pulsed and his bare thighs continue to be conspicuously exposed by sleeked abdomen and belly feathering and by the leg positions (Fig. 9.71(b)). Apparently the ♂'s aim at this stage is to reach a point beneath and behind the ♀ in order to pick at her undertail coverts, prior to picking at her nape (see below). She turns around atop the apex in order to continue to face the advancing ♂ as he rotates around the perch (anti-clockwise in the filmed ♂), in pursuit.

Events to this point may be repeated 1–6 or more times but a most significant additional display phase is often performed by the ♂ to the ♀ clinging to the side of the perch below him —the *Wire-wipe Display*. In this the ♂ stops pecking at the ♀'s bill tip as she clings below, to rotate himself to become head-uppermost in alignment with but directly above the ♀. His wings are held tightly to his body and his yellow flank plumes held as before. He then performs a series of careful ritualised, slow, and exaggerated body-swaying movements by subtly hopping to alternate his foot positions (the upper one becoming the lower and vice versa repeatedly). During this he holds his downward-pointing head and bill out from the body, alternating sides, apparently in order to watch the ♀ while directing his swaying plumes and wires to best effect. The alternate body swaying, at a rate of a single direction sway each 1.5–2 sec, is performed to exhibit the yellow flank plumes and to cause the highly-modified 12 bare plume wires to brush across alternate sides of the ♀'s bill, face, neck, or breast. The body-swaying action is not continuous, like the perpetual motion of a pendulum, but is given a slow rhythmic quality by the ♂ pausing ever-so-briefly at the two extremes of each of his swaying arc, when the broadest area of yellow flank plumes is presented to best advantage before swaying back in the opposite direction (Fig. 9.71(c,d)). The ♂ was seen to sway in each direction up to four times but may do so more often.

While the ♂ is performing the *Wire-wipe Display* into the face of the ♀ she now and then delicately picks or probes his dense yellow flank plumes with closed or open bill. While doing this on one occasion several of the bare 'wires' passing across her face and bill got caught between the back of her closing mandibles. She instantly gaped and slightly shook her head sharply to release the wires as the ♂ continued his swaying display. The ♀ may hop backward and downward away from the displaying ♂ in which case he hops gently backward down with her while continuing *Wire-wipe Display*. The ♀ on occasion flies up to the apex of the display perch during or directly after a *Wire-wipe Display* at which point the ♂ usually rotates to face bill downward immediately prior to flying up after the ♀.

The copulatory sequence may take place after only the briefest of a pursuit phase or after many pursuits up and down the perch with or without one or more *Wire-wipe Displays*. The entire copulatory phase can last more than 120 sec. It always starts with the ♀ perching atop the apex of the display stump. The ♂ approaches her to 'fence' with her bill tip but at this point his breast-shield plumage

Twelve-wired Bird of Paradise *Seleucidis melanoleuca* (Daudin, 1800)

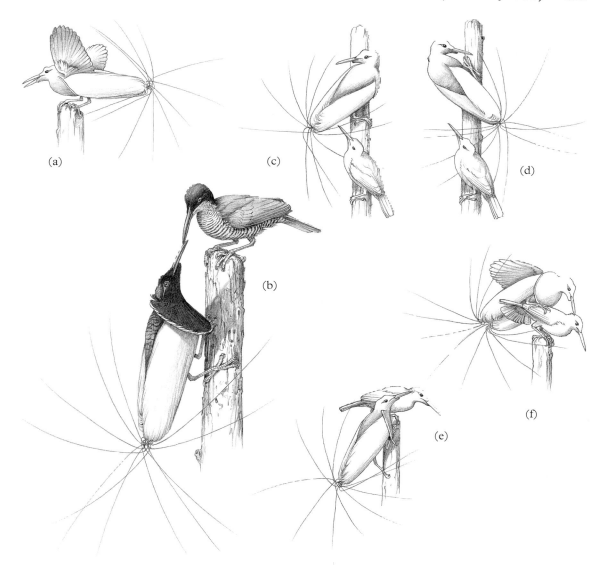

9.71 Some courtship display postures of ad ♂ Twelve-wired Bird: (a) initial advertisement song display, (b) ♂ parrying bill of ♀, (c–d) the two extreme positions of the *Wire-wipe Display* swaying, (e) pre-copulatory pecking at ♀'s nape, (f) copulation. After Frith and Beehler (1997), with kind permission. See text.

is fully extended about his foreparts (Fig. 9.71(b)), his wings are held tightly to his body and the yellow plumes held as before to conceal the leading wing edges and tips, and to expose the bare thighs. In addition to parrying the ♀'s bill tip he repeatedly sharply flicks his head and bill tip downward in a mechanical way. He now more rapidly parries at the ♀'s bill tip as he rotates about the circumference of the perch apex in an attempt to bill her lower flanks. She continues to parry with her bill while rotating backward on the perch apex away from him until she is sufficiently confident to remain still.

At this point the ♂ progressively changes the point of his vigorous but careful parrying, from her bill tip to its base, sides of face, breast, flanks, wings, and alternate sides of her undertail coverts until finally he rapidly alternates from one side of her rear to the other while stretching up and over her upperwings and back to bill her nape (Fig. 9.71(e)). At this point he lowers his breast-shield. If the ♀ permits him to bill her nape she tends to lower her head and neck in response and then usually solicits mounting by very rapidly flicking, or shuddering, her outer primaries. Shortly thereafter the ♂ hops up to mount and copulate for 2–3 sec while flapping his extended wings to maintain position and billing/pecking her nape. At what would appear to be the ♂'s ejaculation his tail is conspicuously fanned and held down to one side and beneath the ♀'s tail which remains in the normal position (Fig. 9.71(f)). He dismounts and flutters down to perch 1–2 m below the ♀ on the display perch or to an adjacent perch. The ♀ remains quietly atop the perch for a min or so but does not flutter her wings or move any other body part in a distinctive way before departing. The ♂ returns to the apex of his perch at her departure to start advertisement singing again.

The above account includes display behaviour documented by Bergman (1957a), Coates (1990), and Frith and Beehler (1997). Most behaviour performed by a single ad ♂ in the NYZS is also described above, except a frequently-repeated, sharp, metallic, single note, accompanied by a wide opening of the mouth for each call, showing the bright green interior (Crandall 1937a). This captive also rapidly opened and closed its wings 10 or 12 times silently with its body held stiffly horizontal with breast plate extended, abdominal feathers compressed, and plumes slightly expanded. While Crandall made no mention of the bare colourfully-pigmented thighs of ad ♂♂ featuring in display, his artist J. Stolper drew the ad ♂ in display with bare thighs conspicuous (Crandall 1937b, fig. 2).

Rand and Gilliard (1967) describe a captive ♂ rapidly vibrating the body while hopping along a perch and giving little thin calls. This bird also hopped along a perch while raising and spreading its wings which enhanced the purplish iridescence, while giving a rather loud, clear *caw* note, raising and lowering the wings with each call.

A displaying ♂ will sometimes attract ♀-plumaged Magnificent Riflebirds (Coates 1990). Near Kakoro, Gulf Province, PNG, an adult ♂ Magnificent Riflebird displayed within 50 m of the display perch of an adult ♂ Twelve-wired. Observations seemed to indicate the calling of one often stimulated calling by the ♂ of the other sp. (BB). This sp. exhibits some displays and call characteristics of both riflebirds and *Paradisaea* plumed birds (see Chapter 5).

Breeding Table 6.5

NEST SITE: a nest found by A. S. Anthony (in Simpson 1907) was in a pandanus. Gilliard (1969) described a nest 3 m above ground in the base of fronds of a pandanus in swampy riverine forest dominated by limbum (black palm), sago palm, and pandanus with a few tall trees, 46 m in from the Sepik R. bank and 15 m from the forest-edge. A second nest, built by the same ♀, was 3.7 m high in a sago palm on dry ground 15 m from the river. Another active nest was 14 m high in a pandanus within 10 m of an ad ♂ calling/display post (BB).

NEST: a shallow egg cup in a bulky deep structure of pandanus bark strippings and vines on a scant foundation of sticks and leaves, lined with rootlets and plant fibres, measuring 200 mm wide and 90 mm deep externally and 90 mm wide and 35 mm deep internally. Within and atop this untidy accumulation is a discrete egg-cup lining of what were once supple vine or tree inflorescence stems up to 1 mm dia that encircle the nest cup. Difficult to see from below. See plate 2b in Gilliard (1969). The 'foundation of sticks' mentioned by Gilliard (1969) may have already been present on the nest site, not placed there by the bird. In view of a lack of sticks in the nest at the AMNH and the lack of them in other bird of paradise nests this sp. probably does not incorporate them. Certainly sticks are not apparent in the photographs of this nest (Gilliard 1955, 1969).

Eggs: the two nests found, with apparently full clutches, both contained a single egg (contrary to the statement that one contained two in Gilliard 1955: 469). Two clutches laid in captivity in Singapore were of single eggs (Khin May Nyunt personal communication) but one laid in Bali in May 1996 was of two (M. Norrie *in litt.*). The egg collected by A. S. Anthony on the Vanapa R. measures 39.2 × 27.1 mm (*contra* 39.2 × 22 mm of Hartert 1910) and that collected by Gilliard on the Sepik 41.1 × 27.3 mm (CF). Ground colour Pale Horn (92) but paler and less buff, more an off-white, marked with longitudinal rufous and greyish broad streaks concentrated mostly about the larger end. Plate 13.

Incubation: observations by Gilliard indicate only the ♀ incubates, no ♂ being seen at nests. The ♀ was most shy and secretive in approaching the nest, keeping to within 1 m of the forest floor until flying steeply upward to it when a couple of metres away. A captive incubation period was 20 days (Khin May Nyunt personal communication).

Nestling care and development: unknown in the wild but in captivity nestling care was exclusively by ♀. The skin of the nestling was still pale and pinkish when half-fledged and it left the nest at 3 weeks old.

Annual cycle   Table 6.5
Display: at least July–Jan. Breeding: peak ♂ gonadal activity is Aug–Dec, ♂♂ collected with enlarged gonads during Jan–Feb and June–Dec. ♀♀ collected with enlarged oocytes during Aug–Sept. Near Port Moresby, PNG, breeding occurs in Jan to late Oct–early Nov and not during the early Nov–early Jan moulting period (Coates 1990). Near Madang, PNG, on 25 Oct a ♀ at a nest made quiet purring sounds suggestive of a nestling's presence (BB). Egg-laying: Jan–Feb, May, Aug–Oct.

Moult: 58 of 170 specimens were moulting involving all months but predominantly May–Dec (Appendix 3). In captivity in New York ad of both sexes took an average of 3 months to complete a moult; specific periods for individual ♂♂ being 7 Apr–15 July, 20 Mar–25 June, 20 May–1 Sept and ♀♀ 1 Apr–1 July and 12 Mar–15 June (Crandall 1937a).

Status and conservation   Table 8.1
Inadequately known, possibly due in part to inhospitable nature of habitat and associated difficulty of travel other than water-bourne; in general, the sp. is said to vary locally from uncommon to common. During annual visits over 1986–96, D. Bishop (*in litt.*) found calling ♂♂ common, if widely separated, along most of the lower 20 km of the Elevara R. from its confluence with the Fly R., above Kiunga, to above the last village thereon. A minimum of three ad ♂♂ were found over each of two study sites, one of 150 and one of 100 ha (Fig. 9.70). A ♂ may range far from his display site(s) each day with no indication of territoriality being noted. ♂♂ do, however, defend their display perches from other ♂♂ (Coates 1990). Given the extensive nature of the range and of the preferred habitat, the sp. is not under any threat.

Knowledge lacking and research priorities
What little is known of this bird suggests a close association with pandanus (for fruit foraging and nesting) such as is also known in larger sicklebills but more study is required to confirm this. It is also possible that it is pandanus fruits that provide the fugitive yellow plume pigmentation. This sp. would strongly merit a full field study of its behaviour and ecology.

Aviculture
The first live bird to be taken to Europe was apparently presented by Signor G. E. Serrati to the King of Italy prior to 1881; it survived a few months (Sharpe 1891). The London Zoo received a live bird in 1881, where it lived for almost a year and another bird lived there for

> 13 yr (Seth-Smith 1923a). The ad ♂ Crandall wrote of lived in the NYZS for almost 23 yr. A bird Alfred Wallace (1869) saw 'on a Dutch steamer ate cockroaches and papaya fruit voraciously'. Guillemard (1886) kept a bird for several weeks, about which he reported: 'wonderfully tame, and will eat out of one's hand. He feeds on the fruit of the Pandanus, on Pawpaw (*Carica papaya*) when it can be obtained, on cockroaches and occasionally on banana.' Captive ad ♂♂ lose the yellow pigmentation of their flank plumes unless fed papaya. Successfully bred at Jurong Birdpark, Singapore, in 1995. The single offspring was raised entirely by the ♀ and lived for a full year (Khin May Nyunt personal communication). In May 1996 a female laid a two-egg clutch at the Taman Burong (bird park), Bali but they were destroyed prior to hatching (M. Norrie *in litt.*).

Other

Skinned specimens lose the rich yellow flank plume colour, hence the sp. epithet *melanoleuca* (black & white). Even the best-preserved skin of an ad ♂ does not do justice to the intensity of yellow in a healthy live wild bird, which, when the ♂ is seen in flight with the deep green forest vegetation as a backdrop, appears as a deep yellow flame. An entirely ♀-plumaged bird received at NYZS in Aug 1929 did not show any sign of ♂ plumage until mid 1934, when > 5 yr old, and had acquired ad ♂ plumage except for chestnut mid-secondaries, rump, uppertail coverts, and tail and a barred abdomen, by 24 June 1935 when > 6 yr old. Full ♂ plumage was acquired during a moult of 20 Mar–June 25 1936, at a minimum age of 7 yr. This period was not influenced by the presence of ad ♂ conspecifics.

Genus *Paradisaea* Linnaeus, 1758

Paradisaea Linnaeus, 1758. *Systema Naturae* ed. 10, **1**, p. 110. Type, by subsequent designation (Gray 1840, p. 39), *P. apoda* Linnaeus.

Synonyms: *Uranornis* Salvadori, 1876. *Annali del Museo Civico di Storia Naturale Genova* **9**, 191. Type, by original designation, *Paradisea rubra* Daudin.

Paradisornis Finsch, 1885. *Zeitschrift für die gesammte Ornithologie* **2**, 385. Type, by monotypy, *Paradisornis rudolphi* Finsch.

Trichoparadisea Meyer, 1893. *Abhandelungen und Berichte des Königl. Zoologischen und Anthropologische-Ethnographischen Museums zu Dresden* **4**(3), 20. Type, by original designation, *Paradisea guilielmi* Cabanis.

The large and colourful plumed birds of paradise consist of **seven species** occupying mainland NG and some of the Western and the Eastern Papuan Is. The three spp. *P. apoda, raggiana, minor* constitute a superspecies. The geographical isolates *P. decora* and *rubra* represent sister forms to the *apoda* superspecies, and *guilielmi* and *rudolphi* constitute distinct outliers to the five core taxa. According to Bock (1963) **skull** characters are most like those of *Cicinnurus*. The seven spp. form a discrete closely-related group within the family as ad ♂♂ all wear grossly-elongated, filamentous, pectoral 'flank' plumes, highly-modified, wire-like, elongate central rectrices, and pale bills, and ♀ plumages are unbarred or predominantly so. Its close relationship with *Cicinnurus* excepted, *Paradisaea* has not been particularly closely aligned with other genera until recent biological and behavioural studies have indicated the possibility that these two genera together with *Semioptera* and *Seleucidis* form a closely-related group (C. Frith 1992; Frith and Frith 1995b; Frith and Cooper 1996; Frith and Beehler 1997). Several intrageneric **hybrids** occur and intergeneric hybridisation is known with *Parotia, Ptiloris, Cicinnurus,* and *Seleucidis*.

Adult ♂♂ are *c.* 10–15% larger than ♀♀. **Bill** almost straight and up to *c.* 20% longer than head. Nostril covered by feathers. **Tail** slightly rounded, not graduated, and central pair of ad ♂ rectrices grossly elongated into fine ribbon-like vaned 'tapes' (*rudolphi*), plastic-like and loosely spiralled and curled feather shaft 'tapes' (*rubra*) or fine 'wires' (all other spp.). **Wings** long. Outermost primary with slightest (*raggiana, minor, apoda, decora*) to no (*rubra, guilielmi*) emargination on inner vane or outermost two primaries there emarginated (*P. rudolphi*). Outer vane of P5–P8 (*raggiana, minor, decora, rubra, rudolphi*) or P3–P9 (*guilielmi, apoda*) emarginated. **Tarsus** length *c*. 24% of wing length. Lower tarsus scuted but more faintly above. Legs and feet less black than all genera except *Cicinnurus, Semioptera*, and *Seleucidis*.

Sexual dimorphism great but less so in *P. rudolphi*. A great tuft of pectoral feathers modified into grossly-elongated, brilliantly-coloured filamentous flank plumes extend well beyond the normal tail tip (but not the central tail wires/ribbons). See Plates 11–12. In ad ♂♂, the crown and nape feathering is thick, short, and felt-like (as in *Cicinnurus magnificus*). Notwithstanding that ad ♀ and juv/imm *P. decora* and *P. rudolphi* show some barring on underparts, the juv/imm of all other *Paradisaea* show no sign of ventral barring. ♀ **plumage** is unique within the sexually-dimorphic members of the subfamily Paradisaeinae in not being barred and thus less cryptic (see Paradisaeinae account for discussion of this).

Nests are built in tree branches and are bulky open cups typical of the Paradisaeinae. **Eggs** are also those most typical of the subfamily and are most like those of *Ptiloris, Cicinnurus, Seleucidis*, and related genera (Plate 13).

Lesser Bird of Paradise *Paradisaea minor* Shaw, 1809

PLATE 11; FIGURES 9.72–3

Other names: Little Emerald Bird of Paradise, *Toffu* or *Burong Papuwa* of plume traders of Tidore and Ternate, E. Ceram and *Shag* or *Shague* by 'the Papuans' (Valentyn in Iredale 1950) but *Sjag* and *Sjage* by Papuans and *Tsjakke* at Sergile, and *Sapaloo* on the 'Papuan Islands' (Stresemann 1954). *Boerong Koening* of the Yapen I. Malays (Mayr and Meyer de Schauensee 1939*a*). *Kioom* of the Victor Emanuel Mts, PNG (Gilliard and LeCroy 1961). *Aramba* (♂) and *Avengenga* (♀), and the sp. *Purat*, of the Wanuma people, Adelbert Ra (T. Pratt *in litt.*; Gilliard and LeCroy 1967). *Yabal* or *Yebey* (♂) and *Pkan* or *Yabal pkan* (♀); and *Angoym sgy* or *Yabal angoym* for what are said to be really old ♂♂ (really dark on face and neck) of the Kalam people (Majnep and Bulmer 1977). *Gomiaor, Komia*, or *Gulu Gomia* of the Huli People (Goldman 1981; Timmer 1993).

The common, widespread, and familiar plumed bird of paradise of N and far W NG. The ♂ is unmistakable with its yellow cape and flank plumes; the ♀ is easily identified by its white breast and belly. Polytypic; 3 subspp.

Description

ADULT ♂: 32 cm, lores, forecrown, ear coverts, malar area, chin, and throat finely feathered, intense iridescent yellowish emerald-green which in some lights may appear jet black, especially about bill base. Remainder of head pale orangy yellow, glossed iridescent silver in some lights extending onto nape and continuing onto whole of mantle at the lower edge of which it overwashes and blends with the Walnut Brown (221B) washed darker Warm Sepia (221A) of back and wings. Remaining upperparts, including wings and tail, pale Warm Sepia washed Maroon (31) on back, rump, and uppertail coverts but wings with an extensive, orangy yellow, 'shoulder' bar and outer edges to greater coverts. Central pair of rectrices grossly elongated, and only basally vaned and coloured as rest of tail, but distally reduced to finely-tapering brown 'wires' Breast Burnt Sienna (132) grading to Walnut Brown on belly, thighs, and undertail coverts with no structurally-discrete upper breast 'cushion'. Ventral surface of tail coverts soft, fluffy, and much elongated. Grossly-elongated, finely-filamental flank plumes bright yellow (55) at their bases, streaked Maroon, fading distally from creamish to dirty white to only slightest of Beige (219D) hue, more so on their dorsal surface. Underwing: coverts dark Walnut Brown, flight feathers Hazel (35) with paler, Vinaceaous Pink, trailing edges. Ventral surface of tail pale, slightly greyish, Hazel. Bill chalky bluish grey, iris deep yellow, legs and feet purplish grey-brown, mouth dull pinkish-flesh. It is said by people of the Upper Kaironk Valley, Schrader Ra, PNG, that a further stage of ♂ plumage attainment is reached by only few individuals that become really dark on the face, neck, and around to the nape. A local name applies to such individuals which rarely occur and of which only one is ever present at a lek. It would appear that this plumage is that of particularly old ad ♂♂.

ADULT ♀: wing significantly shorter than that of ad ♂. Entire head Warm Sepia slightly washed Maroon. Slightly paler rear crown washed orangy yellow grading to Buff-Yellow, washed dull, pale, orangy yellow nape and this overwashing dark Walnut Brown mantle. All remaining upperparts dark Walnut Brown washed Warm Sepia except lesser wing coverts and broad outer edges of greater coverts which are dull Buff-Yellow (53) washed pale orangy yellow. Central tail-feather pair shorter, narrower, and more pointed than rest. Entire underparts cotton-white with slightest wash of pale vinaceous pink (221D), more so on flanks, undertail coverts, and darker lower breast which grades to pale Vinaceous Pink (221C) upper breast and then to the darker throat. Thighs Vinaceous Pink. Underwing and ventral surface of tail as in ad ♂, if slightly darker.

SUBADULT ♂: as ad ♀ with a few feathers of ad ♂ plumage intruding, and central rectrices longer, narrower, and more pointed than rest, to as ad ♂ with a few feathers of ♀ plumage remaining. An individual delivered to the Baiyer River Sanctuary, PNG, as a nestling in Dec 1979 gained the first signs of ad ♂ plumage, consisting of a 'broken yellow colour on the head, green throat and two longer tail feathers' in Mar 1983, and by 1985 had acquired 'half-length yellow flank plumes'. In 1987 it was in three quarter ad plumage and in full ad ♂ plumage in 1988. Thus its transition from ♀ to ad ♂ plumage started at age 4 yr and continued for another 4–5 yr, it being 8–9 yr old when first fully plumaged. This individual had a single white primary in one wing. It was kept in an aviary with other *Paradisaea* spp. (some ad ♂♂) in adjacent aviaries (Aruah and Yaga 1992).

IMMATURE ♂: as ad ♀.

NESTLING–JUVENILE: unknown.

OTHER: partial albino ad ♂ specimen (SMT C 17848) is entirely glossy white but for a pale yellow wash on crown becoming even paler on mantle and wing coverts; a single pale brown tertial, wing, and tail feather (see Frith 1998).

Distribution
Misool I. and the Vogelkop, IJ, and N watershed of NG E as far as the mouth of the Gogol R., the upper Ramu R., and along the N coast of the Huon Penin, PNG, where it meets (and hybridises with) the Raggiana Bird. Penetrates into the interior following major river catchments (e.g. up the Yuat, Jimi, and Baiyer tributaries). From sea level to *c.* 1550 m (Coates 1990).

Systematics, nomenclature, subspecies, weights, and measurements
SYSTEMATICS: four subspp. have been widely accepted (Mayr 1941*a*; Gilliard 1969). We here acknowledge three; the status of *pulchra* requires assessment. Hybridisation: known to

Lesser Bird of Paradise *Paradisaea minor* Shaw, 1809

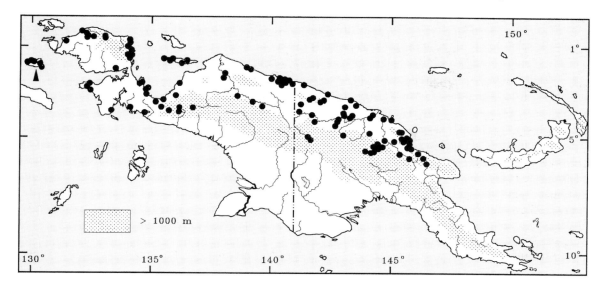

cross with Magnificent Riflebird, Twelve-wired, Magnificent, Raggiana, and Emperor Birds (see Plate 15 and Appendix 1). There is a well-known hybrid zone (with Raggiana) in the upper Ramu, and presumably a similar zone along the westernmost coastal zone of the Huon Penin (E of the mouth of the Gogol R.). The sp. apparently contacts the Greater Bird around Etna Bay, although there do not seem to be any hybrids identified to date—although they are to be expected. There should be a *minor/apoda* hybrid zone between Etna Bay and the Mimika R.

SPECIES NOMENCLATURE AND ETYMOLOGY

Paradisaea minor Shaw, 1809. *General Zoology—Aves*, 7, pt. 2, p. 486 Type locality 'New Guinea', restricted to Dorey, Vogelkop, IJ. Type specimen none designated.

Exploration: this sp. was first brought to the attention of the western world (Spain) by Magellan's crews who returned from the east with footless trade skins (Stresemann 1954).

Etymology: *Paradisaea* = L. *paradisus* = paradise; *minor* = L. smaller, lesser.

SUBSPECIES, WEIGHTS, AND MEASUREMENTS

1. *P. m. minor* Shaw, 1809

Synonym: *P. m. pulchra* Mayr and Meyer de Schauensee, 1939. *Proceedings of the Academy of Natural Sciences of Philadelphia* **91**, 151. Tip, Misol (= Misool I., IJ). Type specimen ANSP 132383.

Range: Misool I. and W NG, eastward in N watershed to the IJ/PNG border, in S watershed to Etna Bay, IJ (at the base of the 'Bird's Neck').

Diagnosis: the availability of 11 specimens from Misool I. allowed Mees (1965) to reassess the validity of *P. m. pulchra*, originally described from only two birds (Mayr and Meyer de Schauensee 1939a). Mees concluded that Misool I. birds did not consistently differ from birds of the adjacent mainland (*P. m. minor*). LeCroy (1981) subsequently presented measurements suggesting that whilst Misool I. birds are a little larger than adjacent mainland birds there is considerable overlap. Our measurements show minimal difference in size between the two forms, birds of Misool I. being no larger than those of the Vogelkop and certainly not even close to as 'large as in *jobiensis*' (*cf.* Gilliard 1969); see Frith and Frith 1997b.

Measurements: wing, ad ♂ ($n = 27$) 180–196 (188), imm ♂ ($n = 23$) 162–194 (179), ad ♀ ($n = 27$) 152–174 (161); tail, ad ♂ ($n = 27$) 116–133 (127), imm ♂ ($n = 23$) 105–130 (123), ad ♀ ($n = 26$) 103–126 (112); tail centrals, ad ♂ (*n* = 23) 420–641 (503), imm ♂ (*n* = 19) 98–127 (116), ad ♀ (*n* = 23) 93–118 (103); bill, ad ♂ (*n* = 27) 34–43 (39), imm ♂ (*n* = 23) 36–41 (38), ad ♀ (*n* = 27) 35–40 (37); tarsus, ad ♂ (*n* = 27) 39–48 (45), imm ♂ (*n* = 22) 39–47 (44), ad ♀ (*n* = 27) 37–45 (40). Weight: ad ♂ (*n* = 12) 185–285 (239), imm ♂ (*n* = 7) 186–242 (216), ad ♀ (*n* = 7) 145–186 (162).

2. *P. m. jobiensis* Rothschild, 1897. *Bulletin of the British Ornithologists' Club* **6**, 46. Jobi I. (= Yapen I.). Type specimen AMNH 678912.

Range: Yapen I., IJ.

Diagnosis: generally larger than other forms, with tarsus a good deal so and flank plumes longer. Flank plume colour as for nominate form.

Measurements: wing, ad ♂ (*n* = 12) 183–210 (200), imm ♂ (*n* = 8) 163–200 (189), ad ♀ (*n* = 1) 172; tail, ad ♂ (*n* = 12) 130–144 (137), imm ♂ (*n* = 8) 124–138 (131), ad ♀ (*n* = 1) 121; tail centrals, ad ♂ (*n* = 9) 353–566 (495), imm ♂ (*n* = 8) 114–130 (122), ad ♀ (*n* = 1) 112; bill, ad ♂ (*n* = 10) 37–41 (39), imm ♂ (*n* = 8) 39–41 (40), ad ♀ (*n* = 1) 36; tarsus, ad ♂ (*n* = 11) 46–49 (48), imm ♂ (*n* = 8) 43–49 (47), ad ♀ (*n* = 1) 42. Weight: ad ♂ (*n* = 3) 293–300 (298), imm ♂ (*n* = 2) 210–250 (230), ad ♀ (*n* = 3) 152–189 (175).

3. *P. m. finschi* Meyer, 1885. *Zeitschrift für die gesammte Ornithologie* **2**, 383. Karan, between Aitape and the mouth of the Sepik R., at long. 142° 30′ E. Type specimen SMT C 8139 (8455, syntype, WW II destroyed).

Range: N PNG, from just E of the PNG border to the Gogol and upper Ramu Rivers.

Diagnosis: on average similarly sized to nominate form but flank plumes averaging proportionately slightly shorter and a brighter orangy yellow. Yellow shoulder marking less extensive than in other forms.

Measurements: wing, ad ♂ (*n* = 27) 180–199 (190), imm ♂ (*n* = 18) 165–193 (179), ad ♀ (*n* = 26) 152–175 (164); tail, ad ♂ (*n* = 26) 124–141 (131), imm ♂ (*n* = 16) 110–131 (124), ad ♀ (*n* = 26) 104–123 (115); tail centrals, ad ♂ (*n* = 24) 430–537 (469), imm ♂ (*n* = 14) 108–131 (121), ad ♀ (*n* = 24) 97–129 (109); bill, ad ♂ (*n* = 27) 36–41 (38), imm ♂ (*n* = 16) 32–39 (37), ad ♀ (*n* = 26) 35–41 (37); tarsus, ad ♂ (*n* = 26) 39–49 (44), imm ♂ (*n* = 16) 41–47 (44), ad ♀ (*n* = 25) 37–47 (41). Weight: ad ♂ (*n* = 21) 183–300 (256), imm ♂ (*n* = 14) 151–268 (205), ad ♀ (*n* = 16) 141–210 (164).

Habitat and habits Table 4.1

Common and widespread in lowland and hill forest, forest-edge, and second-growth. As with the Raggiana Bird, this sp. has adapted to human-altered environments, and is nearly ubiquitous within its range. Usually in parties in the canopy of giant forest trees, but also in the lower vegetation strata and sometimes close to the ground and not rarely in secondary growth and cultivation (Hoogerwerf 1971). Healey (1978*a*) found ad and subad ♂♂ generally restricted to forest and advanced second-growth; by contrast, the ♀-plumaged birds use more habitats and, consequently, the upper altitudinal limit of ad and subad ♂♂ is generally lower than that of ♀-plumaged birds. Healey concluded that this sp. is more dependent upon heavier forest than is the Raggiana. A Doria's Hawk *Megatriorchis doriae* was found to have eaten a Lesser Bird (Rand and Gilliard 1967: 80).

Diet and foraging Table 4.2

Diet thought to be predominantly fruit (Beehler and Pruett-Jones 1983). Valentijn (in Iredale 1950) states that birds eat fruit of the Tschampeda-tree [probably = *Elmerrillia papuana*]. Lesson (in Iredale 1950) noted birds eating the fleshy buds of 'teak-trees' and wrote 'we always found insects in their crops'. Gilliard (1969) states that a number of birds' stomachs contained 'tree and vine fruits, often large and nut-like or small with hard yellow or black pits, and a few insects'. Wild birds at Baiyer River Sanctuary, PNG, took fruits of *Elmerrillia papuana* (= *tsiampaca*) by obtaining bunches of them with acrobatic movements including hanging upside down (Fig. 4.4(b)), then to hold them to a perch with one foot to extract each small arillate fruit with the bill (Cooper in Cooper and Forshaw 1977). Four ad ♂♂ were seen hastily consuming tiny (2–3 mm dia) *Trema orientalis* fruits at Baiyer R., PNG, on 2 June with loud calling but no aggression. Birds also seen to pluck *Omalanthus novoguineensis* fruit and carry it to a branch to hold it with a foot and peck it apart, the handling time for each being 6–10 sec. Birds invariably chased off Brown Cuckoo-Doves *Macropygia amboinensis* when they entered the tree. In Aug, a bird took *Ficus* fruits here in company with five or more *Opopsitta* parrots (BB). See also Appendix 4. Commonly gleans arthropods from small branches, the underside of larger limbs, tree bark, vines, and epiphytes 8–20 m above ground (BB). May join mixed-spp. foraging flocks, such as one in July at Puwani R., W Sepik, PNG, at 200 m, involving a Rufous Babbler flock and a King Bird. Also at Krissa (W Sepik) at 1541 hr in Aug a ♀-plumaged bird was observed hanging upside down to glean tree crevices for insects, 18 m up in sub-canopy, in a flock containing an ad ♂ and ♀-plumaged Twelve-wired Birds, three Jobi Manucodes, and an ad ♂ Pale-billed Sicklebill (Beehler and Beehler 1986).

Vocalisations and other sounds

Figure 9.72

Call descriptions are numerous and varied (Ogilvie-Grant 1905; Ripley 1964; Gilliard 1969; Cooper and Forshaw 1977; Beehler *et al.* 1986; Coates 1990). Overall the songs and calls tend to be typical of those of the *Paradisaea apoda* superspecies, with an advertisement song series of high-pitched, clear, and sharp *wak* notes (Fig. 9.72(a)). A similar but more excited higher-pitched, slightly quicker-delivered, lek convergence vocalisation is usually started with a couple of softer introductory notes (Fig. 9.72(b)). Two displaying ad ♂♂ most commonly gave a loud, somewhat musical, *whick*, interspersed with churrs and growling noises as they moved from limb to limb on the lek (Ripley 1964). Some loud advertising songs of ad ♂♂ similar to those of the Raggiana (but more diverse) whilst others are like those of the Emperor Bird. Series of

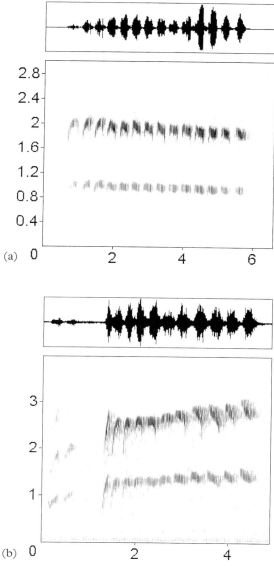

9.72 Sonograms from recordings of ad ♂ Lesser Birds giving lek advertisement song (a) of *wak* notes typical of *Paradisaea* and (b) in a more screeching and rapidly-delivered, excited form in July 1986 at Baiyer R., PNG by K. D. Bishop.

call notes usually start with 3–6 *waik*, *wik*, or *ka* notes followed by a lower-pitched *wok*. As ♂♂ become excited at their lek they may also give a rolling, insistent, nasal series of *werrd*, *werrd*, *werrd*, *werrd* notes. Calls at Baiyer R.,

PNG, in June included the typical advertisement song *wik wong—wau wau* (every 17 sec on average), a descending *kwee kwer kwer kwer kwer*, and a liquid *kwo-dok*. When ♀♀ not present in the lek area ♂♂ relax and give a variety of calls including *ki-kwerk, kwoo-dok, wa wa wa kwi dok bloik, wa wa kyer kyer kwi donk*! (BB). Whitney (1987, fig. 2) noted that a descending series of notes by this sp. is 'quite similar' to the song of the Pale-billed Sicklebill. Coates (1990) observed that most calls are loud, high to medium-pitched, nasal slurs, *whi, uwhi, wha* or *waiy* given as a series of 3–25 or more notes rapidly (up to 4.5 per sec), but sometimes slower (2 notes per sec). Much variation in pitch, tone, and speed occurs between or within a series of calls, more so if birds are excited and counter-singing.

Mating system
Long ago Lesson (1834) noted that birds 'seem to form a harem after the manner of the Gallinaceous birds' and he asked 'Is the Little Emerald Bird of Paradise [the Lesser Bird's name at the time], then, a polygamist?' This sp. is indeed polygynous and the ♂♂ are promiscuous while the ♀♀, once mated at the lek, nest-build, incubate, and provision their offspring unaided (Rand and Gilliard 1967). Leks are traditional, persisting for many years in one or a few adjacent trees. Bergman (in Gyldenstolpe 1955b) reported that birds display in a single lofty tree or occasionally in two immediately-adjacent trees in the forest interior, activity at them being confined to early morning and between 1500 and 1600 hr. He watched up to 12 ad ♂♂ with several ♀-plumaged birds in communal display in a single tree. Gilliard (in Gilliard and LeCroy 1961) found a lek situated at 1433 m 'in the edge of a clump of much-disturbed, second-growth, mid-mountain forest growing on a rounded hill between two garden clearings'. Majnep (in Majnep and Bulmer 1977) states that the tree *Albizia fulva* is often used as a lek. Rand and Gilliard (1967) wrote that local informants reported that some leks are occupied 'for generations'. ♂♂ are known to defoliate the immediate lek area of leaves and tree bark. At Baiyer R., PNG, ♂♂ in June were active at the lek until c. 0830 hr, then dispersed, more or less, and seemed to break up into little parties. Some groups, which included ad ♂♂, seemed to form mini-leks in trees elsewhere, and sang half-heartedly. Although ♂♂ spend little time at the lek (< 1 hr per day), time allotted to extra-lek calling, loafing, and monitoring of lek activity is considerable (several hours at least). Once at 0706 hr, when ♂♂ were displaying at their lek, all birds present simultaneously dropped from sight and 6 sec later a medium-sized falcon passed overhead (BB). Healey (1978a) found an average of about three ad ♂♂ at a Jimi Valley, PNG, lek but recorded that up to 10–20 might be found at some. He noted that some hunters say they have harvested ad ♂ birds at some display trees, or leks, for at least three generations, with birds being shot over 60–100 yr. A lek studied by direct observation during 49 hr over 18 days (2 Jun–4 July) at BRS consisted of a single tree 32 m tall at the edge of a small clearing in forest. Display perches were horizontal or gently-sloping limbs, with some as close as 50 cm to each other. Leaves were removed from the immediate area of these perches by the resident birds. Eight such perches were used daily, the four central ones by individually-identifiable ♂♂ that continuously 'owned' their perch and the others possibly by four other traditional exclusive owners but whose identity was less certain. In addition to traditional perch-owning ♂♂, several other ♂♂, mostly in subad plumage, visited the lek, particularly when intense display activity occurred, and spent much time in forest about it (Beehler 1983d). ♂ attendance at the lek was during 0600–0900 and 1430–1715 hr. Of 26 observed copulations resulting from 99 visits to the lek by ♀♀, most took place during 0645–0715 and 1500–1530 hr. Of these copulations, 25 were performed by the one, centrally-located, ♂ and one by an adjacent ♂. A ♀ once solicited to a peripheral ♂ on the lek but he failed to mount her. Only the three ad ♂♂ most frequently in attendance at the lek received solicitations from visiting ♀♀ (Beehler 1983d).

Lesser Bird of Paradise *Paradisaea minor* Shaw, 1809

Courtship behaviour

As this is the first *Paradisaea* sp. to have displays dealt with herein we adopt standardised terms from LeCroy (1981), Frith (1981), and Beehler (1988) to describe behaviour across most spp. of the genus.

Convergence Display: occurs when the ♂♂ converge from dispersed sites onto their lek perches—usually because ♀♀ have arrived in the vicinity. After loud advertisement singing, ad ♂♂ shake the flank plumes and hop side to side, lower the head, and the body is lowered into a near horizontal position with the neck crooked. Suddenly the wings are opened fully and held out at *c.* 45° to either side of the body, and above the level of the back. The tail is lowered and pulled forward towards and even in front of the feet and perch and the flank plumes spread and erected to arch above the back (Fig. 9.73(a)). After some 10–20 sec, during which the wings are slightly quivered, in this *Upright Wing Pose*, the bird then excitedly hops vigorously back and forth along the display limb with head lowered, wings fully opened horizontally or lower and the flank plumes fully spread and erected, as a loud series of sharp call notes is given. Characteristic of ad ♂♂ is the way in which shorter, pure white, filamentous flank plumes project conspicuously forward either side and over the head (Fig. 9.73(b)) giving an overall appearance of the displaying bird different to congeneric spp. in similar postures.

Static Display: at the apparent climax of the *Convergence Display* the ♂ remains perched still on his own display limb for several sec rubbing or wiping his bill on the perch (Ogilvie-Grant 1905; Bergman in Gyldenstolpe 1955*b*). Gilliard (in LeCroy 1981) describes this posturing in greater detail: 'The shoulders are drawn close to the body and the primaries are extended outward, almost straight out like an oarsman holding oars out of the water. They are held still or a while, and the bird makes no noise except the occasional low *graaa*. Then the wings begin to move up and down, moving up slowly about three inches at their tops, then downward snappily to a point just below the

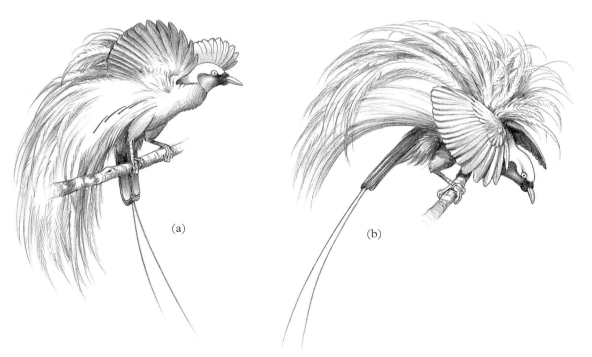

9.73 Some courtship display postures of ad ♂ Lesser Bird: (a) *Upright Wing Pose*, (b) *Static*, or *Flower*, *Display*. See text.

shoulders ... as the performance continues [the flank plumes] begin to rise up behind in a splendid cascade. But the distinctive thing is that certain of the shorter yellow flank plumes are lofted through the opening normally covered by the scapulars and they stick up in random places like separate little golden fountains.' At this point the bird may sometimes almost turn under the perch but no actual upside down posture has been recorded for this '*Flower Display*' (LeCroy 1981).

Copulatory Sequence: this appears to consist of the ♀ approaching the ♂ on his lek perch and crouching submissively in front of him with her back to his breast. Once the ♀ signals her acquiescence to mating, the ♂ rhythmically beats his open wings over and about the ♀ and rocking from side to side for *c.* 20 sec whilst pecking at her and/or she pecking at him before mounting her to copulate (Peckover 1973; Cooper and Forshaw 1977; Coates 1990; A. Aruah *in litt.*).

Gilliard (in Gilliard and LeCroy 1961) made the significant observation that displaying ♂♂ do not thump the wing carpal joint area over the back as do Raggiana ♂♂. ♂♂ perform communal display with or without the presence of ♀-plumaged birds. From the several published accounts, it would appear that the displays and courtship of the Lesser is significantly more simple than that of the following two spp. Gilliard (in Gilliard and LeCroy 1967) observed that two imm ♂♂ in ♀ plumage moved to lek display perches and went through the display motions of ad ♂ birds immediately after the departure of the ad ♂♂ from the lek, but produced less robust calls.

The above account includes display postures described by Ogilvie-Grant (1905) and Crandall (1936).

Breeding Table 6.5

NEST SITE: one was said to be 6 m above ground in the fork of a branch of a slender tree in forest near a native garden (Rand and Gilliard 1967; Gilliard 1969). Another was 'very high up in a tree with thick foliage in a garden and fallow zone' (Bulmer in Gilliard 1969).

NEST: a cup-shaped structure of sticks and vines (Rand and Gilliard 1967). A nest collected by A. E. Pratt at Hambitawuria, W NG at *c.* 457 m was described as twigs lined with black wire-like fibres or rootlets and, on the outside, partially covered with dead leaves. The cup measured 12–13 cm in dia and 8 cm deep (Hartert 1910). These references to sticks and twigs in nests are probably erroneous (see Chapter 6). ♀♀ in captivity started nest-building at an average of 7 days ($n = 6$) prior to egg-laying, total time for most nest construction being 2–5 days (Hundgen *et al.* 1990).

EGGS: clutch usually one, rarely two. Seven eggs average 36.4 (35.2–38.8) × 26.1 (24.3–27.6) mm (CF). An infertile egg of *P. m. finschi*, laid in captivity at Nondugl, PNG, on 30 Jan was creamy-pink and entirely peppered and spotted with fine purplish grey, reddish, and brownish spots. A few more elongate and larger blotches about the larger end and towards the middle of the egg were predominantly purplish grey and over-marked with fine brownish peppering. It measured 37.7 × 27.6 mm and had a medium-glossy surface (Bishop and Frith 1979). Four of six single-egg USA captive clutches averaged 40.2 (40–40.4) × 27.8 (24–32.5) mm (Hundgen *et al.* 1990); ground colour Pale Horn (92) to Pale Pinkish Buff (121D), marked with elongate broad streaks of brown and rufous brown, from the thick end downwards and shorter, deeper, underlying grey ones. See Plate 13. A USA captive ♀ took an average of 22.4 days ($n = 5$) to lay a replacement clutch after previous one removed (Hundgen *et al.* 1990). Three single-egg clutches laid in captivity 3 days after copulation; another was laid 2 days after last observed copulation. Mean weight of four captive-laid eggs, after 18 days of incubation, was 13.3 g (Hundgen *et al.* 1990). A lone captive ♀ subsequently introduced to a ♂ built a nest within 3 days, was seen to copulate 4 and 7 days after introduction of the ♂ and laid an egg on the 12th day after the introduction (Cooper 1995).

INCUBATION: in captivity, ♀♀ spent an average of only 54 min per day off the nest

(Hundgen *et al.* 1990) but this very high attendance level is clearly due to optimal food/water/shelter available. A single USA captive incubation (10 days by the ♀ and the remainder in an incubator) was 18 days and the hatchling weight was 10.4 g (Hundgen *et al.* 1990). A single-egg clutch was fully incubated by the captive ♀ and hatched after 18 days (Laska *et al.* 1992), as did two other captive single-egg clutches (Cooper 1995).

NESTLING CARE: only the ♀ performed nestling care in the wild and captivity.

NESTLING DEVELOPMENT: a single nestling was raised in captivity by hand. For the first 12 days it was fed every 2 hr, during 0600–1800 hr, on small pieces of new-born mice and thereafter also on fruit, 'bird of paradise pellets', and crickets. Pin feathers appeared on its wings by day 4 but feathers did not emerge from tracts on the back and shoulder areas until day 11. By day 15 the young bird preened itself and daily weight gain was *c.* 3–4 g. At 19 days the nestling started perching regularly and at 20 days it started to pick food for itself and was feeding independently at 40 days. It appeared fully feathered at 44 days except on its breast area which did not have pin feathers until 50 days (Hundgen *et al.* 1990). Two captive single-nestling broods departed the nest 18 days after hatching and of these one was still being fed by its parent at 42 days old while the other was encouraged to feed independently by 28 days old (Cooper 1995).

Annual cycle Table 6.5

DISPLAY: leks may be active for up to 7 months a year. Ripley (1964) watched two ad ♂♂ display in an enormous fig tree on 27 July at Bodim, IJ. At Jalan Korea, Nimbokrang, NE coastal IJ, display occurred in the crown of a tall tree on 28 June (B. Poulson personal communication). Gilliard (in Gilliard and LeCroy 1967) observed displays in Mar–Apr in the Adelbert Mts, PNG, but said local informants indicated that July–Aug was the main display season there.

BREEDING: at least July–Feb. Over the sp. range as a whole, ♂♂ were collected with enlarged gonads during most months but more so during Feb–Aug. ♀♀ with enlarged oocytes collected during July–Sept, Nov, and Feb, but mostly in Aug. EGG-LAYING: at least July–Feb. MOULT: 47 of 201 specimens were moulting, with > 25% of each monthly sample showing some moult except for May–Sept when little evident, the peak activity being Nov–Jan (Appendix 3). Birds observed at Baiyer R., PNG, during 8–14 Apr were thought to have completed moult (T. Pratt *in litt.*).

Status and conservation Table 8.1

In spite of generations of hunting of ad ♂♂ the sp. remains common. Healey (1978a) calculated the 'entirely speculative' lower and upper population estimates of 6100 Lesser Birds, at a theoretical sex ratio of one ad ♂ to four ♀-plumaged birds, and 9900 birds, at a ratio of 1:8, over his *c.* 540 km^2 study area. Despite long-term traditional hunting of plumed ♂♂ he concluded that hunting of birds was not greater than the recruitment rate of the population and that this activity posed no threat to the populations. The sp. was very common in the S Adelbert Mts in Mar–Apr 1959 (Gilliard and LeCroy 1967), and moderately common in forests from the base of the Cyclops Mts, to *c.* 1070 m, in IJ during July 1990 (J. Diamond and D. Bishop *in litt.*). During 22–29 Mar 1997 found to be one of the most common birds on Yapen I. Present in lowland, hill, and mountain forests (Simpson 1997).

Knowledge lacking and research priorities

This common sp. is an excellent subject for further field studies, especially of lek behaviour. Particular questions that might be addressed would be lek spacing, establishment of new leks (and demise of old leks), and ♂ mate acquisition behaviour (and mating) away from the lek.

Aviculture

As a result of shipping two birds to England, Alfred Wallace noted that birds of paradise require air and exercise rather than heat. These two birds were in the London Zoo in Apr 1862 (Bartlett in Elliot 1873: 42). A successful USA captive breeding occurred between a ♀ and a ♂ still in ♀ plumage (Laska *et al.* 1992).

Greater Bird of Paradise *Paradisaea apoda* Linnaeus, 1758

PLATE 11; FIGURES 2.3, 9.74–5

Other names: Great Bird of Paradise. *Burong Papua*, *Manu-co-dewata*, *Soffu* or *Sioffu* of Ternate bird traders and *Man-key-aru* of the Amboinese. *Tamuku* and *yau* on the Mimika R. (Goodfellow 1910). *Fanaan* in the Aru Is. (Valentijn via Forster in Iredale 1950, Stresemann 1954). *Iforo* of the Fair language at Bomakia and *Hondo* of the Kombai language, IJ (Rumbiak 1994).

By far the largest but least known of the plumed birds of paradise. The spectacular aggregations of ad ♂♂ in a 'lek' tree on the Aru Is were first made known to the western world by Alfred Russel Wallace (see Fig. 2.3). Females are distinct in being uniformly dark-brown all over. Polytypic; 2 subspp.

Description

ADULT ♂: 43 cm, head and nape as Lesser Bird but yellow nape not extending onto mantle. Mantle and entire upperparts Walnut Brown (221B) washed darker Warm Sepia (221A) on mantle, back, and lesser wing coverts and slightest or orangy yellow wash on greater covert outer edges. Central rectrices grossly elongated being reduced to finely-tapering blackish 'wires', and lacking vanes, except on bases where coloured as rest of tail. Upper breast 'cushion' blackish brown (19) grading at lower border to Warm Sepia and then to Walnut Brown on belly to dark Vinaceous Pink (221C) on thighs, vent, and undertail coverts. Grossly-elongated filamental flank plumes bright yellow (55) at their bases, streaked Maroon (31), fading distally from creamish to dirty white and then this tinted Beige (219D) on the under surface but variably creamy yellow to cream to Beige tips on upper surface. UNDERWING: coverts dark Walnut Brown, flight feathers dark Hazel (35) with paler, Vinaceous Pink, trailing edges. Ventral surface of tail slightly greyish Hazel. Bill chalky pale bluish grey and almost whitish at tip, iris pale yellow, legs and feet purplish grey-brown, mouth colour unrecorded. Alfred Russel Wallace (1869) concluded, from limited specimens, that full ad ♂ plumage is not acquired until 4 years of age. He also thought ♂♂ with longer central tail wires 'very old birds' (in Elliot 1873). Ogilvie-Grant (1915*b*) concluded that wild ♂♂ did not assume ad plumage until their sixth year, or at least 5 years old. Based on his experiences with captives in the UK, Seth-Smith (1923*a*) thought ♂♂ took 8–10 years to attain full plumage. Goodfellow (1910) observed that the rich yellow on the head of ad ♂♂ fades to pale straw colour in captivity, suggesting the possibility this colour is diet based (see Twelve-wired Bird).

ADULT ♀: 35 cm, entire head and upper breast Warm Sepia slightly washed Maroon but slightly-paler rear crown washed orangy yellow and indistinctly-speckled dull orange yellow on nape. Upper breast paler and less densely feathered than ad ♂. Entire upperparts as ad ♂ but slightly darker and lacking orangy yellow wash. Lower breast to undertail coverts as ad ♂ but slightly paler whereas underwing and tail as ad ♂ but slightly darker. Central pair of rectrices shorter, narrower, and more pointed than rest.

SUBADULT ♂: as ad ♀ with a few feathers of ad ♂ plumage intruding, initially about head, to as ad ♂ with a few feathers of ♀ plumage remaining.

IMMATURE ♂: as ad ♀.

Greater Bird of Paradise *Paradisaea apoda* Linnaeus, 1758

NESTLING–JUVENILE: unknown.

Distribution

Aru Is and S NG, from the Timika area (and probably as far W as Etna Bay) E to the Fly/Strickland drainage, PNG, where it meets and hybridises with the Raggiana. Its distribution in the eastern and western extremities of its range are poorly defined and merit scrutiny. Easternmost point of range appears to be Nomad R. station (Mackay 1966), but BB, in a survey of birds of paradise, searched for the sp. there in 1980 without success. Additional eastern verge localities include Ok Tedi, Kiunga, Ningerum, Lake Daviumbu, and (a hybrid population) near Bensbach (Coates 1990; M. LeCroy *in litt*). Ranges from the lowlands to at least 950 m, but does not appear to range as far into the uplands as either the Lesser or Raggiana. A small introduced population existed on Little Tobago I., West Indies, for some 75 yr (Dinsmore 1970b; Emlen and Emlen 1984) before succumbing to habitat stresses and disruptions that were caused by periodic hurricanes.

Systematics, nomenclature, subspecies, weights, and measurements

SYSTEMATICS: considerable confusion exists in the synonymy of this sp. and that of the Raggiana because the latter was long treated as a subsp. of the Greater. The considerable differences in size, ♀ plumage, and ♂ coloration of head, back, and flank plumes support the current treatment. Hybridisation: with Raggiana where they meet; mixed leks that include hybrid ♂♂ have been found in SW PNG, (M. LeCroy and W. Peckover personal communication; D. Bishop *in litt.*). Whether there is assortative mating between the two spp. is unknown. Hybrids are also likely with the Lesser in the Etna Bay region of IJ. See Appendix 1.

SPECIES NOMENCLATURE AND ETYMOLOGY

Paradisaea apoda Linnaeus, 1758. *Systema Naturae* ed. 10, **1**, p. 110. Type locality India (= Aru Is, IJ). Type specimen none designated. Probably the second bird of paradise sp. known to western science (the first being the Lesser).

Etymology: *apoda* = footless or legless (because early trade skins lacked them—see Chapter 2).

SUBSPECIES, WEIGHTS, AND MEASUREMENTS

1. *P. a. apoda* Linnaeus, 1758

Range: Aru Is, IJ.

Measurements: wing, ad ♂ ($n = 16$) 225–240 (232), imm ♂ ($n = 8$) 198–224 (209), ad ♀ ($n = 8$) 194–215 (202); tail, ad ♂ ($n = 16$) 156–175 (165), imm ♂ $n = 8$ 141–161 (151), ad ♀ ($n = 8$) 141–158 (152); tail centrals, ad ♂ ($n = 14$) 500–854 (637), imm ♂ ($n = 8$) 132–160 (146), ad ♀ ($n = 8$) 132–153 (144); bill, ad ♂ ($n = 16$) 38–48 (44), imm ♂ ($n = 8$) 41–48 (44), ad ♀ ($n = 8$) 40–45 (43); tarsus, ad ♂ ($n = 17$) 45–55 (53), imm ♀ ($n = 8$) 45–53 (51), ad ♀ ($n = 8$) 45–53 (49). Weight: ad ♀ ($n = 2$) 170–173 (172).

2. *P. a. novaeguineae* D'Albertis and Salvadori, 1879. *Annali Museo Civico Genova* **14**, 96. Middle Fly R. (300–450 miles upstream). Type specimen MRSN 480 (9634). Wallace (1869) correctly predicted that the Greater might be found on mainland NG.

Range: S NG, from Timika, IJ, eastward to the Fly/Strickland watershed.

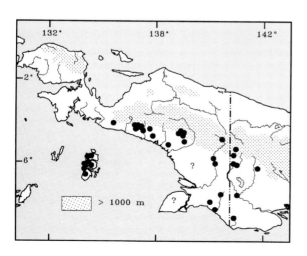

Greater Bird of Paradise *Paradisaea apoda* Linnaeus, 1758

Diagnosis: this mainland subsp. differs from the nominate form in its paler upper breast, more maroon general coloration, and smaller size. Cracraft (1992) found the plumage characters not diagnosably distinct. We find that mainland ad ♀♀ are significantly smaller than nominate ad ♀♀ in both wing and tail length. Moreover, sexual size dimorphism differs between the two populations. As a measure of wing length the ratio ♀/♂ = 88% for nominate and 85% for *novaeguineae*; as a measure of tail length, the ratio ♀/♂ is 95% and 88%, respectively. We therefore retain *novaeguineae*.

Measurements: wing, ad ♂ ($n = 24$) 200–213 (206), imm ♂ ($n = 25$) 180–209 (191), ad ♀ ($n = 26$) 159–188 (173); tail, ad ♂ ($n = 22$) 140–154 (146), imm ♂ ($n = 24$) 130–143 (137), ad ♀ ($n = 26$) 118–142 (128); tail centrals, ad ♂ ($n = 17$) 480–663 (591), imm ♂ ($n = 21$) 125–146 (135), ad ♀ ($n = 25$) 105–132 (119); bill, ad ♂ ($n = 22$) 38–43 (40), imm ♂ ($n = 22$) 38–43 (40), imm ♂ ($n = 22$) 38–42 (40), ad ♀ ($n = 25$) 37–41 (39); tarsus, ad ♂ ($n = 24$) 45–49 (47), imm ♂ ($n = 25$) 45–49 (47). Weight: imm ♂ ($n = 2$) 222–225 (224), ad ♀ ($n = 2$) 170–173 (172).

Habitat and habits Table 4.1

Lowland and hill forest. Mostly frequents the lower canopy of tall forest. Ogilvie-Grant (1915a) noted that mainland birds occupied the lowlands but not near the coast and only up onto the very low foothills. An active and vigorous forager—both for insects and fruits. ♀-plumaged birds frequently encountered in small groups; ad ♂♂ away from the lek are typically seen singly.

Diet and foraging Table 4.2

Probably much like the Raggiana in diet and foraging habit, but observations are few. Wallace (in Elliot 1873) found birds' stomachs full of fruit, noted that birds seek insects 'principally orthoptera', and found one of the largest of Phasmidae almost entire in the gut of an ad ♂. Another early account refers to the red fruits of tall but small-leaved 'Waringha' trees being eaten on the Aru Is (Iredale 1950). Goodfellow (1910) reported seeing 60–70 birds eating fruits in the canopy of high trees on the Aru Is together with several pairs of Glossy-mantled Manucode and a Spotted Catbird. The stomachs of three birds from near Merauke, IJ contained fruit pulp, small seeds, and remains of one large soft insect (Mees 1982). A captive ad ♂ at Macao in 1833 held grasshoppers to the perch to remove legs and wings before eating them (Bennett in Iredale 1950). In remnant forest at Timika, IJ, in Jan 1995, BB observed a ♀-plumaged individual in a mixed flock that included a Twelve-wired Bird. The Greater Bird was 25 m up in a dead tree, hanging inverted and chipping bark from a dead branch. On another occasion at the same site, a ♀-plumaged bird was observed to creep about on a vertical dead stub in the canopy.

Vocalisations and other sounds Figure 9.74

First described by Alfred Wallace (1857), on the Aru Is, as *Wawk-wawk-wawk, wok-wok-wok* which 'resounds through the forest.' Individuals heard calling in the remnant forest in the vicinity of Timika, IJ, produced calls remarkably like those of the Raggiana (BB). Coates (1990) found the typical *wauk* call notes similar to those of the Raggiana but slower, coarser and less urgent-sounding, usually given 4–5 times in immediate succession. The song series of notes is often started by several similar but more pathetic introductory ones (as in King Bird). Dinsmore (1969) described dual calling by two ad ♂♂ with established calling perches 46 m apart in the population of *P. a. apoda* introduced to Little Tobago. In this, each ♂ gave the 'up-3-down' call, which consisted of five *wauk* notes; the first low, then three higher-pitched and louder, and the last 'flattened off' from the preceding three and diagramatically notated by him as _--- _. The entire call lasts *c.* 5 sec. The first and fifth notes are longer than the middle three and a slight pause is given before the last note. One ♂ often gave the last note twice. This call was often preceded by one ♂ giving a series of low, slowly delivered, *wauk* notes. The two ♂♂ gave the 'up-3-down' call in unison or very slightly out of phase so that the second bird's notes sounded like instant echoes of those of the first. Dinsmore (1970a) defined a number of call types as follows. The most common loud call note was a deep *wauk* or *wonk* given by ♂♂ and often associated with activity of the lek. Dinsmore noted two distinct forms of delivery of this note—the *rising call* consisting of four or five deliberate *wauk* notes, repeated at *c.* 1 sec intervals, the first two equal in intensity and the rest increasingly louder and higher. This call was given about once per min, mostly during the early morning near or on the lek but sometimes prior to roosting. The rapid *wauk* call consists of a series of rapid *wauk* notes of one pitch delivered in short bursts of several a sec (Fig. 9.74), each note usually accompanied by short wing-beats involving the primary tips moving sharply only 5 cm from the body and back. The *Wing Pose* call, given as the *Wing Pose* display is performed (see below), is a shrill sliding *eee-ak* note, usually repeated several times but displaying birds may alternate this with the *rapid wauk* call. The *Pump Call* is a series of rapidly-repeated *wauk* notes of the same pitch, delivered faster than in the above calls during the *Pump Display*—the notes often being run together to sound like *wa-wa-wa-wa* and lasting *c.* 10 sec. At the end of the *Pump Call* one or several drawn-out, harsh, nasal, *baa* notes are given as the bird tips its body forward and downward. During the *Dance* phase of display birds produce a slow rhythmic *click* about each sec, sometimes with a faint nasal *bonk* as the wings are raised for another dance step. A call most commonly given after display, as the bird wipes its bill on the perch or tears leaves off the lek foliage, is similar to the *baa* call but is more nasal and abrupt. The last call Dinsmore recorded is one he heard rarely; a rather harsh, guttural *chug'-ich*, *chug'-a* or *chug'-a-la* given by ♂♂ when bouncing along the perches prior to giving the *click* call or as they move from the lek, most often when departing for the evening roost.

Mating system
Polygynous; ♂♂ are promiscuous whilst ♀♀ nest-build, incubate, and provision their offspring unaided. Leks are traditional, persisting for many years (> 20) in the same tree(s) or site (Coates 1990; Gregory 1997). Claud Grant (in Ogilvie-Grant 1915a) noted that a

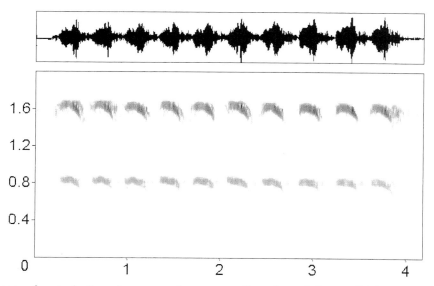

9.74 Sonogram of part of a 9-sec long song, from a recording of an ad Greater Bird giving the lek advertisement song at Kiunga, PNG on 26 Oct 1995 by M. Potts for the BBC. This shows nine *wauk* notes but a song may consist of more and may last *c.* 10 sec.

British Ornithologists' Union Expedition to S IJ observed at least eight ad-plumaged ♂♂ in a lek in single tree and D. Bishop (*in litt.*) has seen a lek consisting of 15 ad ♂♂. Leks of both wild NG birds and of the introduced population on Little Tobago I., West Indies, consist of larger, exposed, horizontal boughs of a large spreading tree, or trees, immediately beneath the canopy. ♂♂ clear leaves and other foliage from the immediate area of their individual display perches. Dinsmore (1970*a*) noted that birds gave harsh nasal calls whilst tearing off pieces of large palm leaves, even landing directly upon leaves then to hang inverted as they tore. Wallace recorded (1869) that during the latter half of Mar 1857 he 'One day got under a tree where a number of the Great Paradise birds were assembled, but they were high up in the thickest of the foliage, and flying and jumping about so continuously that I could get no good view of them.' He subsequently became ill and then (in Apr) depended upon his local assistants for specimens and for information concerning the behaviour of the birds from these hunters and other local people. He was told that 12 or 20 ad ♂♂ assemble in large-crowned trees with exposed spreading branches to 'raise up their wings, stretch out their necks, and elevate their exquisite plumes, keeping them in continual vibration. Between whiles they fly across from branch to branch in great excitement, so that the whole tree is filled with waving plumes in every variety of attitude and motion' (see Fig. 2.3). This oft-quoted passage was the first documentation of group, or lek, ♂ display in a bird of paradise. No naturalist has since published a description of an encounter with the Greater Bird on the Aru Is, although Attenborough (1996) has for mainland birds.

Courtship behaviour

As a result of studying the free-living population of four ad ♂♂ and one subad ♂ at four main display areas on Little Tobago, Dinsmore defined five main phases of display. While quoting him to avoid any possibility of misinterpretation we present his descriptions within the framework of the three major display phases described for the Lesser Bird. It should be noted that Dinsmore uses 'display grounds' for the display tree bough of individual ♂♂ and 'court' for both the lek as a whole and for the display bough of individual ♂♂.

Convergence Display: the upright *Wing Pose* display 'begins with one or more ♂♂ calling from near or on their display grounds. At the approach of another ♂, a ♀, or often in the absence of another *apoda*, the ♂ flies to his court, either directly or with several brief stops at perches on the way, usually giving rapid 'wauk' calls as he flies. Upon arrival at his court, he continues the rapid 'wauk' calls together with wing pose calls. In conjunction with the latter call, the bird extends and holds his wings in a rigid position in front of the body for a few seconds, then continues the rapid '*wauk*' calls. Throughout the performance he holds the plumes erect over the back, tucks the tail forward under the perch, and orients the body perpendicular to the perch' (Fig. 9.75(a)).

'In lengthy displays, which occurred most often when a ♀ came near but did not land on the dance limb, he holds the wing pose for several seconds, then gradually drops the wings to the sides and flaps them rapidly there, accompanied by rapid '*wauk*' calls. Then with a new burst of activity, the bird again gives a wing pose call and posture.' The latter may take a more upright form in which the plumes are not erected (Fig. 9.75(b)). 'This cycle might be repeated every 10–20 sec without interruption with some bouts lasting 30 min or more. With a ♀ present, the calling in this phase seems louder and more excited. The performance apparently functions to attract the ♀ to the court.'

Pump (or *Charging*): in this phase the body is lowered and turned almost in alignment with the perch, wings extended and cupped slightly around the branch, plumes erected almost vertically, and the head and bill pointed down (Fig. 9.75(c)). 'In this position the bird rapidly hops along the court giving the pump call as he bounces up and down.

The vertical motion is exaggerated by flexing the legs, and the tail is moved rapidly sideways with each bounce. The body's motion adds to the splendor of the cascade of plumes above and behind the bird. Often the bird moves from branch to branch while giving the pump but usually returns to the court. A single *Pump* phase lasts at the most 10 sec, but the phase is often repeated several times.'

'This phase often is given as soon as a ♀ arrives at the court. With the body lowered, the plumes are more evident than in the wing pose and thus this phase may serve to keep the ♀ at the court once she has arrived. This phase (which Gilliard called *Charging*) is given most often in ♂ only displays.' It may be performed prior to any *Wing Pose*, or as ♂♂ initially dash about lek perches, and/or between these. At a mainland NG lek it was performed on vertical limbs: ♂♂ in *Wing Pose* suddenly turn rapidly to bill lowermost and hop down the perch about a metre or so then turn bill uppermost and hop back up to their original point on the perch to adopt a static upright *Wing Pose* again but with flank plumes now erected to over the back. This is held rigidly or with occasional odd brief wing fluttering. Suddenly ♂♂ again turn to bill lowermost and hop down their perch again but this time remain at the foot of their descent in inverted pose with plumes absolutely fully erected in the *Static Display*.

Static Display: at the end of the *Pump* (or downward *Charging*) birds perform the *Bow* when 'the body is tipped far forward and down so that the head is below the limb and the back humped. The wings are held out and cupped down around the limb, the tail is brought forward and under the limb opposite the head, and the plumes are held erect over the back' (Fig. 9.75(d)). 'If a call is given in this position, it is the *baa* call. The bow is held rigidly from a few sec to more than a min.' The open cupped wings may be slowly fluttered and the bird may bill-strop the perch. 'This is apparently Gilliard's [and LeCroy's (1981)] "flower" position which he considered to be the climax of the courtship display. ♀♀ weren't necessarily present during this phase, although the bow invariably was given when ♀♀ landed at the court.'

'Often the display ends at this point. The bird gives a few rapid bill-wipes, raises up, and shakes his wings as he lowers his plumes. Frequently the *Wing Pose–Pump–*bow cycle is repeated several times without appreciable interruption although the bow is often left out.' Film obtained by the BBC at a mainland NG lek shows that 4–5, but more often 2, ad ♂♂ may perform the various above movements and static postures of display in uncanny synchrony.

Copulatory Sequence: the *Dance* phase 'usually but not always follows the wing pose–pump–bow sequence'. It usually starts from the bow, or flower, position, when a ♀ is perched directly above the inverted ♂ and often in contact with his tail. From his inverted position the ♂ quickly turns upward to face or hop up to face the ♀ and starts to bounce rhythmically and shuffle back and forth or up and down along the limb with leg flexion exaggerating the vertical motion. 'Both feet are off the branch simultaneously as he bounces along. The *click* call is given instead of the *Pump* call, and the movement in the dance is slow and rhythmic while in the pump the motion is rapid and somewhat jerky. Gilliard (MS field notes) describes a shuffling movement by the birds but doesn't distinguish the distinctive *click* call and rhythmic bouncing of this phase.' Dinsmore saw two variations of the dance, a silent dance without *click* calls and a dance with the plumes lowered and the wings held against the sides. When immediately beneath/beside a ♀ the ♂ hops or bounces on the spot as he then claps the ♀ between his open cupped wings as he bills her nape and/or bill while swaying side-to-side of her (Fig. 9.75(e,f)) all at an increasing tempo, prior to suddenly hopping up to mount and copulate.

This pre-copulatory 'dance may last from a few sec to more than a min but usually takes less than 30 sec. Birds commonly pause briefly to wipe the bill and then continue to dance. At the end of the dance the bird usually bill-wipes several times, shakes his wings, and lowers his plumes.' This is the *Frontal Display* of the Raggiana and other *Paradisaea* spp. but in the Greater the wings are held more widely open and its less developed breast cushion does not feature as it does in the Raggiana.

Dinsmore saw six mountings: 'Five occurred before 08:00, the other at 17:25, and all involved the same ♂. Once a second ♂ was present when the ♀ arrived, but he left before the other two copulated. The mean interval from the time the ♂ first began giving rapid *wauk* calls in response to a ♀'s arrival on the display ground until she landed and remained on the dance limb was 14 min (range 4–30, $n = 6$).'

He described events leading to mounting on these six occasions: 'First the ♂ gave a series of rapid *wauk* calls, flapping his wings against his body and holding wing poses with erected plumes (phase 1). The sight of a ♀ near the display ground apparently triggered this, although I couldn't always see the ♀ until she came very close to the court. The rapid *wauks* continued as the ♀ moved around near the dance tree. While approaching the display ground, she moved quietly but actively from tree to tree.'

'Usually the ♂ drove the ♀ off the court the first few times she landed there; once he did this five times. Eventually she landed at the distal end of the display limb and then moved up to the distal end of the court. Twice the ♂ gave a pump (phase 3) when the ♀ landed at the court, once holding a long bow while facing away from her. More commonly the pump and bow were omitted and the ♂ moved to the proximal end of the dancing surface, turned and danced toward her, usually with wings spread and giving the *click* call, always with the plumes erected. When he reached her, he turned so his body was alongside but at a slight angle to her, their heads close together. He then danced in position, plumes erect, wings spread, and the body bouncing in time with the *click* calls.'

'The ♂ usually obscured the ♀ from view at this point, but once I had a clear view of her. On this occasion I saw that the ♂'s wing next to the ♀ was extended over her body and as he flapped his wings he held her close alongside his body.' This is the wing-clap phase of Raggiana courtship. 'The ♂ also repeatedly rubbed his bill against her bill, bit at her bill, and stretched his head and neck beneath hers and rubbed his bill on the far side of her head, all as he continued to dance. I clearly saw wing-holding and billing only once, but they are probably typical components preceding mounting. After about 20 sec of dancing in contact with the ♀, the ♂ mounted, still flapping his wings. Four times copulation appeared successful while the other two times the ♀ flew before the two could mate. In the dismount the ♀ always flew out from under the ♂ and he dropped to where she had been perched.'

'While the ♀ was on the court before mounting, she sat quietly, usually crouching as the ♂ came toward her. Occasionally after he had mounted she looked up at him, and once she turned her head, rubbed bills, and bit at his throat. I heard no audible calls by the ♀ while she was on the court. The ♀ stayed on the court only 1 to 2 min. After the ♀ flew away, the ♂ stayed near the dance limb for a few min, giving a few nasal calls, preening, and once dancing again briefly. In most cases he left about 10 min after the ♀. The same general sequence took place on several occasions when a ♀ landed at the court and flew before mounting occurred. These five phases, with their calls and movements, seem to form the main features of the courtship display of *apoda*.'

Dinsmore observed a ♂ hanging completely upside down from a display limb on five occasions, for about 5 sec, shortly after display and almost always accompanied by nasal calls but he did not consider this a regular component of courtship. The vast majority of display occurred between 0600 and 0800 hr and between 1400 and 1800 hr (Dinsmore 1970*a*). Film obtained by the BBC in NG confirms that an individual ad ♂ displaying on the lek with other ad ♂♂ may copulate immediately sequentially with the same ♀ or with two different ♀♀. This film includes all behaviour described by Dinsmore.

Greater Bird of Paradise *Paradisaea apoda* Linnaeus, 1758

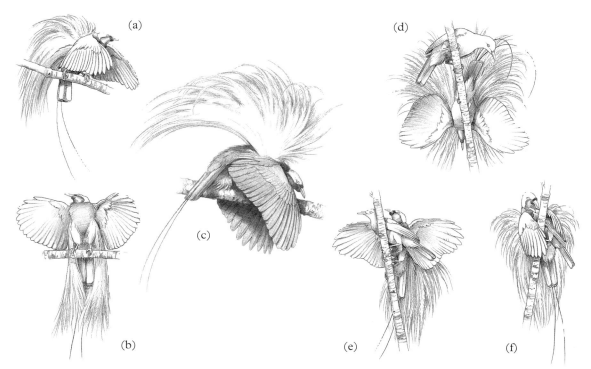

9.75 Some courtship display postures of ad ♂ Greater Bird: (a–b) *Upright Wing Pose*, (c) *Pump* or *Charging Display*, (d) *Static Display*, (e) *Frontal*, or *Dance*, *Display* phase, (f) pre-copulation wing cupping/clapping of ♀. After Attenborough (1996). See text.

Emlen and Emlen (1984) reported a single ♂ (the sole survivor) joining in the group display of the (vaguely similar) oropendola *Psarocolius decumanus* (Icteridae) on Little Tobago.

Breeding   Table 6.5
NEST SITE AND NEST: unknown, although a nest (BMNH N. 193.225) collected at Waboa, Aru Is. in Oct 1925 by W. Frost remains undescribed.

EGGS: Seven BMNH single-egg clutches and a MVZ one are all of Aru Is birds (*P. a. apoda*) and seven of them average 39.3 (35.2–42.8) × 26.5 (25.5–27.3) mm (CF). Ground colour mostly Pale Pinkish Buff (121D), but slightly paler and/or pinker but may be Pale Horn (92) with slightest of pink wash to pale Salmon (6). Marked with broad, elongate, brush-stroke-like, purplish and purplish grey streaks overlaid with similar markings of chestnut and various browns. Markings denser at and around the larger end and all tending to break up into large blotches towards the smaller end where they are sparser.

INCUBATION, NESTLING CARE, AND DEVELOPMENT: unknown but presumably similar to the Lesser, Raggiana, and other *Paradisaea* spp.

Annual cycle   Table 6.5
DISPLAY: of 472 displays seen and heard on Little Tobago only 3% occurred during June–Aug, 12% during Sept–Nov, 45% during Dec–Feb, and 40% during Mar–May. The BBC filmed lek display and copulations during Oct at Kiunga, PNG (Attenborough 1996).
BREEDING: known during Aug–Dec, Mar, and May in NG and would therefore be possible at any time. EGG-LAYING: at least Mar, May, and Aug–Dec in NG. MOULT: 44 of 118 NG specimens were moulting, very predominantly

during Sept–Feb with a single bird moulting during each of Mar–May and Aug and none during June–July (Appendix 3). Valentijn's early (1724–6) account alluded to Aru Is birds moulting during the eastern monsoon (Dec–Apr) and taking 4 months to complete (in Iredale 1950). Wallace (1869) reported Aru Is birds moulting during c. Jan–Feb, showing new flank plumes about Apr, and to be in full plumage in May–June. On Little Tobago, West Indies, birds moulted during Apr–Aug, when little display took place (Dinsmore 1970a). Dinsmore first noted moult in one ad ♂ on 14 Mar; by 5 Apr only one of three ♂♂ still had flank plumes but it started to moult on 16 May. By late Aug new plumes had grown and display activity gradually increased.

Status and conservation   Table 8.1
Common, probably more so in the lower foothills. Coates and Lindgren (1978) observed that in the Ok Tedi, PNG, birds had conspicuously declined where hunting with shotguns had occurred whilst birds remained common and relatively tame where firearms had not been used. Still, this sp. suffered heavy culling during the plume trade and today shows no evidence that there was any long-term impact of that depredation. Common in rolling lowland forest at c. 30 m to the W and N of Kiunga, PNG. Observed there within 1–2 km of the Drimgas road, where Raggiana Birds are regularly seen and a lek of theirs found in Aug 1989 (D. Bishop *in litt.*). During an extensive survey of the Aru Is avifauna J. Diamond and D. Bishop (*in litt.*) found the sp. noticeably more abundant in the N than in the S; being generally moderately common during 28 Mar–12 Apr 1988. They saw no ♂♂ in ad plumage, however, and found birds were hunted regularly for plumes which were sold in Dobo.

Knowledge lacking and research priorities
Little known, but presumably its biology in most respects is similar to that of the Raggiana. It will be important to survey *Paradisaea* populations between Timika and Etna Bay, looking for evidence of interspecific contact and hybridisation. The exact distribution of *raggiana* and *apoda* in the Fly/Strickland drainage needs definition, and a study of mate selection in mixed-spp. leks is required. A complete nesting biology study is a priority.

Aviculture
Pennant noted a living bird was sent to England prior to 1790. Lesson saw two birds in captivity at Amboina (Ambon) in 1828, in the care of a Chinese merchant who fed them on boiled rice and cockroaches. Birds were maintained at NYZS during at least 1935–6 (Crandall 1936).

Raggiana Bird of Paradise *Paradisaea raggiana* Sclater, 1873

PLATE 11; FIGURES 4.4, 9.76–8

Other names: Count Raggi's Bird of Paradise, Raggi's Bird of Paradise, Empress of Germany's Bird of Paradise (*P. r. augustavictoriae*), Marquis Raggi's Bird of Paradise, Grant's Bird of Paradise (*P. r. granti*). *Bounde* of the Kup region, Wahgi Valley, PNG, (Mayr and Gilliard 1954), *Kwa'hua* and *la'kwa* at Lake Kutubu (Schodde and Hitchcock 1968). *Tay* of the Kalam People (Majnep and Bulmer 1977), *Ubiya* of the Huli People (Goldman 1981). *To* of the Fore, *óromo* of the Gimi and *púri* of the Daribi people of the Eastern Highlands, PNG (Diamond 1972). Also *Oromo*, *Pameri* or *Tovoi* of the Gimi people of Crater Mt., PNG (D. Gillison *in litt.*). *Gabi* in the Waria Valley, PNG (W. Cooper *in litt.*).

The classic 'bird of paradise', endemic to Papua New Guinea. The ♂ is unmistakable with the profuse orange or reddish flank plumes, yellow cowl, and green throat. ♀-

Raggiana Bird of Paradise *Paradisaea raggiana* Sclater, 1873

plumaged individuals—seen much more commonly than plumed ♂♂—are identified by a chocolate face contrasting the dull yellow crown and nape. The most common paradisaeid of disturbed habitats in S and E PNG. Polytypic; 4 subspp.

Description

ADULT ♂: 34 cm, head and nape as in Greater Bird but with a narrow yellow collar between green throat and upper breast, and yellow of nape extends onto mantle and overlays the Warm Sepia (221A) to produce an Amber (36) at the lower and outer mantle border. Remaining upperparts, including wing and tail, Warm Sepia washed Maroon (31) on back, rump, and uppertail coverts, except a discrete, pale, orangy yellow wing bar on lesser coverts and deep orangy yellow wash to outer edges of greater coverts. Central pair of rectrices grossly-elongated, reduced to finely-tapering, blackish 'wires', and lacking vanes, except at bases where they are iridescent rich emerald green. Thick, plush, and trim-edged feathers of the upper breast 'cushion' blackish brown (19) with dull suffusion and iridescence of dark coppery bronze grading at lower border into Warm Sepia and then to Walnut Brown (221B) on belly to Vinaceous Pink (221C) on thighs, vent, and undertail coverts. The latter softly filamental and elongated to half tail length. Grossly-elongated and tapering filamental flank plumes Crimson (108) to orangy crimson with buffish white tips on under surface and Ferruginous (41) on upper surface. Underwing: coverts Walnut Brown, flight feathers Hazel (35) with paler, Vinaceous Pink, trailing edges. Ventral surface of tail pale, slightly greyish, Hazel. Bill pale chalky bluish grey, iris deep yellow, legs and feet purplish grey-brown, mouth dull pink flesh. In captivity, the first traces of ad ♂ plumage may be acquired at 4 yr 8 months (Aruah and Yaga 1992); full ad plumage would require at least an additional year or two.

ADULT ♀: 33 cm, foreface, ear coverts, chin, and throat Warm Sepia washed Maroon. Crown, rear and side neck, and narrow foreneck collar dark Buff-Yellow (53) washed dull pale orangy yellow, the latter flecked with Warm Sepia. This collar quickly grades into Light Russet Vinaceous (221D) of remaining underparts, paler at central belly. Underwing and ventral surface of tail as ad ♂ but slightest greyer and darker. Central pair of rectrices shorter, narrower, and more pointed than other tail feathers.

SUBADULT ♂: as ad ♀ with a few feathers of ad ♂ plumage intruding, initially about head, to as ad ♂ with a few feathers of ♀ plumage remaining.

IMMATURE ♂: as ad ♀.

NESTLING–JUVENILE: hatchlings are pink and naked (Dharmakumarsinhji 1943) and subsequently quickly become dark-skinned (Muller 1974). Two-day-old nestlings of a captive brood of two were naked with brownish red back skin and slate bluish grey on the head and wings, and were darker at 4 days old (Rimlinger 1984).

Distribution

S and NE PNG. In the S watershed, ranges W to the Fly/Strickland watershed and marginally into the Trans-Fly region of PNG. In the N watershed, extends W to the upper Ramu R. (in the interior) and coastally to the Gogol R., near

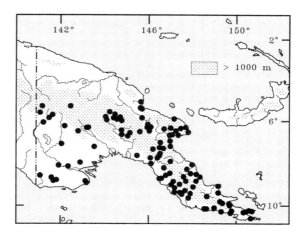

Madang. Ranges from sea level to *c.* 1500 m, and remains common even above 1000 m in many places. Overlaps with *minor* in the N watershed in the W extremity of its range, and with *apoda* in the S (see those spp. accounts).

Systematics, nomenclature, subspecies, weights, and measurements

SYSTEMATICS: long thought to be conspecific with the Greater Bird. The establishment of common names for what earlier ornithologists considered spp., but are now subspp., of *P. raggiana* compounded the confusion. We consider it a member of a superspecies with the Lesser and Greater Birds. As currently constituted, this sp. exhibits considerable regional variation. Hybridisation: wild hybrids are known with the Lesser, Greater, Emperor, and Blue Birds. See Plate 15 and Appendix 1.

SPECIES NOMENCLATURE AND ETYMOLOGY

Paradisaea raggiana Sclater
Paradisea raggiana Sclater, 1873. *Proceedings of the Zoological Society of London*, p. 559. Orangerie Bay. Type specimen MRSN 481 (6064).

Exploration: Italian explorer Luigi D'Albertis obtained the first trade skin of this sp. and wrote in 1880 'If this be a new sp., as I really believe it is, I purpose [*sic*] calling it *Paradisea Raggiana*, after an old and true friend of mine, the Marquis Raggi, of Genoa, a most ardent sportsman and zoologist.' Sclater subsequently named the bird in accord with the wish of D'Albertis.

Etymology: *raggiana* = in honour of the Marquis Raggi.

SUBSPECIES, WEIGHTS AND MEASUREMENTS

1. *P. r. raggiana* Sclater, 1873.
Range: S watershed of SE PNG, from Cloudy Bay to Milne Bay, and into the N watershed as far as Goodenough Bay or Cape Vogel.
Diagnosis: fore-back yellow, fading as it grades into the sepia of the mantle.

Measurements: wing, ad ♂ ($n = 9$) 185–197 (189), imm ♂ ($n = 2$) 183–188 (186), ad ♀ ($n = 5$) 157–172 (164); tail, ad ♂ ($n = 9$) 128–144 (136), imm ♂ ($n = 2$) 128–135 (132), ad ♀ ($n = 5$) 115–125 (121); tail centrals, ad ♂ ($n = 9$) 358–493 (447), imm ♂ ($n = 2$) 125–127 (126), ad ♀ ($n = 5$) 103–125 (114); bill, ad ♂ ($n = 9$) 38–41 (40), imm ♂ ($n = 2$) 40–41 (40), ad ♀ ($n = 4$) 35–38 (37); tarsus, ad ♂ ($n = 8$) 39–46 (42), imm ♂ ($n = 2$) 43–44 (44), ad ♀ ($n = 5$) 37–41 (38). Weight: ad ♂ ($n = 7$) 238–285 (272), imm ♂ ($n = 2$) 225–240 (233), ad ♀ ($n = 5$) 170–215 (184).

2. *P. r. salvadorii* Mayr and Rand, 1935. *American Museum Novitates* **814**, 11. Vanumai, Central Division, Papua, southeast New Guinea. Type specimen AMNH 330366 ♂ ad.
Range: S PNG, from the Fly/Strickland watershed, eastward to Cloudy Bay, and penetrating into interior highland valleys where suitable habitat is available (e.g. the Wahgi and Tari valleys).
Diagnosis: like the nominate form but sepia mantle of both sexes lacking yellow, and the scarlet of the flank plumes less deep. Cracraft (1992) found evidence of clinal variation in characters of *salvadorii*. Very similar in most mean measurements to nominate form but with plume length (a somewhat subjective measure) averaging some 20% shorter. The status of this thinly-differentiated subsp. requires further study to judge its validity.

Measurements: wing, ad ♂ ($n = 41$) 180–198 (188), imm ♂ ($n = 23$) 160–190 (177), ad ♀ ($n = 31$) 150–182 (164); tail, ad ♂ ($n = 42$) 125–154 (135), imm ♂ ($n = 23$) 115–136 (127), ad ♀ ($n = 31$) 110–132 (120); tail centrals, ad ♂ ($n = 38$) 410–527 (463), imm ♂ ($n = 22$) 106–131 (120), ad ♀ ($n = 29$) 100–128 (111); bill, ad ♂ ($n = 42$) 37–41 (39), imm ♂ ($n = 23$) 36–42 (39), ad ♀ ($n = 31$) 35–40 (37); tarsus, ad ♂ ($n = 42$) 39–46 (43), imm ♂ ($n = 22$) 35–41 (43), ad ♀ ($n = 30$) 34–40 (38). Weight: ad ♂ ($n = 21$) 240–295 (266), imm ♂ ($n = 8$) 189–255 (225), ad ♀ ($n = 24$) 135–210 (175).

3. *P. r. intermedia* De Vis, 1894. *Annual Report on British New Guinea, 1893–94*, p. 105. Kumusi R. Type specimen ROM 0050148.

Synonym: *P. r. sororia* Menegaux, 1913. *Revue Française d'Ornithologie* **3**, 50. New Guinea (purchased from the dealer Mantou). Type specimen not located.

Range: N coast of SE PNG, from the lower Mambare R. SE to Collingwood Bay.

Diagnosis: like *salvadorii* but both sexes with yellow mantle/back and yellow streaking down to uppertail coverts. Fractionally smaller than the above two forms but samples of this and the nominate subsp. are too small to be conclusive.

Measurements: wing, ad ♂ ($n = 18$) 178–191 (185), imm ♂ ($n = 4$) 172–181 (175), ad ♀ ($n = 13$) 148–178 (162); tail, ad ♂ ($n = 18$) 127–141 (132), imm ♂ ($n = 4$) 124–133 (128), ad ♀ ($n = 13$) 112–128 (121); tail centrals, ad ♂ ($n = 15$) 384–505 (448), imm ♂ ($n = 4$) 117–129 (121), ad ♀ ($n = 13$) 104–126 (114); bill, ad ♂ ($n = 15$) 36–41 (39), imm ♂ ($n = 3$) 37–40 (39), ad ♀ ($n = 13$) 36–39 (37); tarsus, ad ♂ ($n = 18$) 39–46 (42), imm ♂ ($n = 4$) 41–44 (43), ad ♀ ($n = 12$) 35–44 (38). Weight: ad ♂ ($n = 3$) 234–274 (257), imm ♂ ($n = 1$) 225, ad ♀ ($n = 2$) 163–220 (192).

4. *P. r. augustaevictoriae* Cabanis, 1888. *Journal für Ornithologie* **36**, 119. 'Kaiser Wilhelm's Land', restricted to Finschhafen. Type specimen ZMB 29781 (destroyed in WW II).

Synonyms: *P. granti* North, 1906. *Victorian Naturalist* **22**, 156. 'German New Guinea (?).' Type specimen AM O.14313.

P. apoda subintermedia Rothschild, 1921. *Bulletin of the British Ornithologists' Club* **41**, 138. 'Inland from Huon Gulf.' Type specimen AMNH 678862.

Range: upper Ramu R. and Huon Penin (and interior valleys to the S—the Markham, Wafa, Watut, and Bulolo R.) and also southeastward to the Waria R. watershed and coastally to the lower Mambare.

Diagnosis: an examination of the majority of pertinent specimens (CF) of what was known as *P. r. granti* (Measurements: wing, ad ♂ ($n = 4$) 180–196 (187); tail, ad ♂ ($n = 4$) 126–136 (131); tail centrals, ad ♂ ($n = 3$) 433–471 (457); bill, ad ♂ ($n = 4$) 35–40 (37); tarsus, ad ♂ ($n = 4$) 39–42 (40)) proved confusing owing to considerable variation and lack of localities. Given the now better-defined southeasternmost distribution of *augustaevictoriae* and northwestern-most *intermedia* we concur entirely with Rothschild (1930a) and Mayr (1941a, and in Mayr and Greenway 1962) that *granti* is a changing, intergrading, population geographically between the two (contra Cracraft 1992). Thus, the flank plumes of ad ♂♂ to the NW of the range of '*granti*' are more orange while those to the SE are more scarlet.

P. r. augustaevictoriae is like *intermedia*, with which it hybridises, but flank plumes apricot orange, abdomen/belly paler, and yellow throat collar narrower. Average size very like above three forms. This form hybridises with the Lesser and Emperor Bird—see Plates 11, 15 and Appendix 1.

Measurements: wing, ad ♂ ($n = 37$) 160–194 (184), imm ♂ ($n = 15$) 163–191 (177), ad ♀ ($n = 27$) 150–180 (161); tail, ad ♂ ($n = 37$) 124–136 (131), imm ♂ ($n = 15$) 120–146 (129), ad ♀ ($n = 27$) 108–126 (117); tail centrals, ad ♂ ($n = 36$) 373–523 (451), imm ♂ ($n = 15$) 106–137 (119), ad ♀ ($n = 26$) 100–120 (108); bill, ad ♂ ($n = 36$) 35–40 (38), imm ♂ ($n = 15$) 36–40 (38), ad ♀ ($n = 27$) 35–38 (36); tarsus, ad ♂ ($n = 37$) 39–46 (42), imm ♂ ($n = 15$) 39–45 (42), ad ♀ ($n = 26$) 34–44 (38). Weight: ad ♂ ($n = 16$) 234–300 (276), imm ♂ ($n = 3$) 199–270 (223), ad ♀ ($n = 10$) 133–195 (159).

Habitat and habits   Table 4.1

Lowland, hill, and lower montane forest, second-growth, forest-edge, gardens, and casuarina trees and copses in open deforested sites. Frequents the midstage to canopy of forest and secondary growth. In Varirata National Park, PNG, birds found in all wooded habitats, but fully-plumed ad ♂♂ remained mainly in the forest interior. In semi-original woodlands and in high original forest between Illolo and Ower's Camp, SE PNG (near Varirata), birds fed in upper portions of trees and flocks were infrequently seen streaming through the canopy on apparently well-defined arboreal foraging routes (Gilliard 1950). Near Port Moresby, birds habitually accompany foraging *Pitohui–Pomatostomus* flocks, which may enter teak plantations and forage to within 1 m of the ground (Bell in Cooper and Forshaw 1977). Often seen in the company of other passerines in mixed-spp. flocks sometimes with one or more other birds of paradise, including Trumpet Manucode, Lawes' Parotia, and Magnificent Bird (BB). At 1220 hr on 24 July at Varirata National Park, PNG, three birds were foraging with a Magnificent Bird, a Black Cuckoo-shrike, a Spangled Drongo, and four Rusty Pitohuis (BB). Seen in a mixed-spp. foraging flock with 12 other species on Mt. Bosavi, PNG, in Apr (T. Pratt *in litt.*). In SE coastal PNG, birds were calling at dusk, in mangroves, at the base of a rocky bluff that led up to grass hillocks; all were subad (BB). Of 160 mixed-spp. foraging flocks observed at Brown R., near Port Moresby, PNG, 45 included at least one Raggiana. Of 1196 sightings of the sp., 118 were associated with such flocks. The average number of Raggiana Birds in each mixed-spp. flock was 2.2, SD 1.1 (Bell 1983). Bell found that 89% of his 1196 sightings involved birds in the forest canopy (25–35 m high), 8% in the sub-canopy (8–25 m), 2% in the understorey (1–8 m), and 1% in emergent trees at > 35 m high. A small goshawk, probably *Accipiter cirrhocephalus*, flew into the centre of a lek tree creating an instantaneous 'fire drill' of ♂♂ in attendance. Another goshawk, possibly *A. novaehollandiae* perched in a lek tree to preen and as a result the Raggianas did not return to it during that observation period (BB). ♂♂ approach the lek from below the forest canopy as birds of paradise normally travel (T. Pratt *in litt.*). The typical flight pattern is several convulsive flaps followed by a brief glide (Diamond 1972). Birds frequently use a foot to pin food to a perch or to hold moss while seeking invertebrates.

Diet and foraging Table 4.2

A frugivore that supplements its diet with arthropods gleaned from bark, branches, and leaves of the forest canopy. Simpson (1942) reported that individuals collected had 'pulp of an orange-coloured fruit, called by the natives *varvio*, in its crop, sometimes other fruit, and occasionally a tree grasshopper'. Stomachs of six birds taken from two locations contained 90–100% fruits and 0–10% ants (Schodde 1976). Of 154 foraging events, involving 42 birds, Bell (1984) found that 63% involved feeding on fruits, 12% foraging on tree branches, 8% on a log or tree stump, 7% in leaves, 5% on tree trunks, and 2% on vines. He considered the sp. to eat more arthropods than fruit in its overall diet. Animal foods identified in faecal samples include: crickets/grasshoppers, beetles, spiders, cockroaches, ants, and ad lepidopterans (BB). Numerous birds may utilise a single food tree. Between 0630 and 0830 hr on two consecutive days, 17 ♀-plumaged and one imm ♂ bird visited a fruiting emergent fig at Varirata National Park, PNG, but only once were there two birds present together. Visits to the fig lasted 1–6 min and averaged c. 2 min (Wahlberg 1992). At Varirata National Park in Apr a bird was seen eating *Elmerrillia papuana* fruits with Brown Oriole, Magnificent Riflebird, and Fawn-breasted Bowerbirds in the same tree (BB). Of 227 observations of feeding bouts by birds at 16 different fruiting plants on Mt. Missim, PNG, 36% were upon four *Ficus* spp. (mostly *F. drupacea* and *F. gul*); 33% upon *Omalanthus* sp.; 11% upon *Gastonia* sp., and 7% upon *Dysoxylum* sp. Of the 16 fruit spp., 7 were capsular, 5 drupe/berry, and 4 fig type, but capsules represented 49% (see Fig. 4.4), figs 36%, and drupe/berries only 15% of total feeding bouts (Beehler 1983*a*,*b*)—see Chapter 4. Beehler and Pruett-Jones (1983) estimated that fruit represents 90% of the diet. At Varirata National Park, where there are at least 26 spp. of avian fruit-eaters, two fruiting tree spp. were visited exclusively by birds of paradise, and the Raggiana Bird represented 96% of fruit foraging records at these (Beehler and Dumbacher 1996). This represents one of the most extreme food plant/bird spp. specialisations known to date. At Varirata in July an ad ♂ took 8 *Gulubia* palm fruit in 59 sec, a ♀-plumaged bird took 7 in 57 sec, another 4 in 35 sec, and another a single fruit in 21 sec (BB). A seed capture tray set beneath a Raggiana lek at Varirata National Park over a 6-week period (June–July) collected seeds of > 50 plant spp. Andrew Mack (*in litt.*) records feeding upon 'various Meliaceae, *Piper* sp. and *Ficus* sp. at Crater Mt., PNG. A bird at Mt. Missim, PNG, ate flower buds of *Canarium macadami*, and at Kaisenik, near Wau at 1200 m, PNG, an ad ♂ foraged on *Endospermum labios* fruits in Aug (BB). See Appendix 4 for a list of food plants. Crandall (1932) reports the regurgitation of 'casts' by captive birds, consisting of the shed, empty, inner lining of the gizzard. The significance of this remains obscure and requires confirmation and clarification. ♀-plumaged individuals have often been seen foraging for arthropods by probing and pulling at moss in dead canopy limbs; scansorial gleaning up and down on vertical branches and pecking bark; chipping at bark and lichen and chiselling for up to 81 sec on vertical 30-cm-dia trunk and small limbs of dry *Castanopsis*, hunting sub-bark insects; foraging on 1–3 cm-dia bamboo poles, 3–4 m high in shaded bamboo thicket (BB). Four ♀-plumaged birds were seen feeding low in the forest middle storey by grasping pieces of stick with the bill, breaking them off and holding them to a perch with a foot to probe them for arthropods (Gary Opit personal communication).

Vocalisations and other sounds

Figure 9.76

♂♂ give four distinct vocalisations, each in a specific context (Beehler 1988). The *Advertisement Call* is a loud series of far-carrying cries, the series increasing in intensity, and sometimes also in pitch, the last notes being the longest and coarsest sounding: *wau wau wau Wau Wau WAU WAAUU WAAUU WAAAUUU* (Fig. 9.76(a)). The *High-pitched Call* is a more rapid, higher-pitched, and raspier *wok wok wok wak wak waagh waagh* (Fig. 9.76(b)). There is a considerable amount of variation in these two calls. The first of these is the common call given by ♂♂ away from the lek, but also in the lek in the absence of ♀♀ or display activity. The *Display Call* is given by a ♂ from his lek perch and is a long series of high-pitched, excited, and insistent, keening notes, most commonly given in the presence of ♀♀ or other displaying ♂♂ during lek convergence. One or, more commonly, several ♂♂ give this series while vigorously giving the ('charging') display. This averages 20 sec long and includes c. 25–50 rapidly-delivered notes. It is this call that attracts birds of all ages and sexes to the lek in a 'convergence', and it is the call that fieldworkers listen for to locate a lek in the forest. Finally, the ♂ gives *Single Notes*. The first, given when the lek is quiet, is a single high-pitched note, delivered at irregular intervals. The second, given when ♀♀ are in the lek but activity is minimal, is an occasional quiet honking, growling, cussing, or popping note (Beehler 1988). Diamond (1972) reported that in the Karimui, PNG, area, the advertisement song is 12 or more loud *caws* progressively rising in pitch and decelerating, often concluding with two or more *caws* at lower pitch and longer intervals. Ad ♂♂ noted as vociferous, giving display calls in unison, in early Mar at Wau, PNG, thus vocalising all-year in spite of moult (BB). Vocalisations of ♂♂ at a Varirata National Park lek were quite distinct from those at Wau, the cough note being very rough and birds having an Emperor Bird-like popping note as an ending to their advertisement song. The higher pitched call is also very distinct (BB). A *Bill-click-call* is given in display (see below). A scold note is a very dry, scowling, Trumpet Manucode-like cussing possibly used as conspecific feeding territory call in fruiting trees (BB). Call notes of ♀♀ disturbed near their nest are low-pitched *kss kss kuss* (BB).

Mating system

Polygynous. Promiscuous ♂♂ congregate and display in a communal lek. ♀♀ nest and raise

462 Raggiana Bird of Paradise *Paradisaea raggiana* Sclater, 1873

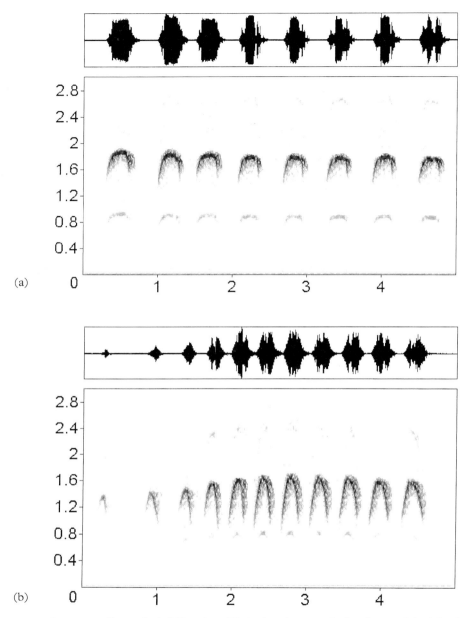

9.76 Sonograms from recordings of ad ♂ Raggiana Birds showing two similar forms of the lek advertisement song of *wau* notes; (a) ending in *waauu waaauuu* and the other, slightly higher-pitched and faster-delivered series, ending in *waagh waagh* notes (b). Sonogram (a) lacks the final note of the song recorded and (b) shows part of what was an intense 11-note song. At Varirata National Park, PNG in Aug. 1987 by B. M. Beehler.

young alone, as Goodfellow (1908) correctly stated. Beehler (1992, unpublished report) found lek calling could be heard as far away as 1 km, and that lek form was similar in both montane and lowland forest suggesting lekking is not ecologically based but is conservative (genetically programmed). He found leks to be long-lived, some being known for > 10 yr. Peckover (1990) states that a lek in Varirata National Park was known to be in continuous use for > 20 yr; this lek was abandoned in 1988/9 (BB). Leks may be formed in any habitat type. Major leks, which may involve several to 10 ad ♂♂, do not appear to be located near significant food resources. Other leks may be temporary, however, (several to c. 14 days) to take advantage of such things as a locally-available, significant fruit harvest. Three leks near Wau, PNG, were all in *Toona sureni* trees (BB). One lek here had three adjacent trees in fruit, *Endospermum labios*, *Ficus* sp., and *Omalanthus novoguineensis*, in which the lekking ♂♂ and their visitors fed, the ♂♂ using them regularly as a 'fast-food' source (BB). Table 9.1 summarises 10 leks found at eight sites by Beehler (1988). All were in the upper portion or top branches of a canopy tree in the forest interior or within a forest remnant. Five were within 100 m of a forest-edge, and five were within large tracts of continuous forest. Five were in a prominent position in local topography, such as on a high ridge crest, and five were in valleys or on sites lacking relief. Nine consisted of a single canopy tree but one included a group of four ♂♂ in one large, dead, vine-covered tree and a pair of actively-displaying ♂♂ in a second, smaller, living tree 70 m from the dead one. At 1553 hr on 14 June at a Varirata National Park lek the ad ♂♂ jostled one another and two then locked claws to tumble out of the lek in battle (BB). Delacour (in Gilliard 1969) noted a successful captive breeding of this sp. between a ♀ and an imm ♂ in ♀ plumage (see also Lesser Bird account). See Fig. 9.78 below.

Courtship behaviour

Display sequences follow a period of advertisement singing by ♂♂ perched in or near the lek, consisting of loud notes or series (see above) given singly, antiphonally, or simultaneously. As soon as one or more ♀♀ approach the lek a sequence of three display phases follows.

Convergence Displays: ♂♂ move from random and peripheral perches to their individual ones within the lek and excitedly give the loud display call whilst charging between perches and whilst raising and fluttering half-opened wings and partly erecting the flank plumes in a vertical (Fig. 9.77(a)), horizontal, or inverted posture (see Frith 1981), during which the upper surface of the wings are struck together (Fig. 9.77(b)). Birds then vigorously beat their half-opened wings as the head and chest is alternately sharply lowered and raised causing the now fully-erected flank plumes to thrash about as the birds call loudly (Fig. 9.77(c)). Behaviour constituting the *Convergence Displays* has previously been described as 'raised half-open wing-flutter', 'vertical wing-beating', 'inversion' and 'wing-beat phase' (Frith 1981).

Static Display: this is given by relatively-immobile and silent ♂♂, each on his particular (usually sloping to near-vertical) perch as ♀♀ move about the lek. ♂♂ lower their heads nearly to invert themselves then to beat the wings together again over the back before opening the wings fully. They then slowly turn the, slightly lowered, head side-to-side whilst simultaneously rhythmically, and repeatedly, sharply striking the carpal area of the wings together whilst giving soft, guttural, growling noises and occasionally clicking the mandibles together and/or wiping the perch with the bill. Whilst in this '*Flower Display*' ♂♂ sometimes descend their perch or stop the wing-striking and head-swaying to remain still in this pose prior to descending. If a ♀ is perched above the ♂ at this point in the display she may descend behind him to peck at his legs or his conspicuously-elongate and elevated undertail coverts or both (Fig. 9.77(d,e)). In so doing the ♀'s head is sometimes within the ♂'s flank plumes and the central pair of tail 'wires' may touch the ♀ and may thus play a tactile role. At this point a ♀ may briefly mount the displaying ♂ but this is a rare occurrence.

Copulatory Sequence: following the inverted *Flower Display* (performed by captive ♂♂ on horizontal perches if sloping/vertical ones are

Table 9.1 Composition and physical setting of ten Raggiana Bird leks at eight locations (five in the Wau area and three the Port Moresby area) in PNG (see Beehler 1988 for further details).

Lek	Elevation (m)	Observation dates	No. of males*	Physical location	Habitat
1	850	29–30 Sept 1975	8+	Subcanopy branches; tall forest tree	Interior of patch of selectively logged hill forest
2	1430	Aug–Oct 1980	4(2)	Limbs and vines in open canopy of dying *Castanopsis acuminatissima* (Fagaceae)	Ridgecrest forest at edge of clearing
3	1250	16 Sept 1980	4 + (2)**	Main lek: a dead tree with canopy vine tangle; second display tree; small, live	Sloping submontane forest in deep ravine
4	1050	14–16 Apr 1975	1(7)	Subcanopy branches, forest tree	Interior of a remnant plot of submontane forest
5	950	Aug 1979–July 1980	2(3)	Canopy limbs of 18 m *Toona sureni* (Meliaceae)	Small patch of regenerating forest surrounded by garden and village
6	850	10 Aug 1980	1(3)	Subcanopy of 33 m forest tree	Original riverine hill forest
7	750	11 July 1986	1(3)	Canopy limbs of hill forest tree	Crest of steep mountainside ridge in forest interior
8	800	9 Aug 1981 28 July–2 Aug 1987	7± 4	Canopy limbs of forest tree	Interior of flat plateau-top forest
9	800	2 Aug 1987	4	Upper limbs of subcanopy forest tree	Flat forest near patch of open savanna
10	550	28 July 1987	5	Canopy limbs of hillforest tree	Prominent ridge crest in forest interior

* Number in parentheses indicates a maximum count of non-resident males to visit the lek when resident males were in display.
** At this lek there were two groups of males: 4 + at the main tree and a pair at the second tree, 70 m distant.

Raggiana Bird of Paradise *Paradisaea raggiana* Sclater, 1873

unavailable—see Fig. 15 in LeCroy 1981) and perch descent the ♂ suddenly turns, bill uppermost, in the full open-wing pose, then to briefly adopt a half-open wing pose. He then ascends his perch in a rhythmic ritualised hopping gait involving several short hops upward followed by two or three shorter downward hops in a bouncing up–down gait, repeated several times, as he pivots the body and head from side to side of the perch without the wing-striking action. This perch ascent is often accompanied by the soft, guttural, growling noise followed by a single bill-click or *Bill-click-call* (see Vocalisations). If a ♀ remains on the perch above the ascending ♂ he then performs the *Frontal Display* which starts with the throat and cushion-like plush breast-shield feathers puffed into a ball-like shape as he approaches her with a bouncing gait to start to sway, increasingly rapidly and vigorously side-to-side while pecking at her bill (Fig. 9.77(f)). As his excitement intensifies he raises his flank plumes and half opens his wings (Fig. 9.77(g)) beating them over the back of the ♀ and continuing to peck at her bill now and then. His bouncing on legs, swaying of body and head, and wing-beating increase in speed until suddenly he mounts the crouching ♀ to copulate for a few sec with flapping wings. His tail is fanned extensively for < 1 sec as he thrusts forward, presumably ejaculating. If the ♀ remains perched after

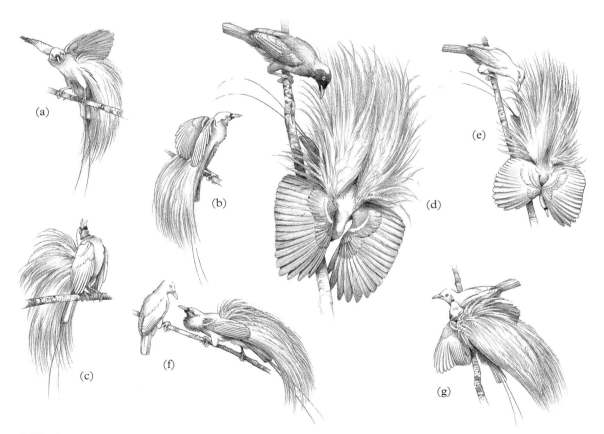

9.77 Some courtship display postures of ad ♂ Raggiana Bird: (a) *Upright Wing Pose* and (b) beating wings above the back, (c) *Wing-beat Phase*, (d) *Static* or *Flower Display* with wing joints apart and (e) together, (f–g) *Frontal*, or *Dance*, *Display*. After Frith (1981), Beehler (1988), and Coates (1990), with kind permission. See text.

mating the ♂ immediately continues *Frontal Display* to her, repeatedly pecking at her bill.

Gilliard (in Mayr and Gilliard 1954) once watched an ad ♂ on his display perch hang suspended upside-down and a ♀ fly to him and land directly above him then to look down at him and the two birds then peck at each other's bill. Margaret Mackay (1981) reported two captive ad ♀♀ (both reported to have laid eggs) performing display postures like those of ad ♂♂. The above account includes display behaviour described by Crandall (1931, 1932), Cooper and Forshaw (1977), LeCroy (1981), Frith (1981), and Beehler (1988).

Table 9.2 summarises activities of birds observed at a Mt. Missim, PNG, lek by Beehler (1988).

Breeding Table 6.5

NEST SITE: Goodfellow (1908) reported one nest under construction in the fork of a branch 3 m above ground on the edge of a small inhabited clearing and another in a similar situation but placed even 'lower down'. A single young was raised in both nests and they were taken live to England together with the ♀ parent from both. A nest containing a young which fledged by 4 Oct was *c.* 2 m high in a mango tree (Watson *et al.* 1962). One nest with a single-egg clutch, 4–5 m above ground in a small *Castanopsis* tree was located near Wau on 7 Dec 1987 (Sakulas 1988). Another nest was found 7 m high in the closed leafy sub-canopy of an 8-m-tall *Ficus* tree on 25 July, at Varirata National Park, PNG (BB). A nest brought to Gilliard (in Mayr and Gilliard 1954) at Kup, Wahgi Valley, with a nestling, was said to have been 11 m up in a casuarina tree standing within a small 'island' of trees in grassland near Kerowagi, at 1524 m. A nest was found 7.5 m above ground in a small *Rhus taitensis* tree and hidden by associated climbing bamboo (Davis and Beehler 1994). On the densely human-populated Wahgi R. plateau, PNG, 'In nearly every case the nests were high up, 50 to 60 feet [15–18 m], in casuarina trees' (Shaw Mayer in Sims 1956).

NEST: Goodfellow (1908) described a nest as a very flimsy structure composed of fibre and dead leaves, 'but twice during the period of incubation the ♀ bird entwined a fresh green creeper negligently about the nest and the fork in which it was placed'. A nest brought to Gilliard (in Mayr and Gilliard 1954) at Kup, Wahgi Valley was a 'flimsy nest of rootlets'. An active nest was a bowl-shaped structure, of *c.* 150 mm external dia and *c.* 110 mm internal dia, of entwined vegetation, including a live orchid and fern, lined with thin fern rhizomes (Davis and Beehler 1994). A nest from which a young fledged was a sparse and flimsy open cup of twisted vine tendrils with two large mango leaves in the bottom measuring 180 mm in dia and 100 mm deep externally and 100 mm dia and 50 mm deep internally (Watson *et al.* 1962). The nest collected with a clutch of 2 (AMNH 17076) in Aug 1908 is 160 mm in total external dia, 80 mm total depth, 120 mm internal dia, and 45 mm internal cup depth. It is substantially of large tree leaves and long narrow leaves and leaf pieces, possibly of gingers or similar plant leaves. A discrete egg-cup lining is of very fine horse-hair-like material (but thicker than the hairlike fungus), dark brown and up to 0.5 mm dia, more than encircling the entire nest cup at its largest dia. This substantial lining is simply sitting directly atop the accumulation of leaves and leaf pieces (CF).

EGGS: clutch usually two ($n = 8$) or one ($n = 5$). 20 eggs average 36.4 (34.3–39) × 25.2 (23.6–27.1) mm (CF). Ground colour Pale Pinkish Buff (121D) to pale Salmon (6) marked with darker browns and purplish greys in typical elongated, brush-stroke-like markings more dense about the larger end (Plate 13). In captivity, the eggs of a two-egg clutch were laid on consecutive days (Muller 1974). A captive ♀ in the USA laid three single-egg clutches over a 12-month period and then two two-egg clutches over the subsequent 12-month period (Rimlinger 1984).

INCUBATION: incubation of a single-egg clutch in the wild 'appeared to last ... 18 days'

Table 9.2 Raggiana Bird activities at a Mount Missim, PNG lek during 80.5 hr of observation between August 1979 and July 1980 by Beehler (1988).

	Time of day															
	0600–0630	0630–0700	0700–0730	0730–0800	0800–0830	0830–0900	1430–1500	1500–1530	1530–1600	1600–1630	1630–1700	1700–1730
No. of days observed	9	19	21	20	19	14	4	8	1	11	12	8
Maximum no. of males (all n)*	5	5	5	5	4	4	3	4	4	4	4	4
Median of maximum counts of males	3	4	4	4	3	3	2	2	3	2	3	3
Median of minimum counts of males	3	3	2	1	1	1	1	2	1	1	2	2
Median no. of males of all counts	3	3	3	3	2	2	2	2	2	2	2	2
Median no. Advertisement Calls**	1	1	2	5	11	7	1	4	3	5	5	8
Median no. Convergence Displays	2	7	8	3	3	1	1	1	1	2	4	1
Median no. Static Displays	0	3	3	2	2	2	0	0	1	1	2	2
Maximum no. of females (all n)	5	6	6	5	5	4	0	1	2	2	2	0
Median of maximum counts of females	3	4	3	1	1	0	0	0	0	0	0	0
No. of copulations	4	15	14	2	0	0	0	0	0	0	0	0

* Includes both perch-holders and visitors.
** Per 15-min block within each 30-min period.

(Goodfellow 1908). Muller (1974) noted that a captive two-egg clutch was not incubated until after the laying of the second egg. Incubation lasted 20 days, with both eggs hatching at different times on the same day. Incubation of another captive clutch of two was 18–19 days (Rimlinger 1984). Some 60 h of observation of parental care of a single-egg clutch in the wild over 9 days prior to its hatching showed that only the ♀ attended the nest, incubating for an overall average of *c.* 75% of daylight time with only heavy continuous rain causing her to spend an even higher proportion of time on the nest (Davis and Beehler 1994).

NESTLING CARE: at one wild nest (contents unknown) only a ♀ was seen, showing great agitation at human presence (Rand and Gilliard 1967). A captive nesting ♀ in India always took grasshoppers to a perch there to hold them by the foot to the perch to remove their legs with the bill before swallowing them, then to fly to her nest to regurgitate them to her nestling. She ate all nestling faeces (Dharmakumarsinhji 1943). From a brood of two, one of each sex, bred at Taronga Zoo, Sydney in 1958 the ♂ offspring lived for 2 yr but the ♀ survived to mate with her father (who was then at least 18 yr old) in Dec 1965. She hatched her single egg only to have her 2-week-old nestling die of exposure during a rain storm. The ♂ concerned died shortly thereafter and the ♀ was then housed with the only remaining ♂ which was estimated at that time to be 28 yr old. After several failed nesting attempts this ♀, at the time 13 yr old, was fertilised by the then *c.* 33-yr-old ♂ and successfully raised two young to independence (Muller 1974). All wild and captive broods were raised by their ♀ parent regurgitating all meals to nestlings (Dharmakumarsinhji 1943, 1944; Muller 1974; Searle 1980; Rimlinger 1984; Davis and Beehler 1994). Observation of parental care of a single-nestling brood in the wild for 110 hr during the first 12 days of nestling life showed the ♀ spent an overall average of *c.* 50% of time in attendance at the nest, this dropping to an average of 33% over the last 4 days. Her absence from the nest did not exceed 45 min until nestling age of over 6 days, after which brooding was mostly performed only during rain. The brooding ♀ often regurgitated a seed before reswallowing it or, if about to leave the nest, carrying it off in the bill. The nestling was fed only arthropods for its first 5 days of life and thereafter was occasionally fed fruits, of the drupe/berry type or the pulpy flesh of figs *Ficus* spp. She fed her young at a rate of mostly slightly over 1 meal per h. Meals at nestling age 1–4 days consisted of an average of 2–3 items per meal but thereafter quickly increased to 4–7 items per meal. The distinct impression was gained that the parent fed herself predominantly, if not exclusively, on fruit whilst raising her offspring predominantly upon arthropods (Davis and Beehler 1994) All nestling faeces were eaten by the ♀ parent.

NESTLING DEVELOPMENT: A captive brood of two had visible feather tracts and primary quills at 12 days and at 18 days were fully feathered, with short primaries and rectrices and one perched on the nest rim. By 20 days old both nestlings had left the nest, could fly, and were three quarters of their mother's size (Muller 1974). In captivity, a single nestling left its nest at 17 days old (Searle 1980) when it could fly quite strongly. The fledgling begged for food by wing quivering but not calling whilst the ♀ gave a soft, disyllabic, 'clucking' note. The fledged young was first noted feeding itself, on papaya, when 43 days old but it may have done so previously. It was still being fed by its parent, by regurgitation, at 75 days old, however. When 4 days old the wing flight feathers of two 2-day-old captive nestlings were visible and in pin (Rimlinger 1984). At 13 days old the nestlings had the head feathers still in pin but those of the wing coverts and flight feathers well out of pin and developed. These young left the nest at 17 days old and were fed by their mother for another 6–8 weeks. By 4 months old they were similar in size and colour, but less yellow on the nape and neck, to their mother. They had dark eyes and took at least 3 yr to attain the completely-yellow iris colour (Rimlinger 1984). A nestling that died in the wild at 12 days old had most feathers still ensheathed, or in pin, with a wing length of 52.5 mm and a weight of 100 g (Davis and Beehler 1994).

Annual cycle   Table 6.5

DISPLAY: Goodfellow (1908) claimed the sp. did not start displaying before Apr in the Owen Stanley Ra. At Varirata National Park, PNG (Sogeri Plateau), lek display peaked during May–Aug. In the Wau Valley, display peak was June–Oct, with activity for each extending for at least a month on either side of the peak. BREEDING: at least Apr–Dec. ♂♂ collected with moderately-enlarged gonads during all months except Nov (sample lacking) but birds with much-enlarged ones only during Apr–Oct and one in Jan. Nesting at Varirata was noted in late July (Beehler and Dumbacher 1990) and Aug; in Wau it was Oct–Nov. EGG-LAYING: July–Nov. MOULT: 22 of 238 specimens were moulting, the vast majority Nov–Mar inclusive with otherwise only a single bird during each of July, Aug, and Sept (Appendix 3).

Status and conservation   Table 8.1

Common, widespread, and flexible in habitat requirements. There is no evidence of current or future threats. At Brown R., PNG, 1196 observations over survey transects gave an average of 1 bird per ha (Bell 1982a).

After a 28-month study over 200 ha on Mt. Missim, PNG, 10 ad ♂♂ were found to be resident and to form two leks (Beehler and Pruett-Jones 1983), or 1 ad ♂ to each 20 ha of forest (see Fig. 9.78). Beehler (1983b) found the average number of individuals visiting a fruiting tree during a single day on Mt. Missim to be 5–10 birds (with 1–6 birds in a single tree simultaneously).

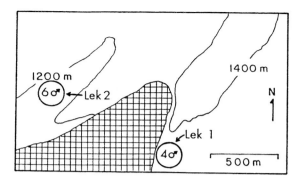

9.78 Display dispersion of ♂ Raggiana Birds on Mt. Missim, PNG. The two circles show the locations of leks, with the number of ♂♂ attending each. Hatching indicates an experimental pine plantation. From Beehler and Pruett-Jones (1983), © Springer-Verlag with kind permission.

Knowledge lacking and research priorities

The status of the well-marked peninsular subspp. requires reassessment, especially in areas of contact between them. In addition, study of mate selection and lek formation in hybrid zones with *minor* and *apoda* should be a priority. No adequate description of a fresh wild nest is available.

Aviculture

Birds were long kept, and bred, in PNG, India, Taronga Zoo, Sydney, Hong Kong, and San Diego Zoo. The sp. was successfully bred from the same ♂ and ♀ at BRS in Aug 1979, and again in 1980, 1981, and 1983. An additional nesting ♀ also raised one offspring in 1983 (Anon 1981, 1984). A hand-raised ♂ lived at BRS for 25 yr and Muller (1974) reports a ♂ at least c. 33 yr old breeding in captivity.

Other

A ♀-plumaged imm ♂ (weight 147 g, iris yellow with green inner ring) caught and ringed on Mt. Missim on 1 Sept 1980 was retrapped in July 1997 in full ad ♂ plumage- at a minimum age of 16 yr 10 months (J. Dumbacher *in litt.*; Anon 1998).

Goldie's Bird of Paradise *Paradisaea decora* Salvin and Godman, 1883

PLATE 12; FIGURE 9.79

Other names: *Siae* (ad ♂) and *Wagolina* (♀-plumaged) on SW Fergusson I. (LeCroy *et al.* 1980).

An aberrant, crimson-plumed *Paradisaea* of hill forests of the D'Entrecasteaux Archipelago, SE PNG. Only member of the genus on these islands, and readily located by distinctive vocalisations. The ♀ shares with the ♀ Blue the obscure ventral barring found in no other members of the genus. Monotypic.

Description

ADULT ♂: 33 (60) cm, lores, anterior forehead, malar area, chin, and throat covered with fine, highly-iridescent, plush, yellowish green to Dark Green (262) feathering that may appear jet black in some lights. Crown and mantle to uppertail coverts Orange Yellow (18), richer on crown, with iridescent white light sheen in some light. Upperwing: coverts pale Orange Yellow, alula, flight feathers, and tertials pinkish washed olive-brown (119B) with variable pale and dull yellow outer edges. Dorsal surface of tail similar to wing but slightly darker. Greatly-elongated central pair of rectrices reduced to fine wire-like shafts except at bases where narrow tapering vanes are blackish with rich Dark Green iridescence. Breast Medium Neutral Grey (184) with strong Lavender (77) suffusion, separated from iridescent green throat by a narrow, sub-obsolete, dull, greyish yellow collar, fading beneath lower breast to more pinkish Beige (219D) vent, thighs, and undertail coverts. Greatly-elongated flank plumes silky brilliant Crimson (108) with white filamental tips on ventral surface and Ferruginous (41) with paler tips on dorsal surface. A discrete subset of innermost flank plumes short, inwardly curving, and jet black with deep violet (172) suffusion and iridescent gloss. Underwing: coverts pale salmon pink (6), flight feathers greyish fawn (27) with broad, paler, salmon pink trailing edges to secondaries and all but outermost two primaries. Ventral surface of tail greyish fawn. Bill chalky blue grey, iris Sulphur Yellow (157), legs and feet purplish brown grey, mouth colour grey.

ADULT ♀: 29 cm, considerably smaller than ♂♂ of all age classes except in bill length. Lores, anterior forehead, malar area, chin, and throat Fuscous (21). Crown, nape, and upper mantle yellowish Buff (24) grading to remaining upperparts of Brownish Olive (29) strongly overwashed with yellowish Buff to a slightly more Cinnamon (123A) wash on rump. Upperwing: coverts and tertials Brownish Olive with yellowish Buff outer edges, alula and flight feathers as ad ♂. Dorsal surface of tail Olive-Brown (28), the more pointed pair with yellowish Buff outer edges, and other rectrices more pinkish washed olive-brown (119B) as in ad ♂. Underwing as ad ♂. Underparts Cinnamon (39) washed pale Amber (36), more so on breast, and finely barred throughout greyish.

SUBADULT ♂: as ad ♀ with a few feathers of ad ♂ plumage intruding, initially about head and central rectrices strongly pointed and longer than rest, to as ad ♂ with a few feathers of ♀ plumage remaining and central tail wires retaining some terminal vanes as elongate spatulate tips (see Plate 12).

IMMATURE ♂: as ad ♀.

NESTLING–JUVENILE: unknown.

Distribution

The Eastern Papuan Is of Fergusson and Normanby, of the D'Entrecasteaux Archipelago, E PNG; forested hills above 300 m to at least 600 m (and probably considerably higher).

Systematics, nomenclature, subspecies, weights, and measurements

SYSTEMATICS: presumed to be a sister form to the *Paradisaea apoda* superspecies, based on

Goldie's Bird of Paradise *Paradisaea decora* Salvin and Godman, 1883

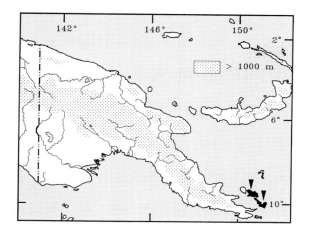

a range of characters (♂ flank plumes, tail wires, head pattern, and ♀ plumage pattern). Distinguished from the superspecies primarily by ♂ breast plumage and ♀ ventral plumage barring. Hybridisation: unrecorded and not expected; sympatric only with monogamous Curl-crested and Trumpet Manucodes.

SPECIES NOMENCLATURE AND ETYMOLOGY
Paradisaea decora Salvin and Godman
Paradisea decora Salvin and Godman, 1883. *Ibis* 1883 (Jan) p. 131. Fergusson I. Type specimen BMNH 834.46.

Etymology: *decora*, from L. *decorus* = becoming or beautiful. The common name honours the original collector Andrew Goldie, a Scottish plant collector who discovered the sp. in 1882.

WEIGHTS AND MEASUREMENTS
Measurements: wing, ad ♂ ($n = 22$) 177–185 (181), imm ♂ ($n = 7$) 164–178 (172), ad ♀ ($n = 8$) 157–165 (160); tail, ad ♂ ($n = 22$) 128–143 (137), imm ♂ ($n = 7$) 134–144 (138), ad ♀ ($n = 8$) 123–131 (128); tail centrals, ad ♂ ($n = 16$) 368–536 (431), imm ♂ ($n = 6$) 116–123 (120), ad ♀ ($n = 8$) 104–112 (109); bill, ad ♂ ($n = 20$) 35–40 (37), imm ♂ ($n = 6$) 35–38 (37), ad ♀ ($n = 8$) 34–39 (36); tarsus, ad ♂ ($n = 21$) 39–43 (41), imm ♂ ($n = 7$) 40–42 (41), ad ♀ ($n = 8$) 36–39 (37). Weight: ad ♂ ($n = 1$) 237. Our measurements indicate dimorphism in average size between the sexes in ♀ plumage, 6% for wing, and 7% for both tail and tarsus, respectively, and the ranges of measurements for each overlap considerably (*contra* LeCroy *et al.* 1980; LeCroy 1981).

Habitat and habits   Table 4.1
Hill forest and edge and fallow upland gardens. Birds visit water-holding tree knot holes, 10 m above ground, to drink and to bathe.

Diet and foraging   Table 4.2
Ad ♂♂ seen leaving their lek trees to feed nearby on the capsular fruits of *Medusanthera laxiflora* at c. 10 m above ground. This plant was common in the forest beneath the lek trees (LeCroy 1981). Birds were twice observed foraging in the forest floor leaf litter, two ad ♂♂ on one occasion and a subad ♂ on another. Local people indicated that ground feeding was not unusual (LeCroy *et al.* 1980). See Appendix 4 for food plants.

Vocalisations and other sounds
The first western naturalist to encounter the bird, Andrew Goldie, described the call as like that of the Raggiana Bird but with the addition of an occasional peculiar shrill whistle. The British anatomist Charles Stonor (1936) observed a captive bird in London and concluded that the sp. had a greater range of call notes than the other *Paradisaea* spp. he knew. LeCroy *et al.* (1980) recorded the following in 4 days observing birds at a lek on SW Fergusson I.: a *Wok-wok* or *wark*, similar to notes of other *Paradisaea* spp. (Fig. 9.79(a)) was usually given in the absence of ♀♀ and was therefore believed to be a ♂–♂ contact call. It could presumably also function, however, to attract ♀♀ to the lek. A soft *Whick-whick* call was also given in the absence of ♀♀ and is possibly a ♂–♂ contact call but a very loud, liquid, ringing version of this two-note call was given when a ♀ was present on the lek or just before a calling ♂ dived out of a lek tree. A low 'growling' call was given by a ♂ starting to display to a ♀. Two ♂♂ displayed simultaneously, often in the presence of a ♀,

they duet by starting with loud ringing metallic *waak* notes given alternately. These notes become increasingly rapid and become a continuous metallic rattling sound described as 'gurgling'. Duetting birds perched 1.5–3.5 m apart facing each other with their plumes raised, or one or both birds may hop up and down on their perches as they call. A relatively-softer call of *wuk wuk whik-whik-whik-whik* (Fig. 9.79(b)) and an explosively-loud, clear and sharp, somewhat *Epimachus* sicklebill-like, *whit whit whit* call (Fig. 9.79(c)) are also known.

Mating system

♂♂ congregate in a few adjacent trees to call and display in lek fashion. As with all other *Paradisaea* spp., ♂♂ presumed to be polygynous, ♀♀ to perform all nesting duties alone. Because a few adjacent trees were used at the only lek studied to date this has been considered to represent an exploded lek (LeCroy et al. 1980; Beehler and Pruett-Jones 1983), but this distinction may be premature since all other plumed *Paradisaea* spp. except *P. rudolphi* form true leks that may involve several trees. The question that remains is whether the birds in the lek are in more or less continual visual contact (true lek) versus only vocal contact (exploded lek). LeCroy et al. (1980) noted that the lek consisted of four main trees, and a fifth used only occasionally, spaced out within a rectangle of c. 92 × 46 m on a steep, south-facing slope just below a ridge crest at 400 m in mature rainforest, with a fairly even canopy of c. 30–33 m high. Display trees included at least three spp., but all had tall straight trunks and rather shallow crowns not

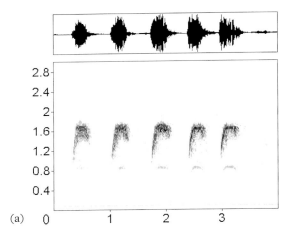

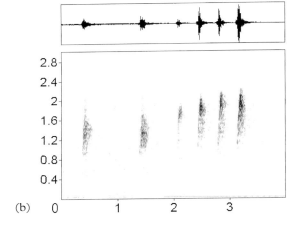

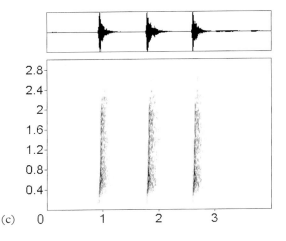

9.79 Sonogram from a recording of an ad ♂ Goldie's Bird giving: (a) typical song of loud *wahk* notes; (b) a softer song of *wuk wuk whik-whik-whik-whik* and (c) *whit whit whit* notes, explosively loud, clear, sharp and somewhat *Epimachus* sicklebill-like. Recorded on Fergusson I. in Nov 1993 by D. Gibbs.

densely foliaged. Display limbs were in the mid-canopy, at 20–23 m above ground, and were larger, open, limbs nearer the tree trunk. Only brief displays were performed on outer branches. Some 8–10 plumed ad ♂♂ displayed in this lek, usually two birds to each tree. No more than two ♀♀ and six unplumed ♂♂ were seen at any one time on the lek (LeCroy *et al.* 1980). On Fergusson and Normanby Is, a casual search suggested leks were 'perhaps 1 mile apart and the birds in a particular arena are able to hear birds in other arenas at least sometimes' along ridges (LeCroy *et al.* 1984). Lek ♂♂ regularly pluck the leaves near their display limbs. On three occasions a ♂ flew upward to take a leafy small branch in the bill, pull it down to his original perch and hold it there by one foot to pluck individual leaves off before releasing it.

Courtship behaviour

LeCroy *et al.* (1980) made observations at a lek over 18 hr from 8 to 11 Nov 1978, from which the following account is based.

Convergence Display: at 0610 hr on 11 Nov three plumed ♂♂ were present at the lek, two of them duetting as the third left the area, together with up to six unplumed birds of which two were thought to be ♀♀ because of their smaller size and quiet behaviour. The four unplumed presumed ♂♂ were actively hopping about perches and 'rowing' their wings. Calling and display by the duetting ad ♂♂ increased as they chased unplumed ♂♂ now and then, but mostly ignored them. During this the display posture mostly adopted was a 'wing pose' similar to that of the Greater Bird (Fig. 9.75(a,c)), but with the wings not held so far forward and more straight down at the sides of the body and moved with a 'rowing' action whilst the tail and body are held in a more horizontal position.

Static Display: at the peak of the dual display and duetting, one ad ♂ stopped posturing and moved away from the main display area, perching quietly during the remainder of the courtship. The display was continued by the remaining plumed ♂. His movements became slow and the display almost static with no audible vocalisation. The unplumed ♂♂ moved in to surround the ♀ closely, and two or three of them started rowing movements of their wings. The remaining plumed ♂♂ did not attempt to drive them away. The ♀ perched quietly near the displaying plumed ♂. The ♀ then left the main display area several times, but she returned almost immediately. Throughout this the plumed ♂ retained his plumes in display position, but a few times he accompanied the ♀.

'While standing near the displaying ♂, the ♀ began soliciting by quivering her wings, held slightly out from the body. The ♂ continued his slow and rhythmic display. Several times in the 5 min that the ♀ solicited, one displaying unplumed ♂ moved in and copulated with the ♀ and once two unplumed ♂♂ in succession copulated with her. These copulations were brief, lasting only a few sec, and there was no preliminary neck rubbing, although once the unplumed ♂ did put his wings down around the body of the ♀ (see below). The ♀ did not leave or stop soliciting after these copulations, but remained near the displaying plumed ♂ (LeCroy *et al.* 1980).

Copulatory Sequence: 'After about 30 min of display, the plumed ♂ began hopping stiffly up and down near the soliciting ♀ (from below we could see that his feet left the limb and came back to the same place). He edged over to the ♀, put his neck and breast on her back and rubbed back and forth. Then he mounted her, brought his wings down around her body and they copulated. While still mounted he rubbed his neck and breast on her back again, and then they copulated again. In all, the copulatory sequence lasted for *c.* 30 sec.'

'After the plumed ♂ had copulated with the ♀, all of the birds remained in the tree and the entire sequence of events was repeated, starting with duetting and joint displays by the two plumed ♂♂. We had no way of knowing whether or not the ♂ that now displayed alone was the one that had displayed alone during the previous display bout. The unplumed ♂♂ were around as before, were chased away on several occasions during the duetting and joint displays by the two plumed ♂♂, but were tolerated during the period of intense display by the single ad ♂. Once again they copulated briefly several times with the soliciting ♀. The displaying plumed ♂ copulated with the ♀ as before, dismounted and moved immediately to the second ♀ and copulated with her—this copulation lasting only a few sec. The first ♀ moved away about 3.5 m, after the ♂ dismounted, and began preening. After the second copulation all birds left the tree. It was then 0715 hr.'

Another display sequence ending in copulation when no unplumed birds were present was observed on 8 Nov, 1200–1220 hr. When the single ♀ appeared in the lek tree two plumed ♂♂ were present. 'She was greeted by low growling by at least one ♂. For the most part the second ♂ sat quietly in the same tree, but twice approached and displayed briefly near the ♀. The main display perch in this tree was a sharply sloping limb. The ♀ sat on a horizontal limb branching from it. The ♂ hopped slowly up and down the sloping limb, zigzagging his body back and forth as he did so, so that he alternately presented back, front and side views to the ♀. The ♀ watched him continuously, sometimes facing him, sometimes peering sideways at him with head cocked. This ♀ appeared quite wary and frequently flew to the outer branches, where the ♂ followed her and displayed. They always quickly came back together to the main display perch. Movements by the ♂ around the ♀ were always slow and deliberate. Several times he gave high intensity *whick-whick* calls in front of the ♀, but generally no calls accompanied the display. After 15 min of display by the ♂ the ♀ began soliciting and the ♂ gradually moved closer to her, hopped up and down on the branch next to her (we did not see neck and breast rubbing on this occasion) and mounted her with his wings down around her body.' Copulation was reported to have lasted *c.* 15 sec (LeCroy *et al.* 1980), before both birds flew from the tree in opposite directions.

We find the repeated mounting of the soliciting ♀ by unplumed ♂♂ in the presence of a courting ad ♂ inexplicable, and suggest that this anomaly requires further assessment in the field to determine its significance and to document whether it is a regular feature of the mating behaviour of *Paradisaea decora*.

Breeding Table 6.5

NEST SITE: people of Normanby and Fergusson Is informed LeCroy *et al.* (1980, 1983) that birds make a hole in a bird's nest fern *Asplenium* sp. to nest but this unlikely nest site for a *Paradisaea* sp. requires confirmation.

NEST, EGGS, INCUBATION, NESTLING CARE, AND DEVELOPMENT: unknown.

Annual cycle Table 6.5

DISPLAY: the only data are from LeCroy *et al.* (1980) for lek displays and matings during the second week of Nov. BREEDING: one ♂ collected in Sept and four in Dec had much-enlarged gonads. EGG-LAYING: unknown. MOULT: 9 of 40 specimens were moulting, involving May and Dec but Jan, Mar–Apr, July. Oct–Nov were not represented by samples (Appendix 3).

Status and conservation Table 8.1

Not uncommon according to earlier collectors and ornithologists but a contemporary assessment of populations, habitat, and potential threats is needed. Goldie (in Sharpe 1891–8 and Gilliard 1969) found ad and imm ♂ birds, away from the lek, in groups of three or four; ♀♀ were encountered as solitary individuals. Albert Meek (in Gilliard 1969) found the sp. 'not rare, but by no means very numerous on the hills of South Fergusson, from about 1500 feet [*c.* 450 m] upwards'. The global range is tiny, but current and expected threats remain low.

Knowledge lacking and research priorities

There is an urgent need for a detailed survey of populations, habitat selection, foraging ecology, and breeding biology of the sp. The systematic position of the sp. could be clarified by molecular studies, if tissue were to be made available. Additional details of lek behaviour of the sp. would be useful, considering the unusual behaviours reported by preliminary field studies.

Aviculture
Unknown in captivity.

Red Bird of Paradise *Paradisaea rubra* Daudin, 1800

PLATE 12; FIGURES 9.80–2

Other names: *Uranornis rubra*, *Burung Kun* (specimen label)

Another isolated insular *Paradisaea*, distinctive for the ♂'s black and curled central tail 'tapes,' and the broad yellow breast-bar of the ♀. The sp., the only *Paradisaea* on Waigeo and Batanta Is. of the Rajah Ampat group, IJ, is easily located because of the loud and persistent vocalisations of the ♂♂. Monotypic.

Description

ADULT ♂: 33 (72) cm, head to just behind the eye, ear coverts, chin, and extensive throat of finely-scaled feathering intensely-iridescent dark emerald-green which may appear jet black in some lights, particularly about bill base. Feathers above each eye slightly elongated and curved to form an erectile cushion-like structure. Remainder of head, nape, upper and central mantle, and modified (mantle-like) wing coverts washed-out, pale Orange-Yellow (18). Sides and lower centre of mantle rich Amber (36) washed yellow with highly-iridescent, white, spun glass-like sheen in some light grading to Russet (34) brown back. Rump and uppertail coverts pale Orange Yellow washed Amber. Dorsal surface of wings, including alula and primary coverts, and dorsal surface of tail rich reddish brown (223A) except central pair of rectrices which are uniquely modified into 3–4 mm-wide, gently-twisting, concave, shiny black 'tapes' of plastic-like appearance and texture and considerable length. Somewhat stiff, pale, Orange-Yellow, upper breast feathers, elongated and pointed to the lower sides, form a small breast-shield of sorts. Beneath this the breast is slightly glossy Warm Sepia (221A) washed Maroon (31) becoming paler and less maroon on the belly (221B) grading to Hazel (35) on vent, thighs, and undertail coverts. Elongate, slightly-stiffened, and downward-curving flank plumes Carmine (8) to Crimson (108), paler towards concealed whitish bases, and variably Warm Sepia to Walnut Brown (221B) on their upper surface with off white filamental tips. Underwing: coverts Hazel, flight feathers likewise but paler. Ventral surface of tail pale greyish Hazel. Bill yellow washed pale green, iris dark reddish brown, legs and feet lead bluish grey, mouth colour unrecorded. Isenberg (1961) stated that a bird he acquired in ♀-plumage did not attain full ad ♂ plumage until at least its fifth year of life. Studies of captive birds (Frith 1976) indicated ad ♂ plumage might require as much as 6 yr to acquire.

ADULT ♀: 30 cm, foreface, to posterior eye, and throat Warm Sepia strongly washed Maroon (feathers above fore-eye elongate and bristly but discernible only on bird in hand). Rear crown, neck, and nape Buff-Yellow (53) washed pale dilute Orange-Yellow this becoming darker on mantle as it overlays Warm Sepia of the lower mantle where it appears Amber. Remaining upperparts, including wing and tail, pale Warm Sepia except lesser coverts and outer edges of greater coverts, which are overwashed orangy yellow to appear amber-buff, and uppertail coverts, which are faintly washed orangy yellow. Upper breast a discrete

Red Bird of Paradise *Paradisaea rubra* Daudin, 1800

Buff-Yellow with orange yellow wash and below this to undertail coverts Walnut Brown washed Hazel, very dark against the upper breast. Underwing and ventral surface of tail as ad ♂. Central pair of rectrices narrower and more pointed than rest.

SUBADULT ♂: as ad ♀ with a few feathers of ad ♂ plumage intruding, to as ad ♂ with a few feathers of ♀ plumage remaining. Central tail feather pair initially simply longer, narrower, and more pointed than rest, to a relatively-short, bare, central feather shaft 'wire' with elongate spatulate tip of vanes that becomes progressively longer and with smaller vane tips.

IMMATURE ♂: as ad ♀ but on average larger.

NESTLING–JUVENILE: hatchling naked and pink skinned, becoming dark bluish grey after 3 days. At 14 days old the bill is horn coloured, the iris dark brown, legs bluish grey, and mouth interior white. See also Nestling development below.

Distribution

Endemic to forests of Waigeo and Batanta Is, in the Western Papuan (Rajah Ampat) group, IJ. Also apparently the two tiny islands of Gemien and Saonek off the W and S coast, respectively, of Waigeo I. Presumably also possibly the much larger Gam I., that lies off the SW coast of Waigeo. From sea level to *c.* 600 m.

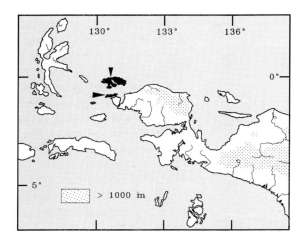

Systematics, nomenclature, subspecies, weights, and measurements

SYSTEMATICS: a highly distinctive *Paradisaea* sp. that appears to be an outlier to the core spp. group that includes the *Paradisaea apoda* superspecies and *P. decora*. Gyldenstolpe (1955b) considered a single ♀ from Saonek I. to be 'much paler above and beneath as well as on the outside of the wings' than ♀ Waigeo and Batanta Is specimens he examined. Hybridisation: unrecorded, the only potential sp. available being the diminutive Wilson's Bird but, given that hybrids between Magnificent and Lesser Birds are known, hybridisation is theoretically possible.

SPECIES NOMENCLATURE AND ETYMOLOGY

Paradisaea rubra Daudin
Paradisea rubra Daudin, 1800. *Traité d'Ornithologie* (Lesson) **2**, p. 271. 'New Guinea' (= Waigeo I.). Type specimen none designated.

Etymology: *rubra* = L. *ruber*, red, ruddy.

WEIGHTS AND MEASUREMENTS

Measurements: wing, ad ♂ (*n* = 27) 169–186 (176), imm ♂ (*n* = 10) 162–178 (170), ad ♀ (*n* = 26) 151–174 (159); tail, ad ♂ (*n* = 27) 114–125 (119), imm ♂ (*n* = 10) 112–126 (119), ad ♀ (*n* = 26) 105–121 (114); tail centrals, ad ♂ (*n* = 23) 478–765 (563), imm ♂ (*n* = 9) 109–127 (117), ad ♀ (*n* = 24) 102–117 (106); bill, ad ♂ (*n* = 26) 31–39 (36), imm ♂ (*n* = 10) 33–38 (35), ad ♀ (*n* = 26) 33–39 (35); tarsus, ad ♂ (*n* = 26) 38–44 (42), imm ♂ (*n* = 10) 40–45 (42), ad ♀ (*n* = 25) 36–44 (39). Weight: ad ♂ (*n* = 7) 158–224 (201), imm ♂ (*n* = 8) 156–212 (195), ad ♀ (*n* = 10) 115–208 (158).

Habitat and habits   Table 4.1

Lowland rainforest and hill forest. Captive birds in Singapore were often seen to sun themselves for long periods by perching in direct sunlight and holding themselves in a peculiar pose with one wing (facing the sun) drooping and the body leaning away from

Red Bird of Paradise *Paradisaea rubra* Daudin, 1800

9.80 Ad ♂ Red Bird in typical sunning posture; the direction of the sun's rays is from top left. From Frith (1976), with kind permission. See also LeCroy (1981, fig. 20).

the sun with the head held high and the bill pointing upward or tilted to one side (Fig. 9.80). Seen in tall (> 35 + m) hill forest on steep limestone on Waigeo I. (D. Bishop *in litt.*)

Diet and foraging   Table 4.2

Alfred Russel Wallace (1869) observed birds on Waigeo 'quite low down; running along a bough searching for insects, almost like a Woodpecker' and eating the fruits of a fig *Ficus* sp. Eleven birds kept in a large and well-foliaged Singapore aviary were often seen to seek insects actively, by pecking off the bark of dead twigs then to probe the soft wood beneath or by carrying pieces of dead bamboo or leaves from the ground to a perch there to hold them to the perch with one foot and tear them apart. Birds also clambered acrobatically about fine outer twigs and foliage of shrubs to snatch insects from leaves. One bird hung upside-down from fine outer twigs to pick and swallow small buds (Frith 1976).

Vocalisations and other sounds   Figure 9.81

Wild birds are vocal in early mornings and late afternoons (B. Poulson, personal communication). Most commonly-heard call given by captive ad ♂♂ was a loud clear repeated *wak* note, birds building up to this with a throaty guttural *work—wok, wak, wak, wak, wak, wak* at *c.* 2–4 notes per sec which then becomes a very loud clear *wok—wau-wau-wau* at *c.* 1–4 notes per sec. Another similar loud but higher-pitched call was a *ca-ca-ca-ca-ca-ca* delivered at *c.* 4–5 notes per sec (Fig. 9.81). An oft-given crow-like call consisted of a coarse and guttural *kaw, kaw, kaw* at 1–2 notes per sec. Also common was a soft pathetic-sounding single *weep* note. A very soft and high-pitched mewing *meew* was not uncommon, as was a single snap of the mandibles, *tick*. Many variations of these calls were heard, mostly of the crow-like call which varied in tone, volume, and duration, but all were clearly defined as one of the above call types. Excitedly-displaying ad ♂♂ frequently produced an odd 'bill-click-call' prior to intense display, by rapidly and repeatedly snapping together the mandibles usually, but not always, starting and ending with a brief vocalisation. The individual *tick* notes sound like the noise produced by smacking one's tongue against the back of the teeth and roof of the mouth. This sound is repeated 4–5 times, being preceded by a high-pitched *beep* and ended by a low guttural *book* note, thus *beep t-t-t-t-t book*: a mechanical sound that lasts little more than 1 sec. The bill-click-call is a display sound, delivered with ritualised body movements and most often given just before intense displays, usually by groups of ♂♂ close together. A continuous bill-clicking

Red Bird of Paradise *Paradisaea rubra* Daudin, 1800

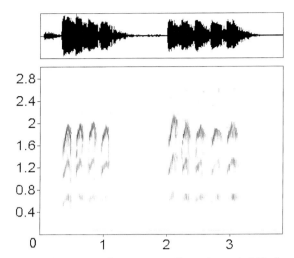

9.81 Sonogram from a recording of an ad ♂ Red Bird giving short, rapid bursts of sharp *ca* notes (typical of *Paradisaea* spp.) in the lek advertisement song in July 1990, Waigeo I., IJ by K. D. Bishop.

without accompanying vocalisations or body movements was also given by a ♂ chasing another in flight. The soft, high-pitched, mewing note described above is given infrequently by displaying ♂♂, as is the single snap of the mandibles.

In flight, birds produce a slight rustling with the wings, but this is not as loud as that of many other birds of paradise.

Mating system

Observations at a single lek on Waigeo I. that was attended by up to 10 ad ♂♂ (Bergman in Gyldenstolpe 1955*b*) indicate this is a polygynous sp. with communally-displaying promiscuous ♂♂, and ♀♀ that doubtless attend nests without assistance. The lek concerned was a single gigantic tree located in a small forest clearing. ♂♂ called and displayed on high dead limbs and other leafless limbs apparently defoliated by them (confirmed by Eastwood 1996: 23). During Sept 1989 and July 1990 D. Bishop (*in litt.*) watched 3–4 ad ♂♂ displaying shortly after dawn in the lower crown of a conspicuously-tall tree in hill-forest on limestone. The birds were shy and wary.

Courtship behaviour

Frith (1976) observed 10 ad ♂♂ and one ♀ captive in a large Singapore aviary over 6 days. Displays were performed on vertical or nearly vertical perches, birds preferring bare branches in an open area. Components of the display sequence are here tentatively ascribed to the three standardised main phases of lekking *Paradisaea* display.

Convergence display: excited ♂♂ hop quickly from branch to branch with fluttering, flicking, and extended wings like a butterfly then to land suddenly on their individual display perches. A ♂ usually begins displaying by perching on a branch near or on the upper part of a vertical limb. He gives the bill-click-call frequently, flicks his outer primaries rapidly forward and back, and occasionally performs perch *Bill-wiping*. Once on the vertical branch, he perches diagonally with bill pointing upward and continues to flick his primaries. As he becomes more excited the wings are slightly spread and quivered (Fig. 9.82(a)); again he wipes his bill and sways his body gently from side to side. He then stops swaying and leans increasingly to one side, the head thus becoming lower, while not only flicking his primaries but actually spreading and shaking his wings, until completely inverted (Fig. 9.82(b)). At this point he increasingly spreads his wings, shaking them very rapidly so that they appear to vibrate, and all the time swaying his body from side to side, making the flank plumes conspicuous, though not raised or spread. He may still bill-wipe the perch briefly; and move his head about in a rather stiff *Head-peering* motion (Fig. 9.82(c)).

Static Display: so far the ♂ may have uttered the soft high-pitched *meew* or more often a single snap of the mandibles. When completely inverted he shakes his wings, fully extended at either side of his body, but does not raise them above his back. He vibrates them rapidly but so slightly that the vibration is only just discernible as he hangs momentarily motionless, save perhaps for a little *Head-peering* as if looking intently at the perch and giving an occasional single mandible snap

(Fig. 9.82(d)). The flank plumes are raised very slightly but not spread and the tail is held only slightly downward, the plumes thus being 7 cm or less above the tip of the tail.

Copulatory Sequence: suddenly the plumage is back in the normal non-display position; the bird hops down the perch, usually directly but sometimes in a spiral, till almost at the base, which (in captivity) is often on the ground, for a distance of *c.* 1.5 m. He then hops once around the branch and back up it to the original spot or higher, rarely ascending in a spiral but usually by hopping to face from one side to the other. He hops down and up the vertical perch with a slow, deliberate, ritualised gait. Having completed the ascent, the bird then performs a *Hopping-on-the-spot Display*, again slowly and deliberately with the body and head moving in a peculiar mechanical fashion. *Hopping-on-the-spot* may be uninterrupted or be interspersed with bouts of *Head-peering* or ritualised *Bill-wiping*. Suddenly he again opens his wings (Fig. 9.82(d) but with head uppermost) and becomes almost motionless except for a very slight rapid vibration of the wings and continuous *Head-peering*. He then begins to sway from side to side very slightly, pointing the bill predominantly upward, with *Head-peering*, and occasionally gives a single bill-click. Finally he augments *Head-peering* by tapping the sides of the very tip of his bill sharply on the perch every few sec while rotating his head (*Head-peering*) with its green plumage conspicuously erected. The bill-tapping produces a *tick* sound clearly audible at 10 m. This completes the entire display, which lasts 45–120 sec, the bird then hopping from his display perch to preen or to feed. Figures in Frith (1976) show the entire sequence of display postures and movements.

We emphasise that the allocation of particular display postures and movements to the three main phases of display above is tentative. As these displays were performed within the confines of a 25.5 × 9.5 × 3.5-m aviary, containing only one ♀, with no copulation observed, it is possible some components of the normal display were performed out of normal sequence or were missing. Observations of wild-displaying and mating birds are required to confirm this.

Various component parts of any one, two, or all three of the above phases of display were performed as a briefer display, the most frequent of them being the *Hopping-on-the-spot* during which the bird would flick his primaries and give a single bill-click or tap the perch or do both.

The *Bill-click-call* is accompanied by ritualised movements that appear to constitute a display of sorts: while the initial *beeb* note is produced the head and bill are thrust downward to the breast; during the bill-clicking the bill is thrust upward; and during the concluding *book* note the head is again thrust downward and then raised to the normal position. While giving the call and associated movements of the head, the bird jerks its body very rapidly and almost imperceptibly, while hopping along the perch and back again a very small distance indeed. The whole action appears mechanical and the sudden jerky body movements are emphasised by the stiff flank plumes.

Crandall (1937a), in observing a NYZS captive ad ♂, made the significant observation that it was not until a downward-sloping (at *c.* 45°) branch was made available that displays were performed. His descriptions of this bird's display include some of the above behaviour and nothing additional to it. He did conclude that 'It seems quite possible that the presence of a slanting perch may be necessary for the full performance'.

Wild ad ♂♂ have been observed to hang upside down quivering their wings and 'stropping' the bill from side to side against a branch and with the red plumes in close contact with the face of a ♀ perched above. This was followed by the ♂ climbing down the vertical display branch for a short distance (B. Poulson *in litt.*). This is a broad description of the display observed in greater detail performed by captives and described above.

Breeding Table 6.5

Information available only from birds in captivity.

480 Red Bird of Paradise *Paradisaea rubra* Daudin, 1800

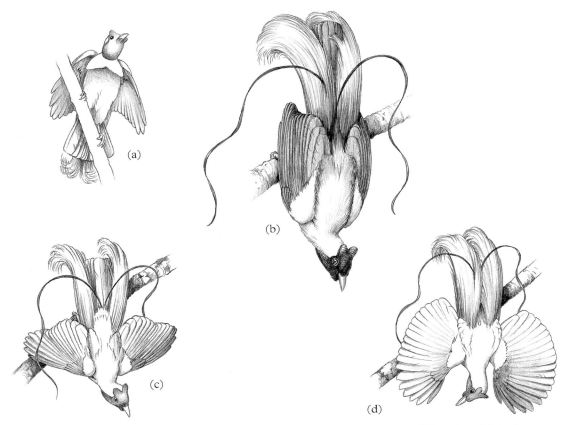

9.82 Some courtship display postures of ad ♂ Red Bird: (a) initial display, or *Wing Pose*, (b) initial inverted posture, (c) open wings inverted posture while *Head-peering*, (d) *Static Display*. After Frith (1976), with kind permission. See text.

NEST SITE AND NEST: unknown. In captivity only the ♀ builds the nest (Isenberg 1961, 1962; Todd and Berry 1980; Hundgen *et al.* 1991; Worth *et al.* 1991).

EGGS: clutch of one or two eggs. Seven BMNH eggs average 36.2 (34.8–39.1) × 24.3 (23.8–25) mm (CF). Ground colour pale buffy-white with palest pinkish wash to Pale Pinkish Buff (121D) or Salmon (6). Three USA captive clutches were of two eggs (Isenberg 1961, 1962). A captive ♀ at Houston Zoo laid an infertile two-egg clutch and a fertile single-egg clutch (Todd and Berry 1980). Mean size of nine captive-laid eggs: 33.3 × 23.0 mm (SD = 1.34 and 1.05, respectively) and mean weight at 14 days old 9.6 g (SD = 0.65, n = 23) (Hundgen *et al.* 1991). Eggs of eight captive-laid two-egg clutches said to be laid at intervals of *c.* 36 hr but five clutches also involved eggs laid on alternate days, two on consecutive days, and one both on same date. Usual laying period in birds of paradise appears to be at *c.* 24 hr intervals.

INCUBATION: in captivity, exclusively by ♀ (Isenberg 1961, 1962; Todd and Berry 1980; Hungen *et al.* 1991; Worth *et al.* 1991). A two-egg clutch laid in the USA by a captive bird took *c.* 14–16 days to hatch (Isenberg 1962). Inconclusive circumstantial evidence suggests that another ♀ USA captive bird took at least 15 days to incubate her single-egg clutch (Todd and Berry 1980). Sixteen two-egg clutches were laid over 3 yr by one ♀ in the NYZS, during Dec–June. Of these 32 eggs, 18 were fertile and 17 hatched. Incubation averaged 16.5 (16–17, $n = 17$) days, with the eggs being brooded for the first 14 days by the ♀ parent and thereafter by an artificial incubator.

NESTLING CARE: a captive ♀ fed her single hatchling exclusively on regurgitated soft fruit for 5 days before then feeding it animal food (Todd and Berry 1980). She continued to eat fruit herself but fed her young only animal food such as cockroaches, crickets, mealworms, and new-born mice, which she dismembered before swallowing them—then to regurgitate them at the nest to her nestling. When the young bird was just over a month old the ♀ again started to feed it fruit, in addition to animal foods. All nestling faeces were eaten at the nest by the ♀ (Todd and Berry 1980). The lack of egg shell remains from repeated captive nestings with successful hatchings by one ♀ suggested she was eating them (Worth *et al*. 1991).

NESTLING DEVELOPMENT: mean weight of 16 captive hatchlings was 8.3 g (SD = 0.64) on the day they hatched. In five hand-raised captive nestlings average daily weight gain during first month of life was 3.4 g (SD = 0.74). Some 3 days after hatching nestling skin colour starts to change from pink to dark bluish grey and wing quills begin to appear. By 14 days old wing quills start to burst from pin and short feathering appears on head and dorsal feather tract, eyes fully open. At 15 days old a tail-less nestling was stretching and preening itself and it left the nest the next day. At 24 days this young bird's tail was *c*. 13 mm long and its plumage brown, slightly darker on the face, and the neck sparsely feathered. When 28 days old golden brown nape feathers like those of the ♀ parent were visible, and a week later the base of the lower mandible and the ear openings were still bare and, although underwing covert feathering was incomplete, the young could fly. It was not seen to feed itself until separated from its mother, however. When 63 days old this juv could only be distinguished from its parent by its darker beak (Todd and Berry 1980). In other hand-raised captive nestlings pin feathers emerged first on wings, both primary and secondary coverts, at average age of 5 days. This was followed by breast at 9 days, back at 11 days, head at 12 days, and tail at 14 days. Eyes opened by 11 days old on average. By some 2 weeks old primary coverts had burst from pin, but secondary coverts were still encased in sheaths. By 3 weeks old nestlings were fully feathered, although breast feathers were still soft and downy. Most of head, back, and wing feathers were russet, but breast was light beige. A narrow line of bare skin ran down centre of breast, and remained visible for *c*. 90 days. Between 40 and 90 days head, nape, and breast gradually changed to a golden colour. By 60 days old the yellow upper breast began to appear and by 125 days this had become distinct and resembled that of ad ♀♀. Also, a patch of dark brownish feathers extended from crown to throat and surrounded the eyes and the cream coloured bill. Nestlings were first seen perched on the edge of their artificial nest bowls at an average of 17.5 (15–20, $n = 4$) days (Todd and Berry 1980).

Annual cycle   Table 6.5
DISPLAY: no information for wild birds additional to that of B. Poulson cited above. BREEDING: ♀♀ with enlarged oocytes in Sept–Oct. EGG-LAYING: no information for wild birds. MOULT: 12 of 55 specimens were moulting involving all months except Feb, Apr, and Aug (Appendix 3). In captivity in NYZS an imm ♂ took *c*. 3 months (3 May–10 Aug) to moult, and a subsequent moult, into its first ad ♂ plumage, took 4 (21 May–25 Sept) months (Crandall 1937*a*). Isenberg (1961) noted that, in captivity in the USA, his ♀ started moulting 2 months before his ad ♂.

Status and conservation   Table 8.1
In 1948, Bergman (in Gyldenstolpe 1955*b*) found the sp. common, and particularly

plentiful about the central village of Waifoi, Waigeo I., and fairly common on Batanta I. Common on Batanta I. in late Aug 1995 (M. Carter *in litt.*). Although the global range of the sp. is small, it appears to be under no threat, especially since there are protected areas on both Batanta and Waigeo. Collecting for the cage bird trade needs monitoring, however.

Knowledge lacking and research priorities
All too little is known of the sp. biology in the wild. It is important to know if it inhabits Gam I. For bibliographic completeness we note that Berger (1956) presented a detailed and thorough description of the feather tract (pterylosis), skeleton, and musculature of this bird based on study of a single ♂ that died in captivity.

Aviculture
Not uncommonly maintained in Asian collections, such as in Singapore but not bred there. Recent significant successes have occurred in the USA (Todd and Berry 1980; Hundgen *et al.* 1991; Worth *et al.* 1991) leading to a pair being loaned to the Vogelpark Walsrode breeding facility at Mallorca, Spain.

Emperor Bird of Paradise *Paradisaea guilielmi* Cabanis, 1888

PLATE 12; FIGURES 5.3, 9.83–4

Other names: Emperor of Germany Bird of Paradise, Emperor of Germany's Bird of Paradise, Emperor's Bird of Paradise, White-plumed Bird of Paradise. *Kina* (♂) and *yagun* (♀) of the Nokopo people of the Finisterre Ra, PNG, (Schmid 1993).

This, the montane Huon Penin isolate of the genus *Paradisaea*, is, like *P. rubra*, an aberrant sister form to the main *Paradisaea* clade, distinguished by the unusual flank plumes, facial mask, courtship displays of the ♂, and the distinctive colour pattern of the ♀. Monotypic.

Description
ADULT ♂: 33 (86) cm, most of head and upper breast highly-iridescent, oily, brilliant Dark Green (262) or greenish yellow subject to light, but some areas, particularly about bill base, may appear jet black. Rear crown, nape, rear neck, and extensive mantle Sulphur Yellow (57) washed Sulfur Yellow (157) with Maroon (31) of feather bases sometimes discernible through yellow of central mantle. Lower back, rump, dorsal surface of tail, and upperwing coverts and tertials dark Warm Sepia (221A) with Maroon wash. Remainder of upperwing and dorsal surface of tail Warm Sepia with central tail feather pair of 'wires' grossly elongated as in Lesser Bird. Yellow cowl extends down to each pectoral to border the lower sides of the green breast. Lower breast dark deep Maroon, belly dark Dusky Brown (19) and vent, thighs, and undertail coverts dark Burnt Umber (22) washed maroon. Grossly-elongated, fine, and sparse filamental flank plumes ventrally and basally pale orangy yellow but elsewhere white except where elongate, maroon, outer upper breast feathers overlap them. Patches of elongate white feathers to either side of the belly appear to expand the flank plumes across underparts to meet almost centrally. Entire underwing and ventral surface of tail dark Warm Sepia. Bill chalky bluish grey, iris dark reddish brown, legs and feet pale purplish brown, mouth chalky blue inside mandibles to pale aqua blue beyond.

ADULT ♀: 31 cm, green head and breast area of ad ♂ is dark Warm Sepia in the ♀ and yellow upperparts and sides of breast of ad ♂ are much more dull owing to the strong influence of underlying and intruding Warm Sepia coloration. Remaining upperparts Warm

Emperor Bird of Paradise *Paradisaea guilielmi* Cabanis, 1888

Sepia slightly paler than in ad ♂. Lower breast and belly Walnut Brown (221B) washed with Vinaceous Pink (221C) and thighs and undertail coverts Warm Sepia. Underwing and tail as ad ♂, if slightly paler.

SUBADULT ♂: as ad ♀ with a few feathers of ad ♂ plumage intruding, to as ad ♂ with a few feathers of ♀ plumage remaining.

IMMATURE ♂: as ad ♀.

NESTLING–JUVENILE: unknown.

Distribution
Throughout the lower mountains and hills of the Huon Penin (Saruwaged, Finisterre, Rawlinson and Cromwell Ra) PNG, from c. 450 to 1500m (mainly 670–1350 m). The Raggiana Bird replaces it at lower altitudes in the E of its range, the Lesser Bird in the W.

Systematics, nomenclature, subspecies, weights, and measurements
SYSTEMATICS: as with *Paradisaea rubra*, this sp. is a geographical isolate that is also a sister form to the main *Paradisaea* clade (*rubra*, *guilielmi*, and *rudolphi* all are systematically isolated taxa that are peripheral to the main *Paradisaea* lineage). Hybridisation: known with Lesser and Raggiana Birds (Plate 15 and Appendix 1).

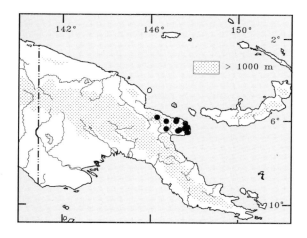

SPECIES NOMENCLATURE AND ETYMOLOGY
Paradisaea guilielmi Cabanis
Paradisea Guilielmi Cabanis, 1888. *Journal für Ornithologie* 36, 119. 'Kaiser Wilhelm's Land', restricted to the Sattelberg, Huon Penin., PNG (Mayr 1941). Type specimens ZMB 29783 (♂), ZMB 29784 (♀).

Exploration: discovered on the Sattelberg by Carl Hunstein in Jan 1888 and named by Jean Louis Cabanis of the University Museum at Berlin.

Etymology: *guilielmi* = Mediaeval L. *Guilielmus* = William, after Willem III Emperor of Germany and King of Prussia (1859–1991) (Jobling 1991).

WEIGHTS AND MEASUREMENTS
Measurements: wing, ad ♂ ($n = 27$) 172–188 (177), imm ♂ ($n = 4$) 163–179 172), ad ♀ ($n = 15$) 151–171 (158); tail, ad ♂ ($n = 27$) 107–121 (114), imm ♂ ($n = 3$) 113–120 (117), ad ♀ ($n = 15$) 101–117 (106); tail centrals, ad ♂ ($n = 17$) 455–693 (557), imm ♂ ($n = 3$) 104–120 (114), ad ♀ ($n = 15$) 95–113 (104); bill, ad ♂ ($n = 27$) 37–43 (41), imm ♂ ($n = 4$) 42–46 (43), ad ♀ ($n = 15$) 38–42 (40); tarsus, ad ♂ ($n = 27$) 42–47 (45), imm ♂ ($n = 4$) 41–46 (44), ad ♀ ($n = 15$) 39–45 (40). Weight: ad ♂ ($n = 6$) 250–265 (256), imm ♂ ($n = 2$) 188–250 (219).

Habitat and habits — Table 4.1
Primary hill forest, including pockets of forest isolated by gardens and other disturbance, where predominantly frequents the midstage to canopy. Seen singly or, particularly in the case of ad ♂♂, in twos or in groups. Said to be unwary and easy to observe.

Diet and foraging — Table 4.2
Fred Shaw Mayer found mostly fruits in birds' stomachs (Gilliard 1969). Assessed as primarily frugivorous (Beehler and Pruett-Jones 1983), but it also takes arthropods. Along the Boana Road in July and Aug, seen to feed upon a *Ficus* sp., a large-stoned drupe (Lauraceae?), and fruits of a rubiaceous tree; also seen to

Emperor Bird of Paradise *Paradisaea guilielmi* Cabanis, 1888

perform much scansorial insect-bark gleaning, actually chipping bark off limbs, 10–15 m up in forest substage (BB). In July, four birds were observed sharing a food tree with Raggiana and Magnificent Birds and white-eyes *Zosterops* sp. (BB). See Appendix 4 for food plants.

Vocalisations and other sounds

Coates (1990) provides some detailed descriptions of calls given by captive ad ♂♂ at Baiyer River Sanctuary, PNG: '(i) A loud series of 6 to 8 notes, the first shorter and lower in pitch, the remainder a repeated high-pitched, rising note given at a rate of one note per 0.8 sec, WHU, WHOW, WHOW, WHOW, WHOW, WHOW (this is an advertising call, given in an upright pose with the head elevated and wings closed). (ii) A loud series of five identical upslurred nasal notes of almost metallic quality, delivered at a rate of one note per 0.6 sec: *WHAI, WHAI, WHAI, WHAI, WHAI* [another advertising call, sometimes given in an apparent duet with a nearby ♂ and also described as *Wuk wau wau wau* ..., Fig. 9.83(a)]. (iii) A very distinctive, fairly rapid series of four or five downward inflected bubbling notes, given at a rate of about one note per 0.5 sec: *WHOP-WHOP-WHOP* ... or *WHOC-WHOC-WHOC* ... (a preliminary display call, given while perching upright with the wings open, head raised, throat expanded and plumes slightly spread); often this call is preceded by a loud rising liquid note, e.g.: *WHICK, WHOP, WHOP, WHOP, WHOP* (duration of call 2.1 sec) [Fig. 9.83(b)]; variations when excited include *WHICK, WHOP, WHOP* (1.1 sec) and *WHICK-WHICK, WHOP* (0.75 sec). (iv) Nasal, upward inflected notes, *whaaiy* or *whi*, given singly or in a series, reminiscent of Raggiana Bird. (v) A variable, quiet, nasal, conversational-sounding note, *whoh, hah, whaw, oh* or *ow*, given singly or repeated several times at intervals of a few seconds (given while moving around the display court, plucking leaves and worrying vegetation, when in the mood to display).'

Another transliteration of the above call (iii) is *whick bau bau bau* ..., the first sharp note upward-inflected and the *bau* notes 'popping' in quality (CF). Wild ♂♂ at a lek gave *pop*, *baupop*, and *bau* notes repeated up to five times, the latter call sometimes being followed by the *pop bau* call. The *pop pop* may be extremely loud, unlike soft introductory *poop* or *pop bau* calls, and when delivered rapidly by excited birds has been likened to the sound

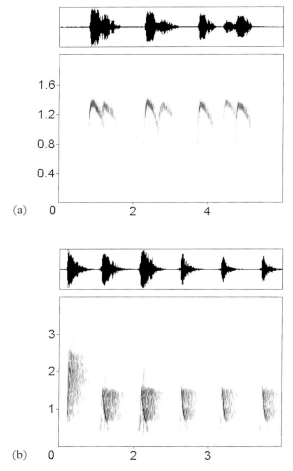

9.83 Sonograms from recordings of ad ♂ Emperor Birds showing (a) a series of *wuk wau wau wau* ... notes typical of *Paradisaea* spp. lek advertisement song but here given in duet by two individuals as is typical of the sp. and (b) the *whick bau bau bau* ... popping notes peculiar to the sp. (see text). Huon Penin in Aug 1986 by K. D. Bishop.

of exploding Chinese firecrackers (Draffan 1978). Some songs can sound rather like that of Lesser Birds. Duetting is typical of lekking ad ♂♂ (see Fig. 9.83(a)) and may involve several of the above calls.

Mating system
Polygynous. Promiscuous ♂♂ call and display communally at a lek consisting of a few adjacent trees in several areas within sight of each other. This represents a true lek. Draffan (1978) wrote 'There were certainly favorite display trees but display was not limited to a particular tree. Rather, I found three areas within my field of observation where displays took place, each consisting of three of four trees.' Harry Bell (*in litt.*) saw several ♂♂ move from one display tree to another and also watched several ♂♂ displaying at the same time in the one tree during Jan at Sialum, PNG. During display, ♂♂ cluster together in close proximity (BB). David Bishop (*in litt.*) found a lek in tall secondary forest within steep-sided hill-forest and considered the birds to have a less-defined lek than other *Paradisaea* spp. He noted birds displaying most intensely from c. 1515 hr onwards and some displaying as low as 3–6 m above the ground. It can be safely assumed that the ♀♀ nest-build and perform all nesting duties alone.

Courtship behaviour
Wagner (1938) described some display postures performed synchronously by two captive ad ♂♂ at Taronga Zoo, Sydney. His editor added a footnote to note that Detzner (in Stresemann 1924) saw five or six ♂♂ in the wild hanging upside down in display side by side in a tree and that they attained their unlikely posture by dropping slowly backwards to hang beneath their perch. This observation appeared so bizarre that it was seriously doubted by Stresemann (1924) and considered questionable by Mayr (1931) and Gilliard (1969). Draffan (1978) added considerably to knowledge of wild birds, however, and confirmed Detzner's report. He repeatedly observed activity at a Sattelberg lek during Jan–May. In general terms the behaviour of the ♂♂ he observed was similar to that of most other lekking *Paradisaea* spp., with up to six ad ♂♂ congregating in one or two adjacent trees to display.

Taking the observations of wild birds by Draffan (1978) and of captives by Crandall (1932), Stonor (1936), and himself into account, Coates (1990) described the displays of this bird, which we divide into phases as follows.

Convergence Display: up to six ad ♂♂ loosely congregate in one or two adjacent lek trees. Following initial loud calling from various locations about a lek ♂♂ congregate and mutually excite themselves with increasingly frequent calls and animated leaping about lek perches. They jump up and down with head and neck stretched upward and their green throat plumage expanded. The wings are then opened and, with the head jerking upwards with each note, several loud preliminary display calls given (Fig. 9.84(a)). This is followed by ♂♂ posturing by keeping the head down, the body more or less horizontal, the wings spread and vibrated rapidly, and the flank plumes slightly spread vertically. Birds move about, bobbing slowly up and down until the body becomes rigid, the head and neck extended forward and slightly downward and the wings suddenly spread and flicked forwards with their upper surfaces towards the head. This pose (Fig. 9.84(b)) is held for a few sec before the wings are suddenly snapped closed and this action repeated about six times at intervals of every few sec. These 'bobbing' and 'wing-flicking' elements of display may be repeated several times.

Static Display: ♂♂ perch still and silently upright for a few moments or may suddenly call loudly before turning head (sometimes tail) first under the display perch to hang completely upside-down with fully-expanded flank plumes. The latter, uniquely, form a lacy white disk with two obvious central patches of bright sulphur yellow. During this display the wings and tail are partly spread and the head turned upward (Fig. 9.84(c)). In this inverted pose ♂♂ constantly move themselves by rocking slowly and twisting the body from side to side, using their legs as a fulcrum. Birds may turn the head gradually to face one direction and then the other or may turn completely around beneath the perch to face the opposite direction.

The *Inverted Display* is mostly performed in silence for up to several min but a bird may call in reply to others. As the ♂ displays

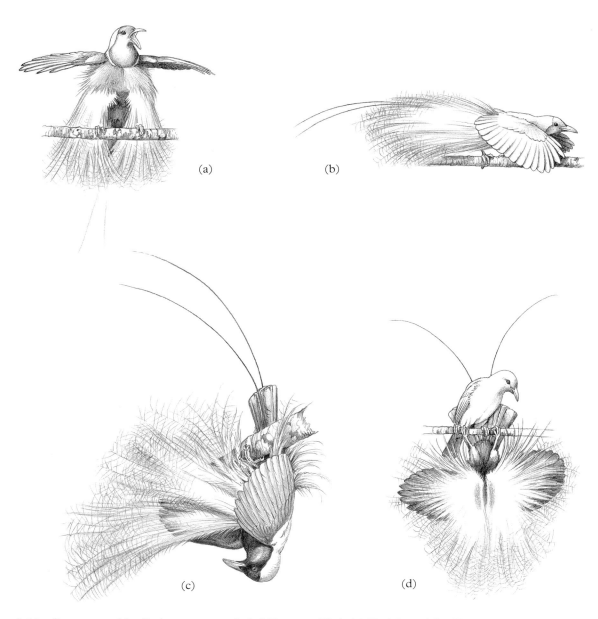

9.84 Some courtship display postures of ad ♂ Emperor Bird: (a) *Upright* and (b) *Horizontal Wing Pose*, (c) '*Static*' or *Inverted Display*, (d) pre-copulatory *Inverted Display*. After Coates (1990), with kind permission. See text.

upside down, a ♀, and, sometimes another ♂, may move along, or fly to, the display limb to perch directly above the inverted ♂ to look directly down at him as he looks up (Fig. 9.84(d)). An inverted displaying ♂ regains a normal upright perched position, at the completion of this phase of display, by rising head first. After this a period of time may be spent hopping about the display perch, or from perch to perch, whilst vibrating the wings and occasionally turning in small circles, or pirouetting. After some 10–15 min of calling and/or leaping about perches, bill-wiping, or plucking adjacent leaves, a ♂ may then repeat the entire performance.

All birds seen by Draffan to display by hanging upside-down from their lek perch reached this position by dropping head first from the normal perched position, as did the captive birds observed by Crandall (1932) and Wagner (1938). The London captives watched by Stonor (1936) and the wild ones seen by Detzner (in Stresemann 1924) inverted tail first however.

Copulatory Sequence: this phase remains unknown in the Emperor Bird.

Breeding Table 6.5
NEST SITE: unknown.

NEST: a nest was collected by C. Keysser from 'inland of the Huon Gulf, S.E. New Guinea' on an unknown date, but apparently during Sept (Gilliard 1969), is a fairly firm deep cup measuring 94 mm deep and 148 mm in dia externally and 72 mm deep and 83 mm in dia internally. It is built on a firm foundation of large broad leaves and is mostly constructed of creeper tendrils and vines which on average measure c. 1.6 mm thick, the heaviest piece being 2.8 mm thick and c. 260 mm long. A few larger vine lengths are beneath the leaf foundation lending support to it. One creeper-like tendril encircling the nest rim bears several fairly large, oval leaves c. 62 × 26 mm, perhaps as a form of nest 'decoration' or 'camouflage'. The egg cup lining consists of very fine tendrils (Frith 1971).

EGGS: clutch of one or two eggs. Five eggs average 36.8 (36.2–38.5) × 25.7 (24.4–27.1) mm (CF). Ground colour from whitish with slightest of pinkish wash to Pale Horn (92), Pale Pinkish Buff (121D), or a washed-out pale Flesh Colour (5), marked with elongate 'brush-stroke-like' streaks of reddish browns and greys. Eggs are thus typical of those of the higher Paradisaeinae.

INCUBATION, NESTLING CARE, AND DEVELOPMENT: unknown.

Annual cycle Table 6.5
DISPLAY: at Sattelberg, most frequent late Jan–early Mar and became markedly less so during Apr–May, with the wet season starting about May. In the Boana area of PNG where the wet season starts about June, ♂♂ have been seen to display in July–Aug. BREEDING: at least Sept–Dec. ♂♂ collected with enlarged gonads in Oct and much-enlarged ones in Nov and Dec. EGG-LAYING: a nest with clutch was found in Sept. MOULT: 11 of 29 specimens were moulting, involving all months except Mar, May, and Sept (Appendix 3). Crandall (1932) noted that two newly-arrived captive ad ♂♂ at the NYZS started moult c. 10–17 Dec and took c. 4 months to complete it. See Plate 3.

Status and conservation Table 8.1
Probably not vulnerable at this time, but given range restricted to Huon Penin, the sp. certainly requires an intensive survey of distribution, habitat selection, and foraging ecology. This should be integrated with an assessment of potential conservation threats.

Knowledge lacking and research priorities
This is another accessible sp. that would permit an excellent field study of ecology and behaviour. Of particular interest would be the behavioural and ecological relationship of this sp. with the Raggiana and Lesser Birds in areas of sympatry. Also of interest are the particulars of lek structure in this sp.—preliminary indications are that *guilielmi* leks are more fluid than those of typical *Paradisaea* sp.

Aviculture
Reported to have been bred in Sydney and offspring to have attained 2 yr of age (Hallstrom 1962). Most detailed knowledge of display postures and movements has been gained from captive birds in Sydney, Australia and BRS, PNG.

Blue Bird of Paradise *Paradisaea rudolphi* (Finsch, 1885)

PLATE 12; FIGURES 5.3, 9.85–7

Other names: *Paradisornis rudolphi*, Prince Rudolph's Blue Bird of Paradise, Archduke Rudolph's Blue Bird of Paradise, Blue Kumul. *Goy* or *Goi* (♂) and *Manga* (♀) in the Kubor region, PNG (Mayr and Gilliard 1954, W. Cooper *in litt.*). *Kongonamu* of the Fore, and *Barimoi* of the Daribi, of PNG (Diamond 1972). *Kabay asdal* of the Kalam people of PNG (Majnep and Bulmer 1977). *Aro Goeabe* or *Aragoiabe* of the Huli of PNG (Goldman 1981). *Ngeowai* at Kompiam, PNG (R. Whiteside personal communication).

This PNG mainland endemic, inhabiting a narrow belt of lower montane forest, is the most aberrant of the *Paradisaea*. Characters that make the sp. unmistakable are the blue wings, broken white eye-ring, the black head and mantle of both sexes, and the lovely blue flank plumes of the ad ♂. Polytypic; 2 subspp.

Description
ADULT ♂: 30 (67) cm, entire head, neck, and mantle glossy jet black with dull, iridescent, bronzed green sheen except on rear crown and nape which are suffused with dark Carmine (8) red, washed Magenta (2) in some light. Eye much enhanced by crescents of silver white feathers immediately above and below, forming a broken eye-ring. Back and rump blackish with dull, iridescent, blue-green (upper back), blue (168), and Indigo Blue (173) sheens. Uppertail coverts blackish cobalt-blue. Dorsal surface of wing and tail variably bright Cerulean and Cobalt blue (67 and 68) on coverts and along inner wing to purplish blue (70) on tertials, slightly darker at leading edge of wing, and flight feathers matte blackish with Cerulean Blue outer edges to all but outer three primaries. Dorsal surface of tail purple blue but central pair grossly-elongated 'ribbons' of narrow, matte, bluish black vanes with paler spatulate tips that in some lights show iridescent blue. Upper breast matte-blackish with slightest of dark, Cobalt blue, iridescent sheen gaining strength on lower breast, greener in some lights, to jet black belly and very dark brownish black thighs and undertail coverts. Grossly-elongated, fine, and sparse filamental flank plumes basally rich dark purple blue, with violet (172) sheen in some lights, to rich Cobalt centrally and variably purplish blue to rich Mauve (75) distally. Upper surface of flank plumes rich rusty Amber (36), giving the bird a different appearance from behind (see Plate 12). Two areas of inner flank plumes form a discrete black and a dark Crimson (108) linear patch either side of the belly. These meet during display to form a continuous bicoloured broad line (see Fig. 9.86). Underwing: coverts variably Cinnamon (123A) to blackish, flight feathers dark greyish (83) with faint bluish wash and paler trailing edges to all but outermost two primaries. Ventral surface of tail dark, greyish, Cobalt blue except the black central 'ribbons' the tips of which are Ultramarine Blue (170A). Bill palest chalky bluish white, almost white, iris dark brown, legs and feet purplish grey, mouth greenish yellow.

ADULT ♀: 30 cm, head and entire upperparts similar to ad ♂ but black areas duller, more very dark brownish black, with only faintest trace of iridescence and no carmine or magenta on head. Underparts unlike ad ♂, being chestnut brown (223A) grading from the blackish throat to dark Cinnamon on remaining underparts, except blackish brown (219) thigh feathering, with slight matte-blackish barring but less so on central belly (absent in some populations). Underwing as in ad ♂ and ventral surface of tail similar to ad ♂ but less bright blue, more greyish. Younger ♀♀ have more extensive and blacker ventral barring and a dark bill.

SUBADULT ♂: as ad ♀ but with a few feathers of ad ♂ plumage intruding into plumage, grading to as ad ♀ but with a few feathers of ♀ plumage remaining.

IMMATURE ♂: as ad ♀, but individuals acquiring progressively longer and narrower pair of central rectrices with age.

NESTLING–JUVENILE: a nestling close to departing the nest, photographed by Don Hadden (*in litt.*), had a blackish, white-tipped, bill, and a pale yellow gape and mouth. The white feathers about the eyes were fully developed. Body plumage was dull sooty black above and rich rufous below, and dull blue on the secondaries and coverts of the wings but the mantle, lesser coverts, and back were dull blackish. Head plumage densely 'plush' and jet black with a smoky brown lustre to it. Lores and forecrown unfeathered. Iris dark brown and legs and feet grey. A nestling found by Simpson (1942) almost exactly resembled the ad ♀. Juvs like ad ♀ but wings darker, bill dark greyish, and abdomen whitish.

Distribution
Endemic to the central cordillera of E and central PNG, where it ranges W to Mt. Hagen, Kompiam (possibly to Maramuni, W of Wabag), and the Tari area of the S Highlands. Western verge of range requires delineation. Ranges from 1100 to 2000 m (mainly 1400–1800 m). The sp. is locally rare or absent, especially on the N slope of SE PNG.

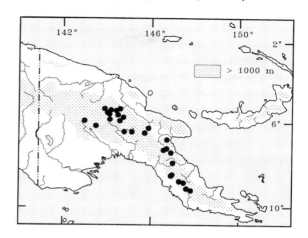

Systematics, nomenclature, subspecies, weights, and measurements

SYSTEMATICS: a highly distinctive member of the genus *Paradisaea* well deserving of placement in the subgenus *Paradisornis*. Without question, this is the most divergent member of the genus, and some would argue this sp. should be placed in its own genus. Hybridisation: known to cross with Lawes' Parotia and the Raggiana Bird (Gilliard 1969; Christidis and Schodde 1993; Frith and Frith 1996*a*). See Plates 14, 15 and Appendix 1.

SPECIES NOMENCLATURE AND ETYMOLOGY
Paradisaea (Paradisornis) rudolphi (Finsch)
Paradisornis Rudolphi Finsch, 1885. *Zeitschrift für die gesammte Ornithologie* **2**, 385, pl. 20. Hufeisengebirge, southeast New Guinea. Type specimen SMT C 8154 (8456), WW II destroyed.

Etymology: *rudolphi* = in honour of Archduke Rudolph (1857–89), then Crown Prince of Austria-Hungary of whom Finsch wrote 'His Imperial and Royal Highness, Crown Prince of Austria, the high and mighty protector of ornithological researches over the entire world'. Rudolph died in the Mayerling Tragedy (he and his princess comitted suicide). This breathtakingly beautiful bird was discovered by Carl Hunstein in 1884.

Blue Bird of Paradise *Paradisaea rudolphi* (Finsch, 1885)

SUBSPECIES, WEIGHTS, AND MEASUREMENTS

Three subspp. have been described since 1951. Comments by Gilliard (1969), Forshaw (in Cooper and Forshaw 1977), and Cracraft (1992) regarding the questionable validity of *P. r. ampla* would suggest that the subsuming of this subsp. into nominate *P. rudolphi* is overdue.

1. *P. r. rudolphi* (Finsch, 1885)

Synonym: *Paradisornis rudolphi hunti* Le Souef, 1907. *Emu* **6**, 119. British New Guinea. Type specimen MV B1132.

P. rudolphi ampla Greenway, 1934. *Proceedings of the New England Zoological Club* **14**, 1. Mt. Misim, Morobe district. Type specimen MCZ 167001. Some measurements of *ampla* are on average marginally smaller than those of the nominate form but with much overlap (Frith and Frith 1997*b*).

Range: southeastern PNG, from the Owen Stanley Ra, NW to the Herzog Ra and the eastern highlands (Okapa).

Measurements: wing, ad ♂ (n = 25) 151–163 (157), imm ♂ (n = 11) 142–157 (153), ad ♀ (n = 16) 142–158 (149); tail, ad ♂ (n = 25) 76–91 (82), imm ♂ (n = 11) 90–99 (93), ad ♀ (n = 16) 90–102 (95); tail centrals, ad ♂ (n = 20) 341–452 (409), imm ♂ (n = 10) 90–102 (95), ad ♀ (n = 15) 90–101 (95); bill, ad ♂ (n = 24) 39–46 (41), imm ♂ (n = 11) 39–42 (41), ad ♀ (n = 15) 38–43 (41); tarsus, ad ♂ (n = 24) 36–42 (40), imm ♂ (n = 11) 36–41 (39), ad ♀ (n = 16) 36–41 (38). Weight: ad ♂ (n = 8) 174–189 (178), imm ♂ (n = 5) 149–174 (163), ad ♀ (n = 1) 157.

2. *P. r. margaritae* Mayr and Gilliard, 1951. *American Museum Novitates* **1524**, 11. Kimil R., 20 miles west-northwest of Nondugl, Wahgi Valley, Central Highlands, Mandated Territory of New Guinea. Type specimen AMNH 348209.

Range: central PNG, W of the range of the nominate form—mainly the Bismarck Ra, Sepik–Wahgi Divide, Kubor Ra, Mt. Karimui westward to Tari, Mt. Giluwe, Mt. Hagen, the Enga Highlands. Western boundary of range not well delineated. Westward populations have not been subspecifically diagnosed (for lack of material). The assumption is that they will be attributable to *margaritae*.

Diagnosis: like nominate form but ♀ with underparts uniformly and narrowly-barred blackish and tail (excluding central pair) and tarsus on average shorter.

Measurements: wing, ad ♂ (n = 5) 151–158 (156), imm ♂ (n = 5) 151–161 (156), ad ♀ (n = 7) 143–159 (148); tail, ad ♂ (n = 5) 76–81 (78), imm ♂ (n = 5) 95–101 (98), ad ♀ (n = 7) 86–94 (89); tail centrals, ad ♂ (n = 3) 436–458 (451), imm ♂ (n = 5) 95–100 (97), ad ♀ (n = 6) 88–96 (92); bill, ad ♂ (n = 4) 39–42 (41), imm ♂ (n = 5) 39–42 (40), ad ♀ (n = 7) 38–42 (40); tarsus, ad ♂ (n = 5) 37–39 (38), imm ♂ (n = 4) 38–40 (39), ad ♀ (n = 7) 36–40 (37). Weight: ad ♂ (n = 1) 158, imm ♂ (n = 2) 166–172 (169), ad ♀ (n = 3) 124–166 (152).

Habitat and habits   Table 4.1

Lower montane forest, forest-edge, and denser, older, secondary growth of fallow gardens. Gilliard (1950) found ♀-plumaged birds numerous in the middle strata of open ridge forest, usually perched solitarily 6–12 m above ground beside small openings in the forest on Mt. Maguli, PNG. Ad ♂♂ found to be resident about limited calling territories based upon relict stands of *Lithocarpus* and *Castanopsis* oak at the forest-edge by Healey (1986), who noted never seeing or hearing the sp. in the depths of montane forest or in secondary growth < 25 yr old. Near Kompiam ad ♂♂ held territories on cleared mountain slopes incorporating islands of secondary and/or primary forest where they sang and displayed during 1994 (R. Whiteside *in litt.*). On Mt. Missim an uncommon inhabitant of oak forest interior (BB). Ad ♂♂ are solitary, usually encountered alone, occasionally in twos, whereas ♀-plumaged birds may be sometimes seen in small aggregations at a fruiting tree where they may also feed side by side with one or more other bird of paradise spp. An ad ♂

frequented, and displayed about, the immediate area of a Lawes' Parotia court and at dusk perched within a few cm of the ad ♂ parotia high in a dead tree (D. Gillison *in litt.*). Both an ad ♂ and a ♀-plumaged bird have been seen moving with a flock of ad ♂ and ♀-plumaged Carola's Parotias (R. Whiteside *in litt.*). Feeding birds have been recorded foraging from the forest canopy level down to within 1–2 m of the ground. An ad ♂ drank from a hollow atop a tree fern trunk stump (R. Whiteside *in litt.*). Diamond (1972) heard and saw ad ♂♂ only in the upperparts of the sp. range, but Pruett-Jones and Pruett-Jones (1986) assessed the distribution of birds of paradise (including *P. rudolphi*) on Mt. Missim and found the ad ♂♂ inhabit the centre of the elevational range of the sp., with young ♂♂ tending to be at the upper and lower verges of the distribution.

Diet and foraging   Table 4.2

Will feed high in the canopy, upon fruits, but more often lower in the forest (Beehler 1978; R. Whiteside *in litt.*). Fruits of *Trema orientalis*, *Schefflera* sp., *Piper* sp., *Planchonella* sp., and *Musa* sp. (wild banana) are eaten (Healey 1986; Hicks and Hicks 1988a: Hopkins 1988). Of 57 observations of birds feeding upon the fruits of 11 plant spp. on Mt. Missim, PNG, 40% were upon *Omalanthus novoguineensis*, 18% upon *Schefflera pachystyla*, and 23% upon four *Ficus* spp. (Beehler 1983a,b). Other fruits eaten less frequently are listed in Appendix 4. Of the 57 observed feedings by fruit class, 47% were upon figs, 30% upon drupes/berries, and 23% upon capsular fruits (see Chapter 4). A subsequent analysis that included additional field data from Thane Pratt found fruits of the plant *Gastonia spectabilis* to be almost as important as *O. novoguineensis* (Beehler 1989a). Of 34 complete foraging visits to fruit-bearing trees, mean visit duration was 4.6 min during which feeding was continuous; 91% lasted 1–10 min, 9% 11–20 min (Pratt and Stiles 1983).

In a study of seed dispersal by birds in forests on Mt. Missim, PNG, a ♀ systematically defended one of three fruiting branches of an epiphytic *Schefflera pachystyla* shrub by chasing off three other bird of paradise spp. (Lawes' Parotia, Superb, and Magnificent Birds). She restricted her movements away from this food resource to its local vicinity where she could presumably be aware of intruders (Pratt 1984). It is possible that ♂-like ♀ plumage in polygynous spp. such as the Blue Bird may favour nest/food resource defence plus species-specific signalling (T. Pratt personal communication). Blue Birds do take fruit in company of conspecifics and other spp., however, and have fed in the same food tree with their own kind and several Carola's Parotias and Superb Birds (R. Whiteside *in litt.*).

Five birds collected from two locations contained 70–100% fruit and 0–30% arthropods, the only identified animal group being cockroaches (Blattidae) (Schodde 1976). Based on stomach samples, faecal samples, and random direct observations of feeding birds, the diet on Mt. Missim was calculated as 85% fruit (Beehler 1983a: 235). Of three faecal samples one was entirely of fruit, one 90% fruit, and one half fruit and half arthropods (including ants, crickets, grasshoppers, a spider, and a sphecoid wasp). A ♀-plumaged bird obtained a large stout spider by tearing into rotten bamboo with its beak (Opit 1975b).

Infrequently seen to forage for arthropods by bark-gleaning and searching vines and creeper foliage at most forest levels (BB). Birds often seen 'hopping along moss-covered branches probing about for insects and on one occasion I saw a well-plumed ♂ bird moving up the trunk of a dead tree in the manner of a tree creeper' (Smyth 1970). Cockroaches and grasshoppers are obtained from mossy tree branches and trunks (Coates 1990).

Vocalisations and   Figure 9.85
other sounds

♂ advertisement songs include a high-pitched, slightly bell-like, upslur followed by an oft-repeated hard downslur *kouwi—carr-carr-carr-carr ...*, a repeated loud, nasal, *quoi*, a series of several upslurred, high-pitched notes on one scale *kwank* or *waah–wah-wha-wha-wha-wha-*

wha reminiscent of the Raggiana Bird call but not as full-bodied (Fig. 9.85(a)) and a single *quaa* or *kwaa* note (Coates 1990). Much like Raggiana advertisement but lower-pitched with notes more *kwank* bell-like in quality (CF). This may be more slowly and deliberately delivered at a lower pitch (Fig. 9.85(b)). Several calls are reminiscent of the Raggiana Bird: a slowly-cadenced plaintive series of downslurred notes that are more nasal and higher-pitched than those of Raggiana—*wahr, wahr, wahr, wahr, wahr, wahr*; and a *we wah wah wah wah wah weh weh wehweh*, the series descending, each note upslurred (BB). Also described is a series of notes similar to Raggiana's but with longer intervals between the notes which are successively lower in pitch and bell-like in quality, this being the origin of the Fore name *Kongon ámu* (Diamond 1972). The typical song note reminded Smyth (1970) of the clang of a hammer on steel when heard from a distance but this quality was lost when heard close up. The display call, given by the ♂ hanging upside down, is ventriloquial and is

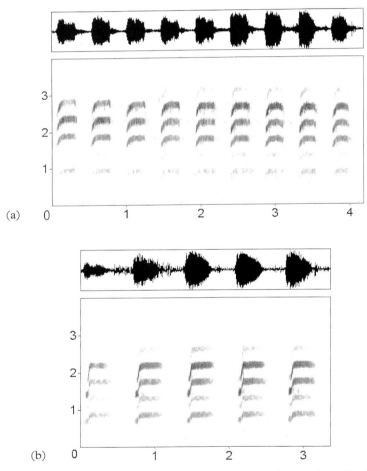

9.85 Sonograms from recordings of an ad ♂ Blue Bird giving two forms of the display perch area advertisement song: (a) the more typical higher-pitched, nasal, ringing honk-like note *kwank* and (b) the lower-pitched song, more deliberate and slowly delivered. Ambua area of the Tari Valley, PNG in Aug 1991 by K. D. Bishop.

well described as a 'continuous, rhythmical song, a mixture of low nasal chitterings and chattering notes interspersed with low *caws*' and often with a strange vibrating twanging note (which is also sometimes used in advertisement song). This becomes a 'continuous fast rhythmic buzzing, as if produced by an electric motor' as ♀ perches close by. If the ♀ hops away this sound is discontinued and is replaced with enthusiastic chittering and chatterings apparently to entice the ♀ back. The buzzing may, however, be performed in the absence of a ♀. The rhythmic element of this call gives it a throbbing quality that appears to be physically felt at very close proximity to a calling bird. ♂♂ moving between singing limbs and other trees give low growling and rolling notes—*brrn brrn*. A blackish-billed juv foraging in substage leaves 10 m above ground on Mt. Missim, PNG, in June gave *skss* notes periodically (BB). A ♀ disturbed at the nest gives a harsh scold like the scold note of Carola's Parotia and Superb Bird and when feeding her nestling a quiet enticement note while vibrating her tail (R. Whiteside *in litt.*).

Mating system

Polygynous but, unlike all other *Paradisaea* spp., each promiscuous ♂ establishes a solitary and dispersed calling station. One ad ♂ at Ubaigubi, PNG, in mid Aug mated with at least three ♀♀ (D. Gillison *in litt.*). The ♀ nest-builds and performs all nest duties alone. See Fig. 9.87. At Mt. Pugent/Trauna Ridge, PNG, at 1800 m, in June an ad ♂ had a regular (but also two others nearby) open, horizontal, sub-canopy calling perch, 18 m high, from which it sang in early mornings with head tilted back to 45° angle and bill opened slightly with a call approximately every 2 min. Another ♂ could be heard singing c. 300 m distant (BB). Ad ♂♂ at Kompiam and at Ubaigubi have numerous advertisement perches within a territory and sing whilst preening and foraging during the early morning (R. Whiteside *in litt.*; M. LeCroy *in litt.*, respectively). ♂♂ at Ubaigubi, PNG, are said to select display trees that are almost directly over a display court of Lawes' Parotia and interact with them (Peckover 1990). Richard Whiteside (*in litt.*) has, however, observed Carola's Parotias possibly associating with two ad ♂ Blue Birds at their respective low display perches and a Magnificent Bird with another.

Courtship behaviour

Ad ♂♂ perform solitary and relatively static courtship displays, often in vegetation only a few metres above ground. They remove leaves from the immediate area of their display perches as do ♂♂ of other *Paradisaea* spp. Display perches are usually well shaded during the mornings, but ad ♂♂ have been observed displaying to ♀♀ at them in direct sunlight (R. Whiteside *in litt.*). Recent observations indicate that although some ad ♂♂ may occasionally display high in the vegetation (up to c. 20 m), they typically display within 1–3 m of the forest floor, on a slim gently to steeply-sloping branch, bamboo, grass, or vine stem with some foliage cover directly above. Smyth (1970) was able to approach a wild ad ♂ to within a few m as it displayed inverted in dense undergrowth at < 2 m above ground, whereas R. Whiteside (*in litt.*) found ♂♂ at display sites extremely nervous and unapproachable. Up to four ♀-plumaged birds at a time were seen attending a displaying ad ♂ but only one at a time perched on the display stem (R. Whiteside *in litt.*). Contrary to statements by Coates (1990), two different ad ♂♂ were seen to face both directions from their display perch while inverted (R. Whiteside 1995, 1998 and *in litt.*).

Displays are mostly performed in mornings between c. 0600 and 0930 + hr, up to midday, and also (but less frequently) during the afternoon. An adult ♂ sings from a perch high in the vegetation usually thereafter descending to a low display perch. In view of a recently-described Blue Bird × Lawes' Parotia hybrid (Frith and Frith 1996*a*), and the observed interaction of the two spp., it is noteworthy that D. Gillison (*in litt.*) has observed an ad ♂ display site immediately above a Lawes'

Parotia court. Moreover, R. Whiteside (*in litt.*) several times saw or heard at least one ad ♂ Blue Bird displaying adjacent to an active Carola's Parotia court, and he observed Carola's Parotia active near display perches of another Blue Bird, so a hybrid between these two spp. would not be surprising.

Coates (1990) witnessed many displays by a captive ad ♂ at Baiyer River Sanctuary, PNG. He noted the ♂'s display activity took place between 1230 and 1730 hr. What follows summarises his observations: from a normal perched position the ♂ lowers himself backward to hang upside down and then spreads his flank plumes with wings held closed. With his head then turned up to one side he starts to call. He jerks his head and constantly moves his plumes to cause shimmering waves of blue and violet over them but the central tail 'ribbons' simply remain curved down to either side (Fig. 9.86(a)). The black central abdomen patch becomes oval in shape and then expands as the bird rhythmically moves his body from his hips. As the display progresses, after several min or more, the ♂'s eyes close to narrow slits (which enlarges the white patches of feathers above and below the eyes) and if a ♀ is now on his display perch the ♂ twists his head and body towards her to direct as much of his display plumage as possible at her. During this part of the display the black abdomen patch may change shape from longer to shorter oval or vice versa (Fig. 9.86(a,b)). If a ♀ remains, at this point the ♂ stiffly draws his body upward by his legs and raises it upward and forward towards the ♀ as the black abdomen patch becomes more ovate, the head and bill point directly up the centre of his body, and the central tail 'ribbons' are rapidly swung from side to side through a wide arc (as in Fig. 9.86(b)). If the ♀ does not then hop along the perch to stand directly above him, the ♂, while remaining inverted, may step along the underside of his perch towards her until directly beneath her.

During numerous inverted courtship displays by ad ♂♂ at Kompiam, PNG, the ♂ never swung or lashed his central pair of rectrices from side to side or side-stepped along the perch towards a watching ♀, but did erect breast feathering and point the bill vertically up at the ♀ towards the end of the buzzing display phase. Sometimes a ♀ would hop away a little at this point. Sometimes the ♂ would stop his buzzing, relax his body, and continue to hang while producing low vibrating *chuck* sounds at which the ♀ would also seem to lose interest. Sometimes display was terminated by the ♂ suddenly righting himself, always head and breast (front) first, to displace the startled ♀. He would then immediately invert and start chattering again or would chase the ♀ (R. Whiteside and M. Feignon *in litt.*).

Copulatory Sequence: as performed by a captive pair at Taronga Zoo, Sydney, consisted of no more than the inverted ♂ suddenly regaining the normal upright perching position immediately beside the ♀ by lifting himself forward (head first) back atop his perch by using his powerful legs and feet, as his plumage is all returned to its usual position, and he stops calling. The ♀ then turns away from him with her head downward in a submissive posture to solicit by flicking her tail several times before the ♂ mounts her to copulate (A. Hiller *in litt.*).

The courtship display, unlike that of the other (lekking) *Paradisaea* spp., lacks a *Convergence Display* and consists only of a *Static Inverted Display* with no ritualised *Copulatory Sequence*. It is possible, however, that future studies may reveal that some form of billing of the ♀ by the ♂ and/or vice versa typically occurs before copulation, in view of what other *Paradisaea* spp. do.

Interestingly, Crandall watched a confirmed (post mortem) ♀ hang upside down from the cage roof wire with fully-extended legs to spread her wings slightly and expand her breast feathers laterally, to project just beyond her wings, and turned slightly forward. She pressed her tail forward between her legs until it was almost horizontal. She did not turn her head upward as much as the ♂ in display, nor close her eyes to any extent. She did, however,

Blue Bird of Paradise *Paradisaea rudolphi* (Finsch, 1885)

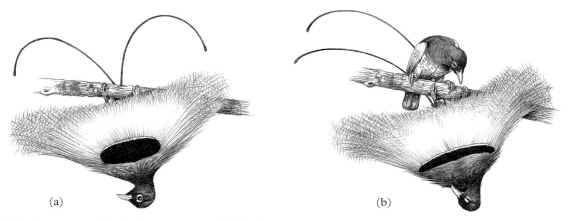

9.86 Some courtship display postures of ad ♂ Blue Bird: (a) *Static Inverted Display*, (b) pre-copulatory *Inverted Display*. After Coates (1990), with kind permission. See text.

vibrate her body back and forward from the hips and occasionally drew herself upward to a horizontal position. In addition, she vibrated her body rapidly up and down by bending her leg joints (Crandall 1932). Thus, this ♀ performed a display much like that of the ad ♂ but held her tail quite differently, moved her body up and down by using her legs, and did not call. The above account includes all other behaviour by captives described by Crandall (1921).

Breeding Table 6.5

NEST SITE: one wild nest was < 4 m above ground in a low tree in thick bush near the top of a ridge (Goodfellow 1926b). A nest found by Simpson (1942) was in the thickest part of 'low, thick scrub on the ridge, and dense masses of bamboos, with larger trees on the slopes'. Another was 19 m above ground in a sub-canopy tree 21 m tall (Pruett-Jones and Pruett-Jones 1988a), and another c. 9 m above ground in the crotch of a *Casuarina oligodon* tree on a steep slope in secondary growth at the edge of temporarily-disused gardens, the nearest mature forest being c. 25 m upslope (Mack 1992). A nest examined at Kompiam, PNG, by R. Whiteside (*in litt.*) was 4–5 m above ground and a nestling was earlier collected and eaten by a local man from a nest at similar height and in the same tree sp.

c. 300 m away and the man thought the same ♀ built both nests.

NEST: composed of 'strips of pandanus-leaves and fibre from the leaves of some palm and with no lining' (Goodfellow 1926b). We examined a photograph of the supposed complete nest in which a nestling of this sp. was delivered to Hadden (1975) which appears to consist of a nest lining of supple, woody, fine tendrils, possibly of vines only. This is the 'open, rather flattish structure composed of sticks' of Hadden (1975) and cited by Cooper and Forshaw (1977) and Coates (1990). We also examined a photograph by R. Whiteside of a nestling about to leave its nest. The nest is dishevelled and displaced in the tree fork as a result of the birds' activities, but was clearly a sparse, deep, circular bowl shape of long, green, supple, epiphytic orchid stems (identification confirmed by specimens) that encircled the structure. The nest appears to have involved few or no leaves. Another nest was an oblong-shaped cup of small pieces of vines and, mostly, the needle-like twigs of casuarina less than 2 mm in dia; measuring externally 150 mm long and 120 mm wide and 80 mm deep and internally 70 × 50 mm and 50 mm deep. Not lined with distinctly different material but with slightly finer examples of the external vegetation (Mack 1992).

EGGS: clutch usually one but may (? rarely) be two eggs (see Nestling Development below). Five eggs average 36.3 (35.2–38.6) × 25.1 (23–27.3) mm (CF). Ground colour Pale Pinkish Buff (121D) but much paler, to a creamy-pinkish or pale Salmon (6) marked with elongate brush-like streaks of lavender greys overlaid with same of dark browns (Hartert 1910) or blotched and spotted with these colours, all more so about the larger end. Generally they resemble eggs of the Raggiana Bird. A single infertile egg laid by a captive ♀ at the Hallstrom Sanctuary, Nondugl, PNG, in Mar measured 36.6 × 27.3 mm, was dully glossed, and pale pinkish fawn with a thin wash of pinkish russet in places about the larger end, apparently as a result of running of the pigment of russet blotches. Thin and slightly-elongate purplish grey blotches predominantly about larger end but also extending over rest of egg sparsely where these markings are smaller and more like spots. These markings overlaid with russet brown blotches that are irregularly shaped and not elongate and are most dense and large on the large end of the egg. Small, fine, russet brown spots occur very sparsely over the rest of the egg (Bishop and Frith 1979). Another single-egg clutch measuring 38.6 × 24.7 mm was described as salmon marked about the larger end with a ring of splotches of cinnamon rufous and tawny, and the smaller end slightly flecked with the same two colours (Mack 1992).

INCUBATION: known to be > 18 days at one wild nest and to be performed only by the ♀ (Pruett-Jones and Pruett-Jones 1988a). Incubation of a single-egg clutch of unknown age was observed for 50 hr over 6 days at one wild nest by Mack (1992), and during one of these days it was possible to confirm only the single individual was involved. The ♀ incubated for 64% of total diurnal observation time, being absent the remainder of time. Incubation bouts averaged 21 min (6–106, min, $n = 47$) and absences 19 min (4–49 min, $n = 43$). Morning incubation bouts (mean = 23 min, $n = 25$) were shorter than afternoon ones (mean = 43 min, $n = 19$). The ♀ was frequently seen to regurgitate seeds whilst incubating. This particular nest and egg was deserted (Mack 1992).

NESTLING CARE: exclusively by ♀. An active wild nest studied for 104 hr over 24 days (27 Sept–20 Oct) containing a nestling was located within the home range of a known ad ♂ but no ♂ was seen near the nest or in close association with the banded ♀ parent at any time. The nesting ♀ was aggressive towards conspecifics and to other bird of paradise spp., including a Superb Bird and Trumpet Manucode that came close to her nest (Pruett-Jones and Pruett-Jones 1988a). Observation during a single morning in Dec at a nest with a well-developed nestling revealed the ♀ parent visited to feed the nestling predominantly fruits (but also a lizard 5 cm long and a green cricket) by regurgitation about once ever half hr (R. Whiteside *in litt.*). A newly-fledged nestling was caught in adjacent forest in mid-Dec as the ad ♀ parent scolded (R. Whiteside *in litt.*).

NESTLING DEVELOPMENT: two nestlings, presumably siblings, 'at first looked grotesquely spiny, owing to the abnormally long grey pinfeathers which covered them all over. Each quill ended in a small tuft of grey down which dropped off with a large part of the quill as they grew, leaving the newly-opened feathers the normal length. The orbital region, and, in fact, the greater part of the head, remained bare long after the rest of the body was covered with feathers' (Goodfellow 1926b).

Annual cycle   Table 6.5

DISPLAY: mostly Apr–late Nov near Kompiam, PNG, but ♂♂ in active moult seen and heard displaying there in Jan (R. Whiteside *in litt.*). BREEDING: Over the range of the sp. may occur at any time of year; general peak seems to be July–Feb. A nest containing a single well-developed nestling near Kompiam, PNG, was attended during Nov. EGG-LAYING: July–Jan and Apr in the wild and therefore possibly any time of year, depending on location. A nest studied on Mt. Missim contained a single-egg clutch on 26 Sept but it disappeared sometime between 13 and 16 Oct (Pruett-Jones and Pruett-Jones 1988a). MOULT: 17 of 54 specimens examined showed moult, involving Jan–June and Sept–Oct inclusive but predominantly Jan–Apr (Appendix 3). An ad ♂ captive at the NYZS observed over 3 consecutive years by Crandall (1932) started to moult in mid June to 7 July and finished it 17–19 weeks later; which he also found to be the case for two other ♂♂ during a single year. A subad ♂ at Kompiam, PNG, started moulting flank plumes in mid Dec (R. Whiteside personal communication).

Status and conservation Table 8.1
Uncommon throughout its range and locally absent in some areas that appear suitable, but given the large geographical distribution, it is difficult to assess the sp. as seriously threatened, although it may legitimately be classified as vulnerable. Very uncommon and locally distributed up to 1920 m in the Wahgi Valley area where 'vast areas' of forest have been lost to agriculture; in the Kubor region 'several small colonies numbering from 10 to perhaps 50 birds still exist at the top fringe of precipitous valleys not yet farmed' (Mayr and Gilliard 1954). This remained true, but of smaller numbers, in 1989 (D. Bishop *in litt.*). Ad ♂♂ were readily found calling from tall trees at the edges of forest in the Kuli area, N slopes of the Kubor Ra in the 1980s (BB). Gyldenstolpe (1955a) pointed out that vast forests removed from the Wahgi Valley for subsistence agriculture denied this sp. much former habitat, birds only appearing in certain remote parts of the Wahgi Divide. He was informed that birds were, however, still fairly numerous along the Kimil R. and on some of the tributaries to the Upper Wahgi R. Information provided by the Kalam people of the upper Kaironk R., near Simbai, PNG, to Bulmer (in Majnep and Bulmer 1977) indicate the possibility this sp. was exterminated from that area by forest clearing from > *c*. 1750 m. Rare in the Jimi Valley, PNG, where ad ♂♂ were hunted for their plumes (Healey 1986). Using a 'feeding bout' as a measure of frugivorous bird abundance on Mt. Missim, PNG, an average of 1–2 Blue Birds were calculated to visit a fruiting tree per day (Beehler 1983b). One ad ♂ and two ad ♀♀ were here radio-tagged and subsequently tracked on 9–13 days over 42–52 days (Pruett-Jones and Pruett-Jones 1988a): the home ranges of all three birds overlapped (Fig. 9.87). One of the radio-tagged ♀♀ was mist-netted within the home range of the ♂ where other ♀♀ were also observed passing through and foraging. Further analysis of selected locations at which the ♂ was recorded giving his advertisement song showed that this 'proclaimed' area was

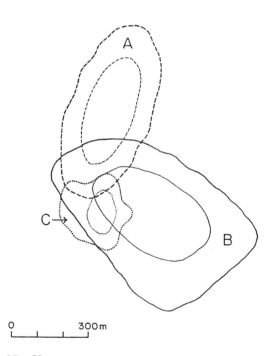

9.87 Home range contours of two ♀♀ (A and B) and one ad ♂ (C) Blue Birds. For each bird the thicker, outer, line represents 95% and the lighter, inner, line 50% probability contours. From Pruett-Jones and Pruett-Jones (1988a), with kind permission.

roughly central within his home range and formed 16% of it. Pruett-Jones and Pruett-Jones (1988a) heard 15 ♂♂ calling from non-overlapping advertisement territories along a 2 km ridge line and 12 ♂♂ along another ridge line 3.5 km long, giving a figure of *c.* 5 ♂♂ per km of suitable ridge line habitat. The 15 ♂♂ on one ridge line vocalised solitarily at a mean distance between centres of 'proclaimed' areas of 220 m (range 160–440 m, SD = 100) and the 12 ♂♂ on the other ridge at a mean distance of 400 m (280–520 m, SD = 80). The advertisement singing areas of neighbouring ♂♂ did not apparently overlap whereas their all-purpose home ranges may have done so. Two solitary ♂♂ were seen and heard calling in remnant forest patches in extensive gardens and associated secondary growth at *c.* 0.5 km apart near Tigibi, E Tari Valley, PNG, on 11 Oct (CF). The sp. is quite plentiful in remnant forest and secondary growth on cleared mountain slopes of the Kaindan area, near Kompiam, PNG (R. Whiteside *in litt.*).

Knowledge lacking and research priorities

The western extremity of the range in the N and S segments of the central cordillera remains to be delineated. Additional observations of ♂ dispersion and territorial behaviour would be useful. Finally, a field survey of the bird's status in the Wahgi Valley would provide data useful to determine its overall conservation status.

Aviculture

An ad ♂ was delivered to Sir William Ingram in the UK in 1907 and a pair was received at the NYZS in 1920 (Crandall 1921). Several hatched in aviaries in New Guinea (Hallstrom 1962). Kept in Taronga Zoo, Sydney during early 1960s where ad ♀♀ nest-built and laid eggs, which were destroyed by the ad ♂♂ before hatching. A ♂ in ♀ plumage received by London Zoo early in the summer of 1933 remained in ♀-plumage until acquiring complete ad ♂ plumage in a single moult 4 yr later, in summer 1937 (Stonor 1938).

Appendix 1

The hybrid birds of paradise

Perhaps no other family of birds (the typical ducks, Anatidae, excepted) has exhibited a propensity for interspecific and intergeneric hybridisation in the wild equal to the birds of paradise. Hybrid birds of paradise have long been the subject of controversy, mainly because most of these remarkable and unusual-looking hybrids are known by only one or two specimens of some antiquity (Fuller 1995). Most hybrid specimens arrived in Europe as commercial trade skins, a saga described in general terms by Gilliard (1969), Fuller (1987, 1995), and Swadling (1996), and summarised in Chapter 2. Given their distinctive appearance (see Plates 14 and 15) it is not surprising that most hybrids were originally described as new species by ornithologists. Whereas many of the supposed hybrids are intermediate in appearance and size between their proposed parental forms, leaving little doubt in the minds of the present authors of their origins, several forms are far less easily attributable to a particular interspecific cross.

The German ornithologist Reichenow (1901) made the first significant step in recognising the hybridisation phenomenon in birds of paradise when he noted that *Janthothorax mirabilis*, which he had originally described as a new species, was perhaps the product of a wild cross between a *Paradisaea* species and a Twelve-wired Bird. This idea was strongly rejected by his contemporaries. Three decades later, Stresemann (1930a,b) suggested that, indeed, Reichenow had been correct with respect to the hybrid nature of '*Janthothorax mirabilis*'.

Stresemann went on to make a complete review of all of the 'aberrant' bird of paradise taxa and presented the notion, revolutionary at the time, that some 17 named forms were of hybrid origin, with many representing hybridisation between genera (Stresemann 1930a,b). He brilliantly deduced, and dogmatically pronounced, the most likely two parental forms in the case of each specimen (see below).

Of course, many ornithologists were incredulous at this hybridisation hypothesis. Significantly, however, the pre-eminent ornithologist, systematist, and evolutionist Ernst Mayr (1941a), himself a student of Stresemann, was wholly convinced of the correctness of his mentor's conclusions. By contrast, Australian ornithologist Tom Iredale (1950) disagreed in the strongest terms with what he saw as 'the hybrid fantasy' and described it as 'the most astonishing debacle in ornithological history'. Iredale (1950) concluded that Stresemann was entirely wrong in his assessment of the hybrids. While Iredale railed and raged against the hybrid theory he never objectively substantiated his contrary view, but mentioned only the odd character for each form by way of an argument. Iredale was not alone in this view, but his was the opinion most widely read.

While some of the forms Stresemann considered to be hybrids were illustrated when originally described (as good species) it was Lilian Medland who first illustrated them all in colour. This she did in Iredale's (her husband's) monograph on the birds of paradise and bowerbirds (Iredale 1950).

Unfortunately Medland was ignorant of birds of paradise in life and as a result her illustrations were at best inadequate and at worst grossly confusing and misleading.

Gilliard (1969), by contrast, simply listed the putative hybrid combinations in his monograph and expressed some dissatisfaction with respect to the provenance of some specimens. Forshaw (in Cooper and Forshaw 1977) did not mention hybrids. Almost no other ornithologist, Fuller (1979, 1995) excepted, has critically assessed Stresemann's perceptive work and many are today unaware of it or even of the remarkable hybrid birds of paradise.

Errol Fuller most recently (1995) resurrected the hybrid debate, suggesting that certain of the so-called 'hybrids' are, in fact, true species whose native homelands have not yet been discovered. He quite rightly questioned the brief and dogmatic way in which the parental origins of most presumed hybrid specimens were established (Stresemann 1930a,b) and the fact is that this assessment had been long and widely accepted without critical analysis. Fuller (1979) did originally impose a limited view of the physical/genetic potential for hybridisation within the family—as only between birds (genera) with a 'closeness of relationship between putative parents' and those species assumed to be polygynous. At the time he complicated his interpretation of plumage characters by considering novel characters in a hybrid to be indicative of immaturity and did not therefore consider the possibility of them being instead the result of hybridisation. He did, however, modify his views considerably (Fuller 1995) and accepted Stresemann's (1930a,b) assessment for many of the forms. Indeed, he now seriously questions the parentage of only four or five of Stresemann's designations. Fuller's remaining doubts for the most part concern forms that we also wonder about (see below). In view of this and because doubts about the origins of some of the hybrid birds have been resurrected (Fuller 1979, 1987, 1995) we review and illustrate them herein. This we do in the light of new comparative morphological and biometrical observations on the specimens in question and those of the putative parent taxa.

It is clear that the cnemophilines, the manucodes, and the King of Saxony Bird excepted, all polygynous genera are capable of hybridising with at least some other polygynous paradisaeids when given the opportunity permitted by microgeographical contact (which excepts the Moluccan genera, as well, of course; see Table 3.5). We believe more hybrid crosses will be documented in decades to come as fieldworkers take an interest in the problem and actively seek hybrids in areas where they might be expected.

Hybrid birds of paradise and the taxonomy of the family

As the most rampant examples of hybridisation, the Raggiana Bird has produced intrageneric crosses with four other species of its genus and the Superb Bird has produced intergeneric crosses with six other species in five genera (see Table 3.5). To ornithologists unfamiliar with this bird family, the wide and diverse array of wild hybrid forms, especially of the intergeneric ones, will come as a surprise and may cause some puzzlement. Of course this situation reflects a closer relationship between the various species of the Paradisaeinae than is expressed in most, if not all, other equivalent avian lineages. Indeed it is conceivable that the polygynous member species of this subfamily may prove genetically closer than any equivalent number of species in any other. This is seen as indicative of relatively-recent and rapid evolution much influenced by extreme sexual selection—so much so that the polygynous species of birds of paradise have been likened to the breeds of domesticated dogs—extremely diverse in morphology and yet fully interfertile (Gilliard 1969).

The common response to this phenomenon is that the present paradisaeinine genera are excessively over-split (Diamond 1972). Lumping genera is, in all probability, an over-simplified solution to a complex problem (see Chapter 3). Genera, so long as they properly delineate monophyletic lineages, are tools of convenience for defining groupings of species taxa in a manner that aids their study. To use as a criterion for generic lumping the existence of hybrids among the species would logically mean combining nine of our paradisaeinine genera into a single genus of 30 species. This obscures more information than it generates.

The hybrid controversy will probably not be entirely laid to rest until molecular techniques are applied to the putative hybrids and their true parental forms. Rapid developments in the application of such technologies to the study of avian systematics, and genetic 'fingerprinting' using DNA amplified from existing avian study skins, suggest the real possibility of definitive conclusions about the parentage of all of the hybrid birds of paradise in the foreseeable future. Such studies are eagerly awaited.

Hybrid birds of paradise, species isolation, and sexual selection

Given the vast number of adult male birds of paradise that reached the western world during each of many years of plume trading and the small number of hybrids, it is clear that hybrids are extremely rare in nature. It has been estimated that the ratio of hybrid to non-hybrid individuals in wild birds of paradise is of the order of one in 20 000 (Mayr 1945) but this must be viewed as no more than an informed guess.

All of the historical hybrids treated by Stresemann (1930*a*,*b*) were in adult male plumage. Not only would these be easier to identify as hybrids, but it may be that hybrid males are more common than females because of chromosomal sex determination in birds and the effects of deleterious recessive genes (Lack 1974). In birds, males are determined by the presence of two active sex chromosomes (XX), whereas females are XY (the Y being inactive). Males are thus better shielded from lethal or sublethal sex-linked genes, or other sex-linked genetic effects that might be brought about by crossing with a different species.

It is widely agreed that hybridisation is prevalent in polygynous birds of paradise because of the influence of court-based mating and the general lack of pair formation in advance of copulation. Such a competitive system would foster hybridisation for two reasons. First, young of species of polygynous court-displaying bird of paradise genera are raised exclusively by their female parent. Nestling females, as a consequence, never see, know, or bond with their male parent. It is thus clear that female mate-choice is either learned through observation, is innate, or is some combination of the two (perhaps the last is the most likely). There is ample room for error in such a system of mate-selection. Second, a male that successfully ascends the mating hierarchy over time develops into a 'mating machine'. Intrasexual competition among males ensures that a successful male must focus on advertisement, display, and copulation. Because sperm are produced in abundance and are nutritionally easy to produce, it is unlikely there is strong selective pressure for sperm conservation (such as strong species recognition cues)—especially given the fact that selectivity is the trait of females in this sort of system. We presume that a male will mate with whatever bird solicits his copulation—another weakness in the system that might allow hybridisation.

We would expect hybridisation to be more likely to occur in situations where females are unable to locate a conspecific male—this would mainly be in range edges, either at the geographical boundary of a species' range, or else at the high or low point in the elevational distribution of a montane-dwelling species. Furthermore, females at either edge might be expected to be young birds, because range edges tend to be suboptimal habitat and young birds generally are forced into suboptimal habitats. Young birds, in turn, might be expected to be more prone to error in mate-choice. This scenario has been recorded by historical hybridisation events between two lekking European game birds. When the Capercaillie *Tetrao urogallus* became relatively rare, owing to a decline in numbers or few birds moving into new areas as pioneers, females were found to be visiting and hybridising at leks of the Black Grouse *Tetrao tetrix* at a far greater frequency than recorded at other times and circumstances (Johnsgard 1983).

Might there be additional reasons for hybridisation? It has recently been suggested that strong sexual selection produced by polygynous court-based regimes might promote females actively to select morphologically novel partners as mates (Christidis and Schodde 1993), which might further foster the cross-species mating pattern. The argument is that (within a court-based species) a novel partner will provide genes that will produce novel offspring which, in turn, will be more attractive to the next generation of females (runaway selection).

We suggest that there are limits to the attraction of novelty to females, especially if it would foster interspecific hybridisation or species introgression. Certainly in birds of paradise there is little evidence that hybrid individuals (offspring of an extremely 'novel' mating event) are anything but extremely rare. The only hybrid populations known are from narrow zones of overlap between closely-related sister forms in the genera *Astrapia* and *Paradisaea*. Even in these, there is little evidence that hybrids receive any benefit from their 'novel' appearance— the two parental forms remain the most common even in these contact zones, and the zones do not appear to be broadening.

A second reason we believe the novel partners hypothesis is not applicable to the paradisaeinine hybridisation phenomenon relates to the strong intergeneric divergence of display site, display behaviour, and display vocalisation. Many known paradisaeinine hybrids are between species whose males are behaviourally disparate. In fact, the most common cross known is between the King and Magnificent Birds, and yet the former displays high in a vine tangle with animated and inverted displays and flights, whereas the latter displays on a cleared terrestrial court while perched on a vertical sapling, with mostly static postures. Vocalisations of the two are very distinct (Figs 9.58 and 9.64). Female selection for a superior male of the appearance of *her own species* presumably contributes to the retention of species integrity (species isolation).

Final evidence against citing the 'novel partners' hypothesis to explain paradisaeinine hybridisation is the fact that adult males within species populations are in fact remarkably uniform in appearance (Beehler 1988; J. Diamond *in litt.*). If females were actively choosing to mate with males of novel appearance, then the opposite situation should prevail. We believe female birds of paradise in fact select to mate specifically with males exhibiting the generally conservative appearance of their species but with subtle but discernible indications of age (survival) and intra-male dominance such as plume length, court advertisement calling frequency, display quality, and lek position. If any evidence for female choice of 'novel partners' does exist it might be found in the Ruff *Philomachus pugnax* (Scolopacidae), a polygynous wader species in which adult males show conspicuous male plumage and behavioural polymorphism (Hogan-Warburg 1966).

The remarkable diversity of known intergeneric hybrid birds of paradise emphasises genetic compatibility and thus the presumed close relationships within the subfamily Paradisaeinae. Moreover, the presently-known range of intergeneric hybrids can leave little doubt that a good many more await discovery. For example, the Superb Bird is presently known to cross with the Western and Carola's

Parotias and presumably has, therefore, some potential to do so with all parotia species within its extensive range and likewise with further members of other genera (see Table 3.5).

In the majority of intergeneric hybrid specimens, characters of both putative parents are readily apparent in near-typical to generally intermediate form as has long been acknowledged to be the case in hybrids (Graves 1996 and references therein). There are a few hybrids, known by only one or two specimens, that appear to exhibit characters of three genera. To account for these it has been suggested that a product of an initial hybridisation, for example a Black Sicklebill × Long-tailed Paradigalla individual, mated with a third species, such as the Arfak Astrapia (to bring about '*Epimachus ellioti*'—see below). There is also, however, a probability that hybrids will deviate significantly from the biparental midpoint as a function of genetic dissimilarity of the parent species. Moreover it is also conceivable that hybrid individuals might exhibit ancestral (atavistic) characters not apparent in the two parent species (Graves 1996).

The hybrid birds of paradise—comparatively illustrated and reviewed

What follows is an annotated list of the known wild hybrid birds of paradise (see Table 3.5). One captive hybridisation is worthy of note here as it represents a combination of species not recorded below but involving two species that do occur sympatrically in the wild. A male Brown Sicklebill was accidentally housed with a female Ribbon-tailed Astrapia at Taronga Zoo, Sydney, and they mated and the female hatched and reared an offspring that died at about 10 days old because of dry conditions (Hallstrom 1962).

The colour plates of hybrids (Plates 14 and 15) are based, in most instances, on numerous detailed photographs of the rare hybrid specimens plus an artistic interpretation, based on W. T. Cooper's knowledge of the parental forms in life (and/or film of them in life) and museum specimens of them as reference. It should be kept in mind that bare parts and iris colours were not recorded for these hybrid specimens by collectors. As a result the colours applied are those thought to be likely in life given the dominant species' plumage characters exhibited in individual cases. Reproductions of the two colour plates of the hybrids herein were sent to various museum curators for direct comparison against the hybrid specimens in their care. Their comments and corrections, and those of other authorities, were then incorporated into the plates by Cooper before reproduction here.

It has been suggested that the few ornithologists aware of the hybrid phenomenon in birds of paradise simply assume Stresemann's conclusions were correct because it is impractical to conduct a critical comparative survey of these rare and widely-scattered specimens (Fuller 1995). During a recent review of the major paradisaeid collections around the world, the opportunity was taken to examine and measure the hybrid specimens (Frith and Frith 1996a,b, 1997). This comparative overview of hybrids represents the most comprehensive review to date, and enables us to provide the first significant comparative analysis of their biometrics together with those of putative parent taxa. Our hybrid accounts are ordered below by their apparent parental forms; we also include the original scientific name attributed to each (when it was thought to constitute a valid new species) as these names are helpful in accessing the original literature about the hybrids. For a fuller synonomy of scientific names attributed to each hybrid form see Fuller (1995).

It must be stressed here that while we have attempted to assign all hybrids to their putative parents in the most parsimonious way this remains highly subjective at this time. Possibilities such as backcrosses and mutants must also be considered as possibilities (see below).

An annotated list of known or presumed wild hybrid birds of paradise

Intrageneric hybrids

Ribbon-tailed Astrapia *Astrapia mayeri* ×
Stephanie's Astrapia *Astrapia stephaniae*
Plate 14
'*Astrarchia barnesi*' of Iredale 1948, *Australian Zoologist* **2**, 160. Type: AM O 37670, Mt. Hagen, PNG. Also known as Barnes' Astrapia or Barnes's Long-tail. At least 12 ad ♂ specimens are known (AM, AMNH, BMNH, CSIRO) from Mt. Hagen and Mt. Giluwe, PNG.
Measurements (ad ♂♂): wing, ($n = 7$) 164–177 (171); tail, ($n = 7$) 101–116 (111); tail centrals, ($n = 7$) 582–839 (727); bill, ($n = 5$) 34–36 (35); tarsus, ($n = 7$) 40–42 (41); weight ad ♂ ($n = 6$) 141–160 (153), ad ♀ ($n = 1$) 121. Sizes are consistently intermediate between the means of the parent taxa but the bill length tends towards that of the larger-billed Stephanie's Astrapia (Frith and Frith 1997b).

This is a hybrid combination that occurs in the relatively-limited area of sympatry between the parent forms. Considerable variation in male hybrid individuals exists, particularly with respect to the amount of black in the central pair of tail feathers. Birds lower in the hybrid zone tend to be more like pure Stephanie's and those at higher elevations more like a pure Ribbon-tailed Astrapia. This pattern of variability is the product of the successful reproduction of hybrid forms, with abundant back-crossing with the more abundant parental form. Hybrids are commonly observed in the field in the zone of overlap, suggesting that there is little assortative mating where both species occur together in the forest. A marked difference in shape of the outer tail feather (narrowly pointed in Ribbon-tailed and bluntly broadened in Stephanie's) probably leaves hybrids, of both sexes and all ages, detectable by the intermediate-shape of their outer rectrices.

Emperor Bird of Paradise *Paradisaea guilielmi* ×
Raggiana Bird of Paradise *Paradisaea raggiana augustaevictoriae*
Plate 15
'*Paradisea maria*' of Reichenow 1894, *Ornithologische Monatsberichte* **2**, 22. Type: ZMB 31049, Finisterre Mts, Astrolabe Bay, PNG. Also known as Frau Reichenow's, or Maria's, Bird of Paradise. At least six ad ♂♂ (AMNH 4, ZMB 2) and a ♀ specimen are known, from Sattelberg and the Finisterre Mts, Huon Penin, PNG.
Measurements: wing, ($n = 6$ ♂♂) 183–189 (185); tail, ($n = 6$) 124–132 (127), tail centrals ($n = 5$) 126–599 (401); plumes ($n = 5$) 150–215 (180); bill, ($n = 5$) 38–42 (41); tarsus, ($n = 6$) 42–47 (45). These measurements are consistently intermediate between the means of the parent taxa but for the tail (being the length of shorter-tailed *P. r. augustaevictoriae*) and the plume length (being longer than the plumes of any known specimen of *P. guilielmi*; Frith and Frith 1997b).

Initially diagnosed as a hybrid by Rothschild (1910) and agreed with by Stresemann (1923). Female AMNH 679107 is said to be a product of this cross combination. It is a large bird but is as pale as typical female *P. r. augustaevictoriae* with a little 'tan' colour in the mantle (which appears the only sign of *P. guilielmi* influence). Otherwise it shows no sign of *P. guilielmi*'s darker throat, face, forecrown, wings, tail, and underparts, and thus it is difficult to confirm its hybrid status.

Emperor Bird of Paradise *Paradisaea guilielmi* ×
Lesser Bird of Paradise *Paradisaea minor finschi*
Not illustrated
'*Paradisea duivenbodei*' of Menegaux 1913, *Revue Francaise d'Ornithologie* **5**, 49. Type: MHN 863. Also Duivenbode's Bird of Paradise. Known from one ad ♂ specimen (in the MHN) alleged to have come from *Près de Yaour, dans la baie de Geelvink* (probably Waar I., Geelvink Bay, IJ) but presumably it was purchased as a trade skin there and is in fact

from the Huon Penin, PNG, home of the Emperor Bird (Stresemann 1930a,b).

Measurements (*n* = 1): wing, 191; tail, 126; tail centrals, 490; plumes, 180; bill, 41; tarsus, 46. The wing length is average and tail length that of small *P. r. augustaevictoriae* size, which are longer than any *P. guilielmi*; bill length is fractionally longer than the largest *P. r. augustaevictoriae* but that of average sized *P. guilielmi*; tarsus length is near maximum length of *P. guilielmi* and a little above average size in both putative parents (Frith and Frith 1997b).

This form is not illustrated as it looks like the previous hybrid except that it has brown uppertail coverts not marked with straw yellow streaking and its flank plumes are more yellow than red as would be expected in view of its parent species.

Raggiana Bird of Paradise *Paradisaea raggiana augustaevictoriae* ×
Raggiana Bird of Paradise *Paradisaea raggiana intermedia*   Not illustrated
Paradisaea granti of North (1906), *Victorian Naturalist* **22**, 156–158. Type specimen AM O. 14313 from 'German New Guinea (?)'. It was subsequently known as *P. apoda subintermedia* (Rothschild 1921a), *P. a. granti* (Mayr 1941a, and in Mayr and Greenway 1962) and *P. r. granti* (Gilliard 1969).

Because the orange-plumed subspecies *P. r. augustaevictoriae* is so different from the various red-plumed subspecies, a mixed population between the former and *P. r. intermedia* was named *P. granti* as a distinct species. As is discussed in the systematics text for the Raggiana Bird (Chapter 9), the population known as *P. r. granti* is in fact a changing, intergrading one between orange-plumed *P. r. augustaevictoriae* to the NW and red-plumed *P. r. intermedia* to the SE. The plumes of these intraspecific hybrids vary from near pure orange to near pure red and with all possible intermediate forms. We also note here that the Raggiana Bird subspecies *P. r. salvadorii* and *P. r. augustaevictoriae* have hybridised with each other in captivity (Muller 1974).

Raggiana Bird of Paradise *Paradisaea raggiana salvadorii* ×
Blue Bird of Paradise *Paradisaea rudolphi margaritae*   Plate 15
'*Paradisea bloodi*' of Iredale 1948, *Australian Zoologist* **11**, 161. Type: AM O 37682, Minyip, Mt. Hagen, PNG. Also known as Captain Blood's, or Blood's Bird of Paradise. Only the one ad ♂ specimen known.

Measurements (*n* = 1): wing, 174; tail, 111; tail centrals, 493; plumes, 125; bill, 41; tarsus, 41. Wing, tail, and plume length are intermediate in size between the means of the larger *P. r. salvadorii* and the smaller *P. r. margaritae*, tail centrals are within the size range of the former but are longer than in any of the latter (Frith and Frith 1997b).

Only Iredale (1948, 1950) failed to accept the parentage of this now undoubted and most spectacular of intrageneric hybrids. The subtle dilute bluish sheen to back, wings, and underparts, flank plume structure and colour, slightly-broadened and spatulate-tipped central tail wires, and the slightest trace of the white feathers (typically encircling the eye of the Blue Bird) leave no room for doubt about this individual's putative parents.

Gilliard suggested that human encroachment upon the altitudinally-restricted habitat of the Blue Bird might have diminished the natural isolating barriers between it and the Raggiana Bird. There may well be an element of truth to this but the number of other *Paradisaea* intrageneric hybrid combinations does indicate that the Blue and Raggiana Birds might cross under entirely natural circumstances. Both species occur together between 1300 and 1600 m on Mt. Missim (Pruett-Jones and Pruett-Jones 1986).

Raggiana Bird of Paradise *Paradisaea raggiana augustaevictoriae* ×
Lesser Bird of Paradise *Paradisaea minor finschi*   Plate 15
'*Paradisea mixta*' of Rothschild 1921, *Bulletin of the British Ornithologists' Club* **41**, 127. Type: AMNH 679102. Also known as Rothschild's Bird of Paradise. At least four ad ♂ specimens

known (AMNH 2, BMNH 1, AM 1), three of them lacking localities and one from c. 110 km S of Madang, PNG.
Measurements (n = 3); wing, 191–194 (193); tail, 133–137 (135); tail centrals, 304–535 (445); plumes, 225–260 (238); bill, 38–40 (39); tarsus, 44–46 (45). Adult ♂♂ of the parent taxa are generally similar in size with some individual *P. m. finschi* being larger than all *P. r. augustaevictoriae* (Frith and Frith 1997*b*).

Even when describing this form as a new species Rothschild (1921*a*) stated it may in fact be a hybrid. Hybrid specimens and typical ones of the putative parent forms indicate that this undoubted hybrid occurs throughout a band *c.* 35 km wide between the parent forms. Lesser Bird genes predominate in the NW, whereas Raggiana Bird genes predominate in the SE. The main hybrid zone appears to lie in the 'Gusap' pass separating the headwaters of the Ramu and Markham Rivers, but also a second zone should be looked for on the northern side of the Finisterre Ra, along the coast S and E of Madang. As with the preceding cross, there appears to be significant introgression, producing considerable variation in individuals of hybrid origin. Those close to being pure Lesser Birds (*P. m. finschi*) are predominantly yellow-plumed and lack the yellow throat collar; those closest to parental Raggiana are predominantly orange-plumed, with a narrow yellow throat collar.

Raggiana Bird of Paradise *Paradisaea raggiana salvadorii* ×
Lesser Bird of Paradise *Paradisaea minor finschi*   Not illustrated
An un-named hybrid previously known from adult male specimens from upper Baiyer Valley, PNG where M. LeCroy (*in litt.*) actually witnessed an apparent copulation between a soliciting ♀ *P. r. salvadorii* and an ad ♂ *P. m. finschi*. As this hybrid combination was clearly elucidated by E. Thomas Gilliard, and given that a common name has been applied to almost every hybrid bird of paradise combination, we suggest it appropriately be informally known as Gilliard's Bird of Paradise.
Measurements: not recorded but as the two parent taxa are extremely similar in size (Frith and Frith 1997*b*) the measurements of the hybrids would be uninformative.

This combination is not illustrated as it looks much like the previous hybrid except for the slightly more reddish flank plumes as would be expected when the red race of Raggiana is one parent.

Greater Bird of Paradise *Paradisaea apoda novaeguineae* ×
Raggiana Bird of Paradise *Paradisaea raggiana salvadorii*   Plate 15
'*Paradisea apoda luptoni*' of Lowe 1923, *Bulletin of the British Ornithologists' Club* **43**, 10. Type: BMNH 1923.3.1.1. Also known as Lupton's Bird of Paradise. Numerous ad ♂ specimens are known in various collections, taken from the Fly R. area of S NG.
Measurements: wing, (n = 4) 197–199 (198); tail, (n = 5) 139–144 (141); tail wires, (n = 5) 500–567 (539); plumes, (n = 4) 190–215 (204); bill, (n = 5) 37–40 (39); tarsus, (n = 3) 44–46 (45). As intermediate between the parent taxa sizes as could be expected, the wing length perfectly so (Frith and Frith 1997*b*).

When described as a new species this variable form was acknowledged by the author as possibly representing 'an interesting instance of hybridism' and that it is 'exactly intermediate' between the Greater and Raggiana Birds (Lowe 1923). Hybrid individuals vary greatly in appearance geographically, with most forms being almost pure Greater in the W and most forms being almost pure Raggiana Bird in the E—a product of repeated back-crosses with the geographically predominant parental form, and evidence that hybrids successfully reproduce. We suspect hybrid swarms form in the centre of the overlap zone, much as is found between *Astrapia stephaniae* and *A. mayeri*. This system would profit from a field study of mate selection by ♀♀ in the hybrid zone.

Magnificent Bird of Paradise *Cicinnurus magnificus* ×
King Bird of Paradise *Cicinnurus regius*
Plate 15

'*Diphyllodes gulielmi III*' of Meyer 1875, *Der Zoologischer Garten* **16**, 29. Type: ZMA 782. Also known as King of Holland's Bird of Paradise, King William III's Bird of Paradise and *Diphyllodes (Rhipidornis) gulielmitertii*. At least 25 ad ♂ specimens are known, in various collections (12 in AMNH alone), predominantly from N coastal NG or unknown localities. This is the same hybrid combination as the next but in this form of it the influence of the Magnificent Bird is more apparent (see Plate 15).

Measurements: wing, ($n = 16$) 107–114 (111); tail, ($n = 16$) 33–38 (35); tail wires, ($n = 15$) 162–212 (196); bill ($n = 14$) 27–30 (28); tarsus, ($n = 15$) 28–31 (30). Wing and tail lengths are shorter than average-sized *C. magnificus*, tail wire length spans the gap between those of shorter *C. regius* and longer *C. magnificus*, and bill and tarsus are generally intermediate in length between those of the parent species. The measurements thus provide strong evidence for the proposed parentage, given the dominant expression of *C. magnificus* genes. The glossy green underparts are more extensive than in the King Bird.

Berlioz (1927) first diagnosed this relatively common hybrid and Meise (1929) agreed with the now widely-accepted parentage. In fact, specimens that approximate this form probably represent *Cicinnurus* hybrid × pure *C. magnificus* back-crosses. Although specimens of this form exhibit characters of both putative parents as clearly as any hybrid, Iredale (1950) scathingly rejected what he saw as the 'hybrid-concoction' for this form.

King Bird of Paradise *Cicinnurus regius* ×
Magnificent Bird of Paradise *Cicinnurus magnificus*
Plate 15

'*Cicinnurus lyogyrus*' of Currie 1900, *Proceedings of the United States National Museum* **22**, 497. Type: USNM 124628 and '*C. goodfellowi*' of Ogilvie-Grant 1907, *Bulletin of the British Ornithologists' Club* **19**, 39. Type: BMNH 1908.5.27.74. Cyclops Mts, 3,000', Humbolt's Bay. Also known as Lyre-tailed King Bird of Paradise, Lyre-tailed King, Lonely Little King or Crimson Bird of Paradise. At least three ad ♂ specimens are known (BMNH, USNM, ZMB) from Humboldt Bay, Cyclops Mts, and unknown localities. This is the same hybrid combination as the last but in this form of it the influence of the King Bird is more apparent (see Plate 15).

Measurements ($n = 2$): wing, 107–113; tail, 33–35; tail wires, 145–168; bill, 25–28; tarsus, 28. Wing and tail length are longer than the average *C. regius* and shorter than the average *C. magnificus* but tail centrals and tarsus lengths are shorter than the shortest of the latter and bill length is roughly intermediate. The measurements thus support the proposed parentage given that *C. regius* genes dominate (Frith and Frith 1997b).

Rothschild (in Stresemann 1930a) argued that because this form exhibits predominantly those characters of the King Bird it is most likely to be a cross between a pure King Bird and a Magnificent × King Bird hybrid. This, then, is the counterpart to the preceding cross. The iridescent green underparts are far less extensive than in the Magnificent Bird, being intermediate in extent in this character between the two parent species, unlike the previous hybrid in which it is more extensive.

The BMNH type of *goodfellowi* clearly shows slight but clear influence of Magnificent Bird genes: at the lower and outer edge of the green breast shield the feather tips are green-*blue* as in *C. magnificus* but unlike in *C. regius*. Also the rear crown and nape feathers in the *goodfellowi* type are short and stubby as in the Magnificent Bird and are not longer and smooth as in the King Bird. In view of this and the measurements of specimens we cannot agree that 'the specimens are so close to *regius* that they could conceivably be merely individual variants' (Fuller 1979).

Intergeneric hybrids

Arfak Astrapia *Astrapia nigra* ×
Black Sicklebill *Epimachus f. fastuosus*

Plate 14. '*Epimachus astrapioides*' of Rothschild 1897b, *Bulletin of the British Ornithologists' Club* 7, 22. Type: AMNH 679119, Dutch NG. Also '*Astrapimachus ellioti*' of Mayr (1941a), Astrapian Sicklebill and Green-breasted Riflebird. One ad ♂ specimen known only, presumably from the Vogelkop, IJ. The name originally applied to this hybrid emphasises that it clearly exhibits characters of both putative parent genera. This is the same hybrid combination as the next but in this form of it the influence of the Arfak Astrapia is more apparent (see Plate 14).

Measurements ($n = 1$): wing, 189; tail, 393; tail centrals, 595; bill, broken; tarsus, 50. Wing length is intermediate between that of the putative parents and in the other characters the size is of smaller *E. f. fastuosus* individuals except the tarsus, which is the average length of the latter but is longer than in any *A. nigra* (Frith and Frith 1997b). Measurements are thus compatible with the proposed parentage.

The neck and nape exhibit some broad scale-like feathers of the Arfak Astrapia type while the upper back has the large, iridescent, blue-green, scale-like feathers of the Black Sicklebill. The crown is purple and not blue as in the Black Sicklebill. The central pair of tail feathers are Arfak Astrapia-like in shape (but slightly narrower and uniform in width) with purple iridescence like this species but with a slight blue wash (of sicklebill influence). Throat and chin are black with deep, blue-green, iridescent gloss. A few dispersed coppery-tipped feathers to the sides of the lower border of the darker throat clearly indicate the fine copper red line found in the Arfak Astrapia. Undertail colour is as in the Arfak Astrapia. There is little doubt about the putative parents of this form, its hybrid origin being widely accepted (Fuller 1995).

Black Sicklebill *Epimachus f. fastuosus* ×
Arfak Astrapia *Astrapia nigra*

Plate 14. '*Epimachus ellioti*' of Ward 1873, *Proceedings of the Zoological Society of London* p. 742. Type: BMNH 1881.5.1.1696, 'Waigeo' (= Vogelkop, IJ). Also known as '*Astrapimachus ellioti*' of Mayr (1941a), Elliot's Bird of Paradise. Two ad ♂ specimens only are known (BMNH and SMT), presumably from the Vogelkop, IJ. This is the same hybrid combination as the last but in this form of it the influence of the Black Sicklebill is more apparent (see Plate 14).

Measurements ($n = 2$—those of the SMT specimen were kindly made for us by S. Eck): wing, 202 and missing; tail, 271, 259; tail centrals, 389, 403; bill, missing and 59; tarsus, 55 and missing. Of the BMNH specimen (second measurement given) both tail measurements are considerably shorter than in both putative parents whereas the bill length is exactly intermediate between the average length of that of both.

This bird raises as many doubts about its parental origins as any other hybrid bird of paradise (Stresemann 1930a,b; Rothschild in Stresemann 1930a,b; Fuller 1979, 1987, 1995); the inferior quality of the BMNH specimen and of illustrations of it exacerbate the doubts. If it is indeed a hybrid, which we believe, the well-developed pectoral fan and filamental flank plumes indicate the Black Sicklebill as one of the parents. The wing length (202) is that of a large Black Sicklebill of the nominate subspecies (188–205) and is merely 9 mm larger than the longest-winged Arfak Astrapia. Both tail measurements of both hybrid specimens are shorter than those of ad ♂♂ of nominate Black Sicklebills and Arfak Astrapias but the measurements of the central rectrices fall within the range of subad Arfak Astrapias and that of the remaining tail within that of subad nominate Black Sicklebills. The bill is intermediate between Black Sicklebill and Arfak Astrapia in length and structure with a relatively small yellow gape lobe at the base of both mandibles. The tarsus length is fractionally larger than the largest of ad ♂ nominate Black Sicklebills (53.5 mm) but is within the range for the species and may possibly reflect a difference in measuring technique (by S. Eck).

Only the three innermost secondaries of each wing are present in the BMNH specimen. The iridescent area of head plumage is as in the Black Sicklebill grading to the burnished greenish yellow patch of neck colour. The latter is not a discrete area as shown in the plate in Elliot's (1873) monograph and reproduced in Fuller (1995), the preparation of the skin having apparently misled the artist.

At the lower breast is a broad, dully-glossed, coppery brown line which would appear indicative of the influence of Arfak Astrapia genes (see Plate 6). The green underparts are apparently of Arfak Astrapia influence but are darker and duller. The tail feather undersides are like, but a little blacker than and lacking the 'barred' appearance of, the Arfak Astrapia.

The upper central tail colour/iridescence is like that of the Arfak Astrapia (deep reddish purple with velvet-like finish) and not like the more harsh 'electric' bluish purple of the Black Sicklebill. The central tail feathers are, however, even in width throughout and end in a sabre-like point and are thus intermediate between the terminally-broadening feathers of Arfak Astrapia and the finely-tapering ones of the Black Sicklebill. The tips of the remaining tail feathers appear perfectly intermediate between pointed sicklebill and 'round-squared' astrapia ones, being conspicuously notched (like those of Long-tailed Paradigallas). The tail length is, however, significantly shorter than might be expected in a Black Sicklebill × Arfak Astrapia cross. This and the paradigalla-like tail feather tip notching possibly, and understandably, influenced Rothschild (in Stresemann 1930a) who suggested that this bird might represent a secondary hybrid combination involving (1) the Black Sicklebill × Long-tailed Paradigalla (known as *Pseudastrapia lobata*—see Plate 14) crossing with an Arfak Astrapia or (2) the Arfak Astrapia × Black Sicklebill (known as *Epimachus astrapioides*—see Plate 14) hybrid subsequently crossing with a Long-tailed Paradigalla.

We can see that, on average, the characters of this form support the Black Sicklebill × Arfak Astrapia parentage proposed by Stresemann (1930a,b) but we can also see why Rothschild (1930a) considered it to show those of three genera. We cannot share Fuller's view that the bird is a possibly extinct pure *Epimachus* but we can see some grounds for his view that if it is a hybrid involving only two genera it could conceivably be a Long-tailed Paradigalla × Black Sicklebill one, which its measurements would not exclude. A reciprocal cross of the following hybrid is also conceivable (M. LeCroy *in litt.*). It is, however, as (if not more) possible this specimen represents an Arfak Astrapia × '*Pseudastrapia lobata*' cross.

Long-tailed Paradigalla *Paradigalla carunculata* ×
Black Sicklebill *Epimachus f. fastuosus*

Plate 14

'*Pseudastrapia lobata*' of Rothschild 1907, *Bulletin of the British Ornithologists' Club* **21**, 25. Type: AMNH 678118, Dutch NG. Also known as False-lobed Long Tail and False-lobed Astrapia. One ad ♂ specimen known only, undoubtedly from the Vogelkop, IJ.

Measurements ($n = 1$): wing, 189; tail, 262; tail centrals, 389; bill, 41; tarsus, 46. Wing, tail centrals, and tarsus are close to, and tail length is perfectly intermediate between, the average size of the putative parents but the bill length is short as in *P. carunculata* (Frith and Frith 1997b). Measurements of the hybrid can thus been seen as well supporting the proposed putative parents if it is accepted that *Epimachus f. fastuosus* genes failed to influence its bill length. The very tip of the upper mandible is missing from the unique specimen and so we made allowance for this, relative to the complete lower mandible, and cannot be significantly in error.

Because the sexually-monomorphic paradigallas were at the time widely assumed to be pair-bonding monogamous birds, Fuller (1979) argued that they were the most unlikely of putative parents for any bird of paradise hybrid. This objection was far less justified than in the case of the other two hybrids believed to involve paradigallas as the present hybrid clearly exhibits paradigalla characters. Moreover, it was subsequently demonstrated that the paradigallas are polygynous, removing Fuller's objection.

Given that it exhibits so little evidence of Black Sicklebill genes, this hybrid bird strikes one as a Long-tailed Paradigalla with several Black Sicklebill tail feathers added—but close examination of the tail base and the gape wattles leaves no doubt it is an authentic specimen. Thus the tail is intermediate between those of the putative parent species. The bill is like that of a Long-tailed Paradigalla but only small gape wattles are present. Iridescent head feathering is restricted to the hybrid's forecrown, as in the Long-tailed Paradigalla. Nape, back, rump, and wings are like the latter species but are slightly paler and more brown. Underparts are much as in the Long-tailed Paradigalla but are slightly paler. Iridescence on the upper central tail feather pair is like that of the Black Sicklebill but is not to the feather edges or on the terminal 10 cm of them which are dull blackish. The next pair of tail feathers are Black Sicklebill-like and the remaining ones (four on bird's left and two on its right) are rather square-ended, centrally notched, and coloured as in the Long-tailed Paradigalla.

In listing type bird specimens at the then Tring Museum, Ernst Hartert (1919) indicated that of *Pseudastrapia lobata* to be an immature bird. In his influential paper Stresemann (1930a) suggested this unique specimen is in its first year plumage, presumably seeing the lack of sicklebill-like pectoral fan plumes as indicative of immaturity. As imm ♂ Black Sicklebills wear the barred ♀ plumage of their species for up to 6 or 7 years we assume Stresemann must have meant the first year of adult plumage (see Plate 9). Given the fully-developed iridescent forehead and upper tail of the specimen we consider this erroneous notwithstanding the obvious dominance of paradigalla (in which immatures are duller versions of the unbarred adults—see Plate 5) genes. Nothing about the specimen is unequivocally indicative of immaturity.

Influenced by Hartert (1919) and Stresemann's statements Fuller (1995) introduced the novel notion that the *Pseudastrapia lobata* specimen represents an immature of the bird known as *Epimachus ellioti* (in Stresemann's and our view a Black Sicklebill × Arfak Astrapia hybrid). In noting the bird is generally blackish, Fuller (1995) stated this is 'just what one might expect in the immature of [*Epimachus*] *ellioti*', given that he does not believe this individual involves the genes of the Arfak Astrapia. This observation would be invalid if the bird is the result of a Black Sicklebill × Arfak Astrapia cross as the immature plumage of both these putative parents of '*ellioti*' include ventral barring. If the bird is, in fact, a Black Sicklebill × Long-tailed Paradigalla hybrid (see below) Fuller's observation is a valid one. It is equally plausible that this bird is in full adult plumage, however, as the paradigalla genes may simply have prevented the development of sicklebill-type pectoral plumes.

We accept as the best available hypothesis Stresemann's original diagnosis that this hybrid is probably the result of a Long-tailed Paradigalla × Black Sicklebill cross but we do have some reservations given the relatively short bill and dilute nature of supposed Black Sicklebill features in general. Perhaps this is the product of a hybrid back-cross with a pure paradigalla.

Long-tailed Paradigalla *Paradigalla carunculata* ×
Superb Bird of Paradise *Lophorina s. superba* Plate 14
'*Loborhamphus nobilis*' of Rothschild 1901, *Bulletin of the British Ornithologists' Club* **12**, 34. Type: AMNH 679113, Dutch NG. Also known as Rothschild's Lobe-billed Bird of Paradise and Noble Lobe-bill. Long known

from two ad ♂ specimens (AMNH) until a third old skin was found at Bogor (BZM 24.478) in 1995 (CF), probably from the Vogelkop, IJ.

Measurements: wing, ($n = 3$) 155–165 (161); tail, ($n = 3$) 103–135 (117); tail centrals, ($n = 2$) 108, 151; bill, ($n = 3$) 37–38 (38); tarsus, ($n = 3$) 39–41 (40). All mean measurements of the hybrids lend strong support for the proposed cross by being close to perfectly intermediate between the respective averages for the putative parent taxa (Frith and Frith 1997*b*).

This is the least adequately documented hybrid bird of paradise. The bill is intermediate in size between that of the Long-tailed Paradigalla and Superb Bird but is perhaps slightly more like the Superb Bird's in shape. Its tail is mildly graduated, like Long-tailed Paradigallas, but the outer rectrices are not more than 5–6 mm shorter than the pair next to the central pair. The shape of the hybrids' outer two primaries is intermediate between that of the putative parents. Their gape wattles are small, being little more than exaggerated gapes of those found in riflebirds and sicklebills—but are clearly suggestive of paradigallas. The shape of the iridescent crown and breast-shield, dense elongate mantle cape, iridescent black central tail feathers, and the iridescent green on throat, sides of face, mantle, and back suggest the Superb Bird. The (1) deep purple of the crown and breast-shield, (2) elongate and iridescent purple-edged, broad-tipped, pectoral, fan-like, outer feathers of an extensive breast-shield, (3) deep iridescent purple on wings, (4) oil green iridescent feathers beneath the breast-shield and (5) the deep claret-plum colour of the underparts are suggestive of Black Sicklebill and Magnificent Riflebird characters!

The above hybrid combination, suggested by Stresemann (1930*a,b*), does seem the most credible, particularly in view of the Superb Bird-like cape and paradigalla-like wattles and tail form. We well understand and to some extent share Fuller's (1979, 1995) doubts, however, as this form is yet another (others being those previously known as *Epimachus ellioti*, *Janthothorax bensbachi* and the next hybrid) appearing to exhibit some characters of at least three bird of paradise genera. If this bird has not been genetically influenced by *Epimachus*, the sicklebill-like, pectoral, fan-shaped feathers have to be seen as the coincidental result of the proposed hybrid combination. On the other hand it might represent a secondary cross. Another possibility not beyond consideration is that of a Superb Bird × an astrapia cross, if one accepts the gape wattles are not influenced by paradigalla genes but merely represent a 'new' character resulting from hybridisation (E. Mayr *in litt.* to Stresemann during 1952).

Black Sicklebill *Epimachus fastuosus atratus* × **Superb Bird of Paradise** *Lophorina superba feminina* Plate 14

Reported as *Epimachus fastuosus atratus* × *Lophorina superba feminina* Junge 1953, *Zoologische Verhandelingen, Leiden* **20**, 62. Type: RMNH 17345, Bobairo, Wissel Lakes, IJ. Also known as the Mysterious Bird of Bobairo (Fuller 1995). Only one ad ♂ specimen known, from Bobairo, near Enarotali on Lake Paniai (Wissel Lakes) Weyland Mts, west IJ.

Measurements ($n = 1$): wing, 168; tail missing; bill 50; tarsus, 41. It is noteworthy that the wing and tarsus length are both exactly intermediate between the average size of those of the putative parents and that the bill is within 5 mm of being perfectly intermediate between them notwithstanding the disparate average size of 78 mm for *E. f. atratus* and 31 mm for *L. s. feminina* (Frith and Frith 1997*b*). These facts add weight to other evidence supporting the above putative parentage for this unique hybrid.

Although less obviously the product of its suggested parents, a close examination of the unique specimen does suggest they are the most likely. This specimen has the small gape normally found in both putative parents. The pectoral fans and filamentous flank plumes clearly express Black Sicklebill genes. Bill length and shape suggest the other parent is the Superb Bird as it would be significantly longer had it been a Magnificent Riflebird. Moreover, it wears a substantial mantle cape of elongate black feathering arising from the lower nape which can only be attributed to the Superb Bird. Gloss on the primaries is as found in the Black Sicklebill. That on the mantle, secondaries, and tertials is certainly, however, far more violet purple than on the latter species (being more like that of the Twelve-wired Bird) which, at present, must be presumed to be an unpredictable result of the influence of the coppery green found on these areas on the Superb Bird. Such changes in what are structural iridescent colours are far from surprising.

A puzzling character of this bird is the fairly strong sheen of glossy, oily green (as in most *Astrapia* spp) over the otherwise Superb/Black Sicklebill-like abdomen and belly. This also occurs in the Twelve-wired × Magnificent Riflebird cross. Perhaps it sometimes results when two birds lacking it hybridise.

It is most unlikely that this specimen represents an undiscovered species (*contra* Fuller 1979) as it comes from an ornithologically well-collected area. We consider Junge's (1953) original diagnosis of a cross between Black Sicklebill and Superb Bird as the most likely explanation, a conclusion even the most sceptical can concede (Fuller 1995). We would make the observation, however, that this bird is not greatly unlike a tail-less Black Sicklebill × astrapia might appear (see Plate 14) and that such a cross (= Black Sicklebill × Splendid Astrapia given the specimen location) might account for the oily green sheen on its underparts and the trace of a narrow coppery red band across the chest side. This bird is then perhaps another showing possible characters of three genera. When collected its gonads measured 7 × 3 mm.

Western Parotia *Parotia sefilata* ×
Superb Bird of Paradise *Lophorina superba*
Plate 14
'*Parotia duivenbodei*' of Rothschild 1900, *Bulletin of the British Ornithologists' Club* **10**, 100. Type: AMNH 679117, Dutch NG. Also known as Duivenbode's Six-wired Bird of Paradise. Two ad ♂ specimens are known (AMNH, MHN), from (probably purchased at) Yaour Aolfe, Geelvink Bay, IJ.

Measurements: wing, ($n = 2$) 154; tail, ($n = 2$) 110, 114; tail centrals, ($n = 2$) 115, 116; bill, ($n = 1$) 34; tarsus, ($n = 2$) 41. Note that in view of the geographical range of *P. sefilata* the subspecies of *L. superba* involved could be the nominate or *niedda* and while we assume it to be the nominate for comparing measurements we would point out that both subspecies are in any event near-identical in all character sizes (Frith and Frith 1997*b*). In all characters except bill length the sizes of larger *P. sefilata* and smaller *L. superba* are mutually exclusive and yet in all characters the two hybrid specimens are almost perfectly intermediate between the average size of the putative parent taxa (Frith and Frith 1997*b*) providing strong support for the proposed cross.

It is generally Superb Bird-like, but with a breast-shield and the shape of the two outermost primaries intermediate in form between those of the putative parents. The two occipital plumes have narrowly-elongate 'ear-tuft' feathers at their bases. The erectile, silver white, narial tuft of the Western Parotia is lacking but there is a dense 'cushion' of black feathers atop the upper mandible base. Behind this is a dished depression and then Superb Bird-like iridescent crown feathering but this confined to an area extending from the posterior edges of the eyes to the rear crown. The discrete line of purple, scale-like, nuchal bar feathers of the Western Parotia is absent. Back and upper tail are as in the Western Parotia, with no sign of the mantle cape of the Superb Bird. Underparts feathering is structurally like the flank 'skirt' plumes of the Western Parotia but is less elongate and with a slight greenish wash.

There can be little doubt this is indeed the result of the Western Parotia hybridising with the Superb Bird. As noted by Fuller (1995), female parotias and those of the Superb Bird are similarly plumaged (see Plates 7 and 10). Superb Birds have, moreover, been observed to encounter parotias aggressively about their courts (M. LeCroy *in litt.*). The present hybrid certainly raises the probability of the geographically widespread Superb Bird hybridising with (all) other sympatric parotia species.

Long-tailed Paradigalla *Paradigalla carunculata* ×
Western Parotia *Parotia sefilata*   Plate 14
'*Loborhamphus ptilorhis*' of Sharpe 1908, *Bulletin of the British Ornithologists' Club* **11**, 67. Type: BMNH 1908.4.10.1. Also known as Sharpe's Lobe-billed Riflebird (a most illogical name if the Magnificent Riflebird is not involved) but Sharpe's Lobe-billed Parotia is more appropriate and should be used if a vernacular is required. One subad ♂ specimen only is known, from unknown locality but 'pretendedly from Dutch NG' (Sharpe 1908) or IJ—could only be from the Vogelkop, IJ.
Measurements ($n = 1$): wing, 181; tail, 148; tail centrals, 164; bill, 42; tarsus, 53. This hybrid's wing is longer than that of the largest (170) *P. sefilata* and is almost as short as the smallest (180) *P. carunculata* specimen. The tail is longer than that of both putative parents while the tail centrals are longer than the average for *P. carunculata*. The bill is longer than that of *P. sefilata* and is within the lower range of *P. carunculata* while the tarsus is longer than any of the latter but is the average size of the former. Thus this unique specimen's measurements are compatible with the suggested cross.

This is one of the most obscure of the hybrids for which there is some certainty of its parentage. The unique specimen has the peculiarly-modified parotia skull shape (see Fig. 1.4) and, while lacking occipital flag plumes, it has the narrowly-pointed elongate 'ear tuft' feathers typically at the base of such occipital plumes. Thus there is no doubt that at least one parent was a parotia. The tail structure suggests the influence of paradigalla genes: for while the Western Parotia's tail is graduated with each feather tip symmetrically pointed, the feather tips in the Long-tailed Paradigalla are broad and strongly notched. This bird's tail feathers are approximately intermediate in form between the two, but tend towards that of the latter bird. The remnant barred underpart plumage is indicative of, but much darker than, imm and subad parotia plumage, but is not like the black unbarred plumages of paradigallas (see Plates 5 and 7) which may, however, account for this hybrid's darker underparts.

On balance, the parentage suggested by Stresemann (1930*a*,*b*) seems correct given its (1) appropriately-lengthened graduated tail, (2) tail feather tip shape, (3) reduced gape wattles, (4) green sheen to upperparts, and (5) relatively-large legs and feet. The only other conceivable option for second putative parent species might be the Arfak Astrapia but in such a cross an even longer tail, finer bill, and smaller legs and feet might be expected to result.

Superb Bird of Paradise *Lophorina superba* ×
Magnificent Bird of Paradise *Cicinnurus magnificus*   Plate 14
'*Lamprothorax wilhelminae*' of Meyer 1894, *Abhandlungen und Berichte des Koniglichen Zoologischen Museums zur Dresden* **5**, 3. Type: SMT 1428C, Arfak, NG. Also known as Wilhelmina's Bird of Paradise and, again illogically, Wilhelmina's Riflebird. Three ad ♂ specimens are known (AMNH, RMNH, SMT), two from Arfak Mts, IJ and one from an unknown locality.
Measurements: wing, ($n = 3$) 129–132 (130); tail, ($n = 3$) 71–72 (72); tail centrals, ($n = 3$) 126–146 (134); bill, ($n = 2$) 32, 35; tarsus, ($n = 3$) 33–34 (33). Because two of these hybrids are said to be from the Vogelkop and all three are similarly sized we use the nominate subspecies of both putative parents for comparisons but acknowledge this may be erroneous. The wing length of this hybrid is that of the smallest (131) *L. superba* and is thus much longer than the much smaller (111–121) *C. magnificus*. Its tail and tail centrals lengths are almost intermediate between those of the respective average sizes of the Superb (99 and 102) and Magnificent (39 and 269) Birds, respectively. Its bill

length is the same or slightly larger than the largest (33) *L. superba* and is longer than all *C. magnificus* while its tarsus length is that of the largest (34) *C. magnificus*. Measurements are thus generally compatible with hybridisation between the putative parent species.

This form is one of the eight that Fuller (1979) considered unlikely to be a hybrid between the proposed putative parent genera. He subsequently underscored his difficulty in demonstrating it to be otherwise, however, by conceding that it 'certainly shows characteristics that might be interpreted as intermediate' and accepting that the 'probability is that this is the hybrid stated' while making the rejoinder that 'these are also features that could have arisen independently' (Fuller 1995). We consider the putative parents correctly diagnosed. We can find no evidence justifying Fuller's (1995) view that the Dresden and Leiden specimens show some signs of immaturity. It would seem likely that in this (and other cases) Fuller is misinterpreting novel plumage characters resulting from the hybridisation as being those of immaturity.

The bill is Superb Bird-like but is longer and less narrow. The crown appearance is generally Superb Bird-like but the feathers are far finer and are a rich, strong, iridescent, pinkish purple (not bluish green) this colour continuing onto the Superb Bird type of nape cape grading into the iridescent coppery green cape and back. Throat, sides of face, and ear coverts are black with strong iridescent gloss of greenish yellow (like the throat of Magnificent Birds) with coppery highlights in some lights. Fine feathers immediately in front of eyes are not green (as in the Magnificent Bird) but are coppery pinkish purple. The wing primaries, secondaries, and coverts are edged with matte blackish. The uppertail is glossy black with a strong pinkish purple sheen (of the Superb Bird) and the central pair of feathers show iridescent 'electric' blue (of the Magnificent Bird's tail 'wires') but with deep, dark, rich, blue to blue-black, highly-glossy highlights.

The breast-shield is like that of the Superb Bird in general appearance but in texture and colour is intermediate between that of the two putative parents, these feathers not being sharply edged and fish-scale-like (as in Superb Birds) nor finely 'hair-like' (as in Magnificent Birds) but intermediate (more like the throat and neck feather structure of the Ribbon-tailed Astrapia). Feather tips of the upper and elongate feathers of the breast-shield edge are tipped a strong iridescent blue but those of the lower border dully-iridescent blue-green.

Carola's Parotia *Parotia carolae* ×
Superb Bird of Paradise *Lophorina superba*
Plate 14

Lophorina superba pseudoparotia of Stresemann 1934, *Ornithologische Monatsberichte* **42**, 144. Type ZMB 3.49.2, Hunsteinspitze. Also known as Stresemann's Bird of Paradise (Frith and Frith 1996, *Journal für Ornithologie* **137**: 515–521). Initially considered to be and treated as a ♀ specimen of Carola's Parotia (Stresemann 1923) and subsequently known as the Superb Bird subspecies *Lophorina superba pseudoparotia* (Stresemann 1934). One unique ♀ specimen known, from Mt. Hunstein, Sepik District, PNG.

Measurements ($n = 1$): wing, 130; tail, 80; tail centrals, 83; bill, 31; tarsus, 37. On geographical grounds we use ♀ *P. c. chrysenia* and ♂ *L. s. addenda* for size comparisons but acknowledge that an undescribed Hunstein Mts subspecies of the latter species could be involved. The unique hybrid specimen's wing length is near intermediate between the averages of the mutually-exclusive size ranges of the putative parent taxa. The lengths of the tail and central rectrices are those of average-sized *L. s. addenda* and thus shorter than any *P. c. chrysenia*. Its bill length is similar to the smallest (29) *P. c. chrysenia* and largest (31) *L. s. addenda*, while its tarsus is as long as the largest *L. s. addenda* (Frith and Frith 1997*b*). These figures support the suggested putative hybrid parentage.

The specimen was collected by Dr J. Bürgers on 7 Mar 1913. It was first identified as a ♀ Carola's Parotia by Stresemann (1923) who changed his mind and described it as a new subspecies of the Superb Bird *Lophorina superba pseudoparotia* (Stresemann 1934). He used the trinomial *pseudoparotia* as a joke upon himself, but his misidentification is understandable now that we know this is a new hybrid combination with parents from the two species Stresemann cited. The unique ♀ specimen clearly exhibits plumage characters and measurements perfectly intermediate between those of the two parent species (Frith and Frith 1996*b*).

Lawes' Parotia *Parotia l. lawesii* ×
Blue Bird of Paradise *Paradisaea rudolphi margaritae* Plate 14
Schodde's Bird of Paradise of Frith and Frith (1996*a*), *Records of the Australian Museum* **48**: 111–116. Only one ad ♀ specimen known (AM O.40100) from Trepikama, Baiyer Valley, PNG collected at *c.* 1616 m..
Measurements (*n* = 1): wing, 146; tail, 93; bill; 37; tarsus, 40. Female wing length is similar in both the putative parents and that of the unique hybrid is smaller than the average of both, while in tail length it is almost exactly that of the shortest *P. l. lawesii* and longest *P. r. margaritae* individuals (both 94). The hybrid's bill length is exactly intermediate between the mutually-exclusive lengths (31–36 and 38–42, respectively) of the putative parent taxa. Its tarsus length is the same as only the largest *P. r. margarita* individual and is slightly smaller than the smallest (44) *P. l. lawesii* individual (Frith and Frith 1997*b*). These figures support the suggested parentage of this hybrid.

The unique specimen was collected by R. Bulmer on 15 Feb 1956 and was thought to be a ♀ Lawes' Parotia until recently recognised to be the result of the above hybrid combination by Richard Schodde (in Christidis and Schodde 1993). The specimen clearly exhibits plumage characters and measurements perfectly intermediate between those of the putative parent species (Frith and Frith 1996*a*).

Magnificent Riflebird *Ptiloris magnificus intercedens* ×
Superb Bird of Paradise *Lophorina superba minor* Plate 15
'*Paryphephorus (Craspedophora) duivenbodei*' of Meyer 1890, *Ibis* p. 420. Type: ZMB C9935 (12990), NW NG. Also Duivenbode's Riflebird. Three ad ♂ specimens are recorded (AMNH, BMNH and ex SMT—the latter type specimen being World War II destroyed), from Deva Deva and Foula which are inland from Yule I., PNG.
Measurements: wing, (*n* = 3) 161–167 (164); tail, (*n* = 2) 94–96 (95); tail centrals, (*n* = 3) 96–98 (97); bill, (*n* = 3) 40–42 (41); tarsus, (*n* = 3) 36–38 (37). Wing, tail, and bill length of the hybrids are, tellingly, near perfectly intermediate between the mutually-exclusive measurements of the putative parent taxa. The lengths of the hybrids' central rectrices are those of average-sized *P. m. intercedens* and their tarsus lengths are those of smaller individuals of this taxon. Thus the sizes of the hybrids represent strong evidence for the results of a crossing of the putative parent taxa.

This hybrid was diagnosed by Meise (1929). It is one of eight forms that Fuller (1979) considered unlikely to be hybrid between the proposed parent genera, noting that the specimens do not show incontrovertible characters of both putative parents and that Gilliard (1969) maintained the latter to be altitudinally isolated (by a mere *c.* 60 m). In fact while the two species are usually isolated by *c.* 800 m of altitude they do in some sites overlap elevationally by as much as 300 m (Beehler *et al.* 1986; Pruett-Jones and Pruett-Jones 1986).

The bill of this form is intermediate between those of the putative parents. Its crown feathering and central pair of tail feathers are as in Magnificent Riflebirds. While there is no real Superb Bird-like mantle cape on existing specimens the mantle feathers are distinctly longer than in the Magnificent Riflebird and are thus intermediate between the putative parent species in form (and colour). More importantly, the Dresden specimen (destroyed during World War II) was painted by J. G. Keulemans (Meyer 1890) and clearly possessed the well-developed nape cape of the Superb Bird. Puzzlement that one specimen should have a cape while two have not has been expressed by Fuller (1995) and we share it.

The hybrid's breast-shield is also intermediate in shape and extent. It does not extend onto the throat and is Superb Bird-like in general shape but lacks the more elongated outer feathers of the latter species while being Magnificent Riflebird-like in colour with black centres to the central feathers. The breast-shield is bordered below with a narrow black line and then an iridescent green one as in the Magnificent Riflebird. This hybrid's underparts are like those of the latter species but are overall slightly darker and with less green dorsally and less obvious 'claret-red' below. Its filamentous flank plumes are also 'intermediate' in that they are typical of the Magnificent Riflebird but lack the longer and more wiry feathers that typically extend beyond the former plumes. This description is contrary to Fuller's initial (1979) view of its characters which he subsequently modified to concede 'perhaps ... is a hybrid after all', particularly as two specimens are from within *c.* 160 km of Port Moresby, capital of PNG (Fuller 1995) where bird collecting was rife. These hybrids vary in central tail feather length. We have no doubt it is a hybrid produced from the above parent taxa.

Twelve-wired Bird of Paradise *Seleucidis melanoleuca* ×
Lesser Bird of Paradise *Paradisaea minor*
Plate 15

'*Paradisea mirabilis*' of Reichenow 1901, *Ornithologische Monatsberichte* **9**, 185. '*Janthothorax mirabilis*' of Rothschild 1903, *Bulletin of the British Ornithologists' Club* **13**, 31. Type: AMNH 679100, near Kaiser-Wilhelmshafen, German NG. Also known as the Wonderful Bird of Paradise. Five ad ♂ specimens are known (4 in AMNH, 1 in BZM), from Stroom gebied, W. Wasami, inland of Sorong, Vogelkop, N NG (1), somewhere in IJ (1) and unknown localities (3).

Measurements: wing, ($n = 5$) 187–195 (191); tail, ($n = 5$) 117–123 (120); tail centrals, ($n = 4$) 148–167 (158); plume, ($n = 4$) 155–250 (214); bill, ($n = 4$) 48–60 (54); tarsus, ($n = 3$) 43–44 (43). As we cannot be sure of the subspecies involved we use measurements for the entire species' populations. The hybrids' wings are slightly longer than the largest (185) Twelve-wired Bird, being the average size of Lesser Birds, while their tails are those of smaller Lesser Birds and their tail centrals far longer than in the former but far shorter than in the latter. Average hybrid bill length is exactly intermediate between the [mutually-exclusive] ones of the putative parent species. The tarsus is as long as that of a large Twelve-wired Bird or a small Lesser Bird. These figures are thus highly compatible with this hybrid combination.

This is another form that Reichenow (1901) thought to be a hybrid but because of its distinctive appearance he decided to describe it as a new species. It is variable in coloration and tail length. The fine iridescent head feathering is structurally like that of *Paradisaea* species. The mustard yellow feathering overlaying the lower back is clearly an expression of the Lesser Bird's genes. The mantle colour is variable, with two AMNH specimens showing the extremes from an almost entirely iridescent deep bluish purple one (as illustrated) to one that is mostly brown. Significantly the legs and feet of the specimens are distinctly pale cinnamon, like 123A of Smithe (1975), but not as pale as in Twelve-wired Bird specimens. The fine iridescent edging to the extreme edge breast feathers is emerald green like the Twelve-wired Bird.

Fuller (1979) found the generally-acknowledged putative parents of this hybrid easier to accept than many others but subsequently suggested that the Magnificent Riflebird might substitute for the Twelve-wired bird as the mate of the Lesser Bird parent. We can see no justification for the suggestion that this is the same cross as the following one. The hybrid is clearly in full adult plumage (*contra* Fuller 1995).

Magnificent Riflebird *Ptiloris m. magnificus* ×
Lesser Bird of Paradise *Paradisaea m. minor*
Plate 15

'*Janthothorax bensbachi*' of Büttikofer 1894, *Notes from the Leyden Museum* **16**, 163. Type: RMNH 1, Arfak Mts, IJ. Also known as Bensbach's Bird of Paradise and Bensbach's Riflebird. Only the one ad ♂ specimen is known.

Measurements ($n = 1$): wing, 201; tail, 132; tail centrals, 224; bill, 50; tarsus, 42. The hybrid's tarsus is just fractionally longer than that of both putative parents. Its tail equals that of the largest nominate Lesser Bird, and its tail centrals are considerably shorter than the shortest (420) ones of that putative parent. In bill length it is perfectly intermediate between the mean bill lengths of the putative parents while its tarsus length is that of the largest nominate Magnificent Riflebird and near-smallest Lesser Bird. These measurements strongly support the case for the putative parentage of this hybrid.

Fuller (1979) found the supposed hybrid characters of this form unconvincing and thus made the suggestion that this unique specimen might represent a 'lost—and possibly extinct—species' and the observation that the specimen 'does not seem quite mature' without explanation (Fuller 1995). Fuller also proposed that a parotia might just as well have mated with a Lesser Bird to produce this form. If this were so the progeny's bill would have been far shorter and, in view of the appearance of other hybrids involving parotias, some sign of a parotia-type breast-shield might also be expected.

The hybrid's primary and broad secondary shape and the physical feel of the wings and tail are strongly reminiscent of those of the Magnificent Riflebird. Its flank plumes are clearly of the *Paradisaea* type but their tips are, significantly, intermediate in form between those of the Lesser Bird and the Magnificent Riflebird. The shape and iridescence of the central pair of tail feathers also suggest a cross between a *Paradisaea* and the Magnificent Riflebird. Moreover the feather bases of the Magnificent Riflebird-like mantle have the conspicuous, pale, concealed centres typical of Lesser Birds. While most characters of this bird lead us to agree with Stresemann's original diagnosis of putative parents, we note that the few pale-tipped feathers on the extreme upper side of the breast and the patchily pale-pigmented legs and feet are curiously suggestive of the Twelve-wired Bird.

Twelve-wired Bird of Paradise *Seleucidis melanoleuca* ×
Magnificent Riflebird *Ptiloris magnificus*
Plate 15

'*Craspedophora mantoui*' of Oustalet 1891, *Le Naturaliste* **13**, 260. Type: MHN 11 932, NG. '*Heteroptilorhis mantoui*' of Sharpe 1898, *Monograph of the Paradisaeidae*. '*Craspedophora bruyni*' of Büttikofer 1895, *Notes from the Leyden Museum* **16**, 161. Also known as Mantou's Riflebird and Bruijn's Riflebird. At least 12 ad ♂ specimens are known (AMNH, BMNH, MHN, RMNH, ZMB, Singapore), from unknown localities save one RMNH specimen from the Arfak Mts, IJ.

Measurements: wing, ($n = 10$) 187–196 (192); tail, ($n = 10$) 85–99 (94); tail centrals, ($n = 8$) 88–104 (97); plume length, ($n = 9$) 50–125 (92); bill, ($n = 10$) 65–70 (68); tarsus, ($n = 5$) 41–45 (43). As the putative parent subspecies concerned are not known, but for one hybrid specimen, we use total species samples for measurement comparisons. The hybrids' wing lengths are longer than those of Twelve-wired Birds and are the lengths of larger Magnificent Riflebirds. Their tail lengths are typical of larger individuals of the former and smaller of the latter species, while the central pair of rectrices are longer than in any of the former and equal the average length of the latter. Significantly, however, their bills are longer than all Magnificent Riflebird's bills and are the length of that of an average-sized Twelve-wired Bird. Tarsus length in the hybrids varies from equal to the average size of both putative parent species to (oddly) slightly larger than the largest individuals of both.

The bills of these hybrids are slightly curved, Twelve-wired Bird-like, but are clearly intermediate between those of the putative parents. The throat- and breast-shield typical of the Magnificent Riflebird has lost its discrete shape and is absent from the throat in this hybrid combination and the iridescent feathers have become more rounded and 'frilled' like some manucode feathers. The upper outer edges of the breast-shield have Twelve-wired Bird-like, but smaller, semi-circular feathers tipped with iridescent blue, washed pinkish purple. Flank plumes are longer than in Magnificent Riflebirds with none of the upward-curving wires of the Twelve-wired Bird (but see below). As noted by Stresemann (1930a), the hybrid legs and feet are intermediate in tone between the pink and black ones of the putative parents, respectively.

Sharpe (1892: xxiv), writing of M. Suchettet's suggestion that this form is a hybrid between the Magnificent Riflebird and the Twelve-wired Bird, noted that 'it is certainly one of the most extraordinary propositions ever conceived in the history of ornithology'. Interestingly, Stresemann (1930a) expressed no surprise at this combination, noting the putative parent species to be 'very closely related' and the riflebird to exhibit the upwardly-curving flank plume wires of (and unique to) the Twelve-wired Bird in a rudimentary or relict form. Of particular note is that ♀-plumaged Magnificent Riflebirds have been seen attending courting ad ♂ Twelve-wired Birds (Coates 1990), and that display sites of these two species have been observed to be situated close to each other (BB). There is general agreement that this is the hybrid proposed.

A painting in Levaillant (1801–6) by Jacques Barraband, who was typically meticulously accurate, shows what is clearly a Twelve-wired Bird with pale flank plumes but black (not yellow) underparts (see Plate 1). The ♀-plumage-like brown wing of this bird suggests it is subad. If such a specimen did exist the possibility of it being of the present hybrid combination must be admitted, the black underparts presumably expressing its Magnificent Riflebird parent.

Magnificent Bird of Paradise *Cicinnurus m. magnificus* ×
Lesser Bird of Paradise *Paradisaea m. minor*
Plate 15

'*Neoparadisea ruysi*' of van Oort 1906, *Notes of the Leyden Museum* **28**, 129. Type: RMNH 480, near Warsembo, W coast of Geelvink Bay. Also known as Ruys' Bird of Paradise. Only the one ad ♂ specimen is known.

Measurements ($n = 1$): wing, 146; tail centrals, 360; bill, 36; tarsus, 38. In all of these measurements the unique hybrid specimen is near perfectly intermediate between the mean size measurements of the two putative parent taxa, thus lending the strongest support for this hybrid combination.

This is another form Fuller initially (1979) considered unlikely to represent a hybrid between the proposed genera; but he subsequently softened his view to concede that 'Hybrids between them could, therefore, turn up in any of the hundreds of square miles they jointly occupy' (Fuller 1995). He stated the unique specimen 'may not be' (Fuller 1979) and then (1995) 'is not that of' a fully-mature individual. We find this interpretation of the specimen doubtful. We consider it to be exhibiting the plumage of a fully-ad ♂. It might conceivably have once worn longer Lesser Bird type flank plumes that were moulted at the time this individual was killed.

We firmly believe the present specimen to be a hybrid in view of its ornithologically well-collected area of origin, its structure, and its appearance. Its obvious *Paradisaea* characters (central tail wires, remnant flank plumes, iridescent facial feathers restricted to lores, chin, and throat, and its wing pigmentation) must indicate the Lesser Bird as one putative parent if only on geographical grounds. Other characters can clearly lead only to the conclusion that the second parent was indeed the Magnificent Bird. The latter has a most distinctive square-backed and flat-crowned skull clearly discernible in the present hybrid specimen. Moreover, the upper surface of the feather vanes of the central tail wires are strongly iridescent blue at their broader bases and iridescent green along their entire narrow length, quite unlike those of the Lesser Bird but very much like those of the Magnificent Bird.

The glossy green underparts typical of the Magnificent Bird have simply become deep purplish blue in the hybrid. Its belly shows the influence of the deep maroon-like colour typical of the Magnificent Bird and its much-reduced flank plumes are predominantly influenced by this colour while showing the slightest influence of the Lesser Bird at their bases where a dirty yellowish intrudes. A lovely subtle, violet purple, glossy sheen on the wings of this hybrid bird in some lights is clearly like that also visible on the Magnificent Bird. In addition, the hybrid's crown feathers are clearly not typical of those of Lesser Birds, but are undeniably strongly influenced by the unique stubbly structure of those found in the Magnificent Bird. The Magnificent Bird also typically exhibits an area of contrastingly-dark feathering curving immediately over its eyes and a visibly-bare area of skin directly behind them—and these characters are present in the hybrid.

Presumed hybrid birds of paradise for which reasonable grounds for doubt remain

In strict scientific terms some uncertainty will continue to be attached to the majority of the above hybrid combinations (are they really hybrids and if so are the suggested putative parents correctly assigned?) until such time as definitive molecular evidence is produced and assessed. There is little if any doubt, however, about the vast majority of the intrageneric hybrids which are generally accepted as such. Of the intergeneric hybrids only the following give some cause for reservation but on balance are likely to be the product of the parent crosses proposed and discussed in the preceding section: Long-tailed Paradigalla × Western Parotia ('*Loborhamphus ptilorhis*'), 1 specimen; Long-tailed Paradigalla × Superb Bird of Paradise ('*Loborhamphus nobilis*'), 3 specimens; Long-tailed Paradigalla × Black Sicklebill ('*Pseudastrapia lobata*'), 1 specimen; Black Sicklebill × Arfak Astrapia ('*Epimachus ellioti*'), 2 specimens and a possible third in Rome; and to a lesser extent the Magnificent Riflebird × Lesser Bird of Paradise ('*Janthothorax bensbachi*'), 1 specimen; and Black Sicklebill × Superb Bird of Paradise (*Epimachus fastuosus atratus* × *Lophorina superba feminina*), 1 specimen.

It is noteworthy that three of these four more doubtful proposed hybrid combinations involve a paradigalla. These three hybrids have conspicuous gapes, not as well developed and extensive as in paradigallas, attributed to the influence of paradigalla genes. On balance the involvement of paradigallas in these hybrids seems more likely than not, but their unspecialised plumage makes their influence in the hybrids' plumage difficult to discern clearly. It is of course possible that the conspicuous gapes in all three of these hybrid forms do not result from the influence of paradigalla genes but merely represent a novel character resulting from hybridisation. This would seem a remarkable coincidence, however, given the biometrics and some other characters being in agreement with the possible involvement of paradigallas. Given that one might expect paradigalla wattles to be reduced in their hybrid progeny, and that the species suggested as second putative parents in all three combinations with Long-tailed Paradigallas are sympatric with the latter geographically-and altitudinally-restricted species, the case for paradigalla involvement is as good as that for any other in all three cases. Nevertheless considerable room for doubt remains.

While one or two undiscovered distinct populations of birds of paradise may remain to be discovered and described, we believe it is unlikely that any would turn out to be from among the list we have determined to be hybrids above. We simply do not believe there are any 'lost' (as distinct from yet-to-be discovered) birds of paradise. That being said, we do believe the hybridisation phenomenon is worthy of much additional exploration, especially in the few instances where hybrids form a portion of a population.

Appendix 2

Exploration of Australasia and the study of birds of paradise—an annotated list

This chronology of exploration is an elaboration of the compilation of that of Melanesia provided by Gilliard (1969: Appendix 2). Although based on Gilliard (1969), we have modified the regions treated and added eastern Australia. We also have included some important field studies of birds of paradise that did not involve the collection of specimens. Citation of resulting publications are given, but only in instances where they have been previously cited in this work. Thus publications merely listing specimens collected (i.e. no biology), no matter how substantial, are not cited below. We include only areas that support populations of birds of paradise (see map below).

It is possible that some names given as the collector (initials of some unknown to us) of specimens may prove to be those of a cabinet collector that obtained the material subsequent to collection. Dates given are those periods when birds of paradise were collected and these may not necessarily be the total period a collector was in any particular location or area. It is possible that in a few cases the date may prove to be other than the collecting date if the latter was not in fact recorded on the specimen and we failed to note this. In some cases we could not determine the collector or precise dates and destination of specimens. Where only the odd bird of paradise was collected, we indicate which species but we do not list all bird of paradise species obtained by major collections.

522 Appendix 2

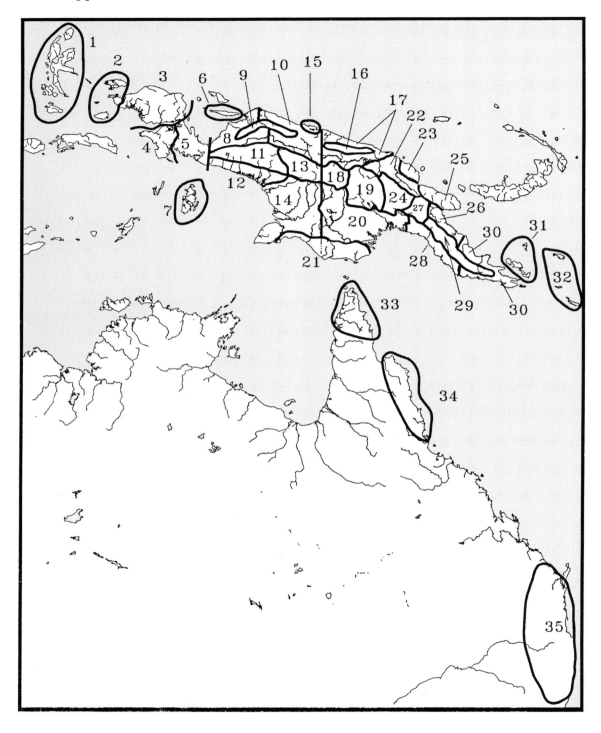

Appendix 2

Map showing the 35 subregions of the world range of the birds of paradise that are used in the following chronology of exploration:

1: Northern Moluccan Islands (Maluku Utara—Halmahera, Bacan, and Obi Is)
2: Western Papuan (Rajah Ampat) Islands—Misool, Waigeo, Batanta, Salawati, and Kofiau Is.
3: Vogelkop (the Bird's Head or Cendrawasih Peninsula)
4: Bomberai and Onin Peninsulas (includes Fakfak and Kumawa Mts)
5: The 'Bird's Neck' Region (including the Wandammen Peninsula and Arguni, Triton, and Etna Bay regions)
6: Yapen Island, Geelvink (Cendrawasih) Bay
7: Aru Islands
8: East coast of Geelvink Bay and Western Meervlakte lowlands (including Mamberamo, Rouffaer, and van Daalen R.)
9: Van Rees and Foya (Gauttier) Ranges of IJ
10: Northern lowlands and Eastern Meervlakte E to Humboldt Bay, IJ
11: Weyland and Nassau Ranges—Western Cordillera of IJ
12: SW Lowlands of IJ (Kapare R, E to the Balim/Lorentz R)
13: Oranje and Star Ranges, IJ (E of the Balim to PNG border: Peg. Jayawijaya)
14: Eastern lowlands of Southern IJ (east of the Balim/Lorentz R)
15: Cyclops Mts, IJ
16: North Coastal Ranges of PNG (Bewani, Torricelli Mountains, and Prince Alexander Ranges)
17: Western and northern Sepik lowlands (including main course of the Sepik and lowlands and foothills west of Chambri Lake), PNG
18: Dap, Hindenburg, and Victor Emanuel Ranges—westernmost high ranges of PNG (east to the Strickland R.)
19: Muller, Hagen, Giluwe, and Central Ranges (west-central highlands of PNG, including Mt Bosavi)
20: Fly platform and Gulf lowlands, PNG
21: Trans-Fly Savannas (Merauke-Bensbach-Oriomo Region), IJ & PNG
22: East Sepik and Ramu lowlands and hills, PNG
23: Adelbert Range and vicinity, PNG
24: Schrader, Bismarck, and Kubor Ranges, Sepik-Wahgi Divide, Jimi R., Wahgi Basin, PNG
25: Huon Peninsula, PNG (including Finisterre, Saruwaged, and Cromwell Ranges)
26: Markham and Watut and Waria drainages, and the Herzog, Kuper, and Bowutu Ranges, PNG
27: Eastern Highlands of PNG (Mt. Karimui, Crater Mt., Mt. Michael, Kratke Ra)
28: Southern Watershed, Papuan Peninsular hills and lowlands
29: Chapman, Wharton, and Owen Stanley Ranges, PNG
30: Southeasternmost lowlands and hills of PNG
31: D'Entrecasteaux and Trobriand Islands, PNG
32: Louisiade Archipelago and small islands of Milne Bay, PNG
33: Cape York Peninsula, Queensland, Australia
34: Cairns/Daintree/Atherton Wet Tropics, N. Queensland, Australia; all involving *P. victoriae* only.
35: Australian Subtropics of S. Queensland (Qld) and New South Wales (NSW); all involving *P. paradiseus* only.

1: Northern Moluccan islands (Maluku Utara—Halmahera, Bacan, and Obi Is)

1858	(Jan–Sept and Oct–Apr) A. R. Wallace; Halmahera and Bacan I. respectively (BMNH).
1858	Charles Allen for A. R. Wallace; Morotai and Halmahera Is (BMNH, MHN).
1860	(late in year) A. R. Wallace; Halmahera I. (BMNH).
1861	(Feb) H. A. Bernstein; Bacan I., Gilolo [= Halmahera I.], (16 Nov) (RMNH, MHN).
1862	(25 July and 17 Aug) H. A. Bernstein; Halmahera and Obi I. respectively (RMNH).
1873	(5 May) A. B. Meyer; Halmahera I., *Lycocorax* (SMT).
1873	Verreaux; Bacan I., *Semioptera* (MHN, USNM).
1873 and 74	(June–July and Dec respectively) A. A. Bruijn; Halmahera I. (RMNH).
1876	(Dec and 1878) Raffray; Halmahera I., *Lycocorax* (MHN).
1883	(Dec) A. A. Bruijn collectors; Halmahera I., *Lycocorax* (MHN, USNM).
1883	(12 Oct and Nov–Dec) F. H. H. Guillemard and Lt R. Powell; Obi Major and Bacan Is, *Lycocorax* and *Semioptera* (BMNH, MCSN).
1890	(Mar) J. F. G.; Halmahera I., *Semioptera* (ZSM).
1891	van Heurn; Halmahera I., *Semioptera* (ZSM).
1891	(May) E. Warburg; Obi I., *Lycocorax* (ZSM).
1892–3	(1 Dec–11 Jan) Dr C. Platen; Bacan I., *Semioptera*, *Lycocorax* (MCZ).
1894	(10 Jan–8 Apr) W. Kukenthal; Halmahera I. (BMNH, ZSM).
1894	(17 Apr–May) W. Kukenthal; Bacan I. (BMNH).
1894	(1 Apr) Ruppell; Halmahera I., *Semioptera* (MV).
1895	(15 May) R. Kukenthal; Obi I., *Lycocorax* (ZSM).
1895	(10 Jan) K. Schluter; Halmahera I., *Semioptera* (SMT).
1896	(Nov) W. Doherty; Halmahera I., Patani region, *Lycocorax*, *Semioptera* (BMNH).
1897	(Aug–Sept) W. Doherty; Bacan and Obi Major Is, *Semioptera*, *Lycocorax* (BMNH, AMNH)
1898	(Apr) Vorderman; Bacan I., *Semioptera* (RMNH).
1899	(18 June–2 July) H. Kuhn; Bacan I., *Semioptera* (BMNH).
1902	(Feb–Sept) Johannes Waterstradt; Bacan, Obi, Halmahera Is, *Lycocorax*, *Semioptera* (BMNH).
1904	(Aug) Schreyer; Obi I., *Lycocorax* (ZSM).
1907	(Feb) T. Barbour; Halmahera I., *Semioptera* (MCZ).
1914	(30 June–2 July) Tarip; Obi I., *Lycocorax* (BZM).
1926	(May and June) W. Goodfellow; SE Halmahera I., *Semioptera* (BMNH).
192?	V. Martom; Bacan I., *Lycocorax* (ZMB).
1928–9	(Dec–Jan) F. Shaw Mayer; Halmahera I., Patani Village, *Lycocorax* (BMNH).
1931	(Apr–June) G. Heinrich; Halmahera I., *Lycocorax*, *Semioptera* (ZMB).
1931	(June–Aug) G. Heinrich; Bacan I., *Lycocorax*, *Semioptera* (ZMB).

1931	(July) W. J. C. Frost; Halmahera I., (BMNH, ROM, PMNH).
1933	(Dec) S. Egawa; Halmahera I., *Lycocorax* (YIO).
1936–8	(Sept, Mar, and June in respective years) H. Abe; Halmahera I., *Lycocorax*, *Semioptera* (YIO).
1936–9	G. Heinrich; Halmahera and Bacan Is, nest and egg of *Lycocorax* (ZMB).
1937	(15 Nov) G. A. Z. DeHaan; Obi I., *Lycocorax* (BZM).
1938	(16–23 Mar) G. A. Z. DeHaan; Morotai I. (BZM).
1938	(28 Apr–29 May) G. A. Z. DeHaan; Kobe, Weda, Ake Loeing, Halmahera I., *Semioptera* (BZM).
1940	(14 Nov) H. Barstra; Obi I., *Lycocorax* (BZM).
1941	(28 Aug) H. Barstra; Obi I., *Lycocorax* (BZM).
1944	(26 Oct) R. H. Knoll; Morotai, *Lycocorax* (CORNELL, ITHACA).
1949	(7 June and 15 Dec) G. A. Z. DeHaan; Halmahera I., *Lycocorax* (BZM).
1950–1	(25 Oct and 14 Sept, May and Oct) G. A. Z. DeHahn; Halmahera I., *Lycocorax*, *Semioptera* (BZM, MCZ).
1952	(6–19 Jan) G. A. L. DeHahn; Ake Wosin, Halmahera I., *Semioptera* (BZM).
1953	(14 June) A. M. R. Wegner; Bira, Obi I., *Semioptera* (BZM).
1953	(June, Aug–Oct) Saan; Obi I., *Lycocorax* (BZM).
1953	(July) Saan; Bacan I., *Semioptera* (BZM).
1953	(July–Nov) G. DeHahn; Bacan I., Aug–Oct and Obi I., July and Nov (BZM)
1953	(7 July) A. M. R. Wegner; Labuhon, Babang, Wajana, Bacan I., *Semioptera* (BZM).
1954	(Sept–Oct) Dr and Mrs S. Dillon Ripley; W Moluccas, Halmahera and Bacan Is (PMNH, BZM), (Ripley 1959).
1981	(Mar–May) P. M. Taylor; Obi I., *Lycocorax* (USNM).
1982	(July–Aug and Nov) Y. Mamou and R. Tatu; Obi I., *Lycocorax* (USNM).
1983	(Aug) K. D. Bishop; discovered and observed *Semioptera* lek on N Halmahera (Bishop 1984).
1984–90	K. D. Bishop; opportunistic and intermittent annual observations of *Lycocorax* and *Semioptera* (Bishop 1992; Coates and Bishop 1997).
1985	(25 Mar) P. M. Taylor; Halmahera I., *Semioptera* (USNM).
1985	(Apr and June) K. D. Bishop; *Semioptera* studied in field (Bishop 1992).
1994	The University of Bristol Expedition to Indonesia involved avian population surveys including *Lycocorax* and *Semioptera* on Halmahera I. (Mackinnon *et al.* 1995; Frith and Poulsen in press).
1994–5	(June–Aug and Feb–May respectively) BirdLife International avian population surveys including *Lycocorax* and *Semioptera* on Halmahera I. (Anon 1995).

2: Western Papuan (Rajah Ampat) Islands — Misool, Waigeo, Batanta, Salawati, and Kofiau Is.

1818–9	(16 Dec–6 Jan) J. R. C. Quoy and J. P. Gaimard; Waigeo I. (MHN).
1823	(6–16 Sept) R. P. Lesson and P. Garnot; Waigeo I., N coast (MHN).
1860	A. R. Wallace and C. Allen; (June and July) Misool I. and (29 June–2 Oct) Waigeo I. (BMNH).
1860	C. E. H. von Rosenberg; Misool and Waigeo Is (BMNH).
1863	H. A. Bernstein; Kofiau I. and (1 Mar–6 May) Waigeo I. (RMNH, MHN).
1865	(8 Apr) H. A. Bernstein; Batanta I.
1867	D. Hokum; Kofiau I. (25 May) Verreaux; Misool I., *C. regius* (MNH).
1868	(11 June and 21 Aug) Misool I. (BMNH).
1869	(27–28 May and 2 July) Batanta I., *P. rubra* (RMNH).
1870	(3 Feb) Salawati I., *S. melanoleuca* (RMNH).
1873	Verreaux; Misool I., *M. atra*; Waigeo I., *P. rubra* (MNH).
1875 and 76	(6–14 Mar 75; Mar 76) O. Beccari; Waigeo I. (MCSN).
1875	O. Beccari; Salawati I. (June) and Kofiau I. (July), (MCSN, ZMB).
1875	(25 June) A. A. Bruijn; Waigeo I. (AMNH).
1875	(May and Nov) collectors for A. A. Bruijn; Salawati and Kofiau Is. (BMNH).
1877	(June) A. A. Bruijn; Batanta I. (MNH, AMNH).
1878	Raffray; Waigeo, Salawati and Batanta Is. (MHN).
1878	(Apr) M. Laglaize; Batanta and Waigeo Is. (MHN).
1879	(July) collectors for A. A. Bruijn; Batanta and Waigeo Is. (BMNH, MHN).
1883	J. J. Labillardiere; Waigeo I. (BMNH).
1883	(2 Sept–27 Oct) R. Powell; Batanta I. (AMNH).
1883	(16–31 Oct, 15 Nov) F. H. H. Guillemard and Lt R. Powell; Waigeo I. (BMNH, MCSN).
1883	R. Powell; Salawati I. (Nov) and Misool I. (Nov–Dec), Samatee (BMNH, AMNH).
1883 and 84	(14 Nov, 21 Dec, 6–13 Jan, Feb) C. Platen; Waigeo I. (ZMB, ZSM).
1892	Schlufer; Waigeo I. (SNHM).
1895	Dr Ad. Struball; Waigeo I., *C. respublica* (ZSM).
1897	Doherty; Waigeo I., *P. rubra* (AMNH).
1899	Th. Hallmann; Waigeo I., *C. respublica* (ZSM).
1900	(Jan–Feb) Heinrich Kühn; Misool I. (BMNH).
1901	R. de Neufrille; Waigeo I. (ZSM).
1902	(Nov) F. H. H. Guillemard; Misool and Salawati Is (MCSN, BMNH).
1902–3	(26 Nov–27 Jan-Apr) J. Waterstradt; Waigeo and Gebe Is (BMNH, ROM, MCZ).
1906	(Nov) W. J. C. Frost; Salawati I. (ROM).
1906	(Nov) W. Goodfellow; Waigeo I., Sassas (BMNH).
1909–10	(17 Dec–27 Jan) L. F. de Beaufort; Waigeo I., Saonek region (observations only).

1910	(2–5 July) van Dedem; Misool I., *C. regius*, *P. minor* (RMNH).	1949	(31 May, June, Oct) S. Bergman; Batanta I., *P. rubra* (BZM), (Gyldenstolpe 1955*b*)
1926	(July–Nov) W. J. C. Frost; Batanta and Salawati Is (ROM, BMNH).	1954–5	(Nov and Jan–Feb) Dr and Mrs S. Dillon Ripley; Misool I. (PMNH).
1929	(6–8 June) W. A. Weber; Waigeo I. (CNHM).	1955	(25 Apr–9 May) J. Khakiaj (for S. Dillon Ripley); Kofiau I. (PMNH).
1930	(Feb and Apr) W. J. C. Frost; Salawati I. (ROM).		
1930	(May and Mar) J. C. Frost; Batanta and Salawati Is (BMNH, AMNH).	1955	(1 Sept) J. Khakiaj (for S. Dillon Ripley); Ajoe I. N of Waigeo I. (PMNH).
1931	(15–21 May) E. Mayr; Waigeo I., *M. atra* (AMNH).	1955	(17–20 Sept) S. D. Ripley; Waigeo I. (PMNH).
1931	(6 May–16 June) Mr and Mrs G. Stein; Waigeo I. (ZMB, BMNH).	1964	(15 June–8 July) E. T. Gilliard and S. Somadikarta with taxidermists Toha and Tojibun; Batanta I., Inland from Wailabet Village (AMNH, BZM), (Greenway 1966).
1934	(Apr–May) W. J. C. Frost; Batanta, Waigeo and Salawati Is (BMNH).		
1936	(25 Nov) E. Jacobson; Waigeo I., *P. rubra* (BZM).	1969	? (24 June) Powell; Misool I., *C. regius* (SAM).
1937	(19–13 Oct) S. Dillon Ripley; Schildpad Is, (Lophon, Kam Wa, Marian, and Jef Fa), (ANSP).	1983	(Oct) J. M. Diamond; Batanta and Salawati Is (Diamond 1985).
		1986	(Jan–Feb) J. M. Diamond and K. D. Bishop; observation survey of avifaunas on Salwati, Misool, Kofiau, Waigeo and Batanta Is.
1937	(Oct–Nov) S. Dillon Ripley; Misool I., vicinity of Tip Village on Kassim R. (ANSP).		
1938	S. Dillon Ripley; Salawati and Sagewin Is (ANSP).		
1938	S. Dillon Ripley; Batanta I., mostly from Mondok, near Yenanas, Aijem and Mt. Beheneo (ANSP).		

3: Vogelkop (the Bird's Head or Cendrawasih Peninsula)

1824	René P. Lesson; Dorey Bay (MHN).
1827	(25 Aug–6 Sept): J. R. C. Quoy and J. P. Gaimard; Dorey Bay (MHN).
1855	R. van Duivenbode accompanied by Prince Ali of Tidore; N foothills of Tamrau Mts via Mega Village.
1858	(11 Apr–29 July) A. R. Wallace; Dorey Bay (BMNH).

1938 — (12 Aug) Denison-Crockett Expedition; Batanta I., *C. respublica* (ANSP)

1938 — (Nov–Dec) J. Khakiaj; Waigeo I., Saonek, Nafer Baai, Mt. Lapon, Linsok (ANS), (Mayr and Meyer de Schauensee 1939*b*).

1948 — (14–15 Sept and 11–13 Oct) S. Bergman; Waigeo and Saonek Is (SMNH).

528 Appendix 2

1858	collectors for A. R. Wallace; Amberbak I. (BMNH).
1861	C. Allen; Sorong area; collecting for Wallace, Allen worked here for *c.* 1 month (BMNH).
1863–8	D. S. Hoedt; Vogelkop and W. Papuan Is (RMNH).
1858–70	C. E. H. von Rosenberg; Vogelkop (RMNH), (Rosenberg 1875).
1871–6	hunters for A. A. Bruijn; Manokwari region (Jan and Apr 1875), Sorong region (Mar and Apr 1875), Mansinam I. (May 1875), mts behind Andai (June 1875), Arfak Mts (Feb 1876), Warmendi, Arfak Mts (25 Jan 1876).
1872	Count L. M. D'Albertis and O. Beccari; W. Vogelkop, Amberbaki region and Arfak Mts (MCSN).
1873	(Mar–July) A. B. Meyer (SMT).
1874	(June) A. A. Bruijn; Arfak Mts, *E. fastuosus* (AMNH).
1874	(July and 10 Dec) W. H. Woelders; Hatam, Arfak Mts (RMNH).
1875	O. Beccari; Sorong region and (4–29 Feb) mts near extreme W of Vogelkop (MCSN).
1875	(Apr–July) O. Beccari; Arfak Mts (MCSN, ZMB, AM, AMNH).
1875	(Apr, May, Dec) A. A. Bruijn; Arfak Mts (BMNH, AMNH, CNHM).
1876	(2–5 Feb–Aug) Bruijn; Arfak Mts (RMNH, MHN).
1876	(8 Feb) Rosenberg; Arfak Mts (RMNH).
1876	(May) Leon Laglaize; Arfak Mts, Woipirboe (MHN, BMNH).
1877	Raffray; Dorey (Feb–Mar) and (Apr–Aug) Amberbaki region, near village of Memiaona (MHN).
1878	A. A. Bruijn; Arfak, *P. minor* (RMNH).
1883	(Nov) F. H. H. Guillemard; Manokwari (MCSN).
1883	(Nov) R. Powell; Manokwari (BMNH).
1884	(May) L. Laglaize; Kafou (MHN).
1887	(?) L. Laglaize; Arfak Mts (MHN).
1896 and 97	(Oct and June) W. Doherty; Manokwari (BMNH).
1899	(Jan) J. M. Dumas; Mt. Moari (BMNH, BZM).
1904	(June) W. Goodfellow; Sekar, N. W. New Guinea (BMNH).
1907	(11 Feb) T. Barbour; near Sorong, *S. melanoleuca* (MCZ).
1908	(Oct) A. E. Pratt; central Arfak Mts, *A. nigra* (ROM).
1909	(Apr) A. E. Pratt; central Arfak Mts, *M. chalybata* (ZMK).
1912	(June–July) K. Gjellerup and Feuilletau de Bruijn; Anggi Lakes (BZM).
1921	(14 Jan) Obdyn; Manokwari region (BZM).
1928	(June–Oct) F. Shaw Mayer; Arfak Mts (BMNH).
1928	Ernst Mayr; (7–14 Apr) Manokwari area and (14 Apr–4 July) Arfak Mts (BMNH), (Hartert 1930).
1929	(Feb) F. Shaw Mayer (AMNH).
1929	Ex-King (then Crown Prince) Leopold of Belgium; Sorong, Manokwari and Anggi Gita, Arfak Mts (BRUSSELS).
1930	(Feb–Mar) W. J. C. Frost; NW NG (BMNH).
1931	(Feb) G. Stein; Manokwari (ZMB).
1931	(July) W. J. C. Frost; NW NG, near Sorong (BMNH).

1934	(Apr–May) W. J. C. Frost; NW NG (BMNH).	1981	John Mackinnon; Tamrau Mts. Observations only.
1937	(15–21 November) S. Dillon Ripley; Manokwari region, Andai, Mansinam I. and Manokwari (ANSP).		
1937	(17 Nov) W. J. C. Frost; Sorong (AMNH).		
1938	S. Dillon Ripley; (22 Jan–7 Feb) Sorong region, coastal islands of Dom, Tsiof, Efman and the Ramoi, and Waron R. and (9 Feb–1 Apr) Tamrau Mts, also revisited by collectors for Ripley in May and Sept (ANSP).		
1939	(18 Mar) S. Bergman; Teminaboem, W Vogelkop (BZM).		
1942–4	Yoko-oji; Manokwari area (YIO).		
1944–5	(24 Sept and 20–21 June) H. N. Russell; Cape Sansoper (CNHM).		
1948–9	(July, Aug, Oct–May) S. Bergman, H. Sjoquist and two assistants from BZM; Sorong region (SMNH, BZM).		
1949	(July, Aug–Oct) S. Bergman; Anggi Lakes, Arfak Mts (SMNH, BZM).		
1961	R. Versteegh; (9 Sept) Kebar Valley and (26 Oct) Ransiki, Vogelkop (BZM), (Gyldenstolpe 1955b).		
1962	(Jan) R. Versteegh, collecting for L. and S. Quate; Kebar Valley (BPBM, AMNH).		
1963	(19 Jan–1 Mar) M. C. Thompson, L. P. Richards, P. Temple; Oransberi. (BPBM).		
1963	(Feb–Mar) A. Hoogerwerf; Manokwari area (BZM).		
1964	(15 July–15 Aug) E. T. Gilliard, P. Suparlan and taxidermists Toha and Tojibun; Tamrau Mts (AMNH), (Gilliard and LeCroy 1970).		

4: Bomberai and Onin Peninsulas (includes Fakfak and Kumawa Mts)

1860	C. E. H. von Rosenberg; Fakfak area (RMNH, BMNH), (Rosenberg 1875).
1896–7	(Dec–Feb) C. Schadler; Sekru (MCZ, RMNH).
1897	(Jan–Feb) W. Doherty; Kapaur (AMNH).
1904	(June) Goodfellow; 'Sekar' (possibly Segaar, Onin Penin), (BMNH).
1948–9	(Dec–Jan) S. Bergman; S coast of Bomberai Penin and Pulu Adi, off S tip of that penin (SMNH), (Gyldenstolpe 1955b).
1981	(Feb and Mar) J. M. Diamond; Fakfak Mts, Buff-tailed Sicklebill (Diamond 1985)
1983	(Aug and Sept) J. M. Diamond; Kumawa Mts (Diamond 1985).
1992	(Aug–Sept) D. Gibbs; Fakfak Mts. Observations of *Paradigalla* sp., Buff-tailed Sicklebill plus six other bird of paradise species (Gibbs 1994).
1995	A. Forsyth and W. Betts; Fakfak Mts. Observations only.

5: The 'Bird's Neck' Region (including the Wandammen Peninsula and Arguni, Triton, and Etna Bay regions)

1828	H. C. Macklot and S. Muller; Lobo, Triton Bay (RMNH).

530 Appendix 2

1873	(May–June) A. B. Meyer; Rubi area at least (SMT).
1896	(July and Aug) Capt. C. Webster; Triton and Etna Bays (BMNH, AMNH).
1896–1900	H. Kuhn; Etna Bay (RMNH, BMNH).
1897	(May–July) W. Doherty; Wandammen Penin and islands off it (BMNH, AMNH).
1904	(4–14 Dec) J. W. R. Koch; Etna Bay (BMNH, RMNH).
1928	(5 to *c.* 26 July) E. Mayr; Wandammen Mts, Mt. Wondiwoi (BMNH), (Hartert 1930).
1930	(Jan) W. J. C. Frost; near Oliphantsburg (AMNH, BMNH).
1930	(Sept) F. Shaw Mayer; Wandammen Bay and Mts (BMNH).
1939	(Aug and Nov) collectors for H. Boschma; Etna Bay (RMNH).
1949	(Aug and Sept) S. Bergman; Wandammen Penin (SMNH), (Gyldenstolpe 1955*b*).
1963	(10 Feb) M. C. Thompson; Oransbari (BPBM).
1983	(Oct) J. M. Diamond; Wandammen Mts (Diamond 1985).

6: Yapen Island, Geelvink (Cendrawasih) Bay

1869	(5 Apr–6 May) H. von Rosenberg (RMNH) (Rosenberg 1875).
1873	(8–29 Apr) A. B. Meyer (SMT).
1874	(Apr) A. A. Bruijn (AMNH).
1875	(4–14 Apr and 11–14 Dec) O. Beccari (MCSN).
1875 and 85	collectors for A. A. Bruijn (BMNH).
1878	(?) L. Laglaize (MHN).
1883	(Nov–Dec) F. H. H. Guillemard and R. Powell (MCSN, BMNH).
1897	(Apr–May) W. Doherty (BMNH, AMNH).
1926	(15 Apr) W. J. C. Frost, *P. minor* (ROM).
1931	(20 Feb–26 Mar) Mr and Mrs G. Stein (ZMB).
1962	(Oct–Nov) N. Wilson and L. P. Richards (BPBM).
1983	(Aug) J. M. Diamond (Diamond 1985).

7: Aru Islands

1857	(8 Jan–2 July) A. R. Wallace; Kobroor I. (BMNH).
1870–2	S. White with J. Cockerill and F. W. Andrews (disposition of specimens and journals unknown (? RMNH, ZMB).
1873	J. Cockerill and son.
1880	J. W. Egan *P. apoda* (AM).
1883	(6 Dec) R. Powell (AMNH).
1887	J. W. Egan (AM).
1893	(June) F. Ferdinand *C. regius* (NMW).
1894	C. G. Forest.
1896	(May–June) Capt. C. Webster; Dobo I. (BMNH).
1897	(Feb) W. Doherty; Dobo I. (BMNH).
1896–7, 1900	(Feb, May, Nov, Dec in 1897 and Aug–Sept 1900) H. Kuhn (BMNH).
1902	E. Lehreyer, *P. apoda* (SMT).
1904	(Feb–Apr) W. Goodfellow; Silbattabatta (BMNH).
1908	(Jan–May) Dr H. Merton; Dobo, Pulau Wamar and Kobroor, *M. atra, C. regius, P. apoda* (ZSM).

1908–9	W. J. C. Frost; live *P. apoda* captured and shipped to Little Tobago I. (BMNH), (see Ingram 1918).	1979	(Oct) J. M. Diamond, Danau Bira, field observations.
1909	(May) W. Stalker, BOU Expedition (BMNH).	1981	(Jan–Feb) J. M. Diamond, Danau Bira, Biri Village, field observations.
1914	(May) W. J. C. Frost; Golili (BMNH).	1986	(10–13 July) B. M. Beehler; Danau Bira.
1925	(Aug) W. J. C. Frost; (BMNH)	1990	(July) J. M. Diamond and K. D. Bishop; observation survey of avifaunas of Rouffaer and van Daalen R.
1927	(May) W. J. C. Frost; Wanumbai (ROM).		
1929	(Mar) W. J. C. Frost; Kobroor I. (BMNH).	1996	A. Mack; Danau Bira.
1930	W. J. C. Frost (destination of specimens unknown).		

9: Van Rees and Foya (Gauttier) Ranges of IJ

1935	(Jan–Feb) Soekarno Wakoea (BZM).
1937	(11 Aug) Pamong; Pradja daerah, *P. apoda* (BZM).
1956	K. August Heise; Dobo I. (Z1ZM).
1964	(25 July–2 Aug) M. Djajasasmita; Dobo I. (BZM).
1988	(Mar–Apr) J. M. Diamond and K. D. Bishop; all major islands. Observations only.

1979	(Oct) J. M. Diamond, Foya Mts.
1981	(Jan–Feb) J. M. Diamond, Foya Mts, *P. carolae*, *D. albertisi* (Diamond 1985).
1994	(Sept–Oct) J. M. Diamond and K. D. Bishop; observation survey of avifauna of Van Rees Mts.

8: East coast of Geelvink Bay and Western Meervlakte lowlands (including Mamberamo, Rouffaer, and van Daalen R.)

10: Northern lowlands and Eastern Meervlakte E to Humboldt Bay, IJ

1896–7	W. Doherty; Waropen, Mt. Kurudy (Oct and Apr–May), (BMNH).
1938	(Oct–Nov) J. P. K. van Eechoud; Mamberamo R. (RMNH, BZM).
1939	(July–Dec) Ch. F. van Krieken; Siriwo R. (Ta R. tributary), E coast Geelvink Bay (BZM).
1962	(Oct–Nov) N. Wilson and L. P. Richards; Nabire (BPBM).
1963	(17 Dec) Boeadi; Nabire (BZM).

1896–7	W. Doherty; Takar (Oct–Nov), Kapur (Dec–Jan), Sarmi (Jan–May) and along Witriwai R., W of Hollandia [= Jayapura], (BMNH).
1897	W. Doherty; Humboldt Bay, Sarmi (AMNH).
1902	(?) F. L. de Beaufort; Hollandia region (BMNH).
1903	(27 Apr) Geschenk; Humboldt Bay (BZM).
1903	(8 Apr–27 June) Dr H. A. Lorentz and F. L. de Beaufort; Lake Sentani and Humboldt Bay (RMNH).

532   Appendix 2

1906	(June–Sept.) W. Goodfellow; Humboldt Bay (BMNH).		R. Wollaston, C. H. B. Grant, G. Chester Shortridge; Nassau Ra, S watershed (BMNH), (Ogilvie-Grant 1915a).
1910	(Apr–May) L. Schultze; Tami, Mundung, and Humboldt Bay (ZMB).	1910	(12 Oct) G. Stein; Weyland Mts (ZMB)
1910	(7 Aug and 29 Dec) K. Gjellerup; Humboldt Bay (BZM).	1912–3	(Aug–Dec, Jan–Mar, Feb–May) BOU, A. F. R. Wollaston Expedition, A. F. R. Wollaston, C. Boden Kloss, Lieut. Van der Water; Nassau Ra, S watershed (BMNH), (Ogilvie-Grant 1915a).
1911	(Jan–Oct) K. Gjellurup; Mamberamo R., Sentani Lake, Humboldt Bay, and Tami R. (BZM).		
1914	(Apr) K. G.; 'Pionierbivak', Mamberamo R. (BZM).	1920–1	(Oct–Jan) Pratt brothers; Mt. Kunupi, Weyland Mts (Oct–Dec); (AMNH, ROM, PMNH, MCZ, ZSM).
1920–1	(Oct–Jan) Pratt Brothers; Wangger R. (Jan) (AMNH, ROM, PMNH, MCZ, ZSM)		
1920	(June–Dec). W. C. van Heurn; Mamberamo and Idenburg R. (ZMB, BZM).	1930	(June–July) F. Shaw Mayer; Weyland Mts, N watershed (BMNH), (Rothschild 1931).
1928	(4 Aug) Capt. C. Webster; Hollandia (CNHM).	1931	(Aug–Dec) Mr and Mrs G. Stein; Weyland Mts, N watershed (ZMB).
1928	(Aug–Oct) E. Mayr; Lake Sentani, Hollandia area (BMNH), (Hartert 1930).	1938	(Oct–Nov) J. P. K. van Eechoud; Wissel Lake region (RMNH, BZM), (Junge 1953).
1938–9	Third Archbold Expedition, R. Archbold, A. L. Rand, W. B. Richardson, L. J. Brass; Hollandia region (AMNH, BZM), (Rand 1942a).	1939	(Aug–Dec) H. Boschma. Wissel Lakes region, NG Expedition K.N.A.G. (RMNH).
1939	(July and Nov) J. P. K. van Eechoud; Pioneer Bivouac, Mamberamo R. (BZM).	1939	(Aug–Nov) C. F. van Krieken; Siriwo R. (BZM).
1945	(3 June) S. G. Jamett; Hollandia (CNHM).	1938–9	(May–May) Third Archbold Expedition, R. Archbold, A. L. Rand, W. B. Richardson, L. J. Brass; Snow Mts (Lake Habbema, Mt. Wilhelmina, Balim Valley); (AMNH, BZM), (Rand 1942a).
1945	(27 Feb–1 Mar) C. G. Sibley; Lake Sentani (MVZ).		
1959	(25–30 June) S. D. Ripley; Tami R., near Humboldt Bay (PMNH).		
		1955	(May) L. Pospisil; Wissel Lakes, *Pteridophora* (PMNH).
		1958	(Mar) S. Bergman; Kadubka, Swart Valley, *Lophorina* (SMNH).

11: Weyland and Nassau Ranges—Western Cordillera of IJ

1876	J. Umlauff, Charles Louis Mts, *P. apoda* (SMT).	1959	(Apr–Aug) R. G. Gardner, P. Matthiessen, M. Rockefeller; Balim Valley (MCZ), additional work in 1962.
1909–11	BOU Expedition, W. Goodfellow, W. Stalker, A. F.		

1960–1	Dr and Mrs S. Dillon Ripley; Nassau Ra, N watershed (Ilaga); (PMNH), (Ripley 1964).
1962	(July–Aug) N. Wilson and H. Clissold; Enarotali; Wissel Lakes (BPBM).
1963	(Mar) B. King (AMNH), field observations.
1963–4	(Dec–Feb) Boeadi, Tjendrawasih 1 Expedition; Nassau to Snow Mts, S watershed; Wissel Lakes and Baliem Valley, Titigina, Homejo, Beoga, Keniri, Komopa, Hitalipa, Nabire and Balim (BZM).
1964	(Jan) Boeadi; Homejo (= Wandai), (BZM).
1980	(25–30 March) B. M. Beehler; Lake Habbema, IJ.; *Macgregoria*, *A. splendidissima* (Beehler 1991a).
1981	B. M. Beehler and B. Finch; Tembagupura; *P. brevicauda*, *A. splendidissima*, *P. apoda* (field observations).
1990	(July) J. M. Diamond and K. D. Bishop; observation survey of avifauna of N slope of Nassau Mts.
1991	(Oct–Nov) M. Toha; Wamena area (BZM).
1994	(14 July) Boeadi; Kigwaki, Tembagapura, *M. pulchra* (BZM).
1997	(Jan–Feb) B. Van Balen; Tembagapura and mine area (surveys).

12: SW Lowlands of IJ (Kapare R. E to the Balim/Lorentz R.)

1909–11	BOU Expedition, W. Goodfellow, W. Stalker, A. F. R. Wollaston, C. H. B. Grant, G. Chester Shortridge; Nassau Ra., S watershed (BMNH).
1912–3	(Aug–Dec, Jan–Mar, Feb–May) BOU, A. F. R. Wollaston Expedition, A. F. R. Wollaston, C. Boden Kloss, Lieut. Van der Water; Nassau Ra., S watershed (BMNH), (Ogilvie-Grant 1915a).
1995	(Jan) B. M. Beehler; Timika, field observations.
1997	(Jan) B. Van Balen; Timika to Tembagapura, field surveys.

13: Oranje and Star Ranges, IJ (E of the Balim to PNG border: Peg. Jayawijaya)

1907 and 9–10	(May–Aug, Sept–Dec, Jan–Mar, respectively) Dr H. A. Lorentz, NG Exped. (RMNH).
1910–1	(July–Mar) A. S. Meek; Mt. Goliath (BMNH).
1911	(Mar) Geschenk; Eilanden R. *P. brevicauda* (BZM).
1912–3	(Jan–Mar) G. Versteeg; Snow Mts, S watershed (RMNH, MCZ), (Junge 1939).
1959	(Apr–Aug) J. J. Staats; Star Mts, Mt. Antares and Sibil Valley; and F. Scharff up to the summit region of Mt. Juliana (RMNH).
1961	L. W. Quate; Sibil Valley at Aliemkop, Star Mts (BPBM).
1976	(Feb–Mar and Oct–Dec) W. Schultz and W. Nelke; Malingdam area c. 120 km E of Wamena (ZIZM).
1980	J. M. Diamond and K. D. Bishop; Star Mts, IJ, field observations.
1987	(3–12 Apr) T. Flannery; Mt. Capella area, Star Mts (CSIRO).
1993	(Mar–Apr) J. M. Diamond and K. D. Bishop; observation survey of avifauna of E Star Mts.

14: Eastern lowlands of Southern IJ (east of the Balim/Lorentz R.)

1920 and 23	(May and 14–21 Sept), T. Jackson; Eilanden R. region, IJ (QM, MCV)
1983	(Sept–Dec) K. D. Bishop; observation avifaunal survey.

15: Cyclops Mts, IJ

1911	(Jan, June, and Oct) K. Gjellerup (BZM).
1928	(Aug–Oct) E. Mayr (BMNH), (Hartert 1930).
1938–9	Third Archbold Expedition, R. Archbold, A. L. Rand, W. B. Richardson, L. J. Brass (AMNH, BZM), (Rand 1942*a*).
1980	(March) B. M. Beehler, field observations.
1990	(July) J. M. Diamond and K. D. Bishop; field observations.

16: North Coastal Ranges of PNG (Bewani, Torricelli Mountains, and Prince Alexander Ranges)

1910	Dr Schlaginhaufen; foothills of Torricelli Mts (SMT).
1953–5	Father O. Shelly (Shellenberger); Wewak area (AMNH).
1954	(Mar) E. T. and M. Gilliard; Wewak region (AMNH), (Gilliard and LeCroy 1966).
1962	W. B. Hitchcock; Prince Alexander Mts (CSIRO).
1965	(July) J. M. Diamond; Mt. Somoro, Torricelli Mts (AMNH).
1966	(July–Sept) J. M. Diamond; Torricelli Mts (AMNH), (Diamond 1968, 1969, 1973, 1985).
1972	(23 Oct and 11–13 Nov) A. beg Mirza; near Wewak, *P. minor* (BPBM).
1981	(2 May) C. Unkau; near Wewak, *C. regius* (PNGM).
1988	(9–13 June), I. J. Mason and L. Seri; Mt. Somoro, Torricelli Mts (CSIRO).
1986	(Sept–Oct) A. Allison; Bewani Mts (BPBM).

17: Western and northern Sepik lowlands (including main course of the Sepik and lowlands and foothills west of Chambri Lake), PNG

1887	C. Hunstein; lower and middle Sepik (ZMB).
1910	L. Schultze-Jena; west Kaiser Wilhelmsland to middle Sepik R. (ZMB).
1910	(July–Sept) L. von Weldenfeld; Berlinhafen [Aitape], (ZSM).
1912–3	(Apr–Dec and Jan–Sept) J. Bürgers often with W. Behrmann, German Sepik Expedition; middle Sepik region and Lordberg, Hunstein and Schrader Ra (ZMB).
1930	(Feb) G. C. Eichhorn; middle Sepik R. region (AMNH).
1937	(4–10 Mar) S. Campbell; Upper Sepik, *P. minor* (AM).
1953–4	(Nov–Mar) E. T. and M. Gilliard; middle Sepik region (AMNH, CSIRO), (Gilliard and LeCroy 1966).
1962	(Nov) W. B. Hitchcock and K. Keith; Sepik District, Maprik, Pagwi, Chambri Lake (CSIRO).
1963	(10 Apr–June) P. Temple; Sepik R. (BPBM).

1965	(Apr–May) W. B. Hitchcock and K. Keith; Amarab and Maprik, Sepik District (CSIRO).	1985–7	A. Murray, Tabubil and environs, field observations.
1966	(Apr–Aug) L. Craven; Sepik and Mt. Hunstein (CSIRO).	1991–7	P. Gregory; Ok Tedi area, upland field observations.
1986	(July–Aug) B. M. & C. H. Beehler; Puwani R., Krissa trail (*D. bruijnii*), (Beehler and Beehler 1986).		
1989	(Sept–Oct) A. Allison, A. Mack, D. Wright; Hunstein Mts (BPBM).		

18: Dap, Hindenburg, and Victor Emanuel Ranges—westernmost high ranges of PNG (east to the Strickland R.)

1936–7	Second Archbold Expedition, A. L. Rand, G. H. H. Tate, L. J. Brass; Dap Mts (AMNH), (Mayr and Rand 1937).
1954	(19 Mar–29 May) E. T. and M. Gilliard, Hindenburg and Victor Emanuel Mts (AMNH, CSIRO), (Gilliard and LeCroy 1961).
1962	(5 Sept) P. Temple; Telefomin, *P. minor* (BPBM).
1966	(Apr) H. L. Bell; Lake Sogolomik, *E. fastuosus* (PNGM).
1968	(21 Oct) A. beg Mirza and R. Traub; Oksapmin, *C. loriae* (BPBM).
1969	(10 July) R. D. Mackay and J. I. Menzies; Olsabip, *C. magnificus* (PNGM).
1970	(19 Feb) A. beg Mirza; Lake Louisa, *A. splendidissima* (BPBM).
1971	(10–22 Mar) A. beg Mirza; Oksapmin (BPBM).
1980	(Mar–Apr) P. Wanga; Bafumin near Telefomin (PNGM).

19: Muller, Hagen, Giluwe, and Central Ranges (west–central highlands of PNG, including Mt. Bosavi)

1938	F. Shaw Mayer; Wabag region. *A. mayeri* type (BMNH).
1943–4	(Oct–July) N. B. Blood; Mt. Hagen area (AM, USNM, BMNH).
1945–7	Captain N. B. Blood; Mt. Hagen region. *P. rudolphi* X *P. raggiana* hybrid, *A. mayeri* and much more (AM).
1946	(Oct–Nov) F. Shaw Mayer; Mt. Hagen (BMNH).
1947	(June) F. Shaw Mayer; Tomba region, Mt. Hagen *C. macgregorii* (BMNH).
1950	(1 July–30 July) E. T. Gilliard, R. Doyle, W. Lamont; Mt. Hagen (AMNH, AM)
1950–1	(Dec–July) F. Shaw Mayer; Tomba region, Mt. Hagen and Mt. Giluwe (BMNH), (Sims 1956).
1952	(11–18 May) E. T. Gilliard and R. Doyle; Mt. Hagen, N flank from Baiyer Valley (AMNH, AM).
1952	(Sept–Oct) F. Shaw Mayer; Loke Wan Tho, Mt. Hagen, Tomba region. Photographs include first of wild bird of paradise at nest (Loke 1957).
1954	(July–Aug) N. Camps and E. Troughton; Wahgi Valley–Mt. Hagen, S watershed (AM).
1955–6	(Jan–Mar) and 1960 (Sept–Dec) R. Bulmer; Mt. Hagen, N slope (AM).

1956	(July) E. T. Gilliard; Mt. Hagen, S watershed (AMNH).	1978	(15 May) M. Koke; Mt. Bosavi, *C. magnificus* (PNGM).
1956	(24 Feb–5 Mar) R. Bulmer; Tari region (AM).	1983	(11 June) R. D. Mackay; Tari Gap, *Pteridophora* (PNGM).
1961	(13–30 Sept and 1–10 Oct) Mr and Mrs R. J. Schodde; Lake Kutubu region (CSIRO), (Schodde and Hitchcock 1968).	1984	(18 Aug) E. Troughton; Tomba, *E. mayeri* (AM).
		1986–9	(Sept–Oct; Oct–Dec; Dec–Feb, respectively) C. B. Frith and D. W. Frith; Tari Gap. See Bibliography.
1961	(July–Sept) Mr and Mrs R. J. Schodde; Mt. Giluwe (CSIRO).	1991	(June–July) M. Laska; Tari, field observations.
1963	(May–June) H. Clissold; Tambul and Mt. Giluwe (BBM, PNGM).	1994	R. Whiteside; Kompiam area, observations of *Lophorina* and *P. rudolphi*, (Whiteside 1995, 1998).
1963	(Aug–Sept) D. V. Holst; Tomba, Mt. Hagen (ZIZM).		
1963	(20 Aug and Dec) Konrad; Mt. Hagen, *Pteridophora* (NMW).		

20: Fly platform and Gulf lowlands, PNG

1963	(27 Nov) J. H. Sedlacek; Mt. Giluwe, *A. mayeri* (BPBM).
1963	(24 Sept) P. Temple; Tari, *P. rudolphi* (BPBM).
1967	(Sept–Dec) P. H. Coleman; Mendi area (BPBM).
1967	(1 Jan and Dec) A. beg Mirza; Mur-Mur Pass and Mt. Giluwe, respectively (BPBM).
1971	(2 and 25–27 July) G. Fennon; Tomba, *C. macgregorii* (PNGM).
1971	(Aug) G. George; Mendi, *A. stephaniae* (PNGM).
1972	(11–17 July) A. beg Mirza; Mt. Giluwe, *A. stephaniae* (BPBM).
1972	(5 June) A. beg Mirza; Mt. Bosavi, *Loboparadisea* (BPBM).
1973	(5 July) H. Clissold; Mt. Bosavi, *C. loriae* (BPBM).
1973	(10–12 July) R. Lossin; Hagen Ra. (AM, PNGM).
1973	(Jan–Mar and May–July) A. beg Mirza; Mt. Giluwe and Mt. Bosavi, respectively (BPBM).

1875–7	Count L. M. D'Albertis; Fly R. (MCSN).
1886	(8–24 Oct) W. W. Froggott; Fly R. *Cicinnurus* (QM).
1889	(17 Dec) Sir William Macgregor; Arowanga, Fly R. (QM).
1920–1	J. Todd Zimmer; Delta region, Kikori (AMNH).
1936–7	(Feb–Jan) Second Archbold Expedition (A. L. Rand, G. H. H. Tate, L. J. Brass); S central NG and middle Fly R. Wassi Kussa area, near Daru I. and lower, middle, and upper Fly R., Palmer R. and Black R. regions (AMNH), (Rand 1942*a*).
1966	H. L. Bell; Balimo, Gulf Province, field observations, (Bell 1967).
1967	(16 Oct) Selbert; Nomad area, *P. raggiana* (PNGM).
1968	H. L. Bell; Ok Tedi area, field observations, (Bell 1969).

1969	(Nov) R. T. Lawson; *C. regius* (PNGM).	1964	(17 June) H. Clissold; Bensbach R., *P. raggiana* (BPBM)
1969	H. L. Bell; Nomad, field observations.	1978	(16 Dec) H. Sakulas; *M. atra* (PNGM)
1977	(31 Mar) L. Walo; Purari R., *C. regius* (PNGM).	1983	(Sept–Dec) K. D. Bishop; observation avifaunal survey.
1980	(4–9 Oct) B. M. Beehler; Nomad area, observations of 5 species.		
1985–7	A. Murray, Tabubil and environs, field observations.		
1991–4	P. Gregory; Ok Tedi area, lowland field observations.		

21: Trans-Fly Savannas (Merauke–Bensbach–Oriomo region), IJ and PNG

1902	A. E. and H. Pratt; Merauke (BMNH).
1904	(22 Aug) J. W. R. Koch; Merauke, *M. atra* (RMNH).
1910	(May) A. S. Meek; Merauke and points W (BMNH).
1914	(Apr) W. J. C. Frost, Bian R., *P. apoda* (ROM).
1920–4	(Sept–Nov) T. Jackson; Merauke and points W (MCV).
1925	Dr Thierfelder; behind Merauke (ZMB).
1933	(Sept–Dec) Dr Hans Nevermann; near Merauke; Toerey, Welbuti (ZMB).
1933–4	First Archbold Expedition, A. L. Rand, G. H. H. Tate, L. J. Brass; S NG. Daru I. and mainland opposite; Wuroi, Dogwa and Oriomo R. (AMNH), (Mayr and Rand 1937).
1951	(2 July) Monsanto; Merauke, *M. atra* (RMNH).
1959–62	A. Hoogerwerf; Merauke region (RMNH).
1964	(Feb–Mar) H. Clissold; Oriomo R. area (BPBM).

22: East Sepik and Ramu lowlands and hills, PNG

1899	(May–June) Ramu Expedition; Bismarck Ra (ZMB).
1898–9	(May–Sept and Jan–Mar) E. Tappenbeck; Ramu R. (ZMB).
1904	(16 Sept) Dr R. Poch; Ramu R. mouth area (NMW).
1929	(13–25 May) Crane Pacific Expedition, W. A. Weber, F. C. Wonder; Marienberg, Arnbot, Keram R. (FMNH).
1945	(30 Jan–11 Feb) C. G. Sibley and C. Howe; Upper Ramu R., Hagen Mts, *P. raggiana*, *P. minor* (MVZ).
1964	(7–25 May) E. T. Gilliard; Aiome region of Upper Ramu R.

23: Adelbert Range and vicinity, PNG

1871–3	Mikloucho Maclay; Bogadjim, Astrolabe Bay (RUSSIA).
1886–7	F. Grabowsky; Samoa Harbour, Hatzfeld Harbour, and Tscherimotsch I. (ZMB).
1887–8	J. S. Kubary; Constantine Harbour (ZSM).
1888–9	Rohde; Kelana and Stephansort (ZMB).
1890–2	B. and H. Geisler; Astrolabe Bay and coasts of Kai Penin (SMT, ZMB).

1890	G. Lauterbach; Gogol R. to source (ZMB).
1891 and 93	(Dec and Mar) S. Fenichel; Astrolabe Bay region (BUDAPEST).
1894–5	B. Hagen; Astrolabe Bay region, chiefly Stephansort (ZMB, ZSM and some to Karlsruhe Museum).
1894	Capt. C. Webster; Stephansort (BMNH).
1894–5	J. S. Kubary; Constantine Harbour (MHN, ZSM).
1896	(4 June–Aug) C. Lauterbach, O. Kersting, E. Tappenbeck, '1 Ramu Expedition'; Stephansort overland to Ramu R. (ZMB).
1896	(Sept) C. Wahnes; *M. atra*, *C. regius* (SMT).
1896	Wolff; Astrolabe Bay, *M. atra*, *M. chalybata*, *P. minor* (SNHM)
1896–7	L. Biro; Berlinhafen, Astrolabe Bay, Erima (ZMB, BUDAPEST).
1897	(Apr) A. B. Meyer; Konstantinhafen, *P. minor* (SMT).
1898–9	Erik Nyman; Astrolabe Bay area, chiefly Stephansort (BMNH).
1898–9 and 01	(16 July, 12 Sept and July, Feb respectively) C. Wahnes; Bongu (SMT).
1900–1	O. Heinroth, '1 German South Sea Expedition of Br. Mencke'; various harbours (ZMB).
1905–6	C. Wahnes; Astrolabe Bay region (BMNH).
1909	G. Duncker; coastal area (ZIZM).
1909–10	C. Schoede; coastal areas; 3 June 1910, Braunschweig Hafen (ZMB).
1912	(23 Mar) R. de Neufrille; Stephansort (ZSM).
1917	(Oct) E. Rowan; Madang, *C. regius* (AM).
1919–20	(Dec and 7 May) W. Potter; Astrolabe Bay and Madang, *P. raggiana* and *M. chalybata* respectively (BMNH).
1924	(7 Sept) R. H. Beck; Madang, *M. chalybata* (AMNH).
1928–9	(Mar–Jan) Mr and Mrs R. Beck; Madang-Bogadjim, upland areas behind Madang (AMNH).
1953 and 56	(Dec and June) E. T. and M. Gilliard; lowlands about Madang (AMNH).
1958	(14–18 June) E. T. Gilliard; Adelbert Mts (AMNH).
1959	(2 Mar–22 Apr) E. T. and M. Gilliard; Adelbert Mts (AMNH), (Gilliard and LeCroy 1967).
1969	(8 May) J. M. Diamond; Sempi, near Madang, *P. minor* (AMNH).
1974	(28 Jan–22 Mar) A. beg Mirza, T. K. Pratt, and G. Opit; Wanuma area (see Pratt 1982*b*), (PNGM).
1974	(May–June) K. Mara and D. Kuro; Sapi Creek, *C. regius*, *P. minor* (PNGM).
1982	(21–26 Aug) P. Kunumai and P. Wanga; Madang, *P. minor* (PNGM).
1985	(Oct), C. B. Frith, D. W. Frith and R. D. Mackay; Mt. Mengam. Adelbert Mts; *L. superba* (Frith and Frith 1988).
1985–8	(several field trips) R. D. Mackay; field observations.

24: Schrader, Bismarck, and Kubor ranges, Sepik–Wahgi Divide, Jimi R., Wahgi Basin, PNG

1947	(Nov–Dec) F. Shaw Mayer; Mt. Wilhelm (BMNH).
1948	(Oct–Dec) F. Shaw Mayer; Mt. Wilhelm (BMNH).

Appendix 2

1948–54	Capt. N. B. Blood; Wahgi Valley and surrounding mts—Wahgi Divide, Hagen, Kubor.	1957	D. Attenborough; Wahgi Valley, Jimi R., Bismarck and Schrader Mts. Films of birds of paradise and c. 20 taken live to UK (Attenborough 1960).
1949	(Apr–June) F. Shaw Mayer; Mt. Wilhelm, N slopes (BMNH).	1959	(June) E. T. Gilliard; Banz area, central Wahgi Valley (AMNH).
1950	(Mar–June) F. Shaw Mayer; Mr Wilhelm, N and S watershed (BMNH).	1959	(May–June) Sixth Archbold Expedition, L. J. Brass, H. M. Van Deusen, J. D. Collins; Mt. Wilhelm (AMNH).
1950	(17 Apr–14 May) E. T. Gilliard and R. Doyle; Nondugl region and adjacent Wahgi Divide (AMNH, AM).	1960	(Feb) R. Bulmer; Schrader Mts, Kaironk R. Valley.
1950	(16 May–1 June) E. T. Gilliard and R. Doyle; Kubor Mts, Omong R., and Mt. O'mar (AMNH).	1963	(June–Sept) W.B. Hitchcock, W. Vink, and R. Pullen; Kubor Mts (CSIRO).
1950	(5–20 June) E. T. Gilliard and R. Doyle; Mt. Wilhelm (AMNH, AM).	1963	(29 June) J. H. Sedlacek; Banz, *P. raggiana* (BPBM).
1950	(Oct–Dec) F. Shaw Mayer; Kubor Mts and Wahgi region near Nondugl and Wahgi Divide (BMNH), (Sims 1956).	1963	(2 July) H. Clissold; Kuta, *P. raggiana* (BPBM).
		1963	(Aug–Sept) W. Vink; Minj area (CSIRO).
		1963	(9 July) J. H. Sedlacek; Karimui, *P. raggiana* (BPBM).
1951	(29 May) N. B. Blood and N. Gyldenstolpe; Jimi R., *Loboparadisea* (SMNH).	1963	(Sept) D. V. Holst; Kubor Ra. and Jimi R. (ZIZM).
1951	(17 Aug–Nov) Count Nils and Lady Gyldenstolpe; Nondugl region and adjacent Wahgi Divide Mts (SMNH), (Gyldenstolpe 1955a).	1963–4	(Sept–Jan, July–Aug): R. Bulmer; Kaironk Valley and adjacent spurs of Bismarck and Schrader Ra. (AMNZ).
1952	(26 Mar–15 July) E. T. and M. Gilliard, R. Doyle, H. Kaltenthaler; Kubor Mts (AMNH, AM).	1964	(21 Apr–7 May) E. T. Gilliard; Schrader Mts, flank above Simbai Valley, watershed of Ramu R., to Mt. Kominjim (AMNH), (Gilliard and LeCroy (1968).
1954(?)–63	F. Shaw Mayer; Wahgi Valley and surrounding mts—Wahgi Divide, Hagen, Kubor Mts.	1964	(19 Aug) A. Ferguson; Goroka area, *A. stephaniae* (PMNH).
1954	N. B. Blood, H. G. Slade, N. Camps, E. Troughton, and R. McFazdean; Wahgi Divide and Jimi Valley (AM).	1964	(Jan–Apr) D. V. Holst; Wurup area (ZIZM).
		1966	(Apr–Dec) G. Jackson; Simbai (PNGM).
1955	(June) R. Bulmer; S central Jimi Valley, from Wahgi Divide to Jimi R.	1967	(Nov) P. H. Colman; Chuave and Kassam Pass areas (BPBM).
1955	(12–28 June) Dr and Mrs R. Bulmer; Jimi R. region (AM).	1967	(26 Nov) A. beg Mirza; Chuave area, *Lophorina* (BPBM).

1968	(1 Sept) J. I. Menzies; near Simbai, *C. macgregorii* (PNGM).	1890	(14 July) W. Schluter; Finschhafen (BMNH).
1969	(July–Aug) K. W. Corbin; A. Ferguson, C. G. Sibley, Trauna Valley, Baiyer R. (PMNH).	1890	(June–Sept, Nov–Dec) C. Lauterbach; vicinity of Finschhafen (ZMB, NMW).
		1891–2	(July and Apr–May, July–Aug) B. Geisler; Bukawasip, Huon Gulf (BMNH, USNM).
1969	(Aug) A. Ferguson; Nondugl, *P. raggiana* (PMNH).	1892	(Mar and July) B. Geisler; 'Finsch Harbour' (BMNH).
1970	(21 Oct) A. beg Mirza; Chuave area, *Lophorina* (BPBM).	1893	W. Schluter; (MHN).
		1893–4	(7 June, Dec and 18 Feb–15 May, 10 Dec, respectively) Kubarg; Konstaninhafen, *M. atra, P. guilielmi* (NMW, ZSM).
1971	(12 June) G. George; Tuman, *P. brevicauda* (PNGM).		
1973–4	C. Healey; Tsuwenkai village, Bismarck Ra., field observations, (see Healey in Bibliography).		
		1894	(Feb) Cotton and Capt. C. Webster; Finisterre Mts, *C. magnificus* (AMNH).
1977	(Aug) K. D. Bishop; Trauna (Lepa) Ridge, field observations, *Loboparadisea*.	1895	(Feb) Cotton and Capt. C. Webster, *C. regius* (AMNH).
		1895	(Aug) W. Schluter; Sattelberg, Finschhafen (RMNH).
1980	(June) B. M. Beehler; Trauna (Lepa) Ridge; *C. loriae, Loboparadisea; P. carolae, Lophorina, P. rudolphi* (Beehler 1991*a*).	1896	(Feb) Cotton and Capt. C. Webster; Simbang (AMNH).
		1898–9	(Jan and Jan–Apr) C. Wahnes; Sattelberg (SMT).
1982	(17 Oct) D. Bruning and R. D. Mackay; Lepa Ridge, Sepik-Wahgi Divide, *Loboparadisea* and *P. carolae* (PNGM).	1898–9	L. Biro; Simbang, Finschhafen, Sattelberg (BUDAPEST).
		1898–9	(Dec and June–Sept, Dec) E. Nyman; Finisterre Mts?, SW slopes? and Sattelberg and Simbang region (BMNH).
1982	(21–26 Aug) P. Kunumai and P. Wanga; Bundi, *P. minor* (PNGM).		
		1901	van Kaemperr; (MHN).

25: Huon Peninsula, PNG (including Finisterre, Saruwaged, and Cromwell Ranges)

		1903	(Aug) C. Wahnes; Sattelberg (SMT).
		1904	(Feb) Capt. C. Webster and Cotton; Finisterre Mts?, SW slopes? (BMNH).
1871–3	(?) M. Maclay; Finisterre Mts, SW aspect (RUSSIA?).	1904–5	(Dec–Jan) Dr R. Poch; Sattelberg and Bussum, Finschhafen (NMW).
1878	(?) Heynemann; *M. keraudrenii* (MHN).	1905–6	(July–Jan) C. Wahnes; Sattelberg and Rawlinson Mts and Sattelberg Ra. (BMNH).
1888	(Jan) C. Hunstein, with S. von Kotze; Rawlinson Mts (ZMB).		
1889	(Oct) Dr Lauterbach; Finschhafen, *M. chalybata* (NMW).	1906	(28 Nov) Holderer; Sattelberg, *M. chalybata* (SNHM).

1908	Dr R. Poch; Mission Keysser (NMW).		Headwaters (Umi R.) of Markham R. in foothills of Finisterre Mts just N of Ramu-Markham Rift Valley (AMNH).
1908–9	(Dec–Mar) R. Neuhauss with missionary C. Keysser; Sattelberg to interior of Kai Penin (ZMB).		
1909	C. Wahnes; Sattelberg (BMNH).	1964	S. O. Grierson and H. M. Van Deusen; Mt. Rawlinson (AMNH).
1909–10	(Apr, May, Aug) Lothar von Wiedenfeld; Heldbach coast, Simbang, Sattelberg (ZSM, ZMB, BRESLAU).	1966	(Aug) R. M. Mitchell; Saruwaged Mts, *A. rothschildi* (BPBM).
		1966	(4 Apr) G., Lippert; Singauwa R. (BPBM).
1910	(Feb–Mar) L. von Wiedenfeld; Finschhafen area (ZSM).	1970	(4 Dec) W. Potter; Rawlinson Mts, *P. magnificus* (BPBM).
1910	Capt. Hang or Haug; Finschhafen (SNHM).	1970	(2 Oct) F. J. Radovsky; 12 miles NE of Lae, Bupu R., *C. regius* (BPBM).
1911	(July, Sept, and Nov) C. Keysser; Kai Penin, in the Saruwaged and Rawlinson Mts (BMNH).	1973	(10 June) R. D. Mackay; Konge, *P. wahnesi* (PNGM).
1912	(23 Mar) R. de Neufrille, *A. stephaniae, P. wahnesi* (ZSM).	1973	(Oct–Nov) R. S. Schodde, and I. J. Mason; Rawlinson-Saruwaged Mts (CSIRO).
1920	(Apr–May) W. Potter; Rawlinson Mts, Singaua, Bulo (BMNH).	1974	(Jan–Feb) T. K. Pratt; Wasu, PNG.
1920–1	(Mar–Apr and May) W. Potter; Musom and Singaua, Huon Gulf (BMNH, AM).	1975	(Sep) B. M. Beehler; Derim, Mt. Bangeta (MCZ).
1928	(19 Oct–1 Dec) Mr and Mrs R. H. Beck; Finisterre Mts, SW slopes in Keku and Mt Tyo regions (AMNH).	1978–91	G. P. Smith; hills and lowlands N and E of Lae (long-term field observations).
		1986–7	C. K. Schmid; Nokopo area, Finisterre Ra, anthropological observations.
1928–9	(8 Dec–27 Apr) E. Mayr; Saruwaged Mts (ZMB).	1992–4	C. Eastwood, Sankwep Road area, regular field observations.
1929	(Jan–Apr) Mr and Mrs R.H. Beck; Saruwaged Mts (AMNH).		

26: Markham and Watut and Waria drainages, and the Herzog, Kuper, and Bowutu Ranges, PNG

1931	(June–July) F. Shaw Mayer; Sattelberg Mts (BMNH, AMNH).
1944	J. F. Cassel; Finschhafen lowlands (CORNELL, ITHACA).
1956	E. T. Gilliard; Finisterre Mts, SW aspect (AMNH).
1959	(13–29 Nov) Sixth Archbold Expedition, H. M. Van Deusen and L. J. Brass;

—?	H. Andechser; Herzog Mts, Sudhong (ZMB).
1910–2	Governor Hahl and 'Klink'; upper Babu R., Batchelor Mt. (ZSM).
1921	(5 May) O'Hay (BMNH).

1929	(7 May–29 June) E. Mayr; Snake R. region (ZMB).	1966	(Apr–Aug) O. R. and J. Wilkes; Mt. Missim, (BPBM).
1932–3	(23 Jan–5 May) H. Stevens; Herzog Mts, Mt. Missim (MCV), (Greenway 1935).	1966	(14 July) R. M. Mitchell; Bulldog Rd., *A. stephaniae* (BPBM).
1936	(Dec) F. Shaw Mayer; Upper Waria R. (BMNH).	1966	(19 July), N. Wilson and R. M. Mitchell; Edie Creek, *E. meyeri* (PNGM).
1950	(8–9 Aug) E. T. Gilliard; Markham R. forest behind Lae (AMNH).	1967	(June–Oct) P. H. Coleman; Wau/Bulolo area (BPBM).
1958	(23 Oct–3 Nov) E. T. Gilliard; Lae region and lower Markham R. (AMNH).	1967	(Aug–Sept) A. beg Mirza; Mt. Missim/Bulolo (BPBM).
1959	(25 Mar–10 Apr) Sixth Archbold Expedition (AMNH).	1967	(June–Aug) A. C. Ziegler; Mt. Kaindi/Bulolo areas (BPBM).
1959	(Apr–May) Sixth Archbold Expedition, L. J. Brass and H. M. Van Deusen; Herzog Mts area (AMNH).	1968	(1 Jan) A. beg Mirza; Edie Creek area, *A. stephaniae* (BPBM).
1959	(1 July) H. W. Clissold; Mt. Kaindi, *C. loriae* (AM).	1968	(12 Nov) M. Nadchadram; Mt. Shungol, Mumeng area (BPBM).
1961	(28 Sept–2 Nov) J. Sedlacek; *C. magnificus* (BBM, USNM).	1969	(Sept–Oct) A. beg Mirza; Watut R. district (BPBM).
1962	(24 Sept–29 Oct) A. H. Miller and W. Z. Lidicker; Herzog Mts and Bulolo basin region (MVZ).	1970	(27 Aug) A. beg Mirza; Bowutu Mts, near Garaina, *Lophorina* (BPBM).
1962	(Jan–May and Sept) H. Clissold; Herzog Mts region (BPBM).	1971	(July–Dec) P. Shanahan; Wau (PNGM).
1963	(Jan–May) H. Clissold; Wau (ZMB, BBM, PNGM).	1972	(Jan and Aug) A. beg Mirza; Bulldog Rd., near Edie Creek, *C. macgregorii* (BPBM).
1963	(5 Mar) J. Sedlacek; Watut Valley, *P. raggiana* (BPBM).	1973	(Oct) I. J. Mason; 12 miles W of Busu R. (CSIRO).
1963	(16 Apr) O. R. Wilkes (BPBM).	1973	(Oct–Nov) R. S. Schodde, I. J. Mason and M. Ainie; Wagau and Mindik areas (CSIRO).
1963	(Mar–Nov) P. Shanahan; Wau-Mt. Missim area (BPBM, PNGM).	1973–5, 77–8	T. K. Pratt; Mt. Missim, Bulldog Rd.
1963	(19 June) P. Shanahan; Coulak, *P. raggiana* (PNGM).	1975–6, 78–80	B. M. Beehler; Mt. Kaindi, Bulldog Rd., Mt. Missim (field studies; MCZ), (see Bibliography).
1963	(11–18 Aug) A. beg Mirza and A. C. Ziegler; 10 km W of Bulolo, *P. lawesii* (BPBM).		
1964	(31 Aug) J. H. Sedlacek; Mt. Kaindi, *A. stephaniae* (BPBM).	1981–3	S. G. Pruett-Jones and M. A. Pruett-Jones; Mt. Missim; *P. lawesii* and other species (see Bibliography).
1966	(18 May) P. Shanahan; Wau, *C. magnificus* (BPBM).		

1984	(17 Feb) A. Engilis; Mt. Missim, *P. lawesii*, *Pteridophora* (BPBM).	1875	(July) Count L. M. D' Albertis; Hall Sound and Yule I. area (MCSN).
1985	(9 Apr) A. Engilis; Mt. Missim, *Lophorina* (BPBM).	1876	(Apr) Dr James; Aleya, SE NG, *M. keraudrenii* (BMNH).

27: Eastern Highlands of PNG (Mt. Karimui, Crater Mt., Mt. Michael, Kratke Ra)

1932	(May–Nov) F. Shaw Mayer; Kratke Mts; Buntibasa district (BMNH).
1940	(May–June) F. Shaw Mayer; Baiyanka, Purarai–Ramu Divide (BMNH).
1959	(Aug–Nov) Sixth Archbold Expedition, L. J. Brass, H. M. Van Deusen, J. D. Collins; Mt. Michael and Kratke Mts (AMNH).
1963	(16 Nov) P. Coleman; Kassam Pass, *M. keraudrenii* (BPBM).
1964	(June–Aug) J. M. Diamond, J. Terborgh; Karimui area (MCZ, AMNH).
1965	(June–Sept) J. M. Diamond; Mt. Karimui area (MCZ, AMNH), (Diamond 1972).
1987	(Aug) W. Peckover, M. LeCroy; Crater Mt. region; Heroana, Ubaigubi, field observations.
1980, 84, 86 91–2, 95	(June–Sept) D. Gillison; Crater Mt. region, Heroana and Ubaigubi areas. Field observations and photography.
1990–6	A. Mack, D. Wright; Crater Mountain Reserve, field observations.

28: Southern Watershed, Papuan Peninsular hills and lowlands

1873–5, 78–9	K. Broadbent, SE NG (QM).
1876–7 +	A. Goldie; Port Moresby region (BMNH).
1878	Carl Hunstein joined A. Goldie; Astrolabe Mts and mts to E. Disposition of specimens unknown.
1882	(26 Apr–27 May) O. Finsch; Port Moresby area, *M. atra* (NMW).
1883	C. Hunstein; Astrolabe Mts *P. lawesii* (GERMANY?).
1884	A. Goldie; Astrolabe Mts (BMNH).
1885	(Apr) W. Lawes; Astrolabe Ra., *P. magnificus* (AM).
1885	(May) H. O. Forbes; Sogeri district (29 Oct.), *Manucodia*; Moroka, 5 Nov, *M. keraudrenii* (MHN).
1891	(24 Aug) Mt. Cameron, *C. magnificus* (MCZ).
1892	(10 Feb) O. Finsch; Port Moresby (ZMB).
1893	(June) F. Ferdinand; Port Moresby, *M. atra* (NMW).
1893	(July) Lix; Nicura, *P. raggiana* (AMNH).
1895	(July–Aug) A. S. Anthony; inland from Mailu I., S watershed (BMNH).
1895–6	A. S. Anthony; SE New Guinea. (BMNH).
1896	(1 Feb) E. Weeks; Brown R., *C. regius* (USNM).
1897	(Mar) Sir William Macgregor; Vanapa R. (QM).
1899	(27 May) A. S. Anthony; Vanapa R., S. E. NG.
1899–1900	(July–Jan) E. Weiske; mts of SE NG, Aroa R.
1900	(June) A. E. Pratt, Mt. Kebea, *Lophorina* (BMNH).

1903	(Jan–Sept) A. E. Pratt and son Harry; mts behind Yule I. (BMNH, SMNS, ROM, AMNH, MCZ, ZMB).	1962	(July–Oct) S. Schodde; Port Moresby, Rigo and Brown R. lowlands. (CSIRO).
1904	(Apr) E. Schreyer; SE NG (SMT).	1963	(Mar–Apr and July) D. V. Holst; Iavarere and Lakekamu, respectively (ZIZM).
1904	(27 Oct–2 Nov) A. S. Meek; Aroa R. (BMNH, AMNH).	1965	(18 June) J. M. Diamond; Brown R., *M. keraudrenii* (PNGM).
1905	(Feb) A. S. Meek; Angabunga R. Owgarra, base of Wharton Ra. (BMNH).	1965	(19 July) R. D. Mackay; Laloki Swamp, *P. raggiana* (BPBM).
1905	(Apr–May) A. S. Meek; Aroa R. region, S watershed, Wharton Ra. (BMNH).	1966	(Jan–Mar) R. S. Schodde; extreme N of S coast area of Gulf District (CSIRO).
1906	(June) W. Stalker; Madeu, Auka, *P. lawesii* (ROM).	1966	(May–June) D. Gosney; Port Moresby area (PNGM).
1908	(18–23 Mar) C. B. Horsburgh and W. Stalker; near Hall Sound, *P. raggiana* (BMNH).	1966	(Aug) R. S. Schodde; Rigo, *M. atra* (CSIRO).
1908	(May–June) W. Stalker; Angabunga R. area (ROM, BMNH, MCZ.	1967	(Nov–Dec) H. L. Bell; Moroka (PNGM).
1908	(Sept) A. E. Pratt; 08° 35′, 147° 03′ (ROM).	1968	(8 Aug) A. Wolf; Laloki, *P. raggiana* (PNGM).
1913	(20 May) A. S. Meek; Aroa R., *C. loriae* (QM).	1968	(3 Oct) M. Keroud and Y. Kans; Brown R. (PNGM).
1920–1	(July and Jan) J. T. Zimmer; SE NG, S watershed (AMNH).	1969	(3 Mar) R. Lossin; Brown R., *M. keraudrenii* (PNGM).
1921	(Sept–Nov) R. Neill and E. P.; Elenagora (BMNH).	1969	(19 May) A. Wolfe; Brown R., *P. magnificus* (PNGM).
c1925	W. Goodfellow; Central Division from coast near Yule I. to 9–10 000 ft in ranges linking Mt. Yule and main Owen Stanley cordillera (BMNH).	1974	(Feb–Apr) K. Mara, J. Iruru, K. Yande, J. Barre, A. Ainie and Harry Sakulas; Port Moresby-Rigo general area (PNGM).
1928	(5 Oct–9 Dec) L. S. Crandall; SE NG, S watershed—40 live birds to NYZS.	1975–7	H. L. Bell, Brown River, field studies of community ecology, including birds of paradise, (see Bell in Bibliography).
1928–9	H. Hamlin; SE NG, S watershed (AMNH).	1975–7	M. Hoyle; Brown R., field work on *C. regius* (PNGM).
1935	(Nov–Dec) F. Shaw Mayer; Deva Deva (BMNH).	1979,82,92–3	B. M. Beehler *et al.*; Lakekamu Basin, PNG, field observations of five species (Beehler *et al.* 1995).
1948	(7–11 Apr) E. T. Gilliard; SE NG, S watershed (AMNH).		

1989–90	B. M. Beehler and J. Dumbacher; Varirata Nat. Park, PNG, field studies of *P. raggiana*.	1897	(July and Oct) Sir William Macgregor; Wharton Ra (QM), (see De Vis 1897*b*).
		1897–8	Emil Weiske; Owen Stanley Mts, S watershed (BMNH).
		1898	(Aug) A. S. Anthony; Mt. Knutsford and Kotoi areas (AMNH).

29: Chapman, Wharton, and Owen Stanley Ranges, PNG

1881–3	W. Denton and sons, S. W. and S. F.; Owen Stanley Mts (CMNH).	1898	Schluter; Hufeisenebirge, *A. stephaniae*, *P. rudolphi* (SNHM).
1884	C. Hunstein; Mt. Maguli, Owen Stanley Mts (SMT).	1902	(Aug) A. E. Pratt and sons; Keba, mts of SE NG *P. rudolphi* (BMNH).
1885	Schneider; Owen Stanley Mts, *P. raggiana* (ZSM).	1903	(Jan–June) A. S. Meek; Aroa R., S watershed, Wharton Ra. (BMNH). *c.* 60 miles inland on 'Manna Manna Mt.'. Meek spent 6 weeks on way to Okuma and 3 months at Bwoidunna.
1889	Sir William Macgregor with G. Belford, K. Kowald and A. P. Goodwin; Musgrave Mts, Mt. Knutsford *C. macgregorii* (QM), (see De Vis 1890).		
1890	(May) W. Goodfellow; Owen Stanley Ra (BMNH).	1905–6	(Nov–Dec and Mar) C. C. Simson; Owen Stanley Mts (BMNH), (Simson 1907).
1891	(26 June–26 July, 2 Aug–25 Sept) Sir William Macgregor and R.E. Guise; Mt. Suckling *P. lawesii*, *M. ater* (QM), (see De Vis 1891).	1908	(Feb–Apr) W. Goodfellow; Moroka District, Owen Stanley Mts (BMNH).
		1909	(July) E. Weiske; Arora R., *C. loriae* (ZMB).
1893	L. Loria; Owen Stanley Mts *C. loriae* (MCSN).	1909	(Mar and May) W. Goodfellow; Bagutana Camp, Owen Stanley Mts and Taora District, respectively (BMNH).
1894–5	R. E. Guise with W. E. Armit; Mt. Dayman (Mt Maneao), (QM).		
1895	(Oct) A. S. Anthony; Eafa district below Mt. Alex and Mt. Bellamy, Owen Stanley Mts (BMNH).	1917	(Sept) A. S. Anthony; Mt. Victoria (ROM).
		1933–4	First Archbold Expedition, R. Archbold, A. L. Rand, G. H. H. Tate, L. J. Brass; SE NG, Mt. Albert Edward, Wharton Ra (AMNH), (Mayr and Rand 1937).
1896	(July–Sept) A. S. Anthony; Mt. Cameron and Mt. Victoria. (BMNH, MHN).		
1896	(Sept–Oct, 18 Nov) Sir William Macgregor; Mt. Scratchley (QM).		
1896	(May and 6–22 Sept) Amedeo Giulianetti; Wharton Ra, *Macgregoria* (QM).	1936	(Oct–Nov) F. Shaw Mayer; upper Waria and Bubu R., *P. lawesii*, *M. pulchra* (BMNH).
		1940–1	(Aug–Dec and Jan) F. Shaw Mayer; Mt. Simpson

1953	(BMNH), (Harrison and Frith 1970). Fourth Archbold Expedition, L. J. Brass, H. M. Van Deusen, G. M. Tate, K. M. Wynn; Mt. Dayman and Cape Vogel (AMNH).	1884 1885 1891	McCormack; Milne Bay, *P. magnificus* (RMNH). (June) O. Finsch; Milne Bay, *P. raggiana* (RMNH). (26 June–26 July, 2 Aug–25 Sept) Sir William Macgregor and R. E. Guise; Collingwood Bay, *M. ater* (QM).
1966	(Mar–Apr) R. S. Schodde; Kukukuku Ra., Aseki area (CSIRO).	1896	(June–July) Sir William Macgregor; Orangerie and Cloudy Bays (QM).
1968	(9 Aug) A. beg Mirza; Smith's Gap, *C. macgregorii* (BPBM).	1896	(Aug–Nov) Sir William Macgregor; Mambare R. (QM), (see De Vis in references).
1969	(Aug) J. M. Diamond; Mt. Albert Edward, (AMNH).		
1969	(June–Sept) W. B. Hitchcock; Agaun area (CSIRO).	1897	(26–30 June) A. S. Meek; Collingwood Bay, *P. raggiana* (BMNH, AMNH, ROM, ZMB).
1969	(June–Aug) R. S. Schodde and B. Kunik; Mt. Wadimana, Dumai Ridge, Agaun, Opanabu, Nowata areas (CSIRO, MV).	1898	(22 Oct) A. S. Meek; Milne Bay, *P. raggiana* (AMNH).
		1899	(Feb–June) A. S. Meek; Collingwood Bay and Milne Bay regions (BMNH).
1970	(Aug–Sept) R. S. Schodde and J. L. McKean; Efogi and Astrolabe Ra., above Barakau (CSIRO, MV).	1905	(Sept) A. S. Meek; Mt. Lamington foothills, Aicora (BMNH).
1978	(Aug–Sept) J. Auki, P. Wanga, A. Keva, J. Iruru; Fane area (PNGM).	c1906	(Jan–May) A. S. Meek; Mambare R., N watershed, Wharton Ra. (BMNH).
1985	(16–17 Mar) A. Engilis; Agaun, Milne Bay Province (BPBM).	1906	(Jan–June) A. F. and G. C. Eichhorn; Mambare R. (ROM).
1985	(24 May and 22 Aug) R. S. Schodde; Efogi (CSIRO).	1907	(Apr–June) A. S. Meek; Kumusi R., N watershed, Owen Stanley Mts a short distance W of Buna and Collingwood Bay (BMNH).
1986–8	B. M. Beehler, H. & M. G. H. Hopkins, I. Burrows, A. Hare; several field trips to Lake Ohma (English Peaks), study of *M. pulchra* (see Beehler in References).	1913	A. F. Eichhorn; Sonbon I., near Samarai (BMNH).
		1918	(Jan–May) A. F. and G. C. Eichhorn; Hydrographer Ra, N watershed (BMNH).
1991–2	(Dec–Jan) B. A. Iova; Laronu Village, Central Prov., field observations.	1928	(15–27 Feb) R. H. Beck; SE NG, Samarai I. and adjacent mainland (AMNH).

30: Southeasternmost lowlands and hills of PNG

1873	Sailors aboard corvette *Vettor Risani*; Orangerie Bay (MCSN).	1963	(30 Aug) P. Shanahan; Popondetta, *M. chalybata* (BPBM).

1963	(Oct) H. Clissold; Popondetta area (BBM, PNGM).	1894	(May and early June) W. E. Armit; Goodenough I. (QM).
1965	(29 Mar) P. Shanahan; Milne Bay, *M. atra* (BPBM).	1894	(May and Sept–Dec): A. S. Meek; Fergusson I. (BMNH).
1966	(7 July) E. Lippert; Mt. Lamington, *M. chalybata* (BPBM).	1895	(Mar–July) A. S. Meek; Kiriwini I.; Trobriand I. (BMNH).
1966	(1–7 July) P. Shanahan; Mt. Lamington, *M. chalybata* (BPBM).	1896	(Nov–Dec) A. S. Meek and W. G. Meek; Goodenough and Fergusson I. (BMNH).
1969	(June–Sept) W. B. Hitchcock; Cape Rodney (CSIRO).	1897	(17–19 Feb, 12–26 May, 4–30 June) A. S. Meek; Fergusson I. (ZMB, BMNH).
1969	(June–Aug) R. S. Schodde and B. Kunik; Amazon Bay, Mt. Wadimana, Opanabu, Nowata areas (CSIRO, MV).	1899	(25–26 May, 10–20 and 23 Aug) A. S. Meek; Goodenough, Normanby and Fergusson Is., respectively (BMNH, SMT).
1969	(2 Sept) A. Ferguson; Alotau, *P. raggiana* (PMNH).	1901	(Aug–Sept) A. S. Meek; D'Entrecasteaux Group, Normanby I. (BMNH).
1982	(3–5 July) Maeva; Nunumeri, Amazon Bay, *P. lawesii*, *C. regius* (PNGM).	1912–3	A. S. Meek; D'Entrecasteaux Group, Goodenough I., Mar–May, and Fergusson I., Dec 1912–10 Jan 1913 (BMNH).
1984	(Apr) G. E. Clapp; Sibium Mts, field observations.		
1987	(July–Aug) R. S. Schodde and L. Christidis; Kokoda to Kumusi R. areas, Oro Province (CSIRO, MV).	1923	A. and G. Eichhorn; Goodenough I. (BMNH).
		1928	(1–7 Nov) H. Hamlin; Trobriand I., *M. comrii* (AMNH, ZMB).
1992–3	A. Fabbro; Alotau area (field observations).	1928	(Nov) H. Hamlin; Fergusson and Goodenough Is (AMNH).

31: D'Entrecasteaux and Trobriand Islands, PNG

		1935	(July and Sept) F. Shaw Mayer; D'Entrecasteaux Group, Fergusson I. (BMNH).
1873–4	R. N. Comrie; Trobriand I. *M. comrii* (BMNH).		
1882	C. Hunstein and A. Goldie; Fergusson and Normanby Is. *P. decora* (BMNH, MHN).	1953	Fourth Archbold Expedition, L. J. Brass, H. M. Van Deusen, K. M. Wynn; D'Entrecasteaux Group, Goodenough I. (AMNH).
1885	O. Finsch; Fergusson I., *M. comrii* (RMNH).	1956	Fifth Archbold Expedition, L. J. Brass and R. F. Peterson, L. Evennett; D'Entrecasteaux Group, Normanby I. 10 Apr–12 May; Fergusson I. 2. May–7 July (AMNH).
1888	(Autumn) B. H. Thomson; d'Entrecasteaux Is, *M. keraudrenii* (BMNH).		
1890	(13–18 Jan) L. Loria and A. Giulianetti; Goodenough I. (MCSN or MRSN?)		
1891	(July) Rickard; Fergusson I. (AM).		

32: Louisiade Archipelago and small islands of Milne Bay, PNG

1956	(9–10 Dec) Fifth Archbold Expedition, R. F. Peterson; Trobriand I. (AMNH).
1976	(March) B. M. Beehler; Goodenough I.; field observations (Beehler 1991a)
1978	(Nov) M. LeCroy et al.; Fergusson I., field observations of *P. decora* (LeCroy et al. 1980).
1980	(2 May) C. Unkau; Goodenough I., *M. comrii* (PNGM).
1980	(July) B. M. Beehler and H. Sakulas; Goodenough I., *M. comrii* (WEI).
1987	(15 Aug) T. Flannery; Goodenough I., *M. comrii* (CSIRO).

32: Louisiade Archipelago and small islands of Milne Bay, PNG

1846–50	John Macgillivray; Louisiade Archipelago (BMNH).
1897	(Mar) A. S. Meek with Gulliver, W. B., H. and T. Barnard; Woodlark and Mura I. (BMNH).
1897	(23 July) A. S. Meek; Basilisk I., *M. atra* (BMNH).
1897	(Aug) A. S. Meek; Louisiade Archipelago, St. Aignan or Misima I. (BMNH).
1898	(Mar–Apr) A. S. Meek; Louisiade Archipelago, Sudest or Tagula I. (BMNH).
1916	(Jan–Mar) A. S. Meek (? with A. and G. Eichhorn at least 21 Feb); Louisiade Archipelago, Sudest or Tagula I. (BMNH).
1928	(27 Sept) H. Hamlin; Samarai I., *P. raggiana, M. atra* (AMNH).
1930?	H. Hamlin; Louisiade Archipelago, 31 July Panopompom, Deboyne I. (AMNH).
1956	Fifth Archbold Expedition, L. J. Brass and R. F. Peterson; Louisiade Archipelago, Misima I. (formerly St. Aignan), 16 July–13 Aug; Sudest I. (15 Aug–25 Sept); Rossel I. 26 Sept–29 Oct (AMNH).
1956	(1–24 Nov) Fifth Archbold Expedition, L. J. Brass and R. F. Peterson; Woodlark I. (AMNH).

33: Cape York Peninsula, Queensland, Australia

Note: collectors often wrote 'Cape York'—a location at the extreme N of Cape York Peninsula, Queensland—when they actually meant the latter.

1884	(Jan–Sept) K. Broadbent; *M. keraudrenii, P. magnificus* (QM).
1894	R. Grant; Cape York, *M. keraudrenii* (AM).
1896	(16 Oct–3 Nov) H. G. Barnard; Somerset, *M. keraudrenii, P. magnificus* (SAM).
1897	(14–23 Jan) H. G. Barnard; Somerset, *P. magnificus* (QM, MV).
1897	(23 Jan) R. Grant; Somerset, *M. keraudrenii* (MV).
1899	(1 Aug) A. S. Meek; Cape York, *P. magnificus* (ROM).
1899–00	(18–21 Dec) C. A. Barnard; Somerset, *M. keraudrenii, P. magnificus* (MV, BMNH).
1902	(3 Dec) R. Jardine; Somerset, *M. keraudrenii* (AMNH).
1910	(29 Nov and 7–17 Dec) H.G. Barnard; Lockerbie, *P. magnificus* (MV).

1911	(Jan–Mar) W. R. McLennan; Paira, Cape York, *M. keraudrenii*, *P. magnificus* (SAM).		Tozer Gap, *M. keraudrenii*, *P. magnificus* (QM).
1911	(6 Mar) W. D. Dodd; Cape York, *M. keraudrenii* (SAM).	1985	(11–16 June) R. S. Schodde; Silver Plains, S. McIlwraith Ra, Qld., *M. keraudrenii* (CSIRO).
1911	(5 Sept) J. P. Rogers; Cape York, *P. magnificus* (AMNH).	1986	(24–29 Nov) G. Ingram; AM/QM Exped., Claudie R., *M. keraudrenii*, *P. magnificus* (AM, QM).
1912	(2 Apr) W. R. McLennan; Cape York, *P. magnificus* (SAM).	1990	(July–Aug) I. J. Mason, J. C. Wombey and N. W. Longmore, E.; McIlwraith Ra, *M. keraudrenii*, *P. magnificus* (CSIRO, QM).
1912–3	(July–Dec and Jan–Aug) R. Kemp; Cape York, *M. keraudrenii* (AMNH).		
1913	(July–Oct) W. R. McLennan; Claudie and Pasco R., Cape York, *P. magnificus* (SAM).	1990	(Nov–Dec) C. and D. Frith; Iron Ra observations (Frith and Frith 1993*d*).
1913	(23 Oct–21 Nov) W. D. Dodd; Cape York, *M. keraudrenii*, *P. magnificus* (SAM).		

34: Cairns/Daintree/Atherton Wet Tropics, N. Queensland, Australia; all involving *P. victoriae* only.

1913	(Nov–Dec) J. A. Kershaw; Claudie R., *M. keraudrenii* (MV).
1914	(2–4 Jan and 8 Mar) J. A. Kershaw and W. R. McLennan, respectively; Claudie R., Cape York, *P. magnificus* (MV, SAM).
1915	J. Thorpe; Cape York, *P. magnificus* (AM).
1916	(5 July) Dr S. Fern; Cape York, *M. keraudrenii* (MV).
1922	(30 Sept–1 Oct) W. R. McLennan; Coen, *P. magnificus* (MV).
1923	(July–Aug) G. H. Wilkins; Olive R., *P. magnificus* (BMNH).
1932	(25 June) P. J. Darlington; McIlwraith Ra, *P. magnificus* (MCZ).
1938	(5 July–1 Sept) G. Scott; Somerset and Rocky Scrub, *M. keraudrenii*, *P. magnificus* (CNHM).
1948	(July–Aug) D. P. Vernon; Rocky Scrub, Peach R., and

1874	(May) Rockingham Bay district (AM).
1878	(Nov) A. Morton; Barnard I. (AM).
1886	(Oct) K. Broadbent; Herbert R. Gorge (QM).
1889	(May and Nov) K. Broadbent; Herberton Scrubs and Bellenden Ker Ra (QM).
1889	(17 July and 7 Nov) R. Grant; Johnstone R. and Herberton (AM).
1896	(Dec) ex Mathews collection; Barnard I. (AMNH).
1899	(6–8 Nov) E. Dhoe; Mt. Sapphire (AMNH, ZSM).
1899	(24 Dec) E. Olive; Bellenden Ker (AMNH).
1899	(Nov and 9–30 Dec) H. C. Robinson; Bellenden Ker (SMT, ZSM).
1901	(11 Nov) W. Day; Boar Pocket near Cairns (MV).

1903	(12 July) R. Grant; Johnstone R. (AM).
1908	(21 June) E. Olive; Johnstone R. (AMNH).
1908	(Nov) G. Sharpe; Atherton and Herberton (MV).
1909	(Nov) G. Sharpe; Herberton Ra. (MV).
1910	(17 July) G. Sharpe; Allumbah (SAM).
1916	(Sept–Oct) H. G. Barnard; Cardwell and Cardwell Ra. (MV).
1922	(12–18 June) T. V. Sherrin; Vine Creek, Ravenshoe (MV).
1925	(Oct–Dec) H. G. Barnard; Kirrima Ra. and Hinchinbrook I. (MV, QM).
1932	(7–9 Apr) P. J. Darlington; Milla Milla (MCZ).
1946	(22–23 July) G. Scott; Puzzle Creek (CNHM).
1948	(13 Sept) D. P. Vernon; Mt. Finnegan (QM).
1956	(Aug) D. P. Vernon; Walsh Camp, near Atherton (QM).
1960	(12 Nov) S. Breeden; Gap Creek, near Cooktown (QM).
1964	(Jan–July) Harold Hall Expeditions; Big Tableland to Bloomfield area, near Cooktown (BMNH).
1965	(17 Aug) J. S. Dorrow; Innisfail area (QM).
1971–2	(3 Apr and 14–21 Sept) F. Crome; Mission Beach and near Mt. Haig, Lamb Ra and Lacy Creek, Mission Beach (CSIRO).
1975	(24 Nov and 3 Dec) 7 km N of Bloomfield R. (QM).
1985	(2–3 July) J. C. Wombey; Mission Beach (CSIRO).
1986	(18 July) L. Christidis; Mt. Walker, S. of Cooktown (CSIRO).
1988	(14 Oct) R. S. Schodde and J. C. Wombey; Atherton (CSIRO).
1989	(Dec) CSIRO; Lake Eacham (CSIRO).
1991	(12 Apr) CSIRO; Lake Eacham (CSIRO).
1993 and 96	(Oct–Dec); Malanda (Frith and Frith 1995*b*, 1998*a*).
1996	Malanda (Frith and Cooper 1996).

35: Australian Subtropics of S Queensland (Qld) and New South Wales (NSW); all involving *P. paradiseus* only.

1870	(Oct) Richmond R., NSW (AM).
1893	(May) F. Ferdinand; Richmond R., Cape Byron, NSW (NMW).
1893	(28 Jan and 5 July) A.P. Goodwin; Lismore, NSW (BMNH).
1896	(11 Oct and Nov) ex Mathews collection; Richmond R., NSW (AMNH, MHN).
1898	(5 and 8 Jan) A. J. Campbell; Richmond R., NSW (MV).
1898	(Oct) S. W. Jackson; Dorrigo Scrubs, NSW (MV).
1899	(10–12 Dec) S. W. Jackson; Booyong, Richmond R., NSW (MV).
1903	(Oct) E. Ashby; Blackall Ra, Qld. (SAM).
1903	(6 Aug) R. Grant; Bellinger R., NSW (MV, ?QM).
1903	(28 Jan) C. J. Wild; North Pine R., Qld. (QM).
1905	(22 Nov) R. Grant; Tweed R., NSW (MV).
1906	(10 Oct) R. Grant; Tweed R., NSW (MV).

1907	R. Grant; Richmond R., NSW (AM).	1940	(19 Sept–10 Oct) L. Macmillan and J. R. Henry; Bunya Mts, Qld. (MV, AMNH).
1907	(5 Sept) McKenzie; North Pine R., Qld. (QM).		
1907	(10 Nov) Kirkpatrick; Emu Vale, Qld. (QM).	1954	(24 Aug) D. M. Parsons; Canunga, Qld. (PMNH).
1907	(Nov) S. Robinson; Tweed R., NSW (SAM).	1964	(22 Sept) D. Dulley; Mt. Glorious, Qld. (QM).
1910	(27 Oct) J. Lamb; Eumundi, Qld. (QM).	1964	(18–30 June) T. Kirkpatrick; The Head, SE Qld. (QM).
1910	(Oct–Nov) S. A. White; Mt. Tamborine and Blackall Ra, Qld. (SAM).	1975	(20 Mar) D. P. Vernon; Dawes Ra, SE Qld. (QM).
		1978	(12–14 July) F. M. Sheldon; Brindle Creek, N. of Kyogle, NSW (PMNH).
1914	(8 Nov) R. Grant; Lismore, NSW (AM).		
1918	(20 Nov) R. Grant; Mymbodoyda, NSW (AM).	1983	(9–16 Apr) R. S. Schodde and J. C. Wombey; Richmond R. area, NSW (CSIRO).
1919	(Aug–Sept) C. M. Hoy; NSW (USNM).		
1919	(3 Oct) S. A. White; Bunya Bunya Mts, Qld. (SAM).	1985	(13 Apr–19 June) J. C. Wombey; Wilson R., NSW and Wide Bay district, Conondale Ra., Qld. (CSIRO).
1919	(Oct–Nov) N. W. Cayley; Bunya Bunya, Qld. (AM).		
1920	(19 Oct) J. Ramsay; Tweed R., NSW (MV).	1985	(15 June) L. Christidis; Murwillumbah, Richmond-Tweed R. district, NSW (CSIRO).
1921	(14 May) S. A. White; Imbil Forest, SE Qld. (SAM).		
1923	(Sept) N. W. Cayley; Bunya Mts, Qld. (AM).	1985	(9 Dec) G. Borgia; Tenterfield Shire, NSW (AM).

Appendix 3

Results of an examination of 4852 museum skin specimens with a locality and date of collection for obvious signs of active moult

Species	Total no. examined	Total no. in moult	Jan	Feb	Mar	Apr	May	June	July	Aug	Sept	Oct	Nov	Dec
Loria's Bird of Paradise	192	59	2	12	9	14	25	2	12	7	7	7	3	2
Crested Bird of Paradise	96	19	0	0	0	0	0	5	53	11	5	11	11	5
Yellow-breasted Bird of Paradise	53	22	0	5	5	9	23	9	0	9	27	14	0	0
Macgregor's Bird of Paradise	35	12	0	8	0	0	0	0	17	8	42	17	8	0
Paradise Crow	103	59	3	2	8	10	10	10	12	19	10	8	5	2
Glossy-mantled Manucode	160	65	12	19	15	9	9	9	5	2	5	6	2	8
Jobi Manucode	61	37	11	5	14	5	11	8	5	11	5	16	5	3
Crinkle-collared Manucode	105	26	4	12	19	8	0	0	15	15	12	4	12	0
Curl-crested Manucode	65	21	0	0	0	10	19	0	0	0	0	0	57	14
Trumpet Manucode	294	101	13	15	10	18	9	7	9	2	1	9	4	4
Long-tailed Paradigalla	9	3	33	0	0	0	33	0	33	0	0	0	0	0
Short-tailed Paradigalla	84	34	27	12	3	0	9	0	12	6	3	9	18	3
Arfak Astrapia	11	5	0	20	0	80	0	0	0	0	0	0	0	0
Splendid Astrapia	142	65	12	12	2	9	6	0	15	3	3	12	22	3
Ribbon-tailed Astrapia	83	28	0	4	0	0	0	0	50	29	7	0	11	0

*% of moulting birds in moult each month

Appendix 3 *continued*

Species	Total no. examined	Total no. in moult	% of moulting birds in moult each month*											
			Jan	Feb	Mar	Apr	May	June	July	Aug	Sept	Oct	Nov	Dec
Stephanie's Astrapia	174	72	3	1	7	38	18	14	7	3	3	1	6	0
Huon Astrapia	44	22	9	18	50	9	0	5	5	0	0	5	0	0
Western Parotia	64	11	9	0	0	27	9	18	36	0	0	0	0	0
Lawes' Parotia	140	48	15	19	19	10	4	13	2	4	2	6	4	2
Wahnes' Parotia	34	16	13	19	31	6	6	0	0	0	0	6	6	13
Carola's Parotia	100	22	18	9	5	0	0	5	0	5	27	5	9	18
King of Saxony Bird of Paradise	139	53	0	4	6	11	19	2	17	8	4	13	15	2
Magnificent Riflebird	213	63	16	14	13	10	6	3	2	10	2	10	5	11
Paradise Riflebird	56	12	17	0	8	25	0	0	0	8	0	25	8	8
Victoria's Riflebird	73	3	0	0	0	33	0	0	0	0	0	0	0	67
Superb Bird of Paradise	304	58	17	21	22	9	7	2	4	2	2	12	2	2
Black Sicklebill	150	47	9	11	0	4	6	23	11	6	6	2	15	6
Brown Sicklebill	145	59	12	10	9	12	9	3	14	2	5	7	15	3
Buff-tailed Sicklebill	73	26	12	31	12	4	4	8	0	12	8	0	0	12
Pale-billed Sicklebill	30	13	0	0	8	0	0	0	0	38	0	46	8	0
Magnificent Bird of Paradise	285	84	6	4	12	7	11	1	10	12	5	6	13	14
Wilson's Bird of paradise	34	8	0	0	0	0	38	63	0	0	0	0	0	0
King Bird of Paradise	294	77	5	13	9	9	8	1	7	4	4	13	8	20
Standardwing Bird of Paradise	105	20	5	15	0	20	10	20	10	5	0	0	10	5
Twelve-wired Bird of Paradise	169	58	7	2	5	4	9	4	9	10	9	28	10	5
Lesser Bird of Paradise	201	47	11	6	23	19	4	6	9	2	2	2	9	6
Greater Bird of Paradise	118	44	9	9	2	2	2	0	0	2	14	23	25	11
Raggiana Bird of Paradise	238	22	27	9	23	0	5	0	5	5	5	0	14	9
Goldie's Bird of Paradise	38	9	0	0	0	0	44	0	0	0	0	0	0	56
Red Bird of Paradise	55	12	17	0	8	0	25	17	0	0	0	17	8	8
Emperor Bird of Paradise	29	11	9	36	0	36	0	0	0	0	0	0	9	9
Blue Bird of Paradise	54	17	12	18	24	12	6	6	0	0	12	12	0	0
	4852	1490												

* to nearest whole % number

Appendix 4

Appendix 4 Summary of records of plant species recorded to be eaten by birds of paradise, from the literature and unpublished so...

Plant	C. loriae	C. mac	L. sericea	M. pulchra	L. pyrr	M. atra	M. job	M. chal	M. comrii	M. keraud	P. carunc	P. brev	A. nigra	A. splend	A. mayeri	A. steph	A. roths	P. sefilata	P. lawesii
Acronychia kaindiensis	+	+													+				
Acronychia sp																			
Aglaia cf. verteeghi																			+
Alocasia brisbanensis																			
Alocasia macrorrhizos																	+		
Alocasia sp																+			
Alphitonia macrocarpa																			
Alphitonia petrei																			
Alpinia singuliflora		+																	
Alpinia tephrochalmys		+													+				
Alpinia sp		+		+															
Anaphalis mariae															+				
*Annona muricata***																			
Apodytes brachystylis																			
Aporusa cf. nigro-punctata																			
Aralia sp															+				
Archontophoenix alexandrae																			
Archontophoenix cunningharniana																			
Ardisia brevipedata																			
Arthrophyllum macrantrum																			
Arytera pauciflora																			
Astelia alpina				+															
Bubbia sp		+																	
Calamus australis																			
Calamus sp												+							+
Callicarpa pentandra								+											
Canarium macadami																			
Canthium cf. brevipes/valetonianum																			
Carica papaya *																			
Carpodetus sp																	+		
Celastrus subspicata																			
Chisocheton cf. lasiocarpus																			
Chisocheton weinlandii						+													+
Chisocheton sp						+													
Cinnamomum oliveri																			
Cissus aristata						+													
Cissus hypoglauca								+											+
Cissus sp														+					
Citronella smythii																			
Citrus sinensis *																			
Gladomyza acrosdera				+															
Coprosma divergens				+															
Coprosma novoguineensis		+													+				
Cryptocarya mackinnoniana																			
Cucurbita sp																	+		
Cupaniopsis dallachyi																			
Cyrtostachys sp																			
Cypholophus sp.	+																		
Dacrycarpus compactus				+															
Decaspermum sp																			
Dimorphanthrea alpinia		+													+				
Diploglotis sp																			
Drimys sp	+																		
Dysoxylum cf. macrothyrsum																			
Dysoxylum cf. pettigrewianum						+													
Dysoxylum rufum																			
Dysoxylum sp						+		+											+
Elaeocarpus largiflorens																			
Eleocarpus sp	+	+															+		
Elmerrillia papuana (= *E. tsiampaca*)						+	+												+
Elmerrillia sp						+													
Endospermum labios																			
Endospermum moluccanum										+									+

Appendix 4

	P. alberti	P. magnif	P. paradis	P. victoria	L. superb	E. fastuo	E. meyeri	D. albert	D. bruijn	C. magnif	C. respubl	C. regius	S. wallac	S. melano	P. minor	P. apoda	P. raggian	P. decora	P. rubra	P. guiliel	P. rudolph
	+																				
																	+				
			+																		
					+	+															
																	+				
			+																		
						+															
																	+				
		+		+																	
			+																		
				+																	
				+	+	+															
			+																		
				+																	
			+																		
			+																		
				+																	
			+														+				
				+														+				
								+													
			+	+																	
				+																	
			+																		
		+								+								+			
		+		+						+								+			+
													+								
		+																			
										+											
			+		+			+		+								+			+
			+																		
			+																		
			+																		
				+	+	+															
				+																	
		+							+		+										
					+																
+																					
		+																	.	.	
								+									+				
		+															+				
			+																		
			+						+								+				
			+	+																	
				+		+		+							+		+				
																	+	+			
																	+				
				+				+									+				

Appendix 4 *continued*

Plant	C. loriae	C. mac	L. sericea	M. pulchra	L. pyrr	M. atra	M. job	M. chal	M. comrii	M. keraud	P. carunc	P. brev	A. nigra	A. splend	A. mayeri	A. steph	A. roths	P. sefilata	P. lawesii
Epipremnum sp																			
Eurya brassii				+															
Eurya tigang																+			
Euodia (Melicope) crispula																			
Fagraea salticola															+				
Fagraea sp															+	+			
Ficus coronata																			
Ficus destruens																			
Ficus drupacea										+									
Ficus gul							+			+									+
Ficus macrophylla																			
Ficus obliqua																			
Ficus odoardi							+			+									+
Ficus sterrocarpa																	+		
Ficus sp		+	+			+	+			+								+	+
Flindersia pimenteliana																			
Freycinetia excelsa																			
Freycinetia inermis																+			
Freycinetia sp														+		+	+		
Garcinia sp		+													+				
Gastonia spectabilis										+									+
Gastonia sp																			+
Glochidion sp																			
Gulubia sp																			
Harretia sp																			
Harpullia sp																	+		+
Hibbertia scandens																			
Horsfieldia sp																			
Ilex sp	+																		
Levieria acuminata																			
Maesa edulis																			
Maoutia sp		+														+		+	
Medinilla markgrafii																+			
Medusanthera laxiflora									+										
Meliaceae sp																			
Musa sp												+				+			
Myristica fragrans																			
Myristica hollrungii																			
Myristica insipida																			
Myristica longipes																			+
Myristica subalulata							+			+									+
Myristica 'variratii'																			
Myristica sp							+			+								+	+
Neolitsea dealbata																			
Nothocnide melastomatifolia																			
Nothofagus grandis																+			
Omolanthus (Homalanthus) novoguineensis		+														+			
Omolanthus sp									+							+			+
Pandanus brosimus																+			
Pandanus conoideus																			
Pandanus lauterbachi																			
Pandanus limbatus									+										
Pandanus sp															+	+			
Perrottetia alpestris												+				+			
Pinanga sp				+															
Piper novae-hollandiae																			
Piper sp						+	+		+										+
Pipturus argenteus																			
Pittosporum pullifolium		+													+				
Pittosporum ramiflorum																+			
Pittosporum rhombifolium																+			
Pittosporum sp	+															+	+		
Planchonella sp	+											+							
Plerandra stahliana																			
Podocarpus neriifolius									+										
Polycias elegans																			
Polycias murrayi																			

	P. alberti	P. magnif	P. paradis	P. victoria	L. superb	E. fastuo	E. meyeri	D. albert	D. bruijn	C. magnif	C. respubl	C. regius	S. wallac	S. melano	P. minor	P. apoda	P. raggian	P. decora	P. rubra	P. guiliel	P. rudolph
														+							
	+			+																	
					+																
	+						+														
	+			+	+																
			+																		
				+																	
																	+				
										+							+			+	
			+																		
			+																		
		+			+					+							+				
	+	+			+	+	+			+				+			+		+	+	+
			+																		
					+	+															
					+	+	+														
	+																				
		+								+							+				+
										+							+				
										+			+								
																	+				
																	+				
			+																	+	
		+	+																		
																+					
			+																		
					+	+	+														
	+			+	+	+															
	+			+	+	+															
																	+				
						+		+			+										+
													+								
																	+				
																	+				
										+							+				
			+																		
	+			+																	
		+		+						+					+		+				+
	+			+		+				+							+				+
					+	+													+		
														+							
		+			+	+	+			+							+				+
	+			+	+	+															
			+	+																	
					+					+							+				+
										+		+								+	
	+			+	+																
	+	+		+	+																
	+																				+
																	+			+	
		+		+													+				
			+																		
				+																	

Appendix 4 (continued)

Plant	C. loriae	C. mac	L. sericea	M. pulchra	L. pyrr	M. atra	M. job	M. chal	M. comrii	M. keraud	P. carunc	P. brev	A. nigra	A. splend	A. mayeri	A. steph	A. roths	P. sefilata	P. lawesii
Pothos longipes																			
Procris sp	+																		
Psychotria sp	+																		
Rapanea involucrata				+															
Rapanea vaccinioides kan cf. Hat																+			
Rapanea sp		+		+															
Rhododendron macgregoriae																+			
Rhododendron sp															+				
Riedelia sp	+	+																	
Rubus sp		+														+			
Sabia pauciflora																			
Sabia sp.																			
*Sambucus canadensis**																+			
Sarcotoechia protracta																			
Schefflera actinophylla																			
Schefflera 'chimbuensis'															+	+			
Schefflera lasiophaera															+	+			
Schefflera pachystyla																			+
Schefflera straminea															+	+			
Schefflera sp		+												+	+	+	+		+
Sericolea pullei	+										+				+				
Sericolea sp															+				
Sloanea cf. aberrans																			
Sloanea australis																			
Sloanea cf. sogerensis										+									+
Sloanea sp																			
Smilax ovata-lanceolata																			
Solanum mauritianum																			
Steganthera gracile		+																	
Stephania japonica																			
Sterculia sp																			+
Sterculiaceae sp																			
Styphelia suaveolens				+												+			
Symplocos cochinchinensis		+		+											+	+	+		
Symplocos sp	+																		
Synima cordieri																			
Synima macrophylla																			
Synoun gladulosum																			
*Syzygium cormiflorum***																			
Syzygium sp		+								+						+			+
Tamarind, native																			
Taro, wild															+				
Tectona grandis																			
Tetrastigma cf. dichotoma		+													+				
Timonius belensis	+	+													+	+			
Timonius singularis																			
Timonius sp																			
Tinospora sp.																			
Toechima monticola																			
Trema orientalis														+		+			
Trichosanthes sp.																+			+
Urtica sp											+					+			
Urticaceae sp															+	+			
Uvaria sp									+										
Vaccinium sp															+	+			
Vitaceae sp																			
Weinmannia trichophora																+			
Xanthomyrtus sp	+	+													+				
Xylopia sp.																			
Zingiberaceae sp							+												
Zygogynum argentia		+																	

Bird species are indicated by the first letter of the genus and an abbreviation of the specific name, in systematic order left to right.
Plants are listed in alphabetical order top to bottom.
* = Exotic plant species; ** = Flowers eaten; See species accounts (Chapter 9) for references.
Addenda: for additional plants eaten by the Paradise Riflebird see Diet and foraging text of that sp. account.

P. alberti	P. magnif	P. paradis	P. victoria	L. superb	E. fastuo	E. meyeri	D. albert	D. bruijn	C. magnif	C. respubl	C. regius	S. wallac	S. melano	P. minor	P. apoda	P. raggian	P. decora	P. rubra	P. guiliel	P. rudolph
			+																	
									+											
				+																
				+	+															
+																				
				+	+	+														
		+																		
																+				
		+																		
		+																		
				+																
																				+
	+			+	+				+							+				+
									+											
		+																		
				+					+							+				
						+			+											
				+																
	+																			
			+																	
+				+																
+																				
				+																
				+																
		+																		
				+																
																+				
		+																		
									+											
+					+															
+																				
		+																		
																+				
		+																		
									+											
				+					+							+				+
				+		+														
+				+																
+				+	+	+														
																+				
			+	+																
																			+	
									+											
																				+

Appendix 4

Appendix 5

Some published recordings of bird of paradise vocalisations

Crouch, H. and Crouch, A. (1980). Papua New Guinea Bird Calls. A cassette tape. Includes Magnificent, King, Raggiana, and Emperor (although the latter not listed on the cover) Birds. PNG Bird Society, PO Box 1598, Boroko, Papua New Guinea.

Curth, H. and Peckover, W. S. (1968). Birds of Paradise and the Sounds of Papua New Guinea. A seven inch 33.3 rpm disc, JPNG-001, that accompanies the 50 pp. book 'Papua New Guinea'. Includes the Brown Sicklebill, Lesser, Raggiana, and Blue Birds. Narrations by J. Leigh. Jacaranda Press Pty., Ltd., 46 Douglas Street, Milton, Queensland 4064, Australia.

Griffin, A. C. M. and Swaby, R. J. (1985). Bird calls of north Queensland Rainforests. A cassette tape. Includes Victoria's Riflebird. A. C. M. Griffin, 'Wirrawilla', CMB 16, Paluma, north Queensland 4816, Australia.

Peckover, W. S. (1965). Birds of Paradise. A seven inch 33.3 rpm disc. Includes the Brown Sicklebill, Lesser, Raggiana, and Blue Birds. Narrations by R. Wilson. Issued with 'The Stamp and Coin Collector' 2 (11), to illustrate article on PNG stamps. H. R. Productions Inc., 17 East 45th Street, New York 10017, USA. November.

Peckover, W. S. (1966). Birds of Paradise. A seven inch 45 rpm disc, Austas Ea 151. Includes the four species as in above 1965 publication. Pty. Ltd., Dubbo, NSW. Privately produced as promotion for PNG stamp albums: not commercially available.

Stewart, D. A. (1994). Voices of Subtropical Rainforests, Australia. A 12 cm diameter compact disc. Includes Paradise Riflebird. Nature Sound, PO Box 256, Mullumbimby, NSW 2482, Australia.

Stewart, D. A. (1996a). Australian Bird Sounds—Lamington National Park. A cassette tape. Includes Paradise Riflebird. Nature Sound, PO Box 256, Mullumbimby, NSW 2482, Australia.

Stewart, D. A. (1996b). Australian Bird Sounds—Queensland's Wet Tropics and Great Barrier Reef: the smaller, or passerine birds. A cassette tape. Includes Victoria's Riflebird. Nature Sound, PO Box 256, Mullumbimby, NSW 2482, Australia.

References: Boswell (1965), Boswell and Kettle (1975), Boswell (1981).

Appendix 6

A brief guide to where and how best to study wild birds of paradise

There is little in nature that can rival one's first glimpse of a plumed male bird of paradise in its native habitat. Although getting to a forest where birds of paradise range through the canopy is beyond the expectation of many ornithologists, the logistics involved grow easier as each year passes.

The most practical (but expensive) means for most people to see a variety of birds of paradise in the field is to join a commercial birding tour to an area with birds of paradise. Speak to the leader and get assurance that the tour will make serious attempts to observe the birds before signing on. Note that 'nature' tours—by contrast to 'birding' tours—often will make only minimal effort to show its clients these elusive birds. It is worth noting that some birding tours are better than others at showing its clients birds of paradise. The reader should determine which tour is most serious about seeing them. That being said, note also that these tours are arduous, requiring long hours, rugged hikes, and a variety of discomforts (rain, biting insects, bone-jolting drives in hot and dusty vans). The tour company will instruct the participants on clothing and equipment needs and other necessary preparations.

Those who wish to look for these birds on their own should expect to see fewer species, but nonetheless the experience will be unique. First, be prepared for the unexpected (delayed flights, broken-down vehicles, bad weather, uncooperative birds). Allow extra time for unforeseen delays. It is necessary to prepare by studying the travel guides that feature jungle venues and trips off the beaten track. In addition, one will need to learn the distributions and habits of the birds in order to plan an adequate itinerary.

Some very general guidance is useful here. Take dull or dark-coloured cotton clothing—long trousers and shorts as well as long-sleeved and short-sleeved shirts. A hat with a visor is recommended for the sun and rain. Footwear should be diverse and should include lightweight and easily-dried boots (they will become wet daily), tennis shoes, and river-runners and/or inexpensive plastic thongs, plus many pairs of socks. Insect repellent and treatment for insect bites/stings are a must. As ever, a small personal medical/ first aid kit should be carried. A raincoat is usually only marginally useful in the hotter lowlands (expect to get wet) but at higher elevations offers important protection against cold and wet afternoons. A small umbrella can always be handy. A small waterproof daypack is also mandatory for your books, binoculars, etc. in the field. When you are afield in Papua New Guinea or Indonesia, be sure to have small change for tipping and for purchase of fresh fruit or guide services. In most of these places, malaria is the greatest health threat and it can best be avoided by taking some sort of prophylaxis and, more importantly, avoiding village areas at dusk and after nightfall—virtually all malaria is transmitted in the village at night—not in the jungle during the day.

Most birds of paradise are wide-ranging through their habitat, so it is a matter of getting into the right habitat and listening for the birds' telltale vocalisations (the birds are heard many times for every time they are seen). Remember that adult males in full nuptial plumage are greatly outnumbered by birds in the dull female plumage. Note also that many species prefer undisturbed forest and are only uncommonly seen at the forest-edge. Lek-breeding species are best observed at an active lek, and these leks can be difficult to locate. Display sites of the solitary species are even more difficult to find. Get help from local village informants—every village usually has one or two men who have an excellent grasp of local nature lore.

It is not necessary to go only to the places birdwatchers typically go to—it's more exciting and more interesting, and one has an opportunity to make a scientific contribution, if you visit a site never before visited by ornithologists. These are readily accessible today and thus are waiting for the more adventurous among us.

Australia

This is where most birdwatchers see wild birds of paradise, simply because this part of Australasia has the greatest ease of access and the most familiar culture and facilities. Birding in Queensland is safer, easier, and cheaper than any other place where birds of paradise live.

Places to observe the Paradise and Victoria's Riflebirds are quickly and easily accessible by car out of Brisbane and Cairns, respectively, and, as ornithologists are relatively numerous, ornithological societies and infrastructure are well developed in these areas, so no more need be said here. The Magnificent Riflebird and Trumpet Manucode are more difficult to get to, but can be found at various locations on Cape York Peninsula. Two of the most convenient involve flying or driving into *Iron Range* or *Bamaga*. If driving, the somewhat unpredictable rainy season (*c.* Dec–June) should be avoided. While birds can be seen within a walk of the *Iron Range* airstrip, this and *Bamaga* are remote places not to be visited without adequate preparation. Be sure to seek local knowledge before departing for the bush.

Papua New Guinea

Without question, Papua New Guinea (PNG) is the destination of choice for ornithologists wishing to see many species of birds of paradise in the wild. With 33 species (nine of them endemic), PNG is the nation that supports more birds of paradise than any other on earth.

Birds of paradise inhabit all forested environments in PNG, thus field observers can find members of the family in virtually every part of the country—not just in selected locales. Some areas, such as the Ambua/Tari region are rightly famous for the Paradisaeidae, but many other lesser-known spots can be equally productive. Whereas a typical lowland forest plot will support five or six species of birds of paradise (Raggiana, King, Twelve-wired, Magnificent Riflebird, and two manucodes) upland forest plots between 1300 and 2000 m may support as many as 10 coexisting species—including many species not found in the lowlands. An observer seeking a wide range of species should work forests at a range of elevations (e.g. sea level, 1500 m, 2200 m, 2700 m), as well as visit sites in both eastern and western PNG.

Papua New Guinea has welcomed tourists for decades, but it is worth planning a trip to PNG carefully, especially if one's specific goal is to see birds of paradise. Although many species can be observed at roadsides and near towns, one may need to invest special effort to find some of the more elusive species. We advise that tourists should make the effort to learn prior to travel about conditions in the areas they plan to visit and that they fully prepare all logistics in advance. Risk of robbery is fairly high, and visitors may not be safe from personal harm unless proper precautions are taken. This advice should in no way discourage the adventurous birdwatcher from seeking birds of paradise but is merely intended to ensure the visitor is well prepared. Although one must remain aware and take precautions in the towns, cities and provincial centres, in the back country (accessible by 'bush' plane) conditions are rustic but the populace will (usually but not inevitably) be cordial, hospitable, and peaceable. Confirm conditions before going to any outstations by speaking to someone in the country who has recently been there. Camping and bushwalking are popular for residents and visitors alike, and are a good way to get to little-visited parts of the country where conditions are pristine, people continue to live the traditional life, and birds of paradise continue to haunt every patch of forest. It is of great importance that the rights of land owners be at all times respected. There are several tourist and backpacking guidebooks to PNG that provide the necessary information to arrange one's own bird of paradise tour. Note, however, that unless one is part of a birding tour, it will be difficult to locate local guides or experts willing or sufficiently knowledgeable to show you a range of birds of paradise. The exception is at Ambua Lodge on the slopes of the Tari Valley, Southern Highlands Province, where the Paradisaeidae is a speciality of ecotourist entertainment and a mecca for wildlife film crews. Below, listed alphabetically, is a selection of recommended sites that support many paradisaeid species. Remember that conditions change and that every effort should be made to obtain fresh information prior to finalising travel plans. The Papua New Guinea Bird Society (P.O. Box 1598, Boroko, N.C.D., PNG) is a source of useful guidance.

Crater Mountain Wildlife Management Area. The vast hill forests of the Crater Mountain reserve are bounded on the north by the imposing Crater Mountain, an extinct Pleistocene volcano. This locally-managed reserve is an excellent destination for the more adventurous ornithologist and bushwalker. Access is from Goroka by bush plane to the villages of Heroana and Haia. Heroana supports a growing ecotourism initiative that provides village accommodation and naturalist/guides for visiting scientists, hikers, and birders. The reserve also boasts an ecological field station in the rugged hill forest a long day's walk from the airstrip at Haia. All of the hill and montane forest species inhabiting central PNG can be found in the reserve (including both Lawes' and Carola's Parotias, Yellow-breasted Bird, and Buff-tailed, Black, and Brown Sicklebills).

Lakekamu Basin. The lowland alluvial forests of the upper Lakekamu drainage are surrounded by hills and mountains, forming a remarkable interior basin that is currently the focus of a locally-managed 'integrated conservation and development' initiative much like that at Crater Mountain. Whereas Crater provides access to the montane and hill forest birds of paradise, Lakekamu is an excellent place to observe lowland forest species in nearly pristine circumstances. Spartan field accommodation is available at Tekadu and Kakoro, villages with airstrips served by flights from Wau and, less frequently, from Kerema or Port Moresby. As with Crater, Lakekamu is a perfect site for field research studies on birds of paradise, and is also an exciting destination for the adventurous birdwatcher (Raggiana, King, Twelve-wired Birds, two manucodes, and the Magnificent Riflebird). The adjacent hill forests support additional species. The most adventurous may wish to attempt the hike from the Basin back to Wau via either the old wartime Bulldog Road (very rugged) or via the Biaru valley (moderate difficulty). Guides are needed for both treks.

Madang. Madang is a lovely seaside town that is home to the Christensen Research Institute. Lodging for visiting bird students and birders is available in town (several hotels), at the Institute, or at an adjacent coastal resort. Although reef diving is the main tourist pursuit in this area, a range of lowland and hill forest birds of paradise can be observed in the remnant forests accessible by jeep from the Institute.

Tari Valley/Ambua Lodge/Tari Gap. This area of the Southern Highlands represents one of the finest sites for finding and observing birds of paradise on earth, mainly because there is ready road access to altitudes of *c.* 1650–3570 m, where no fewer than 13 species have been recorded (Frith and Frith 1992*a*, 1993*a*). Moreover, roads and tracks make the ornithologically-unexplored Muller Range, to the west of Tari township, accessible to the adventurous. The most desirable species here include the Ribbon-tailed Astrapia, King of Saxony, the Blue, Crested, and Loria's Birds, and three sicklebills.

Wau. Nestled in a mid-montane valley an hour's flight north from Port Moresby, Wau is an old gold and timber town that is home to the Wau Ecology Institute, an excellent base for searching for birds of paradise in the surrounding forests of Mount Missim, the Ilaru Range, and Mount Kaindi. Although the town itself has not prospered since the departure of the major gold and timber operations, the mountain forests remain rich with birds of paradise, and the staff of the Ecology Institute can assist both students and birdwatchers in their quest. Hill forest and montane species are well represented.

Varirata National Park, an hour's drive out of Port Moresby, is perched atop the Sogeri Plateau overlooking the Coral Sea. This Park is justly famous for its abundant population of Raggiana Birds, but there are also two manucodes, the eastern population of the Magnificent Riflebird, and the Magnificent Bird present—making the Park an excellent first-stop for the ornithologist visiting PNG.

Irian Jaya, Indonesia

Irian Jaya is, for the naturalist, one of the most exciting places on earth, in part because it is home to 28 species of birds of paradise, five of them endemic. Much of this easternmost province of Indonesia is undeveloped, with vast expanses of forest and rugged mountain ranges—a paradise for bushwalkers and those curious about diverse traditional cultures and wonderful wildlife.

Visitors to Irian Jaya should plan carefully and in advance should make themselves aware of the current conditions as well as government requirements for visas and travel authorisations (*surat jalan*). It is worth the effort, since some of the least-known birds of paradise inhabit this western half of the great island of New Guinea. Below we discuss five possible birding destinations in Irian Jaya.

Jayapura/Sentani/Cyclops Mountains. Most visitors to Irian Jaya will visit the provincial capital of Jayapura. This is a useful base for searching for a range of lowland birds of paradise. The Pale-billed Sicklebill inhabits the lowland forests east of Jayapura (listen for its weird whistled song from roadsides that pass through original or selectively-logged lowland forests). This same habitat will support Lesser and King Birds as well as the western population of the Magnificent Riflebird. As many as four species of manucodes inhabit the lowland and hill forests of the region. The lovely and picturesque Cyclops Mountains rise behind Jayapura and are best visited via a well-known hiking trail that starts at Sentani town. The array of lowland species, plus the Magnificent Bird, can be found in the forests of the Cyclops. The Jayapura/Sentani region is fast-developing and, as more roads penetrate westward and southward into the interior, opportunities arise to visit novel sites for finding birds of paradise.

Wamena/Lake Habbema. Perhaps Irian Jaya's most famous tourist destination is Wamena, in the Grand Valley of the Balim—a spectacular grassy interior highland valley surrounded by high rocky peaks and forested ranges. This is also a good base for students wishing to observe some of Irian's montane birds of paradise. The two-day hike up to Lake Habbema is both spectacular and physically challenging, and provides access to the high montane habitats where one can see Macgregor's and King of Saxony Birds and Splendid Astrapia (the route via Wamena Creek is easier than that via Ibele, and a minimum of five days should be allowed for the entire trek). Birdwatchers have recently located Crested Birds in this area as well (in the upper Ibele Valley). Visiting Wamena will be of interest only to those willing to take to the trail and camp out in the forest—'day trippable' access to birding habitats by automobile apparently remains limited.

Arfak Mountains/Manokwari. The Vogelkop Peninsula of western Irian Jaya is famous for its endemic birds of paradise, and the best base for searching for them is perhaps via the coastal centre of Manokwari. Lowland and hill forest species can be found in the forests around Manokwari (Lesser, King, Magnificent Birds, Magnificent Riflebird, and one or more manucodes), and upland species (Long-tailed Paradigalla, Black Sicklebill, Arfak Astrapia, Western Parotia) can be found at higher elevations in the Arfaks (accessible either by foot or by bush plane into one of the upland airstrips).

Rajah Ampat Islands. Two of Irian Jaya's endemic birds of paradise are found only on Batanta and Waigeo Islands of the Rajah Ampat group—off the west coast of the Vogelkop. Apparently neither island is accessible by air, and thus the intrepid birder must charter a motor-powered prahu (dugout canoe) or other such craft—apparently quite expensive. Waigeo is a rugged and physically-complex island that is bisected by a remarkable inland bay. Batanta is smaller and apparently easier of access. Wilson's Bird inhabits hill forests and is apparently widespread but difficult to observe except at the courts of males. The Red Bird inhabits lowlands and hill forests and is readily found once one gets on the islands.

Timika. The operational base for the huge copper and gold mine owned by Freeport Indonesia is the rapidly-growing town of Timika. Timika has a large airport as well as guest houses and a luxury hotel. The lowland forests around Timika support good populations of birds of paradise—Greater, King, Twelve-wired Birds, Magnificent Riflebird, and two manucodes. All of these species inhabit the remnant forest found abutting the town's airstrip (and accessible on foot from the Sheraton). A network of roads gives access to other forests at differing elevations, but before visiting Timika it is important to inquire about visitor access to these areas. This is a site rarely visited by touring birdwatchers, so there is much to be learned about its potential to the ornithologist.

Maluku Utara, Indonesia

The Northern Moluccas is one of Indonesia's least visited and least accessible regions, so any visit promises both adventure and valuable field observations. Only two paradisaeid species inhabit the region—the Standardwing and the Paradise Crow. Both can be found on Halmahera and Bacan islands, and the Paradise Crow can also be found on nearby Obi Island.

The only ready access to the region is via the daily flight to Ternate Island, a small volcanic outlier of the large, complex, and rainforest-clad island of Halmahera, the logical destination for most visitors. Take a ferry from Ternate to Halmahera, and explore the coastal and hill forests for the two birds of paradise. The Paradise Crow should be readily found in the coastal forest and forest-edge. The Standardwing appears to be most readily found in the rugged hill forests just behind the coast. There is a small guest house or *losmen* at Tanah Batu Putih, 8 km out of Sidangoli, whose owner is a knowledgeable naturalist who can lead you to a nearby Standardwing lek. This and other matters relating to birdwatching in Maluku and Irian Jaya are detailed by Jepson and Ounsted (1997).

Appendix 7

Gazetteer

Most places are located by geographical coordinates. Note that all coordinates are decimal degrees rather than degrees and minutes (hence the decimal place). Thus 2.5S = 2°30′S. To convert from decimal degrees to the more traditional degrees and minutes, take the fractional part of the degree (e.g. 0.5 in the above example) and multiply this by 60. For example, decimal degrees 2.5S = 2 degrees and 0.5 × 60 = 30 minutes. Decimal degrees are useful for digital plotting of localities. Note that those coordinates for which there is greater accuracy of location have more decimal places, those less exact have fewer. Coordinates for major rivers are located not at the mouth, but in the middle section of the river course, between the mouth and the source. Coordinates for mountain ranges are situated in a central point in each range.

Adelbert Ra, PNG, 4.9S, 145.4E
Aiome, PNG, 5.14S, 144.72E
Albany I., Aust., 10.75S, 142.60E
Aleya, Hall Sound, PNG, 8.77S, 146.54E
Ambon, Indonesia, 3.67S, 128.17E
Ambua Lodge, PNG, 5.97S, 143.05E
Ambua Ra, PNG, see Ambua Lodge
Arafura Sea, between Moluccas/IJ and N. Aust., 9S, 136E
Arfak Mts, IJ, 1.2S, 134.0E
Aru Is, S. Moluccas, 6.0S, 134.5E
Aruma Apa-Maguli Ra, PNG, see Mt. Maguli
Atherton Tableland, Aust., 17.5S, 145.45E
Astrolabe Mts, PNG, 9.5S, 147.5E
Awande, near Okapa, PNG, 6.56S, 145.75E

Bacan (Batjan, or Batchian) I, Moluccas, 0.6S, 127.5E
Baiyer R (Sanctuary), Baiyer Valley, PNG, 5.50S, 144.15E
Balim (Baliem) Valley, IJ, 4.0S, 138.9E
Balim (Baliem) Gorge, IJ, 4.5S, 139.3E
Balimo, Fly R., PNG, 8.02S, 142.95E
Barnard Is, Qld., Aust., 17.67S, 146.18E
Barrington Tops, Aust., 32.95S, 151.50E
Batanta I., Rajah Ampat Is., IJ, 0.95S, 130.6E
Bensbach, PNG, 8.87S, 141.73E
Bernhard Camp, Idenburg R., IJ, 3.48S, 139.22E
Bewani Mts or Ra, PNG, 3.2S, 141.5E
Biak I., IJ, 1.0S, 136.0E
Big Scrub, Aust., 28.75S, 153.50E
Big Tableland, near Cooktown, Aust., 15.93S, 145.37E
Binna Burra, Lamington, Aust., 28.18S, 153.18E
Biri, Van Rees Mts., IJ, 2.82S, 137.83E
Bisa I., Moluccas, 1.22S, 127.5E
Bismarck Mts or Ra, PNG, 5.8S, 145.0E
Blackall Ra, Aust., 26.58S, 152.83E
Bloomfield R., N of Cairns, Aust., 16.0S, 145.3E
Boana, Huon Penin, PNG, 6.43S, 146.83E
Bobairo, IJ, just N. of Enarotali, see Enarotali
Bodim, IJ, 2.28S, 138.83E
Boigu I., Aust., 9.28S, 142.17E
Bomakia, Mappi R., SE IJ, [= Boma?, 6.0S, 139.95E]
Bomberai Penin, IJ, 3.2S, 133.2E
Boneno, PNG, 9.88S, 149.39E

Booyong Scrubs, Richmond R. area (see latter), Aust.
Bosavi Mission, PNG, 6.57S, 142.83E
Brisbane, Aust., 27.47S, 153.03E
Brown R., near Port Moresby, PNG, 9.20S, 147.23E
Bulldog Road, near Wau (Abid's Camp), PNG, 7.4S, 146.67E
Bulolo R., near Wau, PNG, 7.25S, 146.67E
Cairns, Aust., 16.93S, 145.77E
Calliope Ra, Aust., 24.35S, 151.0E
Cape Vogel, PNG, 9.7S, 150.0E
Cape York, Aust., 10.68S, 142.53E
Cape York Penin, Aust., 15.0S, 143.0E
Carstensz, IJ—see Mt Carstensz, IJ
Cedar Bay, Aust., 15.82S, 145.32E
Ceram (Seram), S. Moluccas, 3.0S, 129.0E
Charles Louis Ra., IJ, 4.13S, 135.52E
Chimbu Province, PNG, 6.0S, 144.9E
Clarence R., Aust., 29.5S, 153.1E
Claudie R., Cape York Penin., Aust., 12.75S, 143.28E
Cloudy Bay, PNG, 10.17S, 148.67E
Collingwood Bay, PNG, 9.43S, 149.20E
Cooktown, Aust., 15.47S, 145.25E
Crater Mountain Reserve, PNG, 6.6S, 145.2E
Cromwell Mts, Huon Penin, PNG, 6.33S, 147.25E
Cyclops Mts, IJ, 2.5S, 140.6E
Darnley I., Aust., 9.58S, 143.77E
Dawong, Herzog Mts, PNG, 6.87S, 146.77E
D'Entrecasteaux Archipelago, PNG, 9.5S, 150.6E
Deva Deva, PNG, 8.75S, 146.63E
Digul (Digoel) R., PNG, 7.2S, 139.6E
Dinawa (Dilava), NE of Yule I., PNG, 8.54S, 146.97E
Ditschi, Arfak Mts, Vogelkop, IJ, 1.50S, 134.0E
Dobo, Aru Is, S. Moluccas, 5.77S, 134.22E
Dobu I, D'Entrecasteaux Archipelago, PNG, 9.75S, 150.83E
Dokfuma meadow, near Mt. Capella, PNG, 5.17S, 141.10E
Dolok (Frederik Hendrik) I, IJ, 7.9S, 138.6E
Doma Peaks, PNG, 5.9S, 143.13E
Doorman Top, IJ, 3.50S, 137.15E

Dorrigo, Aust., 30.37S, 152.75E
Dorey (Harbour), or Manokwari, IJ, 0.86S 134.08E
East Cape, PNG, 10.28S, 150.72E
Edie Creek, Wau, PNG, 7.31S, 146.68E
Ekuti Ra (Ekuti Divide), PNG, 7.43S, 146.58E
Elevala R., Fly R. basin, PNG, 6.08S, 141.50E
Enarotali, Lake Paniai, IJ, 3.98S, 136.38E
Enga Province, PNG, 5.5S, 143.8E
English Peaks, PNG, 8.75S, 147.47E
Etna Bay, IJ, 3.96S, 134.71E
Fakfak Mts, Onin Penin., IJ, 3.0S, 132.5E
Faralulu, Fergusson I., (see latter) PNG
Fergusson I., PNG, 9.6S, 150.75E
Fly R., PNG, 8.0S, 142.0E
Finisterre Mts or Ra, Huon Penin, PNG, 5.9S, 146.3E
Finschhafen, Huon Penin, PNG, 6.52S, 147.84E
Foula, PNG, see Mafulu
Foya Mts, IJ, 2.58S, 138.75E
Frederik Hendrik, IJ, 8.0S, 138.5E
Gam I., Rajah Ampat Is, IJ, 0.5S, 130.58E
Gamkanora, Halmahera I., IJ, 1.38N, 127.52E
Gebe I., Rajah Ampat Is, IJ, 0.05S, 129.45E
Gebroeders Mts, Weyland Ra, IJ, 3.92S, 136.13E
Geelvink Bay, IJ, 2.0S, 135.0E
Gemien I., IJ, 0.33S, 130.5E
Gilolo I., see Halmahera I.
Giluwe Ra, PNG, 6.05S, 143.88E
Goenong Tobi, (NW NG), ? IJ, (see Büttikofer 1894), not located
Gogol R., PNG, 5.29S, 145.6E
Goodenough Bay, PNG, 9.9S, 150.4E
Goodenough I., PNG, 9.27S, 150.3E
Goroka, PNG, 6.03S, 145.4E
Gunung (Mt.) Gamkanora, Halmahera I., Moluccas, 1.38N, 127.5E
Gunung (Mt.) Ngribou, Arfak Mts (see latter), Vokelkop, IJ
Gusap Pass, PNG, 6.05S, 145.95E
Grand Valley, IJ, see Balim Valley
Great Dividing Ra, Aust., 23.0S, 146.0E
Haia, Purari Basin, PNG, 6.70S, 144.99E
Hall Sound, PNG, 8.83S, 146.60E
Hagen Mts or Ra, PNG, 5.77S, 144.03E

Halmahera I., N. Moluccas, 1.4N, 128.0E
Hambitawuria, IJ, 2.88S, 132.33E
Hatam, Arfak Mts Vogelkop, IJ, 1.09S, 133.97E
Herowana (Heroana), PNG, 6.62S, 145.20E
Herzog Mts, PNG, 6.9S, 146.7E
Hindenberg Mts or Ra, PNG, 5.15S, 141.58E
Hoiyeria, Mt. Hagen (see latter), PNG
Hollandia, IJ, see Jayapura
Holtekong, IJ, 2.63S, 140.78E
Hufeisengebirge, PNG, see Mt. Maguli
Humboldt('s) Bay, IJ, 2.63S, 140.78E
Hunstein Ra, PNG, 4.51S, 142.67E
Hunter R. or Valley, Aust., 32.92S, 151.00E
Huon Gulf, PNG, 6.9S, 147.2E
Huon Penin, PNG, 6.33S, 147.33E
Huon Ra, PNG, see Huon Penin.
Ialibu, PNG, 6.22S, 143.54E
Idenburg (Tatitatu) R., IJ, 3.25S, 139.0E
Ibele Valley, IJ, 3.75S, 139.15E
Ingham, Aust., 18.65S, 146.15E
Ilaga Valley, Snow Mts, IJ, 4.00S, 137.62E
Illolo, see Varirata National Park, PNG
Irian Jaya, Indonesia (western New Guinea, Irian Barat, West Irian, Netherlands New Guinea, Dutch New Guinea)
Iron Ra, Cape York Penin, Aust., 12.78S, 143.30E
Jabogema, N of Lake Habbema (see latter), IJ
Jalan Korea, Nimbokrang, IJ, 2.62S, 140.27E
Jayapura, IJ, 2.57S, 140.73E
Jayawijaya Mts, IJ, 4.5S, 139.5E
Jimi R., PNG, 5.4S, 144.25E
Jobi (Yapen) I., IJ, 1.8S, 136.3E
Kagi, Owen Stanley Ra, PNG, 9.13S, 147.67E
Kaileuna I., PNG, 8.53S, 150.95E
Kaindan, near Kompiam (see latter), PNG
Kaintiba, PNG, 7.50S, 146.00E
Kaironk Valley, Schrader Ra, PNG, 5.25S, 144.47E
Kaisenik, near Wau, PNG, 7.37S, 146.73E
Kaiser [Friedrich] Wilhelmshafen, PNG, harbour in former German New Guinea (= Madang)
Kakoro, Gulf Province, PNG, 7.84S, 146.53E
Kanga, Mambare Valley, PNG, 8.79S, 147.65E

Karan, PNG, 3.3S, 142.5E
Karawari R., Sepik, PNG, 4.72S, 143.43E
Karimui Basin, PNG, 6.5S, 144.7E
Kasiruta I., Moluccas, 0.42S, 127.2E
Karius Ra, PNG, 5.67S, 142.5E
Kemabu Plateau, IJ, 4.0S, 137.25E
Kemp Welch R., PNG, 10.0S, 147.79E
Kerema, PNG, 7.97S, 145.46E
Kerowagi, PNG, 5.88S, 144.84E
Kikita Village, near Tari (see latter), PNG
Kimil R., near Nondugl, Wahgi Valley, PNG, 5.75S, 144.52E
Kiriwina I., PNG, 8.5S, 151.08E
Kiunga, PNG, 6.13S, 141.28E
Kobroor I., Aru Is S. Moluccas, IJ, 6.17S, 134.5E
Kokoda Trail, PNG, 9.2S, 147.7E
Kompiam, PNG, 5.41S, 143.93E
Kratke Mts or Ra, PNG, 6.9S, 145.95E
Krissa (Krisa) Trail/Camp, West Sepik, PNG, 2.76S, 141.34E
Kubor Mts or Ra, PNG, 6.12S, 144.75E
Kuli, Kubor Ra, PNG, 5.92S, 144.46E
Kumawa Mts, IJ, 3.98S, 133.0E
Kumdi, Mount Hagen (see latter) district, PNG
Kumusi R., PNG, 8.53S, 147.95E
Kup, Kubor Mts, PNG, 5.96S, 144.83E
Kuper Ra, PNG, 7.23S, 146.83E
Kurudu I., off Yapen I., IJ, 1.88S, 137.0E
Labuha Village, Bacan I., N. Moluccas, 0.63S, 127.47E
Lae, PNG, 6.73S, 147.0E
Lagaip R., PNG, 5.17S, 142.67E
Lakekamu Basin, PNG, 7.87S, 146.47E
Lake Daviumbu, PNG, 7.58S, 141.3E
Lake Eacham, Atherton Tableland, Aust., 17.28S, 145.63E
Lake Habbema, IJ, 4.15S, 138.65E
Lake Kutubu, PNG, 6.36S, 143.29E
Lake Omha (near English Peaks), PNG, 8.77S, 147.48E
Lake Paniai (Wissel Lakes), IJ, 3.93S, 136.33E
Lakeplain (Meerlakte), IJ, 3.0S, 138.4E
Lake Sogolomik, near Ok Tedi R., PNG, 5.23S, 141.23E
Laloki R., PNG, 9.85S, 147.22E

Lamington National Park, Aust., 28.33S, 153.08E
Lismore, Richmond R., Aust., 28.82S, 153.27E
Lockerbie, Cape York Penin., Aust., 10.8S, 142.47E
Lordberg, Sepik Mts, PNG, 4.92S, 143.0E
Louisiade Archipelago, PNG, 11.5S, 153.5E
Madang, PNG, 5.30S, 145.78E
Mafulu, PNG, 8.52S, 147.02E
Maguli Ra, PNG, see Mt. Maguli
Mai I., Aust., 10.72S, 142.62E
Mailu I., PNG, 10.39S, 149.35E.
Manning R., Aust., 31.92S, 152.0S
Manokwari (Dorey), IJ, 0.86S, 134.08E
Mambare R., PNG, 8.1S, 148.0E
Mamberamo R., IJ, 2.0S, 137.8E
Mapi R., IJ, 7.0S, 139.27
Markham R., PNG, 6.7S, 146.5E
McIlraith Ra, Cape York Penin., Aust., 13.5S, 143.3E
Meervlakte, IJ, see Lakeplain
Mekeo District, PNG, 8.5S, 146.6E
Menyamya, PNG, 7.25S, 146.0E
Merauke, IJ, 8.5S, 140.35E
Milne Bay, PNG, 10.33S, 150.33E
Mimika R., IJ, 4.42S, 136.55E
Minyip (Minyimp), Mt. Hagen, PNG, 5.8S, 144.15E
Mission Beach, Aust., 17.95S, 146.15E
Misool I., Rajah Ampat Is, IJ, 1.9S, 130.0E
Moluccan Is, Indonesia, 1.5S, 127.7E
Moroka, Owen Stanley Mts, PNG, 9.37S, 147.51E
Mountain Creek, near Paluma (see latter), Aust.
Mount Hagen town, PNG, 5.86S, 144.22E
Morotai I., N. Moluccas, 2.3S, 128.3E
Mt. Albert Edward, PNG, 8.42S, 147.33E
Mt. Batchelor, PNG, 8.00S, 147.08E
Mt. Besar, Batanta I., Rajah Ampat Is, IJ, 0.70S, 130.13E
Mt. Bosavi, PNG, 6.6S, 142.86E
Mt. Capella, PNG, 4.99S, 141.08E
Mt. Carstensz (Puncak Jaya), IJ, 4.15S, 137.05E
Mt. Chapman, PNG, 7.98S, 146.97E
Mt. Crater, PNG, see Crater Mountain

Mt. Dayman, PNG, 9.84S, 149.28E
Mt. Elliot, near Townsville, Aust., 19.5S, 146.95E
Mt. Gahavisuka Park, PNG, 6.03S, 145.40E
Mt. Gamkonora, Halmahera, N. Moluccas, 1.38N, 127.52E
Mt. Gigira, S Karius Ra, PNG, 5.92S, 142.70E
Mt. Giluwe, PNG, 6.05S, 143.88E
Mt. Glorious, Aust., 27.5S, 153.0E
Mt. Goliath, Oranje Mts, IJ, 4.69S, 139.83E
Mt. Hagen, PNG, 5.77S, 144.03E
Mt. Horseshoe, PNG, see Mt. Maguli
Mt. Hunstein, Sepik, PNG, 4.51S, 142.67E
Mt. Ialibu, Kubor Ra, PNG, 6.23S, 144.07E
Mt. Ifal, Victor Emanuel Mts, PNG, 4.98S, 141.72E
Mt. Kaindi, PNG, 7.35S, 146.68E
Mt. Karimui, PNG, 6.56S, 144.77E
Mt. Knutsford, PNG, 8.83S, 147.5E
Mt. Kominjim, PNG, 5.22S, 144.52E
Mt. Kourange (Bon Kourangen), Tamrau Ra, Vogelkop, IJ, 5.33E, 132.73E
Mt. Kunupi, Weyland Mts, IJ, 3.87S, 135.52E
Mt. Lewis, Atherton Tableland, Aust., 16.57S, 145.27E
Mt. Liwaro (Leiwaro), PNG, 5.55S, 143.25E
Mt. Maguli (Mt. Horseshoe), Owen Stanley Mts, PNG, 9.33S, 147.69E
Mt. Maneao, PNG, 9.68S, 149.63E
Mt. Menawa, PNG, 3.31S, 141.72E
Mt. Mengam, Adelbert Ra, PNG, 4.87S, 145.32E
Mt. Michael, PNG, 6.41S, 145.32E
Mt. Missim, PNG, 7.28S, 146.78E
Mt. Mura, PNG, 9.88S, 149.39E
Mt. Musgrave, PNG, 8.93S, 147.48E
Mt. Nipa [= Nipa?], PNG, 6.29S, 143.66E
Mt. Nisbit, PNG, 9.17S, 147.83E
Mt. Orata, Kubor Ra, PNG, 6.10S, 144.50E
Mt. Pugent (or Trauna Ridge), PNG, 5.47S, 144.22E
Mt. Scratchley, PNG, 8.75S, 147.47E
Mt. Simpson, PNG, 10.03S, 149.57E
Mt. Somoro, PNG, 3.41S, 142.19E
Mt. Strong, PNG, 7.98S, 146.97E
Mt. Suckling, PNG, 9.72S, 148.97E
Mt. Tafa, PNG, 8.63S, 147.18E
Mt. Victoria, PNG, 8.89S, 147.53E

Mt. Wilhelm, PNG, 5.78S, 145.03E
Mt. Wilhelmina, IJ, see Puncak Trikora
Muller Ra, PNG, 5.4S, 142.3E
Mung R., Kubor Mts, PNG, 6.0S, 144.7E
Murray Pass, PNG, 8.51S, 147.35E
Nade, Fergusson I, PNG, 9.14S, 150.71E
Nanango, Aust., 26.07S, 152.0E
Narirogo, PNG, see Varirata National Park
Naru R., PNG, 5.5S, 145.7E
Nassau Mts or Ra, IJ, 4.0S, 139.5E
Nawawu, Adelbert Ra, PNG, 4.88S, 145.32E
Nawen Hill, near Wewak, PNG, 3.58S, 143.62E
Neneba, upper Mambare R., PNG, 8.69S, 147.54E
Neon (Neowa) Basin, PNG, 8.47S, 147.30E
Newcastle, Aust., 32.92S, 151.77E
Nipa, PNG, 6.17S, 143.47E
Ningerum, PNG, 5.67S, 141.13E
Nomad R., PNG, 6.3S, 142.2E
Normanby I, PNG, 9.9S, 150.9E
North Coastal Ra, consists of the Bewani, Torricelli and Prince Alexander Mts, PNG
Nondugl (Sanctuary), PNG, 5.88S, 144.76E
Obi I., N. Moluccas, 1.50S, 127.67E
Ogeranang, PNG, 6.48S, 147.36E
Okbap, Star Mts, IJ, 4.60S, 140.43E
Ok Ma Road, PNG, see Ok Tedi
Ok Menga, PNG, 5.37S, 141.3E
Ok Om, PNG, 5.0S, 142.0E
Ok Tedi, PNG, 6.07S, 141.13E
Okapa, PNG, 6.56S, 145.75E
Oksapmin, PNG, 5.22S, 142.22E
Onin Penin, IJ, 2.83S, 132.25E
Oranje Mts or Ra, IJ, see Jayawijaya Mts.
Orangerie Bay, PNG, 10.42S, 149.67E
Owen Stanley Mts or Ra, PNG, 8.9S, 147.5E
Ower's Corner (and Camp), Sogeri Plateau, PNG, 9.42S, 147.42E
Paluma, Aust., 19.0S, 146.33E
Paniai (Wissel) Lakes, IJ, 3.93S, 136.33E
Patani, Halmahera, N. Moluccas, 0.3N, 128.67E
Popondetta, PNG, 8.75S, 148.25E
Porgera, PNG, 5.47S, 143.12E
Port Moresby, PNG, 9.45S, 147.2E
Pulau Kimaam, IJ—see Frederik Hendrik (Dolok) I.
Puncak Trikora, IJ, 4.25S, 138.75E
Purari R., PNG, 7.23S, 145.32E
Puwani R., PNG, 3.9S, 141.15E
Putei, near the Purari R., PNG, 7.80S, 146.13E
Ra R., Bacan I., N. Moluccas, 0.62S, 127.43E
Raba Raba, PNG, 9.97S, 149.83E
Rajah Ampat Is, off the W coast of the Vogelkop, IJ, see Batanta, Salawati, Waigeo, Misool Is.
Ramu R., PNG, 4.2S, 144.7E
Rawlinson Ra, Huon Penin, PNG, 6.53S, 147.28E
Rau I., N. Moluccas, 2.2N, 128.1E
Richmond R., Aust., 28.9S, 153.6E
Rocky R., Cape York Penin., Aust., 13.8S, 143.5E
Rockhampton, Aust., 23.38S, 150.53E
Rouna Falls, PNG, 9.42S, 147.37E
Rous, Richmond R. (see latter), Aust.
Rubi, Geelvink Bay, IJ, 3.17S, 134.97E
Saibai I., Aust., 9.43S, 142.67E
Salawati I., IJ, 1.08S, 130.92E
Samarai I., PNG, 10.6S, 150.7E
Saonek I., IJ, 0.5S, 130.8E
Sariba I., PNG, 10.6S, 150.78E
Saruwaged Ra, Huon Penin, PNG, 6.3S, 147.1E
Sattelberg Mts, PNG, 6.5S, 147.8E
Schrader Mts or Ra (Sepik Mts), PNG, 5.3S, 144.4E
Sepik Mts, PNG, see Schrader Mts.
Sepik R., PNG, 4.2S, 144.0E
Sepik-Wahgi Divide, PNG, 5.7S, 144.3E
Sergile, near Sorong (see latter), IJ (see Stresemann 1954: 274)
Setakwa R., IJ, 4.57S, 137.35E
Sewa Bay, Normanby I., PNG, 10.0S, 150.95E
Sidangoli, Halmahera I., N. Moluccas, 0.11N, 127.5E
Sii R., PNG, 7.86S, 146.55E
Silver Plains, near Rocky R. (see latter), Cape York Penin., Aust.
Simbai, PNG, 5.28S, 144.42E
Siwi, Arfak Mts, Vogelkop, IJ, 1.50S, 134.00E

Snow Mts IJ, see Nassau Ra.
Sogeri Plateau, PNG, 9.43S, 147.42E
Sogolomik, Ok Tedi area, PNG, 5.23S, 141.23E
Soliabeda or Soliabedo, PNG, 6.70S, 144.83E
Sorong, Vogelkop, IJ, 0.88S, 131.25E
South Naru, PNG, 5.5S, 145.75E
Star Mts, IJ, 4.6S, 140.3E
Stephansort, Astrolabe Bay, PNG, 5.42S, 145.72E
Strickland R., PNG, 7.0S, 142.0E
Stroom gebied [= river bed], W. Wasimi [= Wasami R], N NG, IJ, (0.78S, 131.40E)]
Styx R. State Forest, Aust., 30.55S, 152.25E
Sudest (Tagula) I, PNG, 11.75S, 153.50E
Sydney, Aust., 33.83S, 151.25E
Tabubil, Ok Tedi valley, PNG, 5.37S, 141.23E
Tagula (Sudest) I., PNG, see Sudest I.
Tamrau Mts, Vogelkop, IJ, 0.6S, 132.1E
Tanah Batu Putih, Halmahera I. (see latter), N. Moluccas
Tanah Merah, Arfak Mts (see latter), Vogelkop, IJ
Tari, PNG, 5.98S, 143.02E
Tari Gap, PNG, 5.09S, 143.16E
Tari Basin or Valley, PNG, see Tari
Taritatu R., IJ, see Idenburg R.
Taua, lower Mamberamo R., IJ
Telefomin, PNG, 5.12S, 141.63E
Tembagapura, IJ, 4.17S, 137.08E
Ternate I., N. Moluccas, 0.80N, 127.40E
Tidore I., N. Moluccas, 0.67N, 127.42E
Tigibi, near Tari (see latter), PNG
Timika, IJ, 4.05S, 136.90E
Tinaroo Scrub, Atherton Tableland, Aust., 17.17S, 145.55E
Tip, Misool I., IJ, 1.82S. 130.08E
Tobelo, Halmahera I., N. Moluccas, 1.72N, 128.00E
Tomba, PNG, 5.78S, 143.97E
Tondon Ra., PNG, 6.13S, 143.53E
Tooloom Scrub, Aust., 28.48S, 152.40E
Torres Strait, Aust., 10E, 143E
Tor R., IJ, 1.98S, 138.97E
Torricelli Ra, PNG, 3.37S, 142.17E
Townsville, Aust., 19.25S, 146.80E

Townsville, Ok Tedi, PNG, see Tabubil
Trauna Ridge, PNG, see Mt. Pugent
Treubbivak, Treub Mts, IJ, 4.47S, 138.58E
Trobriand Is, PNG, see Kaileuna I
Tsuwenkai, Bismarck Ra, PNG, 5.42S, 144.63E
Tufi, PNG, 9.08S, 149.32E
Ubamurai, see Crater Mt., PNG
Ubaigubi, PNG, 6.05S, 145.17E
Uni, Mapi R. (see latter) area, IJ
Utai, Sepik R. drainage, PNG, 3.42S, 141.58E
Utakwa R., IJ, 4.33S, 137.23E
Vanapa R., PNG, 9.12S, 147.12E
Vanimo, PNG, 2.07S, 141.03E
Vanumai, PNG, 8.78S, 146.63E
Van Rees Mts, IJ, 2.58S, 138.25E
Varirata National Park, PNG, 9.44S, 147.36E
Veimauri, PNG, 9.05S, 147.08E
Victor Emanuel Mts, PNG, 5.25S, 141.07E
Vogelkop (Penin.), IJ, 1.6N 133.0E
Wabag, PNG, 5.05S, 143.72E
Wabo, Gulf Province, PNG, 7.00S, 145.07E
Wafa R., PNG, 6.68S, 146.17SE
Wagifa, Fergusson I. (see latter), PNG
Wahgi Divide, PNG, see Sepik–Wahgi Divide
Wahgi R. and Valley, PNG, 5.8S, 144.1S
Waifoi, Waigeo I., Rajah Ampat Is, IJ, 0.13S, 130.75E
Waigeo I, Rajah Ampat Is, IJ, 0.2S, 131.0E
Wamena, IJ, 4.12S, 138.92E
Wandammen Mts, Vogelkop, IJ, 2.7S, 134.6E
Wanoem Bay, Kobroor I., Aru Is, S. Moluccas, 6.03S, 134.30E
Wanuma, Adelbert Mts, PNG, 4.09S, 145.32E
Wara Sii, PNG, see Sii River
Waria R., PNG, 7.87S, 147.13E
Warr I., Geelvink Bay, IJ, 2.01S, 134.75E
Warsembo, W coast of Geelvink Bay (see latter), IJ
Wassi Kussa R. PNG, 9.15S, 142.05E
Wasu, Huon Penin PNG, 5.97S, 147.20E
Watut R., PNG, 6.8S, 146.4E
Watut/Tauri Gap, PNG, 7.85S, 146.25E
Wau (Ecology Institute), Wau Valley, PNG, 7.35S, 146.72E

Weipa, Cape York Penin., Aust., 12.63S, 141.87E
Wewak, PNG, 3.57S, 143.06E
Western Papuan Is, IJ, see Rajah Ampat Is.
Weyland Mts or Ra, IJ, 3.87S, 135.72E
Wissel (Paniai) Lakes, IJ, 3.93S, 136.03E
Woitape, PNG, 8.55S, 147.25E
Wonapi, Jobi (Yapen) I (see latter), IJ
Wondiwoi Mts (Mt. Wondiwoi), IJ, 2.67S, 134.6E
Yalumet, Huon Penin, PNG, 6.07S, 147.02E
Yaour Aolfe, Geelvink Bay (see latter), IJ
Yapen (Japen or Jobi) I, IJ, 1.8S, 136.3E
Yuat R., PNG, 5.1S, 144.1E
Yule I., PNG, 8.28S, 146.53E

Glossary

agonistic behaviour with a primarily aggressive motivation; includes elements of attack & fleeing

alkaloidal primarily plant-derived chemicals rich in nitrogenous basic compounds (typically physiologically active and insoluble in water)

allopatry the distributional phenomenon in which two or more populations inhabit geographically-distinct ranges without overlap (ant. sympatry).

allopreen behaviour in which one bird preens the feathers of another

altricial a developmental term referring to bird species in which the nestlings hatch helpless and dependent on their parent(s) (ant. precocial)

analogous similar in function but not in origin and structure (ant. homologous)

angiosperm seed-bearing plants (of Angiospermae) in which ovules are enclosed in an ovary and develop into fruit once fertilised (cf. gymnosperm)

anthropogenic human-made

antiphonal song sung in two responsive, alternating parts, e.g. duetting.

apterium (pl. **apteria**) patch of skin that is unfeathered

arachnids spiders and their relatives

Araneida the true spiders

archaeology the study of past human cultures

aril the edible covering that envelops some seeds, developing from the stalk of the ovule

arillate possessing an aril

arthropods insects and their allies (spiders, centipedes, etc.)

atavistic showing a resemblance to an earlier form or relative

auctorum of the authors

Australasia a geographical faunal region including Australia, New Guinea, New Zealand, the islands of the SW Pacific, and the Moluccas

autapomorphy a morphological character exhibited by a single lineage—not shared (ant. synapomorphy)

Batesian mimicry when a palatable and vulnerable species resembles a relatively unpalatable species so gaining protection

bathymetric a measuring of ocean or sea depth

biodiversity the diversity of organisms (syn. biological diversity)

biogeography the study of the geographical distribution of life on earth

biological species concept based on the reproductive isolation of the constituent populations from other species

biometrics measurements of morphological features of an animal

biota the plant and animal life of a region

Blattidae the insect family of cockroaches

Cerambycidae the coleopteran insect family of longicorn or longhorn beetles

Cetoniine of the coleopteran (beetle) insect subfamily Cetoniinae, or flower-chafers

Chelisochidae a dermapteran insect family of earwigs

Chilopoda the arthropod order of centipedes

Chrysomelidae the coleopteran insect family of leaf beetles

clade group of species descended from, and including, a common ancestor

cladistic, cladistics a methodology for delineating evolutionary relationships among lineages based on establishment of monophyletic groups (clades), which are practically defined by the presence of uniquely-shared derived characters.

clavicle the collarbone or wishbone in birds

Clubionidae a spider family of mostly small species

cnemophilinine (adj.), **the cnemophilines** (n.) the wide-gaped birds of paradise of the subfamily Cnemophilinae

Coleoptera the insect order of the beetles

convergent, convergence in evolutionary biology, the independent evolution of like forms or like features by two unrelated lineages (e.g. lek behaviour in the birds of paradise and the neotropical fruit-crows is a convergent behaviour)

cordillera, cordilleran (adj.) a linear chain of mountains, usually referring to the principal mountain range of a continent

congener (n.), **congeneric** (adj.) a species of the same genus as one being compared to it

conspecific (adj.) a population of the same species as one being compared to it

corvid (adj.) of the avian family that includes the crows and their allies

culmen the ridge running lengthwise down the dorsal midline of the upper mandible

Curculionidae the coleopteran insect family of weevils

dehiscent with a tendency to split (as in capsular fruits that burst or split open when ripe)

deme a homogeneous population whose members are readily interbreeding

demography, demographics the study of the dynamics (birth/death rates, age/sex ratios and numbers) of populations

Dermaptera the insect order of the earwigs

diaspore exemplified by a winged seed, the reproductive dispersal unit produced by a plant

dichromatic of two colours

dimorphic appearing in two distinct forms

dipteran of the fly insect family Diptera

distally towards the tip of an appendage

DNA–DNA hybridisation a molecular technique used to estimate evolutionary relatedness by comparing base-pair similarity between the nuclear DNA of two species.

ectethmoid a bone in the fore part of the orbital space of the skull—see Fig. 9.1

ectoparasite a parasite that lives externally on its host

Elateridae the coleopteran insect family of click beetles

El Niño (now formally termed **ENSO**) a periodic global climatic phenomenon produced by the establishment of anomalously high surface water temperatures in the central Pacific; typically appearing every 4–6 years, an El Niño produces severe drought conditions in Australasia westward into the Greater Sundas.

endemic present within and confined to a localised area

ephemeral of short duration; transitory (ant. chronic, permanent)

epiphyte a non-parasitic plant that lives entirely upon another plant, but receives its nourishment independent of the supporting plant, as in certain orchids, ferns, mosses, or lichens

ethanol ethyl alcohol; surgical spirits

etymology the origin and development of a word

fiat a decree, order, or sanction

fledging the point in development of a bird when it acquires its first true flight feathers—at which point it usually leaves the nest (and often used to mean this event)

foramen (pl. **foramina**) a small opening, e.g. a hole in bone through which a nerve passes—see Fig. 9.1

Formicidae the hymenopteran insect family of ants

founder effects the genetic divergence evidenced by small isolate populations, postulated to be the product of the small gene pool from which the 'founders' of that population derived

frugivore (n.), **frugivorous** (adj.) a fruit-eater, fruit-eating

genotype the whole genetic constitution of an individual (cf. phenotype)

granivorous seed- or grain-eating

Gryllidae the orthopteran insect family of field crickets

Gryllacrididae the orthopteran insect family of camel crickets

gymnosperm 'naked seed' plants—a major plant lineage that includes the pines, araucarias, podocarps, and other primitive tree families (cf. angiosperm)

haemochromatosis a disorder caused by excess storage of iron, particularly in the liver, recorded for birds of paradise held in captivity (avoidable by provision of low iron diet)

herpetofauna the amphibian and reptilian fauna

heterochrony the differential rate of development in the two sexes

heterospecific of a different species

home range the area occupied by an individual, pair, or group of birds

homogeneous similar in kind or nature

homologous of the same evolutionary and structural derivation (ant. analogous)

inflorescence a flower or flower cluster

insectivorous insect-eating (but commonly used to indicate arthropod-eating)

intrageneric within genus

intergeneric between genera

introgression the incorporation of the genes of one lineage into the lineage of another

iridescent, iridescence a spectrum of colours that shimmer and change due to structural interference

isolating mechanism a feature differing between species (or populations) that tends to prevent cross-mating and so bring about or maintain species isolation/integrity

jugal bar a fine bone that links the quadrate to the base of the maxilla—see Fig. 9.1

katydids cricket-relatives of the orthopteran insect family Tettigoniidae

Kina equivalent of a dollar denomination in Papua New Guinea currency

kinetic full of motion

lacrymal a part of the cranium near the orbit—see Fig. 9.1

land-bridge island an island that has in the past been linked to the mainland (or other island)—usually during a period of lowered sea level

lek (n.), **to lek** (v.), **lekking** (ger.) a place where males traditionally cluster in order to attract and mate with visiting females; also the term for the actual cluster of males in display, and the act or habit of forming in clusters to display.

lineage a group of species allied by common descent

lipid biochemical term for any of a large group of organic compounds that are esters of fatty acids

Lepidoptera the insect order of butterflies and moths

mandible the lower jaw

maxilla the upper jaw

maxillopalatine bony structures on either side of the vomer that extend out and attach to the maxilla, forming part of the palate—see Fig. 9.1

Melanesia, Melanesian the Pacific island region, centring on the island of New Guinea, that includes the Admiralty, Bismarck, and Solomon archipelagoes, and Fiji, Vanuatu, and New Caledonia, home to the dark-skinned island people (cf. Micronesia, Polynesia)

melanin the most common pigment in the feathers of birds, typically appearing black, but also as brown, red-brown, or yellow

microfilaria group of internal nematode parasitic worms that are only partly developed when born

microsympatry the occurrence of two or more populations or species in the same habitat (cf. sympatry)

millinery (trade) of hats and hat trimmings—sold by a milliner

mitochondrial DNA non-nuclear DNA that resides in the mitochondrion

monochromatic possessing a single colour pattern

monogamous mating with only one mate

monomorphic possessing a single form or pattern

monophyletic (adj.), **monophyly** (n.) of a single evolutionary lineage

monotreme the mammalian order including the echidnas and platypus

monotypic of a single type (with no variation); hence a monotypic species includes no subspecific variation; a monotypic genus includes but one species

morphology, morphological the science of form and shape—in ornithological terms: usually refers to external (including coloration) and internal characters

Müllerian mimicry the convergent evolution of similar colour patterns by two or more noxious or unpalatable species that mutually reinforces avoidance effects upon potential predators

Muscidae the dipteran insect family of the house flies

mutualism a mutually-beneficial relationship between two species, maintained by special behaviours or habits

myology the study of musculature; the anatomy of the muscles of an animal

myriapod the arthropod group that includes the centipedes, millipedes, and their relatives

narial of the nares—the anterior or external nares being the nostrils

Newcastle disease an avian disease caused by a world-wide multistrain paramyxovirus that commonly strikes domestic fowl and also representatives of most avian orders

neotropical pertaining to the New World tropics

necropsied a dissected dead specimen, examined to determine cause of death

nectary plant structure that produces nectar

nominate used in this book to refer to that subspecies that bears the same name as the species, e.g. *Paradisaea apoda apoda* is the 'nominate subspecies'

non-eumelanic schizochroism abnormal plumage lacking the black or grey pigmentation normally present in the plumage

nuptial pertaining to courtship and making

occipital pertaining to the large bone at the base of the skull to which the vertebral column is attached; the back part of the head

omnivorous eating a wide range of foods

oocyte(s) immature female germ cell(s) that gives rise to eggs in birds

orbit the cavity on each side of the skull that holds the eye—see Fig. 9.1

Orthoptera the insect order of crickets and their relatives

oscines the higher songbirds (cf. Deutero-Oscines)

ossification the process by which tissue is converted to bone

osteology the study of skeletal anatomy and bones

palatine either of two bones forming the hard palate—see Fig. 9.1

panmictic in which the entire population or species is freely inter-breeding

Papuan pertaining to the island of New Guinea

paradisaeid (n.), **paradisaeinine** (adj.) pertaining to the family of birds of paradise, Paradisaeidae; pertaining to the subfamily of true birds of paradise, Paradisaeinae

parapatry a distributional phenomenon in which the ranges of two species meet but do not overlap (cf. allopatry, sympatry)

pars dorsalis a feather tract that is part of the Pteryla Spinalis located on the back

pars pelvica a feather tract that is part of the Pteryla Spinalis located on the lower back

passerine pertaining to the perching birds, the order Passeriformes

petiole a leaf stalk; a footstalk of a leaf, connecting the blade with the stem

phasmids of the phasmid insect family Phasmatidae, the 'walking sticks'

phyletic of or pertaining to evolutionary lineages (cf. phylogenetic)

phyllodes flattened leafstalks that resemble and function as leaves

phylogeny an evolutionary tree, with branching to indicate evolutionary divergence of lineages

phylogenetic species concept in which each morphologically-diagnosable population is treated as a species (cf. biological species concept)

placental term describing mammals having a placenta (eutherians—of the mammal subclass Eutheria) (cf. marsupials)

polychotomous, polychotomies with multiple branches originating from the same point

polygyny (n.), **polygynous** (adj.) the act of mating with more than one female; the behavioural tendency to mate with more than one female

polyphyletic possessing more than one evolutionary lineage, as in a grouping of species that incorporates taxa that evolutionarily belong to two or more distinct families

polytypic of more than one type, as in a genus that possesses two or more species, or a species that has two or more subspecies

posterior condyle a rounded process of the quadrate bone—see Fig. 9.1

postorbital posterior to the orbit or eye socket—see Fig. 9.1

primaries the major wing flight feathers, which are attached to the carpometatarsus and digital phalanges

protein electrophoresis a laboratory technique that measures the movement of proteins suspended in a solution under the influence of an electrical field; used in ornithology to map variation in protein electro-types in egg albumin as a means of developing biochemical characters for systematic analysis or the study of population genetics

protozoa group of single-celled organisms, including amoebas and ciliates (Kingdom Protista)

proximal, proximally close to the base of an appendage or process (ant. distally)

pterylography the study of feather tracts; mapping of feather tracts of birds

pterylosis the arrangement of contour feathers on the skin of a bird

putative supposed

quadrate the bone at the base of cranium to which the mandible is attached—see Fig. 9.1

radiation the simultaneous evolutionary divergence of several related lineages

rain shadow the relatively dry area to leeward of high ground in the path of rain-bearing winds

rectrices the major tail feathers

Reduviidae the hemipteran insect family of assassin bugs

relict a population of an animal or plant, existing as a remnant of a formerly more widely-distributed lineage, or the last species or population of a formerly speciose lineage

rictal bristles stiff hair-like feathers originating from each side of the gape, at the bill base

rupiah the basic unit of Indonesian currency

sagittal pertaining to the longitudinal vertical plane dividing a bilaterally-symmetrical animal into right and left halves—following the backbone and including the midline of the crown from forehead to occiput

scansorial pertaining to climbing or creeping

scapulars the feathers above a bird's shoulder

scarabeoid an insect relative of the scarab beetles

semi-species one of a member of a super-species group; a well-marked geographical population obviously related to one or more allopatric forms; a distinct population at or below the species level.

sexual dimorphism the presence of differences between the sexes

sexual selection the evolutionary product of competition among members of one sex for reproductive access to members of the other sex, usually resulting in distinctive behavioural and morphological differences between the sexes

Southern Oscillation see El Niño

speciation the process by which two or more populations differentiate to the level of distinct species

Staphylinidae the coleopteran insect family of rove beetles

Stenopelmatidae the orthopteran insect family of sand crickets

subcutaneous beneath the skin

suboscines the lower perching birds, including all families of Passeriformes not included in the oscines

superspecies a monophyletic species-group of two or more distinct but closely-related allopatric forms, each of which has achieved near-species status (semi-species)

sympatry (n.), **sympatric** (adj.) the distributional phenomenon in which two or more populations inhabit the same geographical area (ant. allopatry; cf. microsympatry)

synergy, synergies the combined action of two distinct forces

synonym in biological taxonomy, a scientific name that refers to a subspecies or species that already has a valid name

syrinx the avian organ producing voice or song; an evolutionarily distinct structure from the larynx of mammals

systematics the science of objectively ordering the diversity of life

tarsus (s.), **tarsi** (pl.) in birds, common name for the lower leg bone in birds, technically called the tarsometatarsus, formed by the fusion of the tarsal and metatarsal elements

taxon (s.), **taxa** (pl.) general term for any category used in biological classification (species, genus, etc.); also used in the specific sense to refer to a population or cluster of related populations (a clade)

taxonomy the science of naming and classifying organisms

taxonomist one who practises taxonomy

tectonic plate a distinct section of the earth's crust, one or more of which form each of the earth's continents and major island arcs

Tmesisternine of the coleopteran insect genus *Tmesisternus* (family Cerambycidae)

Tenebrionidae the coleopteran insect family of darkling beetles

Tertiary in geology, the first period of the Cainozoic Era, which followed the Cretaceous period and preceded the Quaternary period

Tettigoniidae the orthopteran insect family of katydids

tibia in birds, common name for the upper leg bone, technically the tibiotarsus, formed by the fusion of the tibia and proximal tarsals

Tipulidae the dipteran insect family of crane flies

trachea the windpipe, leading from the glottis to two bronchi that lead to the lungs

trap-line (n., v.t.) in foraging studies, referring to a pattern of foraging in which an individual visits a series of feeding sites (e.g. fruit-trees) in a regular manner

vicariance the process of geographical speciation in which a once-continuous population is broken into two or more subpopulations, which subsequently differentiate in geographical isolation

vicariant a geographical population that has been isolated from a parent population by some environmental process

Wallace's Line a boundary between the Asian and Australasian vertebrate fauna delineated by A. R. Wallace, which generally follows a north–south trend though the Indonesian archipelago between the islands of Bali and Lombok (in the south) and between Borneo and Sulawesi (in the north)

wattles unfeathered, in some cases brightly-coloured, fleshy flaps or appendages about the head of a bird that are nuptial and/or are species/sex specific recognition marks

zoogeographical pertaining to the geographical distribution of animals

zygomatic process a process of the temporal bone forming part of the zygomatic arch (in cranial anatomy)—see Fig. 9.1

Bibliography

Amadon, D. (1944). The genera of Corvidae and their relationships. *American Museum Novitates*, **1251**, 1–21.

Anon (1972). Third ornithlon results. *New Guinea Bird Society Newsletter*, **74**, 1–5, supplement.

Anon. (1974). September excursion report. *New Guinea Bird Society Newsletter*, **102**, 2.

Anon. (1981). *Baiyer River Sanctuary Papua New Guinea 1975–1980 Report*. Baiyer River Sanctuary, PNG.

Anon. (1984). *Baiyer River Sanctuary Papua New Guinea 1981–1983 Report*. Baiyer River Sanctuary, PNG.

Anon. (1989). Recovery Round-up. *Corella*, **13**, 93–5.

Anon. (1990). *Semioptera wallacii* Gray, 1859 (Aves, Paradisaeidae) conserved as the correct spelling of the generic and specific names. *Bulletin Zoological Nomenclature*, **47**, 169–70.

Anon. (1995). Progress report on the BOU-sponsored Maluku Programme *Ibis*, **137**, 607.

Anon. (1998). Recovery Round-up. *Corella*, **22**, 34–6.

Archbold, R. and Rand, A. L. (1940). *New Guinea Expedition*. McBride, New York.

Armstrong, E. A. (1947). *Bird display and behaviour*. Lindsay Drummond Ltd., London.

Aruah, A. and Yaga, A. (1992). Acquisition of adult male plumage in some birds of paradise at Baiyer River Sanctuary. *Muruk*, **5**, 49–52.

Assink, J. A. and Frankenhuis, M. T. (1981). Report on iron accumulation in livers of birds of paradise at Blijdorp Zoo. Unpublished report, Jan 1981, Rotterdam.

Attenborough, D. (1960). *Quest in paradise*. Lutterworth Press, London.

Attenborough, Sir D. (1996). 'Attenborough in paradise'. 8 April 1996 BBC, Bristol.

Baillie, J. and Groombridge, B. (ed.) (1996). 1996 IUCN red list of threatened animals. IUCN, Gland, Switzerland.

Barker, R. D. and Vestjens, W. J. M. (1990). *The food of Australian birds 2: Passerines*. CSIRO, Melbourne.

Barker, W. R. and Croft, J. R. (1977). The distribution of Macgregor's Bird of Paradise. *Emu*, **77**, 219–22.

Barnard, C. J. (1979). Predation and the evolution of social mimicry in birds. *American Naturalist*, **113**, 613–18.

Barnard, H. G. (1911). Field notes from Cape York. *Emu*, **11**, 17–32.

Beach, J. (1975). Display and mating of the King of Saxony Bird of Paradise. *New Guinea Bird Society Newsletter*, **109**, 1–2.

Beddard, E. E. (1891). On the convoluted trachea of *Manucodia comrii*. *Ibis*, Ser. 6, vol. 3, 512–4.

Beecher, W. J. (1953). A phylogeny of the oscines. *Auk*, **70**, 270–333.

Beehler, B. M. (1978). *Upland birds of northeastern New Guinea*. Wau Ecology Institute Handbook No. 4, Wau Ecology Institute, Wau.

Beehler, B. M. (1980). A comparison of avian foraging at flowering trees in Panama and New Guinea. *Wilson Bulletin*, **92**, 513–19.

Beehler, B. M. (1981). Ecological structuring of New Guinea forest bird communities in New

Guinea. In *Biogeography and ecology in New Guinea*, (ed. J. L. Gressitt), pp. 837–61. Junk, The Hague.

Beehler, B. M. (1983a). *The behavioral ecology of four birds of paradise.* PhD. thesis Princeton University, Princeton.

Beehler, B. M. (1983b). Frugivory and polygamy in birds of paradise. *Auk*, **100**, 1–12.

Beehler, B. M. (1983c). Notes on the behavior and ecology of Macgregor's Bird of Paradise. *Emu*, **83**, 28–30.

Beehler, B. M. (1983d). Lek behavior of the Lesser Bird of Paradise. *Auk*, **100**, 993–5.

Beehler, B. M. (1985a). Adaptive significance of monogamy in the Trumpet Manucode *Manucodia keraudrenii* (Aves, Paradisaeidae). *Ornithological Monographs*, **37**, 83–99.

Beehler, B. M. (1985b). Conservation of New Guinea rainforest birds. In *Conservation of tropical forest birds*, (ed. A. W. Diamond and T. E. Lovejoy), pp. 233–46. ICBP Technical Publications No. 4, Cambridge.

Beehler, B. M. (1987a). Ecology and behavior of the Buff-tailed Sicklebill (Paradisaeidae, *Epimachus albertisi*). *Auk*, **104**, 48–55.

Beehler, B. M. (1987b). Birds of paradise and mating system theory—predictions and observations. *Emu*, **87**, 78–89.

Beehler, B. M. (1988). Lek behavior of the Raggiana Bird of Paradise. *National Geographic Research*, **4**, 343–58.

Beehler, B. M. (1989a). *Patterns of frugivory and the evolution of birds of paradise.* In *Acta XIX Congressus Internationalis Ornithologica*, (ed. H. Ouellet), pp. 816–828. University of Ottawa Press, Ottawa.

Beehler, B. M. (1989b). The birds of paradise. *Scientific American*, **261**, 116–23.

Beehler, B. M. (1991a). *A naturalist in New Guinea.* University of Texas Press, Austin.

Beehler, B. M. (1991b). Papua New Guinea's wildlife and environments—what we don't yet know. In *Conservation and environment in Papua New Guinea: establishing research priorties*, (ed. M. Pearl, B. Beehler, A. Allison, and M. Taylor), pp. 1–10. Wildlife Conservation International, New York.

Beehler, B. M. (1992). Behavioural ecology of the Raggiana Bird of Paradise. Unpublished Final Report for National Geographic Society Grant no. 4026–89.

Beehler, B. M. (ed.) (1993). *A biodiversity analysis for Papua New Guinea*. Volume 2 of the Papua New Guinea Conservation Needs Assessment. Biodiversity Support Program, Washington, D.C.

Beehler, B. M. and Beehler, C. H. (1986). Observations on the ecology and behaviour of the Pale-billed Sicklebill. *Wilson Bulletin*, **98**, 505–15.

Beehler, B. M. and Dumbacher, J. P. (1990). Interesting observations of birds at Varirata National Park, June–July 1989. *Muruk*, **4**, 111–2.

Beehler, B. M. and Dumbacher, J. P. (1996). More examples of fruiting trees visited predominantly by birds of paradise. *Emu*, **96**, 81–8.

Beehler, B. M. and Finch, B. W. (1985). Species-Checklist of the birds of New Guinea. *Australasian Ornithological Monographs*, **1**, 1–127. Royal Australian Ornithologists Union, Melbourne.

Beehler, B. M. and Foster, M. S. (1988). Hotshots, hotspots and female preference in the organization of lek mating systems. *American Naturalist*, **131**, 203–19.

Beehler, B. M. and Pruett-Jones, S. G. (1983). Display dispersion and diet of birds of paradise, a comparison of nine species. *Behavioral Ecology and Sociobiology*, **13**, 229–38.

Beehler, B. M. and Swaby, R. J. (1991). Phylogeny and biogeography of the *Ptiloris* riflebirds (Aves, Paradisaeidae). *Condor*, **93**, 738–45.

Beehler, B. M. Pratt, T. K. and Zimmerman, D. A. (1986). *Birds of New Guinea*. Princeton University Press, Princeton.

Beehler, B. M., Sengo, J. B., Filardi, C., and Merg, K. (1995). Documenting the lowland rainforest avifauna in Papua New Guinea—effects of patchy distributions, survey effort and methodology. *Emu*, **95**, 149–61.

Bell, H. L. (1967). Bird life of the Balimo sub-district, Papua. *Emu*, **76**, 57–79.

Bell, H. L. (1969). Field notes on the birds of Ok Tedi River drainage, New Guinea. *Emu*, **69**, 193–211.

Bell, H. L. (1970). Additions to the avifauna of Goodenough Island, Papua. *Emu*, **70**, 179–82.

Bell, H. L. (1971). Field notes on birds of Mt Albert Edward, Papua. *Emu*, **71**, 13–9.

Bell, H. L. (1977). The vertical distribution of a lowland rain forest bird community in New Guinea. MSc. thesis, University of Papua New Guinea, Port Moresby.

Bell, H. L. (1982*a*). A bird community of lowland rainforest in New Guinea. 1. Composition and density of the avifauna. *Emu*, **82**, 24–41.

Bell, H. L. (1982*b*). A bird community of lowland rainforest in New Guinea. 2. Seasonality. *Emu*, **82**, 65–74.

Bell, H. L. (1982*c*). A bird community of New Guinea lowland rainforest. 3. Vertical distribution of the avifauna. *Emu*, **82**, 143–61.

Bell, H. L. (1982*d*). A bird community of lowland rainforest in New Guinea. 4. Birds of secondary vegetation. *Emu*, **82**, 217–24.

Bell, H. L. (1983). A bird community of lowland rainforest in New Guinea 5. Mixed-species feeding flocks. *Emu*, **82**, 256–75.

Bell, H. L. (1984). A bird community of lowland rainforest in New Guinea. 6. Foraging ecology and community structure of the avifauna. *Emu*, **84**, 142–58.

Belon, P. du Mans. (1555). *L'histoire de la nature des oiseaux*. Paris.

Berger, A. J. (1956). On the anatomy of the Red Bird of Paradise, with comparative remarks on the Corvidae. *Auk*, **73**, 427–47.

Berggy, J. (1978). Bird observations in the Madang Province. *Papua New Guinea Bird Society Newsletter*, **148**, 9–20.

Bergman, S. (1956). On the display and breeding of the King Bird of Paradise, *Cicinnurus regius rex* (Scop) in captivity. *Nova Guinea*, 7, 197–205.

Bergman, S. (1957*a*). *Through primitive New Guinea*. Robert Hale, London.

Bergman, S. (1957*b*). On spelet hos stralparadisfageln *Parotia sefilata* (Pennant). *Fauna Flora*, Upps., **52**, 186–99.

Bergman, S. (1957*c*). On the display of the six-plumed bird of paradise *Parotia sefilata* (Pennant). *Nova Guinea*, new series **8**, 81–6.

Bergman, S. (1957*d*). On the display and breeding of the king bird of paradise, *Cicinnurus regius rex* (Scop.) in captivity. *Avicultural Magazine*, **63**, 115–24.

Bergman, S. (1958). On the display of the Six-plumed Bird of Paradise, *Parotia sefilata* (Pennant). *Avicultural Magazine*, **64**, 3–8.

Bergman, S. (1959). *Min far är kannibal*. Bonnier, Stockholm.

Bergman, S. (1961). *My father is a cannibal*. Hale, London.

Bergman, S. (1968). *Mina paradisfaglar*. Bonnier, Stockholm.

Bergtold, W. H. (1929). Egg weights from egg measurements. *Auk*, **46**, 466–73.

Berlepsch, H. (1911). Die Vögel der Aru-Inseln. *Abhandlungen der Senckenbergischen Naturforschenden Gesellschaft*, **34**, 53–98.

Berlioz, J. (1927). Remarques sur l'hybridarion naturelle chez les oiseaux. *Bulletin de la Société Zoologique de France*, **52**, 393–403.

Bernstein, H. A. (1864*a*). Ueber einen neuen Paradiesvogel und einige andere neue Vogel. *Journal für Ornithologie*, **12**, 401–10.

Bernstein, H. A. (1864*b*). *Natuurkunde Tijdschrift Nederlandische Indië*, **27**, 79.

Beruldsen, G. (1990). Cape York in the Wet. *Australian Bird Watcher*, **13**, 209–17.

Birkhead, T. R. (1991). *The magpies*. Poyser, London.

Bishop, K. D. (1984). Notes on Wallace's Standardwing *Semioptera wallacii*. *Bulletin of the British Ornithologists' Club*, **104**, 118–20.

Bishop, K. D. (1987). Interesting bird observations in Papua New Guinea. *Muruk*, **2**, 52–7.

Bishop, K. D. (1992). The Standardwing Bird of Paradise *Semioptera wallacii* (Paradisaeidae), its ecology, behavior, status and conservation. *Emu*, **92**, 72–8.

Bishop, K. D. and Frith, C. B. (1979). A small collection of eggs of birds-of-paradise at Baiyer River Sanctuary, Papua New Guinea. *Emu*, **79**, 140–1.

Blakers, M. Davies, S. J. J. F., and Reilly, P. N. (1984). *The atlas of Australian birds*. RAOU and Melbourne University Press, Melbourne.

Bleiweiss, R. (1987). Development and evolution of avian racket plumes: fine structure and serial homology of the wire. *Journal of Morphology*, **194**, 23–39.

Bock, W. J. (1963). Relationships between the birds of paradise and bower birds. *Condor*, **65**, 91–125.

Bock, W. J. (1994). History and nomenclature of avian family-group names. *Bulletin of the American Museum of Natural History*, **222**, 1–281.

Boddaert, P. (1783). *Table des planches enlumineez d'histoire naturelle de M. d'Aubenton*. Utrecht.

Boddaert, P. (1874). *Table des planches enluminées d'histoire naturuelle de M. d'Aubenton*. Utrecht (reprint).

Boehm, E. M. (1967). Successful breedings at the Edward Marshall Boehm aviaries in 1966. *Avicultural Magazine*, **73**, 116–20.

Bonaparte, C. L. (1850). Nouvelles espèces ornithologiques. *Comptes Rendus hebdomadaires des séances de L'Académie des Sciences*, **30**, 131–39.

Bonaparte, C. L. (1851). *Conspectus Genera Avium* **1** (1850). Lugduni Bataavorum, E. J. Brill, Leyden.

Bonaparte, C. L. (1853). *Comptes rendus hebdomadaires des séances de L'Academie des Sciences*. **37**, 829.

Borecky, S. R. (1977). The appendicular myology and phylogenetic relationships of the avian 'corvid assemblage'. Ph.D. thesis. University of Pittsburgh, Pittsburgh.

Borgia, G. (1985). Bower quality, number of decorations and mating success of male satin bowerbirds (*Ptilonorhynchus violaceus*): an experimental analysis. *Animal Behaviour*, **33**, 266–71.

Borgia, G. (1986). Sexual selection in bowerbirds. *Scientific American*, **254**, 70–9.

Boswell, J. (1965). A catalogue of tape and gramophone records of Australasian region bird sound. *Emu*, **65**, 65–74.

Boswell, J. (1981). Second supplement to a catalogue of Australasian bird sound. *Emu*, **81**, 223–6.

Boswell, J. and Kettle, R. (1975). A supplement to a catalogue of tape and gramophone records of Australasian bird sound. *Emu*, **75**, 143–6.

Bourke, P. A. and Austin, A. F. (1947). The Atherton Tablelands and its avifauna. *Emu*, **47**, 87–116.

Bradbury, J. W. (1981). The evolution of leks. In *Natural selection and social behavior*, (ed. R. D. Alexander and D. W. Tinkle), pp. 138–69. Chiron Press, New York.

Bradbury, J. W. and Gibson, R. (1983). Leks and mate choice. In *Mate choice*, (ed. P. P. G. Bateson), pp. 109–38. Cambridge University Press, Cambridge.

Bravery, J. A. (1970). The birds of Atherton Shire, Queensland. *Emu*, **70**, 49–63.

Breeden, S. and Breeden, K. (1970). *A natural history of Australia: 1. Tropical Queensland*. Collins, Sydney.

Brooker, M. G. and Brooker, L. C. (1989). Cuckoo hosts in Australia. *Australian Zoological Reviews*, No. 2, 1–67. Royal Zoological Society of New South Wales.

Brown, E. D. and Hopkins, M. G. H. (1998). Tests of disperser specificity between frugivorous birds and rainforest fruits in New Guinea. *Emu*, **98**, in press.

Burger, J. Laska, M., and Gochefeld, M. (1993). Metal concentrations in feathers of birds from Papua New Guinea Guinea forests: evidence of pollution. *Environmental Toxicology and Chemistry*, **12**, 1291–6.

Burnett, J. B. (1998). Biodiversity priority-setting map for Irian Jaya. Conservation International, Washington, DC.

Büttikofer, J. (1894). On two new birds of paradise. *Notes from the Leyden Museum*, **16**, 161–5.

Büttikofer, J. (1895). Einige bemerkungen über neu angekommene paradiesvögel. *Notes from the Leyden Museum*, **17**, 36–40.

Cabanis, J. (1888). Vorläufige Notiz über 2 neue Paradies-Vögel. *Journal für Ornithologie*, **36**, 119.

Campbell, A. J. (1897). Description of the nest and egg of the Rifle Bird. *The Victorian Naturalist*, **13**, 145.

Campbell, A. J. (1901). *Nests and eggs of Australian birds*. 2 volumes. Pawson and Brailsford, Sheffield.

Campbell, R. (1977). Magnificent Bird of Paradise *Diphyllodes magnificus*. *New Guinea Bird Society Newsletter*, **138**, 6.

Cassin, J. (1850). [Description of new species, *Paradisea wilsonii* Cabanis] *Proceedings of the Academy of Natural Sciences of Philadelphia*, (Aug), 67–8.

Cayley, N. (1959). *What bird is that?* 3rd edn. Angus and Robertson, Sydney.

Chalmers, J. and Gill, W. W. (1885). *Work and adventure in New Guinea 1877–85.* Religious Tract Society, London.

Christidis, L. and Boles, W. E. (1994). The taxonomy and species of birds of Australia and its territories. *Royal Australasian Ornithologists' Union Monograph*, **2**, 1–112. RAOU, Melbourne.

Christidis, L. and Schodde, R. (1991). Relationships of Australo-Papuan songbirds—protein evidence. *Ibis*, **133**, 277–85.

Christidis, L. and Schodde, R. (1992). Relationships among the Birds-of-Paradise (Paradisaeidae) and Bowerbirds (Ptilonorhynchidae), protein evidence. *Australian Journal of Zoology*, **40**, 343–53.

Christidis, L. and Schodde, R. (1993). Sexual selection for novel partners, a mechanism for accelerated morphological evolution in the birds-of-paradise (Paradisaeidae). *Bulletin of the British Ornithologists' Club*, **113**, 169–72.

Christidis, L., Leeton, P. R., and Westerman, M. (1996). Were bowerbirds part of the New Zealand fauna? *Proceedings of the National Academy of Science*, **93**, 3898–901.

Church, R. J. (1997). Avian frugivory in a subtropical rainforest: eleven years of observations in Lamington National Park. *Sunbird*, **27**, 85–97.

Clapp, G. E. (1986). Birds of Mount Scratchley summit and environs: 3520 metres asl in south-eastern New Guinea. *Muruk*, **1**, 75–84.

Clench, M. H. (1978). Tracheal elongation in birds-of-paradise. *Condor*, **80**, 423–30.

Clench, M. H. (1985). Body pterylosis of *Atrichornis, Menura*, the 'corvid assemblage,' and other possibly related passerines (Aves: Passeriformes). *Records of the Australian Museum*, **37**, 115–42.

Clench, M. H. (1992). Pterylography of birds-of-paradise and the systematic position of Macgregor's Bird-of-paradise (*Macgregoria pulchra*). *Auk*, **109**, 923–8.

Clusius, C. (1605). *Exoticorum libri decem.* The author, Leiden.

Coates, B. J. (1973a). Birds observed on Mt. Albert Edward, Papua. *New Guinea Bird Society Newsletter*, **84**, 3–7.

Coates, B. J. (1973b). Magnificent rifle birds in display. *New Guinea Bird Society Newsletter*, **87**, 3.

Coates, B. J. (1990). *The birds of Papua New Guinea including the Bismarck Archipelago and Bougainville.* Vol. 2. Dove, Alderley.

Coates, B. J. and Bishop, K. D. (1997). *A guide to the birds of Wallacea.* Dove, Alderley.

Coates, B. J. and Lindgren, E. (1978). *Ok Tedi birds.* Report of a preliminary survey of the avifauna of the Ok Tedi area, Western Province, PNG. Unpublished report prepared for Ok Tedi Environmental Task Force. Ok Tedi Development Co. and Office of Environment and Conservation, PNG.

Coates, B. J. Layton, W. A., and Filewood, L. W. (1970). Efogi 26–28.12.1969: complete annotated list. *New Guinea Bird Society Newsletter*, **50**, 2–3.

Coles, C. (1920). Bird-of-Paradise nesting in captivity. *Emu*, **19**, 244.

Collar, N. J. (1997). Taxonomy and conservation: chicken and egg. *Bulletin of the British Ornithologists' Club*, **117**, 122–36.

Collar, N. J. and Andrew, P. (1988). *Birds to watch: the ICBP world checklist of threatened birds.* ICBP, Cambridge.

Collias, N. E. and Collias, E. C. (1984). *Nest building and bird behavior.* Princeton University Press, Princeton.

Collins, M. (ed.) (1990). *The last rain forest.* Oxford University Press, Oxford.

Collins, N. M. Dayer, J. A. and Whitmore, C. (1991). *The conservation atlas of tropical forests—Asia and the Pacific.* Simon and Schuster, New York.

Cooper, P. (1995). Observations of parent-rearing behaviour in the Lesser Bird of Paradise. *Avicultural Magazine*, **101**, 194–9.

Cooper, W. T. and Forshaw, J. M. (1977). *The birds of paradise and bower birds.* Collins, Sydney.

Cotterell, G. W. (1966). A problem species: *Lamprolia victoriae. Emu*, **66**, 253–66.

Cracraft, J. (1981). Toward a phylogenetic classification of the recent birds of the world (Class Aves). *Auk*, **98**, 681–714.

Cracraft, J. (1992). The species of the birds-of-paradise (Paradisaeidae), applying the phylogenetic species concept to a complex pattern of diversification. *Cladistics*, **8**, 1–43.

Crandall, L. S. (1921). The Blue Bird of Paradise. *Bulletin of the New York Zoological Society*, **24**, 111–3.

Crandall, L. S. (1931). *Paradise quest, a naturalist's experience in New Guinea*. Scribner, New York.

Crandall, L. S. (1932). Notes on certain birds of paradise. *Zoologica*, **11**, 77–87.

Crandall, L. S. (1935). The most beautiful birds in the world. *Bulletin of the New York Zoological Society*, **38**, 147–60.

Crandall, L. S. (1936). Birds of Paradise in display. *Bulletin of the New York Zoological Society*, **39**, 87–103.

Crandall, L. S. (1937a). Further notes on certain Birds of Paradise. *Zoologica*, **22**, 193–5.

Crandall, L. S. (1937b). Position of wires in the display of the Twelve-wired Bird of Paradise. *Zoologica*, **22**, 307–10.

Crandall, L. S. (1938). Display of the Magnificent Rifle Bird. *Bulletin of the New York Zoological Society*, **41**, 43–4.

Crandall, L. S. (1940). Notes on the display forms of Wahnes' Six-plumed Bird of Paradise. *Zoologica*, **25**, 257–9.

Crandall, L. S. (1941). Description of an egg of the Long-tailed Bird of Paradise. *Zoologica*, **26**, 47–8.

Crandall, L. S. (1946a). Further notes on display forms of the Long-tailed Bird of Paradise, *Epimachus meyeri meyeri* Finsch. *Zoologica*, **31**, 9–10.

Crandall, L. S. (1946b). A curious display form of a curious bird. *Animal Kingdom*, **49**, 108–10.

Crandall, L. S. and Leister, C. W. (1937). Display of the Magnificent Rifle Bird. *Zoologica*, **22**, 311–4.

Currie, R. P. (1900). [*Cicinnurus Lyogyrus*, n. sp.] *Proceedings of the United States National Museum*, **22**, 497–9.

Cuvier, G. (1817). *Le règne animal*, **1**, 407–8 (Dec 1816). Deterville, Paris.

D'Albertis, L. M. (1880). *New Guinea, what I did and what I saw*. 2 volumes. Sampson Low, London.

D'Albertis, L. M. and Salvadori, T. (1879). Catalogo degli uccelli raccolti da L. M. D'Albertis. durante la 2a e 3a esplorazione del Fiume Fly negli anni 1876 e 1877. *Annal Museo Civico Genova*, **14**, series 1, 21–147.

Darwin, C. (1871). *The descent of man, and selection in relation to sex*. Murray, London.

Daudin, F. M. (1800). *Traite elementarie et complet d'ornithologie, ou histoire naturelle des oiseaux* (2). L'Imprimerie de Bertrand, Paris.

Davis, W. E. and Beehler, B. M. (1994). Nesting behavior of a Raggiana Bird of Paradise. *Wilson Bulletin*, **106**, 522–30.

Delacour, J. (1963). Notes on Austral and southern Pacific birds. *Avicultural Magazine*, **69**, 227–38.

De Vis, C. W. (1890). Report on birds from British New Guinea. *Annual Report for British New Guinea*, 1888–1889, Appendix C.

De Vis, C. W. (1891). Description of new species. *Cnemophilus Macgregorii*. *Ibis*, Ser. 6, vol. 3, 25–41.

De Vis, C. W. (1894). *Annual Report for British New Guinea* 1893–1894. Report on ornithological specimens collected in British New Guinea. *Annual Report for British New Guinea*, 1st. July 1893 to 30 June 1894, 99–105.

De Vis, C. W. (1897a). Description of a new bird of paradise from British New Guinea. *Ibis*, Ser. 7, vol. 3, 250–2.

De Vis, C. W. (1897b). Diagnoses of thirty-six new or little-known birds from British New Guinea. *Ibis*, Ser. 7, vol. 11, 390.

Dharmakumarsinhji, Prince K. S. (1943). Notes on the breeding of the Empress of Germany's Bird of Paradise in captivity. *Zoologica*, **28**, 139–44.

Dharmakumarsinhji, Prince, K. S. (1944). Notes on the breeding of the Empress of Germany's bird of paradise in captivity. *Avicultural Magazine*, **9**, 109–16.

Diamond, J. M. (1968). Search for birds in northern New Guinea. *Explorers' Journal*, **46**, 210–23.

Diamond, J. M. (1969). Preliminary results of an ornithological exploration of the North Coastal Range, New Guinea. *American Museum Novitates*, **2362**, 1–57.

Diamond, J. M. (1972). *Avifauna of the Eastern Highlands of New Guinea*. Publication of the Nuttall Ornithological Club No. 12. Cambridge, Massachusetts.

Diamond, J. M. (1973). Distributional ecology of New Guinea birds. *Science*, **179**, 759–69.

Diamond, J. M. (1981). *Epimachus bruijnii*, the Lowland Sickle-billed Bird-of-Paradise. *Emu*, **81**, 82–6.

Diamond, J. M. (1982). Mimicry of friarbirds by orioles. *Auk*, **99**, 187–96.

Diamond, J. M. (1985). New distributional records and taxa from the outlying mountain ranges of New Guinea. *Emu*, **85**, 65–91.

Diamond, J. M. (1986*a*). Biology of birds of paradise and bower birds. *Annual Review of Ecology and Systematics*, **17**, 17–37.

Diamond, J. M. (1986*b*). The design of a nature reserve system for Indonesian New Guinea. In *Conservation biology*, (ed. M. E. Soute), pp. 485–503. Sinauer Sunderland.

Diamond, J. M. (1987). Flocks of brown and black New Guinea birds, a bicoloured mixed-species foraging association. *Emu*, **87**, 201–11.

Diamond, J. M. (1991). Borrowed sexual ornaments. *Nature*, **349**, 105.

Diamond, J. M. (1992). Rubbish birds are poisonous. *Nature*, **360**, 19.

Dinsmore, J. J. (1967). Ecology and behaviour of the Greater Bird-of-Paradise on Little Tobago Island. M.S. thesis. Madison University, Wisconsin.

Dinsmore, J. J. (1969). Dual calling by birds of paradise. *Auk*, **86**, 139–40.

Dinsmore, J. J. (1970*a*). Courtship behaviour of the Greater Bird of Paradise. *Auk*, **87**, 305–21.

Dinsmore, J. J. (1970*b*). History and natural history of *Paradisaea apoda* on Little Tobago Island, West Indies. *Caribbean Journal of Science*, **10**, 93–100.

Dorst, J. (1973*a*). Structure des plumes de parure des Paradisaeides pereticulierement des plumes generatrices d'interferences. *Compte Rendu des Séances de l'Academie des Science, Paris*, **276**, D, 1441–8.

Dorst, J. (1973*b*). Precisions sur la structure des plumes generatrices d'interferences et de celles d'apparence veloutee chez les paradisiers. *L'Oiseau et La Revue Francaise D'Ornithologie*, **44**, 138–44.

Dorst, J. Gastaldi, G. Hagege, R. and Jacquemart, J. (1974). Differents aspects des barbules de quelques Paradisaerides observes sur coupes en microscopie electronique. Relations avec les phenomenes d'interferences. *Compte Rendu des Séances de l'Academie des Science, Paris*, **738**, D, 285–90.

Doughty, R. W. (1975). *Feather fashions and bird preservation*. University of California Press, Berkeley.

Draffan, B. (1977). List of some birds seen at Ogeranang Airstrip 1600 metres Pindiu Sub District 18–19.5.77. *New Guinea Bird Society Newsletter*, **133**, 8.

Draffan, R. D. W. (1978). Group display of the Emperor of Germany Bird-of-Paradise *Paradisaea guilielmi* in the wild. *Emu*, **78**, 157–9.

Drew, R. A. I. (1988). Amino acid increases in fruit infested by fruit flies of the family Tephritidae. *Zoological Journal of the Linnean Society*, **93**, 107–12.

Dumbacher, J. P. Beehler, B. M. Spande, T. F. Garraffo, H. M. and Daly, J. W. (1992). Homobatrachotoxin in the genus *Pitohui*: chemical defense in birds? *Science*, **258**, 799–801.

Eastwood, C. (1989). Recent observations July–September 1988. *Muruk*, **4**, 25–37.

Eastwood, C. (1996). A trip to Irian Jaya. *Muruk*, **8**, 12–23.

Edwards, G. (1750). *A natural history of birds III*. London.

Egerton, P. de M. G. (1872). Catalogue of fossil fishes in the collection of the Earl of Enniskillen and Sir Philip Grey Egerton. 2 volumes.

Ehrlich, P. and Ehrlich, A. (1981). *Extinction*. Random House, New York.

Elliot, D. G. (1871). Review of the genus *Ptiloris*, Swainson. *Proceedings of the Zoological Society of London*, **1871**, 580–4.

Elliot, D. G. (1873). *A monograph of the Paradiseidae*. The author, London.

Ellis, S. T. (1966). Agonistic behavior in the male Starling. *Wilson Bulletin*, **78**, 208–24.

Emlen, J. T. and Emlen, V. M. (1984). Misdirected displays by a solitary bird of paradise in an oropendola nesting colony. *Wilson Bulletin*, **96**, 482–3.

Emlen, S. T. and Oring, L. (1977). Ecology, sexual selection, and the evolution of mating systems. *Science*, **197**, 215–23.

Everett, M. (1978). *The birds of paradise*. Putnam's, New York.

Everitt, C. (1962). The Paradise Rifle-bird. *Avicultural Magazine*, **68**, 95–7.

Everitt, C. (1965). Breeding the Magnificent bird of paradise. *Avicultural Magazine*, **71**, 146–8.

Everitt, C. (1973). *Birds of the Edward Marshall Boehm aviaries*. Boehm, Trenton.

Feare, C. (1984). *The starling*. Oxford University Press, Oxford.

Filewood, L. W. C. and Peckover, W. S. (1978). Scientific names used in Birds of New Guinea and Tropical Australia and the Handbook of New Guinea Birds. *Wildlife in Papua New Guinea*, **78**, 12.

Finch, B. W. (1983). Birds of the Vanapa-Veimauri Kanosia-Cape Suckling regions. *Papua New Guinea Bird Society Newsletter*, **199–200**, 17–40.

Finch, B. W. and Lowry, J. (1980). Observations. *Papua New Guinea Bird Society Newsletter*, **167–68**, 39–40.

Finsch, O. and Meyer, A. B. (1885). Vögel von Neu Guinea zuneist aus der Alpenregion am Südostabhange des Owen Stanley-Gebirges (Hufeisengebirge 7000–8000′ hoch), gesammelt von Karl Hunstein. *Zeitschrift für die gesammte Ornithologie*, **2**, 369–91.

Fisher, J. (1938). The Orange-wattled bird of paradise, (*Macgregoria pulchra*). *Avicultural Magazine*, Ser. 5, vol. 3, 65–6.

Fisher, R. A. (1958). *The genetical theory of natural selection*. Dover, New York.

Flannery, T. (1995a). *Mammals of New Guinea*. Reed, Sydney.

Flannery, T. (1995b). *Mammals of the South-West Pacific & Moluccan Islands*. Reed, Sydney.

Flynn, E. (1960). *My wicked, wicked ways*. William Heinemann, London.

Foerster, F. (1906). Two new birds of paradise. *Ornithological Tracts*, **32**, 1–3. Hazell, Watson and Viney, London.

Foerster, F. and Rothschild, W. (1906). *Two new birds of paradise*: 2. Tring Museum, Tring.

Forbes, W. A. (1882a). Notes on a peculiarity in the trachea of the Twelve-wired Bird-of-paradise (*Seleucides nigra*). *Proceedings of the Zoological Society of London*, **1882**, 333–5.

Forbes, W. A. (1882b). On the convoluted trachea of two species of Manucode with remarks on similar structures in other birds. *Proceedings of the Zoological Society of London*, **1882**, 353.

Forrest, T. (1781). *A voyage to New Guinea and the Moluccas 1774–1776*. [1969 edition introduced by D. K. Bassett. Oxford University Press].

Forster, J. R. (1781). *Zoologica Indica selecta*. Gebauer, Halle.

Frankenhuis, M. T. van Eyk, H. G. Assink, J. A. and Zwart, P. (1989). Iron storage in livers of birds of paradise. In *Proceedings of the 2nd European Symposium on avian medicine and surgery*, pp. 92–6. Dutch Assoc. Avian Vets, Utrecht.

Friedmann, H. (1934). The display of Wallace's Standard-wing Bird of Paradise in captivity. *Scientific Monthly*, **39**, 52–5.

Friedmann, H. (1935). Die Balz von *Semioptera wallacei halmaherae* in Gefangenschaft. *Journal für Ornithologie*, **83**, 283–6.

Frith, C. (1968). Some displays of Queen Carola's Parotia. *Avicultural Magazine*, **74**, 85–90.

Frith, C. B. (1970). The nest and nestling of the short-tailed paradigalla *Paradigalla brevicauda* (Paradisaeidae). *Bulletin of the British Ornithologists' Club*, **90**, 122–4.

Frith, C. B. (1971). Some undescribed nests and eggs of New Guinea birds. *Bulletin of the British Ornithologists' Club*, **91**, 46–9.

Frith, C. B. (1974). Observations on Wilson's Bird of Paradise. *Avicultural Magazine*, **80**, 207–12.

Frith, C. B. (1976). Displays of the Red Bird of Paradise *Paradisaea rubra* and their significance, with a discussion on displays and systematics of other Paradisaeidae. *Emu*, **76**, 69–78.

Frith, C. B. (1977). Some birds of paradise skins in a Singapore collection. *New Guinea Bird Society Newsletter*, **128**, 8–9.

Frith, C. B. (1979). Ornithological literature of the Papuan Subregion 1915 to 1976 an annotated bibliography. *Bulletin of the American Museum of Natural History*, **164**, 379–465.

Frith, C. B. (1981). Displays of Count Raggi's Bird-of-Paradise *Paradisaea raggiana* and congeneric species. *Emu*, **81**, 193–201.

Frith, C. B. (1985). *Birds of Paradise* and *Silktail*, In *A dictionary of birds*, (ed. R. Campbell and E. Lack), pp. 55–6, 538. Poyser, Carlton.

Frith, C. B. (1987). An undescribed plumage of Loria's Bird of Paradise *Loria loriae*. *Bulletin of the British Ornithologists' Club*, **107**, 177–80.

Frith, C. B. (1992). Standardwing Bird of Paradise *Semioptera wallacii* displays and relationships, with comparative observations on displays of other Paradisaeidae. *Emu*, **92**, 79–86.

Frith, C. B. (1994a). The status and distribution of the Trumpet Manucode *Manucodia keraudrenii* (Paradisaeidae) in Australia. *Australian Bird Watcher*, **15**, 218–24.

Frith, C. B. (1994b). Adaptive significance of tracheal elongation in manucodes (Paradisaeidae). *Condor*, **96**, 552–5.

Frith, C. B. (1994c). Range extension of the Splendid Astrapia *Astrapia splendidissima*, a sighting of an *A. mayeri* × *A. stephaniae* hybrid, or an unidentified *Astrapia* sp. (Paradisaeidae)? *Muruk*, 7, 49–52.

Frith, C. B. (1994d). Egg laying at long intervals in bowerbirds (Ptilonorhynchidae). *Emu*, **94**, 60–1.

Frith, C. B. (1996). Further notes on little-known plumages of the Crested and Loria's Birds of Paradise *Cnemophilus macgregorii* and *C. loriae*. *Bulletin of the British Ornithologists' Club*, **116**, 247–51.

Frith, C. B. (1997). Huia (*Heteralocha acutirostris*, Callaeidae) -like sexual bill dimorphism in some birds of paradise (Paradisaeidae) and its significance. *Notornis*, **44**, 177–84.

Frith, C. B. (1998). Aberrant plumages in birds of paradise (Paradisaeidae). *Memoirs of the Queensland Museum*, **42**, in press.

Frith, C. B. and Beehler, B. M. (1997). Courtship and mating behaviour of the Twelve-wired Bird of Paradise *Seleucidis melanoleuca*. *Emu*, **97**, 133–40.

Frith, C. B. and Coles, D. (1976). Additional notes on displays of Queen Carola's Bird of Paradise. *Avicultural Magazine*, **82**, 52–3.

Frith, C. B. and Cooper, W. T. (1996). Courtship display and mating of Victoria's Riflebird *Ptiloris victoriae* (Paradisaeidae) with notes on the courtship of congeneric species. *Emu*, **96**, 102–13.

Frith, C. B. and Frith, D. W. (1979). Leaf-eating by Birds-of-paradise and Bower birds. *Sunbird*, **10**, 21–3.

Frith, C. B. and Frith, D. W. (1981). Displays of Lawes's Parotia *Parotia lawesii* (Paradisaeidae), with reference to those of congeneric species and their evolution. *Emu*, **81**, 227–38.

Frith, C. B. and Frith, D. W. (1985). Seasonality of insect abundance in an Australian upland tropical rainforest. *Australian Journal of Ecology*, **10**, 31–42.

Frith, C. B. and Frith, D. W. (1990a). Nesting biology and relationships of the Lesser Melampitta *Melampitta lugubris*. *Emu*, **90**, 65–73.

Frith, C. B. and Frith, D. W. (1990b). Archbold's Bowerbird *Archboldia papuensis* (Ptilonorhynchidae) uses plumes from King of Saxony Bird of Paradise *Pteridophora alberti* (Paradisaeidae) as bower decoration. *Emu*, **90**, 136–7.

Frith, C. B. and Frith, D. W. (1990c). Discovery of the King of Saxony Bird of Paradise *Pteridophora alberti*, nest, egg and nestling with notes on parental care. *Bulletin of the British Ornithologists' Club*, **110**, 160–4.

Frith, C. B. and Frith, D. W. (1992a). Annotated list of birds in Western Tari Gap, Southern Highlands, Papua, New Guinea, with some nidification notes. *Australian Bird Watcher*, **14**, 262–76.

Frith, C. B. and Frith, D. W. (1992b). The nesting biology of the Short-tailed Paradigalla *Paradigalla brevicauda* (Paradisaeidae). *Ibis*, **134**, 77–82.

Frith, C. B. and Frith, D. W. (1993a). Results of a preliminary highland bird banding study at Tari Gap, Southern Highlands Province, Papua New Guinea. *Corella*, **17**, 5–21.

Frith, C. B. and Frith, D. W. (1993b). The nesting biology of the Ribbon-tailed Astrapia *Astrapia mayeri* (Paradisaeidae). *Emu*, **93**, 12–22.

Frith, C. B. and Frith, D. W. (1993c). Nidification of the Crested Bird of Paradise *Cnemophilus macgregorii* and a review of its biology and systematics. *Emu*, **93**, 23–33.

Frith, C. B. and Frith, D. W. (1993d). Notes on birds found nesting at Iron Range, Cape York Peninsula, November–December 1990. *Sunbird*, **23**, 44–58.

Frith, C. B. and Frith, D. W. (1993e). Courtship display of the Tooth-billed Bowerbird *Scenopoeetes dentirostris* and its behavioural and systematic significance. *Emu*, **93**, 129–36.

Frith, C. B. and Frith, D. W. (1994a). The nesting biology of Archbold's Bowerbird *Archboldia papuensis* and a review of that of other bowerbirds (Ptilonorhynchidae). *Ibis*, **136**, 153–60.

Frith, C. B. and Frith, D. W. (1994*b*). Discovery of nests and an egg of Loria's Bird of Paradise *Cnemophilus loriae*. *Bulletin of the British Ornithologists' Club*, **114**, 182–92.

Frith, C. B. and Frith, D. W. (1994*c*). Courts and seasonal activities at them by male Tooth-billed Bowerbirds *Scenopoeetes dentirostris* (Ptilonorhynchidae). *Memoirs of the Queensland Museum*, **37**, 121–45.

Frith, C. B. and Frith, D. W. (1995*a*). Court site constancy, dispersion, male survival and court ownership in the male Tooth-billed Bowerbird, *Scenopoeetes dentirostris* (Ptilonorhynchidae). *Emu*, **95**, 84–98.

Frith, C. B. and Frith, D. W. (1995*b*). Notes on the nesting biology and diet of Victoria's Riflebird *Ptiloris victoriae*. *Emu*, **95**, 162–74.

Frith, C. B. and Frith, D. W. (1996*a*). Description of the unique *Parotia lawesii* × *Paradisaea rudolphi* hybrid bird of paradise (Paradisaeidae). *Records of the Australian Museum*, **48**, 111–16.

Frith, C. B. and Frith, D. W. (1996*b*). The unique type specimen of the bird of paradise *Lophorina superba pseudoparotia* Stresemann 1934 (Paradisaeidae), a hybrid of *Lophorina suberba* × *Parotia carolae*. *Journal für Ornithologie*, **137**, 515–21.

Frith, C. B. and Frith, D. W. (1997*a*). The taxonomic status of *Paradigalla carunculata intermedia* (Paradisaeidae) with notes on the other *Paradigalla* taxa. *Bulletin of the British Ornithologists' Club*, **117**, 38–48.

Frith, C. B. and Frith, D. W. (1997*b*). Biometrics of the birds of paradise (Aves: Paradisaeidae): with observations on variation and sexual dimorphism. *Memoirs of the Queensland Museum*, **42**, 159–212.

Frith, C. B. and Frith, D. W. (1997*c*). Courtship display and mating of the King of Saxony Bird of Paradise *Pteridophora alberti* (Paradisaeidae) in New Guinea with comment on their taxonomic significance. *Emu* **97**, 185–93.

Frith, C. B. and Frith, D. W. (1998*a*). Additional notes on the nesting biology of Victoria's Riflebird *Ptiloris victoriae* (Paradisaeidae). *Emu*, **98**, in press.

Frith, C. B. and Frith, D. W. (1998*b*). Golden Bowerbird nesting biology. *Emu*, **98**, in press.

Frith, C. B. and Harrison, C. J. O. (1989). An undescribed plumage of the Crested Bird of Paradise *Cnemophilus macgregorii*. *Bulletin of the British Ornithologists' Club*, **109**, 137–9.

Frith, C. B. and McGuire, M. (1996). Visual evidence of vocal avian mimicry by male Tooth-billed Bowerbirds *Scenopoeetes dentrirostris* (Ptilonorhynchidae). *Emu*, **96**, 12–6.

Frith, C. B. and Poulsen, M. K. (in press). Distribution and status of the Paradise Crow *Lycocorax pyrrhopterus* and Standardwing Bird of Paradise *Semioptera wallacii* (Paradiseaidae), with notes on biology and nidification. *Emu*.

Frith, C. B., Frith, D. W. and Watling, D. (1989). Notes on the nesting, parental care, and taxonomy of the Silktail (*Lamprolia victoriae*) of Fiji. *Notornis*, **36**, 96–8.

Frith, C. B., Gibbs, D., and Turner K. (1995). The taxonomic status of populations of Archbold's Bowerbird *Archboldia papuensis* in New Guinea. *Bulletin of the British Ornithologists' Club*, **115**, 109–14.

Frith, C. B., Borgia, G., and Frith, D. W. (1996). Courts and courtship behaviour of Archbold's Bowerbird *Archboldia papuensis* in Papua New Guinea. *Ibis*, **136**, 153–60.

Frith, C. B., Frith, D. W., and Jansen, A. (1997). The nesting biology of the Chowchilla *Orthonyx spaldingii* (Orthonychidae). *Emu*, **97**, 18–30.

Frith, D. W. (1984). Foraging ecology of birds in an upland tropical rainforest in North Queensland. *Australian Wildlife Research*, **11**, 325–47.

Frith, D. W. and Frith, C. B. (1988). Courtship display and mating of the Superb Bird of Paradise *Lophorhina superba*. *Emu*, **88** 183–8.

Frith, D. W. and Frith, C. B. (1990). Seasonality of litter invertebrate populations in an Australian upland tropical rainforest. *Biotropica*, **22**, 181–91.

Frith, D. W. and Frith, C. B. (1991). Say it with bowers. *Wildlife Conservation*, **94**, 74–83.

Frith, D. W. and Frith, C. B. (1996). *Cape York Peninsula—a natural history*. Reed, Sydney.

Fritts, T. H. (1988). The brown tree snake, *Boiga irregularis*, a threat to Pacific islands. U.S. Department of Interior, Fish and Wildlife Service, Biological Report, **88**, 1–36.

Frost, W. J. C. (1930). The nesting habits of the King Bird of Paradise. *Avicultural Magazine*, Ser. 5, vol. 8, 33–5.

Fuller, E. (1979). Hybridization among the Paradisaeidae. *Bulletin of the British Ornithologists' Club*, **99**, 145–52.

Fuller, E. (1987). *Extinct birds*. Penguin, London.

Fuller, E. (1995). *The lost birds of paradise*. Swan Hill Press, Shewsbury.

Gesner, C. (1669). *Vogel-Buch*. Wilhelm Serlins, Frankfurt am Mayn. Reprint 1981, Schlütersche Verlagshandlung, Hannover.

Gibbs, D. (1994). Undescribed taxa and new records from the Fakfak Mountains, Irian Jaya. *Bulletin of the British Ornithologists' Club*, **114**, 4–11.

Gill, F. B. (1990). *Ornithology*. Freeman, New York.

Gilliard E. T. (1950). Notes on birds of south-eastern Papua. *American Museum Novitates*, **1453**, 1–40.

Gilliard, E. T. (1953). New Guinea's rare birds and stone age men. *National Geographic Magazine*, **103**, 421–88.

Gilliard, E. T. (1955). The land of the head-hunters. *National Geographic Magazine*, **108**, 437–86.

Gilliard, E. T. (1956). The systematics of the New Guinea Manucode *Manucodia ater*. *American Museum Novitates*, **1770**, 1–13.

Gilliard, E. T. (1957). Coronation in Katmandu. *National Geographic Magazine*, **113**, 139–52.

Gilliard, E. T. (1958). Feathered dancers of Little Tobago. *National Geographic Magazine*, **114**, 428–40.

Gilliard, E. T. (1961). Four new birds from the mountains of central New Guinea. *American Museum Novitates*, **2031**, 3–7.

Gilliard, E. T. (1969). *Birds of paradise and bower birds*. Weidenfeld and Nicolson, London,

Gilliard, E. T. and LeCroy, M. (1961). Birds of the Victor Emmanuel and Hindenburg Mountains, New Guinea—Results of the American Museum of Natural History Expedition to New Guinea in 1954. *Bulletin of the American Museum of Natural History*, **123**, 1–86.

Gilliard, E. T. and LeCroy, M. (1966). Birds of the Middle Sepick Region, New Guinea. Results of the American Museum of Natural History expedition to New Guinea in 1953–1954. *Bulletin of the American Museum of Natural History*, **132**, 247–75.

Gilliard, E. T. and LeCroy, M. (1967). Annotated list of birds of the Adelbert mountains, New Guinea. Results of the 1959 Gilliard expedition. *Bulletin of the American Museum of Natural History*, **138**, 51–82.

Gilliard, E. T. and LeCroy, M. (1968). Birds of the Schrader Mountain region, New Guinea. Results of the American Museum of Natural History expedition to New Guinea in 1964. *American Museum Novitates*, **2343**, 1–41.

Gilliard, E. T. and LeCroy, M. (1970). Notes on birds from the Tamrau Mountains, New Guinea. *American Museum Novitates*, **2420**, 1–28.

Ginn, H. G. and Melville, D. S. (1983). *Moult in birds*. British Trust for Ornithology, Tring.

Gmelin, J. F. (1788). *Systema naturae per regna tria naturae*. 1—Part 1 401–2; 468. Georg. Emanuel Beer, Lipsiae.

Goldman (1981). *Talk never dies*. Ph.D. thesis. University of London.

Goode, J. (1977). *Rape of the Fly*. Nelson and Brown, Melbourne.

Goodfellow, W. (1908). [Account of expedition to British New Guinea]. *Bulletin of the British Ornithologists' Club*, **23**, 35–9.

Goodfellow, W. (1910). Notes on Birds of Paradise. *Avicultural Magazine*, Ser. 3, vol. 1, 277–86.

Goodfellow, W. (1926a). Princess Stephanie's Bird of Paradise, (*Astrachia stephaniae*). *Avicultural Magazine*, **4**, series 4, 197–202.

Goodfellow, W. (1926b). Remarks on his recent journey in Papua New Guinea and on the birds of paradise met with. *Bulletin of the British Ornithologists' Club*, **46**, 58–9.

Goodfellow, W. (1927). Wallace's Bird of Paradise (*Semioptera wallacei*). *Avicultural Magazine*, 5th series, **4**, 57–65.

Goodwin, A. P. (1890). Notes on the paradise-birds of British New Guinea. *Ibis*, Ser. 6, vol. 2, 150–6.

Goodwin, D. (1983). *Pigeons and doves of the world*. 3rd edn. British Museum of Natural History, London.

Goodwin, D. (1986). *Crows of the world*. British Museum of Natural History, London.

Gould, J. (1840–8). *The birds of Australia*. The author, London.

Gould, J. (1850). Description of *Ptiloris victoriae*, Gould. *Proceedings of the Zoological Society of London*, **1850**, 111.

Gould, J. (1865). *Handbook to the birds of Australia*. Vol. 1. The author, London.

Gould, J. (1869). *The birds of Australia; supplement*. The author, London.

Gould, J. and Sharpe, R. B. (1875–88). *The birds of New Guinea*. Vol. 1. Sotheran, London.

Graves, G. R. (1995). Sequence of plumage evolution in the Standardwing Bird of Paradise. *Wilson Bulletin*, **107**, 371–3.

Graves, G. R. (1996). Hybrid wood warblers, *Dendroica striata* × *Dendroica castanea* (Aves: Fringillidae: Tribe Parulini) and the diagnostic predictability of avain hybrid phenotypes. *Proceedings of the Biological Society of Washington*, **109**, 373–90.

Gray (1840). A list of the genera of birds. ed. 1, add. and err. p. 1. Taylor, London [*Craspedophora* gen. n.].

Gray, G. R. (1859a). Letter received from A. R. Wallace; minutes of the Society meeting. *Proceedings of the Zoological Society of London*, **1859**, 129–30.

Gray, G. R. (1859b). List of birds lately sent by Mr A. R. Wallace from Dorey or Dorery, New Guinea. *Proceedings of the Zoological Society of London*, **1859**, 153–9.

Gray, G. R. (1859c). Zoological Society (report of meeting). *Literary Gazette* (new series), **39**, 406.

Gray, G. R. (1870). *Hand list of birds* **2**, 17. British Museum, London.

Greenwalt, C. H. (1962). Dimensional relationships for flying animals. *Smithsonian Miscellaneous Collections*, **144**, 1–46.

Greenway, J. C. Jr. (1934). Description of four new subspecies of birds from the Huon Gulf region, New Guinea. *Proceedings of the New England Zoological Club*, **14**, 1–3.

Greenway, J. C. Jr. (1935). Birds from the coastal range between the Markham and the Waria Rivers, northeastern New Guinea. *Proceedings of the New England Zoological Club*, **14**, 15–106.

Greenway, J. C. Jr. (1942). A new manucode bird of paradise. *Proceedings of the New England Zoological Club*, **19**, 51–2.

Greenway, J. C. Jr. (1966). Birds collected on Batanta, off Western New Guinea, by E. Thomas Gilliard in 1964. *American Museum Novitates*, **22**, 1–27.

Greenway, J. C. Jr. (1967). *Extinct and vanishing birds of the world*. Dover, New York.

Gregory, P. (1995). Further studies of the birds of the Ok Tedi area, Western Province, Papua New Guinea. *Muruk*, **7**, 1–38.

Gregory, P. (1997). Range extensions and unusual sightings from Western Province, Papua New Guinea. *Bulletin of the British Ornithologist's Club*, **117**, 304–11.

Gruson, E. S. (1976). *A checklist of the birds of the world*. Collins, London.

Guillemard, F. H. H. (1886). *The cruise of the Marchesa to Kamschatka and New Guinea with notices of Formosa, Liu-kiu, and various islands of the Malay Archipelago*. 2 volumes. Murray, London.

Gyldenstolpe, N. (1955a). Notes on a collection of birds made in the Western Highlands, Central New Guinea, 1951. *Arkiv för Zoologi*, **8**, 1–181.

Gyldenstolpe, N. (1955b). Birds collected by Dr. Sten Bergman during his expedition to Dutch New Guinea 1948–1949. *Arkiv för Zoologi*, **8**, 183–397.

Haan, J. A. B. de (1920). On a doubling of the central tail-feathers in a bird of paradise. *Ibis*, Ser. 11, vol. 2, 720–2.

Hadden, D. (1975). Birds seen in the Tari area from 1 August to 14 August 1975. *New Guinea Bird Society Newsletter*, **113**, 8–9.

Hallstrom, E. (1959). Some breeding results in the Hallstrom collection. *Avicultural Magazine*, **65**, 77–80.

Hallstrom, E. (1962). Some breeding results in the Hallstrom collection. *Avicultural Magazine*, **68**, 46–8.

Hamilton, W. D. and Zuk, M. (1982). Heritable true fitness and bright birds: a role for parasites? *Science*, **218**, 384–7.

Harrison, C. J. O. (1964). Open-billed probing by the Princess Stephanie Bird of Paradise. *Condor*, **66**, 162–3.

Harrison, C. J. O. (1971). Further notes on eggs of New Guinea birds. *Emu*, **71**, 85–6.

Harrison, C. J. O. and Frith, C. B. (1970). Nests and eggs of some New Guinea birds. *Emu*, **70**, 173–8.

Harrison, C. J. O. and Walters, M. P. (1973). Use of nests of other species by the trumpetbird. *Emu*, **73**, 189–90.

Harrison, J. M. (1964). Plumage. In *A new dictionary of birds*, (ed. A. L. Thomson), pp. 639–43. Nelson, London.

Hartert, E. (1903). The birds of the Obi Group, Central Moluccas. *Novitates Zoologicae*, **10**, 1–38.

Hartert, E. (1910). On the eggs of the Paradisaeidae. *Novitates Zoologicae*, **17**, 484–91.

Hartert, E. (1919). Types of birds in the Tring Museum. *Novitates Zoologicae*, **25**, 129.

Hartert, E. (1930). List of the birds collected by Ernst Mayr. *Novitates Zoologicae*, **36**, 27–128.

Healey, C. J. (1975). A further note on the display and mating of *Pteridophora alberti*. *New Guinea Bird Society Newsletter*, **110**, 6–7.

Healey, C. J. (1976). Sympatry in *Parotia lawesii* and *P. carolae*. *Emu*, **76**, 85.

Healey, C. J. (1978a). Effects of human activity on *Paradisea minor* in the Jimi Valley, New Guinea. *Emu*, **78**, 149–55.

Healey, C. J. (1978b). Communal display of Princess Stephanie's Astrapia *Astrapia stephaniae* (Paradisaeidae). *Emu*, **78**, 197–200.

Healey, C. J. (1980). Display of Queen Carola's Parotia *Parotia carolae* (Paradisaeidae). *Papua New Guinea Bird Society Newsletter*, **163–164**, 6–9.

Healey, C. J. (1986). Men and birds in the Jimi Valley. The impact of man on Birds of Paradise in the Papua New Guinea Highlands. *Muruk*, **1**, 34–71.

Healey, C. J. (1990). *Maring hunters and traders*. University of California Press, Berkeley.

Heaney, W. (1982). The changing role of birds of paradise plumes in bridewealth in the Wahgi Valley. In *Traditional conservation in Papua New Guinea, implications for today* (ed. L. Morauta, J. Pernetta, and W. Heaney). Institute of Applied Social and Economic Research. Monograph **16**, 227–31, Boroko.

Heather, B. D. (1977). The Vanua Levu Silktail (*Lamprolia victoriae klenschmidti*), a preliminary look at its status and habits. *Notornis*, **24**, 94–128.

Heatwole, H. (1987). Major components and distributions of the terrestrial fauna. In *Fauna of Australia Vol. 1A, General Articles* (ed. G. R. Dyne), pp. 101–35. Bureau of Flora and Fauna, Canberra.

Heinroth, O. (1918). [*Phonygammus neumanni*]. *Journal für Ornithologie*, **4**, 488–9.

Heinrich, G. (1956). Biologische Aufzeichnungen uber Vogel von Halmahera und Batjan. *Journal für Ornithologie*, **97**, 31–40.

Hermann, J. (1783). *Tabula affinitatum animalium*. Treuttel, Argentorati.

Hicks, J. H. and Hicks, R. K. (1988). Display of Loria's Bird of Paradise. *Muruk*, **3**, 52.

Hicks, R. K. (1988a). Feeding observations at a fruiting *Pipturus*. *Muruk*, **3**, 15.

Hicks, R. (1988b). Feeding observations of female Crested Bird of Paradise. *Muruk*, **3**, 15.

Hicks, R. K. and Hicks, J. H. (1988a). Observations of birds feeding in a fruiting *Planchonella*. *Muruk*, **3**, 10–1.

Hicks, R. K. and Hicks, J. H. (1988b). Feeding observations of Short-tailed Paradigalla. *Muruk*, **3**, 14.

Hides, J. G. (1936). *Papuan wonderland*. Blackie, London.

Hogan-Warburg, A. J. (1966). Social behaviour of the ruff, *Philomachus pugnax* (L.). *Ardea*, **54**, 109–229.

Holmes, G. (1973). The bird species diversity of some subtropical Australian forests. B.Sc. (Hons.) thesis. University of New England, Armidale.

Hoogerwerf, A. (1971). On a collection of birds from the Vogelkop, near Manokwari north-western New Guinea (continued). *Emu*, **71**, 73–83.

Hooper, N. (1972). Display of the Magnificent Rifle-bird. *Australian Bird Watcher*, **4**, 134–5.

Hope, F. W. (1838). *The Coleopterist's manual* ... Part 2: 165. Bohn, London.

Hope G. S., Peterson, J. A., Radok, U, and Allison, I. (1976). *The equatorial glaciers of New Guinea*. Balkema, Rotterdam.

Hopkins, H. C. F. (1988). Some feeding records for Birds of Paradise. *Muruk*, **3**, 12–3.

Hopkins, H. C. F. (1992). Some records of birds feeding on flowers and fruits in montane forest, near Myola, Oro Province, and Tari Gap, Southern Highlands Province, P.N.G. *Muruk*, **5**, 86–9.

Horsbrugh, C. B. (1909). A journey to British New Guinea in search of Birds of Paradise. *Ibis*, Ser. 9, vol. 3, 197–213.

Howe, H. F. (1979). Fear and frugivory. *American Naturalist*, **114**, 925–31.

Howe, R. W. (1984). Local dynamics of bird assemblages in small forest habitat islands in Australia and North America. *Ecology*, **65**, 1585–601.

Howe, R. W. (1986). Bird distributions in forest islands in north-eastern ed. New South Wales. In *The dynamic partnership: birds and plants in Southern Australia*, (eds H. A. Ford and D.C. Paton), pp. 119–29. The Flora and Fauna of South Australia Handbooks Committee, Adelaide.

Hoyle, M. A. (1975). Observations on birds of paradise/bower birds. *New Guinea Bird Society Newsletter*, **110**, 6.

Hudon, J. and Brush, A. H. (1990). Carotenoids produce flush in the Elegant Tern plumage. *Condor*, **92**, 798–801.

Hundgen, K. and Bruning, D. (1988). Propagation techniques for Birds of Paradise at the New York Zoological Park. *AAZPA 1990 Annual Conference Proceedings*, 14–20.

Hundgen, K., Sheppared, C., Bruning, D., Hutchins, M., Worth, W., and Laska, M. (1990). Management and breeding of the Lesser Bird of Paradise *Paradisaea minor* at the New York Zoological Park. *AAZPA 1990 Annual Conference Proceedings*, 199–207.

Hundgen, K., Hutchins, M., Sheppard, C., Bruning, D. and Worth, W. (1991). Management and breeding of the Red Bird of Paradise (*Paradisaea rubra*) at the New York Zoological Park. *International Zoo Yearbook*, **30**, 192–9.

Ingram, C. (1913). Birds of paradise in the West Indies. *Avicultural Magazine*, Ser. 3, vol. 5, 35–41.

Ingram, C. (1956). Birds of paradise in the West Indies. *Country Life*, **119**, 482.

Ingram, W. (1907). On the display of the King Bird-of-Paradise (*Cicinnurus regius*). *Ibis*, Ser. 9, vol. 1, 224–9.

Ingram, W. (1911). The acclimatization of the Greater Bird of Paradise (*Paradisea apoda*) in the West Indies. *Avicultural Magazine*, **18**, 142–7.

Ingram, W. (1917). The Great Bird of Paradise on the island of Little Tobago (*Paradisea apoda*). *Avicultural Magazine*, **23**, 341–51.

Ingram, W. (1918). Birds of paradise on Little Tobago, West Indies. *Avicultural Magazine*, **24**, 279–80.

Iredale, T. (1922). [Proposal of new genus name *Mathewsiella*]. *Bulletin of the British Ornithologists' Club*, **43**, 39.

Iredale, T. (1948). A check list of the birds of paradise and bowerbirds. *Australian Zoologist*, **11**, 161–89.

Iredale, T. (1950). *Birds of paradise and bowerbirds*. Georgian House, Melbourne.

Irwin, R. E. (1996). The phylogenetic content of avian courtship display and song evolution. In *Phylogenies and the comparative method in animal behaviour*, (ed. E. P. Martins), pp. 234–52. Oxford University Press, New York.

Isenberg, A. H. (1961). Nesting of the Red bird of paradise. *Avicultural Magazine*, **67**, 43–4.

Isenberg, A. H. (1962). Further notes on the breeding of the Red bird of paradise. *Avicultural Magazine*, **68**, 48.

IUCN (1996). *1996 IUCN red list of threatened animals*. IUCN, Gland.

Jackson, C. E. (1993). *Great bird paintings of the world—the old masters*. Antique Collectors' Club, Woodbridge.

Jackson, C. E. (1994). *Great bird paintings of the world—the eighteenth century*. Antique Collectors' Club, Woodbridge.

Jackson, S. W. (1907). *Egg collecting and bird life of Australia. Catalogue and data of the Jacksonian Oological Collection*. Sydney. F. W. White, Printer, Melbourne.

Jackson, S. W. (1909). In the Barron River Valley, north Queensland. *Emu*, **8**, 233–83.

Jeikowski, H. and Stephan, B. (1972). Uber die schmuckfedernim Flugel von *Semioptera wallacei*. *Journal für Ornithologie*, **113**, 86–90.

Jamieson, I. G. and Spencer, H. G. (1996). The bill and foraging behaviour of the Huia (*Heteraclocha acutirostris*): were they unique? *Notornis*, **43**, 14–8.

Jepson, P. and Ounsted, R. (Ed.) (1997). *Birding Indonesia*. Periplus Editions, Singapore.

Jobling, J. A. (1991). *A dictionary of scientific bird names*. Oxford University Press, Oxford.

Johnas, W. (1932). Die Balz dere *Parotia sefilata* (L.). *Ornithologische Monatsberichte*, **40**, 38–41.

Johns, R. J. (1977). *The vegetation of Papua New Guinea. Part 1: an introduction to the vegetation*. Papua New Guinea, Forestry College, Bulolo.

Johns, R. (1977). *The vegetation of Papua New Guinea. Training Manual No. 10*. Papua New Guinea Forestry College, Bulolo.

Johnsgard, P. A. (1965). *Handbook of waterfowl behaviour*. Constable, London.

Johnsgard, P. A. (1983). *The grouse of the world*. Croom Helm, London.

Johnsgard, P. A. (1994). *Arena birds*. Smithsonian Institution Press, Washington.

Johnston, G. R. and Richards, S. J. (1994). Notes on birds observed in the Western Province during July 1993. *Muruk*, **6**, 9.

Jones, D. N., Dekker, R. W. R. J., and Roselaar, C. S. (1995). *The megapodes*. Oxford University Press, Oxford.

Joyce, R. B. (1971). *Sir William MacGregor*. Oxford University Press, London.

Junge, G. C. A. (1939). The birds of South New Guinea. Part 2. Passeres. *Nova Guinea*, **3**, 1–94.

Junge, G. C. A. (1953). Zoological results of the Dutch New Guinea Expedition, 1939, No 5. The birds. *Zoologische Verhandelingen*, **20**, 1–77.

Kemp, A. (1995). *The hornbills*. Oxford University Press, Oxford.

Kikkawa, J, Monteith, G. B., and Ingram, G. (1981). Cape York Peninsula: major region of faunal interchange. In *Ecological biogeography of Australia*, (ed. A. Keast), pp. 1695–742. Junk, Hague.

King, B. (1979). New distributional records and field notes for some New Guinea birds. *Emu*, **79**, 146–8.

Kinghorn, J. R. (1939). A new genus and species of bird of paradise. *Australian Zoologist*, **9**, 295–6.

Kleinschmidt, O. (1897). Beschreibung eines neuen Paradies-vogels. *Ornithologische Monatsberichte*, **5**, 46–8.

Knox, A. G. and Walters, M. (1992). Under the skin: the bird collections of the Natural History Museum. *Bulletin of the British Ornithologist's Club Centenary Supplement*, **112A**, 169–90.

Kramer, G. (1930). Bewegungsstudien an Vögeln des Berliner Zoologischen Gartens. *Journal für Ornithologie*, **78**, 257–68.

Kuroda, M. (1943). An apparently new species of bird of paradise of the genus *Astrapia*. *Bulletin of the Biogeographical Society of Japan*, **13**, 33–7.

Kwapena, N. (1985). *The ecology and conservation of six species of birds of paradise in Papua New Guinea*. Biological Resources Management, Port Moresby.

Lack (1965). *The life of the robin*. Witherby, London.

Lack, D. (1968). *Ecological adaptations for breeding in birds*. Chapman and Hall, London.

Lack, D. (1971). *Ecological isolation in birds*. Blackwell, London.

Lack, D. (1974). *Evolution illustrated by waterfowl*. Blackwell, Oxford.

Lambert, F. R. (1994). Notes on the avifauna of Bacan, Kasiruta and Obi, north Moluccas. *Kukila*, **7**, 1–9.

Lambert, F. and Young, D. (1989). Some recent bird observations from Halmahera. *Kukila*, **4**, 30–3.

Lambley, P. (1990). Observations on the feeding habits of the Huon Astrapia *Astrapia rothschildi*. *Muruk*, **4**, 75.

Laska, M. S., Hutchins, M. Sheppard, C., Worth, W., and Bruning, D. (1992). Reproduction by captive unplumed male Lesser Bird of Paradise *Paradisaea minor*, evidence for an alternative mating strategy? *Emu*, **92**, 108–11.

Latham, J. (1790). *Index ornithologicus, sive systema ornithologine*. Vol. 1. Leigh and Sotheby, London.

Lawton, J. H. and May, R. M. (eds.). (1995). *Extinction Rates*. Oxford University Press, Oxford.

Layton, A. (1971). Observations. *New Guinea Bird Society Newsletter*, **68**, 1.

Lea, A. H. and Gray, J. T. (1936). The food of Australian birds. *Emu*, **36**, 335–47.

LeCroy, M. (1981). The genus *Paradisaea*—display and evolution. *American Museum Novitates*, **2714**, 1–52.

LeCroy, M. (1983). The spelling of *Semioptera wallacii* (Paradisaeidae). *Bulletin of the British Ornithologists' Club*, **103**, 144–5.

LeCroy, M. (1988). *Semioptera wallacii* Gray, 1859 (Aves, Paradisaeidae): proposed confirmation as correct spelling. *Bulletin of Zoological Nomenclature*, **45**, 212–3.

LeCroy, M. and Bock, W. (1989). Comments on the proposed conservation of the spelling *Semioptera wallacii* Gray, 1859 (Aves, Paradisaeidae), (2). *Bulletin of Zoological Nomenclature*, **46**, 49.

LeCroy, M., Kulupi, A. and Peckover, W. S. (1980). Goldie's Bird of Paradise display, natural history and traditional relationships of people to the bird. *Wilson Bulletin*, **92**, 289–301.

LeCroy, M., Peckover, W. S., Kulupi A., and Manseima, J. (1984). Bird observations on Normanby and Ferguson, D'Entrecasteaux islands-Papua New Guinea. *Wildlife in Papua New Guinea*, **831**, 1–7.

Lenz, N. (1994). Mating behaviour and sexual competition in the Regent Bowerbird *Sericulus chrysocephalus*. *Emu*, **94**, 263–72.

Le Souef, D. (1907). Description of a new bird of paradise. *Emu*, **6**, 119–20. [*Paradisornis rudolphi hunti*].

Lesson, R. P. (1830). *Voyage autour du monde ... sur ... la Coquille pendant ... 1822–25 ... par M. L. I. Duperry. Zoologie*. Vol. II, Paris.

Lesson, R. P. (1831). *Traité d'ornithologie ...* Levrault, Paris.

Lesson, R. P. (1834–5). *Histoire naturelle de oiseaux des paradis et des epimaques*. Arthus Bertrand, Paris.

Lesson, R. P. and Garnot, P. (1826a). Description d'une nouvelle espèce de Cassican (*Barita Keraudrenii*). *Bulletin Scientifiques Naturelles (Férussac)* **8**, 110–1.

Lesson, R. P. and Garnot, P. (1826b). *Voyage autour du monde ... sur ... la Coquille pendant ... 1822–25 ... par M. L. I. Duperry. Zoologie*. Vol. I. Paris.

Levaillant, F. (1801–6). *Historie naturelle des oiseaux de paradis et des rolliers*. Chez Denne le jeune [et] Perlet, Paris.

Levaillant, F. (1807). *Histoire naturelle des Promerops, et des Guêpiers*. Le Jeune, Paris.

Linnaeus, C. (1758). *Systema naturae per regna tria naturae*. **1**, 110. Laurentii Salvii.

Linsley, M. D. (1995). Some bird records from Obi, Maluku. *Kukila*, **7**, 142–51.

Loke, Wan Tho (1957). *A company of birds*. Michael Joseph, London.

Lowe, P. R. (1923). [Description of '*Paradisea apoda luptoni*']. *Bulletin of the British Ornithologists' Club*, **43**, 110–1.

Macgillivray, J. (1852). *Narrative of the voyage of H.M.S. Rattlesnake, commanded by the late Captain Owen Stanley, during the years 1846–1850*. London.

MacGillivray, W. (1914). Notes on some north Queensland birds. *Emu*, **13**, 132–86.

MacGillivray W. (1918). Ornithologist in North Queensland Part 3. *Emu*, **17**, 180–212.

Mack, A. L. (1992). The nest, egg and incubating behaviour of a Blue Bird of Paradise *Paradisaea rudolphi*. *Emu*, **92**, 244–6.

Mack, A. L. and Wright, D. D. (1996). Notes on occurrence and feeding of birds at Crater Mountain Biological Research Station, Papua New Guinea. *Emu*, **96**, 89–101.

Mackay, M. (1981). Display behaviour by female birds of paradise in captivity. *Papua New Guinea Bird Society Newsletter*, **185–86**, 5.

Mackay, M. (1984). Short notes on a bird of paradise and a bower bird. *Papua New Guinea Bird Society Newsletter*, **209**, 4–5.

Mackay, M. D. (1990). The egg of Wahnes' Parotia *Parotia wahnesi* (Paradisaeidae). *Emu*, **90**, 269.

Mackay, R. D. (1966). Men and birds of Nomad River. *New Guinea Bird Society Newsletter*, **12**, 1.

Mackay, R. D. (1987). *Papua New Guinea birds*. Brown and Associates, Bathurst.

Mackay, R. D. (1990). Variation in the display of the Magnificent Riflebird *Ptiloris magnificus*. *Muruk*, **4**, 65–6.

Mackay, R. D. and Mackay, M. (1974). Observations. *New Guinea Bird Society Newsletter*, **103**, 2–3.

Mackinnon, J. L., Fuller, R., Harper, M. E., Hugh-Jones, T., Knowles-Leak, R., Raham, D., Robb, D., and Vermeulen, J. (1995). Halmahera '94, a University of Bristol expedition to Indonesia (Final Report). Unpublished.

Majnep, I. S. and Bulmer, R. (1977). *Birds of my Kalem country*. Auckland University Press, Auckland.

Mani, M. S. (1990). *Fundamentals of high altitude biology*. Aspect Publications, London.

Manson-Bahr, P. H. (1935). Remarks on the displays of birds of paradise. *Bulletin of the British Ornithologists' Club*, **56**, 63–8.

Marchant, S. (1986). Long laying intervals. *Auk* **103**, 247.

Marshall, A. J. (1954). *Bower-birds their displays and breeding cycles—a preliminary statement*. Oxford University Press, Oxford.

Mathews, G. M. (1915). Additions and corrections to my list of the birds of Australia. *The Austral Avian Record*, **2**, 133.

Mathews, G. M. (1917). New subspecies and notes on species. *The Austral Avian Record*, **3**, 72.

Mathews, G. M. (1922). Additions and corrections. *The Austral Avian Record*, **5**, 1–9.

Mathews, G. M. (1923). Additions and corrections to my lists of the birds of Australia. *The Austral Avian Record*, **5**, 42.

Mathews, G. M. (1930). *Systema avium Australasianarum—a systematic list of the birds of the Australasian region*. 2 volumes. British Ornithologists' Union, London.

Mathews, G. M. (1925–7). *The birds of Australia*. Vol. 12. The author, London.

Mayr, E. (1930a). *Loboparadisea sericea aurora* subsp. nova. *Ornithologische Monatsberichte*, **38**, 147–8.

Mayr, E. (1930b). Die Unterarten des Kragenparadiesvogels (*Lophorina superba*). *Ornithologische Monatsberichte*, **38**, 178–80.

Mayr, E. (1931). Die Vogel des Saruwaged-und Herzoggebirges (No-Neuguinea). *Mitteilungen aus dem Zoologisches Museum in Berlin*, **17**, 649–52; 710–2; 720.

Mayr, E. (1936). New subspecies of birds from the New Guinea region. *American Museum Novitates*, **869**, 1–4.

Mayr, E. (1941a). *List of New Guinea birds—a systematic and faunal list of the birds of New Guinea and adjacent islands*. American Museum of Natural History, New York.

Mayr, E. (1941b). The origin and history of the bird fauna of Polynesia. *Proceedings of the Sixth Pacific Science Congress 1939*, **4**, 197–216.

Mayr, E. (1942). *Systematics and the origin of species*. Columbia University Press, New York.

Mayr, E. (1945). Birds of Paradise. *Natural History*, **54**, 264–76.

Mayr, E. and Gilliard, E. T. (1951). New species and subspecies of birds from the highlands of New Guinea. *American Museum Novitates*, **1524**, 1–15.

Mayr, E. and Gilliard, E. T. (1952). The Ribbon-tailed Bird of Paradise (*Astrapia mayeri*) and its allies. *American Museum Novitates*, **1551**, 1–13.

Mayr, E. and Gilliard, E. T. (1954). Birds of central New Guinea—Results of the American Museum of Natural History expeditions to New Guinea in 1950 and 1952. *Bulletin of the American Museum of Natural History*, **103**, 354–62.

Mayr, E. and Greenway, J. C. Jr. (Ed.) (1962). *Check-list of birds of the world*, Volume 15. Cambridge, Massachusetts.

Mayr, E. and Meyer de Schauensee, R. M. (1939a). Zoological results of the Denison-Crockett South Pacific expedition for the Academy of Natural Sciences of Philadelphia, 1937–1938. Part 4—Birds from northwest New Guinea. *Proceedings of the Academy of Natural Sciences of Philadelphia*, **91**, 97–144.

Mayr, E. and Meyer de Schauensee, R. M. (1939b). Zoological results of the Denison-Crockett South Pacific Expedition for the Academy of Natural Sciences of Philedelphia, 1937–38. Part 5—Birds from the western Papuan Islands. *Proceedings of the Academy of Natural Sciences of Philadelphia*, **91**, 145–63.

Mayr, E. and Rand, A. L. (1935). Results of the Archbold Expeditions. 6. Twenty-four apparently undescribed birds from New Guinea and the D'Entracasteaux Archipelago. *American Museum Novitates*, **814**, 1–17.

Mayr, E. and Rand, A. L. (1937). Results of the Archbold Expeditions. 14. Birds of the 1933–1934 Papuan Expedition. *Bulletin of the American Museum of Natural History*, **73**, 1–248.

McAllan, I. A. W. and Bruce, M. D. (1988). The birds of New South Wales—A working list. 70. Biocon Research Group in association with the New South Wales Bird Atlassers, Turramurra, NSW 2074, Australia.

McAlpine, D. K. (1979). The correct name and authorship for Wallace's Standard Wing (Passeriformes, Paradisaeidae). *Bulletin of the British Ornithologists' Club*, **99**, 108–10.

McAlpine, J. R., Keig, G. and Short, K. (1975). Climatic tables for Papua New Guinea. CSIRO, Australian division land use research technical papers 37, Canberra.

McAlpine, J. R., Keig, G, and Falls, R. (1983). *Climate of Papua New Guinea*. CSIRO and Australian National University Press, Canberra.

McGill, A. R. (1951). Proceedings of the annual congress of the R.A.O.U., Sydney, 1950. *Emu*, **50**, 240–50.

Meek, A. S. (1913). *A naturalist in cannibal land*. Unwin, London.

Mees, G. F. (1964). Notes on two small collections of birds from New Guinea. *Zoologische Verhandelingen*, **66**, 1–37.

Mees, G. F. (1965). The avifauna of Misool. *Nova Guinea, Zoology*, **31**, 139–203.

Mees, G. F. (1982). Birds from the lowlands of southern New Guinea (Merauke and Koembe). *Zoologische Verhandelingen*, **191**, 3–188.

Meise, W. (1929). Verzeichnis der Typen des Staatlichen Museums für Tierkunde in Dresden. 2 Teil., Voegel 1. *Abhandlung und Berichte der Museum für Tierkunde und Volkkunde, Dresden*, **17**, for 1927–9, Dec. 30, 1–22.

Melville, D. (1979). Ornithological notes on a visit to Irian Jaya. *Papua New Guinea Bird Society Newsletter*, **161**, 3–22.

Menegaux, A. (1913). Description de deux ouveaux paradisiers (*Paradisea duivenbodei* et *P. Raggiana sororia*). *Revue Française d'Ornithologie*, **37**, 49–51.

Merton, D. V., Morris, R. B., and Atkinson, I. A. E. (1984). Lek-behaviour in a parrot: the Kakapo *Strigops habroptilus* of New Zealand. *Ibis*, **126**, 277–83.

Meyer, A. B. (1873). Ueber einen neuen Paradiesvogel von Neu-Guinea. *Journal für Ornithologie*, **21** (October), 405–6.

Meyer, A. B. (1875a). [Description of *Diphyllodes gulielmi III*]. *Proceedings of the Zoological Society of London*, **1875**, 30–1.

Meyer, A. B (1875b). *Diphyllodes (Paradisea) Gulielmi III*, v. v. Mussch., ein neuer Paradiesvogel. *Der Zoologische Garten*, **16**, 29–31.

Meyer, A. B. (1885). [*Manucodia rubiensis*] *Zeitschrift für die gesammte Ornithologie*, **2**, 374.

Meyer, A. B. (1890). Notes on birds from Papuan Region, with descriptions of some new species. *Ibis*, Ser. 6, vol. 2, 412–24.

Meyer, A. B. (1893). [List of Paradise Birds]. *Abhandelungen und Berichte des Königlichen Zoologischen und Anthropologisch-Ethnographischen Museums zu Dresden*, **4**, 15.

Meyer, A. B. (1894a). [*Parotia carolae*, sp. n.]. *Bulletin of the British Ornithologists' Club*, **4**, 6.

Meyer, A. B. (1894b). [*Pteridophora alberti*, sp. n.]. *Bulletin of the British Ornithologists' Club*, **4**, 11–2.

Meyer, A. B. (1894c). *Abhandlungen und Berichte des Koniglichen Zoologischen und Anthropologisch-Ethnographischen) Museums zu Dresden*, **5**, 3.

Milkovský, J. (1989). Comments on the proposed conservation of the spelling *Semioptera wallacii* Gray, 1859 (Aves, Paradisaeidae), (1). *Bulletin of Zoological Nomenclature*, **46**, 49.

Montbeillard, G. de (1775) In G. L. L. Buffon's *Histoire naturelle des oiseaux*. Vol. 3. Paris.

Moorhouse, R. J. (1996). The extraordinary bill dimorphism of the Huia (*Heteraclocha acutirostris*): sexual selection or intersexual competition? *Notornis*, **43**, 19–34.

Morris, A., McGill, A., and Holmes, G. (1981). *Handlist of birds in New South Wales*. New South Wales Field Ornithologists' Club, Sydney.

Morrison-Scott, T. (1936). [Display of *Lophorina superba minor*]. *Proceedings of the Zoological Society of London*, **1936**, 809.

Moresby, Captain J. (1876). *New Guinea and Polynesia—discoveries and surveys in New Guinea and the D'Entrecasteaux Islands, a cruise in Polynesia and visits to the pearl shelling stations in Torres Straits of H.M.S. Basilisk*. John Murray, London.

Moynihan, M. (1968). Social mimicry: character convergence versus character displacement. *Evolution*, **22**, 315–31.

Muller, K. A. (1974). Rearing Count Raggi's bird of paradise, *Paradisaea raggiana* at Taronga Zoo, Sydney. *International Zoo Yearbook*, **14**, 102–5.

Murphy, R. C. and Amadon, D. (1966). In Memoriam: E. Thomas Gilliard. *Auk*, **83**, 416–422.

Myers, N. and Simon, J. L. (1994). *Scarcity or abundance—a debate on the environment*. Norton, New York.

Neumann, O. (1922). Neue Formen aus dem papuanischen und polynesischen Inselreich. *Verhandlungem der Ornithologischen Gesellschaft in Bayern*, **15**, 234–6.

Neumann, O. (1932). *Lophorina superba sphinx* nov. subsp. *Ornithologische Monatsberichte*, **40**, 121–2.

Norris, A. Y. (1964). Observations on some birds of the Tooloom Scrub, northen N.S.W. *Emu*, **63**, 404–12.

North, A. J. (1892). Note on the nidification of *Manucodia comrii*, Sclater (Comrie's Manucode). Records of the Australian Museum, **2**, 32.

North, A. J. (1901–4). Nests and eggs of birds found breeding in Australia and Tasmania. *Special Catalogue of the Australian Museum*, **1**, 1–36.

North, A. J. (1906). Description of a new bird of paradise. *Victorian Naturalist*, **22** 156–8.

Nunn, G. B. and Cracraft, J. (1996). Phylogenetic relationships among the major lineages of the birds-of-paradise (Paradisaeidae) using mitochondrial DNA gene sequences. *Molecular Phylogenetics and Evolution*, **5**, 445–59.

Ogilvie-Grant, W. R. (1904). [On the trachea of *Phonygama purpureoviolacea*]. *Bulletin of the British Ornithologists' Club*, **14**, 40–1.

Ogilvie-Grant, W. R. (1905). On the display of the Lesser Bird-of-Paradise (*Paradisea minor*). *Ibis*, Ser. 8, vol. 5, 429–40.

Ogilvie-Grant, W. R. (1907). *Cicinnurus goodfellowi* sp. n. *Bulletin of the British Ornithologists' Club*, **19**, 39–40.

Ogilvie-Grant, W. R. (1912). On the eggs of certain birds-of-paradise. *Ibis*, Ser. 9, vol. 6, 112–8.

Ogilvie-Grant, W. R. (1913). [*Paradigalla intermedia*, sp. n.]. *Bulletin of the British Ornithologists' Club*, **31**, 99–106.

Ogilvie-Grant, W. R. (1915a). Report on the birds collected by the British Ornithologists' Union Expedition and the Wollaston Expedition in Dutch New Guinea. *Ibis*, Jubilee Supplement No. 2, 1–329.

Ogilvie-Grant, W. R. (1915b). [Note on ♂ *Paradisaea apoda noveaguineae* adult plumage aquisition]. *Bulletin of the British Ornithologists' Club*, **36**, 41.

Ogilvy, W. M. B. (1916). In Kaiser Wilhelm's Land—hunting the birds-of-paradise—and avoiding the cannibals. *The Murray Pioneer* (newspaper) of 3 Feb 1916, p. 7.

Olson, S. L. (1980). *Lamprolia* as part of a South Pacific radiation of Monarchine flycatchers. *Notornis*, **27**, 7–10.

Olson, S. L., Parkes, K. C., Clench, M. H., and Borecky, S. R. (1983). The affinities of the New Zealand passerine genus *Turnagra*. *Notornis*, **30**, 319–36.

Oort, E. D, van (1906). On a new bird of paradise. *Notes of the Leyden Museum*, **28**, 129–30.

Oort, E. D, van (1909). Birds from southwestern and southern New Guinea. *Nova Guinea*, **9**, 51–107.

Oort, E. D. van. (1915). On a new bird of paradise from central New Guinea, *Falcinellus meyeri albicans*. *Zoologische Mededlingen*, **1**, 228.

Opit, G. (1975a). Display of Magnificent Rifle Bird. *New Guinea Bird Society Newsletter*, **113**, 15.

Opit, G. (1975b). Observations [along Kokoda Trail]. *New Guinea Bird Society Newsletter*, **115**, 4–5.

Oustalet, E. (1880). D'une Espece nouvelle de Paradisier (*Drepanornis bruijnii*). *Annales des Sciences Naturelles*, Ser. 6, vol. 9, 1.

Oustalet, E. (1891). Description de deux espèces nouvelles d'Oiseaux. Appartenant aux familles des Paradiseidae et des Trogonidae. *Le Naturaliste*, **13**, 260–1.

Paine, J. R. (1991). *IUCN directory of protected areas in Oceania*. IUCN, Gland.

Parer, D. (1981). *Voices in the forest*. Australian Broadcasting Commission, Melbourne.

Parker, S. A. (1963). Nesting of the Paradise Crow *Lycocorax pyrrhopterus* (Bonaparte) and the Spangled Drongo, *Dicrurus hottentottus* (Linn.) in the Moluccas. *Bulletin of the British Ornithologists' Club*, **83**, 126–7.

Pavesi, P. (1874). Intorno ad una nuova forma di trachea di Manucodia. *Annali del Museo Civico di Storia Naturale di Genova*, **6**, 315–24.

Pavesi, P. (1876). Studi anatomici sopra alcini uccelli. *Annali del Museo Civico di Storia Naturale di Genova*, **9** (1876–1877), 66–77.

Peckover, W. S. (1973). Behavioural similarities of birds of paradise and bowerbirds to Lyrebirds and scrub-birds. *Proceedings of the 1972 Papua and New Guinea Scientific Society*, **24**, 10–20.

Peckover, W. S. (1985). Seed dispersal of *Amorphophallus paeoniifolius* by birds of paradise in Papua New Guinea. *Aroideana*, **8**, 70–1.

Peckover, W. S. (1990). *Papua New Guinea birds of paradise*. Brown, Carina.

Peckover, W. S. (1995). Moult in birds of paradise. *Muruk*, 7, 115–6.

Peckover, W. S. and Filewood L. W. C. (1976). *Birds of New Guinea and tropical Australia*. Reed, Sydney, Australia.

Peckover, W. S. and George, G. G. (1992). Obituary: 'Masta Pisin'—the bird man of New Guinea, Fred Shaw Mayer M. B. E. 1899–1989. *Emu*, **92**, 250–54.

Pennant, T. (1781). *Specimen faunulae indicae*. In Forster's *Zoologica Indica Selecta* … Gebauer, Halle.

Pennant, T. (1790–91). *Indian zoology*. 2nd edn. Faulder, London.

Perrins, C. M. (1985). Clutch-size. In *A dictionary of birds*, (ed. R. Campbell), pp. 91–4. Poyser, Carlton.

Pigafetta, A. (1524–34). *Le Voyage et navigation faict par les Espaignolz es Isles des Mollucques*. Paris.

Pizzey, G. (1980). *A field guide to the birds of Australia*. Collins, Sydney.

Pratt, A. E. (1906). *Two years among New Guinea cannibals*. Seeley, London.

Pratt, T. K. (1982a). Biogeography of birds in New Guinea. *Monographiae Biologicae*, **42**, 815–36.

Pratt, T. K. (1982b). Additions to the avifauna of the Adelbert Range, Papua New Guinea. *Emu*, **82**, 117–25.

Pratt, T. K. (1983). *Seed dispersal in a montane forest in Papua New Guinea*. Ph.D. thesis. Rutgers University, New Brunswick, New Jersey.

Pratt, T. K. (1984). Examples of tropical frugivores defending fruit-bearing plants. *Condor*, **86** 123–9.

Pratt, T. K. and Stiles, E. W. (1983). How long fruit-eating birds stay in the plants where they feed, implications for seed dispersal. *American Naturalist*, **122**, 797–805.

Pratt, T. K. and Stiles, E. W. (1985). The influence of fruit size and structure on composition of frugivore assemblages in New Guinea. *Biotropica*, **17**, 314–21.

Preston, F. W. (1974). The volume of an egg. *Auk*, **91**, 132–8.

Price, D. and Nielsen, L. (1991). Bird list for Karawari Lodge and area, East Sepik Province. *Muruk*, **5**, 23–4.

Pruett-Jones, S. G. (1985). The evolution of lek mating behavior in Lawes's Parotia (Aves, *Parotia lawesii*). Ph.D. thesis. University of California, Berkeley.

Pruett-Jones, S. G., and Pruett-Jones, M. A. (1986). Altitudinal distribution and seasonal activity patterns of birds of paradise. *National Geographic Research*, **2**, 87–105.

Pruett-Jones, S. G. and Pruett-Jones, M. A. (1988a). A promiscuous mating system in the Blue Bird of Paradise *Paradisaea rudolphi*. *Ibis*, **130**, 373–7.

Pruett-Jones, S. G. and Pruett-Jones, M. A. (1988b). The use of court objects by Lawes' Parotia. *Condor*, **90**, 538–45.

Pruett-Jones, S. G. and Pruett-Jones, M. A. (1990). Sexual selection through female choice in Lawes' Parotia, a lek-mating bird of paradise. *Evolution*, **44**, 486–501.

Pruett-Jones, S. G., Pruett-Jones, M. A., and Jones, H. I. (1990). Parasites and sexual selection in birds of paradise. *American Zoologist*, **30**, 287–98.

Prum, R. O. (1990). Phylogenetic analysis of the evolution of display behaviour in Neotropical manakins (Aves: Pipridae). *Ethology*, **84**, 202–31.

Pycraft, W. P. (1907). Contributions to the osteology of birds. Part IX. Tyranni; Hirundines; Muscicapae, Lanii, and Gymnorhines. *Proceedings of the Zoological Society of London*, **1907**, 352–79.

Ramsay, E. P. (1885). Contributions to the zoology of New Guinea. Notes on birds from Mount Astrolabe, with descriptions of two new species. *Proceedings of the Linnaean Society of New South Wales*, **10**, 242–4.

Ramsay, J. (1919). Notes on birds observed in the upper Clarence River District, N.S.W., Sept.–Dec., 1918. *Emu*, **19**, 2–9.

Rand, A. L. (1938). Results of the Archbold Expeditions. No. 22. On the breeding habits of some birds of paradise in the wild. *American Museum Novitates*, **993**, 1–8.

Rand, A. L. (1940a). Results of the Archbold Expeditions. No. 26. Breeding habits of the birds of paradise, *Macgregoria* and *Diphyllodes*. *American Museum Novitates*, **1073**, 1–14.

Rand, A. L. (1940b). Courtship of the Magnificent Bird of Paradise. *Natural History Magazine*, **45**, 55.

Rand, A. L. (1942a). Results of the Archbold Expeditions No. 42. Birds of the 1936–1937 New Guinea Expedition. *Bulletin of the American Museum of Natural History*, **79**, 289–366.

Rand, A. L. (1942b). Results of the Archbold Expeditions No. 43. Birds of the 1938–1939 New Guinea Expedition. *Bulletin of the American Museum of Natural History*, **797**, 425–516.

Rand, A. L. and Gilliard, E. T. (1967). *Handbook of New Guinea birds*. Weidenfeld and Nicolson, London.

Reichenow, A. (1901). Ein merkwurdiger Paradiesvogel. *Ornithologische Monatsberichte*, **9**, 185–6.

Reichenow, A. (1918). *Phonygammus neumanni* sp. n.. *Journal für Ornithologie*, **66**, 438.

Rensch, B. (1933). Zoologische Systematik und Artbildungsproblem. *Verhandlungen der Deutschen Zoologischen Gesellschaft*, **6**, 19–83.

Ricklefs, R. (1980). Commentary. *Auk*, **97**, 476–7.

Rimlinger, D. (1984). The Empress of Germany's Bird of Paradise. *Zoonooz*, **57**, 10–4.

Ripley, S. D. (1950). Strange courtship of birds of paradise. *National Geographic Magazine*, **97**, 247–78.

Ripley, S. D. (1957). The display of the Sicklebilled Bird of Paradise. *Condor*, **59**, 207.

Ripley, S. D. (1959). Birds from Djailolo, Halmahera. *Postilla*, **41**, 1–8.

Ripley, S. D. (1964). A systematic and ecological study of birds of New Guinea. *Yale Peabody Museum Bulletin*, **19**, 1–87.

Rosenberg, C. B. H. von. (1875). *Reistochten naar de Geelvinkbaai op Nieuw-Guinea in de jaren 1869 en 1870*. Martinus Nijhoff, 'S Gravenhage.

Rosenthal, A. M. (1956). Plumes save U.S. face in Nepal. *The New York Times*, newspaper May 2 1956.

Rothschild, M. (1983). *Dear Lord Rothschild*. Balaban, Glenside.

Rothschild, W. (1895). A new bird of paradise [*Astrapia splendidissima*]. *Novitates Zoologicae*, **2**, 59–60.

Rothschild, W. (1896a). [*Loboparadisea sericea*. sp. n.]. *Bulletin of the British Ornithologists' Club*, **6**, 15–6.

Rothschild, W. (1896b). [*C. r. coccineifrons*] *Novitates Zoologicae*, **3**, 10.

Rothschild, Lord, W. (1897a). Exhibition of skins of *Paradisea minor* and allies, *P. m. finschi*, and *P. minor jobiensis*, subsp. n.. *Bulletin of the British Ornithologists' Club*, **6**, 45–6.

Rothschild, Lord, W. (1897b). [*Epimachus astrapioides* sp. n.]. *Bulletin of the British Ornithologists' Club*, 7, 22–3.

Rothschild, Lord, W. (1898). Paradiseidae. *Das Tierreich* **2** Lieferung 1–52. Verlag von R. Friedlander und Sohn, Berlin.

Rothschild, Lord, W. (1901). On a new genus of Paradiseidae [*Loborhamphus nobilis* sp. n]. *Bulletin of the British Ornithologists' Club*, **12**, 34.

Rothschild, Lord, W. (1903). [Exhibition of *Janthothorax mirabilis*] *Bulletin of the British Ornithologists' Club*, **13**, 31.

Rothschild, Lord, W. (1907a). Description of new bird-of-paradise (*Lophorina minor latipennis*). *Bulletin of the British Ornithologists' Club*, **19**, 92–3.

Rothschild, Lord, W. (1907b). [Exhibition of *Pseudastrapia lobata*, sp. n., *Loborhamphus nobilis*, and *Janthothorax mirabilis*]. *Bulletin of the British Ornithologists' Club*, **21**, 25.

Rothschild, Lord, W. (1909). [*Parotia duivenbodei* sp. n.]. *Bulletin of the British Ornithologists' Club*, **10**, 100.

Rothschild, Lord, W. (1910). [Exhibition of *Parotia carolae meeki* subsp. n.] *Bulletin of the British Ornithologists' Club*, **27**, 35–6.

Rothschild, Lord, W. (1921a). On *Paradisaea apoda granti* and *Paradisea mixta* n. sp [*P. m. finschi* × *P. a. augustaevictoriae* = *P. r. granti*] *Bulletin of the British Ornithologists' Club*, **41**, 127.

Rothschild, Lord, W. (1921b). [*Paradisea apoda subintermedia*, subsp. n.]. *Bulletin of the British Ornithologists' Club*, **41**, 138–9.

Rothschild, Lord W. (1930a). Notes on the preceding article of Dr. Stresemann. *Novitates Zoologicae*, **36**, 16–17.

Rothschild, Lord, W. (1930b). Exhibition of eggs of the Paradise-Crow (*Lycocrax pyrrhopterus pyrrhopterus*) and *Phonygammus keraudrenii*. *Bulletin of the British Ornithologists' Club*, **51**, 9.

Rothschild, Lord, W. (1930c). Notes on Paradisaeidae with a list of the species, subspecies, and hybrids exhibited at the Seventh International Ornithological Congress. *Proceedings of the Seventh International Ornithological Congress*, Amsterdam 1930, 285–9.

Rothschild, Lord, W. (1931). On a collection of birds made by Mr. F. Shaw Mayer in the Weyland Mts., Dutch New Guinea, in 1930. *Novitates Zoologicae*, **36**, 250–76.

Rothschild, Lord, W. (1932). On a new bird of paradise. *Annals and Magazine of Natural History*, **10**, 126. [*Lophorina feminina lehunti* [= *L. superba feminina*]]

Rothschild, Lord, W. (1936). [*D. alberitisi inversa*] *Mitteilungen aus dem Zoologischen Museum in Berlin*, **21**, 188.

Rothschild, Lord, W. and Hartert, E. (1903). Notes on Papuan birds VII. Paradiseidae. *Novitates Zoologicae*, **10**, 65–89.

Rothschild, Lord, W. and Hartert E. (1911). Preliminary descriptions of some new birds from central New Guinea. *Novitates Zoologicae*, **18**, 159–60.

Rothschild, Lord, W. and Hartert, E. (1929). Note on *Manucodia ater ater* and a new subspecies [*M. a. subalter*]. *Bulletin of the British Ornithologists' Club*, **49**, 109–10.

Rumbiak, A. M. (1984). Observations on the trade in birds of paradise in Bomakia District of Kouh, Region of Merauke. Fellowship studies in Irian Jaya. Report of Cenderwasih University student investigations. Sponsored by the World Wildlife Fund WWF/IUCN conservation programme in Irian Jaya. Project 1528. Mapia mambruk, Universitias Cenderwasih, Jayapura.

Saenger, P., Specht, M. M., Specht, R. L., and Chapman, V. J. (1977). Mangal and coastal saltmarsh communities in Australasia. In *Ecosystems of the world. 1 Wet coastal ecosystems*, (ed. V.J. Chapman), pp. 293–345. Elsevier, Amsterdam.

Safford, R. J. and Smart, L. M. (1996). The continuing presence of Macgregor's Bird of Paradise on Mount Albert Edward, Papua New Guinea. *Bulletin of the British Ornithologists' Club*, **116**, 186–8.

Sakulas, H. (1988). Breeding record of Raggiana Bird of Paradise, *Paradisaea raggiana*. *Muruk*, **3**, 57.

Salvadori, T. (1875). [*Manucodia jobiensis* sp. n.]. *Annali del Museo Civico di Storia Naturale di Genova* [1876], **7**, 969–70.

Salvadori, T. (1876). *Annali del Museo Civico di Storia Naturale di Genova*, **9**, 191–2.

Salvadori, T. (1880–2). *Ornitologia della Papuasia e delle Molucche* Stamperia Reale Della Ditta G.B. Paravia e. Comp., Torino.

Salvadori, T. (1894). Viaggio di Lamberto nella Papuasia Orientale. XII. Caratteri de cinque specie nuove di uccelli della Nuova Guinea Orientale-meridionale raccolti da L. Loria. *Annali del Museo Civico di Storia Naturale di Genova*, Ser. 2a, vol. 14, 150–2.

Salvadori, T. (1896). *Annali del Museo Civico di Storia Naturale di Genova*, Ser. 2a, vol. 16, 110–1. [Descriptions of two new species, *Manucodia orientalis* Salvadori and *Diphyllodes zanthoptera* Salvadori]

Salvin, O. and Godman, F. O. (1883). Discovery of a new bird of paradise. *Ibis*, Ser. 5, vol. 1, 131; 199–202.

Schemske, D. W. (1983). Limits to specialization and coevolution in plant–animal mutualisms. In *Coevolution*, (ed. M. H. Nitecki), pp. 67–109. University of Chicago Press, Chicago.

Schlegel, H. (1863). [*Lycocorax morotensis* Schlegel, sp. n.]. *Ibis*, **5**, 119–20.

Schlüter, W. (1911). *Seleucidis ignotus auripennis* Schlut., subsp. nov. aus Deutsch-Neuguinea. *Falco*, **7**, 2–4.

Schmid, C. K. (1993). Birds of Nokopo. *Muruk*, **6**, 1–61.

Schodde, R. (1973). General problems of fauna conservation in relation to the conservation of vegetation in New Guinea. *Nature Conservation in the Pacific*, **10**, 123–44.

Schodde, R. (1975). *Interim list of Australian songbirds. Passerines*. Royal Australasian Ornithologists' Union, Melbourne.

Schodde, R. (1976). Evolution in the birds-of-paradise and bowerbirds, a resynthesis. In *Proceedings of the 16th International Ornithological Congress*, (eds. H. J. Frith and J. H. Calaby) pp. 137–49. Australian Academy of Science, Canberra.

Schodde, R. and Hitchcock, W. B. (1968). Contributions to Papuasian Ornithology. **1.**. Report on the birds of the Lake Kutabu area, Territory of Papua New Guinea. Division of Wildlife Research Technical Paper **13**, 1–73. CSIRO, Melbourne.

Schodde, R. and Hitchcock, W. B. (1972). Birds. In *Encyclopedia of Papua New Guinea*, (ed. P. A. Ryan), pp. 67–86. Melbourne University Press, Melbourne.

Schodde, R. and Mason, I. J. (1974). Further observations on *Parotia wahnesi* and *P. lawesii* (Paradisaeidae). *Emu*, **74**, 200–1.

Schodde, R. and Mason, I. J. (1980). *Nocturnal birds of Australia*. Lansdowne Editions, Melbourne.

Schodde, R. and McKean, J. L. (1972). Distribution and taxonomic status of *Parotia lawesii helenae* De Vis. *Emu*, **72**, 113.

Schodde, R. and McKean, J. L. (1973). The species of the genus *Parotia* (Paradisaeidae) and their relationships. *Emu*, **73**, 145–56.

Schodde, R. and Tidemann, S. (1988). *The Reader's Digest book of Australian birds*. Reader's Digest, Sydney.

Schodde, R., Van Tets, G. F., Champion, C. R., and Hope, G. S. (1975). Observations on birds at glacial altitudes on the Carstensz Massif, Western New Guinea. *Emu*, **75**, 65–72.

Schönwetter, M. (1944). Die Eier der Paradiesvogel. *Beiträge zur Fortpflanzungsbiologie der Vögel*, **20**, 1–18.

Sclater, P. L. (1857). Notes on the birds in the museum of the Academy of Natural Sciences of Philadelphia, and other collections in the United States of America. *Proceedings of the Zoological Society of London*, **1857**, 1–8.

Sclater, P. L. (1873). [Notes concerning birds collected by D'Albertis]. *Proceedings of the Zoological Society of London*, **1873**, 557–60.

Sclater, P. L. (1876). On birds collected by Dr. Comrie on the south-east coast of New Guinea during the survey of HMS Basilisk [*Manucodia comrii*, sp. n.]. *Proceedings of the Zoological Society of London*, **1876**, 459–61.

Sclater, P. L. (1883). [*Drepanornis albertisi cervinicauda*, subsp. n.]. *Proceedings of the Zoological Society of London*, **1883**, 578.

Sclater, P. L. (1891). Remarks on Macgregor's Paradise-bird, *Cnemophilus macgregori*. *Ibis*, **3**, Sixth series, 414–5.

Scopoli (1786). *Deliciae florae et faunae Insubricae* part 2. Ticini, Monasterii S. Salvatoris. [*C. r. rex*]

Searle, K. C. (1980). Breeding Count Raggi's Bird of Paradise at Hong Kong. *International Zoo Yearbook*, **20**, 210–14.

Seba, A. (1734). *Locupletissimi rerum naturalium thesauri accurata descriptio*. Vol. 1. Amstelaedami.

Selander, R. K. (1966). Sexual dimorphism and differential niche utilization in birds. *Condor*, **68**, 113–51.

Selander, R. K. (1972). Sexual selection and dimorphism in birds. In *Sexual selection and the descent of man*, (ed. B. Campbell), pp. 180–230. Heinemann, London.

Selous E. C. (1927). *Realities of bird life*. Constable, London.

Seth-Smith, D. (1923a). The Birds of Paradise and Bower Birds. *Avicultural Magazine*, Ser. 4, vol. 1, 41–60.

Seth-Smith, D. (1923*b*). On the display of the Magnificent Bird of Paradise *Diphyllodes magnifica hunsteinii*. Proceedings of the Zoological Society of London, **1923**, 609–13.

Seth-Smith, D. (1936). Display posture of *Lophorina superba minor*. Proceedings of the Zoological Society of London, **1936**, 807–8.

Sharland, M. (1977). Display of Victoria's Riflebird. *The Bird Observer*, **550**, 69–70.

Sharpe, R. B. (1877). *Catalogue of the birds in the British Museum. Vol. 3, Passeriformes*. British Museum, London.

Sharpe, R. B. (1882). Contributions to the Ornithology of New Guinea, Part 8. On collections made by Mr. A. Goldie and Mr. Charles Hunstein. *Journal of the Linnean Society, Zoology*, **16**, 422–47.

Sharpe, R. B. (1894). [List of bird of paradise species]. *Bulletin of the British Ornithologists' Club*, **4**, 12–5.

Sharpe, R. B. (1891–8). *Monograph of the Paradisaeidae, or Birds of Paradise, and Ptilonorhynchidae, or Bower-Birds*. Parts 1–8. H. Sotheran and Co., London.

Sharpe, R. B. (1899–1909). *Hand-list of the genera and species of birds in the British Museum*. 5 volumes. British Museum (Natural History), London.

Sharpe, R. B. (1908). [*Lophoramphus ptiloris* sp. n.] *Bulletin of the British Ornithologists' Club*, **11**, 67.

Shaw G. (1809). *General zoology or systematic natural history*. 7, Pt. 2 (Aves), 478–504.

Shaw Mayer, F. W. and Peckover, W. S. (1991). Eggs of hybrid Shaw Mayer's Bird of Paradise: the Ribbontail Astrapia mayeri × ? *Emu*, **91**, 189.

Sibley, C. G. (1996). *Birds of the world*. Version 2.0. Thayer Birding Software, Cincinnati.

Sibley, C. G. and Ahlquist, J. E. (1985). The phylogeny and classification of the Australo-Papuan passerine birds. *Emu*, **85**, 1–14.

Sibley, C. G. and Ahquist, J. E. (1987). The Lesser Melampitta is a bird of paradise. *Emu*, **87**, 66–8.

Sibley, C. G. and Ahlquist, J. E. (1990). *Phylogeny and classification of birds: a study in molecular evolution*. Yale University Press, New Haven, Connecticut.

Sibley, C. G. and Monroe, B. L. Jr. (1990). *Distribution and taxonomy of birds of the world*. Yale University Press, New Haven.

Sibley, C. G., Ahlquist, J. E., and Monroe, B. L. Jr. (1988). A classification of the living birds of the world based on DNA–DNA hybridization studies. *Auk*, **105**, 409–23.

Sims, R. W. (1956). Birds collected by Mr. F. Shaw-Mayer in the central highlands of New Guinea 1950–1951. *Bulletin of the British Museum (Natural History)*, **3**, 389–438.

Simson, C. C. (1907). On the habits of the birds-of-paradise and bower-birds of British New Guinea. *Ibis*, Ser. 9, vol. 1, 380–7.

Simpson, C. C. (1942). Across the Owen Stanley Range. *Victorian Naturalist*, **59**, 98–104.

Simpson, D. (1997). Japen Island. *Papua New Guinea Bird Society Newsletter*, **290**, 6–7.

Skutch, A. F. (1949). Do tropical birds rear as many young as they can nourish? *Ibis*, **91**, 430–55.

Skutch, A. F. (1976). *Parent birds and their young*. University of Texas Press, Austin.

Skutch, A. F. (1987). *Helpers at birds' nests*. University of Iowa Press, Iowa.

Smithe, F. B. (1975). *Naturalist's color guide*. American Museum of Natural History, New York.

Smyth, H. (1970). Hand-rearing and observing birds of paradise. *Avicultural Magazine*, **76**, 67–70.

Snow, B. K. and Snow, D. W. (1979). The Ochre-bellied Flycatcher and the evolution of lek behaviour. *Condor*, **81**, 286–92.

Snow, D. W. (1976*a*). The relationship between climate and annual cycles in the Cotingidae. *Ibis*, **118**, 366–401.

Snow, D. W. (1976*b*). *The web of adaptation: bird studies in American tropics*. Collins, London.

Snow, D. (1982). *The cotingas*. British Museum (Natural History), London.

Snow, D. W. (1997). Should the biological be superceded by the phylogeneric species concept? *Bulletin of the British Ornithologists' Club*, **117**, 110–21.

Somadikarta, S. (1984). Polyrectricyly. *Bulletin of the British Ornithologists' Club*, **104**, 60–1.

Sonnerat, P. (1776). *Voyage a la Nouvelle Guinée*. Ruault, Paris.

Stein, G. H. W. (1936). Beitrage zur Biologie papuanischer Vogel. *Journal für Ornithologie*, **84**, 21–57.

Stephan, B. (1967). Die Schmuckfedern im Flugel von *Semioptera wallacei. Journal für Ornithologie*, **108**, 47–50.

Stewart, D. (1996). *Australian bird sounds: Lamington National Park, rainforests of the subtropics.* Nature Sounds, Mullumbimby.

Stonor, C. R. (1936). The evolution and mutual relationships of some members of the Paradisidae. *Proceedings of the Zoological Society of London*, **1936**, 1177–85.

Stonor, C. R. (1937). On the systematic position of the Ptilonorhynchidae. *Proceedings of the Zoological Society of London* B, **107**, 425–90.

Stonor, C. R. (1938). Some features of the variation of the birds of paradise. *Proceedings of the Zoological Society of London* B, **108**, 417–81.

Stonor, C. R. (1939). A new species of paradise bird of the genus *Astrapia. Bulletin of the British Ornithologists' Club*, **59**, 57–61.

Stonor, C. R. (1940). *Courtship and display among birds.* Country Life, London.

Storr, G. M. (1953). Birds of the Cooktown and Laura districts, North Queensland. *Emu*, **53**, 225–48.

Storr, G. M. (1984). Revised list of Queensland birds. *Records of the Western Australian Museum*, Supplement No. **19**, 1–189.

Strahan, R. (1996). *Finches, bowerbirds and other passerines of Australia.* Angus and Robertson, Sydney.

Stresemann, E. (1922). Neue Formen dem papuanischen gebiet. *Journal für Ornithologie*, **70**, 405–8.

Stresemann, E. (1923). Dr Burger's ornithologische Ausbeute in Stromgebiet des Sepik. Ein Beitrag zur Kenntnis der Vogelwelt Neuguineas. *Archiv für Naturgeschichte*, **89** (7), 1–96; (8), 1–92.

Stresemann, E. (1924). Neue Beitrage zur Ornithologie Deutsch-Newguinea. *Journal für Ornithologie*, **72**, 424–8.

Stresemann, E. (1930*a*). Welche Paradiesvogelarten der Literatur sind Hybriden ursprungs? *Novitates Zoologicae*, **36**, 6–15.

Stresemann, E. (1930*b*). Welche Paradiesvogelarten der Literatur sind Hybriden ursprungs? *Proceedings of the Seventh International Ornithological Congress*, Amsterdam, 284.

Stresemann, E. (1931). Ueber die Balz von *Parotia sefilata* (L.). *Ornithologische Monatsberichte*, **39**, 4–6.

Stresemann, E. (1934). Vier neue Unterarten von Paradiesvogeln. *Ornithologische Monatsberichte* **42**, 144–7.

Stresemann, E. (1954). Die Entdeckungsgeschichte der Paradiesvogel. *Journal für Ornithologie*, **95**, 263–91.

Swadling, P. (1996). *Plumes from paradise.* Papua New Guinea National Museum and Robert Brown, Boroko.

Swainson, W. (1825). On the characters and natural affinities of several new birds from Australasia; including some observations on the columbidae. *The Zoological Journal*, **1**, 463–84.

Swofford, D. L. (1985). *PAUP: phylogenetic analysis using parsimony.* Version 2.4. Illinois Natural History Survey, Champaign, Illinois.

Swofford, D. L. (1986). *CONTREE (consensus tree program).* Illinois Natural History Survey, Champaign, Illinois. pp. 1–7.

Temple, P. (1962). *Nawoki—the New Zealand expedition to New Guinea's highest mountains.* London, Dent.

Templeton, M. T. (1992). Birds of Nanango, south-east Queensland. *Sunbird*, **22**, 87–100.

Terborgh, J. W. and Diamond, J. M. (1970). Niche overlap in feeding assemblages of New Guinea birds. *Wilson Bulletin*, **82**, 29–52.

Thair, S. and Thair, M. (1977). Report on display of Magnificent Bird of Paradise. *New Guinea Bird Society Newsletter*, **128**, 13.

Timmer, J. (1993). Inclined to be authentic: altered contexts and body decoration in a Huli society, Southern Highlands Province, Papua New Guinea. M. A. thesis. University of Amsterdam.

Timmis, W. H. (1968). Breeding of the Superb Bird of Paradise at Chester Zoo. *Avicultural Magazine*, **74**, 170–2.

Timmis, W. H. (1970). Breeding of the Superb Bird of Paradise, *Lophorina superba*, at Chester Zoo. *International Zoo Yearbook*, **10**, 102–4.

Timmis, W. H. (1972). Notes on the display and nest building of the Sickle-billed Bird of Paradise. *International Zoo Yearbook*, **12**, 190–2.

Todd, W. and Berry, R. J. (1980). Breeding the Red Bird of Paradise. *International Zoo Yearbook*, **20**, 206–11

Tolhurst, L. (1989). Extension of the known range of Splendid Astrapia *Astrapia splendidissima*. *Muruk*, **4**, 20.

Tweddle, D., Eccles, D. H., Frith, C. B., Fryer, G., Jackson, P. B. N., Lewis, D. S. C. and Lowe-McConnell, R. H. (1998). Cichlid spawning structures—bowers or nests? *Environmental Biology of Fishes*, in press.

Valentijn, F. (1724–6). *Oud en Nieuw Oost-Indiën*. III, Dordrecht.

Varghese, T. (1977). *Eimeria paradisaeai* sp. n. and *Isospora raggianai* sp, n. from the raggiana bird of paradise (*Paradisaea raggiana* Sclater) from Papua New Guinea. *Journal of Parasitology*, **63**, 887–9.

Vieillot, L. P. (1816). *Analyse d'une nouvelle ornithologie élémentaire*. Deterville, Paris.

Vieillot, L. P. (1819). *Nouveau dictionnaire d'histoire naturelle. Nouvelle edn.* **28**, 165–8. Chez Deterville, Libraire, Paris.

Vigors, N. A. (1825). Observations on the natural affinities that connect the orders and families of birds. *Transactions of the Linnean Society of London*, **14**, 395–517.

Wagner, H. O. (1938). Beobachtungen ueber die Balz des Paradiesvogels *Paradisaea guilielmi*. Cab. *Journal für Ornithologie*, **86**, 550–3.

Wahlberg, N. (1990). Display of the Glossy-mantled Manucode *Manucodia atra*. *Muruk*, **4**, 64.

Wahlberg, N. (1992). Observations of birds feeding in a fruiting fig *Ficus* sp. in Varirata National Park. *Muruk*, **5**, 109–10.

Wallace, A. R. (1857). On the Great Bird of Paradise *Paradisea apoda*, Linn.; Burong Mati (Dead Bird) of the Malays. *Annals and Magazine of Natural History*, **20**, 377–87.

Wallace, A. R. (1862). Narrative of search after birds of paradise. *Proceedings of Zoological Society of London*. **1862**, 153–61.

Wallace, A. R. (1869). *The Malay Archipelago*. Macmillan, London.

Ward, E. (1873). Description of a new Bird of Paradise of the genus *Epimachus*. *Proceedings of the Zoological Society of London*, **1873**, 742–3.

Watson, J. D., Wheeler, W. R. and Whitbourne, E. (1962). With the RAOU in Papua New Guinea, October 1960. *Emu*, **62**, 31–50, 67–98.

Wheelwright, N. T. and Orians, G. H. (1982). Seed dispersal by animals: contrasts with pollen dispersal, problems of terminology, and constraints on coevolution. *American Naturalist*, **119**, 402–13.

White, C. M. N. and Bruce, M. D. (1986). *The Birds of Wallacea (Sulawesi, The Moluccas and Lesser Sunda Islands, Indoneasia)*. British Ornithologists' Union Checklist No. 7. British Ornithologists' Union, London.

White, J.P. and O'Connell, J. F. (1982). *A prehistory of Australia, New Guinea and Sahul*. Academic Press, Sydney.

Whiteside, R. (1995). Notes on the display behaviour of a fully-plumed male Blue Bird of Paradise *Paradisaea rudolphi*. *Muruk*, **7**, 71–3.

Whiteside, R. (1998). The Blue Bird of Paradise *Paradisaea rudolphi*—display and behaviour in wild birds. *Australian Bird Watcher*, **17**. In press.

Whiteside, R. and Feignan, M. (1998). Displays and associated behaviour of wild Superb Birds of Paradise *Lophorina superba*. *Australian Bird Watcher*, **17**. In press.

Whitney, B. M. (1987). The Pale-billed Sicklebill *Epimachus bruijnii* in Papua New Guinea. *Emu*, **87**, 244–6.

Williams, S., Pearson, R., and Burnett, S. (1993). Vertebrate fauna of three mountain tops in the Townsville region, north Queensland: Mount Cleveland, Mount Elliot and Mount Halifax. *Memoirs of the Queensland Museum*, **33**, 379–87.

Williams, T. D. (1995). *The penguins*. Oxford University Press, Oxford.

Willis, E. O. (1972). Behaviour of the Plain-brown Wood-creeper. *Wilson Bulletin*, **84**, 377–420.

Winterbottom, J. M. (1928a). The display of Wilson's Bird of Paradise (*Schlegelia wilsoni*). *Ibis*, **4**, 319–20.

Winterbottom, J. M. (1928b). Polygamy in birds of paradise. *American Naturalist*, **62**, 380–3.

Worth, W., Hutchins M., Sheppard, C., Bruning, D., Gonzaliz, J and McMamara, T. (1991). Hand-rearing, growth, and development of the Red Bird of Paradise (*Paradisaea rubra*) at the New York Zoological Park. *Zoo Biology*, **10**, 17–33.

Yealland, J. J. (1969). Breeding of Princess Stephanie's Bird of Paradise at London Zoo. *Avicultural Magazine*, **75**, 50–1.

Zahavi, A. (1975). Mate selection—a selection for a handicap. *Journal of Theoretical Biology*, **53**, 205–14.

Index

Subfamily, genera, and species accounts are denoted by bold type.

Particular species of birds of paradise can be located either by English name (e.g. Bird of Paradise, Blue), by genus (e.g. *Paradisaea rudolphi*), or by species scientific name (e.g. *rudolphi, Paradisaea*).

Subspecies trinomials and synonyms are listed by those particular names (e.g. *pulchra, Paradisaea minor*; or *mirabilis, Janthothorax*). For synonyms and out-of-use names, the index lists where the name is cited in the text rather than providing a cross-reference.

abbreviations xvii–xviii
addenda, Lophorina superba 348
adelberti, Manucodia keraudrenii 232
adornment, human 143–44, 356
age at first breeding 97
Air Niugini 152
alberti, Pteridophora **305–14**, Plates 10, 13
alberti, Ptiloris magnificus 318
albertisi, Drepanornis **377–85**, Plates 9, 13
albertisi, Drepanornis albertisi 379
albicans, Epimachus meyeri 368
albinism 254
alboundata, Astrapia 274
altera, Manucodia atra 214
allopreening 349
Ambua Lodge 563
Ambua Range, PNG Plate 3
amethystina, Cnemophilus loriae 178
ampla, Paradisaea rudolphi 490
anatomy of birds of paradise 6, 24
annual cycles 100, 139–42
apoda, Paradisaea 9, 42, 43, **448–56**, Plate 11
apoda, Paradisaea apoda 449
Archbold Expeditions 38
Archboldia papuensis 145, 314
Arfak Mountains 565
aruensis, Manucodia keraudrenii 231
Astrapia 60, Plate 6
astrapias 60, Plate 6
Astrapia **249–50**
  alboundata 274
  mayeri **257–65**, Plates 3, 6, 13
  nigra **250–53**, Plate 6
  recondita 259
  rothschildi **273–76**, Plates 6, 13
  splendidissima **253–57**, Plate 6
  stephaniae **266–73**, Plate 6

Astrapia, Arfak **250–53**, Plate 6
  Barnes' Plate 14
  Black 250
  False-lobed Long Tail 509, Plate 14
  Huon **273–76**, Plates 6, 13
  Princess Stephanie 266
  Ribbon-tailed **257–65**, Plates 3, 6, 13
  Rothschild's 273
  Splendid **253–57**, Plate 6
  Stephanie's **266–73**, Plate 6
astrapioides, Epimachus 508
Astrarchia 249, 267
Astrarchia, barnesi 504
ater, Manucodia 211
atra, Manucodia **211–17**, Plates 5, 13
atra, Manucodia atra 213
atratus, Epimachus fastuosus 360
Attenborough, Sir David vii, ix–x, 46
Audebert 4
augustaevictoriae, Paradisaea raggiana Plate 11
augustaevictoriae, Paradisaea raggiana 459
auripennis, Seleucidis melanoleuca 430
aurora, Loboparadisea sericea 192
Australian Bird and Bat-Banding Scheme xix
aviaries 39
aviculture 39–41, 166

Bacan (Batjan) Island 565
Baiyer River Sanctuary 40, 41
Bamaga 562
Barraband, J. 31, Plate 1
Beecher 48
Beehler, Bruce M. i
Beehler, Carol v
bensbachi, Janthothorax 517
Bergman, Sten 40

berlepschi, Parotia carolae 300
bibliography 580–606
bill morphology 83
biogeography of the birds of paradise 66–77
biological species concept 66
biometrics 10–12
Bird of Paradise, Archduke Rudolph's Blue 488
   Arfak 250
   Arfak Six-wired 277
   Bare-headed Little King 401
   Bensbach's Plate 15
   Black Saber-tailed 357
   Black Sickle-billed 357
   Black-and-Gold 183
   Black-billed Sicklebill 377
   Blue 73, 76, **488–98**, Plates 1, 12
   Bobairo Plate 14
   Brown Sickle-billed 366
   Bruijn's 385
   Captain Blood's Plate 15
   Count Raggi's 456
   Crested **183–89**, Plates 3, 4
   D'Albertis's 377
   Duivenbode's 504
   Duivenbode's Six-wired Plate 14
   Elliot's Plate 14
   Emperor **482–88**, Plates 3, 12
   Emperor of Germany 482
   Empress of Germany's 456
   Enamelled 305
   Frau Reichenow's Plate 15
   Goldie's **470–75**, Plate 12
   Great 448
   Greater 9, 42, 43, **448–56**, Plate 11
   Greater Six-plumed 277
   Huon 273
   Incomparable 250
   King 9, **407–16**, Plate 10
   King of Holland's Plate 15
   King of Saxony **305–14**, Plates 10, 13
   Lawes's Six-wired 283
   Lesser **439–48**, Plates 1, 13
   Lesser Superb 345
   Little Emerald 439
   Little King 407
   Loria's **176–82**, Plates 4, 13
   Lupton's Plate 15
   Lyre-tailed King 507, Plate 15
   Macgregor's 9, 59, **198–204**, Plates 5, 13
   Magnificent **391–401**, Plates 10, 13
   Meyer's Sickle-billed 366
   Multi-crested 183
   Orange-wattled 198
   Prince Rudolph's Blue 488
   Princess Stephanie 266
   Queen Carola's Six-wired 298
   Queen of Saxony's 298
   Raggi's 456
   Raggiana **456–70**, Plates 11, 13
   Red **475–82**, Plates 1, 12
   Ribbon-tailed 257
   Rothschild's Plate 15
   Rothschild's Lobe-billed Plate 14
   Ruys' Plate 15
   Schodde's Plate 14
   Shaw Mayer's 257
   Shield-billed 190
   Sickle-crested 183
   Splendid 253
   Standardwing **417–27**, Plate 12
   Stresemann's Plate 14
   Superb 24, 77, **345–55**, Plates 10, 13
   Twelve-wired 24, **428–38**, Plates 1, 8, 13
   Wahnes' Six-wired 292
   Waigeu 401
   Wallace's 417
   Wallace's Standard Wing 417
   Wattle-billed 190
   White-billed 385
   White-billed Sicklebill 385
   White-plumed 482
   Wilhelmina's Plate 14
   Wilson's **401–6**, Plate 10
   Wonderful Plate 15
   Yellow-breasted **190–94**, Plate 4
bird–plant interactions 93
birds of paradise, exploration for 521–51
   plumed 61, Plate 11
   typical paradisaea Plate 11
   where to find 561–65
bloodi, Epimachus meyeri 368
bloodi, Paradisaea 505
Bock, W. 48
body size 7, 9
Boehm, Edward 40
Boiga irregularis 97
bowerbirds (Ptilonorhynchidae) 52, 53, 145, 314
brevicauda, Paradigalla **244–49**, Plates 3, 5, 13
bride-wealth, traditional 144
Bristlehead (*Pityriasis*) 49
brood parasitism 239
brown-and-black flocks 95–6
bruijnii, Drepanornis 73, **385–90**, Plate 9
bruijnii, Epimachus 385
bürgersi, Pteridophora alberti 307
butcherbirds, Australian 51

Calastrapia 249
carolae, Parotia 77, **298–303**, Plate 7
carolae, Parotia carolae 300
carolinae, Macgregoria pulchra 200
carunculata, Paradigalla **242–43**, Plate 5
casts (of stomach) 92
causes of mortality 97
cervinicauda, Drepanornis albertisi 379
chalcothorax, Parotia carolae 300
chalybata, Manucodia **220–24**, Plate 5
character matrix 56–8
character state 57–8
characteristics of birds of paradise 5, 8
Charmosyna papou 261
chrysenia, Parotia carolae 301
chrysopterus, Cicinnurus magnificus 393
Cicinnurus 13, **390**
   lyogyrus 507
   magnificus **391–401**, Plates 10, 13
   regius 9, **407–16**, Plate 10
   respublica **401–6**, Plate 10
CITES 154

cladistic analysis   53–5
claudia, Craspedophora magnifica   318
claudii, Cicinnurus regius   410
clelandiae, Parotia carolae   301
clutch size   129
Cnemophilinae   59, 67, **174–75**, Plate 4
Cnemophilus   9, **175–76**
    loriae   **176–82**, Plates 4, 13
    macgregorii   **183–89**, Plates 3, 4
    mariae   176
coccineifrons, Cicinnurus regius   410
coin collecting   151
colour, in display   118
    other body   24–5
comrii, Manucodia   **224–29**, Plate 5
comrii, Manucodia comrii   226
connectens, Lophorina superba   348
conservation   27–8, 154–67
    aviculture and   166
    enforcement of legislation and   165–6
    forest reserves and   164–5
    grassroots education and   165
    impact of El Niño   159
    impact of exotic predators   158
    impact of global change   158–9
    impact of human population   158
conservation action   163–6
contact zone (of *Astrapia* spp.)   76
CONTREE   55
Convention on the International Trade of Endangered
        Species   154
Cooper, William T.   i, ix, 45
Corvoidea   49
court display   105
courtship display   41–2, 117–22
    body movements in   120–1
    communal   115
    signals   117
    solitary   115
    surprise elements in   12
Courtship feeding   349
Cracraft, J.   63, 66
Cracticus, quoyi   238
Craspedophora, mantoui   517
Crater Mountain Wildlife Management Area   563
Crow, Paradise   9, 48, **205–10**, Plates 3, 5
crows, true   22, 50–1
cryptorhynchus, Cicinnurus regius   410
Cyathea tree fern   Plate 2
Cyclops Mountains   564

Dacrycarpus compactus   79, 82, 91, 201, 202, 203, Plate 2
Darwin, Charles   32
decora, Paradisaea   **470–75**, Plate 12
decoration, human   143–4
Diamond, Jared M.   43, 63, 74, 82
diamondi, Manucodia keraudrenii   232
Dicrurus   238
diet   82–9
    adult vs. nestling   92–3
dimorphism, plumage   14, 22, 103
    sexual   13–19, 102–4
    size   102–3
    vocal   103

Dinsmore, James   43
Diphyllodes   390
    gulielmi III   507
    magnificus   391
    respublica   401
discovery of birds of paradise   521–51
    history of   4, 29–46, 37–38, 44, 260
display site, advertisement of   114
    modification of   13
display sites   109–14
distributional anomalies   73
    ecology   99–100
    patterns   68
distributions, Australian   67
distributions of birds of paradise   xxi
Drepanornis   **376–7**
    albertisi   **377–385**, Plates 9, 13
    bruijnii   73, **385–90**, Plate 9
ducalis, Astrapia stephaniae   268
duet   223, 227
duivenbodei, Parotia   512
duivenbodei, Paryphephorus (Craspedophora)   515
dyotti, Ptiloris paradisea   335

ecological biogeography   74
ecology   26, 78–101
egg, colour of   130
    incubation and hatching of   132–5
    laying interval   132
    shape of   130
    weight of   130
eggs   129, Plate 13
Elcano   30
Elliot, D. G.   45
ellioti, Epimachus   508
elliotsmithi, Astrapia splendidissima   255
environment   78–9
Epimachus   **356–7**
Epimachus albertisi   377
    astrapioides   508
    bruijnii   385
    ellioti   503
    fastuosus   7, **357–65**, Plate 9
    meyeri   **366–75**, Plates 9, 13
Eulacestoma nigropectus   19
evolution of the birds of paradise   47–77
exhibita, Parotia lawesii   284
exploration for birds of paradise   35–9
    of Australasia   **521–55**
extra, Cicinnurus magnificus   394

fastosus, Epimachus   357
fastuosus, Epimachus   **357–65**, Plate 9
fastuosus, Epimachus fastuosus   360
feather, modification of   23, 408
feather form and function   22
feminina, Astrapia stephaniae   268
feminina, Lophorina superba   348, Plate 10
field study   42–5
fig (*Ficus*)   89, 91, 92, 193, 234, 239, 279
finding birds of paradise   560–65
finschi, Paradisaea minor   442
flagbird   60

flocks, black and brown   95–6
Flynn, Errol   29
food plants   94, 554
foraging ecology   95–6
foraging, terrestrial   369
forest, midmontane   Plate 2
Forshaw, J. M.   45
Frith, Clifford B.   i
Frith, Dawn W.   v, ix
frugivorous specialisation   92–3
frugivory   86–9, 554–9
fruit, paradisaeid consumption of   554–9
fruit foraging   89–92, 93
   morphology   87
   preferences   90
fruit, capsular   87–8
   drupe   87–8
   fig   87–8
fuscior, Parotia lawesii   284

Gamkanora (Halmahera)   Plate 3
gazetteer, ornithological   566–72
geisleri, Drepanornis albertisi   379
Gilliard, E. Thomas   v, 42, 123, 146
glossary   573–9
Gould, John   45, Plate 1
gouldii, Manucodia keraudrenii   234
granti, Paradisaea raggiana   459, 505
guilielmi, Paradisaea   **482–8**, Plates 3, 12
Guillemard, F.   32
gulielmi III, Diphyllodes   507
gymnorhynchus, Cicinnurus regius   410

Habbema, Lake   Plate 2
habitat loss   154–7
   selection   79–82
haemochromatosis   93, 166
Hallstrom, Sir Edward   40
hallstromi, Pteridophora alberti   307
Halmahera Island   565
halmaherae, Semioptera wallacii   419
handling and processing food   91
helenae, Parotia lawesii   285, Plate 7
helios, Astrapia splendidissima   255
heterochrony   96–7
historical biogeography   69–74
Huli people   30, 146, 150
human cultures, birds of paradise in   143–53
human traditions, birds of paradise in   143–53
hunsteini, Cicinnurus magnificus   394
hunsteinii, Manucodia keraudrenii   233
hunti, Paradisaea rudolphi   490
hunting birds of paradise   147–9, 157–8
hybrid birds of paradise   Plates 14, 15
hybrid forms   **499–520**
hybridisation   28, 74–6

ignota, Seleucides   428
ignotus, Seleucides   428
illustrations of birds of paradise   45–6
inexpectata, Cnemophilus loriae   178

Ingram, Sir William   39
insectivory   85–6, 88
intercedens, Ptiloris magnificus   318
intermedia, Paradigalla carunculata   245
intermedia, Paradisaea raggiana   459, Plate 11
intermedius, Cicinnurus magnificus   393
inversa, Drepanornis albertisi   379
Iron Range, Australia   562, Plate 3

jamesii, Manucodia keraudrenii   232
Janthothorax bensbachi   517
   mirabilis   499
Jayapura   564
jobiensis, Manucodia   **217–20**, Plate 5
jobiensis, Paradisaea minor   442

Kalam people   150
keraudrenii, Manucodia   76, **229–40**, Plates 3, 5, 13
keraudrenii, Manucodia keraudrenii   231
Keulemans, John Gerrard   45, Plate 1
Kissaba   305
kuboriensis, Cnemophilus macgregorii   185

Lake Habbema   564
Lakekamu Basin   563
Lamprothorax, wilhelminae   513
land-bridge   72
latipennis, Lophorina superba   348
lawesi, Parotia   24, **283–92**, Plates 7, 13
lawesii, Parotia lawesii   284, Plate 7
LeCroy, Mary   v, 43
legislation, conservation   34–5
lehunti, Lophorina superba   348
lek   104
   behaviour   104
   display   106
Lesson, R.   31, 36, 48
Levaillant, F.   31, Plate 1
Linnaeus   30, 48
listing, taxonomic   20–2
Little Tobago Island   39–40
lobata, Pseudastrapia   509
Loboparadisea   9, **189**
Loboparadisea sericea   **190–4**, Plate 4
Loborhamphus ptilorhis   513
Loborhamphus nobilis   510
Long Tail, Rothschild's   273
longevity   96
Lophorina   13, 60, **345**
Lophorina superba   13, 24, 77, **345–55**, Plates 10, 13
   superba pseudoparotia   514
loriae, Cnemophilus   **176–82**, Plates 4, 13
loriae, Cnemophilus loriae   178
luptoni, Paradisaea apoda   506
Lycocorax   9, **204–5**
   pyrrhopterus   9, 48, **205–210**, Plates 3, 5
lyogyrus, Cicinnurus   507

macgregori, Xanthomelus   184
Macgregoria   9, 60, **197**
   pulchra   59, 60, **198–204**, Plates 5, 13

macgregorii, Cnemophilus **183–9**, Plates 3, 4
macgregorii, Cnemophilus macgregorii 185
macnicolli, Taeniaparadisea 259
Madang 563
magic, birds of paradise and 149
magnificus, Cicinnurus **391–401**, Plates 10, 13
magnificus, Cicinnurus magnificus 393
magnificus, Diphyllodes 391
magnificus, Ptiloris **315–27**, Plates 8, 13
magnificus, Ptiloris magnificus 318
Mahendra, King, of Nepal 147
male maturation 107
male–male aggression in lek 116–17
Maluku Utara 565
mantoui, Craspedophora 517
Manucode 229
   Allied 217
   Black 211
   Crinkle-breasted 220
   Crinkle-collared **220–4**, Plate 5
   Curl-breasted 224
   Curl-crested 9, **224–9**, Plate 5
   Glossy 211
   Glossy-mantled **211–17**, Plates 5, 13
   Green 220
   Green-breasted 220
   Jobi **217–20**, Plate 5
   Trumpet 25, 76, **229–40**, Plates 3, 5, 13
manucodes 76, Plate 5
Manucodia 59, **210–11**
   atra **211–17**, Plates 5, 13
   chalybata **220–224**, Plate 5
   chalybatus 220
   comrii 9, **224–9**, Plate 5
   jobiensis **217–20**, Plate 5
   keraudrenii 25, 76, **229–40**, Plates 3, 5, 13
   orientalis 221
manucodiine lineage 59
map of key localities xxviii, xxix
margaritae, Paradisaea rudolphi 490
mate-choice 104
mating strategy 107
mating system, monogamous 105
   polygynous 105
mating systems 26, 105–9
   ecology and evolution of 107–9
mayeri, Astrapia **257–65**, Plates 3, 6, 13
Mayr, Ernst 36, 65
measurements xii
meeki, Parotia carolae 300
megarhynchus, Epimachus meyeri 369
Megatriorchis doriae 97
Melampitta, Lesser 6–7, 49–50, 55, 67
   Greater 50
melanoleuca, Seleucidis **428–38**, Plates 1, 8, 13
melanoleuca, Seleucidis melanoleuca 430
Melidectes, belfordi 261
Melidectes, rufocrissalis 353
meyeri, Epimachus **366–75**, Plates 9, 13
meyeri, Epimachus meyeri 368
migration 240
millinery trade 154, 157, 344
minor, Lophorina superba 348
minor, Paradisaea **439–48**, Plates 1, 13

minor, Paradisaea minor 441
mirabilis, Janthothorax 49
mirabilis, Paradisea 516
mixed flocks 95–6, 219, 279, 319–20, 327, 349, 460
mixta, Paradisea 505
Moluccan distributions 67
Moresby, Captain J. 32
morotensis, Lycocorax pyrrhopterus 207, Plate 5
moult, seasonal timing of 552–3
museum collections 35–6
mythology, birds of paradise and 149

Nagore River Plate 2
Neoparadisea, ruysi 518
Nepal, Royal Court of 147
nest decoration 128–9
nest sites 126–7
nest-building 128
nesting 26–7
   parental investment in 137
nesting biology 123–42
nesting ecology 98
nestling, care of 135
   development of 135
   diet of 138
   growth of 135
   post-fledging care of 138
nests 127–8
neumanni, Manucodia keraudrenii 232
New Guinea, administration of 35
New York Zoological Society 39
New Zealand wattlebirds 52
niedda, Lophorina superba 348
nigra, Astrapia **250–3**, Plate 6
nobilis, Loborhamphus 510
nomenclature xxii
Nondugl aviaries 40
novaeguineae, Paradisaea apoda 449
Nunn and Cracraft 65

obiensis, Lycocorax pyrrhopterus 206, Plates 5, 13
origins of the birds of paradise 48, 66
orioles 51
over-water differentiation 73

pandanus 90, 320, 354, 369, Plate 2
Pandanus tectorius Plate 2
paradigallas 60, Plate 5
Paradigalla 241
   brevicauda **244–9**, Plates 3, 5, 13
   carunculata **242–3**, Plate 5
Paradigalla, Long-tailed **242–3**, Plate 5
   Short-tailed **244–9**, Plates 3, 5, 13
Paradisaea **438–9**, Plate 11
   apoda 9, 42, 43, **448–56**, Plate 11
   apoda luptoni 506
   bloodi 505
   decora **470–5**, Plate 12
   granti 505
   guilielmi **482–8**, Plates 3, 12
   maria 504

Paradisaea (cont.)
   minor **439–48**, Plates 1, 11, 13
   raggiana **456–70**, Plates 11, 13
   rubra **475–82**, Plates 1, 12
   rudolphi 73, 76, **488–98**, Plates 1, 12
Paradisaeidae 6, **171–3**
Paradisaeinae 6, **195–6**, Plates 5–13
Paradise Bird, Scale-breasted 315
   Twelve-wired 428
Paradisea, duivenbodei 504
   mirabilis 516
   mixta 505
   sexipennis 279
   wilsoni 402
paradiseus, Ptiloris **327–34**, Plate 8
Paradisornis 61, 438
   rudolphi 488
parallel variation 77
parasites 97–8
   brood 139
Parer, David 45
parotias 60, Plate 7
Parotia, Arfak 277
   Arfak Six-wired 277
   Carola's 77, **298–303**, Plate 7
   Huon 292
   Parotia, Lawes' 24, **283–92**, Plates 7, 13
   Lawes' Six-wired 283
   Wahnes' **292–7**, Plate 7
   Western **277–82**, Plate 7
Parotia 13, **277**, Plate 7
   carolae 77, **298–303**, Plate 7
   duivenbodei 512
   helenae 283
   lawesii 24, **283–92**, Plates 7, 13
   sefilata **277–82**, Plate 7
   wahnesi **292–7**, Plate 7
Paryphephorus (*Craspedophora*), *duivenbodei* 515
PAUP 55
philately 151–2
Philemon 238
Philomachus pugnax 502
Phonygammus keraudrenii 229
phylogenetic reconstruction 53–62
phylogenetic species concept 66
phylogeny 59, 62–5
Pitohui dichrous 101
   kirhocephalus 101
Pityriasis (Bristlehead) 49
plant species, paradisaeid use of 554–9
plumage xxi
   mimicry 100–1
plume trade, the 34–5, 447
plumed birds of paradise 61, Plate 11
Pomatostomus isidorei 101
population biology 96–8
predation 97, 204, 344, 442, 460
proportions 7, 9, 13
Pseudastrapia, lobata 509
pseudoparotia, Lophorina superba 349, 514
Pteridophora 13, 60, **304–5**
   alberti 13, **305–14**, Plates 10, 13
Ptilonorhynchidae 52, 53
ptilorhis, Loborhamphus 513

Ptiloris 60, **314–15**
Ptiloris magnificus **315–27**, Plates 8, 13
   paradiseus **327–34**, Plate 8
   victoriae **334–45**, 13, 92 Plate 8
pulchra, Macgregoria 59, 60, **198–204**, Plates 5, 13
pulchra, Macgregoria pulchra 200
pulchra, Paradisaea minor 441
purpureoviolacea, Manucodia keraudrenii 233
pyrrhopterus, Lycocorax 9, 48, **205–10**, Plates 3, 5
pyrrhopterus, Lycocorax pyrrhopterus 206

queenslandica, Ptiloris paradisea 329

raggiana, Paradisaea **456–70**, Plates 11, 13
raggiana, Paradisaea raggiana Plates 11, 458
rainforest regeneration 100
Rajah Ampat Islands 565
ranging habits 98–9
Rattlesnake, H. M. S. 31–2
recondita, Astrapia 259
recordings of birds of paradise 560
references 580–606
regius, Cicinnurus **407–16**, Plate 10
regius, Cicinnurus regius 409
regurgitation 92
reproductive behaviour 102–22
   investment 103–4
research on birds of paradise 521–51
respublica, Cicinnurus **401–6**, Plate 10
respublica, Diphyllodes 401
rex, Cicinnurus regius 409
Rifle Bird 327
Riflebird, Albert 315
   Duivenbode's Plate 15
   Lesser 334
   Magnificent **315–27**, Plates 8, 13
   Mantou's Plate 15
   Paradise **327–34**, Plate 8
   Queen Victoria's 334
   Sharpe's Lobe-billed Plate 14
   Victoria's **334–45**, 13, 92 Plate 8
riflebirds 60, Plate 8
ring-species 71
Rosenberg, C. B. H. von 32
Rothschild, Lord Walter 36
rothschildi, Astrapia **273–6**, Plates 6, 13
rothschildi, Cicinnurus magnificus 393
Rouna Falls Plate 2
rubiensis, Manucodia 219
rubra, Paradisaea **475–82**, Plates 1, 12
rudolphi, Paradisaea 73, 76, **488–98**, Plates 1, 12
rudolphi, Paradisaea rudolphi 490
rudolphi, Paradisornis 488
Ruff 502
runaway selection 104
ruysi, Neoparadisea 518

sabretails 61, Plate 9
Salvadori, T. 33
salvadorii, Paradisaea raggiana 458
sanguineus, Cnemophilus macgregorii 185

Schefflera 90, 261, Plate 2
Schodde, R. 63, 74
search for birds of paradise 521–51
seasonality, annual breeding 139–42
seed dispersal 100
sefilata, Parotia **277–82**, Plate 7
selection, sexual 102–4
Seleucidis 427
  melanoleuca 24, **428–38**, Plates 1, 8, 13
Semioptera 417
  wallacii **417–27**, Plate 12
Sentani 564
sericea, Loboparadisea **190–94**, Plate 4
sericea, Loboparadisea sericea 191
sexipennis, Paradisea 279
Sharpe, R. Bowdler 42, 45, 48, 123, Plate 1
Sibley and Ahlquist 47, 49, 60
Sicklebill, Astrapian Plate 14
  Black 7, **357–65**, Plate 9
  Black-billed 377
  Brown **366–75**, Plates 9, 13
  Buff-tailed **377–85**, Plates 9, 13
  Greater 357
  Lowland 385
  Pale-billed 73, **385–90**, Plate 9
  Short-tailed 377
sicklebills 61, Plate 9
sickletails 61, Plate 10
silktail, the 52
similis, Cicinnurus regius 410
sing-sing 144, 153
site guide for birds of paradise 560–65
skull modification 24
Sonnerat 30–1
sonograms xxiii
speciation 69–73
  time-scale of 73
speciation of birds of paradise 69–77
speciosus, Falcinellus 359
sphinx, Lophorina superba 349
spinturnix, Cicinnurus regius 410
splendidissima, Astrapia **253–7**, Plate 6
splendidissima, Astrapia splendidissima 255
stamps, birds of paradise on 152
Standardwing, Wallace's 417
starlings 51
stephaniae, Astrapia **266–73**, Plate 6
stephaniae, Astrapia stephaniae 268
Stonor, C. R. 48
stresemanni, Epimachus fastuosus 360
striatus, Falcinellus 359
study skins 39
subaltera, Manucodia atra 213
subintermedia, Paradisaea raggiana 459
subspecies xxii
sunning 476–7
superba, Lophorina 24, 77, **345–55**, Plates 10, 13
superba, Lophorina superba 347, Plate 10

symbols, birds of paradise as modern 150–3
systematic relationships 47, 53–66

Taeniaparadisea 249
Taeniaparadisea macnicolli 259
tape recordings of birds of paradise 560
Tari Gap 261, 563, Plates 2, 3
Tari Valley 563
terms, definition of scientific 573–9
Ternate Island 565
threatened species 159–63
threats to birds of paradise 154–55
Timika 565
Timonius 91
topography of a bird xxv–xxvi
Torres Strait 67, 72
trachea, modified 25, 211, 221, 225, 427
trade, black market 157
trade of birds of paradise 157–8
trade skins 29
traditional culture and birds of paradise 27
Transilvano, Massimiliano 30
Trichoparadisea 438
Trikora, Puncak Plate 2
trobriandi, Manucodia comrii 226
Trumpetbird 229

ultimus, Epimachus fastuosus 360
Uranornis 438
Uranornis rubra 475

Varirata National Park 564
victoriae, Ptiloris **334–45**, 13, 92 Plate 8
Vieillot and Audebert 4
Vieillot, L. J. B. 31
vocal dialects 76
vocalisation 25
  during courtship 117–18
vocalisations xxiii, 560
vulnerability of species 160–3

wahnesi, Parotia **292–7**, Plate 7
Wallace, Alfred Russel 32, 48, 317, 365, 477
wallacii, Semioptera **417–27**, Plate 12
wallacii, Semioptera wallacii 419
Wamena 564
Wau 564
wide-gaped birds of paradise 174, Plate 4
wilhelminae, Lamprothorax 513
wilsoni, Paradisea 402
wood-swallows 51

yorki, Craspedophora magnifica 319